Table of Atomic Masses

Element	Symbol	Atomic Number	Atomic Mass	Element	Symbol	Atomic Number	Atomic Mass
Actinium	Ac	89	227.0278	Neon	Ne	10	20.179(6)
Aluminum	Al	13	26.981539(5)	Neptunium	Np	93	237.0482
Americium	Am	95	243.0614	Nickel	Ni	28	58.6934(2)
Antimony	Sb	51	121.757(3)	Niobium	Nb	41	92.90638(2)
Argon	Ar	18	39.948(1)	Nitrogen	N	7	14.00674(7)
Arsenic	As	33	74.92159(2)	Nobelium	No	102	259.1009
Astatine	At	85	209.9871	Osmium	Os	76	190.2(1)
Barium	Ba	56	137.327(7)	Oxygen	O	8	15.9994(3)
Berkelium	Bk	97	247.0703	Palladium	Pd	46	106.42(1)
Beryllium	Be	4	9.012182(3)	Phosphorus	P	15	30.973762(4)
Bismuth	Bi	83	208.98037(3)	Platinum	Pt	78	195.08(3)
Boron	B	5	10.811(5)	Plutonium	Pu	94	244.0642
Bromine	Br	35	79.904(1)	Polonium	Po	84	208.9824
Cadmium	Cd	48	112.411(8)	Potassium	K	19	39.0983(1)
Calcium	Ca	20	40.078(4)	Praseodymium	Pr	59	140.90765(3)
Californium	Cf	98	251.0796	Promethium	Pm	61	144.9127
Carbon	C	6	12.011(1)	Protactinium	Pa	91	231.03588(2)
Cerium	Ce	58	140.115(4)	Radium	Ra	88	226.0254
Cesium	Cs	55	132.90543(5)	Radon	Rn	86	222.0176
Chlorine	Cl	17	35.4527(9)	Rhenium	Re	75	186.207(1)
Chromium	Cr	24	51.9961(6)	Rhodium	Rh	45	102.90550(3)
Cobalt	Co	27	58.93320(1)	Rubidium	Rb	37	85.4678(3)
Copper	Cu	29	63.546(3)	Ruthenium	Ru	44	101.07(2)
Curium	Cm	96	247.0703	Samarium	Sm	62	150.36(3)
Dysprosium	Dy	66	162.50(3)	Scandium	Sc	21	44.955910(9)
Einsteinium	Es	99	252.083	Selenium	Se	34	78.96(3)
Erbium	Er	68	167.26(3)	Silicon	Si	14	28.0855(3)
Europium	Eu	63	151.965(9)	Silver	Ag	47	107.8682(2)
Fermium	Fm	100	257.0951	Sodium	Na	11	22.989768(6)
Fluorine	F	9	18.9984032.9	Strontium	Sr	38	87.62(1)
Francium	Fr	87	233.0197	Sulfur	S	16	32.066(6)
Gadolinium	Gd	64	157.25(3)	Tantalum	Ta	73	180.9479(1)
Gallium	Ga	31	69.723(4)	Technetium	Tc	43	98.9072
Germanium	Ge	32	72.61(2)	Tellurium	Te	52	127.60(3)
Gold	Au	79	196.96654(3)	Terbium	Tb	65	158.92534(3)
Hafnium	Hf	72	178.49(2)	Thallium	Tl	81	204.3833(2)
Helium	He	2	4.002602(2)	Thorium	Th	90	232.0381(1)
Holmium	Ho	67	164.93032(3)	Thulium	Tm	69	168.9342(3)
Hydrogen	H	1	1.00794(7)	Tin	Sn	50	118.710(7)
Indium	In	49	114.82(1)	Titanium	Ti	22	47.88(3)
Iodine	I	53	126.90447(3)	Tungsten	W	74	183.85(3)
Iridium	Ir	77	192.22(3)	Unnilennium[a, b]	Une	109	(266)
Iron	Fe	26	55.847(3)	Unnilhexium	Unh	106	263.118
Krypton	Kr	36	83.80(1)	Unniloctium	Uno	108	(265)
Lanthanum	La	57	138.9055(2)	Unnilpentium	Unp	105	262.114
Lawrencium	Lr	103	262.11	Unnilquadium	Unq	104	261.11
Lead	Pb	82	207.2(1)	Unnilseptium	Uns	107	262.12
Lithium	Li	3	6.941(2)	Uranium	U	92	238.0289(1)
Lutetium	Lu	71	174.967(1)	Vanadium	V	23	50.9415(1)
Magnesium	Mg	12	24.3050(6)	Xenon	Xe	54	131.29(2)
Manganese	Mn	25	54.93805(1)	Ytterbium	Yb	70	173.04(3)
Mendelevium	Md	101	258.10	Yttrium	Y	39	88.90585(2)
Mercury	Hg	80	200.59(2)	Zinc	Zn	30	65.39(2)
Molybdenum	Mo	42	95.94(1)	Zirconium	Zr	40	91.224(2)
Neodymium	Nd	60	144.24(3)				

[b]These unnil-names are temporary. Official names and symbols for these elements have not been agreed upon.

BASIC CONCEPTS OF CHEMISTRY

BASIC CONCEPTS OF CHEMISTRY

FOURTH EDITION

Leo J. Malone
Saint Louis University

JOHN WILEY & SONS, INC.
New York • Chichester • Brisbane • Toronto • Singapore

Acquisitions Editor: Nedah Rose
Development Editor: Johnna Barto
Marketing Manager: Catherine Faduska
Production Editor: Marcia Craig
Design: Hudson River Studio
Manufacturing Manager: Andrea Price
Photo Researcher: Lisa Passmore
Illustration: Ed Starr

This book was set in 10/12 New Century Schoolbook by Interactive Composition Corporation and printed and bound by Von Hoffmann Press. The cover was printed by Phoenix Printing Co.

Library of Congress Cataloging in Publication Data:

Malone, Leo J., 1938–
    Basic concepts of chemistry / Leo J. Malone. — 4th ed.
        p.   cm.
    Includes bibliographical references.
    ISBN 0-471-53590-7
    1. Chemistry.   I. Title.
QD31.2.M344   1994
540—dc20   93–38490
                                                                                                     CIP

ISBN 0-471-53590-7

Printed in the United States of America

10   9   8   7   6   5   4   3   2   1

PREFACE

Basic Concepts of Chemistry began life in 1981 as a text directed at students in a preparatory chemistry course. That is, it was intended for those planning to proceed to a main-sequence chemistry course, but who had little or no background in chemistry or who had a significant interruption of their studies. The acceptance of previous editions, however, indicates that this text has served a much broader mission. In addition to preparatory chemistry, it has found extensive use in one-semester, general-purpose courses. Such courses usually include a mix of students. For some of these students, it is a prelude to a main-sequence course; for others, a semester of organic-biochemistry may follow to satisfy allied health professions requirements. Still others are there to satisfy a science requirement. The Fourth Edition has been rewritten with this broader appeal in mind. Yet this text continues to emphasize the quantitative nature of real chemistry. Thus it will continue to satisfy the needs of those intending to continue on in chemistry and other sciences.

In preparing the Fourth Edition, the author and publisher obtained considerable input about how chemistry should be presented and what is needed by instructors at this level. As a result of our research, we have built on the strengths of previous editions to present what may be described as a "learning text." It is meant to be of maximum aid to the many dedicated teachers who believe that chemistry is a fascinating subject that serves as the basis of all sciences and technologies as well as to their students who have little or no background in chemistry.

We set out to accomplish three major goals with the Fourth Edition. They have been addressed by incorporating successful features of previous editions along with new, innovative approaches used for the first time in this edition. A discussion of these goals follows.

GOALS AND FEATURES OF THE FOURTH EDITION

To accomplish the goals of the Fourth Edition, the successful features of previous editions have been enhanced with some innovative pedagogical elements. Called "navigational devices," these learning aids are highlighted by marginal notations to help guide students in understanding basic chemical concepts.

Goal I. To introduce chemistry as a live, relevant science.

Presenting subject matter in a clear and readable style and a logical, smoothly flowing sequence helps achieve this goal.

- *A Friendly, Nonintimidating Writing Style.* The discussions in this text take nothing for granted. It is assumed that this is the first contact the students will have had with almost all concepts; thus a careful discussion is required. This edition continues and expands on the use of simple, understandable analogies that relate abstract concepts to concrete models. For example, in Chapter 5 the placement of electrons in orbitals is compared to the placement of students in dorm rooms. When possible, current topics of interest are used as examples. A case in point is found in Chapter 14 where chlorine is used as an example of a catalyst in a reaction that destroys ozone.

- *A Logical Sequence of Topics.* The sequence of chapters in the text allows a logical step-by-step development of the science. There have been some significant modifications in this edition. The new prologue covers scientific method and the study of chemistry. Chapter 1 proceeds directly to the measurements used in chemistry so that Chapters 2, 3, and 4 follow each other smoothly. An introduction to the periodic table and chemical nomenclature has been combined and moved forward to an earlier chapter (Chapter 4). Many have suggested that students benefit from an early introduction to the language of chemistry. After this chapter, two paths of development of chemistry are equally logical. Chapters on electron structure (Chapter 5) and bonding (Chapter 6) come first, followed by a comprehensive review of Chapters 4–6. However, if instructors prefer, they can proceed from Chapter 4 directly to the two quantitative chapters (Chapter 7 on the mole and Chapter 8 on equations and stoichiometry) followed by a comprehensive review of Chapters 4, 7, and 8. There is no mention of the content of Chapters 7 and 8 in Chapters 5 and 6, so either sequence works well. Another change involves moving nuclear chemistry to later in the book: Chapter 15. However, it can still be covered immediately after Chapter 3, if so desired.

- *Smooth Transitions Between Topics.* Sometimes, one hears the complaint that chemistry is difficult to learn because it is a collection of unrelated facts and concepts. In this edition, we have addressed that concern in two ways. First, each chapter begins with an improved introductory section, labeled as **Setting the Stage**, that sets up the purpose of the chapter by reference to current concerns—such as the greenhouse effect (Chapter 6), or the relevance of the supernova of 1987 (Chapter 3), or the nature of lightning (Chapter 13). Within the chapters, we have made a major effort to smooth the flow between topics by connecting each section to the next. A brief paragraph (labeled in the margin as **Looking Ahead**) lets the student know what comes next, why, and how it relates to the topic just covered.

● *The Use of Color*. Chemistry is indeed a colorful science and its use helps bring the subject to life. The colorful elements, minerals, and chemical reactions can now be shown as they actually appear. Relevant chemicals and chemical reactions are emphasized rather than pictures of chemical companies and manufacturing processes. Color is also used extensively in the illustrations, which adds to their appeal.

Goal II. To encourage not only learning but critical thinking.

Some effective ways of learning chemistry are by summarizing and recitation, critical questioning, and working problems.

● *Frequent Summarizing and Recitation*. The learning of chemistry requires reflection. This edition includes short summaries of two or three closely related sections within each chapter (labeled in the margin as **Looking Back**). The object is to encourage the student to pause, reflect, and mentally gather in the main points before proceeding to recitation and problem reinforcement. As in previous editions, the chapter concludes with a review that summarizes the whole chapter and is labeled in the margin as **Putting It Together**. This summary is often presented in a unique way by use of tables, diagrams, or flow charts. Key terms defined in the chapter are now used in the context of a discussion and are shown in bold type. In addition, a complete Glossary of Terms appears in Appendix G for easy reference.

After the paragraph labeled **Looking Back**, a section called *Learning Check* follows—which is referred to in the margin as **Checking It Out**. The first part of the *Learning Check* summarizes recent topics with key words left out. By filling in the blanks, the student aids learning by recitation.

● *Critical Questioning*. The introductory section (**Setting the Stage**) ends with a series of questions that are to be addressed in the chapter that follows. These questions are labeled in the margin as **Formulating Some Questions**. In addition, other questions in the margins encourage critical thinking about the topic being discussed. Not just simple questions—such as "What is Charles's law?" that obviously is answered in the text—these questions are more thought-provoking. For example: "What keeps a hot-air balloon suspended in the air?" and "Why doesn't a water bug sink?" and "How do our bodies keep cool?"

● *Working Problems*. Real chemistry requires problem-solving skills. This text helps develop these skills in four steps.

1. Example problems (labeled **Working It Out**) are carefully worked out in an easy-to-follow manner. Each step in a solution is explained or diagrammed. Most problems have a section labeled *Procedure* in which a strategy is outlined and a *Solution* where the strategy is carried out. Example problems have descriptive headings in this edition.

2. Shortly after presentation of an example problem, the concept is reinforced with similar problems provided in the *Learning Check*. Worked-out answers to these problems in the *Learning Check* are

provided at the end of the chapter for easy reference. A few simple chapter-end problems are referenced in the *Learning Check* for additional practice.

3. Numerous chapter-end problems are listed by topic and range from the simplest to more complex applications. The hardest are indicated by an asterisk. Many new problems that illustrate practical applications of a concept have been added to this edition. About 60 percent of the answers are provided in Appendix H. Many of the quantitative problems also include worked-out solutions.

   In addition to the chapter-end problems that support a specific topic or concept, this edition includes a section of uncategorized problems titled *General Questions* where students must decide for themselves which concepts are being tested. Concepts from previous chapters may also be needed for these more challenging problems. Many of these problems require the application of several concepts.

4. *Comprehensive Review Tests* are available that integrate the content of three relevant chapters. These tests are found after Chapters 3, 6, 8, 11, and 14. They include multiple-choice questions that help test the material learned in recitation.

Goal III. To understand and provide help with the math anxieties that can distract students of chemistry.

The greatest fear of many of the students taking chemistry for the first time is the math involved. This text has been and remains sympathetic and encouraging to those students. It is understood that many, if not most, students require some preparation or review of mathematical concepts used in introductory chemistry. Extensive end-of-book appendixes (A–F) supplement Chapter 1 on measurements. These appendixes provide not only discussion but worked-out examples and sample problems (with answers provided) in the areas of basic arithmetic, algebra, and scientific notation. In addition, there are separate appendixes on logarithms, graphs, and calculators. These also provide sample problems to test one's understanding. Students have used these appendixes extensively, and they have helped reduce what can become an exaggerated fear of the math involved in this discipline. An acknowledgment of this difficulty and the extensive help provided are unique to this text.

Major efforts have been made throughout the book to ensure the needed math is meaningful. For example, the use of percent is illustrated in some problems concerning the composition of alloys in Chapter 2. Conversion factors between the metric and English systems are illustrated early in Chapter 1. In these and other cases the reviews in the appendixes are referred to in the margin.

SUPPLEMENTS

- *Study Guide/Solutions Manual* is available to accompany this text. In the Study Guide, related sections within a chapter that correspond to a *Learning Check* have been grouped for review, discussion, and testing. In this manner, the Study Guide can be put to use before the chapter is completed. Topics of current interest are included in the Study Guide, such as the demise of the dinosaurs, the greenhouse effect, and the reason for the "ozone hole." Selected solutions to text problems are also included in this edition. These solutions are to the quantitative problems that have answers but no solutions in the text.

- *Instructor's Manual,* by Leo J. Malone, St. Louis University, includes chapter objectives, teaching hints, and solutions to all text problems.

- *Transparencies.* 110 full-color figures from text, resized and edited for classroom use.

- *Test Bank,* by Stanley Grenda, University of Nevada, Las Vegas, contains approximately 1,000 test items consisting of multiple-choice, short answer, and critical thinking questions. These questions are also available as a computerized test bank.

- *Experiments in Basic Chemistry,* by Steven Murov and Brian Stedjee, Modesto Junior College, contains 26 experiments that parallel text organization and provides learning objectives, discussion sections outlining each experiment, easy-to-follow procedures, post-lab questions, additional exercises, and answers to pre-lab questions.

- *Instructor's Manual for Experiments in Basic Chemistry,* written by the lab manual authors, this instructor's manual contains answers to post-lab questions, list of chemicals needed, suggestions for other experiments, as well as suggestions for experiment set-ups.

ACKNOWLEDGMENTS

A number of professional colleagues contributed comments and suggestions that helped in the development of the Fourth Edition. I am grateful to the following chemistry instructors who either evaluated the Third Edition or reviewed manuscript for the Fourth Edition:

Barbara C. Andrews	Central Piedmont Community College
Hal Bender	Clackamas Community College
Paul Bowie	College of the Desert
Bill W. Callaway	Rose State College
Jerry A. Driscoll	University of Utah
Stanley Grenda	University of Nevada, Las Vegas
Leslie N. Kinsland	University of Southwestern Louisiana
Steven Murov	Modesto Junior College
Raymond O'Donnell	SUNY–Oswego
Larry Olsen	Arizona State University
James A. Petrich	San Antonio College
Karen Pressprich	Indiana University
Barbara Rainard	Community College of Allegheny County
George Schenk	Wayne State University
Walter Volland	Bellevue Community College (WA)
Trudie Wagner	Vincennes University

Leo J. Malone
St. Louis

CONTENTS

(page 5)

(page 84)

(page 261)

(page 350)

(page 470)

(page 557)

BASIC CONCEPTS OF CHEMISTRY

Fire is powerful and awesome. Of all the animals on Earth, only humans have dared to tame and use this force of nature.

INTRODUCTION TO THE STUDY OF CHEMISTRY

A. The Mystery of Fire

B. The Scientific Method

C. The Study of Chemistry and Using This Textbook

A. THE MYSTERY OF FIRE

Fire—the powerful and mystical force of nature! Out of control, it causes destruction, pain, and even death. For these reasons, we fear it. Under control, however, it heats our homes, cooks our meals, and transforms ores into metals. For these reasons, we welcome it. It is easy to understand how this awesome force was the source of so much mystery and speculation throughout the ages. For example, in Greek mythology there existed a god named Prometheus. He supposedly gave animals the special tools they needed to survive in a hostile world. Humans received the gift of fire. Legend has it that Prometheus went too far because this was a tool reserved for the gods themselves. As might be expected, the other gods proceeded to punish him for his unacceptable generosity.

Evidence indicates that fire has been used by humans for at least four hundred thousand years. It is difficult to imagine how our ancient ancestors could have managed without it. Humans do not have sharp night vision like raccoons, but fire brought light to the long, dark night. We have no protective fur like the deer, but fire lessened the chill of winter. We do not have sharp teeth or powerful jaws like the lion, but fire rendered meat tender. Humans are not as strong or as powerful as the other large animals, but fire repels even the most ferocious of beasts. It seems reasonable to suggest that the taming of fire was one of the most monumental events in the history of the human race. The use of fire made our species dominant over all others. It is truly the tool of the gods.

Let's fast-forward in time to near the end of the Stone Age, about ten thousand years ago. In the Stone Age, weapons and utensils were fashioned from rocks and a few chunks of native copper metal that were found in nature. Copper was superior to stone because it could easily be shaped into fine points and sharp blades by pounding. Unfortunately, native copper was quite rare. But about seven thousand years ago this changed. Anthropologists speculate that some resident of ancient Persia found copper metal in the ashes of a hot charcoal fire. The free copper had not been there before so it must have come from a green stone called malachite (see Figure P-1), which probably lined the fire pit.

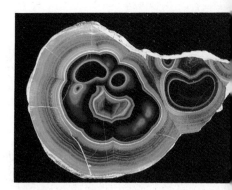

Figure P-1
Malachite. Malachite is a copper ore.

1

Imagine the commotion that this discovery must have caused. A stone could be transformed by hot coals into a valuable metal! Fire was the key that launched the human population into the age of metals. The recovery of metals from their ores is now a branch of chemical science called *metallurgy*. The ancient Persians must have considered this discovery a dramatic example of the magic of fire.

Other civilizations used chemistry in various ways. About 3000 B.C., the Egyptians learned how to dye cloth and embalm their dead through the use of certain chemicals found in nature. They were very good at what they did. In fact, we can still determine from ancient mummies the cause of death and even diseases the person may have had. The Egyptians were good chemists, but they had no idea why any of these procedures worked. Every chemical process they used was discovered by accident.

Around 400 B.C., while the more mystical Greeks were speculating about their various gods, philosophers were trying to understand and describe nature. These great thinkers argued about why things occurred in the world around them, but they were not inclined (or able) to check out their ideas by experimentation or to put them to practical use. At the time, however, people believed that there were four basic elements of nature—earth, air, water, and fire. Of these, fire was obviously the most mysterious. It was the transforming element; that is, it had the capacity to change one substance into another (e.g., certain rocks into metals). We now call such transformations chemical reactions. Fire itself is simply the hot, glowing gases associated with certain chemical reactions. If fire is a result of an ongoing chemical transformation, then it is reasonable to suggest that chemistry and many significant advances in the human race are very much related.

The early centuries of the Middle Ages (A.D. 500–1600) are usually referred to as the Dark Ages in Europe because of the lack of art and literature and the decline of central governments. In fact, the civilizations that Egypt, Greece, and Rome had previously built disappeared. Chemistry, however, began to grow during this period, especially in the area of experimentation. Chemistry was considered a combination of magic and art rather than a science. Many of those who practiced chemistry in Europe were known as *alchemists*. Some of these alchemists were simply con artists who tried to convince greedy kings that they could transform cheaper metals such as lead and zinc into gold. Gold was thought to be the perfect metal. Such a task was impossible, of course, so many of these alchemists met a drastic fate for their lack of success. However, all was not lost. Many important laboratory procedures such as distillation and crystallization were discovered. Alchemists also prepared many previously unknown chemicals, which we now know as elements and compounds.

Modern chemistry has its foundation in the late 1700s when the use of the analytical balance became widespread. Chemistry then became a quantitative science in which theories had to be correlated with the results of direct laboratory experimentation. From these experiments and observations came the modern atomic theory, first proposed by John Dalton around 1803. This theory, in a slightly modified form, is still the basis of our understanding of nature today. Dalton's theory gave chemistry the solid base from which it could serve humanity on an impressive scale. Actually, most of our understanding of chemistry has evolved in the past 100 years. In a way, this makes chemistry a very young science. However, if we mark the beginning of chemistry with the use of fire, it is also the oldest science.

From the ancient Persians five millennia ago, to the Egyptians, to the alchemists of the Middle Ages, various cultures have stumbled on assorted chemical procedures. In many cases, these were used to improve the quality of life. With the exception of the Greek philosophers, there was little attention given to why a certain process worked. The "why" is very important. In fact, the tremendous explosion of scientific knowledge and applications in the past two hundred years can be attributed to how science is now approached. This is called the scientific method, which we will discuss in the next section.

▼ Looking Ahead

B. THE SCIENTIFIC METHOD

The first step in the scientific method involves *making observations and gathering data*. As an example, imagine that we are the first to make a simple observation about nature—"the sun rises in the east and sets in the west." This never seems to vary and, as far as we can tell from history, it has always been so. In other words, our scientific observation is strictly *reproducible*. So now we ask "Why?" We are ready for a hypothesis. A **hypothesis** *is a tentative explanation of observations*. The first plausible hypothesis to explain our observations was advanced by Claudius Ptolemy, a Greek philosopher, in A.D 150. He suggested that the sun, as well as the rest of the universe, revolves around Earth from east to west. That made sense. It certainly explained the observation. In fact this concept became an article of religious faith in much of the western world. However, the hypothesis of Ptolemy did not explain other observations known at the time, which included the movement of the planets across the sky and the phases of the moon.

Sometimes new or contradictory evidence means a hypothesis, just like an old car, must either receive a major overhaul or be discarded entirely. In 1543, a new hypothesis was proposed. Nicolaus Copernicus explained all of the observations about the sun, moon, and planets by suggesting that Earth and the planets revolve around the sun. Even though this hypothesis explained the mysteries of the heavenly bodies, it was considered extremely radical and even heretical at the time. (It was believed that God made Earth the center of the universe.) In 1609, a Venetian scientist by the name of Galileo Galilei turned a telescope that he had built up to the heavens and provided almost unquestionable proof that Copernicus was correct. Galileo is sometimes credited with the beginning of the modern scientific method because he provided direct experimental data in support of a concept. The hypothesis had withstood the challenge of experiments and achieved the status of a theory. A **the-**

ory is a well-established hypothesis. A theory should predict the results of future experiments or observations.

The next part of this story comes in 1684, when an English scientist named Sir Isaac Newton stated a law that governs the motion of planets around the sun. A **law** *is a concise scientific statement of fact to which no exceptions are known*. Newton's law of universal gravitation states that planets remain in stationary orbits around the sun. (See Figure P-2.)

In summary, these were the steps that led to a law of nature.

1   Reproducible observations (the sun rises in the east)

2   A hypothesis advanced by Ptolemy and then a better one by Copernicus

3   Experimental data gathered by Galileo in support of the Copernican hypothesis and eventual acceptance of the hypothesis as a theory

4   The statement by Newton of a universal law based on the theory

Variations on the scientific method serve us well today as we pursue an urgent search for cures of diseases. An example follows.

▶ ──────────────────────────────

The Scientific Method in Action

The healing power of plants and plant extracts has been known for centuries. For example, it was known that an extract of the willow tree relieved pain. (We now know that it contains a drug very closely related to aspirin.) This is the observation that starts us on our journey to new drugs. An obvious hypothesis comes from this observation, namely, that there are many other useful drugs among the plants and soils of the world. We should be able to find them. There are several recent discoveries that support this hypothesis. For example, the rosy periwinkle is a

Figure P-2
The Solar System. The Copernican theory became the basis of a natural law of the universe.

common tropical plant not too different from thousands of other tropical plants except that this one saves lives. The innocent-looking plant contains a powerful chemical called vincristine, which can cure childhood leukemia. Another relatively new drug called taxol has been extracted from the Pacific yew tree. Initial results indicate that taxol can be effective in treating ovarian cancer and possibly breast cancer. In fact, the current top twenty best-selling medicines in the United States originated from plants and other natural sources. These drugs treat conditions such as high blood pressure, cancer, glaucoma, and malaria. The search for effective drugs from natural sources is especially prominent today. One chemical company in the United States randomly tests 3000 plant extracts a year for anti-cancer, anti-arthritic, and anti-AIDS activity. Still, only a small fraction of the 250,000 known species have been tested. Since the greatest variety of plants, molds, and fungi are found in tropical forests, these species are receiving the most attention. The introduction of a new medicine from a plant involves the following steps.

Somewhere in this rain forest, a chemical may exist that cures a dreaded disease.

1 *Collection of materials.* "Chemical prospectors" scour the backwoods of the United States and the tropical forests such as those in Costa Rica collecting and labeling samples of leaves, barks, and roots. Soil samples containing fungi and molds are also collected and carefully labeled.

2 *Testing of activity.* Scientists at several large chemical and pharmaceutical companies make extracts of the sample in the laboratory. These extracts are run through a series of chemical tests to identify if there is any antidisease activity among the chemicals in the extract. Recent advances in these laboratory procedures are astounding and allow large-scale testing that was not possible even a few years ago. If there is antidisease activity, it is considered a "hit" and the extract is taken to the next step.

3 *Isolation and identification of the active ingredient.* The next painstaking task is to separate the one chemical among the "soup" of chemicals present that has the desired activity. Once that's done, the particular structure of the active chemical must be determined. A hypothesis is then advanced about what part of the structure is important and how the chemical works. The hypothesis is tested by attempting to make other more effective drugs (or ones with fewer side effects) based on the chemical's structure. It is then determined whether the chemical or a modified version of it is worth further testing (see Figure P-3).

4 *Testing on animals.* If the chemical is considered promising, it is now ready to be tested on animals. This is usually done in government and university labs under strictly controlled conditions. Scientists study toxicity, side effects, and the chemical's activity against the particular disease for which it is being tested. If, after careful study, the chemical is considered both effective and safe, it is ready for the next step.

5 *Testing on humans.* The final step is the careful testing on humans in a series of clinical trials carefully monitored by the U.S. Food and

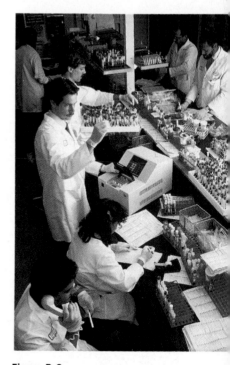

Figure P-3
A Pharmaceutical Laboratory.
Labs like this screen chemicals for drugs that may cure diseases.

Drug Administration. Effectiveness, dosage, and long-term side effects are carefully recorded and evaluated. This process usually takes years, but if all is successful, the drug becomes available to the general public for use.

If a chemical with the desired activity is randomly discovered, only about one in a thousand may actually find its way into general use. Still, the process works. Many chemicals active against cancer and even AIDS (one has been isolated from the mulberry tree) are now in the pipeline for testing. There is some urgency in all of this. Not only are we anxious to cure specific diseases, but the tropical forests that contain the most diverse plants are disappearing at an alarming rate. In any case, nature is certainly our most important chemical laboratory.

At the current time, we have many hypotheses about why specific chemicals are effective in treating a cancer or AIDS. Someday soon we may be able to gather these hypotheses into a theory that will explain the nature of these diseases so that an effective cure will follow.

The scientific method has produced cures for many diseases that have plagued the human race for thousands of years. Eventually it will lead to a cure or vaccine for AIDS. Our modern scientific methods also produce materials for space travel, all sorts of plastics and synthetic fibers, microchips for computers, and processes for genetic engineering. Just a century ago, everything was made out of stone, wood, metal, or natural fibers (wool, cotton, and silk). Our modern society could hardly function on those materials alone.

Looking Ahead ▼

One good thing about the study of chemistry is that, once we establish a law, we can count on it. Laws of nature never change, but our theories sometimes change or are modified. Social scientists have a somewhat tougher task in this regard. It is difficult to make laws about human nature because there is so much variability among people and people are always changing. If we establish that copper is a brownish-red metal, we can be assured that it will always be so, regardless of whether we find the copper in the state of Minnesota or on the planet Mars. That's how sure we are of our facts.

The study of chemistry does require certain skills and study habits. In some ways, it is like the study of math—repetitious problem solving to master a skill. In other ways, it is like the study of a language—basic memorizing and recitation. And it is also like the study of philosophy— the quiet contemplation of concepts. In the next section, we will discuss some specific recommendations about how this science can be mastered and how this text can help.

C. THE STUDY OF CHEMISTRY AND USING THIS TEXTBOOK

Chemistry is *the* fundamental natural science. This is not just an idle boast. Chemistry is concerned with the basic structure and properties of all matter, be it a huge star or a microscopic virus. Biology, physics, geology, as well as all branches of engineering and medicine, are based on an understanding of the chemical substances of which nature is composed. Chemistry is the beginning point in the course of studies that eventually produces all scientists, engineers, and physicians. But it is also important for all responsible citizens. Our environment is very fragile—more fragile than we realized just a few years ago. Many of the chemicals that make life easier also affect our surroundings. Control of air, water, and land pollution needs as much attention from citizens and scientists as the invention of new materials did in the 20th century (see Figure P-4). We all have a big stake in the future, so it is reasonable that chemistry is a prerequisite not only for courses of study but also for life, especially in these complex times. What follows is a brief list of academic self-disciplines and skills needed in the study of chemistry, and how this text can help.

Figure P-4
Air Pollution. Our environment, especially the air we breathe, is very fragile.

1. Time Management

Chemistry is not unlike basketball or piano—it requires lots of practice. It is not only a question of putting in the time but *how* and *when* you put in the time. One does not wait until the night before the big game to first practice jump shots or the night before the recital to first practice the sonata. A master schedule should be prepared, with chemistry (as well as all other subjects) receiving a regularly scheduled study time. One study period should follow as soon as possible after the lecture. In setting up your schedule, it is wise to select shorter but frequent study periods. Two separate one-hour sessions are more effective than one two-hour period. After a length of time, the mind tends to wander into areas not directly related to chemistry (e.g., baseball). Reserve the night before an exam for a comprehensive review and a decent night's sleep. Then is not the time to break new ground.

How This Textbook Can Help

The understanding of chemistry is greatly enhanced by proceeding slowly, mastering one concept at a time. In the text, each chapter is divided into two or three parts followed by a brief summary entitled **Looking Back.** Each of these parts is about right for one or two study sessions. The brief summary is followed by a *Learning Check* section, called **Checking It Out,** where you are asked to fill in the blanks in summarizing statements and do a few problems that reinforce the concepts just covered. To rush through the whole chapter and then work a few problems often leads to frustration. The following are examples of **Looking Back** and **Working It Out** sections from Chapter 3.

Most of our world, from a speck of dust to the mightiest mountain, is composed of just a few of the 88 elements occurring in nature. Elements are made up of extremely tiny particles called atoms. The atoms of one element are alike, but they are different from those of another element. Atoms may bond together to form molecules. Molecules are the fundamental unit of some elements and of a specific type of compound. In molecules of given compound the atoms are arranged in a unique manner.

▲ Looking Back

Learning Check A

◄ Checking It Out

A–1. Fill in the blanks.

The symbol for the element nitrogen is _____ and that of copper is _____. The symbol Ni represents the element _____. The fundamental particle of an element is called an _____. When these fundamenetal particles are covalently bonded to each other, they form _____. The formula of a compound that contains two atoms of carbon, seven atoms of hydrogen, and one atom of nitrogen is written as _____. If two compounds have the same formula, they must have different _____ formulas.

A–2. How many atoms of each element are present in a molecule of acetominophen (a pain reliever), whose formula is $C_8H_9NO_2$?

Additional Examples: Problems 3–5, 3–6, 3–12, 3–15.

Before doing problems in the *Learning Check,* carefully work through the step-by-step examples, which are labeled as **Working It Out**. It is important that you do this with a pencil and not just by reading the example. Write out the example yourself. There is a strong connection between writing and learning in chemistry. Sometimes learning chemistry requires that you put your feet on the desk, lean back, and think. More often it requires that you put your feet on the floor, lean over, and write.

Working It Out●

EXAMPLE 3–1

Calculating the Number of Particles is an Isotope

How many protons, neutrons, and electrons are present in $^{90}_{38}Sr$?

Solution

$$\text{atomic number} = \text{number of protons} = \underline{38}$$
$$\text{number of neutrons} = \text{mass number} - \text{number of protons}$$
$$90 - 38 = \underline{52}$$

In a neutral atom, the number of protons and the number of electrons are the same.

$$\text{number of electrons} = \underline{\underline{38}}$$

EXAMPLE 3–2

Calculating the Atomic Mass of an Element From Percent Distribution of Isotopes

In nature, the element boron occurs as 19.9% ^{10}B and 80.1% ^{11}B. If the isotopic mass of ^{10}B is 10.013 amu and that of ^{11}B is 11.009 amu, what is the atomic mass of boron?

After you have completed the entire chapter, you are ready for a comprehensive summary. The final section of each chapter provides such a review. Don't ignore this section. It should help you put the concepts in the chapter into a logical progression. For that reason, we label the review **Putting It Together**. The important terms defined in the chapter are used in a discussion format in this section but are shown in boldface type. Often we can use tables or flowcharts to summarize the whole chapter so that the review is not just a repetitive rehash. The following is an example of **Putting It Together** from Chapter 3.

C H A P T E R   R E V I E W

▲▼ **Putting It Together**

All of the various forms of nature around us, from the simple elements in the air to the complex compounds of living systems, are composed of only a few basic forms of matter called elements. Each element has a name and a unique one- or two-letter **symbol.**

It has been less than two hundred years since John Dalton's atomic theory introduced the concept that elements are composed of fundamental particles called **atoms.** It has been less than twenty years since we have been able to produce images of these atoms with a microscope. Most elements are present in nature as aggregates of individual atoms. In some elements, however, two or more atoms are combined by **covalent bonds** to produce basic units called **molecules. Molecular compounds** are also composed of molecules although, in this case, the atoms of at least two different elements are involved. Each compound has a unique arranement of atoms in the molecular unit.

There is another type of compound, however. Matter can be positively or negatively charged with **electrostatic forces** between the charges. Atoms can become electrically charged to form **ions.** Groups of atoms that are covalently bonded together can also have charges and are known as **polyatomic ions.** An **ionic compound,** is composed of **cations** (positive ions) and **anions** (negative ions). The **formula** of a molecular compound represents the actual number of atoms of each element present in a molecular unit. The formula of an ionic compound shows the type and number of ions in a **formula unit.** A formula unit which relfects the fact that the positive charge is balanced by the negative charge. The four most common ways that we find atoms in nature is summarized as follows.

2. Attendance

The importance of regular class attendance cannot be overemphasized. Any college teacher knows there is a direct correlation between understanding and attendance, whether it is required or not. This text is a great ally, but what material your instructor covers and in what depth can be discovered only in class. Nor can you sense your instructor's emphasis or benefit from his or her problem-solving hints by reading someone else's notes. The key to what will be on the exams is found in the lecture. *You need to be there.*

How This Textbook Can Help

Before you go to bed the night before a chemistry class, put this book on your dresser so that you will be reminded to go to class. Or perhaps you could put it under your pillow. Whatever it takes, do it.

3. Asking Questions

Few students feel confident enough to ask questions in class. This is natural enough—we all tend to think our questions might be "dumb." Nevertheless, you still need to ask the questions. If you hesitate to ask questions in class or the class is too big, take advantage of help sessions or your instructor's office hours. In many cases, instructors are a significantly underutilized resource during their office hours. You may discover that, in the office, the instructor is really human and helpful. The bottom line is that you owe yourself answers and understanding. Do what you have to do.

How This Textbook Can Help

One of the goals of this text is to promote critical thinking by formulating questions. Each chapter begins with an important section labeled **Setting the Stage.** Read this carefully. Our intent is for you to gain a sense of the relevance of the topic that we are about to discuss. Toward the end of this introduction, we state some questions that pertain to the chapter labeled as **Formulating Some Questions.** Occasionally,

I n late February of 1987, an astronomer on a desolate mountaintop in Chile looked out into the clear night sky through a huge telescope. He planned nothing more than a routine study of the skies but he was in for the surprise of his life. A bright new star had appeared, one that hadn't been there the night before. This wasn't a normal star burning steadily like our sun. It was a star in its death throes undergoing a catastrophic explosion known as a supernova. It was to be the brightest supernova seen on Earth in almost four hundred years. While this dramatic event signaled the end of a star, it also meant the creation of vast quantities of elements. Scientists now believe that heavier elements originate only in these violent infernos in space.

◀ **Setting the Stage**

For billions of years, stars have been churning, boiling, and eventually undergoing catastrophic explosions such as the one seen in 1987. Blasted by the force of exploding suns, new-born elements and some simple compounds of the elements have drifted through cold, dark space as bits of dust and clouds of gases. About four and a half billion years

Formulating ▶ Some Questions

tion and origin of these elements. For example, it is now universally accepted that elements are composed of very small fundamental units called atoms. The existence of the atom brings up many additional questions that we will discuss in this chapter. How did the concept of the atom come to be accepted? If atoms are the fundamental unit of an element, what is the fundamental unit of a compound? What distinguishes one compound from another? Are the atoms of an element exactly the same? How do the atoms of one element differ from those of another? What determines the mass of an atom? Before we address these questions, however, we have more to say about the number of elements, their distribution, and how we name them.

within the chapter, there is a question in the margin. Take at least a few seconds to think about the question, and relate it to what is being discussed. Our goal is for you also to formulate questions. In other words, the author and your instructor hope to pique your curiosity about nature. Once you get started asking questions, you will get answers. Every time we get an answer, however, it just seems to generate more questions. That is the nature of science.

4. Perseverance

Everyone hopes to start off with an A on the first test or quiz. However, few do that, including many who are subsequently very successful in the study of chemistry. If you are disappointed with an exam grade, reanalyze your study habits, make adjustments, and try again. Don't expect better results by doing the same thing. Remember to use your instructor as your primary resource for advice in this matter. Perhaps your problem is not the material but "test anxiety." This is a very real phenomenon and has to be acknowledged. Most colleges have a counseling center that can help you overcome this problem. Sometimes your instructor can give you helpful hints on taking chemistry tests so that you won't have the fear of "freezing up." If you are ultimately not successful in the study of chemistry, at least you can say that you gave it your best try.

How This Textbook Can Help

Your instructor will assign some Exercises provided at the end of each chapter to help check your understanding of basic concepts. Here are some Examples of exercises from Chapter 3.

E X E R C I S E S

NAMES AND SYMBOLS OF THE ELEMENTS

3–1 Write the symbols of the following elements. Try to do this without reference to a table of the elements.
(a) bromine   (d) tin
(b) oxygen   (e) sodium
(c) lead   (f) sulfur

3–2 The following elements all have symbols that symbols are C, Ca, Cd, Ce, Cf, Cl, Cm, Co, Cr, Cs, and Cu. Match the symbol with the element and then check with the table of elements inside the front cover.

3–3 The names of six elements begin with the letter B. What are their names and symbols?

3–4 The names of eight elements begin with the letter S. What are their names and symbols?

3–5 Using the table inside the front cover of the text, write the symbols for the following elements.
(a) barium   (d) platinum
(b) neon   (e) manganese
(c) cesium   (f) tungsten

MOLECULAR COMPOUNDS AND FORMULAS

3–8 Which of the following are formulas of elements rather than compounds?
(a) P_4O_{10} (d) S_8
(b) Br_2 (e) MgO
(c) F_2O (f) P_4

3–9 Name the elements in the previous problem.
(d) $C_9H_8O_4$ (aspirin)
(e) $Al_2(SO_4)_3$
(f) $(NH_4)_2CO_3$

GENERAL QUESTIONS

3–21 Describe the difference between a molecular and an ionic compound. Of the two types of compounds discussed, is a stone more likely to be a molecular or an ionic compound? Is a liquid more likely to be a molecular or an ionic compound?

3–22 Write the symbol of the element, mass number, atomic number, and electrical charge given the following information. Refer to a table of the elements.

Other than knowing the material, one of the best ways to prepare for a test is to practice taking exams. A comprehensive **Review Test** follows every three chapters. The test includes multiple-choice questions as well as problems. A sample from the **Review Test** for Chapter 3 follows.

REVIEW TEST ON CHAPTERS 1-3

MULTIPLE CHOICE

The following multiple-choice questions have one correct answer.

1. Which of the following relates to the precision of a measurement?
   (a) the number of significant figures
   (b) the location of the decimal point

PROBLEMS

Carry out the following calculations.

1. Round off the following numbers to three significant figures.
   (a) 173.8   (b) 0.0023158   (c) 18,420

2. Which of the following is not an exact relationship?
   (a) 1 mile = 5280 ft   (c) 10^3 m = 1 km
   (b) 0.62 mile = 1 km   (d) 4 qt = 1 gal

3. How many significant figures are in the measurment 0.0880 m?
   (a) 5   (b) 4
   (c) 3   (d) 2

Carry out the following calculations. Express your answer to the proper decimal place or number of significant figures.

2. 1.09 cm + 0.078 cm + 16.0021 cm

3. 4.2 in. × 16.33 in.

The answers to all of these questions and the solutions to all of the problems are given in the back of the book (Appendix H). Use these review tests as practice for an exam. Take the test within a specific time allotment, as if it were an exam in class. A Study Guide that provides additional worked-out examples and tests is available with the text.

5. Study Skills

a. *Memorization.*   Most likely, there will be some material that your instructor will ask you to commit to memory. Chemistry has a vocabulary of its own, and, in this respect, it must be approached as a foreign language. Sometimes definitions must be memorized before understanding occurs. For example, a student may approach an instructor with the plea, "I don't understand the concept of oxidation." The instructor may reply, "What is the definition of oxidation?" When the student says that he doesn't know, the instructor may suggest to the student, "First, go memorize the definition of oxidation. Then I bet I can help you understand the concept in a matter of minutes."

How This Textbook Can Help
After first reading through the chapter-end review, go over each term in boldface type and recite in your own words its meaning or definition. If necessary, return to the chapter text or the glossary (Appendix G) for the definition.

b. *Reading Ahead.*   Even in high school, football and basketball opponents are scouted before the game. Every team likes to know what they are up against and what lies ahead. In studying science, the equivalent to scouting is reading ahead. Even if you do not grasp the concepts, reading ahead will give you a feeling for some of the material that you will be discussing in the next lecture or two. If you know something is coming that seems confusing, you will be more alert when the concept is discussed. Reading ahead is also a time saver. When you know that certain definitions or tables are in the book, you can save notetaking time.

How This Textbook Can Help

A complaint commonly heard about the study of chemistry is that it sometimes seems like an accumulation of isolated facts and concepts. Not so! There are logical progressions of chemical information, with each concept building on the previous one. This text has a unique feature that should help you read ahead. You may have already noticed the connectors between sections in the prologue. This also occurs in the text, with a short paragraph that ties two sections together. Consider an example from Chapter 3. The following paragraph (labeled **Looking Ahead**) connects the discussion of two types of chemical substances called compounds.

> *The type of compound that we have discussed is composed of molecules and tends to be the "soft" part of nature—gases, liquids, and solids that melt at low temperatures. What about rocks and the minerals? What are they composed of if not molecules? Another type of compound, called ionic, consists of solids that are hard and have high melting points.*

▼Looking Ahead

c. *Notetaking.* There has been a lot written about notetaking as a necessary skill. A useful suggestion is to leave about one-third of the page as a blank margin when you take notes in class. This can be used for adding any thoughts, notes, or questions later and for summarizing in the review (see Figure P-5). Order on paper is important in an orderly science such as chemistry. Neat and easy-to-read notes, tests, and papers seem to correlate with good performance. It is not a coincidence.

We are now ready to continue the study of chemistry. It is a fascinating subject. We hope you enjoy it. Please let the text and the instructor help you to get as much out of it as you possibly can.

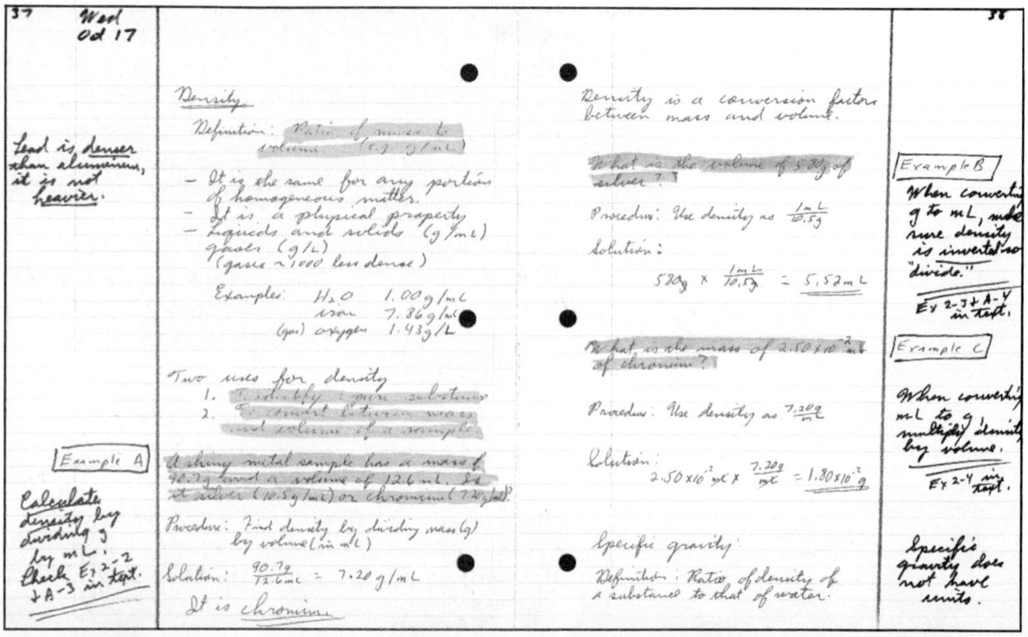

Figure P-5
Chemistry Notes.
Good note taking is essential in the study of chemistry.

To land this large aircraft safely, the pilot must observe many measurements displayed in the cockpit. Chemistry is a quantitative science that also requires careful measurements. The measurements used in Chemistry are the subjects of this chapter.

MEASUREMENTS IN CHEMISTRY

An airline pilot carefully scans the instruments in the cockpit, which report air speed, altitude, current position, and outside temperature. A carpenter plots the dimensions of a roof insuring close fits for the supporting studs. An accountant computes the bottom line. A painter carefully gauges the mix of colors. All of these people are busy making or reading measurements of one sort or another. Hardly any profession can escape the need for measuring, and chemistry is certainly no exception. Chemistry is a science based on the natural laws of matter. Many of these laws originate from reproducible quantitative measurements. Therefore, one of our first major topics in the text concerns these measurements—what they mean, how they are expressed, and how they are manipulated.

◀ **Setting The Stage**

Measurements, of course, imply the use of mathematics. For many this "**M**" word brings on an acute sense of panic. If that is a problem for you, our hope is to significantly reduce such fear in this chapter (with the help of the math review appendixes). If you are not in at least fair physical shape, playing a game of basketball or baseball is not fun—it is exhausting and probably frustrating. Likewise, doing chemistry can be frustrating if you are not in at least fair mathematical shape. Even though only one year of algebra is required for this course, it is likely that most students need a little more mathematical conditioning. For most, a simple review will do, and that is provided in the text. At appropriate points in this chapter, you will be referred to a specific appendix in the back of the book for review. Review appendixes include basic arithmetical operations (Appendix A), basic algebra operations (Appendix B), and scientific notation (Appendix C). Also, Appendix F can aid you in the use of calculators for the mathematical operations found in the text.

Formulating Some Questions ▶

Everyone is busy measuring something, but what exactly is a measurement? *A* **measurement** *determines the quantity, dimensions, or extent of something, usually in comparison to a specific unit. A* **unit** *is a definite quantity adapted as a standard of measurement.* Thus, a measurement (e.g., 1.23 meters) consists of two parts: a numerical quantity (1.23) followed by a specific unit (meters). In the first three sections of this chapter, we will address questions about the numerical quantity. How many numbers should we write? How are numerical values manipulated in mathematical operations? How can very large and very small numbers be conveniently expressed? In the last three sections, we turn our attention to the units used to express various measurements. What units do we use in chemistry to express dimensions, weight, and volume? How can we convert between units of measurement (e.g., miles to kilometers) that are directly proportional? Finally, how do we change temperature measurements from one scale to another?

1-1   THE NUMERICAL VALUE OF A MEASUREMENT

Where do they get those numbers?

Have you ever wondered how we know how many people are present at an outdoor assembly such as a rock concert? Usually, there is someone around who is considered an expert at estimating the size of crowds. Suppose the expert estimates the crowd present at a certain rock concert at 12,000 people. Eight people leave. Should the estimate be changed to 11,992 people? Of course not. The original number was not intended to mean *exactly* 12,000 people. In fact, the three zeros in the number 12,000 were not actually "measured" numbers but simply indicate the magnitude of the number, in other words, to locate the decimal point. In this example, a considerable number of people would have to leave the concert in order for the estimate to be changed to 11,000. In the original number (12,000), only the 1 and 2 are considered significant. *In a measurement, a* **significant figure or digit** *is a digit that is either reliably known or estimated.* For example, in the number 12,000 we can assume the 1 is reliable and reproducible from any number of estimates, but the 2 is uncertain. Other experts may estimate the crowd at 13,000 or 11,000. The zeros are not significant since they actually have no numerical meaning. In our example, there are two significant figures, the 1 and the 2. *The number of significant figures or digits in a measurement is simply the number of measured digits and refers to the precision of the measurement.* **Precision** *relates to the degree of reproducibility of the measurement.* Notice that the original estimate had an uncertainty (reproducibility) of ± 1000.

Now let us have the same crowd at the concert seated in the bleachers of a stadium instead of milling about. In this case, a more precise estimate is possible since the exact capacity of the stadium is known,

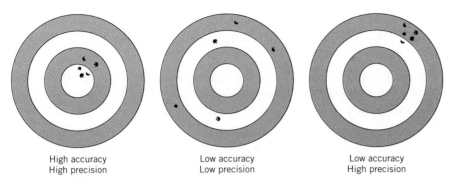

High accuracy
High precision

Low accuracy
Low precision

Low accuracy
High precision

Figure 1–1
Precision and Accuracy

which means that the experts can get a better idea of the number of people present. The crowd is now estimated at between 12,400 and 12,600, or an average of 12,500. This is a measurement with three significant figures. The 1 and 2 are now reliable, but the third significant figure, the 5, is estimated. However, the extra significant figure means that the uncertainty is now reduced to ± 100. The more significant figures in a measurement, the more precise it is. If the crowd went through a turnstile before entering the stadium, an even more precise number could be given. Notice that the significant figure farthest to the right in a measurement is estimated. (As you may guess, people doing chemistry are far more interested in the degree of precision of a measurement than those interested in the size of crowds at rock concerts.)

Participants in the sport of riflery (target shooting) are judged on two points: how close the bullet holes are to each other (the pattern), and how close the pattern is to the center of the target known as the bull's-eye. The *precision* of the contestant's shooting is measured by the tightness of the pattern. How close the pattern is to the bull's-eye is a measure of the shooter's *accuracy*. **Accuracy** *in a measurement refers to how close the measurement is to the true value*. Usually, the more precise the measurement, the more accurate it is—but not always. In our example, if a certain competitor has a faulty sight on the rifle, the shots may be close together (precise) but off-center (inaccurate). (See Figure 1-1.)

Accuracy in measurements depends on how carefully the instrument of measurement has been *calibrated* (compared to the standard). For example, if we attempt to measure length with a ruler that has the first two inches broken off, we obviously will not obtain accurate readings. We would need to recalibrate the ruler so that it starts at zero.

It would be easy to determine the number of significant figures in a measurement if it were not for the number zero. Unfortunately, zero serves two functions: as a reliable or estimated digit, or simply as a filler to locate the decimal point (such as the zeros in the 12,000 people). Since the zeros look alike in both cases, it is important for us to know when a zero is significant or when it is there simply to locate the decimal point. The following rules can be used to tell us about zero.

Is a precise measurement always an accurate measurement?

1 When a zero is between other digits, it is significant (e.g., 709 has three significant figures).

2 Zeros to the right of a digit and to the right of the decimal point are significant (e.g., 8.0 has two significant figures as do 7.9 and 8.1; 5.700 has four significant figures).

How do you evaluate zero as a significant figure?

3 Zeros to the left of the first digit are not significant (e.g., 0.078 and 0.0078 both have two significant figures; 0.04060 has four significant figures).

4 Zeros to the left of an implied decimal point may or may not be significant. In most cases, they are not (e.g., the crowd of 12,000 has two significant figures, and 6600 also has two). What if the zero in a number such as 890 is actually an estimated digit and thus significant? This is a tough question. Some texts use a line over the zero to indicate that it is significant (i.e., 89$\overline{0}$), and others simply place a decimal point after the zero (i.e., 890.). As we will see in the next section, there is a solution to this dilemma. In most problems used in this text, measurements are expressed to three significant figures. Therefore, in calculations where other numbers have three significant figures we will assume that numbers such as 890 also have three significant figures.

EXAMPLE 1–1

Evaluating Zero as a Significant Figure

Working It Out ●

How many significant figures are in the following measurements?
(a) 1508 cm   (b) 300.0 ft   (c) 20.003 lb   (d) 0.00705 gal

Answers
(a) four (rule 1)   (b) four (rule 2) *Note:* Since the zero to the right of the decimal point is significant, the other two zeros are also significant because they lie between significant figures.   (c) five (rule 1)   (d) three (rules 1 and 3)

Looking Ahead ▼

We can determine the degree of precision of a single measurement by counting the number of significant figures. Quite often, however, we need to add, subtract, multiply, or divide two or more measurements with varying degrees of precision. How do we express the result of such calculations? We will examine the guidelines for these operations in the next section.

1-2 SIGNIFICANT FIGURES AND MATHEMATICAL OPERATIONS

How much can you trust a calculator?

The hand calculator is a phenomenal invention. For under 10 dollars we can have a small computer that instantaneously carries out any calculation that we require. Impressive as calculators are, they don't know everything about expressing answers. For example, 7.8 divided by 2.3 reads 3.33913043, etc., on the display. However, if the original numbers were actually measurements with specific units (e.g., 7.8 lb and 2.3 ft), the calculator gives us an answer that is ridiculous—it is much too precise. The calculator assumed that our measurements were more exact than the two significant figures shown. That is, it assumed that each measurement had a string of significant zeros after the 8 and the 3. Since the calculator doesn't give us the answer to the proper number of significant figures, we must know how to do it. There are two sets of rules: one for addition and subtraction and one for multiplication and division. We will discuss the rules for addition and subtraction first.

When numbers are added or subtracted, the answer cannot be expressed to any more decimal places than the fewest present for any one measurement. This rule is illustrated by the following summation:

$$\begin{array}{r} 10.6801 \\ 0.473 \\ \underline{1.32} \\ 12.4731 = 12.47 \end{array}$$

The calculator does not give the answer to the proper number of decimal places.

Note that the "31" in the answer cannot be expressed because those decimal places were not known for the third number, 1.32. It would not be mathematically honest to include any number beyond the 7 since the 7 in the answer is itself estimated. In other words, our degree of uncertainty in this summation is ±0.01, meaning that the true value can be between 12.46 and 12.48.

The rules for rounding off a number are as follows (the examples are all to be rounded off to three significant figures):

1   If the digit to be dropped is less than 5, simply drop that digit (e.g., 12.44 is rounded off to 12.4).

2   If the digit to be dropped is 5 or greater, increase the preceding digit by one (e.g., 0.3568 is rounded off to 0.357, and 13.65 is rounded off to 13.7). Note that the number 12.448 is rounded off to 12.45 if it is to be expressed to four significant figures or two decimal places, and to 12.4 if it is to be expressed to three significant figures or one decimal place.

Recall our original example of a crowd of 12,000 people at a rock concert. Since the three zeros were not significant, subtracting 8 or 80 would not change the number. If 800 were subtracted, however, the estimate does round off to 11,000 as shown below.

$$\begin{array}{r} 12,000 \\ \underline{-8} \\ 11,992 = \underline{12,000} \end{array} \qquad \begin{array}{r} 12,000 \\ \underline{-80} \\ 11,920 = \underline{12,000} \end{array} \qquad \begin{array}{r} 12,000 \\ \underline{-800} \\ 11,200 = \underline{11,000} \end{array}$$

EXAMPLE 1–2

Expressing Summations and Subtractions to the Proper Decimal Place

Carry out the following calculations, rounding off the answer to the proper decimal place.

● **Working It Out**

$$\begin{array}{r} 7.56 \\ +\ 0.375 \\ \underline{+14.2203} \\ 22.1553 = \underline{22.16} \end{array} \qquad \begin{array}{r} 14,000 \\ +\quad 580 \\ \underline{+\quad 75} \\ 14,655 = \underline{15,000} \end{array} \qquad \begin{array}{r} 0.0327 \\ \underline{-0.00068} \\ 0.03202 = \underline{0.0320} \end{array}$$

In multiplication and division, we consider the number of significant figures in the answer rather than the number of decimal places. *The answer is expressed with the same number of significant figures as the multiplier, dividend, or divisor with the least number of significant figures.* In other words, the answer can be only as precise as the least precise part of the calculation (i.e., the chain is only as strong as its weakest link).

Working It Out ●

EXAMPLE 1-3

Expressing Multiplications and Divisions to the Proper Number of Significant Figures

Carry out the following calculations. Assume the numbers represent measurements so that the answer should be rounded off to the proper number of significant figures or decimal place.

(a) 2.34×3.225

The answer on the calculator reads 7.5465. Since the first multiplier has three significant figures and the second has four, the answer should be expressed to three significant figures. The answer is rounded off to 7.55.

(b) $\dfrac{11.688}{4.0}$

The answer shown on the calculator is 2.922 but should be rounded off to two significant figures. The answer is 2.9.

(c) $(0.56 \times 11.73) + 22.34$

In cases where we must use both rules, carry out the exercise in parenthesis first, round off to the proper number of significant figures, and then add and finally round off to the proper decimal place.

$$(0.56 \times 11.73) = 6.6 \quad 6.6 + 22.34 = 28.9$$

The precision of a measurement is indicated by the number of significant figures. Some numbers that we use in calculations are not measurements, however, but are *exact* definitions. For example, there are exactly 12 inches in one foot and four quarts in one gallon. Since these are exact, they can be considered to have unlimited significant figures when used in a calculation. Thus exact definitions are ignored when determining the number of significant figures in an answer. This can be somewhat confusing until we get used to exact numbers, but in the calculations in this chapter, we will remind you when one is being used.

How many significant figures are in a defined relationship?

Looking Ahead ▼

Some aspects of the world of chemistry are either extremely small or unbelievably large. In fact, some of the dimensions of our world require the use of such huge numbers that it is difficult to comprehend the size. One such number is

$$602,000,000,000,000,000,000,000$$

With 21 nonsignificant zeros, this number is very awkward as written. In the next section, we will discuss scientific notation, which is a convenient and compact way to express nonsignificant zeros.

1-3 EXPRESSING LARGE AND SMALL NUMBERS: SCIENTIFIC NOTATION

When numbers require the use of so many nonsignificant zeros that one has to carefully count them, we express the numbers in scientific notation. In **scientific notation,** *a number is expressed with one nonzero digit to the left of the decimal point multiplied by 10 raised to a given power. The* **exponent** *is the appropriate power to which 10 is raised.* Following are some powers of 10 and their equivalent numbers.

Refer to Appendix C for additional help and Appendix F for discussion of calculators.

$$10^0 = 1$$

$$10^1 = 10$$

$$10^2 = 10 \times 10 = 100$$

$$10^3 = 10 \times 10 \times 10 = 1000$$

$$10^4 = 10 \times 10 \times 10 \times 10 = 10,000 \quad \text{etc.}$$

etc.

$$10^{-1} = \frac{1}{10^1} = 0.1$$

$$10^{-2} = \frac{1}{10^2} = 0.01$$

$$10^{-3} = \frac{1}{10^3} = 0.001$$

$$10^{-4} = \frac{1}{10^4} = 0.0001 \quad \text{etc.}$$

The huge number written at the end of the last section is shown in scientific notation as follows.

$$\underline{6.02} \times 10^{23} \quad \swarrow \text{Exponent}$$

Notice that the number (6.02) expresses the proper precision of three significant figures. This expression means that 6.02 is multiplied by 10^{23} (which is 1 followed by 23 zeros). Other sciences also use scientific notation (see Figure 1-2).

In the following exercises, we will give examples of how numbers are expressed in scientific notation and how scientific notation is handled in multiplication and division. These examples can serve as a brief review,

The planet Saturn is 10^{12} meters away at its closest approach to Earth.

2 meters tall

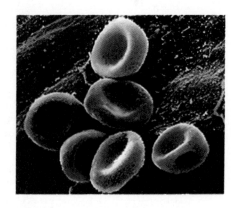

The diameter of one of these cells is about 10^{-6} meters.

Figure 1–2
Powers of Ten In astronomy, extremely large numbers are encountered. In cell biology, very small numbers are used.

but if further practice is needed, see Appendix C for additional discussion on adding, squaring, and taking square roots of numbers in scientific notation. Appendix F includes discussion of manipulation of scientific notation with calculators. Before we consider the examples, we can see how scientific notation can remove the ambiguity of numbers such as 12,000 where the zeros may or may not be significant. Notice that by expressing the number in scientific notation we can make it clear whether one or more of the zeros are actually significant.

*How can you know when a
zero is significant?*

$$1.2 \times 10^4 \quad \text{has two significant figures}$$
$$1.20 \times 10^4 \quad \text{has three significant figures}$$
$$1.200 \times 10^4 \quad \text{has four significant figures}$$

EXAMPLE 1–4

Changing Regular Numbers to Scientific Notation

Express each of the following numbers in scientific notation (one digit to the left of the decimal point).
(a) 47,500 (b) 5,030,000 (c) 0.0023 (d) 0.0000470

Solution

(a) The number 47,500 can be factored as $4.75 \times 10,000$. Since $10,000 = 10^4$, the number can be expressed as

$$\underline{4.75 \times 10^4}$$

A more practical way to transform this number to scientific notation is to count to the left from the old implied decimal point to where you wish to put the new decimal point. The number of places counted to the *left* will be the positive exponent of 10.

$$\overset{4 \quad 3 \quad 2 \quad 1}{4 \; 7 \; 5 \; 0 \; 0} = 4.75 \times 10^4$$

(b)
$$\overset{6 \quad 5 \quad 4 \quad 3 \quad 2 \quad 1}{5, \; 0 \; 3 \; 0, \; 0 \; 0 \; 0} = \underline{5.03 \times 10^6}$$

(c) 0.0023 can be factored into 2.3×0.001. Since $0.001 = 10^{-3}$, the number can be expressed as
$$\underline{2.3 \times 10^{-3}}$$

A more practical way is to count to the right from the old decimal point to where you wish to put the new decimal point. The number of places counted to the *right* will be the negative exponent of 10.

$$\overset{1 \quad 2 \quad 3}{0. \; 0 \; 0 \; 2 \; 3} = 2.3 \times 10^{-3}$$

(d)
$$\overset{1 \quad 2 \quad 3 \quad 4 \quad 5}{0. \; 0 \; 0 \; 0 \; 0 \; 4 \; 7 \; 0} = \underline{4.70 \times 10^{-5}}$$

EXAMPLE 1-5

Multiplication and Division of Numbers Expressed in Scientific Notation

Carry out the following operations. Express the answer to the proper number of significant figures.
(a) $(8.25 \times 10^{-5}) \times (5.422 \times 10^{-3})$
(b) $(4.68 \times 10^{16}) \div (9.1 \times 10^{-5})$

(a) Procedure

Group the digits and the powers of 10. Multiply the digits and add the powers of 10.

Solution

$(8.25 \times 5.442) \times (10^{-5} \times 10^{-3}) = 44.9 \times 10^{-8}$
$$= \underline{4.49 \times 10^{-7}} \text{ (three significant figures)}$$

(b) Procedure

Group the digits and the powers of 10. Divide the exponents and subtract the powers of 10.

Solution

$$\frac{4.68 \times 10^{16}}{9.1 \times 10^{-5}} = \frac{4.68}{9.1} \times \frac{10^{16}}{10^{-5}}$$
$$= 0.51 \times 10^{-16-(-5)}$$
$$= 0.51 \times 10^{21}$$
$$= \underline{5.1 \times 10^{20}} \text{ (two significant figures)}$$

Good science is honest and reliable. To be reliable, the numerical part of a measurement must be expressed to the proper precision, which is shown by the correct number of significant figures. One problem is that the role of zero can be easily misinterpreted. Thus we must take special care in evaluating this number as a significant figure. Calculators rarely give us answers that are expressed to the proper precision, so we need to express the result of mathematical operations very carefully according to the rules of addition or multiplication. Finally, we have prepared ourselves for the convenient use of very large or small numbers with scientific notation. We also need to skillfully manipulate these numbers in mathematical operations. These mathematical tools are as fundamental to science as a hammer is to a carpenter.

▲Looking Back

◀Checking It Out

Learning Check A

A-1. Fill in the blanks.
In a measurement, a significant figure is a number that is either known or _____ . The number of significant figures in a measurement relates to its _____ . How close the number is to the true value is referred to as the _____ of the measurement. In a measurement, zero is a significant figure when it is _____ other digits or to the _____ of a digit and to the _____ of the

decimal point. In addition and subtraction of numbers with various degrees of precision, the answer is expressed to the same decimal place as the number with the _____ number of _____ _____. In multiplication and division, the answer is expressed with the same number of significant figures as the number with the _____ number of _____ _____. Awkwardly large or small values are expressed in _____ _____. These numbers consist of a number with _____ digit to the left of the decimal multiplied by 10 raised to a power. The power of 10 is known as an _____ .

A-2.   How many significant figures are in the following measurements?
(a) 2.33 ft   (b) 40.01 lb   (c) 2.30 L   (d) 10,200 km
(e) 0.020 g

A-3.   Round off the following numbers to three significant figures:
(a) 23.44   (b) 483,550   (c) 0.02203

A-4.   Carry out the following operations and express the answer properly:
(a) 19.63 + 0.366   (b) 0.200 × 12.765
(c) (12.45 − 11.65) × 2.68

A-5.   Express the following numbers in scientific notation:
(a) 456,000,000 to four significant figures   (b) 0.000340

A-6.   Carry out the following operations and express the answer properly:
(a) $(9.41 \times 10^{12}) \times (2.7722 \times 10^{-5})$ (b) $\dfrac{4.856 \times 10^{10})}{(0.020 \times 10^{4})}$

Additional examples: Problems 1-4, 1-6, 1-10, 1-14, 1-18, and 1-28.

Looking Ahead ▼

Having examined the numerical part of a measurement, we are now prepared to look at the other part—the units that are used. A unit is a determined quantity compared to a standard. For example, in the United States, we use the measurement of length called the **foot.** *The standard for this unit was originally the length of a certain English king's foot. Fortunately, we have more reliable and precise standards today than some dead king's body part. In the next section we will examine some of the units used in chemistry.*

1-4   MEASUREMENT OF MASS, LENGTH, AND VOLUME

In the United States, a trophy bass out of a Midwestern lake tips the scale at 9 lb 6 oz. A highly recruited basketball player for a men's college team tops out at 6 ft 11 in. We may need to add one quart and one pint of oil to our car. All of these units were inherited from England when this country was founded. A major problem with these units is that they lack any systematic relationship between units (see Table 1-1). Thus we

TABLE 1-1 RELATIONSHIPS AMONG ENGLISH UNITS

Length	Volume
12 inches (in.) = 1 foot (ft)	2 pints (pt) = 1 quart (qt)
3 ft = 1 yard (yd)	4 qt = 1 gallon (gal)
1760 yd = 1 mile (mi)	42 gal = 1 barrel (bbl)

Mass
16 ounces (oz) = 1 pound (lb)
2000 lb = 1 ton

often need to use two units (e.g., feet and inches) to report only one measurement. Our monetary system is an exception because it is based on the decimal system. Therefore, only one unit (dollars and decimal fractions of dollars) is needed to show a typical student's dismal financial condition (e.g., $11.98). The old English system required two units (pounds and shillings).

Most of the world and the sciences use the metric system of measurement for length, volume, and mass. In this system, units are conveniently related by multiples of 10. Since 1975, there have been plans to convert to the metric system in the United States, but action has been implemented in fits and starts. Complete conversion may not occur in this lifetime, but most citizens are becoming more familar with this system. (See Figure 1-3.) Metric units also form the basis of the SI system, after the French *Systeme International* (International System). The SI units that will concern us are listed in Table 1-2. There are other SI units that designate measurements that are not used in this text. The basic units of the SI system have very precisely defined standards based on certain precisely known properties of matter and light. For example, the unit of one meter is defined as the distance light travels in a vacuum

What is the advantage of the metric system?

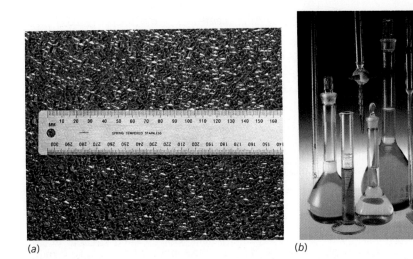

(a)   (b)   (c)

Figure 1–3
Length, Volume, and Mass These properties of a quantity of matter are measured with common laboratory equipment. (a) metric ruler. (b) volume: a graduated cylinder, burettes, volumetric flasks, and burette. (c) an electric balance.

TABLE 1-2 SI UNITS		
Measurement	Unit	Symbol
Mass	kilogram	kg
Length	meter	m
Time	second	s
Temperature	kelvin	K
Quantity	mole	mol
Energy	joule	J
Pressure	pascal	Pa

in 1/299,792,458th of a second. Obviously, this is extremely precise, but when we are aiming a space ship at a planet billions of miles away (e.g., Neptune), we need a great deal of precision in our units. Although some English units were formerly based on a king's anatomy, in modern times many have been redefined more precisely based on a corresponding metric unit. For example, one inch is now defined as exactly equal to 2.54 centimeters (cm).

Use of metric units is simplified by their exact relationships by powers of 10. This is illustrated in Table 1-4 by use of the more common prefixes listed in Table 1-3.

TABLE 1-3 PREFIXES USED IN THE METRIC SYSTEM					
Prefix	Symbol	Relation to Basic Unit	Prefix	Symbol	Relation to Basic Unit
tera-	T	10^{12}	deci-	d	10^{-1}
giga-	G	10^{9}	centi-	c	10^{-2}
mega-	M	10^{6}	milli-	m	10^{-3}
kilo-	k	10^{3}	micro-	μ*	10^{-6}
hecto-	h	10^{2}	nano-	n	10^{-9}
deca-	da	10^{1}	pico-	p	10^{-12}

P.U. A Greek letter, mu.

TABLE 1-4 RELATIONSHIPS AMONG METRIC UNITS USING COMMON PREFIXES								
Mass Unit	Symbol	Relation to Basic Unit	Volume Unit	Symbol	Relation to Basic Unit	Length Unit	Symbol	Relation to Basic Unit
kilogram	kg	10^{3} g	kiloliter	kL	10^{3} L	kilometer	km	10^{3} m
decigram	dg	10^{-1} g	deciliter	dL	10^{-1} L	decimeter	dm	10^{-1} m
centigram	cg	10^{-2} g	centiliter	cL	10^{-2} L	centimeter	cm	10^{-2} m
milligram	mg	10^{-3} g	milliliter	mL	10^{-3} L	millimeter	mm	10^{-3} m
microgram	μg	10^{-6} g				nanometer	nm	10^{-9} m

There is one other convenient feature of the metric system. There is an exact relationship between length and volume. **Volume** *is the space that a given quantity of matter occupies*. The SI unit for volume is the cubic meter (m^3). Since this is a rather large volume for typical laboratory situations, the metric unit, known as the liter, is used. One liter is defined as the exact volume of one cubic decimeter (i.e., 1 L = 1 dm^3). On a smaller scale, one milliliter is the exact volume of one cubic centimeter (1 mL = 1 cm^3 = 1 cc). Thus the units milliliter and cubic centimeter can be used interchangeably when expressing volume (see Figure 1-4).

The basic metric unit of mass is the *gram*, but the SI unit is the *kilogram* (kg) which is equal to 1000 grams (g). **Mass** *is the quantity of matter that a sample contains*. The terms "mass" and "weight" are often used interchangeably, but they actually refer to different concepts. **Weight** *is a measure of the attraction of gravity for the sample*. An astronaut has the same mass on the moon as on Earth. Mass is the same anywhere in the universe. An astronaut who weighs 170 lb on Earth, however, weighs only about 29 lb on the moon. In Earth orbit, where the effect of gravity is counteracted, the astronaut is "weightless" and floats free. You can lose all your weight by going into orbit, but obviously your body (mass) is still there. However, since our relevant universe is confined mostly to the surface of the Earth, we often use weight as a measure of mass. In this text, we will use the term "mass," as it is the more scientific term.

Does weight mean the same as mass?

Several relationships between the metric and English systems are listed in Table 1-5. (See also Figures 1-5 and 1-6.) The relationships *within* systems (e.g., Tables 1-1 and 1-4) are always defined and exact

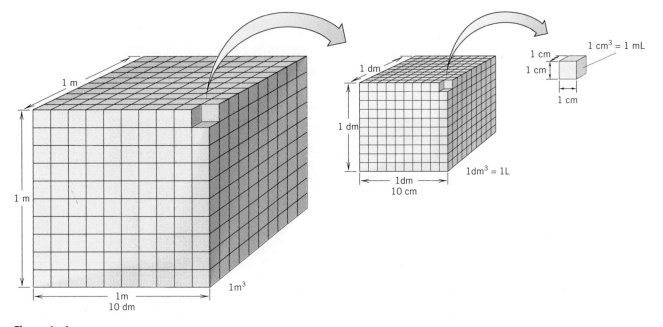

Figure 1–4
Volume and Length. The SI unit for volume is 1 m^3. More practical units for the laboratory are 1 dm^3 (1 L) and 1 cm^3 (1 mL).

**TABLE 1-5 THE RELATIONSHIP BETWEEN
ENGLISH AND METRIC UNITS**

	English	Metric Equivalent
Length	1.00 in.	2.54 cm (exact)
	1.00 mi	1.61 km
Mass	1.00 lb	454 g
	2.20 lb	1.00 kg
Volume	1.06 qt	1.00 L
	1.00 gal	3.78 L
Time	1.00 s	1.00 s

Figure 1–5
Metric Units The use of metric units in this country is becoming increasingly evident.

Figure 1–6
Comparison of Metric and English Units (*a*) 1 meter and 1 yard, (*b*) 1 quart and 1 liter and (*c*) 1 kilogram and 1 pound.

(a)

(b)   (c)

The astronaut is weightless in earth orbit but still has mass.

numbers. The relationships *between* systems, however, are not necessarily exact, so they can be expressed to various degrees of precision. Most of the relationships shown are known to many more than the three significant figures given.

At the present time, most natives of the United States are more comfortable with the familiar English system than they are with the metric system. However, it is important to relate the two systems of measurement so as to have a feeling for the units used in chemistry. Natives of other countries studying in the United States are faced with the opposite chore—relating the more familiar metric to the English system. In the next section we will be converting from one system of measurement to another. Since most of our problems can be considered conversion problems, this is a convenient time to introduce an effective problem-solving tool. That is the subject of the next section.

▼**Looking Ahead**

1-5 CONVERSION OF UNITS BY THE FACTOR-LABEL METHOD

When using certain electrical appliances such as an electric razor, the alternating current (AC) available through the outlet cannot be used. An electric razor requires direct current (DC). The large plug inserted in the outlet is actually an *electrical converter* that changes the current from what we are given (AC) to what is required (DC). Most of the calculations that we do in chemistry require a similar device. In this case, however, we use a *mathematical converter* that changes a unit of measurement that is given to one that is required. These mathematical converters are known as conversion factors. *A* **conversion factor** *is a relationship between two units or quantities expressed in fractional form.* *The* **factor-label method** *(also called* **dimensional analysis**) *converts from one unit to another by the use of conversion factors.* For example, a conversion factor can be constructed from the exact relationship within the metric system.

$$1000 \text{ m} = 1 \text{ km}$$

This can be expressed in either of two fractional forms.

$$\textbf{(1)} \ \frac{1 \text{ km}}{1000 \text{ m}} \text{ and } \textbf{(2)} \ \frac{1000 \text{ m}}{1 \text{ km}} \left(\text{or simply} \ \frac{1000 \text{ m}}{\text{km}} \right)$$

These fractional relationships read as "one kilometer per 1000 meters" and "1000 meters per one kilometer." The latter fraction is usually simplified to "1000 meters per kilometer." When just a unit (no number) is read or written in the denominator, it is assumed to be 1 of that unit. *Factors that relate a quantity in a certain unit to 1 of another unit are sometimes referred to as* **unit factors.** Unit factors are written without the 1 in the denominator in this text. It should be understood, however, that the 1 in the numerator or implied in the denominator can be considered exact in calculations.

Two other conversion factors can be constructed from an equality between English and metric units. For example, the exact relationship between inches (in.) and centimeters (cm) is

$$1.00 \text{ in.} = 2.54 \text{ cm}$$

which can be expressed in fractional form as

$$\textbf{(3)} \ \frac{1 \text{ in.}}{2.54 \text{ cm}} \text{ and } \textbf{(4)} \ \frac{2.54 \text{ cm}}{\text{in.}}$$

Each of the four fractions can be used to convert one unit into the other. The general procedure for a one-step conversion by the factor-label method is

(what's given) × (conversion factor) = (what's requested)

In most conversions, "what's given" and "what's requested" both have one unit. The conversion factor has two: the requested unit in the numerator and the given unit in the denominator.

How do we get from here to there in a problem?

$$\text{A (given unit)} \times \frac{\text{B (requested unit)}}{\text{C (given unit)}} = \text{D (requested unit)}$$

When combined in the calculation, the given units will appear in both the numerator and the denominator and thus cancel just like any numerical quantity. The requested unit survives. This is illustrated as follows, where A, B, C, and D represent the numerical parts of the measurements or definitions.

$$\frac{\text{A (} \cancel{\text{given unit}} \text{)} \times \text{B (requested unit)}}{\text{C (} \cancel{\text{given unit}} \text{)}} = \text{D (requested unit)}$$

The key to using the factor-label method is to select the right conversion factor to do the job. A one-step conversion can be carried out whenever we have a direct relationship between the given and requested units. Notice that the units are treated just like the numerical quantities; that is, they are multiplied, divided, or canceled in the course of the calculation. This method is a very orderly and logical approach to problem solving. Most of the problems in the text are discussed and illustrated by the factor-label method, so it is well worth the investment in time to adjust to this method if it is not already familiar to you.

We will now illustrate some examples of how the factor-label method is used in some one-step conversions. The first problem (Example 1-6) will put to use one of the two conversion factors that we constructed within the metric system, and the next (Example 1-7) will use one of the two conversion factors between the English and metric systems. In the examples, we proceed carefully through three steps.

1 From what is given and what is requested, decide what the conversion factor must do (e.g., convert in. to cm). You can express the procedure in shorthand (e.g., in→cm).

2 Find the proper relationship between units from a table (if you don't already know it). Express the relationship as a conversion factor so that the given unit is in the denominator and the requested or new unit is in the numerator.

How do we organize conversion problems?

3 Put the problem together. Make sure the proper units cancel and your answer has the unit or units requested. Carefully do the numerical part of the problem, making sure to express your answer to the proper number of significant figures.

EXAMPLE 1-6

Converting Meters to Kilometers

●**Working It Out**

Convert 0.468 m to (a) km and (b) mm.

(a) Procedure

1 In shorthand, the conversion is

$$m \longrightarrow km$$

2 From Table 1-4, we find the proper relationship is 10^3 m = 1 km. When expressed as the proper conversion factor, *km* is in the numerator (requested) and *m* is in the denominator (given). The conversion factor is

$$\frac{1 \text{ km}}{10^3 \text{ m}}$$

Solution

$$0.468 \text{ m} \times \frac{1 \text{ km}}{10^3 \text{ m}} = 0.468 \times 10^{-3} \text{ km} = \underline{\underline{4.68 \times 10^{-4} \text{ km}}}$$

(b) *Procedure*

1 In shorthand, the conversion is

$$m \longrightarrow mm$$

2 From Table 1-4, we find the proper relationship is 1 mm = 10^{-3} m. When shown as the proper conversion factor, *mm* is in the numerator (requested) and *m* is in the denominator (given). The conversion factor is

$$\frac{1 \text{ mm}}{10^{-3} \text{ mm}}$$

Solution

$$0.468 \text{ m} \times \frac{1 \text{ mm}}{10^{-3} \text{ m}} = 0.468 \times 10^3 \text{ mm} = \underline{\underline{468 \text{ mm}}}$$

EXAMPLE 1-7

Converting Centimeters to Inches

Convert 825 cm to in.

Procedure

1 In shorthand, the conversion is

$$cm \rightarrow in.$$

2 From Table 1-5, we find the proper relationship is 1.00 in. = 2.54 cm. Expressed as a conversion factor, *in.* is in the numerator (requested) and *cm* is in the denominator (given). The conversion factor is

$$\frac{1 \text{ in.}}{2.54 \text{ cm}}$$

Solution

$$825 \text{ cm} \times \frac{1 \text{ in.}}{2.54 \text{ cm}} = \underline{\underline{325 \text{ in.}}}$$

EXAMPLE 1-8

Converting Quarts to Liters

Convert 6.85 qt to L.

Procedure

1 In shorthand, the conversion is

$$qt \rightarrow L$$

2 From Table 1-5, we find the proper relationship is 1 L = 1.06 qt. Expressed as a conversion factor, *L* is in the numerator (requested) and *qt* is in the denominator (given). The conversion factor is

$$\frac{1 \text{ L}}{1.06 \text{ qt}}$$

Solution

$$6.85 \; \cancel{qt} \times \frac{1 \; L}{1.06 \; \cancel{qt}} = \underline{\underline{6.46 \; L}}$$

What would have happened if we had inverted the conversion factor? In that case, the units would have served as a red flag indicating that a mistake had been made. The units of the answer would have been qt^2/L, which is obviously not correct.

$$6.85 \; qt \times \frac{1.06 \; qt}{1 \; L} = 7.26 \; \frac{qt^2}{L} \; ??????$$

Let's return to the analogy of the electrical converter for the electric razor. Suppose we still need DC current from an AC outlet, but we do not have an AC-DC converter. What we do have, however, is an AC-BC converter and a BC-DC converter. Obviously, we can still get the job done using both converters, but the conversion now requires two steps (i.e., AC → BC → DC). Likewise, in many of our chemistry conversion problems a single conversion factor is not available, and we then have to use two or more. The one-step conversions that have been worked so far are also analogous to a direct, nonstop airline flight between your home city and your destination. Multistep conversions are analogous to the situation in which a nonstop flight is not available and you have to make the flight in several steps before reaching your destination. Each step in the flight is a separate journey, but it gets you closer to your ultimate destination. For multistep conversion problems, you must plan a step-by-step path from your origin (what's given) to your destination (what's requested). Each step along the path toward the answer requires a separate conversion factor.

For example, let us consider a problem requiring a conversion between two hypothetical units of quantity—number of apples and boxes of apples. Let us assume we know that exactly six apples can be placed in each sack and exactly four sacks make up one box. The problem is, "How many boxes of apples can be prepared with 254 apples?" Since we don't have a direct relationship between apples and boxes we need a "game plan." First, we can convert number of apples to number of sacks (conversion **1**) and then number of sacks to number of boxes (conversion **2**). The game plan and problem is set up and completed as follows:

$$
\begin{array}{ccc}
(\mathbf{1}) & (\mathbf{2}) \\
\text{apples} \rightarrow \text{sacks} \rightarrow \text{boxes}
\end{array}
$$

$$254 \; \cancel{\text{apples}} \times \overset{(\mathbf{1})}{\frac{1 \; \text{sack}}{6 \; \cancel{\text{apples}}}} \times \overset{(\mathbf{2})}{\frac{1 \; \text{box}}{4 \; \cancel{\text{sacks}}}} = \frac{254}{6 \times 4} \; \text{boxes} = 10.6 \; \text{boxes}$$

Notice that the two conversion factors represent exact relationships, so they do not affect the number of significant figures in the answer.

We are now ready to work through some real multistep conversions between units of measurement.

EXAMPLE 1-9

Converting Milligrams to Kilograms

Convert 4978 mg to kg.

Procedure

1   A conversion between mg and kg is not directly available. It will be necessary to make a two-step conversion by (a) converting *mg* to *g* and then (b) *g* to *kg*. In shorthand, the conversion is shown as

$$\text{(a) \quad (b)}$$
$$mg \rightarrow g \rightarrow kg$$

2   From Table 1-4, we find the conversion factors needed are 1 mg = 10^{-3} g and 1 kg = 10^3 g. In conversion (a), *mg* is converted to *g* so we need *g* in the numerator (requested) and *mg* in the denominator (given). In conversion (b), *g* is then converted to *kg* so we need *kg* in the numerator (requested) and *g* in the denominator (given). The two conversion factors are

$$\text{(a) } \frac{10^{-3}\,g}{mg} \text{ and (b) } \frac{1\,kg}{10^3\,g}$$

Solution

$$4978\,\text{mg} \times \frac{10^{-3}\,g}{mg} \times \frac{1\,kg}{10^3\,g} = 4978 \times 10^{-6}\,kg = \underline{\underline{4.978 \times 10^{-3}\,kg}}$$

EXAMPLE 1-10

Converting Liters to Gallons

Convert 9.85 L to gal.

Procedure

1   A two-step conversion is needed. *L* can be converted into *qt* (a) and *qt* then converted into *gal* (b). In shorthand, the conversions are expressed as

$$\text{(a) \qquad (b)}$$
$$L \longrightarrow qt \longrightarrow gal$$

2   The two relationships needed are 1.06 qt = 1.00 L (from Table 1-5) for conversion (a), and 4 qt = 1 gal (from Table 1-1) for conversion (b). The two relationships properly expressed as conversion factors are

$$\text{(a) } \frac{1.06\,qt}{L} \text{ and (b) } \frac{1\,gal}{4\,qt}$$

Solution

$$\text{(a) \qquad (b)}$$
$$9.85\,L \times \frac{1.06\,qt}{L} \times \frac{1\,gal^*}{4\,qt} = \underline{\underline{2.61\,gal}}$$

*An exact number does not limit the number of significant figures in the answer.

EXAMPLE 1-11

Converting Miles Per Hour to Meters Per Minute

Convert 55 mi/hr to m/min.

Procedure

1 In this case, we can convert *mi/hr* to *km/hr* (a) and then *km/hr* to *m/hr* (b). It is then necessary to change the units of the denominator from *hr* to *min* (c). In shorthand, the conversions are expressed as

$$\underset{(a)}{\frac{mi}{hr}} \longrightarrow \underset{(b)}{\frac{km}{hr}} \longrightarrow \underset{(c)}{\frac{m}{hr}} \longrightarrow \frac{m}{min}$$

2 The needed relationships are 1.00 mi = 1.61 km (from Table 1-5), 10^3 m = 1 km (from Table 1-4), and 60 min = 1 hr. The relationships expressed as conversion factors are

$$(a)\ \frac{1.61\ km}{mi} \quad (b)\ \frac{10^3\ m}{km} \quad (c)\ \frac{1\ hr}{60\ min}$$

Notice that factor (c) has the requested unit (*min*) in the denominator and the given unit (*hr*) in the numerator. This is because we are changing a denominator in the original unit.

Solution

$$55\ \frac{\cancel{mi}}{\cancel{hr}} \times \underset{(a)}{\frac{1.61\ \cancel{km}}{\cancel{mi}}} \times \underset{(b)}{\frac{10^3\ m}{\cancel{km}}} \times \underset{(c)}{\frac{1\ \cancel{hr}}{60\ min}} = \underline{\underline{1.5 \times 10^3\ m/min}}$$

The SI system of measurements, and the metric units from which they are derived, presents a large variety of units that are based on precisely defined standards. These serve as the universally accepted units of science. In fact, some of the English system of units are now defined by reference to their metric equivalents. Conversion among units can be accomplished using the various relationships among the units. This is most conveniently done using the factor-label method. In this method, relationships between units provide the conversion factors that allow us to change units of a measurement.

▲ **Looking Back**

Learning Check B

◀ **Checking It Out**

B-1 Fill in the blanks.

The basic metric unit of length is the _____, of volume is the _____, and of mass is the _____. Other units relate to a basic unit by powers of 10. For example, a quantity of 10 grams is the same as _____ mg and _____ kg. One cubic centimeter in the metric system is the volume of _____ _____. Changing units by the factor-label method requires the use of _____ factors that are constructed from relationships between units. In a typical problem, the conversion factor has the unit of what's given in the _____ and the units of what's requested in the _____ .

B-2   Write a relationship in factor form from Tables 1-1, 1-3, and 1-5 that would make the following conversions:
(a) feet to yards
(b) decimeters to meters
(c) gallons to liters
(d) pounds to kilograms

B-3   How many miles are in 25.0 kilometers?

B-4   If the unit price of chili is $0.180/oz, what is the cost of a 12-oz can of chili? How many ounces can you buy at this rate for $1.75?

B-5   Convert 12.0 feet to centimeters.

B-6   A certain size of nail costs $1.25/lb. What is the cost of 3.25 kg of these nails?

Additional examples: Problems 1-36, 1-38, 1-43, 1-44, 1-45, 1-53, 1-54.

Looking Ahead ▼

In the final part of this chapter, we examine one additional measurement, which is temperature. U.S. natives are again at a disadvantage because we use a temperature scale different from that of science and most of the world. However, we can use our conversion-problem skills to relate the two scales.

1-6   MEASUREMENT OF TEMPERATURE

What is "hot"?

When someone says an item is "hot," we can interpret this in several ways. It could mean that this item is something we should have, that it looks great, or that it has a somewhat higher temperature than "warm." In this section, we will be concerned with the latter concept. "Hot" in science refers to temperature. **Temperature** _is a measure of the intensity of heat of a substance._ A **thermometer** _is a device that measures temperature._ The thermometer scale with which we are most familiar in the United States is the **Fahrenheit** scale (°F), but the **Celsius** scale (°C) is commonly used elsewhere and in science. Many U.S. television news shows once broadcast the temperature in both scales, but, unfortunately, this practice is losing popularity.

Thermometer scales are established by reference to the freezing point and boiling point of pure water. These two temperatures are constant and unchanging (under constant air pressure). Also, when pure water is freezing or boiling, the temperature remains constant. We can take advantage of these facts to compare the two temperature scales and establish a relationship between them. In Figure 1-7, the temperature of an ice and water mixture is shown to be exactly 0°C. This temperature was originally established by definition and corresponds to ex-

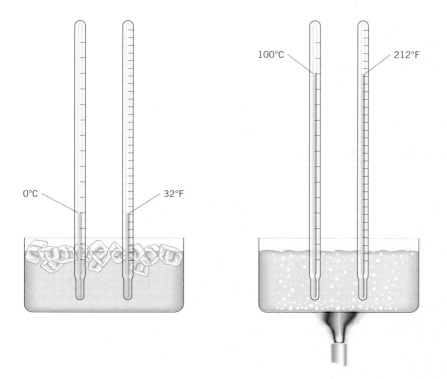

Figure 1–7
THE TEMPERATURE SCALES. The freezing and boiling points of water are used to calibrate the temperature scales.

actly 32°F on the Fahrenheit thermometer. The boiling point of pure water is exactly 100°C, which corresponds to 212°F.

On the Celsius scale, there are 100 equal divisions between these two temperatures, whereas on the Fahrenheit scale, there are 212 − 32 = 180 equal divisions between the two temperatures. Thus we have the following relationship between the scale divisions:

$$100 \text{ C div.} = 180 \text{ F div.}$$

This relationship can be used to construct *exact* conversion factors between an equivalent number of Celsius and Fahrenheit degrees.

$$\frac{100 \text{ C div.}}{180 \text{ F div.}} = \frac{1 \text{ C div.}}{1.80 \text{ F div.}} \text{ and } \frac{1.80 \text{ F div.}}{\text{C div.}}$$

To convert the Celsius temperature [$t(\text{C})$] to Fahrenheit temperature [$t(\text{F})$],

1 Multiply the Celsius temperature by the proper conversion factor (1.8°F/1°C) to convert the equivalent number of Fahrenheit degrees.

2 Add 32°F to this number so that both scales start at the same point (the freezing point of water).

$$t(\text{F}) = \left[t(\text{C}) \times \frac{1.8°\text{F}}{1°\text{C}} \right] + 32°\text{F} = [t(\text{C}) \times 1.8] + 32$$

To convert the Fahrenheit temperature to the Celsius temperature,

1   Subtract 32°F from the Fahrenheit temperature so that both scales start at the same point.

2   Multiply the number by the proper conversion factor (1°C/1.8°F) to convert to the equivalent number of Celsius degrees.

$$t(\text{C}) = [t(\text{F}) - 32°\text{F}] \times \frac{1°\text{C}}{1.8°\text{F}} = \frac{[t(\text{F}) - 32]}{1.8}$$

EXAMPLE 1-12

Converting Between Fahrenheit and Celsius

A person with a cold has a fever of 102°F. What would be the reading on a Celsius thermometer?

Solution

$$t(\text{C}) = \frac{[t(\text{F}) - 32]}{1.8} = \frac{(102 - 32)}{1.8} = \underline{\underline{39°\text{C}}}$$

EXAMPLE 1-13

Converting Between Celsius and Fahrenheit

On a cold winter day the temperature is −10.0°C. What is the reading on the Fahrenheit scale?

Solution

$$t(\text{F}) = [t(\text{C}) \times 1.8] + 32$$
$$t(\text{F}) = (1.8 \times -10.0) + 32.0 = -18.0 + 32.0 = \underline{14.0°\text{F}}$$

The SI temperature unit is called the **kelvin (K)**. The zero on the Kelvin scale is theoretically the lowest possible temperature (the temperature at which the heat energy is zero). This corresponds to −273°C (or more precisely −273.15°C). Since the Kelvin scale also has exactly 100 divisions between the freezing point and the boiling point of water, the magnitude of a kelvin and a Celsius degree is the same. Thus we have the following simple relationship between the two scales. The temperature in kelvins is represented by $T(\text{K})$, and $t(\text{C})$ represents the Celsius temperature.

$$T(\text{K}) = [t(\text{C}) + 273]$$

Thus the freezing point of water is 0°C or 273 K, and the boiling point is 100°C or 373 K. We will use the Kelvin scale more in later chapters.

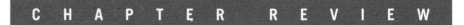

CHAPTER REVIEW

Putting It Together ▲▼

Chemistry, probably more than any other science, is a science of **measurements**. Certainly, that is not all there is—many other concepts are also integral to the science, but handling measurements properly is the highest priority. That is why we need to address this subject very early in the text. Our first item of business was to examine the numerical part of a measurement as to the significance of the numbers that are reported. The numerical value has two qualities: **precision** (number of **significant figures**) and **accuracy,** which are summarized as follows:

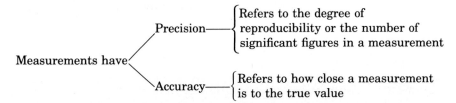

Measurements have

Precision——{ Refers to the degree of reproducibility or the number of significant figures in a measurement

Accuracy——{ Refers to how close a measurement is to the true value

Perhaps a greater challenge than just understanding the precision of one measurement is to handle measurements of varying precision in mathematical operations. The rules are summarized as follows:

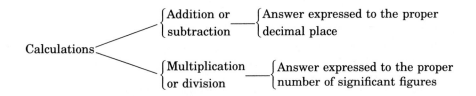

Calculations

{ Addition or subtraction ——{ Answer expressed to the proper decimal place

{ Multiplication or division ——{ Answer expressed to the proper number of significant figures

We won't get too far into the text before we encounter some extremely small and extremely large numbers. Writing large numbers of zeros is cumbersome. **Scientific notation**, however, allows us to write a number in an easy-to-read manner, that is, as a number between one and ten multiplied by ten raised to a power indicated by the **exponent**. Some of the ambiguities of zero as a significant figure are removed when the number is expressed in scientific notation. Remember, help is available in Appendix C if you are uneasy with scientific notation and using it in calculations.

Next, we turned our attention to units of measurements. Most of us have to become familiar with the SI system and the metric system on which it is based, specifically with regard to units of length (meters), **mass** or **weight** (grams), and **volume** (liters). Compared to the rather unsystematic English system, the metric system has the advantage of decimal relationships between units.

Because we need to convert between systems of measurement, we then introduced our main problem-solving technique—the **factor-label method,** also called **dimensional analysis**. In this, we change between units of measurement using **conversion factors** (usually **unit factors**), which are constructed from relationships between units. When a direct relationship is available between what's given and what's requested, conversion can be accomplished in a one-step calculation. If not, the calculation is more involved and requires a game plan that outlines step-by-step conversions from given to requested. This problem-solving method is the language of this text. If you do not follow this method, it is hard to take full advantage of the worked-out examples in this and future chapters.

The final topic of this chapter concerned the measurement of **temperature** using **thermometers.** By comparing the number of divisions between the freezing point and boiling point of water, we can set up conversion factors between the two temperature scales, **Fahrenheit** (°F) and **Celsius** (°C). A third system, closely related to the Celsius scale, is the **Kelvin** scale. Using an algebraic approach, we are able to convert between Fahrenheit and Celsius temperatures, as well as between Celsius and Kelvin.

E X E R C I S E S

Throughout the text, answers to all problems except those with the problem number in color are given in Appendix H. The more difficult problems are marked with an asterisk.

SIGNIFICANT FIGURES

1-1 Which of the following measurements is the most precise?
(a) 75.2 gal   (c) 75.22 gal
(b) 74.212 gal   (d) 75 gal

1-2 How can a measurement be precise but not accurate?

1-3 The actual length of a certain plank is 26.782 in. Which of the following measurements is the most precise and which is the most accurate?
(a) 26.5 in.   (c) 26.202 in.
(b) 26.8 in.   (d) 26.98 in.

1-4 How many significant figures are in each of the following measurements?
(a) 7030 g   (e) 4002 m
(b) 4.0 kg   (f) 0.060 hr
(c) 4.01 lb   (g) 8200 km
(d) 0.01 ft   (h) 0.00705 yd

1-5 How many significant figures are in each of the following measurements?
(a) 0.045 in.   (e) 7.060 qt
(b) 405 ft   (f) 2.0010 yd
(c) 0.340 cm   (g) 0.0080 in.
(d) 21.0 m   (h) 2200 lb

SIGNIFICANT FIGURES IN MATHEMATICAL OPERATIONS

1-6 Round off each of the following numbers to three significant figures.
(a) 15.9994   (e) 87,550
(b) 1.0080   (f) 0.027225
(c) 0.6654   (g) 301.4
(d) 4885

1-7 Round off each of the following numbers to two significant figures.
(a) 115   (e) 55.6
(b) 27.678   (f) 0.0396
(c) 37,500   (g) 1,557,000
(d) 0.47322   (h) 321

1-8 Express the following fractions in decimal form to three significant figures.
(a) $\dfrac{1}{4}$ (b) $\dfrac{4}{5}$ (c) $\dfrac{5}{3}$ (d) $\dfrac{7}{6}$

1-9 Express the following fractions in decimal form to three significant figures.
(a) $\dfrac{2}{3}$ (b) $\dfrac{2}{5}$ (c) $\dfrac{5}{8}$ (d) $\dfrac{13}{4}$

1-10 Carry out each of the following operations. Assume that the numbers represent measurements so the answer is expressed to the proper decimal place.
(a) 14.72 + 0.611 + 173
(b) 0.062 + 11.38 + 1.4578
(c) 1600 − 4 + 700
(d) 47 + 0.91 − 0.286
(e) 0.125 + 0.71

1-11 Carry out each of the following operations. Assume that the numbers represent measurements so the answer is expressed to the proper decimal place.
(a) 0.013 + 0.7217 + 0.04
(b) 15.3 + 1.12 − 3.377
(c) 35.48 − 4 + 0.04
(d) 337 + 0.8 − 12.0

1-12 A container holds 32.8 qt of water. The following portions of water are then added to the container: 0.12 qt, 3.7 qt, and 1.266 qt. What is the new volume of water?

1-13 A container holds 3760 lb (three significant figures) of sand. The following portions are added to the container: 1.8 lb, 32 lb, and 13.55 lb. What is the final mass of the sand?

1–14 Carry out the following calculations. Express your answer to the proper number of significant figures. Specify units.
(a) 40.0 cm x 3.0 cm
(b) 179 ft x 2.20 ft
(c) $\dfrac{4.386 \text{ cm}^2}{2 \text{ cm}}$ (d) $\dfrac{(14.65 \text{ in.} \times 0.32 \text{ in.})}{2.00 \text{ in.}}$

1–15 Carry out the following calculations. Express your answer to the proper number of significant figures. Specify units.
(a) $\dfrac{243 \text{ m}^2}{0.05 \text{ m}}$

(b) 3.0 ft x 472 ft (d) $\dfrac{(1.84 \text{ yd x } 42.8 \text{ yd})}{0.8 \text{ yd}}$

(c) 0.0575 in. x 21.0 in.

*1–16 Carry out the following calculations. Express your answer to the proper number of significant figures or decimal places.
(a) $\left(\dfrac{146}{2.3}\right) + 75.0$ (b) $(157 - 112) \text{ x } 25.6$
(c) $(12.688 - 10.0) \text{ x } (7.85 + 2.666)$

*1–17 Carry out the following calculations. Express your answer to the proper number of significant figures or decimal places.
(a) $(67.43 \times 0.44) - 23.456$
(b) $(0.22 + 12.451 + 1.782) \times 0.876$
(c) $(1.20 \times 0.8842) + (7.332 \times 0.0580)$

SCIENTIFIC NOTATION

1–18 Express the following numbers in scientific notation (one digit to the left of the decimal point).
(a) 157   (e) 0.0349
(b) 0.157   (f) 32,000
(c) 0.0300   (g) 32 billion
(d) 40,000,000 (two   (h) 0.000771
     significant figures) (i) 2340

1–19 Express the following numbers in scientific notation (one digit to the left of the decimal point).
(a) 423,000   (f) 82,000,000 (three
(b) 433.8   significant figures)
(c) 0.0020   (g) 75 trillion
(d) 880   (h) 0.00000106
(e) 0.00008

1–20 Using scientific notation, express the number 87,000,000 to (a) one significant figure, (b) two significant figures, and (c) three significant figures.

1–21 Using scientific notation, express the number 23,600 to (a) one significant figure, (b) two significant figures, (c) three significant figures, and (d) four significant figures.

1–22 Express the following as "ordinary" numbers.
(a) 4.76×10^{-4} (d) 0.489×10^6
(b) 6.55×10^3 (e) 475×10^{-2}
(c) 788×10^{-5} (f) 0.0034×10^{-3}

1–23 Express the following as "ordinary" numbers.
(a) 64×10^{-3} (c) 0.022×10^4
(b) 8.34×10^3 (d) 0.342×10^{-2}

1–24 Change the following numbers to scientific notation (one digit to the left of the decimal point).
(a) 489×10^{-6} (d) 571×10^{-4}
(b) 0.456×10^{-4} (e) 4975×10^5
(c) 0.0078×10^6 (f) 0.030×10^{-2}

1–25 Change the following numbers to scientific notation.
(a) 0.078×10^{-8} (d) 280.0×10^8
(b) 72000×10^{-5} (e) 0.000690×10^{-10}
(c) 3450×10^{16} (f) 0.0023×10^6

1–26 Carry out each of the following operations. Assume that the numbers represent measurements so the answer is expressed to the proper decimal place.
(a) $(1.82 \times 10^{-4}) + (0.037 \times 10^{-4})$
$+ (14.11 \times 10^{-4})$
(b) $(13.7 \times 10^6) - (2.31 \times 10^6)$
$+ (116.28 \times 10^5)$
(c) $(0.61 \times 10^{-6}) + (0.11 \times 10^{-4})$
$+ (0.0232 \times 10^{-3})$
(d) $(372 \times 10^{12}) + (1200 \times 10^{10})$
$- (0.18 \times 10^{15})$

1–27 Carry out each of the following operations. Assume that the numbers represent measurements so the answer is expressed to the proper decimal place.
(a) $(1.42 \times 10^{-10}) + (0.17 \times 10^{-10})$
$- (0.009 \times 10^{-10})$
(b) $(146 \times 10^8) + (0.723 \times 10^{10}) + (11 \times 10^8)$
(c) $(1.48 \times 10^{-7}) + (2911 \times 10^{-9})$
$+ (0.6318 \times 10^{-6})$
(d) $(299 \times 10^{10}) + (823 \times 10^8) + (0.75 \times 10^{11})$

1–28 Carry out each of the following operations. Assume that the numbers represent measurements so the answer is expressed to the proper number of significant figures.
(a) $(149 \times 10^6) \times (0.21 \times 10^3)$
(b) $\dfrac{(0.371 \times 10^{14})}{(2 \times 10^4)}$

(c) $(6 \times 10^6) \times (6 \times 10^6)$

(d) $(0.1186 \times 10^6) \times (12 \times 10^{-5})$

(e) $\dfrac{(18.21 \times 10^{-10})}{(0.0712 \times 10^6)}$

1–29 Carry out each of the following operations. Assume that the numbers represent measurements so the answer is expressed to the proper number of significant figures.

(a) $(76.0 \times 10^7) \times (0.6 \times 10^8)$

(b) $(7 \times 10^{-5}) \times (7.0 \times 10^{-5})$

(c) $\dfrac{(0.786 \times 10^{-7})}{(0.47 \times 10^7)}$

(d) $\dfrac{(3798 \times 10^{18})}{(0.00301 \times 10^{12})}$

(e) $(0.06000 \times 10^{18}) \times (84921 \times 10^{-9})$

LENGTH, VOLUME, AND MASS IN THE METRIC SYSTEM

1–30 Write the proper prefix, unit, and symbol for the following. Refer to Tables 1-2 and 1-3.

(a) 10^{-3} L

(b) 10^2 g

(c) 10^{-9} J

(d) $\dfrac{1}{100}$ m

(e) 10^{-6} g

(f) 10^{-1} Pa

1–31 Write the proper prefix, unit, and symbol for the following. Refer to Tables 1-2 and 1-3.

(a) 10^3 m

(b) 10^{-3} g

(c) $\dfrac{1}{1000}$ L

(d) 10^3 s

(e) 10^{-9} m

(f) $\dfrac{1}{1000}$ mol

1–32 Complete the following table.

	mm	cm	m	km
(Example)	108	10.8	0.108	1.08×10^{-4}
(a)	7.2×10^3	___	___	___
(b)	___	___	56.4	___
(c)	___	___	___	0.250

1–33 Complete the following table.

	mg	g	kg
(a)	8.9×10^3	___	___
(b)	___	25.7	___
(c)	___	___	1.25

1-34 Complete the following table.

	mL	L	kL
(a)	___	___	6.8
(b)	___	0.786	___
(c)	4452	___	___

CONVERSIONS BETWEEN UNITS OF MEASUREMENT

1–35 Which of the following are "exact" relationships?

(a) 12 = 1 doz

(b) 1 gal = 3.78 L

(c) 3 ft = 1 yd

(d) 1.06 qt = 1 L

(e) 10^3 m = 1 km

(f) 454 g = 1 lb

1–36 Write a relationship in factor form that would be used in making the following conversions. Refer to Table 1-3.

(a) mg to g

(b) m to km

(c) cL to L

(d) mm to km (two factors)

1–37 Write a relationship in factor form that would be used in making the following conversions. Refer to Table 1-3.

(a) kL to L

(b) Mg to g

(c) kg to Mg (two factors)

(d) cg to hg (two factors)

1–38 Write a relationship in factor form that would be used in making the following conversions. Refer to Tables 1-1 and 1-5.

(a) in. to ft

(b) in. to cm

(c) mi to ft

(d) L to qt

(e) pt to L (two steps)

1–39 Write a relationship in factor form that would be used in making the following conversions. Refer to Tables 1-1 and 1-5.

(a) qt to gal

(b) kg to lb

(c) gal to L

(d) ft to km (three steps)

1–40 Complete the following table.

	mi	ft	m	km
(a)	___	___	7.8×10^3	___
(b)	0.450	___	___	___
(c)	___	8.98×10^3	___	___
(d)	___	___	___	6.78

1–41 Complete the following table.

	gal	qt	L
(a)	6.78	_____	_____
(b)	_____	670	_____
(c)	_____	_____	7.68×10^3

1–42 Complete the following table.

	lb	g	kg
(a)	_____	_____	0.780
(b)	_____	985	_____
(c)	16.0	_____	_____

1–43 If a person has a mass of 122 lb, what is her mass in kilograms?

1–44 A punter on a professional football team averaged 28.0 m per kick. What is his average in yards? Should he be kept on the team?

1–45 If a student drinks a 12-oz (0.375 qt) can of soda, what volume did she drink in liters?

1–46 A prospective basketball player is 6 ft 10 1/2 in. tall and weighs 212 lb. What is his height in meters and his weight in kilograms?

1–47 Gasoline is sold by the liter in Europe. How many gallons does a 55.0-L gas tank hold?

1–48 If the length of a football field is changed from 100 yd to 100 m, will the field be longer or shorter than the current field? How many yards would a "first and ten" be on the metric field?

1–49 Bourbon used to be sold by the "fifth" (one fifth of a gallon). A bottle now contains 750 mL. Which is greater?

1–50 If the speed limit is 65.0 mi/hr, what is the speed limit in km/hr?

1–51 Mount Everest is 29,028 ft in elevation. How high is this in kilometers?

1–52 It is 525 mi from St. Louis to Detroit. How far is this in kilometers?

1–53 Gasoline sold as low as $0.899 per gallon in 1993. What was the cost per liter? What did it cost to fill an 80.0-L tank?

1–54 Using the price of gas from the preceding problem, how much does it cost to drive 551 mi if your car averages 21.0 mi/gal? How much does it cost to drive 482 km?

1–55 Using the information from the two preceding problems, how many kilometers can you drive for $45.00?

1–56 An aspirin contains 0.324 g (5.00 grains) of aspirin. How many pounds of aspirin are in a 500-aspirin bottle?

1–57 A hamburger in Canada sells for $2.55 (Canadian dollars). That seems expensive to a U.S. resident until we realize that the exchange rate is $1.26 Canadian per one United States dollar. What is the cost in United States dollars?

1–58 A certain type of nail costs $0.95/lb. If there are 145 nails per pound, how many nails can you purchase for $2.50?

1–59 Another type of nail costs $0.92/lb, and there are 185 nails per pound. What is the cost of 5670 nails?

1–60 If an automobile gets 24.5 miles to the gallon of gasoline and gasoline costs $1.22/gal, what would it cost to drive 350 km?

1–61 If a train travels at a speed of 85 mi/hr, how many hours does it take to travel 17,000 ft? How many yards can it travel in 37 min?

1–62 If grapes sell for $1.15/lb and there are 255 grapes per pound, how many grapes can you buy for $5.15?

*1–63 At a speed of 35 mi/hr, how many centimeters do you travel per second?

*1–64 The planet Jupiter is about 4.0×10^8 mi from Earth. If radio signals travel at the speed of light, which is 3.0×10^{10} cm/s, how long would it take a radio command from Earth to reach a spacecraft passing Jupiter?

TEMPERATURE

1–65 The temperature of the water around a nuclear reactor core is about 300°C. What is this temperature in degrees Fahrenheit?

1–66 The temperature on a comfortable day is 76°F. What is this temperature in degrees Celsius?

1–67 The lowest possible temperature is −273°C and is referred to as absolute zero. What is this temperature in degrees Fahrenheit?

1–68 Mercury thermometers cannot be used in cold arctic climates because mercury freezes at −39°C. What is this temperature in degrees Fahrenheit?

1–69 The coldest temperature recorded on Earth was −110°F. What is this temperature in degrees Celsius?

1–70 A hot day in the Midwest of the United States is 35.0°C. What is this in degrees Fahrenheit?

1–71 Convert the following Kelvin temperatures to degrees Celsius.
(a) 175 K   (d) 225 K
(b) 295 K   (e) 873 K
(c) 300 K

1–72 Convert the following temperatures to the Kelvin scale.
(a) 47°C   (d) −12°C
(b) 23°C   (e) 65°F
(c) −73°C   (f) −20°F

1–73 Make the following temperature conversions.
(a) 37°C to K   (d) 127 K to °F
(b) 135°C to K   (e) 100°F to K
(c) 205 K to °C   (f) −25°C to K

*1–74 At what temperature are the Celsius and Fahrenheit scales numerically equal?

GENERAL PROBLEMS

1–75 Carry out the following calculations. Express the answer to the proper number of significant figures.
(a) $\dfrac{12.61 + 0.22 + 0.037}{0.04}$
(b) $0.333 \text{ g} \times (23.60 + 1.2) \text{ cm}$
(c) $\dfrac{6.286 \text{ g}}{(13.68 - 12.48) \text{ mL}}$
(d) $\dfrac{(44.35 + 0.03 + 0.057)}{(22.35 - 20.018)}$

1–76 Write in factor form the two relationships needed to convert the following:
(a) mg to lb   (c) hm to mi
(b) L to pt   (d) cm to ft

1–77 Convert 5.34×10^{10} nanograms to pounds.

1–78 Convert 7.88×10^{-4} megaliters to gallons.

1–79 If gold costs \$345/oz, what is the cost of 1.00 kg of gold? (Metals are traded as "troy" ounces. There are exactly 12 troy ounces per troy pound and one troy pound is equal to 373 g, to three significant figures.)

1–80 Construct unit factors from the following information: A 82.3-doz quantity of oranges weighs 247 lb.
(a) What is the mass per dozen oranges?
(b) How many dozen oranges are there per pound?

1–81 A unit of length in horse racing is the "furlong." The height of a horse is measured in "hands." There are exactly 8 furlongs per mile, and one hand is exactly 4 inches. How many hands are there in 12.0 furlongs? (Express the answer in standard scientific notation.)

1–82 The unit price of groceries is sometimes listed in cost per ounce. Which of the two has the smallest cost per ounce: 16 oz of baked beans costing \$1.45 or 26 oz costing \$2.10?

1–83 Low-tar cigarettes contain 11.0 mg of tar per cigarette. If all the tar gets into the lungs, how many packages of cigarettes (20 cigarettes per package) would have to be smoked to produce 0.500 lb of tar? If a person smoked two packages per day, how many years would it take to accumulate 0.500 lb of tar?

1–84 An automobile engine has a volume of 306 in.³. What is this volume in liters?

1–85 A U.S. quarter has a mass of 5.70 g. How many dollars is one pound of quarters worth?

1–86 The surface of the sun is at a temperature of about 3.0×10^{7} °C. What is this temperature in °F? In kelvins?

SOLUTIONS TO LEARNING CHECKS

A–1 estimated, precision, accuracy, between, right, left, least, decimal places, least, significant figures, scientific notation, one, exponent

A–2 (a) 3   (b) 4   (c) 3   (d) 3   (e) 2

A–3 (a) 23.4 (b) 484,000 (c) 0.0220

A–4 (a) 20.00 (b) 2.55 (c) 2

A–5 (a) 4.560×10^8 (b) 3.40×10^{-4}

A–6 (a) 2.61×10^8 (b) 2.4×10^8

B–1 meter, liter, gram, 10^4 mg, 10^{-2} kg, 1 mL, conversion, denominator, numerator

B–2 (a) $\dfrac{1 \text{ yd}}{3 \text{ ft}}$ (b) $\dfrac{10^{-1} \text{ m}}{\text{dm}}$ (c) $\dfrac{3.78 \text{ L}}{\text{gal}}$ (d) $\dfrac{1 \text{ kg}}{2.20 \text{ lb}}$

B–3 $25.0 \ \cancel{\text{km}} \times \dfrac{1 \text{ mi}}{1.61 \ \cancel{\text{km}}} = \underline{\underline{15.5 \text{ mi}}}$

B–4 $12 \ \cancel{\text{oz}} \times \dfrac{\$0.180}{\cancel{\text{oz}}} = \underline{\underline{\$2.16}}$ $\$1.75 \times \dfrac{1 \text{ oz}}{\cancel{\$}0.180} = \underline{\underline{9.72 \text{ oz}}}$

B–5 $12.0 \ \cancel{\text{ft}} \times \dfrac{12 \ \cancel{\text{in.}}}{\cancel{\text{ft}}} \times \dfrac{2.54 \text{ cm}}{\cancel{\text{in.}}} = \underline{\underline{366 \text{ cm}}}$

B–6 $3.25 \ \cancel{\text{kg}} \times \dfrac{2.20 \ \cancel{\text{lb}}}{\cancel{\text{kg}}} \times \dfrac{\$1.25}{\cancel{\text{lb}}} = \underline{\underline{\$8.94}}$

In 1989, Voyager swept past the cold planet Neptune, lying four billion miles from Earth. Its instruments detected the presence of the same elements and compounds that are present on Earth. Elements and compounds are the subjects of this chapter.

CHAPTER 2

MATTER, CHANGES, AND ENERGY

In 1989, the Voyager spacecraft swept silently by the giant planet Neptune, lying more than four billion miles from Earth. This lonely craft had started its fantastic journey to gather information on the outer planets back in 1979. At the time, Voyager was viewing the planet farthest from the sun. The sensitive instruments on the spacecraft radioed back huge amounts of information about this cold, distant world. As expected, Neptune was composed of the same types of "stuff" found on Earth. The nature of this stuff (which we call matter) is a domain of chemistry. **Chemistry** *can be defined as the study of matter and the changes it undergoes.* **Matter** *is defined as anything that has mass and occupies space.* Matter surrounds us. Obviously, rock is matter, but so is water, and so is the invisible air that we breathe. The oceans, the moon, and the faraway stars are all composed of matter. The forms of matter that we see are often changing. Trees grow, die, and decay. Rocks weather, crumble, and form the fertile soil of the plains or deposits in the oceans. The changes that matter undergoes are another concern of chemistry. But there is even more. *In addition to matter, the universe is also composed of energy.* When a log burns in the fireplace, it is obvious that a change in matter has occurred. The log seems to have disappeared, leaving a small pile of ash. But the burning of the log warms

◀**Setting The Stage**

us—it has given off heat. The heat that was liberated by the burning process is a form of energy. Energy is a more abstract concept than matter. It can't be weighed and it doesn't have shape, form, or dimensions. But energy can be measured, and it does interact with matter (e.g., it warms us, starts our car, and makes trees grow). Since energy is an integral part of the changes that matter undergoes, it is also important in the study of chemistry. Thus the purpose of this chapter is to examine in detail the two key words in the definition of chemistry: "matter" and "changes."

At first glance, the world around us seems so complex. For example, there are billions of people on Earth, so we simplify our understanding of this population by grouping it into general categories such as nationality, age, sex, and so forth. There are also literally millions of different types of matter around us. Can our understanding of matter be simplified by grouping? We will see that the answer is "yes." But how do we describe matter? How can we distinguish between the many types of matter (called substances)? Finally, how does energy relate to changes in matter? In the first section, we will find that matter can be divided into two general categories.

Formulating Some Questions ▶

2–1   TYPES OF MATTER: ELEMENTS AND COMPOUNDS

How elementary can we make this?

All of the matter that we see around us—the oceans, the earth, the air, and even the moon, sun, and stars—are composed of fewer than 90 unique types of matter. These substances are called the elements. *Because it cannot be broken down into simpler substances, an* **element** *is the most basic form of matter that exists under ordinary conditions.* A few elements are found on Earth in their free state. The shiny gold in a ring, the life-supporting oxygen in air, and the sturdy but light aluminum in a soda can are all examples of elements. The elements are listed inside the front cover of this text. In the next chapter, we will discuss the names, symbols, and abundances of these elements.

Obviously, there are millions of substances around other than the free elements. Elements can combine with other elements to form compounds. **A compound** *is a substance composed of two or more elements that are chemically combined.* Most familiar substances are actually compounds. For example, water, table salt, and sugar are compounds. Water is composed of the elements oxygen and hydrogen; table salt is made up of sodium and chlorine; and sugar consists of carbon, hydrogen, and oxygen. Although relatively few free elements occur in nature, various chemical combinations of these elements make up millions of known compounds. (See Figure 2-1.) Every element and compound can be identified by its unique properties. **Properties** *describe the particular characteristics or traits of a substance.* Elements and compounds can be referred to as pure substances. **Pure substances** *have definite compositions and definite, unchanging properties.*

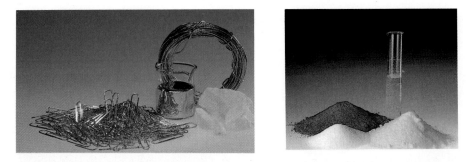

Figure 2–1
Elements and Compounds Iron (the paper clips), liquid mercury, copper, and sulfur are examples of elements (*left*). White sodium chloride, blue copper sulfate hydrate, yellow sodium chromate, and water are examples of compounds (*right*).

We sometimes identify human beings by two kinds of properties: physical (e.g., tall and thin) and emotional (e.g., quiet and reflective). Likewise, there are two kinds of properties used to identify a particular element or compound: physical and chemical. We will first discuss physical properties and then proceed to chemical properties.

▼ **Looking Ahead**

2–2 PHYSICAL PROPERTIES OF MATTER

The physical properties of people can be found on their driver's licenses. Height, weight, sex, and color of eyes are considered constant and unchanging, although weight, unfortunately, may change. Like an individual, **the physical properties** of a substance are those that can be observed without changing the substance into another substance. Odor and color are two physical properties that a substance may or may not display. Another physical property is the physical state. The three **physical states** of matter are *solid, liquid, and gas*. In a later chapter, we will discuss how all elements and compounds are composed of very small particles that can be imaged only with the most powerful microscopes known. In the solid state, the particles remain close together in relatively fixed positions. Movement of the particles is confined mostly to vibrations about these positions. Because of the fixed positions of the fundamental particles, **solids** *have a definite shape and a definite volume*. The fundamental particles in liquids are also close together but are free to move past each other. Liquids flow and take the shape of the lower part of a container. Thus, **liquids** *have a definite volume but not a definite shape*. In gases, the particles are in random motion and move freely in all three dimensions. Gases fill a container uniformly. Thus, **gases** *have neither a definite volume nor a definite shape*. (See Figure 2-2.)

We are already familiar with many examples of all three physical states. Ice, rock, salt, and steel are substances that exist as solids; water, gasoline, and alcohol exist as liquids; ammonia, natural gas, and the

How do we tell one substance from another?

Figure 2–2
The Three Physical States of Matter (*a*) Solids have a definite shape and volume.
(*b*) liquids have a definite volume but an indefinite shape. (*c*) Gases have an
indefinite shape and volume.

What does temperature have to do with physical state?

components of air are present in nature as gases. Whether a particular
element or compound is a solid, liquid, or gas depends on the nature of
the substance and on the temperature. For example, at low temperatures, liquid water freezes to form a solid, and at high temperatures, liquid water boils to form a gas (vapor). If the temperature is very low
(−196°C), even the gases that form our atmosphere condense to the liquid state. In fact, the temperature at which a pure substance changes to
another physical state is a definite, unchanging physical property of
that substance. *The temperature at which a particular element or compound changes from the solid state to the liquid state is known as its*
melting point. (When the reverse change occurs, this same temperature is referred to as the **freezing point.**) At some higher temperature
the liquid begins to boil. *The temperature at which bubbles of the vapor
begin to form in the liquid and escape to the surroundings is known as
the* **boiling point.** (The reverse change occurs at the same temperature,
which is then called the **condensation point.**) The boiling point of a

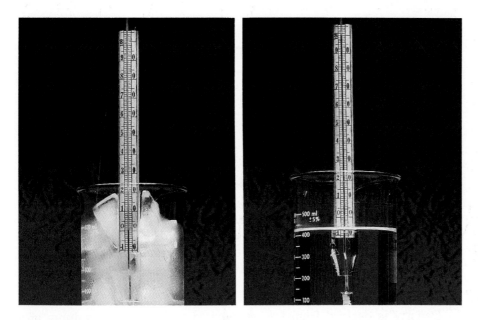

Figure 2–3
Melting and Boiling Points Water melts at 0°C and boils at 100°C at sea level.

liquid is affected by the atmospheric pressure, however. Boiling point temperatures are usually listed as the boiling point of the liquid at average sea-level atmospheric pressure. (See Figure 2-3.)

When a liquid freezes or boils, it undergoes a change to another physical state. It is still the same substance, however, so the change is described as physical. **A physical change** *in a substance does not involve a change in the composition of the substance but is simply a change in physical state or dimensions.* Liquid water, ice, and steam are all physical states of the same compound.

Odor, color, physical state, and melting and boiling points are all important physical properties. One other physical property that is almost always included in a listing of properties is density (or specific gravity). We will discuss density in the next section.

▼Looking Ahead

2–3 A PHYSICAL PROPERTY: DENSITY

A styrofoam coffee cup is "light" but a lead car battery is "heavy." Actually, by themselves these terms don't mean much unless we compare an equal volume of each. The volume and mass of a substance are variable properties, depending on the size of the sample. A more useful physical property that does not depend on the amount present is density. **Density** *is the ratio of the mass (usually in grams) to the volume (usually in milliliters or liters).* The density of a pure substance is a constant property that can be used to identify the particular element or compound. The densities of several liquids and solids are listed in Table 2-1. (Be-

Is gold heavy or dense?

TABLE 2–1   DENSITY (AT 20°C)

Substance (Liquid)	Density (g/mL)	Substance (Solid)	Density(g/mL)
Ethyl alcohol	0.790	Aluminum	2.70
Gasoline (a mixture)	≈0.67 (variable)	Gold	19.3
Carbon tetrachloride	1.60	Ice	0.92 (0°C)
Kerosene	0.82	Lead	11.3
Water	1.00	Lithium	0.53
Mercury	13.6	Magnesium	1.74
		Table salt	2.16

Gold is one of the densest substances known.

cause the volume of liquids and solids expands slightly as the temperature rises, densities are usually given at a specific temperature. In this case, 20°C is the reference temperature.) Because 1 mL is the same as 1 cm³, density is also expressed as g/cm³. The densities of gases are discussed in Chapter 9.

Density is derived from two measurements, mass and volume of a particular sample. It makes no difference how large the sample is, the density will be the same. The calculation of density from the two measurements is discussed in the following two examples and illustrated in Figure 2-4.

Working It Out ●

EXAMPLE 2–1

Calculation of Density

A sample of a pure substance is found to have a mass of 47.5 g. As shown in Figure 2-4, a quantity of water has a volume of 12.5 mL. When the substance is added to the water, the volume reads 31.8 mL. The difference in volume is the volume of the substance. What is the density?

Solution
Refer to Figure 2-4.

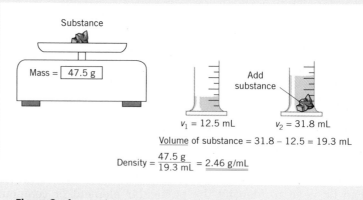

Substance

Mass = 47.5 g

Add substance

v_1 = 12.5 mL v_2 = 31.8 mL

Volume of substance = 31.8 − 12.5 = 19.3 mL

Density = $\dfrac{47.5\ \text{g}}{19.3\ \text{mL}}$ = 2.46 g/mL

Figure 2–4
Measurements of Mass and Volume for Density

EXAMPLE 2–2

Using Density to Identify a Pure Substance

A young woman was interested in purchasing a sample of pure gold having a mass of 8.99 g. Being wise, she wished to confirm that it was actually gold before she paid for it. With a quick test using a graduated cylinder like that shown in Figure 2-4, she found that the "gold" had a volume of 0.796 mL. Was the substance gold?

●Working It Out

Procedure
From the volume and the mass, the density can be calculated and compared with that of pure gold. (See Table 2-1.)

Solution

$$\text{density} = \frac{8.99 \text{ g}}{0.796 \text{ mL}} = 11.3 \text{ g/mL}$$

The sample was *not* gold. It apparently was lead that had been dipped in gold paint.

Density is not only an important physical property but also a conversion factor that relates mass to volume of a particular sample. If the density is known, a given mass can be converted to an equivalent volume and vice versa. These two conversions are illustrated in the following examples.

EXAMPLE 2–3

Converting Mass to Volume

What is the volume in mL occupied by 485 g of table salt?

●Working It Out

Procedure
1 In shorthand, the conversion is

$$\text{g} \longrightarrow \text{mL}$$

2 The density of table salt is given in Table 2-1 as 2.16 g/mL. Expressed as a conversion factor to convert *g* to *mL*, the relationship is

$$\frac{1 \text{ mL}}{2.16 \text{ g}}$$

Solution

$$485 \text{ g} \times \frac{1 \text{ mL}}{2.16 \text{ g}} = \underline{\underline{225 \text{ mL}}}$$

EXAMPLE 2–4

Converting Volume to Mass

What is the mass in grams of 1.52 L of kerosene?

Procedure
1 A two-step conversion is necessary. In the first step (a), *L* is converted

into *mL,* and in the second step (b) *mL* is converted into *g.* In short-hand, the conversion is shown as

$$\overset{(a)}{L \longrightarrow} \overset{(b)}{mL \longrightarrow} g$$

2   The proper relationships are 1 mL = 10^{-3} L (from Table 1-4), and 0.82 g/mL (from Table 2-1). Expressed as proper conversion factors, the relationships are

$$(a) \frac{1 \text{ mL}}{10^{-3}\text{L}} \text{ and (b) } \frac{0.82 \text{ g}}{\text{mL}}$$

Solution

$$1.52 \text{ L} \times \frac{1 \text{ mL}}{10^{-3} \text{ L}} \times \frac{0.82 \text{ g}}{\text{mL}} = \underline{\underline{1.2 \times 10^3 \text{ g}}}$$

Specific gravity is related to density. **Specific gravity** *is the ratio of the mass of a substance to the mass of an equal volume of water under the same conditions.* This definition can be expressed as follows:

$$\text{specific gravity} = \frac{\text{density of a substance}}{\text{density of water}}$$

The density of water at 4°C is 1.00 g/mL. (This was the original definition of the gram in the metric system.) Thus the specific gravity of water is exactly 1 at that temperature. Fortunately, the specific gravity of water changes very little at higher temperatures (i.e., 0.998 at 20°C), so we use a value of 1 in most calculations. The specific gravity of mercury is calculated as follows:

$$\text{density of mercury} = 13.6 \text{ g/mL}$$

$$\text{density of water} = 1.00 \text{ g/mL}$$

$$\text{specific gravity} = \frac{\text{density of mercury}}{\text{density of water}}$$

$$= \frac{13.6 \text{ g/mL}}{1.00 \text{ g/mL}} = 13.6$$

Notice that when we use units of g/mL or g/cm^3 for density, the specific gravity is numerically the same as the density except that it is expressed without units.

Elements and compounds are described by a list of properties. Density is a physical property that is almost always included in such a list. For example, zinc and sulfur are elements and zinc sulfide is a compound. (See Figure 2–5.) Although zinc sulfide is composed of zinc and sulfur chemically combined, it has properties that are entirely unique and distinct from the two elements from which it is formed.

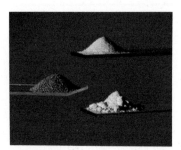

Figure 2-5
Physical Properties Sulfur (*bottom*), zinc (*middle*), and zinc sulfide (*top*) each have distinct and different physical properties.

Looking Back ▲

There is a lot more to us than can be described on our driver's license. Besides our physical description, we also have personality traits that, taken as a whole, are unique to each of us. Our personality includes how we interact with others. Likewise, pure substances have chemical properties that relate to how the element or compound interacts with other substances. We will discuss chemical properties next.

2–4 CHEMICAL PROPERTIES OF MATTER

When heat is applied to ice, it melts. When the water is cooled, it refreezes to ice. All that has changed is the physical state of the water. When we heat a raw egg, it solidifies. When we cool the egg, it stays solid. Obviously, the egg is not the same—it has undergone some profound change. When iron rusts, vegetation decays, and wood burns, these substances have been transformed into one or more other substances. These processes all describe chemical changes. *When a **chemical change** occurs in a substance, the substance is transformed into other substances. The **chemical properties** of a pure substance relate to its tendency to undergo chemical changes.* A chemical property of the element iron is its ability to react with oxygen from the air to form rust (a compound composed of iron and oxygen). In some cases, chemical properties can be stated in the negative. For example, a chemical property of the element gold is that it does *not* rust or tarnish. In the previous section, the physical properties of zinc, sulfur, and zinc sulfide were listed. The chemical properties of these three pure substances can also serve as fingerprints for identification. Some chemical properties are listed in Figure 2-6.

How is a physical change different from a chemical change?

The universe consists of two components: matter and energy. Matter is composed of elements, the basic forms of matter. Although a few elements do exist in the free state, most elements combine in countless ways to form millions of compounds. Each element and compound has a unique combination of physical properties (color, physical state, density, etc.) and chemical properties (tendency to interact with other substances). These properties allow further groupings and classifications of pure substances, as we will find throughout the text.

▲ **Looking Back**

Learning Check A

◀ **Checking It Out**

A–1. Fill in the blanks.
Matter has mass and occupies _____. Two types of matter are elements and _____. These are known as _____ substances, and they have unique _____. The ratio of the mass and the volume of a sample is known as its _____ and is a _____ property. A _____ change transforms one substance into another.

A–2. Calcium, an element, is a dull, gray solid that melts at 839°C and has a specific gravity of 1.54. When it is placed in water, bubbles form, and the calcium slowly dissolves in the water. When the water is evaporated, elemental calcium is not recovered. Which are the physical properties of calcium? Which is a chemical property?

A–3.   A sample of a given pure liquid has a mass of 254 g and a volume of 159 mL. What might the liquid be? Refer to Table 2-1.

A–4.   What is the volume of 178 g of aluminum?

Additional Examples: Problems 2-3, 2-10, 2-12, 2-17, 2-19, 2-23, 2-26.

Looking Ahead ▼

Pure substances can be identified by their properties. In fact, most of our surroundings are mixtures of various compounds and, in some cases, free elements. The subject of the next section is what happens when pure substances are mixed.

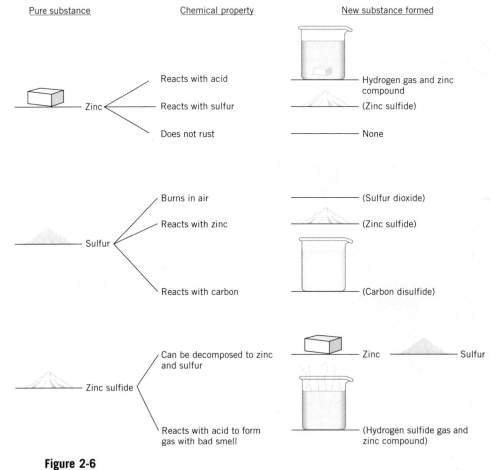

Figure 2-6
Chemical Properties and Changes

2–5 MIXTURES OF ELEMENTS AND COMPOUNDS

You can't mix oil and water! You've probably heard that often enough. On the other hand, when you place alcohol in water, both liquids disperse into each other and no boundary between the two liquids is apparent. (See Figure 2-7.) Oil and water form a heterogeneous mixture. **A heterogeneous mixture** *is a nonuniform mixture containing two or more phases with definite boundaries between the phases.* **A phase** *is one physical state (solid, liquid, or gas) with distinct boundaries and uniform properties.* Alcohol and water form a homogeneous mixture. **A homogeneous mixture** *is the same throughout and contains only one phase.* We will discuss heterogeneous mixtures first.

How do things mix?

Notice in Figure 2-7a that the oil floats on top of the water because the oil (≈0.90 g/mL) is less dense than water (1.00 g/mL). Compare this to the homogeneous mixture formed when alcohol and water are mixed, as shown in Figure 2-7b.

Why does oil float on water?

Besides oil and water (two liquid phases), an obvious heterogeneous mixture is a handful of soil from the backyard. If we look closely, we see bits of sand, some black matter, and perhaps pieces of vegetation. One can easily detect several solid phases with the naked eye. Other examples of heterogeneous mixtures are carbonated beverages (liquid and gas) and dirty water (liquid and solid). Heterogeneous mixtures can often be separated into their homogeneous components by simple laboratory procedures. For example, suspended solid matter can be removed from water by *filtration.* (See Figure 2-8.) When the dirty water is passed through the filter, the suspended matter remains on the filter paper and the liquid phase passes through. In the purification of water for drinking purposes, the first step is the removal of suspended particulate matter.

Another example of a heterogeneous mixture is the mixture of two solid elements, zinc and sulfur, shown in Figure 2-9. Both solid phases are visible, and, if necessary, we could separate the two components. If

(a)   (b)   (c)   (d)

Figure 2–7
Mixtures water and oil (*a*) form a heterogeneous mixture (*b*). Water and alcohol (*c*) form a homogeneous mixture (*d*).

Figure 2–8
Filtration A heterogeneous mixture of a solid and liquid can be separated by filtration

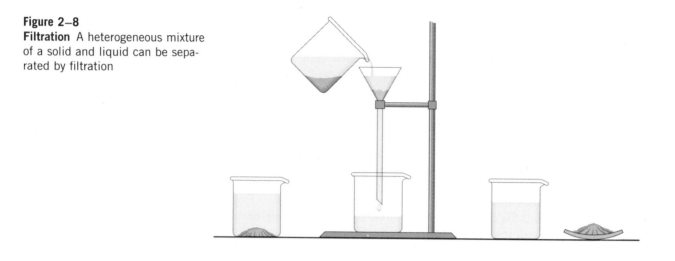

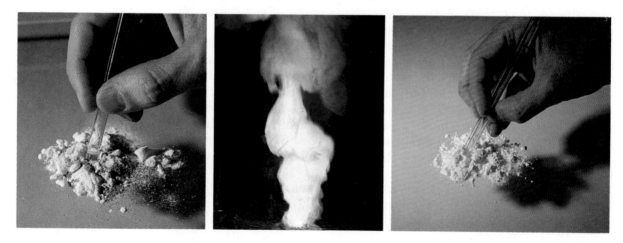

Figure 2–9
Formation of Zinc Sulfide In this chemical reaction, a heterogeneous mixture of elements (*left*) is changed into a pure compound (*right*).

this mixture is ignited with a hot flame, however, a vigorous chemical reaction occurs, forming the compound zinc sulfide. Through the magic of chemistry, the mixture of elements was transformed into one pure substance. *It is important to note that the zinc sulfide is no longer considered a mixture even though it was made from a mixture.* After the chemical reaction, it would take some very involved chemical procedures to decompose the homogeneous compound back to its elements.

Formation of a compound from a mixture is much like baking a cake. We mix eggs, flour, and other ingredients and then heat the mixture. What comes out is a unique substance that no longer looks like, tastes like, or smells like the mixture we put in the oven. The mix of ingredients has been transformed into a new homogeneous substance (we hope).

Sometimes heterogeneous mixtures cannot be detected with the naked eye. For example, creamy salad dressing and fog both appear uniform at first glance. However, if we were to magnify each, the truth would become apparent. The salad dressing has little droplets of oil suspended in the vinegar (two liquid phases), and the fog has droplets of water suspended in the air (liquid and gas phases).

In heterogeneous mixtures, portions of each component are large enough to be detected, although some magnification may be necessary. In homogeneous mixtures, the components disperse uniformly into each other. In this case, there is no detectable boundary between components. When solid table salt is added to water, it forms a solution. Although all homogeneous mixtures are technically solutions, the word **solution** *usually refers to homogeneous mixtures with one liquid phase.* No amount of magnification would reveal pieces of solid salt in the water. Thus components of a solution cannot be separated by filtration. In the case of the dissolved table salt, a laboratory procedure called *distillation* would be necessary. (See Figure 2-10.) In distillation, the water is boiled away from the solution and then retrieved by condensation through a water-cooled tube. When all of the water has boiled away, the solid table salt remains behind in the distilling flask.

You can't tell the difference between a glass of salt water and pure water by inspection. They look exactly the same. You would have to taste the two to know which is which. In fact, *since both solutions and pure substances (elements and compounds) are homogeneous matter, one must examine the physical properties to distinguish between the two.* Mixtures have properties that vary with the proportion of the components. Elements and compounds have definite and unchanging properties. A simple example of a variable property is the taste and color of a cup of coffee. The more coffee that is dissolved in the water, the stronger the taste and the darker the solution.

How can you tell the difference between pure water and a solution of water?

Figure 2–10
Distillation A homogeneous mixture of a solid dissolved in a liquid can be separated by distillation.

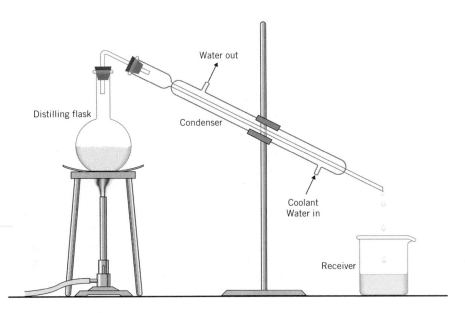

Water out

Distilling flask

Condenser

Coolant
Water in

Receiver

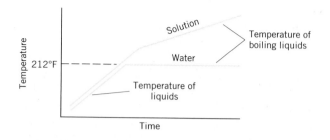

Figure 2–11
Boiling Temperatures of Pure Water and a Solution When pure water boils, the temperature remains constant; when a solution boils the temperature slowly rises.

Now consider the properties of two compounds alone: table salt and water. Solid table salt (sodium chloride) melts at 801°C and water melts at 0°C (32°F). A solution of salt in water freezes anywhere from −18°C to just under 0°C, depending on the amount of salt dissolved. Not only does the presence of dissolved salt change the melting point and boiling point of the water, but it changes the *way* it melts and boils. For example, if pure water is boiled at average sea-level atmospheric pressure until it is all gone, we would notice that the temperature remains a steady 100°C (212°F) from start to finish. When a solution of sodium chloride boils, however, the temperature of the boiling solution slowly rises as the liquid boils away. (See Figure 2-11.)

Density is another physical property that is different for a solution compared to the pure liquid component. For example, "battery acid" is a solution of a compound, sulfuric acid, in water. Its density is greater than pure water. The more sulfuric acid present, the denser the solution. *The density of a liquid can be determined with a device called a* **hydrometer.** The hydrometer tube, as shown in Figure 2-12, is exactly balanced with weights in the bottom so that its level in pure water reads exactly 1.00 g/mL. The scale is calibrated (the divisions previously determined and checked) in such a manner that the density of a liquid is read directly from the scale at the surface of the liquid. For battery acid, the denser the solution, the higher the charge in the battery.

Figure 2–12
A Hydrometer This device is used to measure the density of a liquid.

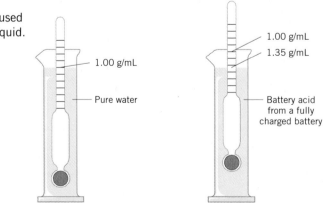

Many of the metals that we use in our daily lives are actually alloys. An **alloy** *is a homogeneous mixture of metallic elements with one solid phase.* Although an alloy is considered a *solid solution,* it is made by mixing the metals in the molten state and then allowing the liquid solution to cool and resolidify. Pure gold (24 K) is a comparatively soft element and is easily deformed. It is made stronger by mixing with other elements. For example, 18 K gold is 75% (by mass) gold with the rest silver and/or copper. Stainless steel is a mixture of three elements: 80% iron, 12% chromium, and 8% nickel. The percent composition of these alloys can be used to convert between the mass of the alloy and the mass of the component. The use of percent as a conversion factor is illustrated in the following two examples. Further exercises on the use of percent are found in Appendixes A and B.

EXAMPLE 2–5

Conversion of Total Mass to Mass of a Component

● **Working It Out**

Manganese steel is very strong and finds use as railroad rails. It is composed of 87% iron, 12% manganese, and 1% carbon. What is the mass of each of the three elements in a 253 kg sample of manganese steel?

Procedure
Percent means parts per 100. In this case, a conversion factor can be constructed from each percent relating kg of a component to kg of the steel as follows:

$$87\% \text{ iron} = \frac{87 \text{ kg iron}}{100 \text{ kg steel}}; \ 12\% \text{ manganese} = \frac{12 \text{ kg manganese}}{100 \text{ kg steel}};$$

$$1\% \text{ carbon} = \frac{1 \text{ kg carbon}}{100 \text{ kg steel}}$$

Each of these conversion factors can now be used as written to convert kg of steel to kg of a component.

Solution

$$253 \text{ kg steel} \times \frac{87 \text{ kg iron}}{100 \text{ kg steel}} = \underline{220 \text{ kg iron}}$$

$$253 \text{ kg steel} \times \frac{12 \text{ kg manganese}}{100 \text{ kg steel}} = \underline{30 \text{ kg manganese}}$$

$$253 \text{ kg steel} - 220 \text{ kg iron} - 30 \text{ kg manganese} = \underline{3 \text{ kg carbon}}$$

EXAMPLE 2–6

Conversion of Mass of a Component to Total Mass

A sample of brass is composed of 72% copper and the remainder zinc. What mass of brass can be made from 25 kg of zinc?

Procedure
The percent composition of zinc is 100% − 72% = 28% zinc.

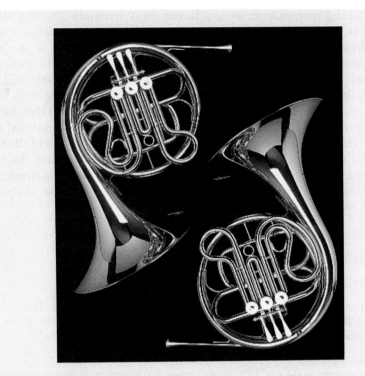

The French horn is made of brass which is a homogeneous solid mixture of copper and zinc.

The conversion factor and its reciprocal are

$$\frac{28 \text{ kg zinc}}{100 \text{ kg brass}} \quad \text{and} \quad \frac{100 \text{ kg brass}}{28 \text{ kg zinc}}$$

The latter factor can now be used to convert kg zinc to kg brass.

Solution

$$25 \text{ kg zinc} \times \frac{100 \text{ kg brass}}{28 \text{ kg zinc}} = \underline{\underline{89 \text{ kg brass}}}$$

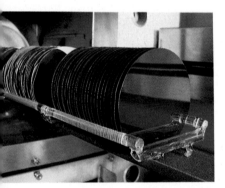

Figure 2–13
Silicon for Computer Chips The silicon used to make computer chips must be "ultrapure."

Actually, everything is a mixture, at least to some extent, since complete purity may be impossible. What we call "fresh drinking water" is hardly pure. It contains dissolved gases and solids. Even rainwater contains some dissolved gases from the air. What we consider pure may depend on the application. For example, if a sample of matter is composed of 99% of one element or compound, it may be considered pure for most purposes. On the other hand, the element silicon must be "super pure" to be used in computer chips. In this case, it is composed of more than 99.9999% silicon. (See Figure 2-13.)

Looking Back ▲

Very little of our surroundings can actually be classified as "pure." Most of what we see is a mixture of the pure substances known as elements and compounds. Sometimes pure substances mix uniformly with no boundaries between components, but sometimes the pure substances form more obvious mixtures with boundaries. Even when the presence of a mixture is not obvious, however, it has properties that are different than the pure component substances. Physical properties distinguish between a mixture and a pure substance.

◀ Checking It Out

Learning Check B

B–1. Fill in the blanks.
Pure substances can mix in either of two ways. If the mixture is composed of more than one phase, it is _____. If there are two liquid phases, the liquid on top has the _____ density. If they mix in such a way that only one phase is formed, the mixture is _____. Mixtures that result in one liquid phase are usually referred to as _____. These mixtures are distinguished from pure substances by their variable properties such as _____ and _____ points and density. Density of liquids is measured with a device called a _____. Mixtures of metals that result in one solid phase are referred to as _____. The percent composition of these mixtures relates the mass of a _____ to the total mass.

B–2. Carbon tetrachloride and water form a heterogeneous mixture. Which liquid is on the top? Refer to Table 2-1.

B–3. 14 K gold is 58.0% gold. What is the mass of pure gold in 4.00 oz of 14 K gold? What mass of 14 K gold can be made from 4.00 oz of pure gold?

B–4. The density of an alloy of aluminum and mercury is 5.40 g/mL. Which element is the principal metal in the alloy? Refer to Table 2-1.

Additional Examples: Problems 2-42, 2-49, 2-51, 2-55.

Matter ranges from complex heterogeneous mixtures, like a handful of soil, to the most basic form of pure matter—an element. We are now ready to examine the second component of the universe, which is energy. The change that matter undergoes in a chemical reaction is accompanied by a change in energy. We look next at the relationship of energy to chemical changes.

▼ Looking Ahead

2–6 ENERGY CHANGES IN CHEMICAL REACTIONS

Some days we just don't have any energy. In other words, we would prefer not to do any work. Actually, this is almost exactly the definition of energy. **Energy** *is the capacity or the ability to do work.* Just as matter has more than one physical state, energy has more than one form. Most of the energy on Earth originates from the sun. Deep in the interior of this star, transformations of elements occur that liberate a form of energy called *nuclear energy.* This energy, however, changes in the

What is the origin of our energy?

Almost all of the energy on Earth originates from the sun.

sun into *light* or *radiant energy* that then travels through space to illuminate Earth. Light energy from the sun warms our planet and shines on the surface vegetation, where it is converted into *chemical energy* by a process called photosynthesis. Chemical energy is stored in the energy-rich compounds that make up the bulk of the vegetation. When logs from a tree are burned, the chemical energy is released in the form of *heat energy.* In a similar process, the metabolism of food in our bodies releases the energy to keep us alive. In the burning or metabolism process, energy-poor compounds are produced, and they recycle to the environment. In the production of electrical power, the heat energy is used to produce steam that turns a turbine. The movement of the turbine is *mechanical energy.* The mechanical energy powers a generator that converts the mechanical energy into *electrical energy.* Other conversions between energy forms are possible. For example, in the chemical change that occurs in a car battery, chemical energy is converted directly into electrical energy. (See Figure 2-14.)

Chemical changes may be accompanied by either the release or the absorption of heat energy. *When a chemical reaction releases heat, the reaction is said to be* **exothermic.** *When a chemical reaction absorbs heat, the reaction is said to be* **endothermic.** Combustion (burning) is a common example of an exothermic chemical reaction. An example of an endothermic process involves an "instant cold pack." When the compounds ammonium nitrate (a solid) and water are brought together in a plastic bag, a solution is formed. The endothermic solution process causes enough cooling to make an ice pack useful for treating sprains and minor aches. (See Figure 2-15.)

In addition to the *forms* of energy, there are two *types* of energy. These depend on whether the energy is available but not being used or

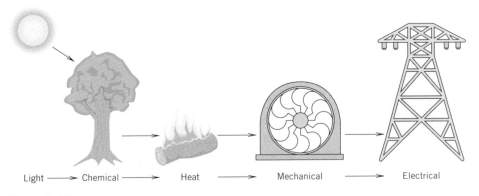

Light ⟶ Chemical ⟶ Heat ⟶ Mechanical ⟶ Electrical

Figure 2–14
Energy Energy is neither created nor destroyed but can be transformed.

is actually in use. **Potential energy** *is energy that is available because of position or composition.* For example, a weight suspended above the ground has energy available because of its position and the attraction of gravity for the weight. Water stored behind a dam (Figure 2-16), a compressed spring, and a stretched rubber band all have potential energy. The chemical energy stored in the compounds of a tree log is also classified as potential energy. *Energy resulting from motion is known as* **kinetic energy.** A moving baseball, a speeding train, and water flowing down a spillway from a dam (Figure 2-16) all have kinetic energy.

Although there are different forms and types of energy that are important in chemistry, heat energy has the most relevance to chemical changes. Perhaps the most obvious thing about heat energy is that it causes changes in the temperature of matter. Heat energy, just like matter, can be quantified. In other words, we can answer the question, How much heat? The quantification of heat energy and how heat changes the temperature of various substances is the next subject.

Can energy be stored?

▼**Looking Ahead**

Figure 2–15
Instant Cold Pack When capsules of ammonium nitrate are broken and mixed with water, a cooling effect results.

Figure 2–16
Potential and Kinetic Energy
Water stored behind a dam has potential energy. The water flowing over the spillway has kinetic energy.

2–7 TEMPERATURE CHANGE AND SPECIFIC HEAT

How does heat affect a substance?

It seems to take forever for a pot of water to get hot, but a heavy iron skillet over the same fire heats up quickly. Obviously, the same amount of heat changed the temperature of the skillet more than it did that of the water. A physical property of matter that relates temperature change to the amount of heat is known as specific heat capacity, or simply specific heat. **Specific heat** *is defined as the amount of heat required to raise the temperature of one gram of a substance one degree Celsius (or Kelvin)*. Like density, specific heat is the same for any amount of a substance, and it can be used to identify a pure substance.

Units of heat energy are related to the specific heat of water. *The* **calorie** *was originally defined as the amount of heat energy required to raise the temperature of one gram of water from 14.5°C to 15.5°C*. For the most part, the unit of heat energy used in chemistry is the SI unit called the **joule.** *The calorie is now defined in terms of the joule.*

$$1 \text{ cal} = 4.184 \text{ joule (J)} \quad \text{(exactly)}$$

From the definition of the calorie, it is apparent that the specific heat of water is

$$1.00 \frac{\text{cal}}{\text{g} \cdot {}^{\circ}\text{C}} = 4.184 \frac{\text{J}}{\text{g} \cdot {}^{\circ}\text{C}}$$

The units read: calories (or joules) per gram per degree Celsius. The degree Celsius unit represents a *temperature change* and not a specific temperature reading. A change in temperature is represented as Δt. A formula that can be used to calculate the specific heat is

$$\text{specific heat} = \frac{\text{amount of heat energy (J or cal)}}{\text{mass (g)} \times \Delta t \, ({}^{\circ}\text{C})}$$

The specific heats of several pure substances are listed in Table 2-2.

TABLE 2–2   SPECIFIC HEATS

Substance	Specific Heat [cal/(g · °C)]	Specific Heat [J/(g · °C)]
Water	1.00	4.18
Ice	0.492	2.06
Aluminum (Al)	0.214	0.895
Gold (Au)	0.031	0.13
Copper (Cu)	0.092	0.38
Zinc (Zn)	0.093	0.39
Iron (Fe)	0.106	0.444

The following examples illustrate the calculation of specific heat and the use of specific heat to calculate temperature change.

EXAMPLE 2–7

The Calculation of Specific Heat

It takes 62.8 J to raise a 125-g quantity of silver 0.714°C. What is the specific heat of silver?

● **Working It Out**

Procedure
The specific heat is calculated by substituting the appropriate quantities in the formula.

Solution

$$\text{specific heat} = \frac{62.8 \text{ J}}{125 \text{ g} \times 0.714°C} = \underline{\underline{0.704 \frac{\text{J}}{\text{g} \cdot °C}}}$$

EXAMPLE 2–8

The Calculation of Temperature Change

If 1.22 kJ of heat is added to 50.0 g of water at 25.0°C, what is the final temperature of the water? $\left(\text{The specific heat of water is } 4.184 \frac{\text{J}}{\text{g} \cdot °C} \right).$

Procedure
(a) Convert kJ to J.
(b) Solve for Δt and substitute the known values in the formula for specific heat.
(c) Find the final temperature of the water.

Solution

(a) $1.22 \cancel{\text{kJ}} \times 10^3 \frac{\text{J}}{\cancel{\text{kJ}}} = 1.22 \times 10^3 \text{ J}$

(b) $\text{specific heat} = \dfrac{\text{heat energy}}{\text{mass (g)} \times \Delta t(°C)}$

$\Delta t(°C) = \dfrac{\text{heat energy}}{\text{specific heat} \times \text{mass}} =$

$\dfrac{1.22 \times 10^3 \cancel{\text{J}}}{4.184 \dfrac{\cancel{\text{J}}}{\cancel{\text{g}} \cdot °C} \times 50.0 \cancel{\text{g}}} = 5.83°C = t(C) = 25.0 + 5.8 = \underline{\underline{30.8°C}}$

Why does it take so long to heat water?

The reason it takes so long to heat water is that it has a comparatively high specific heat. Notice in Table 2-2 that it is almost ten times higher than that of the iron in the skillet. Thus, in a calculation similar to the one in Example 2-8, the same amount of heat will raise the temperature of the iron almost ten degrees for every one degree for the same amount of water.

The health of this planet Earth depends on the high specific heat of water. For example, the sun warms the waters of the Gulf of Mexico. The Gulf Stream carries this warm water across the Atlantic Ocean all the way to England and the Scandinavian countries. There, the heat absorbed in the warm Gulf Stream is released and warms what would otherwise be uninhabitable countries. The island of Greenland is not warmed by the Gulf Stream and is almost entirely covered with ice the year around. That would be the fate of England were it not for the ability of water to absorb and retain vast quantities of heat energy.

The amount of energy that we obtain from food can also be measured and expressed in calories. The nutritional calorie is actually one *kilocalorie* (10^3 cal) as defined above. To distinguish between the two calories, the "c" in calorie is capitalized in the nutritional calorie (1 Cal). The Calorie content of a portion of food is determined by burning the dried food in such a way that the heat released is used to heat water. The temperature change of the water is then converted to the amount of heat. This is illustrated in Example 2-9.

EXAMPLE 2–9

The Calories in a Piece of Cake

Working It Out ●

A piece of cake is dried and burned so that all of the heat energy released heats some water. If 3.15 L of water is heated a total of 75.0°C, how many Calories does the cake contain?

Procedure
(a) Convert the volume of water to the mass of water using the density.
(b) Use the formula for specific heat to find the amount of heat energy.

$$\text{amount of heat energy} = \text{mass} \times \Delta t \times \text{specific heat}$$

(c) Convert calories to Calories.

Solution

(a) $3.15 \text{ L} \times \dfrac{10^3 \text{ mL}}{\text{L}} \times 1.00 \dfrac{\text{g}}{\text{mL}} = 3.15 \times 10^3 \text{ g}$

(b) $3.15 \times 10^3 \text{ g} \times 75.0°C \times 1.00 \dfrac{\text{cal}}{\text{g} \cdot °C} = 236 \times 10^3 \text{ cal}$

(c) $236 \times 10^3 \text{ cal} \times \dfrac{1 \text{ Cal}}{10^3 \text{ cal}} = \underline{\underline{236 \text{ Cal}}}$

Looking Back ▲

Mowing the grass is a lot of work. It takes energy. Fortunately, we usually have a good reserve of chemical energy from recent meals that we can put to use. As we metabolize the food, an exothermic reaction takes place when the chemical energy is converted into heat energy to power our bodies. Some of the heat energy is then converted into the mechanical

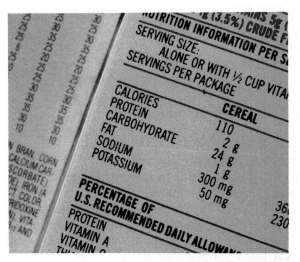

Many people are very conscious of the energy content (calories) of the food they eat.

energy of a moving lawn mower. These are conversions between forms of energy and types of energy. Chemical, heat, and mechanical are forms of energy, while potential (as in stored chemical energy) and kinetic (as in a moving lawn mower) are types of energy. If we get too hot from all of this effort, we may want to take a cool shower. We are aided in this endeavor by the high specific heat of water. As the cool water flows over our skin, the water absorbs the heat and becomes warmer and we become cooler.

◄Checking It Out

Learning Check C

C–1. Fill in the blanks.
Energy is defined as the ability to do _____. Most of the energy used on Earth arrives from the sun in the form of _____ energy. Another form of energy that we rely on to run our appliances is _____ energy. Chemical energy is a type of _____ energy that is stored in compounds. When this energy is released as heat in a chemical change or reaction, the reaction is said to be _____. When heat is added to matter, it causes an _____ in the temperature. The two units of heat energy that are used are the _____ and the _____. The specific heat of a substance is a _____ property. It is defined as the amount of heat required to raise one _____ of the substance _____ degree Celsius.

C–2. A 10.0-g sample of a certain metal is at 25°C. When 87.4 J of heat energy is added to the sample, the temperature increases to 48°C. Identify the metal from Table 2-2.

C–3. A 255-g quantity of gold is heated from 28 to 100°C. How many kilojoules of heat energy were added to the sample? (Refer to Table 2-2.)

Additional Examples: Problems 2-57, 2-60, 2-63, 2-64, 2-67, 2-69.

Looking Ahead ▼

The universe is composed of matter and energy. These two components of nature may seem totally independent. But are they really completely separate phenomena? For a long time it was thought so. A scientist named Albert Einstein first suggested that they are related. The conservation of mass and energy and the relationship between the two is our final topic of the chapter.

2–8   CONSERVATION OF MASS AND ENERGY

What happens to a log when it burns?

Only two centuries ago, scientists were still puzzled when wood burned leaving behind only a small portion of the mass in the form of ashes. At that time, the involvement of gases in chemical reactions was not fully understood. The apparent mass loss was explained by a popular theory, but we now know that the mass does not actually disappear. Most of the solid compounds of the wood have simply been transformed in the combustion process into gaseous compounds that are carried away in the atmosphere. The mass of the wood plus the mass of the oxygen from the air equals the mass of the ashes plus the mass of the gaseous combustion products. (See Figure 2-17.) Chemical changes illustrate the **law of conservation of mass,** *which states that matter is neither created nor destroyed in a chemical reaction.* This law is very important in chemistry and is the basis for quantitative relationships that will be discussed in Chapter 6.

In a previous section of this chapter, we also indicted that energy undergoes changes from one form to another, as illustrated in Figure 2-14. In all of these changes, the total energy remains the same. Energy changes are subject to the same law as matter changes in chemical reactions. **The law of conservation of energy** *states that energy cannot be created or destroyed but only transformed from one form to another.* In the production of electrical energy in Figure 2-13, only about 35% of the chemical energy is eventually transformed into electrical energy. The rest of the energy is lost as heat energy in the various transformations. The total energy remains constant, however.

What does mass have to do with energy?

For our purposes in chemistry, the laws of conservation of mass and energy are entirely valid and can be applied to matter and its changes. In actual fact, the two laws turn out to be less exact than was originally thought. In 1905, Albert Einstein proposed the now well-known relationship between mass and energy:

$E = mc^2$, *where E = energy, m = mass, and c = speed of light*

This amazing theory tells us that the two components of the universe, matter and energy, are actually convertible into each other. In fact, en-

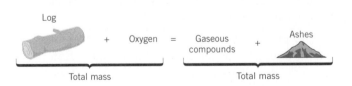

Log   +   Oxygen   =   Gaseous compounds   +   Ashes

Total mass   Total mass

Figure 2-17
Combustion of Wood In a chemical reaction, mass is conserved.

ergy that is released in any process originates from a loss of mass. The equation tells us, however, that a small amount of mass converts into a tremendous amount of energy. In fact, the energy that is involved in a chemical process results from a mass loss far too small to be detected by our most sensitive instruments. Therefore, mass changes in chemical reactions can be ignored. Nuclear processes, however, such as occur in the sun and in nuclear power reactors, do involve significant and measurable mass losses, which result in the release of vast amounts of energy. Nuclear energy is discussed in Chapter 15.

C H A P T E R   R E V I E W

Chemistry is a science concerned with the two components of the universe, **matter** and **energy.** Matter is composed of a few basic substances called **elements.** Actually, we don't find many of the free elements on Earth because they are chemically combined to form **compounds.** Both elements and compounds are known as **pure substances.** Elements and compounds are distinguished by their properties. They display **physical properties** and undergo **physical changes.** An important physical property of a substance is its **physical state: solid, liquid,** or **gas.** A physical change takes place when the substance changes to a different physical state. The temperatures at which an element or compound changes to a different **phase** are known as the **melting point** of a solid (or **freezing point** of a liquid) and **boiling point** of a liquid (or **condensation point** of a gas).

▲▼ Putting It Together

An important physical property of a substance is its **density** or **specific gravity.** Density relates the mass of a substance to its volume; thus it is independent of the size of the sample. For this reason, it can be used as an identifying property as well as a conversion factor between mass and volume of a sample. When density is expressed in units of g/mL or g/cm³, specific gravity has the same numerical quantity but is expressed without units. A pure substance also has **chemical properties** that relate to the **chemical changes** it undergoes.

	Physical	Chemical
Property	Describes properties that do not result in a change into another substance	Describes the ability of a substance to undergo various types of chemical reactions
Change	Changes in dimensions or physical state	Changes into another substance

Mixtures of elements and compounds may form either **homogeneous** or **heterogeneous mixtures.** In heterogeneous mixtures of liquids, the less dense liquid floats on the other liquid. Heterogeneous mixtures of solids and liquids can be separated by filtration. Two types of homogeneous mixtures are **solutions,** which are composed of one liquid phase, and **alloys,** which are composed of one solid phase. The components of a solution can be separated by distillation. Aqueous (water) solutions and pure water often look identical. Pure water, however, has definite and unchanging properties. The properties of a solution vary according to the proportions of the mixture. Among those properties that vary for a solution is density. A **hydrometer** can be used to measure the density of a solution as well as that of a pure liquid. In the diagram below, we start with the most complex type of impure matter and proceed to the simplest form of homogeneous matter, an element.

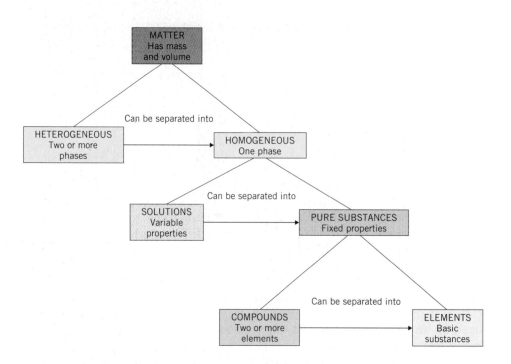

The other component of the universe, energy, is also intimately involved in chemical processes. There are several *forms* of energy such as chemical, heat, and electrical energy. There are also two *types* of energy: **potential energy** and **kinetic energy.** Forms of energy can be converted into each other. When chemical energy is converted into heat energy, the chemical change is said to be **exothermic.** When heat energy is converted into chemical energy, the reaction is said to be **endothermic.**

How much the temperature of a substance is affected by a certain amount of heat is a physical property known as **specific heat.** Specific heat is a measure of the amount of heat in **calories** or **joules** (or kcal or kJ) that will raise the temperature of one gram of the substance one degree Celsius.

In all of the transformations discussed in this chapter, whether between forms of energy or between matter in chemical changes, both mass and energy are conserved. These facts are stated in the laws of **conservation of mass** and **conservation of energy.** In fact, these two laws are more accurately stated as the law of conservation of mass *and* energy. Einstein's law indicates that one does convert to the other. However, the interchange between mass and energy in chemical changes is negligible.

E X E R C I S E S

ELEMENTS AND COMPOUNDS

2–1 List the elements that you think occur in the free state on the Earth. (Refer to the list of elements inside front cover.)

2–2 What is the difference between an element and a compound?

2–3 Identify the following as either elements or compounds. Refer to the list of elements inside the front cover if necessary.
(a) carbon monoxide (d) titanium
(b) hydrogen (e) potash
(c) iron (f) sodium bicarbonate

2–4 Identify the following as either elements or compounds. Refer to the list of elements inside the front cover if necessary.
(a) ammonia (d) mercury
(b) hydrogen peroxide (e) stannous fluoride
(c) aspirin (f) uranium

2–5 Two centuries ago, water was thought to be an element. Why do you think scientists decided that it is a compound?

2–6 A given compound can be decomposed into two other substances. Are the two substances both elements?

PHYSICAL AND CHEMICAL PROPERTIES AND CHANGES

2–7 Which of the following describes the liquid phase?
(a) It has a definite shape and a definite volume.
(b) It has a definite shape but not a definite volume.
(c) It has a definite volume but not a definite shape.
(d) It has neither a definite shape nor a definite volume.

2–8 Which state of matter is compressible? Why?

2–9 A fluid is a substance that flows and can be poured. Which state or states can be classified as fluids?

2–10 Identify the following as either a physical or a chemical property.
(a) Diamond is one of the hardest known substances.
(b) Carbon monoxide is a poisonous gas.
(c) Soap is slippery.
(d) Silver tarnishes.
(e) Gold does not rust.
(f) Carbon dioxide freezes at $-78°C$.
(g) Tin is a shiny, gray metal.
(h) Sulfur burns in air.
(i) Aluminum has a low density.

2–11 Identify the following as either a physical or a chemical property.
(a) Sodium burns in the presence of chlorine gas.
(b) Mercury is a liquid at room temperature.
(c) Water boils at 100°C at average sea-level pressure.
(d) Limestone gives off carbon dioxide when heated.
(e) Hydrogen sulfide has a pungent odor.

2–12 Identify the following as a physical or a chemical change.
(a) the frying of an egg
(b) the vaporization of dry ice
(c) the boiling of water
(d) the burning of gasoline
(e) the breaking of glass

2–13 Identify the following as a physical or a chemical change.
(a) the souring of milk
(b) the fermentation of apple cider
(c) the compression of a spring
(d) the grinding of a stone

2–14 Can an element be distinguished from a compound by examination of its physical properties only? Explain.

2–15   When table sugar is heated to its melting point, it bubbles and turns black. When it cools, it remains a black solid. Describe the changes.

2–16   Aluminum metal melts at 660°C and burns in oxygen to form aluminum oxide. Describe a physical property and change and a chemical property and change.

2–17   A pure substance is a green solid. When heated, it gives off a colorless gas and leaves a brown, shiny solid that melts at 1083°C. The shiny solid cannot be decomposed to simpler substances, but the gas can. List all of the properties given and tell whether they are chemical or physical. Tell whether each substance is a compound or an element.

2–18   A pure substance is a greenish-yellow, pungent gas that condenses to a liquid at −35°C. It undergoes a chemical reaction with a given substance to form a white solid that melts at 801°C and a brown, corrosive liquid with a specific gravity of 3.12. The white solid can be decomposed to simpler substances, but the gas and the liquid cannot. List all of the properties given and tell whether they are chemical or physical. Tell whether each substance is a compound or an element.

DENSITY

2–19   A handful of sand has a mass of 208 g and displaces a volume of 80.0 mL. What is its density?

2–20   A 125-g quantity of iron has a volume of 15.8 mL. What is its density?

2–21   A given liquid has a volume of 0.657 L and a mass of 1064 g. What might the liquid be? (Refer to Table 2-1.)

2–22   What is the volume in mL occupied by 285 g of mercury? (See Table 2-1.)

2–23   What is the mass of 671 mL of table salt? (See Table 2-1.)

2–24   What is the mass of 1.00 L of gasoline? (See Table 2-1.)

2–25   What is the mass of 1.50 L of gold? (See Table 2-1.)

2–26   What is the volume in milliliters occupied by 1.00 kg of carbon tetrachloride?

2–27   The volume of liquid in a graduated cylinder is 14.00 mL. When a certain solid is added, the volume reads 92.45 mL. The mass of the solid is 136.5 g. What might the solid be? (Refer to Table 2-1.)

2–28   Pumice is a volcanic rock that contains many trapped air bubbles. A 155-g sample is found to have a volume of 163 mL. What is the density of pumice? What is the volume of a 4.56-kg sample? Will pumice float or sink in water? In ethyl alcohol? (Refer to Table 2-1.)

2–29   The density of diamond is 3.51 g/mL. What is the volume of the Hope diamond if it has a mass of 44.0 carats (1 carat = 0.200 g)?

2–30   A small box is filled with liquid mercury. The dimensions of the box are 3.00 cm wide, 8.50 cm long, and 6.00 cm high. What is the mass of the mercury in the box? (1.00 mL = 1.00 cm³)

2–31   A 125-mL flask has a mass of 32.5 g when empty. When the flask is completely filled with a certain liquid, it weighs 143.5 g. What is the density of the liquid?

2–32   Which has a greater volume, 1 kg of lead or 1 kg of gold?

2–33   Which has the greater mass, 1 L of gasoline or 1 L of water?

2–34   What is the difference between density and specific gravity?

*****2–35**   What is the mass of 1 gal of carbon tetrachloride in grams? In pounds?

*****2–36**   Calculate the density of water in pounds per cubic foot (lb/ft³)?

*****2–37**   In certain stars, matter is tremendously compressed. In some cases the density is as high as 2.0×10^7 g/mL. A tablespoon full of this matter is about 4.5 mL. What is the mass of this tablespoon of star matter in pounds?

MIXTURES AND PURE SUBSTANCES

2–38   Carbon dioxide is not a mixture of carbon and oxygen. Explain.

2–39   List the following "waters" in order of increasing purity: ocean water, rainwater, and drinking water. Explain.

2–40   Iron is attracted to a magnet, but iron compounds are not. How could you use this information to tell whether a mixture of iron and sulfur forms a compound when heated?

2–41 When a teaspoon of solid sugar is dissolved in a glass of liquid water, what phase or phases are present after mixing?
(a) liquid only
(b) still solid and liquid
(c) solid only

2–42 Identify the following as homogeneous or heterogeneous matter.
(a) gasoline   (e) a new nail
(b) dirt   (f) vinegar
(c) smog   (g) aerosol spray
(d) alcohol   (h) air

2–43 Identify the following as homogeneous or heterogeneous matter.
(a) a cloud   (d) bourbon
(b) dry ice   (e) natural gas
(c) whipped cream   (f) a grapefruit

2–44 In which physical state or states does each of the substances listed in Problem 2-42 exist?

2–45 In which physical state or states does each of the substances listed in Problem 2-43 exist?

2–46 Carbon tetrachloride and kerosene mix, but neither mixes with water. How can water be used to keep the carbon tetrachloride and kerosene apart? Which liquid is on the top? (Refer to Table 2-1.)

2–47 How could liquid mercury be used to tell whether a certain sample of metal is lead or gold? (Refer to Table 2-1.)

2–48 Why does ice float on water? (Refer to Table 2-1.) Is ice water homogeneous or heterogeneous matter? Pure or a mixture?

2–49 Tell whether each of the following properties describes a heterogeneous mixture, a solution (homogeneous mixture), a compound, or an element.
(a) a homogeneous liquid that, when boiled away, leaves a solid residue
(b) a cloudy liquid that after a time seems more cloudy toward the bottom
(c) a uniform red solid that has a definite, sharp melting point and cannot be decomposed into simpler substances
(d) a colorless liquid that boils at one unchanging temperature and can be decomposed into simpler substances
(e) a liquid that first boils at one temperature

but as the heating continues, boils at slowly increasing temperatures. (There is only one liquid phase.)

2–50 Tell whether each of the following properties describes a heterogeneous mixture, a solution (homogeneous mixture), a compound, or an element.
(a) a nonuniform powder that, when heated, first turns mushy and then continues to melt as the temperature rises.
(b) a colored gas that can be decomposed into a solid and another gas (The entire sample of gas seems to have the same chemical properties.)
(c) a sample of colorless gas, only part of which reacts with hot copper

2–51 A sample of bronze is made by mixing 85 kg of molten tin with 942 kg of molten copper. What is the mass percent tin in this bronze?

2–52 A sample of brass weighs 22.8 lb. It contains 14.8 lb of copper and the rest zinc. What is the mass percent zinc in this brass?

2–53 A U.S. "nickel" is actually only 25% nickel. What mass of the element nickel is in 255 kg of "nickels"?

2–54 10 K gold is only 42% gold. What mass of gold is in 186 g of 10 K gold?

2–55 Duriron is used to make pipes and kettles. It contains 14% silicon, with the remainder iron. What mass of duriron can be made from 122 lb of iron?

2–56 A sample of a gold alloy is 7% copper and 10% silver in addition to the gold. What mass of the gold alloy can be made from 175 kg of pure gold?

MASS AND ENERGY

2–57 From your own experiences, tell whether the following processes are exothermic or endothermic.
(a) decay of grass clippings
(b) melting of ice
(c) change in an egg when it is fried
(d) condensation of steam
(e) curing of freshly poured cement

2–58 A car battery can be recharged after the engine starts. Trace the different energy conversions from gasoline to the battery.

2–59   Windmills are used to generate electricity. What are all of the different forms of energy involved in generation of electricity by this method?

2–60   Identify the principal type of energy (kinetic or potential) exhibited by each of the following:
(a) a car parked on a hill
(b) a train traveling 60 mi/hr
(c) chemical energy
(d) an uncoiling spring
(e) a falling brick

2–61   Identify the following as having either potential or kinetic energy or both.
(a) an arrow in a fully extended bow
(b) a baseball traveling high in the air
(c) two magnets held apart
(d) a chair on the fourth floor of a building

2–62   When you apply your brakes to a moving car, the car loses kinetic energy. What happens to the lost energy?

2–63   When a person plays on a swing, at what point in the movement is kinetic energy the greatest? At what point is potential energy the greatest? Assume that once started, the person will swing to the same height each time without an additional push. At what point is the total of the kinetic energy and the potential energy greatest?

SPECIFIC HEAT

2–64   It took 73.2 J of heat to raise the temperature of 10.0 g of a substance 8.58°C. What is the specific heat of the substance?

2–65   When 365 g of a certain pure metal cooled from 100°C to 95°C, it liberated 56.6 cal. Identify the metal from among those listed in Table 2-2.

2–66   A 10.0-g sample of a metal requires 22.4 J of heat to raise the temperature from 37.0°C to 39.5°C. Identify the metal from those listed in Table 2-2.

2–67   If 150 cal of heat energy is added to 50.0 g of copper at 25°C, what is the final temperature of the copper? Compare this temperature rise with that of 50.0 g of water initally at 25°C. (Refer to Table 2-2.)

2–68   How many joules are evolved if 43.5 g of aluminum is cooled by 13 Celsius degrees? (Refer to Table 2-2.)

2–69   What mass of iron is needed to absorb 16.0 cal if the temperature of the sample rises from 25°C to 58°C?

2–70   When 486 g of zinc absorbs 265 J of heat energy, what is the rise in temperature of the metal?

2–71   Given 12.0-g samples of iron, gold, and water. Calculate the temperature rise that would occur when 50.0 J of heat is added to each.

2–72   A 10.0-g sample of water cools 2.00°C. What mass of aluminum is required to undergo the same temperature change?

2–73   A can of diet soda contains 1.00 Cal (1.00 kcal) of heat energy. If this energy was transferred to 50.0 g of water at 25°C, what would be the final temperature?

2–74   If 50.0 g of aluminum at 100.0°C is allowed to cool to 35.0°C, how many joules are evolved?

*2–75   In the preceding problem assume that the hot aluminum was added to water originally at 30.0°C. What mass of water was present if the final temperature of the water was 35.0°C? (Hint: The heat energy lost by the aluminum is equal to the heat energy gained by the water.)

*2–76   If 50.0 g of water at 75.0°C is added to 75.0 g of water at 42.0°C, what is the final temperature?

*2–77   If 100.0 g of a metal at 100.0°C is added to 100.0 g of water at 25.0°C, the final temperature is 31.3°C. What is the specific heat of the metal? Identify the metal from Table 2-2.

GENERAL QUESTIONS

2–78   A 22-mL quantity of liquid A has a mass of 19 g. A 35-mL quantity of liquid B has a mass of 31 g. If they form a heterogeneous mixture, explain what happens.

2–79   The volume of water in a graduated cylinder reads 25.5 mL. What does the volume read when a 25.0-g quantity of pure nickel is added to the cylinder? The density of nickel is 8.91 g/mL.

2–80   A solution of table sugar in water is 14% by mass sugar. The density of the solution is 1.06 g/mL. What is the mass of sugar in 100 mL of the solution?

2–81   Battery acid is 35% sulfuric acid in water and has a density of 1.29 g/mL. What is the mass of sulfuric acid in 1.00 L of battery acid?

2–82   An alloy is prepared by mixing 50.0 mL of gold with 50.0 mL of aluminum. What is the mass percent gold in the alloy?

2–83   When 215 J of heat is added to a 25.0-g sample of a given substance, the temperature increases from 25°C to 91°C. What is the volume of the sample? (Refer to Tables 2-1 and 2-2.)

2–84   25.0 mL of a given pure substance has a mass of 67.5 g. How much heat in kilojoules (kJ) is required to heat this sample from 15°C to 88°C? (Refer to Tables 2-1 and 2-2.)

2–85   When a log burns, the ashes have less mass than the log. When zinc reacts with sulfur, the zinc sulfide has the same mass as the combined mass of zinc and sulfur. When iron burns in air, the compound formed has more mass than the original iron. Explain each reaction in terms of the law of conservation of mass.

SOLUTIONS TO LEARNING CHECKS

A–1   space, compound, pure, properties, density, physical, chemical

A–2   Physical—dull, gray solid, melting point of 839°C, specific gravity of 1.54
Chemical—reacts with water

A–3 $\dfrac{254 \text{ g}}{159 \text{ mL}} = 1.60$ g/mL (carbon tetrachloride)

A-4   178 g aluminum $\times \dfrac{1 \text{ mL}}{2.70 \text{ g aluminum}} = 65.9$ mL

B–1   heterogeneous, lesser, homogeneous, solutions, melting, boiling, hydrometer, alloys, component

B–2   Water floats on top of carbon tetrachloride.

B–3   4.00 oz 14 K gold $\times \dfrac{58.0 \text{ oz pure gold}}{100 \text{ oz 14 K gold}} = 2.32$ oz pure gold

4.00 oz pure gold $\times \dfrac{100 \text{ oz 14 K gold}}{58.0 \text{ oz pure gold}} = 6.90$ oz of 14 K gold

B–4   Aluminum (The density of the alloy is closer to that of aluminum.)

C–1   work, radiant or light, electrical, potential, exothermic, increase, calorie, joule, physical, gram, one

C–2 $\Delta t = 48 - 25 = 23°C$

specific heat $= \dfrac{87.4 \text{ J}}{10.0 \text{ g} \cdot 23°C} = 0.38 \dfrac{\text{J}}{\text{g} \cdot °C}$ (copper)

C–3 $\Delta t = 100 - 28 = 72°C$

heat energy = specific heat $\times$ mass $\times \Delta t =$

$0.13 \dfrac{\text{J}}{\text{g} \cdot °C} \times 255 \text{ g} \times 72°C = 2.4 \times 10^3$ J

$2.4 \times 10^3 \text{ J} \times \dfrac{1 \text{ kJ}}{10^3 \text{ J}} = 2.4$ kJ

Elements are being created in the intensely hot interiors of stars such as the one shown here in the gaseous nebula in Serpens. All known substances in earth are composed of these elements.

ELEMENTS, COMPOUNDS, AND THEIR COMPOSITION

In late February of 1987, an astronomer on a desolate mountaintop in Chile looked out into the clear night sky through a huge telescope. He planned nothing more than a routine study of the skies, but he was in for the surprise of his life. A bright new star had appeared, one that hadn't been there the night before. This wasn't a normal star burning steadily like our sun. It was a star in its death throes undergoing a catastrophic explosion known as a supernova. It was to be the brightest supernova seen on Earth in almost four hundred years. While this dramatic event signaled the end of a star, it also meant the creation of vast quantities of elements. Scientists now believe that heavier elements originate only in these violent infernos in space.

◄ **Setting The Stage**

For billions of years, stars have been churning, boiling, and eventually undergoing catastrophic explosions such as the one seen in 1987. Blasted by the force of exploding suns, new-born elements and some simple compounds of the elements have drifted through cold, dark space as bits of dust and clouds of gases. About four and a half billion years ago, a large cloud containing hydrogen and some heavier elements produced by previous supernovas condensed to form a sun with nine planets. The third planet out from this sun is called Earth. On Earth, there arose living systems composed of these elements. We—and all of nature as we know it—are, in fact, composed of "stardust."

In the past one hundred years, the miracle of the human mind, with its boundless curiosity, has unraveled many mysteries about the composition and origin of these elements. For example, it is now universally accepted that elements are composed of very small fundamental units

Formulating
Some Questions ▶

called atoms. The existence of the atom brings up many additional questions that we will discuss in this chapter. How did the concept of the atom come to be accepted? If atoms are the fundamental unit of an element, what is the fundamental unit of a compound? What distinguishes one compound from another? Are the atoms of an element exactly the same? How do the atoms of one element differ from those of another? What determines the mass of an atom? Before we address these questions, however, we have more to say about the number of elements, their distribution, and how we name them.

3–1 NAMES AND SYMBOLS OF THE ELEMENTS

There are now 109 known elements but, of these, only 88 occur in nature in measureable amounts. The other 21 are products of the nuclear age, as they were synthesized in laboratories or nuclear reactors. These 21 elements spontaneously change into other elements by radioactive decay. Of the 88 elements found naturally, only 10 constitute over 99% of Earth's crust (the outer portion including the oceans and the atmosphere). About 93% of the mass of our bodies is composed of only three elements—carbon, hydrogen, and oxygen. (See Table 3-1.)

The names of elements come from many sources. Some are derived from Greek, Latin, or German words for a color—for example, bismuth (white mass),. iridium (rainbow), rubidium (deep red), and chlorine (greenish yellow). Some relate to the locality where the element was discovered (e.g., germanium, francium, and californium). Four elements (yttrium, erbium, terbium, and ytterbium) are named after a town in Sweden (Ytterby). Other elements honor noted scientists (e.g., einsteinium, fermium, and curium) or mythological figures (e.g, plutonium, uranium, titanium, mercury, and promethium [the fire-giver]). Many of the oldest known elements have names with obscure origins. Elements

Our universe is composed of less than 90 elements.

TABLE 3–1   DISTRIBUTION OF THE ELEMENTS IN THE EARTH'S CRUST* AND HUMAN BODY

Earth's Crust				Human Body	
Element	Weight Percent	Element	Weight Percent	Element	Weight Percent
Oxygen	49.1	Magnesium	1.9	Oxygen	64.6
Silicon	26.1	Hydrogen	0.88	Carbon	18.0
Aluminum	7.5	Titanium	0.58	Hydrogen	10.0
Iron	4.7	Chlorine	0.19	Nitrogen	3.1
Calciuim	3.4	Carbon	0.09	Calcium	1.9
Sodium	2.6	Sulfur	0.06	Phosphorus	1.1
Potassium	2.4	All others	0.50	Chlorine	0.9
				All others	0.9

*Includes the oceans and the atmosphere.

TABLE 3–2 SOME COMMON ELEMENTS

Element	Symbol	Element	Symbol
Aluminum	Al	Iodine	I
Bromine	Br	Magnesium	Mg
Calcium	Ca	Nickel	Ni
Carbon	C	Nitrogen	N
Chlorine	Cl	Oxygen	O
Chromium	Cr	Phosphorus	P
Fluorine	F	Silicon	Si
Helium	He	Sulfur	S
Hydrogen	H	Zinc	Zn

bers. Permanent names can be assigned only by the International Union of Pure and Applied Chemistry (IUPAC). All of the elements are listed inside the front cover of this book.

An element can be designated by a symbol that is shorthand for its full name. *In most cases, the first one or two letters of the name are used as the element's* **symbol.** When an element has a two-letter symbol, the first is capitalized but the second is not. The symbols of the elements are listed along with the names inside the front cover. Some common elements whose symbols are derived from their English names are shown in Table 3-2. Each of the last six elements has a three-letter symbol.

Among the elements listed there are 11 whose symbols do not seem to relate to their names. These elements have symbols are derived from their original Latin names (except wolfram, which is German) and are listed in Table 3-3.

TABLE 3–3 ELEMENTS WITH SYMBOLS FROM EARLIER NAMES

Element	Symbol	Former Name
Antimony	Sb	Stibium
Copper	Cu	Cuprum
Gold	Au	Aaurum
Iron	Fe	Ferrum
Lead	Pb	Plumbum
Mercury	Hg	Hydragyrum
Potassium	K	Kalium
Silver	Ag	Argentum
Sodium	Na	Natrium
Tin	Sn	Stannum
Tungsten	W	Wolfram

How does one recognize the name of a compound?

The names of compounds are usually based on the elements of which they are composed, and most contain two words. Carbon dioxide, sodium sulfite, and silver nitrate all refer to specific compounds. Notice that the second word ends in *ide, ite,* or *ate.* A few compounds have three words, such as sodium hydrogen carbonate. A number of compounds have common names with one word, such as water, ammonia, lye, and methane. We will talk more about the names of compounds in the next chapter.

Looking Ahead ▼

In the next section, we will try to visualize the tiny atoms of an element.

3–2   COMPOSITION OF AN ELEMENT AND ATOMIC THEORY

Two centuries ago, the most knowledgeable scientists thought that a sample of an element such as copper could be divided (theoretically) into infinitely smaller pieces without changing its nature. In other words, they believed that matter was continuous. In 1803, however, an English scientist named John Dalton (1766-1844) proposed a completely different theory of matter. His ideas are now known as *atomic theory.* The major conclusions of atomic theory are as follows:

1   Matter is composed of small indivisible particles called atoms.

2   Atoms of the same element are identical and have the same properties.

3   Chemical compounds are composed of atoms of different elements combined in small whole-number ratios.

4   Chemical reactions are merely the rearrangement of atoms into different combinations.

Dalton's theory is the basis of our current view of matter. Thus we may define an **atom** as *the smallest fundamental particle of an element that has the properties of that element.*

Actually, Dalton wasn't the first to propose such a theory. Over two thousand years earlier, a Greek philosopher named Democritus had suggested that matter was not continuous but existed as basic particles. The proposals of Democritus were simply the product of his mind, however. Dalton's theory was a brilliant and logical explanation of many quantitative experimental observations and laws that were known at the time. The law of conservation of mass (Section 2-8) was among the laws that were explained by atomic theory. Since chemical reactions are simply the shuffling of atoms, none are lost or gained and the mass remains unchanged.

How can we be so sure about the atomic theory?

Why are we so sure that Dalton was right? Besides the overwhelming amount of circumstantial evidence, we now have direct proof. In recent years, a highly sophisticated instrument called the scanning tunneling microscope has produced images of atoms of the heavier elements. Although these images are somewhat fuzzy, they indicate that

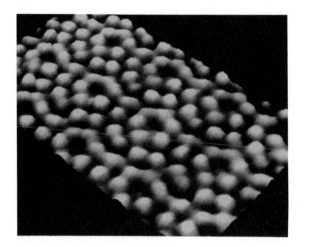

Figure 3–1
Atoms and Marbles The fuzzy spheres in the photograph are individual atoms of silicon. They stack together like the marbles in the glass container.

an element such as (silicon) is composed of spherical atoms packed closely together, just as you would find in a container of marbles all of the same size. (See Figure 3-1.)

When we look at a small piece of copper wire, it is hard to imagine that it is not continuous. This is because it is so difficult to comprehend the small size of the atom. Since the diameter of a typical atom is on the order of 10^{-8} cm, it would take about 10^{16} (ten quadrillion) atoms to appear as a tiny speck. The piece of copper wire is like a brick wall: from a distance it looks completely featureless, but up close we would notice that it is actually composed of closely packed basic units (bricks).

Most of the matter around us is not composed of separate atoms. In fact, atoms are usually joined together with atoms of the same element or other elements. We will examine this phenomenon next.

▼Looking Ahead

3–3 ELEMENTS AND COMPOUNDS COMPOSED OF MOLECULES, AND THEIR FORMULAS

The air that we breathe is composed of three elements plus traces of other gases. These elements are nitrogen (78%), oxygen (21%), and argon (less than 1%). Let's assume that we can magnify a sample of air so that we can see the atoms of these elements. We would notice that argon exists as individual atoms, but that the atoms of nitrogen and oxygen are bonded together in pairs. (See Figure 3-2.) *The force that bonds two atoms together is known as a* **covalent bond.** *A basic unit of two or more atoms held together by covalent bonds is known as a* **molecule**.

Can an element be molecular?

Hydrogen, fluorine, chlorine, bromine, and ıodine in their elemental form also exist as diatomic (two-atom) molecules under normal tempera-

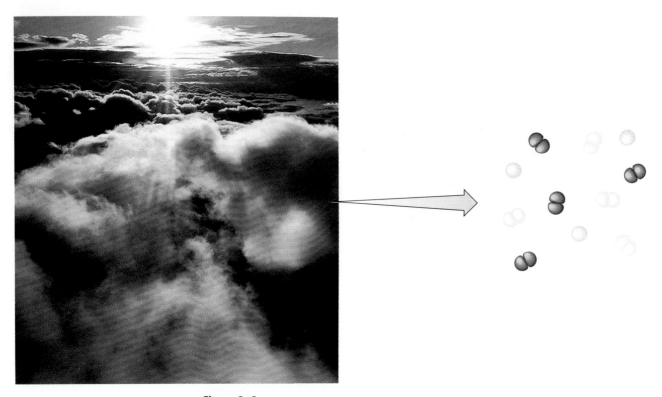

Figure 3–2
Atoms and Molecules Our atmosphere is composed mostly of nitrogen and oxygen molecules, and a small amount of argon atoms.

ture conditions. A form of elemental phosphorus consists of molecules composed of four atoms, and a form of sulfur consists of molecules composed of eight atoms. There is also a second form of elemental oxygen, known as ozone, which is composed of three atoms.

In elements composed of molecules, the atoms are of the same type. Atoms of different elements also combine to form molecules. A **molecular compound** *is composed of molecules with atoms of two or more different elements.* If it were possible to magnify a droplet of water so as to visualize its basic units, we would see that it is an example of a molecular compound. Each molecule of water is composed of two atoms of hydrogen joined by covalent bonds to one atom of oxygen. (See Figure 3-3.) Molecules can contain as few as two atoms or, in the cases of the complex molecules on which life is based, millions of atoms.

A compound is represented by using the symbols for the elements of which it is composed. This is called the **formula** *of the compound or element* (if it is composed of molecules). The familiar formula for water is therefore

$$H_2O$$

This, of course, is pronounced "Haitch-two-Oh." Note that the 2 is written as a subscript, indicating that the molecule has two hydrogen atoms. When there is only one atom of a given element present (e.g., oxygen), a subscript of "1" is assumed but not shown.

Figure 3–3
A Water Molecule A water molecule is composed of two atoms of hydrogen and one atom of oxygen.

What makes one compound different from another? The answer is that each chemical compound has a unique formula or arrangement of atoms in its molecules. For example, there is another compound composed of just hydrogen and oxygen, but it has the formula H_2O_2. Its name is hydrogen peroxide, and its properties are distinctly different from water (H_2O). The formulas of other well-known compounds are $C_{12}H_{22}O_{11}$ (table sugar, also known as sucrose), $C_9H_8O_4$ (aspirin), NH_3 (ammonia), and CH_4 (methane). Elements composed of molecules are also referred to by formulas such as I_2 (iodine), O_2 (oxygen), O_3 (ozone), and P_4 (phosphorus).

Sometimes two or more compounds may share the same chemical formula. What then makes them unique? In this case, their difference stems from the arrangement of the atoms within the molecule. For example, ethyl alcohol and dimethyl ether both have the formula C_2H_6O. The difference in the two compounds lies in the order of the bonded atoms.

Do all compounds have unique formulas?

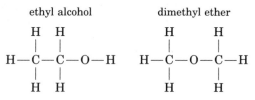

Notice that the alcohol has a C—C—O order and the ether a C—O—C order. (The dashes between atoms represent chemical bonds.) The difference in the arrangement has a profound effect on the properties of these two compounds. Ingestion of alcohol causes intoxication, while a similar amount of ether may cause death. *Formulas that show*

the order and arrangement of specific atoms are known as **structural formulas.**

Most of our world, from a speck of dust to the mightiest mountain, is composed of just a few of the 88 elements occurring in nature. Elements are made up of extremely tiny particles called atoms. The atoms of one element are alike, but they are different from those of another element. Atoms may bond together to form molecules. Molecules are the fundamental unit of some elements and of a specific type of compound. Just as we designated an element by a symbol, we designate a compound by a formula. In molecules of a given compound, the atoms are arranged in a unique manner.

Learning Check A

A–1. Fill in the blanks.
The symbol for the element nitrogen is _____ and that of copper is _____ . The symbol Ni represents the element _____ . The fundamental particle of an element is called an _____ . When these fundamental particles are covalently bonded to each other, they form _____ . The formula of a compound that contains two atoms of carbon, seven atoms of hydrogen, and one atom of nitrogen is written as _____ . If two compounds have the same formula, they must have different _____ formulas.

A–2. How many atoms of each element are present in a molecule of acetominophen (a pain reliever), whose formula is $C_8H_9NO_2$?

Additional Examples: Problems 3-5, 3-6, 3-12, 3-15.

The type of compound that we have discussed is composed of molecules and tends to be the "soft" part of nature—gases, liquids, and solids that melt at low temperatures. What about rocks and the minerals? What are they composed of if not molecules? Another type of compound, called ionic, consists of solids that are hard and have high melting points.

3–4   ELECTRICAL NATURE OF MATTER: COMPOUNDS COMPOSED OF IONS

Electricity is an awesome and powerful force. It is essential in our modern world but still should be treated with respect because of its danger. All matter is electrical in nature. Just shuffling across a rug can create a stinging spark of electricity. Lightning is the same phenomenon but on a much grander scale. (See Figure 3-4.) Sparks and lightning occur because there are two types of electrostatic charge—positive and negative. **Electrostatic forces** *exist between the charges. There is a force of attraction between opposite charges and a force of repulsion between like*

Figure 3–4
Lightning The phenomenal bolts of electricity, known as lightning, are a result of electrostatic forces.

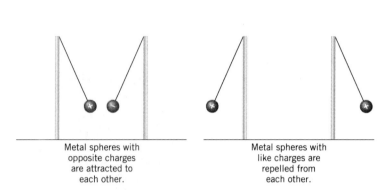

Metal spheres with opposite charges are attracted to each other.

Metal spheres with like charges are repelled from each other.

Figure 3–5
Electrostatic Forces Opposite charges attract; like charges repel.

charges. (See Figure 3-5.) Lightning is a result of this force of attraction. It is a flow of negative charge between the ground that has become negatively charged and a positively charged cloud (or sometimes vice versa). Air normally separates the two charges, but when the attraction becomes too great, the positive and negative charges seek each other through the bolt of lightning.

Atoms can also achieve an electrostatic charge. *When atoms become electrostatically charged, they become* **ions.** *Positively charged ions are known as* **cations**, *and negatively charged ions are called* **anions.** Ordinary table salt is a compound named sodium chloride. In sodium chloride, the sodium is a cation with a single positive charge and the chlorine is an anion with a single negative charge. This is illustrated with a + or a − as a superscript to the right of the symbol of the element.

What does lightning have to do with the atom?

$$Na^+ \text{ and } Cl^-$$

Since the sodium cations and the chlorine anions are oppositely charged, the ions are held together by electrostatic forces of attraction. In Figure 3-6, the ions in sodium chloride are shown as they would appear if sufficient magnification were possible. Note that each cation (one of the smaller spheres) is attached not just to one anion. In fact, each ion is surrounded by six oppositely charged ions. This situation is much different from that of the previously discussed molecular compounds, in which atoms bonded together to form discrete groups of molecules.

Compounds consisting of ions are known as **ionic compounds**. *The electrostatic forces holding the ions together are known as* **ionic bonds.** The ions in ionic compounds are locked tightly in their positions by the strong electrostatic attractions. This results in solid compounds that are almost all hard and rigid, with high melting points. The principal components of most rocks and minerals are ionic compounds.

The formula of sodium chloride is

NaCl

This represents the simplest ratio of cations to anions present (in this case, one to one). This ratio reflects the fact that the two ions have equal and opposite charges and that the negative charge balances the positive

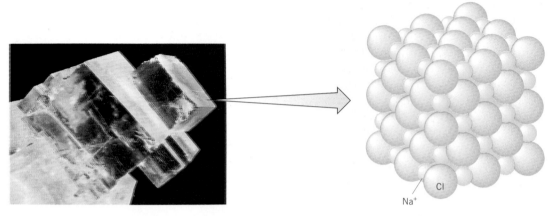

Figure 3–6
The Ions in Sodium Chloride An ionic compound such as sodium chloride exists as an arrangement of charged particles.

What is the difference between a molecular unit and a formula unit?

charge. Notice that the charges are not displayed in the formula. *The simplest whole number ratio of ions in an ionic compound is referred to as a* **formula unit.** Ions may also have multiple charges. Calcium bromide is a compound composed of Ca^{2+} cations and Br^- anions. Its formula is

$$\mathbf{CaBr_2}$$

The white cliffs of Dover along the British coast are composed of calcium carbonate, which is an ionic compound.

An aluminum atom attains a +3 charge to become an aluminum cation represented as Al^{3+}, and a sulfur atom attains a −2 charge to become a sulfur anion represented as S^{2-}. *Groups of atoms that are covalently bonded to each other may also be cations or anions, and they are known as* **polyatomic ions.** Examples are the nitrate anion (NO_3^-), the sulfate anion (SO_4^{2-}), and the ammonium cation (NH_4^+). *In ionic compounds, cations and anions always exist together in a ratio such that the total positive charge is balanced by the total negative charge.* When more than one polyatomic ion is in a formula unit, parentheses and a subscript are used. When there is only one polyatomic ion, no parentheses are used. Barium perchlorate (a compound containing one Ba^{2+} and two ClO_4^-) is represented as in (a), and calcium carbonate (a compound containing one Ca^{2+} and one CO_3^{2-}) is represented as in (b).

(a) **Ba(ClO$_4$)$_2$**   (b) **CaCO$_3$**

We will spend more time on writing and naming ionic compounds in the next chapter.

The matter that we breathe, eat, swim in, or walk on is composed of either individual atoms, molecules, or oppositely charged ions. The formula of a compound represents either a discrete molecular unit, in the case of molecules, or a ratio of ions, in the case of ionic compounds.

▲Looking Back

Learning Check B

◀Checking It Out

B–1. Fill in the blanks.
Compounds are composed of either _____ or _____. The formula of an ionic compound does not represent discrete molecules but represents the smallest whole number ratio of _____ to

_____.

B–2. Write the formula of a compound containing three Na^+ cations and one PO_4^{3-} anion, and a compound containing two NH_4^+ cations and one S^{2-} anion.

Additional Examples: Problems 3-19, 3-21, 3-23, 3-25.

Having identified the atom as the smallest building block of nature, we are now ready to look at the structure of the atom itself. We will find that the atom is composed of even smaller and more fundamental particles. We will begin our journey into the atom in the next section.

▼Looking Ahead

3–5   COMPOSITION OF THE ATOM

It was only about a hundred years ago that scientists began to realize that atoms were not hard, featureless spheres. Beginning in the late 1880s and continuing today, the mysteries and complexities of the atom have been slowly discovered. Ingenious experiments of scientists by the names of Thomson, Rutherford, Becquerel, Curie, and Roentgen, among others, contributed to the current nuclear model of the atom. Much evidence indicates that, if we could ever peer into the atom through a microscope, what follows is what we would see.

In a journey within the confines of the atom, we would first encounter a very small particle called an **electron.** The electron was the first subatomic particle to be identified. In 1897, J. J. Thomson characterized the electron by proving that it had a negative electrical charge (assigned a value of -1) and was common to the atoms of all elements. Matter was electrical in nature because atoms themselves contained electrically charged particles. However, since atoms were known to be neutral, Thomson proposed that it must also contain positive charge to counterbalance the negative charge. From this information, Thomson is usually credited with suggesting the "plum pudding" model of the atom, which explained the facts known at the time. (Plum pudding was a popular English dessert.) In this model, the positive charge was thought to be diffuse and evenly distributed throughout the volume of the atom (analogous to pudding). The negative particles (electrons) would be distributed throughout the atom (analogous to raisins in the pudding). (See Figure 3-7.)

The next major development occurred in 1911. Ernest Rutherford in England conducted experiments meant to support the accepted model of Thomson. His results, however, could be explained only by a radically different model. When students in his laboratories bombarded a thin foil of gold with fast-moving alpha particles (positively charged helium ions), they expected these very small particles to pass right through the atoms of the gold with very little effect, like bullets through a stick of butter. They did not expect the atom to contain anything massive enough to affect the path of the alpha particles. Instead, they found that some of the alpha particles were deflected by large angles, and a few were even deflected right back toward the source. After repeated experiments, Rutherford reluctantly concluded that the plum pudding model must be wrong. He realized that there could be only one explanation for the results of his experiments: The atom must contain a small, dense core containing most of the mass of the atom and all of the positive charge. When alpha particles had a close encounter with the core, they were deflected. "Direct hits" would reflect the alpha particle back toward the source. (See Figure 3-8.) This core was called the **nucleus.** Most of the volume of the atom is actually empty space occupied by the very small electrons. To get an idea about proportions, imagine a nucleus expanded to the size of a softball. In this case, the radius of the atom would extend for about one mile.

Later experiments would show that the nucleus is composed of particles called **nucleons.** There are two types of nucleons: **protons,** which have a positive charge (assigned a value of $+1$, equal and opposite to that of an electron), and **neutrons,** which do not carry a charge. (See Figure 3-9.) The data on the three particles in the atom are summarized in Table 3-4. The proton and neutron have roughly the same mass, which is about "1 amu" (1.67×10^{-24} g). The amu (atomic mass unit) is a convenient unit for the masses of individual atoms and subatomic particles. This unit will be defined more precisely in the next section.

For some time, it was thought that the atom was composed of just these three particles. As experimental procedures became more elabo-

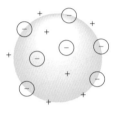

Figure 3–7
The Plum Pudding Model
Electrons were thought to be like tiny particles distributed in a positively charged medium.

What is wrong with Thomson's model?

What is an atom made of?

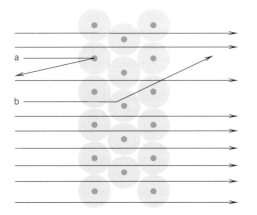

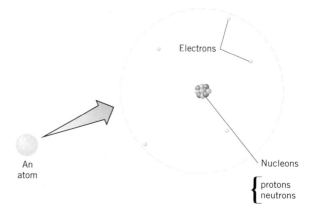

Figure 3–8
Rutherford's Experiment To account for the deflection of some of the positively charged alpha particles, Rutherford proposed that the atom is mostly empty space with the positive charge located in a small, dense core. When an alpha particle encountered a nucleus (a and b) electrostatic repulsions would either deflect it or repel it back.

Figure 3–9
The Composition of the Atom The atom is composed of electrons, neutrons, and protons.

		Electrical		
TABLE 3–4	**ATOMIC PARTICLES**			
Name	Symbol	Electrical Charge	Mass (amu)	Mass (g)
Electron	e	-1	0.000549	9.110×10^{-28}
Proton	p	$+1$	1.00728	1.673×10^{-24}
Neutron	n	0	1.00867	1.675×10^{-24}

rate and sophisticated, however, the picture became more complicated. It now appears that the three particles are themselves composed of various combinations of even more fundamental particles called quarks. So, of what are atoms composed? The answer is actually quite complex. Fortunately for us, the three-particle model of the atom still meets the needs of the chemist.

If we say that the atom is composed of three fundamental particles, we will see how they can be used to define the identity of an element and the relative mass of its atoms. With this goal in mind, we will continue our exploration inside the atom in the next section.

▼Looking Ahead

3–6 ATOMIC NUMBER, MASS NUMBER, AND ATOMIC MASS

We can now take a closer look at an atom of the element copper. We find that a typical atom is composed of a nucleus containing a total of 63 nucleons, of which 29 are protons and 34 are neutrons. The atom also con-

Are all atoms of a certain element the same?

tains 29 electrons that exactly balance the positive charge of the protons. Thus the atom is neutral. *The number of protons in the nucleus (which is equal to the total positive charge) is referred to as the atom's* **atomic number.** *The total number of nucleons is called the* **mass number.** Therefore, this particular copper atom has an atomic number of 29 and a mass number of 63. If we were to examine a number of copper atoms, we would soon come upon one that is different from the one we have just described. This atom of copper also has an atomic number of 29 but a mass number of 65, so it must have 36 neutrons. *Atoms having the same atomic number but different mass numbers are known as* **isotopes.**

What distinguishes one element from another?

If we were to look at the atoms of other elements, we would find the same phenomenon—almost all elements exist in nature as a mixture of isotopes. We also would notice that different elements have different atomic numbers. *Indeed, it is the atomic number that distinguishes one element from another.* Any atom with an atomic number of 29, regardless of any other consideration, is an atom of copper. If the atomic number is 28, the element is nickel; if it is 30, the element is zinc.

Isotopes of elements are identified by the mass number in the upper left-hand corner of the symbol of the element. Sometimes, the atomic number of the element is represented in the lower left-hand corner. This is strictly a convenience, of course, since the atomic number determines the identity of the element. The two isotopes of copper are written as follows

Mass number (number of nucleons)
(29 protons and 34 neutrons)

(29 protons and 36 neutrons)

$^{63}_{29}Cu$ $^{65}_{29}Cu$

Atomic number (number of protons)
(29 protons)

Isotopic notation can be used to calculate the number of particles in an isotope, as we will see in Example 3-1.

Working It Out ●

EXAMPLE 3–1

Calculating the Number of Particles in an Isotope

How many protons, neutrons, and electrons are present in $^{90}_{38}Sr$?

Solution

$$\text{atomic number} = \text{number of protons} = \underline{\underline{38}}$$
$$\text{number of neutrons} = \text{mass number} - \text{number of protons}$$
$$90 - 38 = \underline{\underline{52}}$$

In a neutral atom, the number of protons and the number of electrons are the same.

$$\text{number of electrons} = \underline{\underline{38}}$$

In Chapter 7, an important consideration will be how the mass of one element compares to another. The mass of the electrons is extremely small compared to the masses of the protons and neutrons, so it is not included in the mass of an isotope. Thus the mass number of an isotope is a convenient but rather imprecise measure of its mass. It is imprecise not only because electrons are not included, but because the proton and neutron do not have exactly the same mass. A more precise measure of the mass of one isotope relative to another is known as the isotopic mass. *The **isotopic mass** of an isotope is determined by comparison to a standard, ^{12}C, which is defined as having a mass of exactly 12 atomic mass units (amu). Therefore, one **atomic mass unit** is a mass of exactly 1/12 of the mass of ^{12}C.* By comparison with ^{12}C, the isotopic mass of ^{10}B is 10.013 amu and that of ^{11}B is 11.009 amu. Since boron, as well as most other naturally occurring elements, is found in nature as a mixture of isotopes, the atomic mass of the element reflects this mixture. The **atomic mass** *of an element is obtained from the weighted average of the atomic masses of all isotopes present in nature.* A weighted average relates the isotopic mass of each isotope present to its percentage occurrence. It can be considered as the isotopic mass of an "average atom." Example 3-2 illustrates how atomic mass relates to the distribution of isotopes.

What is an "average" atom?

EXAMPLE 3–2

Calculating the Atomic Mass of an Element from Percent Distribution of Isotopes

In nature, the element boron occurs as 19.9% ^{10}B and 80.1% ^{11}B. If the isotopic mass of ^{10}B is 10.013 amu and that of ^{11}B is 11.009 amu, what is the atomic mass of boron?

● **Working It Out**

Procedure
Find the contribution of each isotope toward the atomic mass by multiplying the percent in decimal form by the isotopic mass.

Solution

$$^{10}B \quad 0.199 \times 10.013 \text{ amu} = \quad 1.99 \text{ amu}$$
$$^{11}B \quad 0.801 \times 11.009 \text{ amu} = \quad \underline{8.82 \text{ amu}}$$
$$\text{Atomic mass of boron} = \underline{10.81 \text{ amu}}$$

Before we conclude this section, we need to discuss how ions differ from neutral atoms. A cation such as Na^+ has a positive electrical charge because there is an unequal number of electrons and protons. A cation has fewer electrons than protons, and an anion has more electrons than protons. In both cases, it is the electrons that are out of balance, not the protons. For example, the Na^+ cation has 11 protons (its atomic number) in its nucleus but only 10 electrons ($+11 - 10 = +1$), and the Cl^- anion has 17 protons in its nucleus and 18 electrons ($+17 - 18 = -1$). The Ca^{2+} cation has 20 protons and 18 electrons, leaving an ion with a charge of +2. The charge of a polyatomic ion is also due to the imbalance of electrons. For example, in the NO_3^- ion, there is a total of 31 protons in the four nuclei, but there are 32 electrons, giving a net charge on the ion of −1.

What is the difference between an atom and an ion?

Looking Back▲

Dalton wasn't completely right when he suggested that the atoms of an element are identical. But Dalton didn't know about isotopes. Isotopes of a particular element are atoms with the same atomic number but different mass numbers (that is, different numbers of neutrons). To establish a more precise measure of isotopic mass, the mass of the isotope is compared to a standard with a defined mass. This also allows a more precise measure of the weighted average of all of the isotopes of an element occurring in nature. This is known as the atomic mass of the element. Finally, we see that, compared to their parent atoms, ions have more or less electrons than they have protons.

Checking It Out▶

Learning Check C

C–1. Fill in the blanks:
An atom is composed of negatively charged _____, positively charged _____, and _____. The nucleus contains _____ and _____, with both particles having a mass of about _____ amu. The atomic number refers to the number of _____ in the nucleus, and the mass number refers to the number of _____. Isotopes of an element have the same _____ number but different _____ numbers. The isotopic mass of an isotope is its mass compared to _____. The atomic mass is the _____ _____ of all of the naturally occurring isotopes of the element. A cation contains more _____ than _____.

C–2. How many protons and neutrons are in each of the three isotopes of oxygen: $^{16}_{8}O$, $^{17}_{8}O$, $^{18}_{8}O$?

C–3. Naturally occurring lead is composed of four isotopes: 1.40% ^{204}Pb (203.97 amu), 24.10% ^{206}Pb (205.97 amu), 22.10% ^{207}Pb (206.98 amu), and 52.40% ^{208}Pb (207.98 amu). What is the atomic mass of lead?

C–4. How many protons and how many electrons are in the N^{3-} ion?

Additional Examples: Problems 3-30, 3-33, 3-41, 3-44.

CHAPTER REVIEW

Putting It Together ▲▼

All of the various forms of nature around us, from the simple elements in the air to the complex compounds of living systems, are composed of only a few basic forms of matter called elements. Each element has a name and a unique one- or two-letter **symbol.**

It has been less than 200 years since John Dalton's atomic theory introduced the concept that elements are composed of fundamental particles called **atoms.** It has been less than 20 years since we have been able to produce images of these atoms with a special microscope. Most elements are present in nature as aggregates of individual atoms. In some elements, however, two or more atoms are combined by **covalent bonds** to produce basic units called **molecules. Molecular compounds** are also composed of molecules, although, in this case, the atoms of at least two different elements are involved. Each compound has a unique arrangement of atoms in the molecular unit. These are sometimes conveniently represented by **structural formulas.**

There is another type of compound, however. Matter can be positively or negatively charged, with **electrostatic forces** between the charges. Atoms can become electrically charged to form **ions.** Groups of atoms that are covalently bonded together can also have charges and are known as **polyatomic ions.** An **ionic compound** is composed of **cations** (positive ions) and **anions** (negative ions). The **formula** of a molecular compound represents the actual number of atoms of each element present in a molecular unit. The formula of an ionic compound shows the type and number of ions in a **formula unit.** A formula unit represents the smallest whole-number ratio of cations and anions, which reflects the fact that the positive charge is balanced by the negative charge. The four most common ways that we find atoms in nature is summarized as follows:

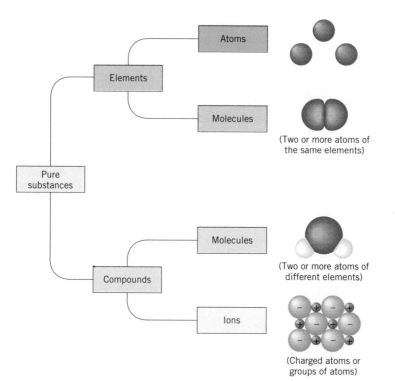

The atom is the unique particle that characterizes an element. It is composed of more basic particles called **electrons** and **nucleons.** There are two types of nucleons, **protons** and **neutrons.** The relative charges and masses of these three particles are summarized in Table 3-4.

Rather than being a hard sphere, the atom is mostly empty space occupied by the negatively charged electrons. The protons and neutrons are located in a small dense core called the **nucleus.** The number of protons in an atom is known as its **atomic number** (symbolized by Z in the following diagram), which distinguishes the atoms of one element from those of another. The total number of nucleons in an atom is known as its **mass number** (symbolized by A in the diagram). Atoms of the same element may have different mass numbers and are known as **isotopes** of that element. An atom is neutral because it has the same number of electrons as protons. When the number of electrons is greater or less than the number of protons, an ion results (the charge, if present, is symbolized by Q in the diagram). This information for a particular isotope is illustrated as follows:

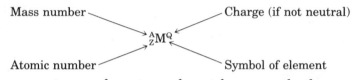

Because protons and neutrons do not have exactly the same mass, the mass number is not an exact measure of the comparative masses of isotopes. A more precise measure of mass is the **isotopic mass.** This is obtained by comparing the mass of the particular isotope with the mass of ^{12}C, which is defined as having a mass of exactly 12 **atomic mass units (amu).** The **atomic mass** of an element is the weighted average of all of the naturally occurring isotopes found in nature.

E X E R C I S E S

NAMES AND SYMBOLS OF THE ELEMENTS

3–1   Write the symbols of the following elements. Try to do this without reference to a table of the elements.
(a) bromine   (d) tin
(b) oxygen   (e) sodium
(c) lead   (f) sulfur

3–2   The following elements all have symbols that begin with the letter C: cadmium, calcium, californium, carbon, cerium, cesium, chlorine, chromium, cobalt, copper, and curium. The symbols are C, Ca, Cd, Ce, Cf, Cl, Cm, Co, Cr, Cs, and Cu. Match the symbol with the element and then check with the table of elements inside the front cover.

3–3   The names of six elements begin with the letter B. What are their names and symbols?

3–4   The names of eight elements begin with the letter S. What are their names and symbols?

3–5   Using the table inside the front cover of the text, write the symbols for the following elements.

(a) barium   (d) platinum
(b) neon   (e) manganese
(c) cesium   (f) tungsten

3–6   Write the elements corresponding to the following symbols. Try to do this without reference to a table of the elements.
(a) S   (d) N
(b) K   (e) Mg
(c) Fe   (f) Al

3–7   Using the table, give the elements corresponding to the following symbols.
(a) B   (e) Cl
(b) Bi   (f) Hg
(c) Ge   (g) Be
(d) U   (h) As

MOLECULAR COMPOUNDS AND FORMULAS

3–8   Which of the following are formulas of elements rather than compounds?
(a) P_4O_{10} (d) S_8
(b) Br_2 (e) MgO
(c) F_2O (f) P_4

3–9 Name the elements in the previous problem.

3–10 Which of the following is the formula of a diatomic element? Which is the formula of a diatomic compound?
(a) NO_2 (d) $(NH_4)_2S$
(b) CO (e) N_2
(c) K_2O (f) CO_2

3–11 Name all of the elements in the compounds of the previous problem.

3–12 Determine the number of atoms of each element in the formulas of the following compounds.
(a) $C_6H_4Cl_2$
(b) C_2H_5OH (ethyl alcohol)
(c) $CuSO_4 \cdot 9H_2O$ (H_2O's are parts of a single formula unit)
(d) $C_9H_8O_4$ (aspirin)
(e) $Al_2(SO_4)_3$
(f) $(NH_4)_2CO_3$

3–13 What is the total number of atoms in each molecule or formula unit for the compounds listed in Probem 3-12?

3–14 How many carbon atoms are in each molecule or formula unit of the following compounds?
(a) C_8H_{18} (octane in gasoline)   (c) $Fe(C_2O_4)_2$
(b) $NaC_7H_4O_3NS$ (saccharin)   (d) $Al_2(CO_3)_3$

3–15 Write the formulas of the following molecular compounds.
(a) sulfur dioxide (one sulfur and two oxygens)
(b) carbon dioxide (one carbon and two oxygens)
(c) sulfuric acid (two hydrogens, one sulfur, and four oxygens)
(d) acetylene (two carbons and two hydrogens)

3–16 Write the formulas of the following molecular compounds.
(a) phosphorus trichloride (one phosphorus and three chlorines)
(b) naphthalene (ten carbons and eight hydrogens)
(c) dibromine trioxide (two bromines and three oxygens)

3–17 Neon gas is composed of individual atoms, and chlorine gas is composed of Cl_2 molecules. Describe how these two elements appear on the atomic or molecular level. In the gaseous state, the particles are comparatively far apart.

IONS AND IONIC COMPOUNDS

3–18 The gaseous compound HF contains covalent bonds, and the compound KF contains ionic bonds. Describe how the basic particles of these two compounds appear.

3–19 Write the formulas of the following ionic compounds.
(a) calcium perchlorate (one calcium and two ClO_4^- ions)
(b) ammonium phosphate (three NH_4^+ ions and one PO_4^{3-} ion)
(c) iron (II) sulfate (one iron and one SO_4^{2-} ion)

3–20 Write the formulas of the following ionic compounds.
(a) calcium hypochlorite (one Ca^{2+} ion and two ClO^- ions)
(b) magnesium phosphate (three Mg^{2+} ions and two PO_4^{3-} ions)
(c) chromium (III) oxalate (two chromiums and three $C_2O_4^{2-}$ ions)

3–21 An ionic compound is composed of an Fe^{2+} ion and one anion. Which of the following could be the other ion?
(a) F^- (b) Ca^{2+} (c) S^{2-} (d) N^{3-}

3–22 An ionic compound is composed of two ClO_4^- ions and one cation. Which of the following could be the other ion?
(a) SO_4^{2-} (b) Ni^{2+} (c) Al^{3+} (d) Na^+

3–23 An ionic compound is composed of one SO_3^{2-} and two cations. Which of the following could be the cations?
(a) I^- (b) Ba^{2+} (c) Fe^{3+} (d) Li^+

3–24 An ionic compound is composed of two Al^{3+} ions and three anions. Which of the following could be the anions?
(a) S^{2-} (b) Cl^- (c) Sr^{2+} (d) N^{3-}

3–25 Write the formulas of the compounds in problems 3-21 and 3-23.

3–26 Write the formulas of the compounds in problems 3-22 and 3-24.

COMPOSITION OF THE ATOM

3–27 Which of the following were not part of Dalton's atomic theory?
(a) Atoms are the basic building blocks of nature.
(b) Atoms are composed of electrons, neutrons, and protons.

(c) Atoms are reshuffled in chemical reactions.
(d) The atoms of an element are identical.
(e) Different isotopes can exist for the same element.

3-28 Which of the following describes a neutron?
(a) +1 charge, mass number 1 amu
(b) +1 charge, mass number 0 amu
(c) 0 charge, mass number 1 amu
(d) −1 charge, mass number 0 amu

3-29 Which of the following describes an electron?
(a) +1 charge, mass number 1 amu
(b) +1 charge, mass number 0 amu
(c) −1 charge, mass number 1 amu
(d) −1 charge, mass number 0 amu

3-30 Give the mass numbers and atomic numbers of the following isotopes. Refer to the table of the elements.
(a) ^{193}Au   (c) ^{118}I
(b) ^{132}Te   (d) ^{39}Cl

3-31 Give the numbers of protons, neutrons and electrons in each of the following isotopes. Refer to the table of the elements.
(a) ^{45}Sc   (c) ^{223}Fr
(b) ^{232}Th   (d) ^{90}Sr

3-32 Three isotopes of uranium are ^{234}U, ^{235}U, and ^{238}U. How many protons, neutrons, and electrons are in each isotope?

3-33 Using the table of elements inside the front cover, complete the following table for each isotope.

	Isotope Symbol	Atomic Number	Mass Number	Subatomic Particles		
				Protons	Neutrons	Electrons
molybdenum-96	$^{96}_{42}$Mo	42	96	42	54	42
(a)	$^{?}_{?}$Ag				61	
(b)		14			14	
(c)			39		20	
(d) cerium-140						
(e)				26	30	
(f)		50	110			
(g)	$^{118}_{?}$I					
(h) mercury-?					116	

3-34 Using the table of elements inside the front cover, complete the following table for neutral isotopes.

	Isotope Symbol	Atomic Number	Mass Number	Subatomic Particles		
				Protons	Neutrons	Electrons
(b) tungsten-?			184			
(b)					12	11
(c)	$^{200}_{?}$At					
(d)	$^{?}_{?}$Pm				87	
(e)			109	46		
(f)			48			23
(g)		21			29	

3-35 What are the total number of protons and the total number of electrons in each of the following ions?
(a) K^+ (d) NO_2^-
(b) Br^- (e) Al^{3+}
(c) S^{2-} (f) NH_4^+

3-36 What are the total number of protons and the total number of electrons in each of the following ions?
(a) Sr^{2+} (d) NO^+
(b) P^{3-} (e) SO_3^{2-}
(c) V^{3+}

ATOMIC MASS

3-37 Using the periodic table, determine the atomic number and the atomic mass of each of the following elements.
(a) Re   (b) Co   (c) Br   (d) Si

3-38 About 75% of a U.S. "nickel" is an element with an atomic mass of 63.546 amu. What is the element?

3-39 White gold is an alloy of gold containing an element with an atomic mass of 106.4 amu. What is the element?

3-40 The elements O, N, Si, and Ca are among several that are composed *primarily* of one isotope. Using the Table of Atomic Masses inside the front cover, write the atomic number and mass number of the principal isotope of each of these elements.

3-41 A given element has a mass 5.81 times that of ^{12}C. What is the atomic mass of the element? What is the element?

3-42 The atomic mass of a given element is about 3 1/3 times that of ^{12}C. Give the atomic mass, the name, and the symbol of the element.

3-43 Bromine is composed of 50.5% ^{79}Br and 49.5% ^{81}Br. The isotopic mass of ^{79}Br is 78.92 amu and that of ^{81}Br is 80.92 amu. What is the atomic mass of the element?

3-44 Silicon occurs in nature as a mixture of three isotopes: ^{28}Si (27.98 amu), ^{29}Si (28.98 amu), and ^{30}Si (29.97 amu). The mixture is 92.21% ^{28}Si, 4.70% ^{29}Si, and 3.09% ^{30}Si. Calculate the atomic mass of naturally occurring silicon.

3-45 Naturally occurring Cu is 69.09% ^{63}Cu (62.96 amu). The only other isotope is ^{65}Cu (64.96 amu). What is the atomic mass of copper?

*3-46 Chlorine occurs in nature as a mixture of ^{35}Cl and ^{37}Cl. If the isotopic mass of ^{35}Cl is approximately 35.0 amu and that of ^{37}Cl is 37.0 amu, and the atomic mass of the mixture as it occurs in nature is 35.5 amu, what is the proportion of the two isotopes?

*3-47 The atomic mass of the element gallium is 69.72 amu. If it is composed of two isotopes, ^{69}Ga (68.926 amu) and ^{71}Ga (70.925 amu), what is the percent of ^{69}Ga?

GENERAL QUESTIONS

3-48 Describe the difference between a molecular and an ionic compound. Of the two types of compounds discussed, is a stone more likely to be a molecular or an ionic compound? Is a liquid more likely to be a molecular or an ionic compound?

3-49 Write the symbol, mass number, atomic number, and electrical charge of the element given the following information. Refer to a table of the elements.
  (a) An ion of Sr contains 36 electrons and 52 neutrons.
  (b) An ion contains 24 protons, 28 neutrons, and 21 electrons.
  (c) An ion contains 36 electrons and 45 neutrons and has a −2 charge.
  (d) An ion of nitrogen contains 7 neutrons and 10 electrons.
  (e) An ion contains 54 electrons and 139 nucleons and has a +3 charge.

3-50 Write the symbol, mass number, atomic number, and electrical charge of the element given the following information. Refer to the table of the elements.
  (a) An ion of Sn contains 68 neutrons and 48 electrons.
  (b) An ion contains 204 nucleons and 78 electrons and has a +3 charge.
  (c) An ion contains 45 neutrons and 36 electrons and has a −1 charge.
  (d) An ion of aluminum has 14 neutrons and a +3 charge.

3-51 An isotope of iodine has a mass number that is 10 amu less than two-thirds the mass number of an isotope of thallium. The total mass number of the two isotopes is 340 amu. What is the mass number of each isotope? (Hint: There are two equations and two unknowns.)

3-52 An isotope of gallium has a mass number that is 22 amu more than one-fourth the mass number of an isotope of osmium. The osmium isotope is 122 amu heavier than the gallium isotope. What is the mass number of each isotope? (Hint: There are two equations and two unknowns.)

3-53 A given element is composed of 57.5% of an isotope with an isotopic mass of 120.90 amu and the remainder an isotope with an isotopic mass of 122.90 amu. What is the atomic mass of the element? What is the element? How many electrons are in a cation of this element if it has a charge of "3"? How many neutrons are in each of the two isotopes of this element? What percent of the isotopic mass of each isotope is due to neutrons?

3-54 A given isotope has a mass number of 196, and 60.2% of the nucleons are neutrons. How many electrons are in a cation of this element if it has a charge of "2"?

3-55 A given isotope has a mass number of 206. The isotope has 51.2% more neutrons than protons. What is the element?

3-56 A given molecular compound is composed of one atom of nitrogen and one atom of another element. The mass of nitrogen accounts for 46.7% of the mass of one molecule. What is the other element? What is the formula of the compound? This molecule can lose one electron to form a polyatomic ion. How many electrons are in this ion?

3–57 A given molecular compound is composed of one atom of carbon and two atoms of another element. The mass of carbon accounts for 15.8% of the mass of one molceule. What is the other element? What is the formula of the compound?

3–58 If the isotopic mass of ^{12}C is defined as exactly 8 instead of 12, what would be the atomic mass of the following elements to three significant figures? Assume that the elements have the same masses relative to each other as before;

that is, hydrogen still has a mass about one-twelfth that of carbon.
(a) H (b) N (c) Na (d) Ca

3–59 Assume that the isotopic mass of ^{12}C is defined as exactly 10 and that the atomic mass of an element is 43.3 amu on this basis. What is the element?

3–60 Assume that the isotopic mass of ^{12}C is defined as exactly 20 instead of 12 and that the atomic mass of an element is 212.7 amu on this basis. What is the element?

S O L U T I O N S   T O   L E A R N I N G   C H E C K S

A–1 N, Cu, nickel, atom, molecules, C_2H_7N, structural

A–2 eight carbons, nine hydrogens, one nitrogen, and two oxygens

B–1 molecules, ions, cations, anions

B–2 Na_3PO_4, $(NH_4)_2S$

C–1 electrons, protons, neutrons, protons, neutrons, one, protons, nucleons, atomic, mass, ^{12}C, weighted average, protons, electrons.

C–2 ^{16}O (8 p, 8 n) ^{17}O (8 p, 9 n) ^{18}O (8p, 10 n)

C–3
$$0.0140 \times 203.97 = 2.86 \,(^{204}Pb)$$
$$0.2410 \times 205.97 = 49.64 \,(^{206}Pb)$$
$$0.2210 \times 206.98 = 45.74 \,(^{207}Pb)$$
$$\underline{0.5240 \times 207.98 = 108.98}(^{208}Pb) \text{ (calculation good to only one decimal}$$
$$\text{place)}$$
$$207.22 = 207.2 \text{ amu}$$

C–4 There are <u>seven protons</u> in a nitrogen atom or a nitrogen ion.

$$\text{charge} = \text{number of protons} - \text{number of electrons}$$
$$\text{number of electrons} = \text{number of protons} - \text{charge} = 7 - (-3)$$
$$\underline{\text{number of electrons} = 10}$$

MULTIPLE CHOICE

The following multiple-choice questions have one correct answer.

1. Which of the following relates to the precision of a measurement?
   (a) the number of significant figures
   (b) the location of the decimal point
   (c) the closeness to the actual value
   (d) the number of zeros that are significant

2. Which of the following is not an exact relationship?
   (a) 1 mile = 5280 ft
   (b) 0.62 mile = 1 km
   (c) 10^3 m = 1 km
   (d) 4 qt = 1 gal

3. How many significant figures are in the measurement 0.0880 m?
   (a) 5   (b) 4   (c) 3   (d) 2

4. The number 785×10^{-8} expressed in scientific notation is
   (a) 7.85×10^{-10} (c) 7.85×10^{-6}
   (b) 78.5×10^{-7} (d) 0.785×10^{-6}

5. One kiloliter is the same as
   (a) 10^6 mL   (c) 10^3 mL
   (b) 10^{-3} L   (d) 10 L

6. The temperature of a sample of matter is a measure of
   (a) its density
   (b) its specific heat
   (c) its physical state
   (d) the intensity of heat

7. Which of the following is the lowest possible temperature?
   (a) 0 K   (c) 0°F
   (b) 0°C   (d) −273 K

8. A substance completely fills any size container. The substance is a
   (a) gas
   (b) solid
   (c) liquid

9. A substance has a distinct melting point and density. It cannot be decomposed into chemically simpler substances. The substance is a
   (a) heterogeneous mixture
   (b) solution
   (c) compound
   (d) element

10. A sample of a given liquid has uniform properties, but the temperature at which the liquid boils slowly changes as it boils away. The liquid sample is a
    (a) heterogeneous mixture
    (b) solution
    (c) compound
    (d) element

11. Which of the following is a chemical property of iodine?
    (a) It melts at 114°C.
    (b) It has a density of 4.94 g/mL.
    (c) When heated, it forms a purple vapor.
    (d) It forms a compound with sodium.

12. A charged car battery has _____ energy that can be converted into _____ energy.
    (a) chemical, electrical
    (b) heat, mechanical
    (c) light, electrical
    (d) mechanical, chemical

13. Which of the following has the highest density?
    (a) a truckload of feathers
    (b) an ounce of gold
    (c) a cup of water
    (d) a gallon of gasoline

14. Which of the following is the symbol for silicon?
    (a) Se   (b) S   (c) Si   (d) Sc

15. Lead dioxide is the name of
    (a) two elements
    (c) a compound
    (b) a mixture
    (d) a solution

16. Which of the following has a charge of +1 and a mass of 1 amu?
    (a) a neutron
    (b) an electron
    (c) a proton
    (d) a helium nucleus

*Answers to all questions and problems in review tests are given in Appendix H.

17. Isotopes have
    (a) the same mass number and atomic number
    (b) the same mass number but different atomic numbers
    (c) different mass numbers and atomic numbers
    (d) different mass numbers but the same atomic number

18. Which of the following describes an isotope with a mass number of 99 that contains 56 neutrons in its nucleus?
    (a) $^{99}_{56}Ba$
    (b) $^{43}_{56}Ba$
    (c) $^{99}_{43}Tc$
    (d) $^{56}_{43}Tc$
    (e) $^{155}_{99}Es$

19. Which of the following isotopes is used as the standard for atomic mass?
    (a) ^{12}C (c) ^{13}C
    (b) ^{16}O (d) ^{1}H

20. Which of the following is a diatomic molecule of a compound?
    (a) NO   (c) H_2
    (b) NH_4Cl (d) CO_2

21. Which of the following is not a basic particle of an element?
    (a) an atom   (b) a molecule   (c) an ion

22. Which of the following is a polyatomic cation?
    (a) CO   (b) K^+ (c) SO_4^{2-} (d) NH_4^+

23. Which would be the electrical charge on a sulfur atom containing 18 electrons?
    (a) 2−   (b) 1−   (c) 0   (d) 2+

24. One formula unit of $Al_2(CO_3)_3$ contains
    (a) nine oxygen atoms
    (b) three oxygen atoms
    (c) six aluminum atoms
    (d) six carbon atoms

25. Which would be the formula of an ionic compound made up of Ba^{2+} ions and Cl^- ions?
    (a) Ba_2Cl
    (b) $BaCl_2$
    (c) BaCl
    (d) $BaCl_3$

PROBLEMS

Carry out the following calculations.

1. Round off the following numbers to three significant figures.
   (a) 173.8
   (b) 0.0023158
   (c) 18,420

Carry out the following calculations. Express your answer to the proper decimal place or number of significant figures.

2. 1.09 cm + 0.078 cm + 16.0021 cm

3. 4.2 in. × 16.33 in.

4. 398 g ÷ 31 mL

5. 0.892×10^4 cm × 0.0022×10^6 cm

6. 7784×10^{-6} kg ÷ 0.56 kg/mL

Express the answers to the following problems in scientific notation. Assume that the numbers are measurements so that the answer is expressed to the proper decimal place or number of significant figures.

7. $(92.8 \times 10^6) \times (1.3 \times 10^4)$

8. $(0.4887 \times 10^{-4}) \div (0.03 \times 10^6)$

9. $(4.0 \times 10^6)^2$

Make the following conversions.

10. 189 cm to in.

11. 43 qt to L

12. 197 mi/hr to km/hr

13. 6.74 lb/ft to g/cm

Solve the following problems.

14. A cube of metal measures 2.2 cm on a side. It has a mass of 47.68 g. What is its density?

15. What is the volume of a 678-g sample of a substance that has a density of 11.3 g/mL?

16. The density of liquid iron is 7.05 g/mL. How many mL of iron must be mixed with liquid chromium to make 1.00 kg of an alloy that is 72.0% by mass iron?

17. What is the temperature in degrees Celsius if the reading is 112°F?

18. The temperature of a given liquid increases by 12.0°C if 69.0 joules of heat is added to a 20.0-g sample. What is the specific heat of the liquid?

19. The atomic mass of an element is 10.15 times that of the defined standard. What is the atomic mass of the element? What are the name and symbol of the element?

20. The atoms of a given element are distributed between two isotopes: 65.0% of the atoms have a mass number of 116, and the remainder have a mass number of 112. Using mass number as an approximation of isotopic mass, calculate the atomic mass of the element to tenths of an atomic mass unit.

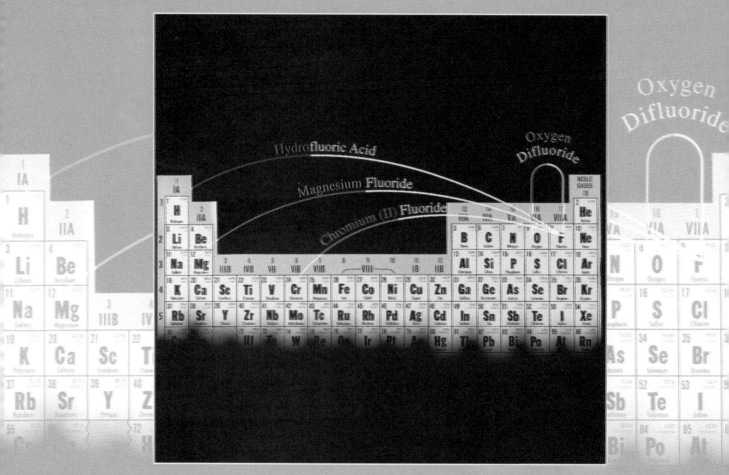

The periodic table helps the chemist systematize the properties of the elements. It is also used for the naming of compounds formed by the elements. The origin of this table and its use in naming compounds are the subjects of this chapter.

THE PERIODIC TABLE AND CHEMICAL NOMENCLATURE

Two things may strike the casual and uninformed observer about chemists and students of chemistry: (1) they seem to glance frequently at an ever-present wall poster containing the symbols of the elements, and (2) they seem to have their own language. The observer is correct on both counts.

◀ **Setting The Stage**

The wall poster is obviously important since it hangs in almost every chemical laboratory and classroom. Known as the *periodic table*, it is as necessary to a chemist as a globe of the world is to a geographer. The location of a city on a globe tells us a wealth of information about climate and culture. Likewise, the location of a specific element in the periodic table gives us important information about the nature of the element.

Chemists have their own vocabulary for the 10 million or so known compounds. Why is it necessary for us to learn the language of chemistry? The answer is the same as why you learn any language other than your native tongue—you can communicate directly with others who speak the same language. The vocabulary of chemistry is known as *chemical nomenclature*. Like learning any other tongue, learning chemical nomenclature requires some memorization.

These two topics, the periodic table and chemical nomenclature, may at first seem unrelated. However, the periodic table groups the elements into two broad categories, and the naming of compounds relates to these classifications. The periodic table displays this information in an easy-to-read manner that simplifies the rules of chemical nomenclature.

Basically, the main function of the periodic table is to provide some order in what otherwise would be total chaos. Many elements have similar properties and can be grouped accordingly. But what kinds of groupings of elements are possible? Why and how did the periodic table originate? What information about the elements does it so conveniently display? How can this information be used in naming compounds? These are some of the questions that we will answer in this chapter.

Formulating Some Questions ▶

4–1   GROUPING THE ELEMENTS: METALS AND NONMETALS, NOBLE METALS AND ACTIVE METALS

Can elements be divided into two groups?

To some extent, all science is simplified by classifying and grouping information. If certain characteristics are common to a number of samples, then those samples can be grouped. In biology, living systems can be grouped into flora (trees and plants) and fauna (animals). In geology, the surface of the earth can be divided into the oceans and the continents. In chemistry, it has been known for hundreds of years that elements can be grouped into one of two broad classifications: *metals and nonmetals.* **Metals** are generally hard, lustrous elements that are *ductile* (can be drawn into wires) and *malleable* (can be pounded into thin sheets). (See Figure 4-1.) We also know they readily conduct electricity and heat. Metals form the strong framework on which our modern society is built. The discovery and use of metals over five thousand years

(a)   (b)

Figure 4 – 1
Two Properties of Metals Metals can be pressed into thin sheets (a) or drawn into wires (b).

Figure 4–2
Nonmetals The bottle on the left contains liquid bromine and its vapor. The flask in the back contains pale green chlorine gas. Solid iodine is in the flask on the right. Powdered red phosphorous is in the dish in the middle, and black powdered graphite (carbon) is in the watch glass in front. Lumps of yellow sulfur are shown in the front.

ago moved civilization beyond the Stone Age. The second type of element is noted by their *lack* of metallic properties. These are the nonmetals. **Nonmetals** are generally gases or soft solids that do not conduct electricity. (See Figure 4-2.) There are some notable exceptions to these general properties, however. The division between metallic and nonmetallic properties is not sharp, so some elements have intermediate properties. There are also examples of very hard nonmetals and very soft metals. For example, a form of the nonmetal carbon (diamond) is one of the hardest substances known. Mercury, a metal, is a liquid at room temperature. Still, almost everyone has a general idea of what a metal is like. In addition to these physical properties, there are some very important chemical differences between metals and nonmetals, which we will explore in a later chapter.

Classifying the elements doesn't stop with the division of elements into these two groups. We find that all metals are not the same, so further classification is possible. It's like classifying the human race into two genders, men and women, but then finding that they can be further subdivided into personality types (e.g., extroverts and introverts). The first thing we note about metals is that some are chemically unreactive. That is, elements such as copper, silver, and gold are very resistant to the chemical reactions of corrosion and rust. These are the metals of coins and jewelry, not only because of their comparative rarity and beauty, but because of this chemical inertness. For this reason, they are known as the *noble metals*. Gold and silver coins on the ocean bottom, deposited from ships that foundered hundreds of years ago, can be easily polished to their original luster. (See Figure 4-3.) Other metals are very

Figure 4–3
One Type of Metal With a little polish, these gold and silver coins regained their original luster after three centuries at the bottom of the ocean.

much different. They are extremely reactive with air and water. In fact, metals such as lithium, sodium, and potassium must be stored under oil because they react violently (to the point of explosion) with water. These metals are among those known as the *active metals*. Thus, copper, silver, and gold can be placed into one family of metals and lithium, sodium, and potassium into another. Similar relationships among other elements were also noticed and appropriate groupings were made.

Looking Ahead ▼

Somehow, it just seemed that there should be more order among the elements than the simple groupings discussed above would indicate. Indeed there is. The full scale of relationships among the elements became apparent with the introduction of the periodic table. We explore this fantastic tool in the next section.

4–2 THE PERIODIC TABLE

Although early scientists had long known that certain elements could be classified into families, it was only a little more than a century ago that more relationships were discovered. The modern periodic table is shown in Figure 4-4. The earliest version of this table was introduced in 1869 by Dimitri Mendeleev of Russia. Lothar Meyer of Germany indepen-

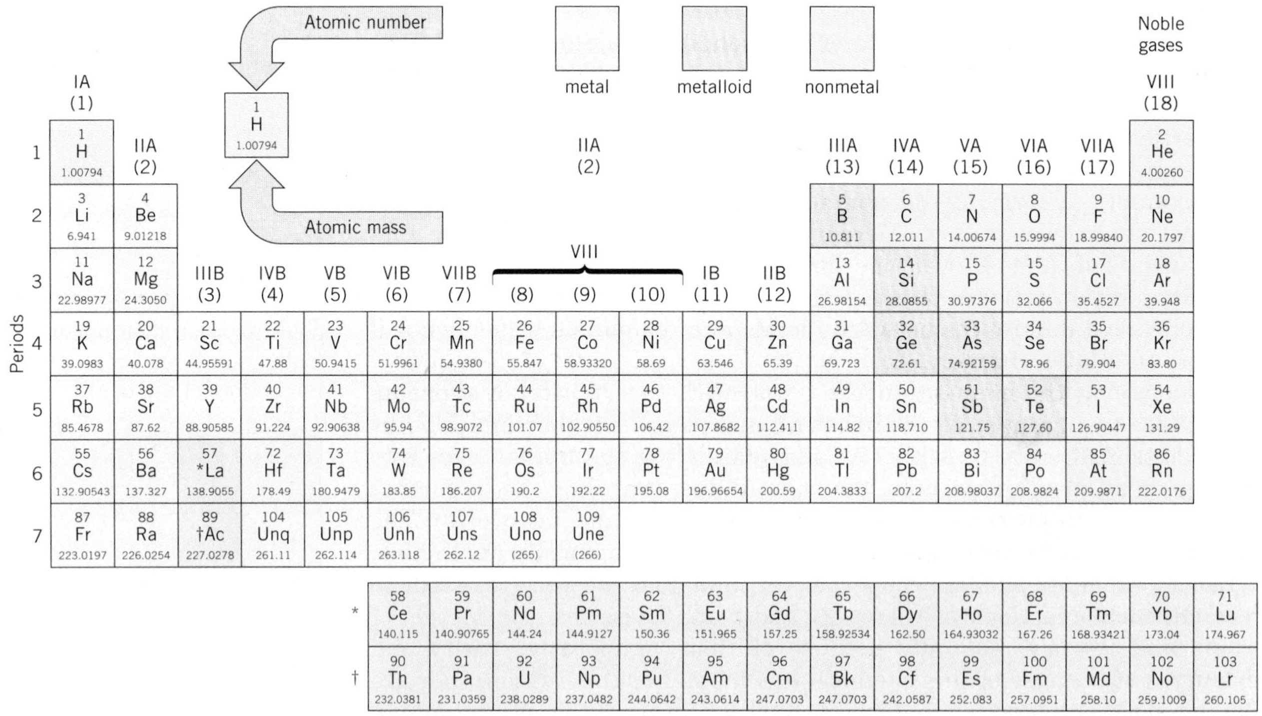

Figure 4–4
The Periodic Table

dently presented a similar table in 1870. When these two scientists arranged the elements in order of increasing atomic masses (atomic numbers were still unknown), they observed that elements in families appeared at regular (periodic) intervals. The periodic table was constructed so that elements in the same family (i.e., Li, Na, and K) fell into vertical columns. At first this did not always happen. Sometimes the next heaviest element did not seem to fit in a certain family. Mendeleev solved this problem by placing the element in a family with similar properties, even if he had to leave a space or two blank. For example, a space was left under silicon and above tin for what Mendeleev suggested was a yet undiscovered element. Mendeleev called the missing element "eka-silicon." Later, an element, germanium, was discovered that had properties intermediate between silicon and tin, as predicted by the location of the blank space.

Did all of the elements "fit" into the original periodic table?

A second problem involved some misfits when the elements were ordered according to atomic mass. For example, notice that tellurium (#52) is heavier than iodine (#53). But Mendeleev realized that iodine clearly belonged under bromine and tellurium under selenium, and not vice versa. Mendeleev simply reversed the order, suggesting that perhaps the atomic masses reported were in error. (That was known to happen sometimes.) We now know that this problem does not occur when the elements are listed in order of increasing atomic number instead of atomic mass. This method of ordering conveniently displays the **periodic law,** which states that *the properties of elements are periodic functions of their atomic numbers.*

We will now look at the periodic table in more detail. Even more classification is possible. For example, we will find that families of elements can be further grouped into categories that are much like extended families.

▼Looking Ahead

4–3 PERIODS AND GROUPS IN THE PERIODIC TABLE

So far, our main emphasis concerning the periodic table has been on the vertical columns, which contain the families of elements. In fact, there are common characteristics in the horizontal rows as well. *Horizontal rows of elements in the table are called* **periods.** Each period ends with a member of the family of elements called the **noble gases.** These elements, like the noble metals, are chemically unreactive and are composed of individual atoms. The first period contains only two elements, hydrogen and helium. The second and third contain 8 each, the fourth and fifth contain 18 each, the sixth 32, and the seventh 23. (The seventh would also contain 32 if there were enough elements.)

Families of elements fall into vertical columns called **groups.** Each group is designated by a number at the top of the group. The most commonly used label employs Roman numerals followed by an A or a B. Another method, which eventually may be accepted, numbers the groups 1 through 18. It is not clear at this time which method will win out, or if some alternative will yet be proposed and universally accepted. The periodic tables used in the text display both numbering systems. In the discussion, however, we will use the traditional method involving Roman numerals along with the letters A and B.

The groups of elements can be classified even further into four main categories of elements.

What are some ways that elements can be classified or grouped?

1 The Main Group or Representative Elements (Groups IA-VIIA)

Most of the familiar elements that we will discuss and use as examples in this text are representative elements. For example, the three main elements of life—carbon, oxygen, and hydrogen—are in this category. One group of the representative elements includes the highly reactive metals that we discussed earlier. Notice that lithium, sodium, and potassium are found in Group IA. This family of elements (except for hydrogen) is known as the **alkali metals.** Another group of metals that are also chemically reactive is found in Group IIA and are known as the **alkaline earth metals.** Group VIIA are all nonmetals and are known as the **halogens.** Group VIA elements are known as the **chalcogens.** The other representative element groups (IIIA through VA) are not generally referred to by a family name. In a later chapter, we will discuss in more detail some of the physical and chemical properties of the representative elements.

2 The Noble or Inert Gases (Group VIII)

These elements form few chemical compounds. In fact, helium, neon, and argon do not form any compounds. They all exist as individual atoms in nature.

3 The Transition Metals (Group B Elements)

These metals include many of the familiar structural metals such as iron and chromium as well as the noble metals, copper, silver, and gold (Group IB) that we discussed earlier.

4 The Inner Transition Metals

The 14 metals between lanthanum (#57) and hafnium (#72) are known as the **lanthanides** or **rare earths,** and the 14 metals between actinium (#89) and unnilquadium (#104) are known as the **actinides.**

Looking Ahead ▼

We are now ready to see how the periodic table gives us some very important information. For example, the position of an element on the periodic table indicates whether it is a metal or nonmetal and whether it is a gas or a solid at room temperature.

4–4   PUTTING THE PERIODIC TABLE TO USE

In our initial discussion of the grouping of elements, we discussed metals and nonmetals. One can easily classify an element as a metal or nonmetal by a glance at the periodic table. The heavy stair-step line in

Figure 4-4 separates the two groups. Metals (about 80% of the elements) are on the left of the line, and the nonmetals are on the right. When hydrogen is displayed as an IA element, however, it appears to the left of the line but is definitely a nonmetal. *Many elements on the borderline have properties that are intermediate between metals and nonmetals and are sometimes referred to as either* **metalloids** *or* **semimetals.** These elements are indicated in Figure 4-4. For example, silicon is a brittle solid, typical of nonmetals, but conducts a limited amount of electricity (a semiconductor). Conduction of electricity is a metallic property. In any case, we will use only the two broad classifications (metal and nonmetal) in nomenclature discussed later in the chapter.

Our next question concerns how the three physical states of matter (solids, liquids, and gases) are represented among the elements in the periodic table. We must be cautious, however, as to the temperature conditions we define. If the temperature is low enough, all elements exist as solids (except helium); if it is high enough, all elements are in the gaseous state. On Triton, a moon of the planet Neptune, the temperature is −236°C (37 K). The atmosphere is very thin because most substances that are gases under Earth conditions are solids or liquids under Triton conditions. On the outer part of the Sun, however, the temperature is 50,000°C so only gases exist. Thus we must come to some agreement as to a reference temperature to define the physical state of an element. **Room temperature,** which is defined as exactly **25°C,** *is the standard reference temperature* used to describe physical state. At this temperature, all three physical states are found among the elements on the periodic table. Fortunately, except for hydrogen, the gaseous elements are all found at the extreme right top of the table (nitrogen, oxygen, fluorine, and chlorine) and in the right-hand vertical column (the noble gases). There are only two liquids: a metal, mercury, and a nonmetal, bromine. All other elements are solids. Two solids, gallium and cesium, melt to become liquids at slightly above the reference temperature (29°C), however. (See Figure 4-5.)

What temperature do we use as a reference for physical state?

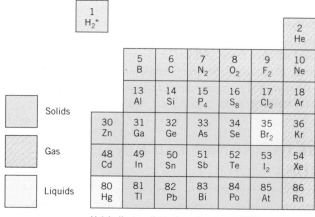

H_2* indicates diatomic molecules at 25°C

Figure 4–5
The Physical States of the Elements

The noble gases, Group VIII, all exist as individual atoms. Many of the other nonmetals, however, exist as molecules. Except for hydrogen, these are all located on the right-hand side of the table. Group VIIA naturally occurring elements exist as diatomic molecules. Other diatomic elements are hydrogen, nitrogen, and oxygen. The formula for the most common form of phosphorus is P_4 and the most common form of sulfur is S_8.

Looking Back ▲

No element is entirely unique. Each element shares certain chemical characteristics with other elements, which allows us to group them together. The periodic table is an ingenious display of these groupings of elements. The table makes it possible to quickly identify the elements that belong to a particular group, such as the alkali metals or the halogens. By being able to locate a certain element on the periodic table, a number of characteristics common to that group or location can be assumed.

Checking It Out ▶

Learning Check A

A–1. Fill in the blanks.
Of the two general classifications of elements, nickel is a _____ and sulfur is a _____ . Some borderline elements such as germanium are sometimes referred to as _____ . In the periodic table, elements are ordered in increasing _____ _____ so that _____ of elements fall into vertical columns. Of the four general categories of elements, calcium is a _____ element, nickel is a _____ _____ , and xenon is a _____ _____ . An element in Group IIA, such as calcium, is also known as an _____ _____ metal. A solid nonmetal that is composed of diatomic molecules is in the group known as the _____ . Metals are all solids except for _____ .

Additional Examples: Problems 4-5, 4-10, 4-12, and 4-14

Looking Back ▲

We are now ready to shift gears somewhat as we delve into the naming of compounds. Chemical compounds can be roughly divided into two groups: organic and inorganic. Organic compounds are composed principally of carbon, hydrogen, and oxygen. These are the compounds of life and will not be discussed at this time. All other compounds are called inorganic compounds, and these will be the subject of this chapter. The identity of an element as a metal or a nonmetal is important in chemical nomenclature, so we will need to keep a periodic table handy.

4–5   NAMING METAL-NONMETAL BINARY COMPOUNDS

What is the chemical difference between metals and nonmetals?

One of the most important chemical properties of a metal is its tendency to form positive ions (cations). Most nonmetals, on the other hand, tend to form negative ions (anions). (The noble gases form neither cations nor anions.) Inorganic compounds composed of just two elements—metal cations and nonmetal anions—will be referred to as *metal-nonmetal* **bi-**

TABLE 4–1 MONATOMIC IONS OF THE REPRESENTATIVE ELEMENTS

IA	IIA	IIIA	IVA	VA	VIA	VIIA
Hydride H^-						
Lithium Li^+	Beryllium Be^{2+}		Carbide C^{4-}	Nitride N^{3-}	Oxide O^{2-}	Fluoride F^-
Sodium Na^+	Magnesium Mg^{2+}	Aluminum Al^{3+}		Phosphide P^{3-}	Sulfide S^{2-}	Chloride Cl^-
Potassium K^+	Calcium Ca^{2+}				Selenide Se^{2-}	Bromide Br^-
Rubidium Rb^+	Strontium Sr^{2+}				Telluride Te^{2-}	Iodide I^-
Cesium Cs^+	Barium Ba^{2+}					

nary compounds. Metals behave in two ways. Some metals form cations with only one charge (e.g., Ca^{2+}). Others form cations with two or more charges (e.g., Fe^{2+} and Fe^{3+}). The first type of compound we will consider are those in which the metal forms ions with only one charge.

Metals with Ions of Only One Charge

Representative element metals in Group IA form $+1$ cations exclusively. Likewise, the metals in Group IIA form $+2$ ions exclusively. Aluminum in Group IIIA forms a $+3$ ion exclusively, but other members of this group also form a $+1$ ion. (See Table 4-1.) (There are also several transition metals that form only one cation charge, but they are not included in this discussion.) When present in metal-nonmetal binary compounds, the nonmetals form one type of anion. Hydrogen and Group VIIA form -1 ions, Group VIA elements form -2 ions, and N and P in Group VA form -3 ions. (See Table 4-1.)

In both naming and writing the formula for a binary ionic compound, the metal comes first and the nonmetal second. The unchanged English name of the metal is used. (If a metal cation is named alone, the word "ion" is also included to distinguish it from the free metal.) The name of the anion includes only the English root plus *ide*. For example, chlor*ide* is derived from <u>chlor</u>ine and oxide from <u>oxy</u>gen. Table 4-1 includes all of the monatomic anions and cations of the representative elements and their names.

Which element is named and written first?

EXAMPLE 4–1

● Working It Out

Naming Metal-Nonmetal Binary Compounds (Metals with Ions of Only One Charge)

Name the following binary ionic compounds: KCl, Li_2S, and Mg_3N_2.

Answer

KCl	potassium chloride
Li_2S	lithium sulfide
Mg_3N_2	magnesium nitride

EXAMPLE 4–2

Writing Formulas of Metal-Nonmetal Binary Compounds (Metals with Ions of Only One Charge)

Write the formulas for the following binary metal-nonmetal compounds: aluminum fluoride, calcium selenide, and potassium phosphide.

Answer

Aluminum is in Group IIIA, so it forms a cation with a +3 charge exclusively. Fluorine is in Group VIIA, so it forms an anion with a -1 charge. Since the positive charge is balanced by the negative charge, it is obvious that we need three F^{1-} anions to balance one Al^{3+} ion [e.g., $+3 + (\underline{3} \times -1) = 0$]. Therefore, the formula is written as

$$Al^{3+} + 3(F^-) = \underline{\underline{AlF_3}}$$

Another convenient way to establish the formula is to write the ions with their appropriate charges side by side. The numerical value of the charge on the cation becomes the subscript on the anion and vice versa. (The number "1" is understood instead of written in.) This is known as the *cross-charge* method.

$$Al^{(3+)} \quad F^{(1-)} = \underline{\underline{AlF_3}}$$

Calcium is in Group IIA, so it forms a +2 cation. Selenium is in Group VIA, so it forms a -2 anion. Notice that, by exchanging values of the charge, we first indicate a formula of Ca_2Se_2. This is not a correct representation, however. Ionic compounds should be expressed with the simplest whole numbers for subscripts. Therefore, the proper formula is written as CaSe.

$$Ca^{(2+)} \quad Se^{(2-)} = Ca_2Se_2 = \underline{CaSe}$$

Potassium is in Group IA, so it has a +1 charge in compounds. Phosphorus is in Group VA, so it has a -3 charge in compounds.

$$K^{(1+)} \quad P^{(3-)} = \underline{\underline{K_3P}}$$

Metals with Ions of More Than One Charge

Except for Groups IA, IIA, and aluminum, other representative metals and most transition metals can form more than one cation. Therefore, a name such as iron chloride would be ambiguous since there are two iron chlorides, $FeCl_2$ and $FeCl_3$. An even more extreme case is that of manganese oxide—there are five different compounds (MnO, Mn_3O_4, Mn_2O_3, MnO_2, and Mn_2O_7). To distinguish among these compounds, the charge on the metal ion follows the name of the metal in Roman numerals and in parentheses. This is referred to as the **Stock method.** Therefore, the two chlorides of iron are named iron(II) chloride ($FeCl_2$) and iron(III)

Why not include (I) after potassium?

chloride ($FeCl_3$). In the **classical method** of naming these compounds, $FeCl_2$ would be named *ferrous chloride*, and $FeCl_3$ would be named *ferric chloride*. In this method, the name of the metal or its Latin root is used along with *ous* for the lower charged ion or *ic* for the higher charged ion. This method is not used in this text so it will not be pursued further.

In order to use the Stock method, it is necessary to determine the charge on the metal by working backward from the known charge on the anion. For example, in a compound with the formula FeS, we can establish from Table 4-1 that the charge on the S is -2. Therefore, the charge on the one Fe must be $+2$. The compound is named iron(II) sulfide.

EXAMPLE 4–3

Naming Metal-Nonmetal Binary Compounds (Metals with Ions of More Than One Charge)

Name the following compounds: SnO_2, Co_2S_3, CrF_6

● Working It Out

Answer
The positive charges must be balanced by the negative charges (add to zero). Therefore, we can set up an equation as follows:

(number of metal atoms) × (positive charge on metal)
+ (number of nonmetal atoms) × (negative charge on nonmetal)
= net charge (zero)

For SnO_2, the equation is

$$[1 \times (Sn\ charge)] + [2 \times (O\ charge)] = 0$$

Substituting the known charge on the oxygen,

$$Sn + (2 \times -2) = 0$$

$$Sn - 4 = 0$$

$$Sn = +4\ (IV)$$

The name of the compound is, therefore, tin(IV) oxide.

For Co_2S_3, we can construct the following equation:

$$[2 \times (Co\ charge)] + [3 \times (S\ charge)] = 0$$

The charge on a Group VIA nonmetal is -2.

$$2\ Co + (3 \times -2) = 0$$

$$2\ Co = +6$$

$$Co = +3\ (III)$$

The name of the compound is cobalt(III) sulfide.

For CrF_6, the equation is

$$(Cr\ charge) + (6 \times F\ charge) = 0$$

The charge on F, a Group VIIA nonmetal, is -1.

$$Cr + (6 \times -1) = 0$$

$$Cr = +6\ (VI)$$

The name of the compound is chromium(VI) fluoride.

EXAMPLE 4–4

Writing Formulas of Metal-Nonmetal Binary Compounds (Metals with Ions of More Than One Charge)

Write the formulas for lead(IV) oxide, iron(II) chloride, chromium(III) sulfide, and manganese(VII) oxide.

Answer

Lead(IV) oxide

If lead has a +4 charge, two O^{2-} ions are needed to form a neutral compound.

$$Pb^{4+} \quad O^{2-} = Pb_2O_4 = \underline{\underline{PbO_2}}$$

Iron(II) chloride

Two chlorides are needed to balance the +2 iron.

$$Fe^{2+} \quad Cl^{-} = \underline{\underline{FeCl_2}}$$

Chromium(III) sulfide

$$Cr^{3+} \quad S^{2-} = \underline{\underline{Cr_2S_3}}$$

Manganese(VII) oxide

$$Mn^{7+} \quad O^{2-} = \underline{\underline{Mn_2O_7}}$$

Looking Ahead ▼

Monatomic ions are the result of an imbalance between the number of electrons and the number of protons in the nucleus of the atom. Cations have fewer electrons than protons, and anions have more electrons than protons. Two or more atoms that are chemically combined with covalent bonds may also have an imbalance of electrons and protons leading to a charged species called a polyatomic ion. How compounds containing these ions are named is our next topic.

4–6 NAMING COMPOUNDS WITH POLYATOMIC IONS

Most of us are somewhat familiar with names of polyatomic ions. We use bicarbonates and carbonates for indigestion, as well as sulfites and nitrites to preserve foods. A list of some common polyatomic ions is given in Table 4-2. Notice that all but one (NH_4^+, ammonium) are anions.

Most of the compounds containing polyatomic ions are ionic, as were the compounds discussed in the previous section. Thus we follow essentially the same rules as before. That is, the metal is written and named first. If the metal forms more than one cation, the charge on the cation is shown in parentheses. The polyatomic anion is then named or written.

There is some systematization possible that will help in learning Table 4-2. In many cases, *the anions are composed of oxygen and one other element.* Thus these anions are called **oxyanions.** When there are

TABLE 4–2   POLYATOMIC IONS

Ion	Name	Ion	Name
$C_2H_3O_2^-$	acetate	HSO_3^-	hydrogen sulfite or bisulfite
NH_4^+	ammonium	OH^-	hydroxide
CO_3^-	carbonate	ClO^-	hypochlorite
ClO_3^-	chlorate	NO_3^-	nitrate
ClO_2^-	chlorite	NO_2^-	nitrite
CrO_4^{2-}	chromate	$C_2O_4^{2-}$	oxalate
CN^-	cyanide	ClO_4^-	perchlorate
$Cr_2O_7^{2-}$	dichromate	MnO_4^-	permanganate
HCO_3^-	hydrogen carbonate or bicarbonate	PO_4^{3-}	phosphate
		SO_4^{2-}	sulfate
HSO_4^-	hydrogen sulfate or bisulfate	SO_3^{2-}	sulfite

two oxyanions of the same element (e.g., SO_3^{2-} and SO_4^{2-}), they, of course, have different names. The anion with the fewer number of oxygens uses the root of the element plus *ite*. The one with the higher number uses the root plus *ate*.

$$SO_3^{2-} \quad \text{sul}\underline{\text{fite}}$$

$$SO_4^{2-} \quad \text{sul}\underline{\text{fate}}$$

There are four oxyanions containing Cl. The middle two are named as before. The one with one less oxygen than the chlorite has a prefix of *hypo*. The one with one more oxygen than chlorate has a prefix of *per*.

$$ClO^- \quad \text{hypochlorite}$$

$$ClO_2^- \quad \text{chlorite}$$

$$ClO_3^- \quad \text{chlorate}$$

$$ClO_4^- \quad \text{perchlorate}$$

Most of the ionic compounds that we have just named are also referred to as salts. **A salt** *is an ionic compound formed by the combination of a cation with an anion.* (Cations combined with hydroxide or oxide form a class of compounds that are not considered salts and are discussed in a later chapter.) For example, potassium nitrate is a salt composed of K^+ and NO_3^- ions, and calcium sulfate is a salt composed of Ca^{2+} and SO_4^{2-} ions. Ordinary table salt is NaCl, composed of Na^+ and Cl^- ions.

EXAMPLE 4–5

Naming Compounds with Polyatomic Ions

Name the following compounds: K_2CO_3 and $Fe_2(SO_4)_3$.

●**Working It Out**

Answer

K_2CO_3: The cation is K^+ (Group IA). The charge is not included in the name because it forms a +1 ion only. The anion is CO_3^{2-} (the carbonate ion).

<u>potassium carbonate</u>

$Fe_2(SO_4)_3$: The charge on the Fe cation can be determined from the charge on the SO_4^{2-} (sulfate) ion.

$$2\ Fe + 3\ SO_4^{2-} = 0$$

$$2\ Fe + 3\ (-2) = 0$$

$$2\ Fe = +6$$

$$Fe = +3$$

<u>iron(III) sulfate</u>

EXAMPLE 4–6

Writing Formulas of Metal-Polyatomic Anion Compounds

Give the formulas for barium acetate, ammonium sulfate, thallium(III) nitrate, and manganese(III) phosphate.

Answer
Barium acetate
    Barium is in Group IIA, so it has a +2 charge. Acetate is the $C_2H_3O_2^-$ ion.

$$Ba^{2+} \quad C_2H_3O_2^{1-} = \underline{Ba(C_2H_3O_2)_2}$$

(If more than one polyatomic ion is in the formula, enclose the ion in parentheses.)
Ammonium sulfate
    From Table 4-2, ammonium = NH_4^+, sulfate = SO_4^{2-}.

$$NH_4^{1-} \quad SO_4^{2-} \qquad = \underline{(NH_4)_2SO_4}$$

Thallium(III) nitrate $Tl^{3+} \quad NO_3^{1-} \qquad = \underline{Tl(NO_3)_3}$

Manganese(III) phosphate $Mn^{3+} \quad PO_4^{3-} \qquad = Mn_3(PO_4)_3 = \underline{\underline{MnPO_4}}$

EXAMPLE 4–7

Naming Compounds by Analogy

Name the following compounds: $NaBrO_4$ and $Cu(IO)_2$.

Answer
$NaBrO_4$: The compound is composed of Na^+ and BrO_4^- ions. Both Br and Cl are in Group VIIA, so they may form analogous ions. Since the name of the ClO_4^- ion is the perchlorate ion, it is logical to assume that BrO_4^- is named the *perbromate* ion. The name of the compound is

<u>sodium perbromate</u>

$Cu(IO)_2$: The anion is analogous to the ClO^- (hypochlorite) ion since both

I and Cl are also in Group VIIA. This is the hypoiodite ion (IO^-). The Cu cation must then have a $+2$ charge.

copper(II) hypoiodite

For the most part, the compounds we have named so far are ionic compounds composed of a metal cation and a nonmetal anion or a polyatomic anion. Nonmetals, however, also form neutral molecular compounds when bonded to other nonmetals. When a compound is composed of two nonmetals, the rules for nomenclature are somewhat different, and that is discussed next.

▼**Looking Ahead**

4–7 NAMING NONMETAL-NONMETAL BINARY COMPOUNDS

In the case of metal-nonmetal compounds we name and write the metal first. What happens when a binary compound is composed of two nonmetals? In this case, we generally name and write the nonmetal that is closer to being a metal first—that is, the nonmetal closer to the metal-nonmetal border in the periodic table (further down or further to the left). Thus we write CO_2 rather than O_2C but OF_2 rather than F_2O. In cases where both elements are equidistant from the border, Cl is written before O (e.g., Cl_2O), and in the order S, N, then Br (e.g., S_4N_4 and NBr_3). When hydrogen is combined with nonmetals in Group VIA and VIIA (e.g., H_2O and HF), it is written first. When combined with other nonmetals (Groups IIIA, IVA, and VA) and metals, however, it is written second (e.g., NH_3 and CH_4.) In some of these latter compounds, hydrogen is actually more metallic, but the formulas were written that way long before any attempts were made to systematize nomenclature.*

The more metallic element is also named first using its English name. The less metallic is named second using its English root plus *ide* as discussed before. If more than one compound of the same two nonmetals exist, the number of atoms of each element present in the compound is indicated by the use of Greek prefixes (see Table 4-3). Table 4-4

Chemical names are familiar to us in many common drugs and cleansers.

TABLE 4–3 GREEK PREFIXES

Number	Prefix	Number	Prefix	Number	Prefix
1	mono	5	penta	8	octa
2	di	6	hexa	9	nona
3	tri	7	hepta	10	deca
4	tetra				

*With few exceptions, in organic compounds containing C, H, and other elements, the C is written first followed by H and then other elements that are present.

Nitrogen dioxide contributes to the brownish haze in polluted air.

TABLE 4–4 THE OXIDES OF NITROGEN	
Formula	Name
N_2O	dinitrogen monoxide (sometimes referred to as nitrous oxide)
NO	nitrogen monoxide (sometimes referred to as nitric oxide)
N_2O_3	dinitrogen trioxide
NO_2	nitrogen dioxide
N_2O_4	dinitrogen tetroxide*
N_2O_5	dinitrogen pentoxide*

*The "a" is often omitted from tetra and penta for ease in pronunciation.

illustrates the nomenclature of nonmetal-nonmetal compounds with the six oxides of nitrogen. Notice that if there is only one atom of the nonmetal written first, *mono* is not used. However, if there is only one of the second nonmetal, *mono* is used. (Notice that the *o* in mon*o* is dropped in monoxide for ease in pronunciation.) The Stock method is rarely applied to the naming of nonmetal-nonmetal compounds because it can be ambiguous in some cases. For example, both NO_2 and N_2O_4 could be named nitrogen(IV) oxide.

Several of these compounds are known only by their common names, such as water (H_2O), methane (CH_4), and ammonia (NH_3).

Are the rules always followed?

According to the rules, compounds such as TiO_2 and UF_6 should be named by the Stock method—for example, titanium(IV) oxide and uranium(VI) fluoride, respectively. Sometimes, however, we hear them named in the same manner as nonmetal-nonmetal binary compounds (i.e., titanium dioxide and uranium hexafluoride.) The rationale for the later names is that when the charge on the metal exceeds +3, the compound has properties more typical of a molecular nonmetal-nonmetal binary compound than an ionic one. For example, UF_6 is a liquid at room temperature, whereas ionic compounds are all solids under these conditions. In any case, in this text, we will identify all metal-nonmetal binary compounds by the Stock method regardless of their properties and confine the use of Greek prefixes to the nonmetal-nonmetal compounds.

Looking Ahead ▼

Our final topic on nomenclature concerns those compounds in which most of the anions listed in Table 4-1 and 4-2 are combined with hydrogen. Since hydrogen is not a metal, it does not form a cation in compounds but forms molecular compounds instead. Most of these molecular hydrogen compounds have special properties, especially when dissolved in water. These common properties earn them a special classification called acids.

4–8   NAMING ACIDS

When hydrogen is combined with an anion such as Cl^-, the formula of the resulting compound is HCl. The fact that HCl is a gas and not a hard solid at room temperature indicates that HCl is molecular rather than ionic. When dissolved in water, however, the HCl is ionized by the water molecules to form H^+ ions and Cl^- ions. This ionization is illustrated as

$$HCl \xrightarrow{H_2O} H^+ + Cl^-$$

Most of the hydrogen compounds formed from the anions in Tables 4-1 and 4-2 behave in a similar manner, at least to some extent. *This common property of forming H^+ in aqueous solution is a property of a class of compounds called* **acids.** Acids are important enough to earn their own nomenclature. The chemical nature of acids is discussed in more detail in Chapter 12.

The acids formed from the anions listed in Table 4-1 *are composed of just two elements, so they are called* **binary acids.** These compounds can be named in two ways. In the pure state, the hydrogen is named like a metal with only one charge (+1). That is, HCl is named *hydrogen chloride* and H_2S is named *hydrogen sulfide.* When dissolved in water, however, these compounds are generally referred to by their acid names. The acid name is obtained by dropping the word hydrogen, adding the prefix *hydro* to the anion root, and changing the *ide* ending to *ic* followed by the word *acid.* Both types of names are illustrated in Table 4-5.

The following hydrogen compounds of anions listed in Table 4-1 are not generally considered to be binary acids: H_2O, NH_3, CH_4, and PH_3.

The acids formed by combination of hydrogen with most of the polyatomic anions in Table 4-2 are known as **oxyacids** *because they are formed from oxyanions.* To name an oxyacid, we use the root of the anion to form the name of the acid. If the name of the oxyanion ends in *ate,* it is changed to *ic* followed by the word *acid.* If the name of the anion ends in *ite,* it is changed to *ous* plus the word *acid.* Most hydrogen compounds of oxyanions do not exist in the pure state as do the binary acids. Generally, only the acid name is used in the naming of these compounds. For example, HNO_3 is called nitric acid and not hydrogen nitrate. Development of the acid name from the anion name is shown for some anions in Table 4-6.

If the anion does not contain oxygen, it is obviously not an oxyanion. Acids formed from these anions are usually named in the same manner as the binary acids. For example, the anion cyanide (CN^-) forms an acid (HCN) that is named *hydrocyanic acid.*

TABLE 4–5 BINARY ACIDS

Anion	Formula of Acid	Compound Name	Acid Name
Cl^-	HCl	hydrogen chloride	hydrochloric acid
F^-	HF	hydrogen fluoride	hydrofluoric acid
I^-	HI	hydrogen iodide	hydroiodic acid
S^{2-}	H_2S	hydrogen sulfide	hydrosulfuric acid

TABLE 4–6	OXYACIDS		
Anion	Name of Anion	Formula of Acid	Name of Acid
$C_2H_3O_2^-$	acetate	$HC_2H_3O_2$	acetic acid
CO_3^{2-}	carbonate	H_2CO_3	carbonic acid
NO_3^-	nitrate	HNO_3	nitric acid
PO_4^{3-}	phosphate	H_3PO_4	phosphoric acid
ClO_2^-	chlorite	$HClO_2$	chlorous acid
ClO_4^-	perchlorate	$HClO_4$	perchloric acid
SO_3^{2-}	sulfite	H_2SO_3	sulfurous acid
SO_4^{2-}	sulfate	H_2SO_4	sulfuric acid

Working It Out ●

EXAMPLE 4–8

Naming Acids

Name the following acids: H_2Se, $H_2C_2O_4$, and $HClO$.

Answer

H_2Se	hydroselenic acid
$H_2C_2O_4$	oxalic acid
$HClO$	hypochlorous acid

EXAMPLE 4–9

Writing Formulas of Acids

Give formulas for the following: permanganic acid, dichromic acid, and acetic acid.

Answer

permanganic acid	$HMnO_4$
dichromic acid	$H_2Cr_2O_7$
acetic acid	$HC_2H_3O_2$

Looking Back ▲

Our first important use of the periodic table is to aid us in the language of chemistry known as chemical nomenclature. The identification of an element as a metal or a nonmetal determines how we name its compounds. There are four major classifications of compounds that we have identified in this chapter: (1) metal-nonmetal binary compounds, (2) metal-polyatomic ion compounds, (3) nonmetal-nonmetal binary com-

pounds, and (4) acids. In addition to these classifications, we also noticed that we represent the charge on a metal in its name if it can form more than one charge; otherwise we do not.

Learning Check B

◀Checking It Out

B–1. Fill in the blanks.
Given the following four elements: K, Se, Al, and Br.
The metals are _____ and the nonmetals are _____ . The charges that these elements have in binary ionic compounds are K = _____ , Se = _____ , Al = _____ , and Br = _____ . The formula of a compound formed between K and Se is _____ and its name is _____ . The formula of a compound formed between Al and Br is _____ and its name is _____ . The formula of a compound formed between Se and Br is $SeBr_2$ and its name is _____ . An acid formed from the Se ion has the formula _____ and its name is _____ .

B–2 Fill in the blanks.
Given the following ions: Ca^{2+}, Au^{3+}, MnO_4^-, and CrO_4^{2-}.
The formula of a compound formed between Ca^{2+} and MnO_4^- is _____ and its name is _____ . The formula of a compound formed between Au^{3+} and CrO_4^{2-} is _____ and its name is _____ . An acid formed from the CrO_4^{2-} ion has the formula _____ and its name is _____ .

B–3. Three fluorine compounds formed by Group IIIA elements are BF_3, AlF_3, and TlF_3. All three are named somewhat differently. Name the compounds.

B–4. Give the chemical names for Br_2O and OF_2. Why is oxygen named first in OF_2 and second in Br_2O?

Additional Examples: Problems 4-18, 4-20, 4-22, 4-34, 4-36, 4-43, 4-45, 4-47, 4-48.

C H A P T E R   R E V I E W

There is no more important time-saving device for the chemist than the **periodic table,** which demonstrates in table form the **periodic law** for the elements. Horizontal rows are known as **periods,** and vertical columns are known as **groups.** Although there are four categories of elements in the table, the category that we will emphasize in the text is the **main group** or **representative elements.** This category includes some named groups known as the **alkali metals,** the **alkaline earth metals,** the **chalcogens,** and the **halogens.** The other three categories

▲▼Putting It Together

are the **noble** or **inert gases,** the **transition metals,** and the **inner transition metals.** The latter category includes the **lanthanides** and the **actinides.** These named groups are summarized in the chart that follows.

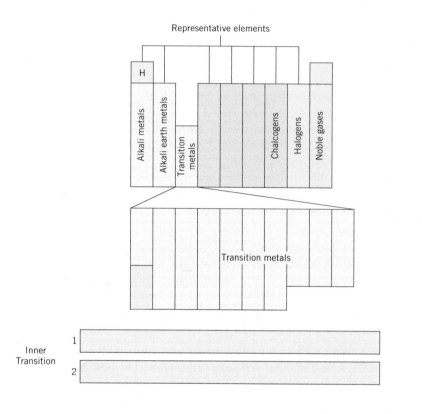

One important function of the periodic table is that it shows a clear boundary between elements that are classified as **metals** and **nonmetals.** Some metals and nonmetals have intermediate characteristics and may be referred to as **metalloids.** All three physical states are found among the elements at the reference temperature of 25°C, known as **room temperature.** Some of the nonmetals also exist as molecules rather than individual atoms at the reference temperature.

Our first important use of the periodic table is to aid us in one aspect of **chemical nomenclature.** For example, binary compounds containing a metal and a nonmetal are named differently than those composed of two nonmetals. Also, all metals form cations but some form only one charge and others form more than one. In using the **Stock method** of nomenclature, we need the periodic table to tell us which metals are in the latter group. In addition to the binary compounds, we discussed the naming of **salts** containing **oxyanions.** Finally, a class of hydrogen compounds called **acids** (both **binary** and **oxyacids**) was discussed as a special group of compounds. The naming of compounds is summarized by the chart that follows on the next page.

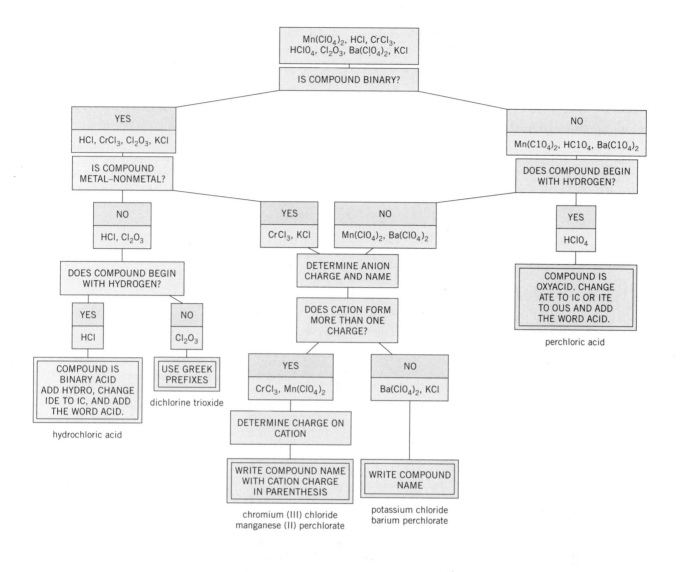

$Mn(ClO_4)_2$, HCl, $CrCl_3$, $HClO_4$, Cl_2O_3, $Ba(ClO_4)_2$, KCl

IS COMPOUND BINARY?

YES — HCl, $CrCl_3$, Cl_2O_3, KCl

NO — $Mn(ClO_4)_2$, $HClO_4$, $Ba(ClO_4)_2$

IS COMPOUND METAL–NONMETAL?

DOES COMPOUND BEGIN WITH HYDROGEN?

NO — HCl, Cl_2O_3

YES — $CrCl_3$, KCl

NO — $Mn(ClO_4)_2$, $Ba(ClO_4)_2$

YES — $HClO_4$

DOES COMPOUND BEGIN WITH HYDROGEN?

DETERMINE ANION CHARGE AND NAME

COMPOUND IS OXYACID. CHANGE ATE TO IC OR ITE TO OUS AND ADD THE WORD ACID.

perchloric acid

YES — HCl

NO — Cl_2O_3

DOES CATION FORM MORE THAN ONE CHARGE?

COMPOUND IS BINARY ACID ADD HYDRO, CHANGE IDE TO IC, AND ADD THE WORD ACID.

hydrochloric acid

USE GREEK PREFIXES

dichlorine trioxide

YES — $CrCl_3$, $Mn(ClO_4)_2$

NO — $Ba(ClO_4)_2$, KCl

DETERMINE CHARGE ON CATION

WRITE COMPOUND NAME

WRITE COMPOUND NAME WITH CATION CHARGE IN PARENTHESIS

chromium (III) chloride
manganese (II) perchlorate

potassium chloride
barium perchlorate

E X E R C I S E S

THE PERIODIC TABLE

4–1 How is an active metal different from a noble metal?

4–2 How many elements would be in the seventh period if it were complete?

4–3 Which of the following elements are halogens?
   (a) O_2
   (b) P_4
   (c) I_2
   (d) N_2
   (e) Li
   (f) H_2
   (g) Br_2

4–4 Which of the following elements are alkaline earth metals?
   (a) Sr
   (b) C
   (c) B
   (d) Be
   (e) Na
   (f) K

4–5 Classify the following elements into one of the four main categories of elements.
   (a) Fe
   (b) Te
   (c) Pm
   (d) La
   (e) Xe
   (f) H
   (g) In

4–6   Classify the following elements into one of the four main categories of elements.
(a) Se
(b) Ti
(c) Ni
(d) Sr
(e) Zn
(f) I
(g) Er

4–7   Which of the following elements are transition metals?
(a) In
(b) U
(c) Ca
(d) Xe
(e) Pd
(f) Tl
(g) Ag

PHYSICAL STATES OF THE ELEMENTS

4–8   What is the most common physical state of the elements at room temperature? Which are more common, metals or nonmetals?

4–9   Which metals, if any, are gases at room temperature? Which metals, if any, are liquids? Which nonmetals, if any, are liquids at room temperature?

4–10   Referring to Figure 4-5, tell which of the following are gases at room temperature.
(a) Ne
(b) S
(c) B
(d) Cl
(e) Br
(f) N
(g) Na

4–11   Referring to Figure 4-5, tell which of the following elements exist as diatomic molecules under normal conditions.
(a) N
(b) C
(c) Ar
(d) F
(e) H
(f) B
(g) Xe
(h) Hg

4–12   Referring to Figure 4-4, tell which of the following elements are metals.
(a) Ru
(b) Sn
(c) Hf
(d) Te
(e) Ar
(f) B
(g) Se
(h) W

4–13   Which, if any, of the elements in Problem 4-12 can be classified as a metalloid?

4–14   Identify the following elements using the information in Figures 4-4 and 4-5.
(a) a nonmetal, monatomic gas in the third period
(b) a transition metal that is a liquid
(c) a diatomic gas in group VA
(d) the second metal in the second period
(e) the only member of a group that is a metal

4–15   Identify the following elements using the information in Figures 4-4 and 4-5.
(a) a nonmetal, diatomic liquid
(b) the last element in the third period
(c) a nonmetal, diatomic solid
(d) the only member of a group that is a nonmetal

4–16   Assume that elements with atomic numbers greater than 109 will eventually be made. If so, what would be the atomic number and group number of the next nonmetal? (Assume that the border between metals and nonmetals continues as before.)

4–17   What is the atomic number of the metal farthest to the right in the periodic table that would appear after element number 109?

METAL-NONMETAL BINARY COMPOUNDS

4–18   Name the following compounds.
(a) LiF
(b) BaTe
(c) Sr_3N_2
(d) BaH_2
(e) $AlCl_3$

4–19   Name the following compounds.
(a) CaI_2
(b) FrF
(c) BeSe
(d) Mg_3P_2
(e) RaS

4–20   Give formulas for the following compounds.
(a) rubidium selenide
(b) strontium hydride
(c) radium oxide
(d) aluminum carbide
(e) beryllium fluoride

4–21   Give formulas for the following compounds.
(a) potassium hydride
(b) cesium sulfide
(c) potassium phosphide
(d) barium telluride

4–22   Name the following compounds using the Stock method.
(a) Bi_2O_5
(b) SnS
(c) SnS_2
(d) Cu_2Te
(e) TiC

4–23   Name the following compounds using the Stock method.
(a) CrI_3
(b) $TiCl_4$
(c) IrO_4
(d) MnH_2
(e) $NiCl_2$

4–24 Give formulas for the following compounds.
(a) copper(I) sulfide
(b) vanadium(III) oxide
(c) gold(I) bromide
(d) nickel(II) carbide
(e) chromium(VI) oxide

4–25 Give formulas for the following compounds.
(a) yttrium(III) hydride
(b) lead(IV) chloride
(c) bismuth(V) fluoride
(d) palladium(II) selenide

4–26 From the magnitude of the charges on the metals, predict which of the compounds in Problems 4-22 and 4-24 may be molecular compounds.

4–27 From the magnitude of the charges on the metals, predict which of the compounds in Problems 4-23 and 4-25 may be molecular compounds.

COMPOUNDS WITH POLYATOMIC IONS

4–28 Which of the following is the chlorate ion?
(a) ClO_2^- (d) Cl_3O^-
(b) ClO_4^- (e) ClO_3^+
(c) ClO_3^-

4–29 Which of the following ions have a -2 charge?
(a) sulfate   (d) carbonate
(b) nitrite   (e) sulfite
(c) chlorite   (f) phosphate

4–30 What are the name and formula of the most common polyatomic cation?

4–31 Which of the following oxyanions contain four oxygen atoms?
(a) nitrate   (e) phosphate
(b) permanganate   (f) oxalate
(c) perchlorate   (g) carbonate
(d) sulfite

4–32 Name the following compounds. Use the Stock method where appropriate.
(a) $CrSO_4$ (e) $(NH_4)_2CO_3$
(b) $Al_2(SO_3)_3$ (f) NH_4NO_3
(c) $Fe(CN)_2$ (g) $Bi(OH)_3$
(d) $RbHCO_3$

4–33 Name the following compounds. Use the Stock method where appropriate.
(a) $Na_2C_2O_4$ (c) $Fe_2(CO_3)_3$
(b) $CaCrO_4$ (d) $Cu(OH)_2$

4–34 Give formulas for the following compounds.
(a) magnesium permanganate
(b) cobalt(II) cyanide
(c) strontium hydroxide
(d) thallium(I) sulfite
(e) indium(III) bisulfate
(f) iron(III) oxalate
(g) ammonium dichromate
(h) mercury(I) acetate (The mercury(I) ion exists as $Hg_2{}^{2+}$.)

4–35 Give formulas for the following compounds.
(a) zirconium(IV) phosphate
(b) sodium cyanide
(c) thallium(I) nitrite
(d) nickel(II) hydroxide
(e) radium hydrogen sulfate
(f) beryllium phosphate
(g) chromium(III) hypochlorite

4–36 Complete the following table. Write the appropriate anion at the top and the appropriate cation to the left. Write the formulas and names in the other blanks.

Cation/anion	HSO_3^-	_____	_____
NH_4^+	NH_4HSO_3 ammonium bisulfite	_____ _____ _____	_____
_____ _____	_____ _____	$CoTe$ _____ (name)	_____
_____ _____	_____ _____	_____ _____	_____ (formula) aluminum phosphate

4–37 Complete the following table. Write the appropriate anion at the top and the appropriate cation to the left. Write the formulas and names in the other blanks.

Cation/anion	_____	$C_2O_4{}^{2-}$	_____
_____	_____ (formula) thallium(I) hydroxide	_____ _____	_____
Sr^{2+}	_____ _____	_____ _____	_____ _____
_____ _____	_____ _____	_____ _____	TiN _____ (name)

4–38 Give the systematic name for each of the following.

Common Name	Formula
(a) table salt	NaCl
(b) baking soda	$NaHCO_3$
(c) marble or limestone	$CaCO_3$
(d) lye	NaOH
(e) chili saltpeter	$NaNO_3$
(f) sal ammoniac	NH_4Cl
(g) alumina	$Al2O_3$
(h) slaked lime	$Ca(OH)_2$
(i) caustic potash	KOH

4–39 The perzenate ion has the formula $XeO_6{}^{4-}$. Write formulas of compounds of perzenate with the following.
(a) calcium   (c) aluminum
(b) potassium

*4–40 Name the following compounds. In these compounds, an ion is involved that is not in Table 4-2. However, the name can be determined by reference to other ions of the central element or from ions in Table 4-2 in which the central atom is in the same group.
(a) PH_4F (d) $CaSiO_3$
(b) KBrO   (e) $AlPO_3$
(c) $Co(IO_3)_3$ (f) $CrMoO_4$

NONMETAL-NONMETAL BINARY COMPOUNDS

4–41 The following pairs of elements combine to form binary compounds. Which element should be written and named first?
(a) Si and S   (d) Kr and F
(b) F and I   (e) H and F
(c) H and Se   (f) H and As

4–42 The following pairs of elements combine to make binary compounds. Which element should be written and named first?
(a) S and P   (c) O and Br
(b) O and S   (d) As and Cl

4–43 Name the following:
(a) CS_2 (e) SO_3
(b) BF_3 (f) Cl_2O
(c) P_4O_{10} (g) PCl_5
(d) Br_2O_3 (h) SF_6

4–44 Name the following:
(a) PF_3 (d) AsF_5
(b) I_2O_3 (e) $SeCl_4$
(c) ClO_2 (f) SiH_4

4–45 Write formulas for the following:
(a) tetraphosphorus hexoxide
(b) carbon tetrachloride
(c) iodine trifluoride
(d) dichlorine heptoxide
(e) sulfur hexafluoride
(f) xenon dioxide

4–46 Write formulas for the following:
(a) xenon trioxide
(b) sulfur dichloride
(c) dibromine monoxide
(d) carbon disulfide
(e) diboron hexahydride (also known as diborane)

ACIDS

4–47 Name the following acids.
(a) HCl   (d) $HMnO_4$
(b) HNO_3 (e) HIO_4
(c) HClO   (f) HBr

4–48 Write formulas for the following acids.
(a) hydrocyanic acid   (d) carbonic acid
(b) hydroselenic acid   (e) hydroiodic acid
(c) chlorous acid   (f) acetic acid

4–49 Write formulas for the following acids.
(a) oxalic acid   (c) dichromic acid
(b) nitrous acid   (d) phosphoric acid

*4–50 Refer to the ions in Problems 4-39 and 4-40. Write the acid names for the following:
(a) HBrO   (d) $HMoO_4$
(b) HIO_3 (e) H_4XeO_6
(c) H_3PO_3

*4–51 Write the formulas and the names of the acids formed from the arsenite $(AsO_3{}^{3-})$ ion and the arsenate $(AsO_4{}^{3-})$ ion.

GENERAL PROBLEMS

4–52 The halogen (A) with the lowest atomic number forms a compound with another halogen (X) that is a liquid at room temperature. The compound has the formula A_5X or XA_5. Write the correct formula with the actual elemental symbols and the name.

4–53 A metal that has only a +2 ion and is the third member of the group forms a compound with a nonmetal that has a −2 ion and is in the same period. What are the formula and name of the compound?

4–54 The only gas in a certain group forms a compound with a metal that has only a +3 ion. The compound contains one ion of each element. What are the formula and name of the compound? What are the formula and name of the compound the gas forms with a Ti^{2+} ion?

4–55 An alkali metal in the fourth period forms a compound with the phosphide ion. What are the formula and name of the compound?

4–56 A transition metal ion with a charge of +2 has 25 electrons. It forms a compound with a nonmetal that has only a −1 ion. The anion has 36 electrons. What are the formula and name of the compound?

4–57 The lightest element forms a compound with a certain metal in the third period that has a +2 ion and with a nonmetal in the same period

that has a −2 ion. What are the formulas and names of the two compounds?

4–58 The thiosulfate ion has the formula $S_2O_3{}^{2-}$. What are the formula and name of the compound formed between the thiosulfate ion and a Rb ion; an Al ion; a Ni^{2+} ion; and a Ti^{4+} ion? What are the formula and name of the acid formed from the thiosulfate ion?

4–59 Name the following compounds: NiI_2, H_3PO_4, $Sr(ClO_3)_2$, H_2Te, As_2O_3, Sb_2O_3, and SnC_2O_4.

4–60 Name the following compounds: SiO_2, SnO_2, MgO, $Pb_3(PO_4)_2$, $HClO_2$, $BaSO_4$, and HI.

4–61 Give formulas for the following compounds: tin(II) hypochlorite, chromic acid, xenon hexafluoride, barium nitride, hydrofluoric acid, iron(III) carbide, and lithium phosphate.

SOLUTION TO LEARNING CHECKS

A–1 metal, nonmetal, metalloids, atomic number, groups, representative, transition metal, noble gas, alkaline earth, halogens, mercury.

B–1 K and Al, Se and Br, +1, −2, +3, −1, K_2Se, potassium selenide, $AlBr_3$, aluminum bromide, selenium dibromide. H_2Se, hyderoselenic acid.

B–2 $Ca(MnO_4)_2$, calcium permanganate, $Au_2(CrO_4)_3$, gold(III) chromate, H_2CrO_4, chromic acid.

B–3 BF_3—boron trifluoride (molecular compound from two nonmetals)
AlF_3—aluminum fluoride (metal has one charge)
TlF_3—thallium (III) fluoride (more than one type of metal ion)

B–4 Br_2O—dibromine oxide
OF_2—oxygen difluoride
Br is more metallic than O, but O is more metallic than F

Fireworks produce spectacular colors. Each color is caused by the presence of a specific element. The emission of color by the atoms of hot gaseous elements is a topic of this chapter.

MODERN ATOMIC THEORY

Bursts of bright silver, streaks of green, yellow, red, and blue explode across the sky. How we love it! The Fourth of July, the closing ceremony of the Olympics, the celebration of a centennial are all highlighted by spectacular displays of fireworks. But what causes such exciting colors? Actually, it has been known for hundreds of years that certain elements in fireworks produce specific colors. Strontium imparts a red color, barium green, copper blue, and sodium yellow. In fact, the hot gaseous atoms of all elements emit specific colors that can be used to identify elements like fingerprints can be used to identify individuals. The study of the colors of these *emission spectra* is what opened the door to a deeper understanding of the nature of the atom.

◄Setting The Stage

Often in science, when a theory is developed to explain one phenomenon, other mysteries are explained as well. This happened with the theory that explained the emission spectra of the elements. The theory, which emphasized the electrons in the atom, led to an understanding of the theoretical basis of the periodic table. As we explained in Chapter 4, this marvelous table displays elements that are chemically related in vertical columns called groups. Although the existence of chemically related elements has been known for almost two hundred years, only in the past sixty years or so have scientists had a feeling for *why* elements are chemically related.

A theory of the atom that emphasized the electrons was first advanced in 1913 by a student of Lord Rutherford named Niels Bohr. This had the immediate effect of explaining the emission spectrum of hydrogen, the simplest of the elements. More significant to our purposes, however, is that Bohr's theory eventually led to a new, improved theory that explained and predicted chemical similarities among certain elements.

Formulating ▶ Some Questions

In this chapter, we will attempt to answer several important questions. What is the nature of light? How does one color differ from another? How does an atom emit light? What is the relationship between electrons and the periodic table? What trends in atomic properties can be predicted by the periodic table? Our first goal is to examine the nature of light so that we can appreciate how atoms emit specific colors.

5–1   THE EMISSION SPECTRA OF THE ELEMENTS

What are the characteristics of light?

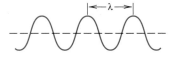

We all sense that the glowing blob of light in the daytime sky, which we call the sun, is very hot. Actually, any substance glows if it is heated to a high enough temperature. The tungsten filament in an incandescent lightbulb emits a bright light as a result of heat generated by the flow of electricity. Light, which is also known as **electromagnetic radiation,** *is a form of energy like heat or electricity.* Light travels through space in waves much like the waves moving across a pond or a lake. Light waves travel at a given velocity and also have properties called wavelength and amplitude. All light waves travel at the same velocity in a vacuum, which is at the phenomenal rate of 3.0×10^{10} cm/s (186,000 mi/s). The amplitude of a wave refers to its height, which in turn relates to the intensity of light. The amplitude is not important to our discussion so it will not be mentioned further. *The* **wavelength** *of light is the distance between two adjacent peaks in the wave.* Wavelength is very important because it relates to the energy of light. Wavelength is designated by the Greek letter λ (lambda).

Let us return to the white light emitted from a tungsten filament in a lightbulb. We can analyze this light by passing it through a glass prism. Since all colors of light do not have the same velocity through a medium such as glass, the white light is separated by the prism into its component colors. (See Figure 5-1.) The white light separates into a continuous range of colors that we associate with a rainbow. *Since one color blends gradually into another, this is known as a* **continuous spectrum.** A rainbow after a rainstorm is a continuous spectrum caused by the separation of sunlight into its component colors by raindrops in the air.

Each color of light in a rainbow has a specific wavelength. Violet light, on one end of the spectrum, has the shortest wavelength, and red, on the other end of the spectrum, has the longest. Important to our consideration of light, however, is that the energy of light is inversely proportional to its wavelength. That is,

$$E \propto \frac{1}{\lambda} \qquad E = \frac{hc}{\lambda}$$

where h is a constant of proportionality known as Planck's constant, and c is the velocity of light in a vacuum, which is also a constant.

Due to the inverse relationship, the larger the value of λ, the lower the energy. Thus red light, with the longest wavelength, has the lowest

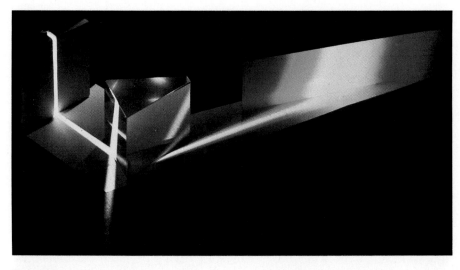

Figure 5–1
The Spectrum of Incandescent Light. When a narrow beam of light from an ordinary lightbulb is passed through a glass prism, the white light is found to contain all of the colors of the visible spectrum.

energy in the visible spectrum, whereas violet, with the smallest wavelength, has the highest energy. The visible part of the spectrum has wavelengths between 400 nm (1 nm = 10^{-9} m) to about 800 nm. The spectrum of electromagnetic radiation extends in both directions well beyond the wavelengths of visible light. Light that extends beyond violet (*ultraviolet light*) is not visible but is highly energetic. Indeed, it is powerful enough to damage living organisms. Our atmosphere shields us from the more powerful ultraviolet light from the sun. Light with extremely small wavelengths and thus extremely high energies includes X-rays and gamma-rays. Only small amounts of this radiation can be tolerated by living systems. On the other side of the visible spectrum is light with wavelengths longer than red (*infrared light*). This direction extends into the extremely long wavelengths of radio and TV waves. (See Figure 5-2.)

When elements are heated to the point where the hot gaseous atoms begin to emit light, the results are quite different from the continuous spectrum emitted by a lightbulb or the sun. In these cases, only definite or discrete colors are produced. Since the colors are discrete, the ener-

What is the difference between red light and blue light?

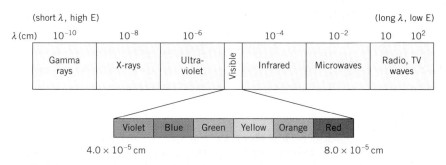

Figure 5–2
The Spectrum of Light

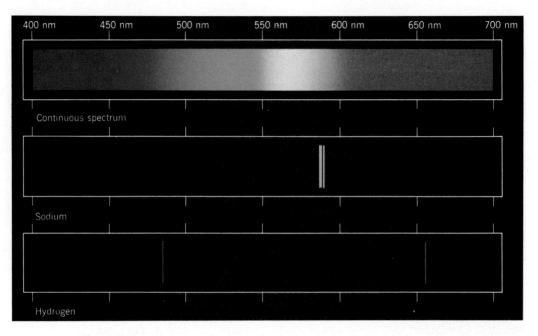

Figure 5–3
The Continuous Spectrum and the Emission Spectra of Sodium and Hydrogen in the Visible Range The light from an incandescent light (*top*) is continuous, whereas the atomic spectra of sodium (*middle*) and hydrogen (*bottom*) are composed of discrete colors.

gies emitted from the atoms are discrete. The spectrum of each element (called its atomic emission spectrum) is unique to that element and is the reason specific elements are used in fireworks. (See Figure 5-3.) *Since only certain colors are produced by the atoms of hot gaseous elements, their atomic emission spectra are referred to as* **discrete** *or* **line spectra.** In Figure 5-4, the spectrum of hydrogen in the visible range is

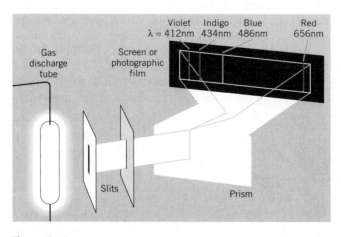

Figure 5–4
Production of the Line Spectrum of Hydrogen The light from excited hydrogen atoms is passed through a glass prism. The prism divides the light into four discrete colors in the visible range.

shown along with the wavelengths of light of the four lines that appear. Energy is supplied to the hydrogen gas by an electrical discharge similar to that in the familiar neon light.

Hydrogen, the simplest of elements with one proton and one electron, has the most orderly atomic spectrum, as shown in Figure 5-4. The unique symmetry of the converging lines of the hydrogen spectrum made this the most interesting of all of the elements. It is this spectrum that Niels Bohr set out to explain with a straightforward theory.

▼Looking Ahead

5–2   A MODEL FOR THE ELECTRONS IN THE ATOM

By 1913, the nuclear theory of the atom proposed by Lord Rutherford was generally accepted by scientists. In that year, Bohr set out to develop this theory further in order to explain the origin of the discrete spectrum of hydrogen. His main goal was to explain how the electron in the hydrogen atom was responsible for the energy emitted in the form of light. Bohr did not realize that his theory would lay the foundation for the explanation of other phenomena such as the periodicity of elements and the way atoms bond to each other.

First, Bohr proposed that the electron in the hydrogen atom revolves around the nucleus in a stable, circular orbit. The electrostatic attraction of the negative electron for the positive proton keeps the electron in a stable orbit. His theory of the hydrogen atom is analogous to the orbiting of the planets around the sun and thus is often referred to as a model. *A* **model** *is a description or analogy used to help visualize a phenomenon.* Bohr's model does not seem so revolutionary at first glance, but classical physics prohibits such a model. Classical physics, which was all that was known at the time, stated that a charged particle (the electron) orbiting around another charged particle (the proton) would lose energy and spin into the nucleus. Bohr sidestepped this problem by postulating (suggesting without proof as a necessary condition) that classical physics did not apply in the small dimensions of the atom. Eventually, he would be proved correct.

Second, Bohr suggested that there are several orbits where the electron may be located. The orbits available to the electron are said to be **quantized,** *which means that they are at definite, or discrete, distances from the nucleus.* Since the energy of a particular orbit is a function of the distance from the nucleus, the energy of an electron is also quantized. Thus *the discrete orbits available to an electron in a hydrogen atom are referred to as* **energy levels.** (See Figure 5-5.)

The quantized energy levels in the hydrogen atom can be compared with a stairway in your home. In Figure 5-6 you can see that between floor A and floor B there are five steps that are analogous to energy levels in an atom. Since you cannot stand (with both feet together) between steps, we can say that each step represents a discrete or *quantized* amount of energy. On the other hand, a ramp between the two floors represents a *continuous* change in energy. In this case, all energy levels are possible between floors.

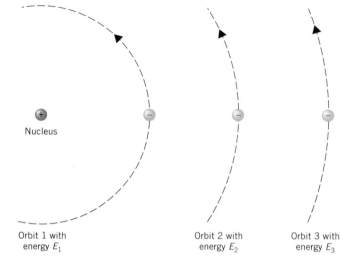

Figure 5–5
Bohr's Model of the Hydrogen Atom In this model, the electron can exist only in definite or discrete energy states

Nucleus

Orbit 1 with energy E_1

Orbit 2 with energy E_2

Orbit 3 with energy E_3

Under normal conditions, *the single electron in a hydrogen atom will occupy the lowest energy level, which is the orbit closest to the nucleus. This is called the* **ground state.** When energy is supplied to a hydrogen atom, however, as when it is heated, the electron can absorb the appropriate amount of this energy to "jump" to a higher energy level. Because of its new position in a higher energy level, the electron now has potential energy just like a weight suspended above the ground. *Energy levels higher than the ground state are called* **excited states.** Bohr suggested that the light energy that is emitted from a hydrogen atom originates from the transition of an electron from an excited state down to lower excited states or back to the ground state. Since energy levels are quantized, the difference in energy between any two levels is quantized. Thus, when an electron falls back to a lower energy level, it must emit a discrete amount of energy. Since this energy is emitted as light, the light would have a discrete energy, a discrete wavelength, and a discrete color (if the light is in the visible part of the spectrum). (See Figure 5-7.)

Why does an atom give off light of a specific color?

This is the qualitative explanation of the discrete spectrum. The real significance of Bohr's model is that he was able to calculate the expected wavelength of light in the hydrogen spectrum. The experimental values listed in Figure 5-4 correspond beautifully with those computed by Bohr

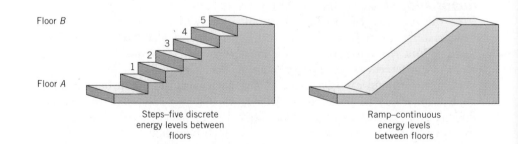

Floor *B*

Floor *A*

Steps–five discrete energy levels between floors

Ramp–continuous energy levels between floors

Figure 5–6
Discrete Versus Continuous Energy Levels The steps represent discrete energy levels; the ramp, continuous energy levels.

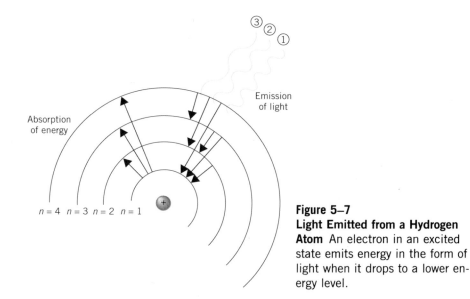

Figure 5–7
Light Emitted from a Hydrogen Atom An electron in an excited state emits energy in the form of light when it drops to a lower energy level.

from his theoretical calculations. Although Bohr's model was specifically for the hydrogen atom, the discrete spectra of other elements result from similar considerations.

Because of the simplicity of Bohr's model of the hydrogen atom, modern scientists still use this picture in certain situations. In the modern theory of the atom, however, we do not view electrons as simple particles with definite velocities and energies orbiting around a nucleus. Due to discoveries about the nature of matter, we now describe the electron as having properties of a particle (mass) *and* of light (wave nature). *The result is a complex mathematical approach to the electrons in the atom, known as* **wave mechanics.** According to the wave mechanical model of

Is Bohr's model still accepted?

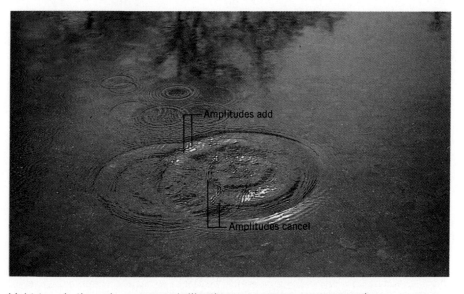

Light travels through space much like the waves move across a pond.

the atom, the electron has a certain probability of existing in a given region of space. Because of its high velocity, we can't tell exactly where it is. This is analogous to a fastball in baseball. The faster it is thrown, the less the hitter knows about its location (and the less likely that the ball will be hit). Our main interest in the wave mechanical model is that it tells us more about the electrons in atoms, and this information will give us an understanding of the periodicity of elements.

Comparing Bohr's model to the wave mechanical model is like comparing a Model T Ford to a present-day automobile with fuel injection and other modern conveniences. The Model T was very simple and easy to repair. On the other hand, it was hard to drive and had to be repaired a lot. The modern automobile does everything except drive itself, but only a skilled mechanic dares to tamper with anything beneath the hood. Fortunately, however, we don't need to understand either the workings of a modern car or the theoretical basis of the wave mechanical model to put both to use.

Looking Ahead ▼

Bohr suggested that atoms have discrete energy levels. The wave mechanical model goes much further, suggesting that energy levels can be further subdivided. Eventually, we will find that these subdivisions of energy levels give us the theoretical basis of the periodic table.

5–3   SHELLS AND SUBSHELLS

Although Bohr and the wave mechanical model describe only the energy of the one electron in the hydrogen atom, it is assumed *that electrons in the atoms of all other elements occupy energy levels like those of hydrogen.* As with hydrogen, the level closest to the nucleus is the lowest in energy. Each successive level out from the nucleus is higher in energy and has a progressively higher capacity for electrons. The capacity of an energy level is determined from the formula

$$2n^2$$

where *n is an integer that designates each energy level and is referred to as the* **principal quantum number.** For the first energy level, $n = 1$; for the second energy level, $n = 2$; and so forth. The capacity of the first four energy levels is given in Table 5-1.

Bohr's model described the electron only in terms of the principal quantum number. The wave mechanical model carries us much further.

TABLE 5–1   SHELLS

Quantum Number Designation (*n*)	Shell Capacity ($2n^2$)
1	$2 \times 1^2 = 2$
2	$2 \times 2^2 = 8$
3	$2 \times 3^2 = 18$
4	$2 \times 4^2 = 32$

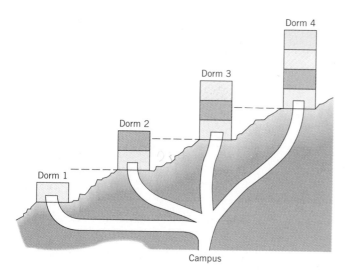

Figure 5–8
Floors within a Dorm. The floors represent different subdivisions of energy within each dorm, analogous to the subshells within each shell.

Each principal energy level, which is known as a **shell,** *has one or more* **subshells.** To understand this, let us consider an analogy. Assume that there are four dormitories located near a campus, with successive dormitories located farther and farther up a hill. Obviously, the dorm lowest on the hill is the lowest in energy since it requires the least amount of energy to reach. In Figure 5-8, we have labeled these four dorms 1 through 4. Each dorm is analogous to a shell (principal energy level) in an atom. But there is more to getting to a room in a dorm than reaching the entrance to the building. Chances are good that the elevators are full, stuck on the top floor, or completely out of order. Thus a student has to use the stairs. Therefore, which floor a student's room is on is also an important energy consideration. In our analogy, the first dorm lowest on the hill has only one floor, the second has two, the third has three, and so on. (This unusual situation is for the sake of the analogy.) Each floor represents a different subdivision of energy *within* each dorm.

How can an electron shell be subdivided?

Just as the dorms have one or more floors, each shell has one or more subshells. The subshells found in the known elements are labeled in order of increasing energy.

$$s < p < d < f$$
increasing energy
$\longrightarrow$

For each shell, the number of subshells is equal to the value of its quantum number n. Thus, analogous to the dorms, the first shell ($n = 1$) has an s subshell. The second shell ($n = 2$) has two subshells, an s and a p. The third shell ($n = 3$) has three (s, p, and d) and the fourth shell ($n = 4$) has four (s, p, d, and f). Each type of subshell has a different electron capacity. This information has been summarized in Table 5-2, which indicates how the total electron capacity of the first four shells is distributed among the subshells.

In Figure 5-9, the first four shells are shown in order of increasing energy, analogous to the dorms shown in Figure 5-7.

TABLE 5–2   SHELLS AND SUBLEVELS

Shell	1	2		3			4			
Subshell	s	s	p	s	p	d	s	p	d	f
Subshell capacity	2	2	6	2	6	10	2	6	10	14
Shell capacity	2		8		18			32		

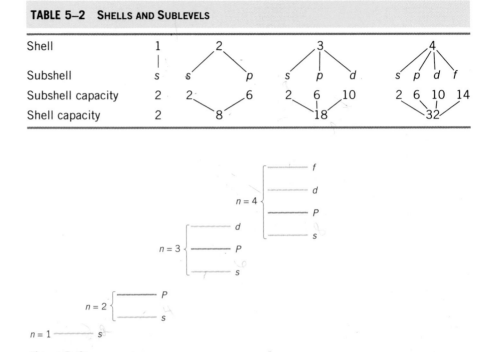

Figure 5–9
Energy Diagram of the First Four Shells Each successive shell contains one additional subshell.

Looking Ahead ▼

To locate a student in a particular dorm on a certain floor, more information is needed. Since each floor probably has more than one room, we need to know the location of the room. Just as students reside in regions of space called rooms, electrons reside in regions of space called orbitals. We will discuss orbitals in the next section.

5–4   ORBITALS

How can an electron subshell be subdivided?

A subshell is composed of one or more orbitals. *An **orbital** is the region of space where there is the highest probability of finding a particular electron.* Each orbital can hold two electrons. An *s* subshell is composed of one orbital. This is analogous to a certain floor having only one room. Thus an *s* subshell, being composed of one orbital, has a capacity of two electrons. A *p* subshell is composed of three orbitals, however, so it has a capacity of six electrons (two per orbital). A *d* subshell is composed of five orbitals, and an *f* subshell is composed of seven orbitals.

Consider the third shell ($n = 3$). It is composed of three subshells (s, p, and d). The 18-electron capacity of the third shell is distributed as follows: two in the one *s* orbital, six in the three *p* orbitals, and ten in the five *d* orbitals. This is analogous to a dorm having one room on the first floor, three on the second, and five on the third. (See Figure 5-10.)

Figure 5–10
Dorm Rooms and Orbitals. The rooms represent different regions of space on each floor of a dorm, analogous to the orbitals of each subshell.

The energy of an electron is determined only by the shell and subshell, under normal conditions. This means that all of the three different orbitals in a given p subshell (e.g., the $3p$) have the same energy. In the dorm analogy, once you have reached a particular floor, it doesn't matter whether you go to the room on the right, left, or straight ahead. The same amount of energy is required in each case.

Each type of orbital (s, p, d, or f) has a particular shape. By shape we refer to the dimensions of the region of highest probability of finding the particular electron or electrons. Thus, when we represent a sphere, the electron will have at least a 90% probability of being within that volume. Despite Bohr's model, scientists no longer picture an electron as orbiting the nucleus in some designated path within the orbital. It just exists within that region.

The s subshell is composed of one orbital that has a perfectly spherical shape. There is an equal probability of finding the one or two electrons in an s orbital in any direction from the nucleus. The highest probability of finding a $2s$ electron is in a spherical volume farther from the nucleus than a $1s$ orbital. (See Figure 5-11.)

The p subshell is composed of three orbitals. Each orbital has two regions of high probability called "lobes" on either side of the nucleus. A p orbital is shaped like a weird baseball bat with two fat ends. In a three-coordinate graph, if we define x as the axis along which the two lobes of

What is the difference between two types of orbitals?

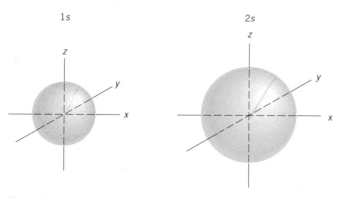

Figure 5–11
s **Orbitals.** The *s* orbitals have a spherical shape. The region of highest probability extends further from the nucleus for a 2*s* orbital compared to a 1*s*.

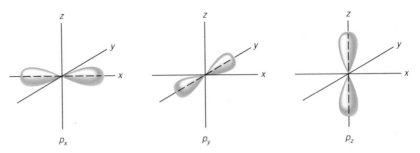

Figure 5–12
p Orbitals Each p orbital has two lobes.

a p orbital are directed, we can designate that p orbital as the p_x orbital. The other two p orbitals are each at a 90° angle to the p_x orbital and to each other. They are referred to as the p_y and the p_z orbitals. (See Figure 5-12.)

The d subshell is composed of five orbitals that are more complex. A d orbital is usually drawn with four lobes. The one or two electrons in a d orbital may exist in regions of space *between* the axes (d_{xy}, d_{xz}, d_{yz}) or *along* the axes ($d_{x^2-y^2}$, d_{z^2}). (See Figure 5-13.)

The f orbitals are even more complex. As we will see in the next chapter, electrons are intimately involved in bonding. However, since electrons in f orbitals are rarely involved in bonding, their shapes usually are not important to the chemist.

Looking Back ▲

The relationship between the light emitted by hot gaseous atoms and the electrons in the atom was the key that unlocked some hard-core mysteries of the elements. Niels Bohr was responsible for bringing the word "quantum" into frequent usage, at least in the scientific world. After Bohr's model came wave mechanics and the description of the electron in terms of subshells and orbitals as well as shells. Ultimately, all electrons in an atom exist in a specific orbital. The subshell defines the shape of a particular orbital. The shell defines the distance from the nucleus of the highest electron probability.

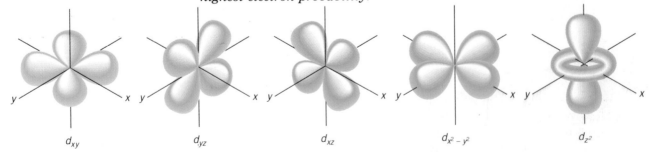

Figure 5–13
d Orbitals Except for the d_{z^2}, the d orbitals have four lobes.

Learning Check A

A–1. Fill in the blanks.

Light has a _____ nature. The wavelength of light is _____ proportional to the energy. Therefore, in the visible spectrum, red light has the _____ wavelength and the _____ energy. When the hot gaseous atoms of an element emit light, it is called a _____ spectrum. Niels Bohr suggested that the orbits of an electron were _____. The lowest energy state is the _____ state, and the higher states are the _____ states. Wave mechanics tell us that a principal energy level, called a _____, can be subdivided into _____. The four types of these subdivisions, in order of increasing energy, are the _____, _____, _____, and _____. Electrons exist in regions of space called _____. The *s* subshell has _____, the *p* _____, the *d* _____ and the *f* _____ orbitals. Each orbital can hold _____ electrons. The shape of the *s* orbital is described as _____.

A–2. Consider the $n = 3$ shell.
(a) What is its capacity?
(b) What subshells are present?
(c) What are the capacities of each subshell?
(d) How many orbitals are present in each subshell?

Additional Examples: Problems 5-7, 5-10, 5-11, 5-13, and 5-15.

We are now ready to apply the information about shells, subshells, and orbitals to the electrons in atoms of elements other than hydrogen. Although calculations were for the hydrogen atom only, we will assume that atoms with more than one electron also have "hydrogen-like" orbitals. In the next section, we will confine our discussion to the shells and subshells occupied by the electrons. In a later section, we will include the orbitals in the total picture.

5–5 ELECTRON CONFIGURATION OF THE ELEMENTS

We will assume that electrons are something like students residing in dorms. Each student is assigned to a room on a specific floor in a specific dorm. Likewise, all electrons in an atom can be assigned to a specific shell and a specific subshell. *The designation of all of the electrons in an atom into specific shells and subshells is known as the element's* **electron configuration.** The assignment of electrons is simple. *Electrons will go into the shells and subshells of the lowest energy.* This is known as the **Aufbau principle.** It is like assigning students to dorm rooms. The most desirable room available will be on a low floor in a dorm as low as possible on the hill.

A shell is designated by the principal quantum number (n), the subshell by the appropriate letter, and the number of electrons in that subshell by the appropriate superscript number. For example, the existence of three electrons in the p subshell of the fourth shell ($n = 4$) is shown as follows:

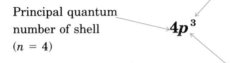

Number of electrons in the subshell (3)

Principal quantum number of shell ($n = 4$)

$4p^3$

Subshell (p)

We begin with the ground state of the one electron in the simplest and first element, hydrogen. The electron is in the $n = 1$ shell, which has an s subshell. The electron configuration for H is

$$1s^1$$

The next element, helium, has two electrons. The $1s$ subshell has a capacity of two electrons, so He has the configuration

$$1s^2$$

Next comes Li with three electrons. The first two fill the first shell, but the third is assigned to the second shell ($n = 2$). The second shell has two subshells, the s and the p, but the s is lower in energy so it fills first. The next element is Be. Its four electrons are also assigned to the $1s$ and $2s$ subshells. The electron configurations of Li and Be are

$$\text{Li} \quad 1s^2 2s^1 \quad \text{Be} \quad 1s^2 2s^2$$

The next element is B. Four of its five electrons have the same configuration as Be, but the fifth electron begins the filling of the next subshell, the $2p$. The next five elements, C through Ne, complete the filling of the $2p$ subshell.

B $1s^2 2s^2 2p^1$	N $1s^2 2s^2 2p^3$	F $1s^2 2s^2 2p^5$
C $1s^2 2s^2 2p^2$	O $1s^2 2s^2 2p^4$	Ne $1s^2 2s^2 2p^6$

With the element neon, the second shell is full. We now continue with the element sodium, which has 11 electrons. The first 10 fill the first and second shells, so the 11th electron is assigned to the third shell. The third shell has three subshells, the s, p, and d. The lowest in energy is the s subshell, so the 11th electron is in the $3s$ subshell. At this point, it becomes somewhat tedious to write all of the filled subshells. A shorthand method of writing configurations is to represent all of the filled subshells of a noble gas by the symbol of that noble gas in brackets (e.g., [Ne]). Thus, in the following electron configurations, $[\text{Ne}] = 1s^2 2s^2 2p^6$. The electron configurations of the next eight elements after neon are shown as follows:

Na $[\text{Ne}]3s^1$	P $[\text{Ne}]3s^2 3p^3$
Mg $[\text{Ne}]3s^2$	S $[\text{Ne}]3s^2 3p^4$
Al $[\text{Ne}]3s^2 3p^1$	Cl $[\text{Ne}]3s^2 3p^5$
Si $[\text{Ne}]3s^2 3p^2$	Ar $[\text{Ne}]3s^2 3p^6$

Notice that the electron configuration of Li is similar to Na (i.e., [He] $2s^1$ and [Ne] $3s^1$), Be to Mg, B to Al, and so forth. Notice also that these pairs of elements are in the same groups in the periodic table. This is very significant. In fact, we can make a general statement for this observation. *The electron configuration of the elements is a periodic property.* Elements in the same group have the same outer subshell electron configurations in successively higher shells. These similar electron configurations are a major reason that the elements in a group have similar properties. The existence of subshells is a result of the study of wave mechanics. By placing electrons in these subshells, we have actually developed the theoretical basis for the existence of the periodic table. Now, however, let us continue with the electron configurations of elements beyond argon.

What does electron configuration have to do with the periodic table?

The next element after Ar presents a problem. The third subshell in the third shell ($3d$) is still available, so at first we may be inclined to assign the 19th electron in K to the $3d$ subshell. However, notice that K is under Na and Li in the periodic table. The latter two elements have their last electron in an s subshell. If the basis of the periodic table is correct, then the location of K (under Na) indicates that it also should have its last electron in the $4s$ subshell. This is indeed the case. Using noble gas shorthand notation, the configuration of K is

$$[Ar]\ 4s^1$$

To understand how the $4s$ fills before the $3d$, let us return to the analogy of the dorms as shown in Figure 5-8. Although dorm 4 is at a higher level than dorm 3, note that not all floors in dorm 3 are lower than those in dorm 4. In fact, it is easier to proceed farther up the hill to occupy the lowest floor in dorm 4 than to go all the way up to the third floor in dorm 3. In the case of the electrons in atoms, a similar phenomenon is true. For a neutral atom, the $4s$ subshell is lower in energy than the $3d$ subshell; thus $4s$ fills first. After it is filled, the $3d$ begins filling, as in the element Sc. The next nine elements after Sc complete the filling of the $3d$ subshell, which has a capacity of 10 electrons. After the $3d$, the $4p$ begins filling as in the element Ga. The $4p$ subshell is filled at the next noble gas, Kr.

Why does the $4s$ subshell fill before the $3d$?

K [Ar] $4s^1$

Ca [Ar] $4s^2$

Sc [Ar] $4s^2 3d^1$

$\vdots$

Zn [Ar] $4s^2 3d^{10}$

Ga [Ar] $4s^2 3d^{10} 4p^1$

$\vdots$

Kr [Ar] $4s^2 3d^{10} 4p^6$

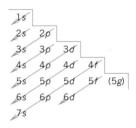

Figure 5–14
Order of Filling of Subshells The arrows indicate the normal order of filling of subshells.

Are there exceptions to the normal order of filling of subshells?

At this point, the order of filling of subshells becomes more variable. A scheme may be helpful. In Figure 5-14, the subshells in each shell are written starting with the 1s at the top. All of the *s* subshells are written in a vertical column, as are the other types of subshells. A stairstep line is then drawn on the right. A diagonal arrow is inserted through each corner. The top arrow points to the first subshell filled (the 1s). The second points to the second subshell filled (the 2s). The third points to the next two (the 2p followed by the 3s). The fourth points to the next two (the 3p followed by the 4s). The next arrow tells us that the order of filling of the next three subshells is 3d followed by 4p followed by 5s. The scheme continues in a similar manner for subshells of higher shells.

Unfortunately, the electron configurations of many elements are not always exactly as we would predict by following the usual order of filling. For example, there are two notable exceptions among the transition elements where the *d* subshells are filling. The actual electron configuration of chromium (#24) and copper (#29) are

$$\text{Cr} \quad [\text{Ar}]\, 4s^1 3d^5 \qquad \text{Cu} \quad [\text{Ar}]\, 4s^1 3d^{10}$$

If the normal order had been followed, we would predict the configuration of Cr to be $[\text{Ar}]\, 4s^2 3d^4$ and Cu to be $[\text{Ar}]\, 4s^2 3d^9$. These exceptions are explained on the basis of a particular stability for a half-filled and a completely filled subshell (the 3d).

There are many other exceptions, especially where *d* and *f* subshells are concerned. *We should be aware that such exceptions exist, but it is not important or necessary to know them all.*

Notice that Figure 5-14 predicts the filling of the 4f before the 5d. However, La and Ac do not follow this order. The 57th electron of La and the 89th electron of Ac go into the 5d and 6d subshells, respectively rather than the 4f and 5f. Their correct configurations are

$$\text{La}\ [\text{Xe}]\, 6s^2 5d^1 \qquad \text{Ac}\ [\text{Rn}]\, 7s^2 6d^1$$

Thus these two elements are listed in the periodic table under Sc and Y, which also have one electron in a *d* subshell. The 58th electron of Ce and the 90th electron of Th, however, reside in the 4f and 5f subshells, respectively. These two electron configurations are

$$\text{Ce}\ [\text{Xe}]\, 6s^2 5d^1 4f^1 \qquad \text{Th}\ [\text{Rn}]\, 7s^2 6d^1 5f^1$$

Working It Out ●

EXAMPLE 5–1

The Electron Configuration of Elements

Write the electron configuration of (a) iron and (b) bismuth

Solution
(a) Iron
From the table inside the front cover, we find that iron has an atomic number of 26, which means that the neutral atom has 26 electrons. Write the subshells in the order of filling plus a running summation of the total number of electrons involved with respect to iron. Use Figure 5-14 to determine the order of filling.

Subshell	Number of Electrons in Subshell	Total Number of Electrons
1s	2	2
2s	2	4
2p	6	10
3s	2	12
3p	6	18 [Ar]
4s	2	20
3d	6	26

Iron has six electrons past the filled 4s subshell, but it does not completely fill the 3d subshell, which could accommodate 10 electrons. The complete electron configuration of iron is

$$1s^2 2s^2 2p^6 3s^2 3p^6 4s^2 3d^6$$

or, if we use noble gas shorthand

$$[\text{Ar}]\, 4s^2 3d^6$$

(b) Bismuth

Bismuth is element number 83, which means that we must assign 83 electrons to shells and subshells. We can continue the table from part (a), starting at the 3d subshell.

Subshell	Number of Electrons in Subshell	Total Number of Electrons
3d	10	30
4p	6	36 [Kr]
5s	2	38
4d	10	48
5p	6	54 [Xe]
6s	2	56
4f	14	70
5d	10	80
6p	3	83

The electron configuration of bismuth using the noble gas shorthand that begins with the noble gas xenon (with 54 electrons) is

$$[\text{Xe}]\, 6s^2 4f^{14} 5d^{10} 6p^3$$

The energy of an electron in an atom is determined by its shell and subshell. The particular orbitals within a subshell all have the same energy. However, the distribution of electrons in the orbitals of a particular subshell accounts for certain properties of elements. We will include information about the electron distribution in orbitals in the next section.

▼ **Looking Ahead**

5–6   ORBITAL DIAGRAMS OF THE ELEMENTS

Subshells are composed of one or more orbitals, which are regions of space in which electrons reside. Each orbital can hold a maximum of two electrons. In the following scheme, we will represent an individual orbital by a box. Therefore, an *s* subshell will have one box, a *p* subshell three, a *d* subshell five, and an *f* subshell seven. Electrons will be represented by arrows.

As mentioned earlier, an electron has a dual nature. In some respects it has properties of a wave, and in other respects it has properties of a particle. One particle property is that the electron behaves like a charged particle spinning in either a clockwise or counterclockwise direction. A spinning charged particle is like a tiny magnet. What if there are two electrons in the same orbital? The **Pauli Exclusion principle** *states that no two electrons in the same orbital can have the same spin.*

How can two electrons occupy the same orbital?

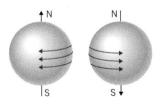

We will represent the electrons in orbitals by means of orbital diagrams. The **orbital diagram** *of an element represents the orbitals in a subshell as boxes and its electrons as arrows. The spin of an electron is indicated by the direction of the arrow pointing either up or down. Two electrons with opposite spins in the same orbital are said to be paired.* Thus a doubly occupied orbital is represented as follows:

We will now expand on the electron configuration of the first five elements by including their orbital diagrams.

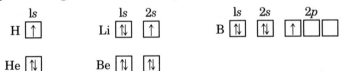

Before going on, we need to consider the placement of the sixth electron in carbon. There are three possibilities. Does it pair with the first 2*p* electron in the same orbital, have opposite spin in a different orbital, or go into a different orbital with the same spin? We have one more rule to guide us. **Hund's rule** *states that electrons occupy separate orbitals in the same subshell with parallel spins.* At least part of this rule is understandable. Since electrons have the same charge, they will repel each other to different regions of space. Electrons "want their space," so they prefer separate orbitals rather than pairing in the same orbital. Pairing occurs when separate empty orbitals in the same subshell are not available. With Hund's rule in mind, we can now write the orbital diagrams of the next five elements.

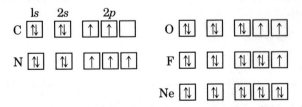

A similar phenomenon occurs with elements that have electrons occupying *d* or *f* orbitals. For example, Mn (#25) has the following electron configuration and orbital diagram.

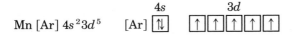

Mn [Ar] $4s^2 3d^5$

Although we will not pursue the topic in this text, orbital diagrams give us important information about the magnetic properties of elements. Orbital diagrams are also relevant to the types of bonds that a particular element forms.

EXAMPLE 5–2

The Electron Configurations and Orbital Diagrams

Determine the electron configuration and write the orbital diagram of the electrons beyond the noble gas for the elements S, Cr, and As.

● **Working It Out**

Solution

First, use the table inside the front cover to find the atomic number, which equals the number of electrons in a neutral atom. Then, use Figure 5-14 to determine the electron configuration of the element as in Example 5-1. Finally, show the orbital diagram for the outer orbitals beyond the noble gas core. Apply the Pauli principle and Hund's rule in positioning the arrows.

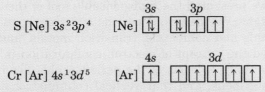

S [Ne] $3s^2 3p^4$

Cr [Ar] $4s^1 3d^5$

Remember that Cr is an exception to the normal order of filling; it has six unpaired electrons. This exception is due to the stability of the half-filled $3d$ subshell.

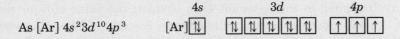

As [Ar] $4s^2 3d^{10} 4p^3$

Wave mechanics tell us that each electron in an atom can be assigned to a shell and a subshell according to the Aufbau principle. Using this method, we see that the electron configuration is a periodic property. As it turns out, this is the ultimate periodic property because it explains many of the others. Following the Pauli principle and Hund's rule, we can further define the atom by assigning electrons to orbitals in a subshell. Electrons in orbitals are illustrated with orbital diagrams.

▲ **Looking Back**

Learning Check B

◀ **Checking It Out**

B–1. Fill in the blanks.
The electron configuration $5d^5$ indicates that there are _____ electrons in the _____ subshell in the _____ shell. "Electrons go into the lowest energy subshell" is a statement of the _____ principle. "Electrons in the same orbital must be paired" is a statement of the _____ principle. "Electrons go into separate orbitals with parallel spins" is a statement of _____ _____.

B–2.   What is the electron configuration of element Ni (#28), Sr (#38), and Pb (#82)? Use the noble gas shorthand notation.

B–3.   Show the electron configuration and orbital diagram of all of the electrons beyond the previous noble gas for Nb (#41).

Additional Examples: Problems 5-19, 5-20, 5-22, 5-24, and 5-25.

Looking Ahead ▼

About 70 years after the modern periodic table was first displayed, the theoretical basis of its existence was explained in terms of electron configurations. Why not let the periodic table tell us an element's electron configuration? If we can do this, we will not need to memorize the order of filling of subshells or even use a scheme such as the one in Figure 5-14.

5–7   ELECTRON CONFIGURATION AND THE PERIODIC TABLE

In Chapter 4, we presented the indispensable tool of the chemist, which is the periodic table. We described periods and groups, and located gases and solids, metals and nonmetals, transition metals, and other metals. As we developed the concept of electron configuration in Section 5-5, it became obvious that the basis of the periodic table lies in the consecutive filling of subshells, and that vertical columns consist of elements with the same outermost subshell configuration in different shells. We can now put this knowledge to work for us and let the periodic table tell us the electron configuration of a specific element. In Figure 5-15, we have included in the space for each element the value of n for the shell of the outermost subshell configuration. A study of this table and some practice will allow us to predict configurations with only the periodic table as a reference.

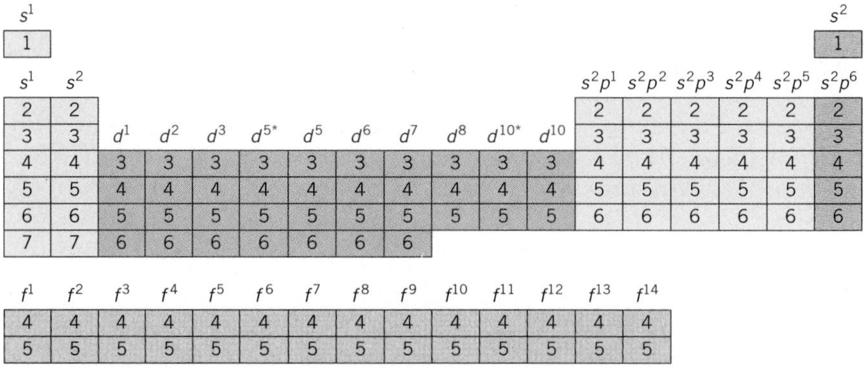

*This has an ns^1 configuration rather than ns^2.

Figure 5–15
Electron Configuration and the Periodic Table. The electron configuration of an element can be determined from its position in the periodic table. The value of "n" shown in each box is the shell of the outermost subshell.

As we will see in the next chapter, the electrons beyond the last noble gas are the ones commonly involved in the formation of chemical bonds. To focus on these electrons, we use the noble gas shorthand in expressing electron configurations. We will now relate electron configurations with groups and categories of elements discussed in Chapter 4. Once again, the four categories and some of the groups within a category are listed. After the group number, the general electron configuration for that group is noted. For example, [NG] ns^1 means a noble gas configuration followed by one electron in the n shell and the s subshell.

General Configuration: [NG] ns^1

Specific configuration for potassium (K): [Ar] $4s^1$

Representative Elements: [NG] $ns^x np^y$

IA. [NG] ns^1　　Except for hydrogen, these are the *alkali metals*. All have one electron beyond a noble gas configuration. Notice that the numbering for the s subshell begins at 1 for hydrogen and is consecutive down the table.

What is characteristic about the electron configuration of alkali metals?

IIA. [NG] ns^2　　These are the *alkaline earth metals*. All have two electrons beyond a noble gas. The numbering begins at 2 in this column, with Be, because He, which has a $1s^2$ configuration, is located to the far right with the noble gases.

IIIA. [NG] $ns^2 np^1$　　In numbering these elements, note that the p subshell begins at the $2p$ for B through Ne, and each p subshell is numbered consecutively down the table. The fourth and fifth period elements Ga (#31) and In (#49) also have filled d subshells beyond the noble gas configuration. In the sixth period, Tl (#81) has a filled $6d$ and $4f$ beyond the noble gas. This is true for the other representative elements in these three periods. For example, the electron configuration of In (#49) begins with the previous noble gas Kr, fills the $5s$ and the $4d$, and has one electron in the $5p$. That is,

In [Kr] $5s^2 4d^{10} 5p^1$

IVA. [NG] $ns^2 np^2$
VA. [NG] $ns^2 np^3$
VIA. [NG] $ns^2 np^4$　　These elements are known as the *chalcogens*. They are all nonmetals except for Po.

VIIA. [NG] $ns^2 np^5$　　These elements are known as the *halogens*. The halogens are all nonmetals and one electron short of having a noble gas configuration.

Noble Gases: [NG] $ns^2 np^6$

The noble gas elements are significant because they rarely form chemical bonds. These elements have full outer s and p subshells except for He, which has a full s subshell. The fourth and fifth period noble gases, Kr and Xe, also have filled $3d$ and $4d$ subshells, respectively, and Rn has a filled $4f$ and $5d$ in addition to the filled outer $6s$ and $6p$ subshells.

Transition Metals: [NG] $ns^2 (n - 1)d^x$

Group IIIB begins the filling of a d subshell. In this group, there are some exceptions to the normal order of filling. Note that group VIB has the configuration [NG] $ns^1(n - 1)d^5$, and group IB has the configuration [NG] $ns^1(n - 1)d^{10}$. The first transition element, Sc, begins the filling of the $3d$ subshell. Each d subshell is numbered consecutively beginning with 3 at the top. Hf (#72) through Hg (#80) also have a filled $4f$ subshell. It is assumed that Unq (#104) through the last element (#109) have filled $5f$ subshells, although little experimental information confirms this at the present.

Inner Transition Metals: [NG] $ns^2(n - 1)d^1(n - 2)f^x$

There are two series of inner transition elements: the *lanthanides*, which have the general configuration [Xe] $6s^2 5d^1 4f^x$; and the *actinides*, which have the general configuration [Rn] $7s^2 6d^1 5f^x$.

Working It Out ●

EXAMPLE 5–3

Electron Configurations of V and Pb from the Periodic Table

Write the electron configuration for (a) vanadium and (b) lead.

Solution

(a) Inside the front cover of the book, note that vanadium (V) has atomic number 23. Use the periodic table inside the front cover to locate V in the periodic table shown in Figure 5-15. The outermost subshell is $3d$, and since V is the third element in the $3d$ series, its subshell configuration is $3d^3$. If we write out all subshells preceding the $3s$ subshell, the complete electron configuration for V is

$$1s^2 2s^2 2p^6 3s^2 3p^6 4s^2 3d^3$$

or, starting with the previous noble gas

$$[Ar] 4s^2 3d^3$$

(b) Lead (Pb) has atomic number 82. Using the periodic table inside the front cover, locate Pb and the noble gas before it, which is Xe. Find Pb in Figure 5-15. Notice that it has two electrons in a $6p$ subshell. Using the noble gas shorthand, insert [Xe] to start the electron configuration. Notice that between Xe and the $6p$ subshell the $6s$, $4f$, and $5d$ have filled. The electron configuration of Pb is

$$[Xe] 6s^2 4f^{14} 5d^{10} 6p^2$$

EXAMPLE 5–4

The Identity of an Element from Its Electron Configuration

What element has the electron configuration [Kr] $5s^2 4d^{10} 5p^5$?

Solution
In Figure 5-15, locate p^5 in the upper right of the table. Using the periodic table inside the back cover, locate Kr. The element after Kr in the p^5 column in Figure 5-15 has an atomic number 53. In the list of elements in the front cover, atomic number 53 corresponds to the element <u>iodine</u>.

The following three characteristics of atoms vary in a regular periodic manner related to electron configurations: the size (radius) of the atoms, the energy it takes to remove an electron, and the energy released when an atom adds an electron. The latter two properties relate to the topic of bonding presented in the next chapter. The first characteristic that we will consider is the size of neutral atoms.

▼**Looking Ahead**

5–8 PERIODIC TRENDS

Perhaps one of the most fundamental trends in the periodic table concerns the size or radius of the atom of an element. The **radius** *of an atom is the distance from the nucleus to the outermost electrons.* It is not an easy task to measure the radius of an atom. There are both experimental and theoretical problems (i.e., electrons do not have a fixed distance from the nucleus). Despite all of these difficulties, consistent values for the radii of neutral atoms have been compiled. Some generally accepted values are shown in Figure 5-16 for the first three rows of representative elements. Two units of measurement are most often used for such small distances—the **nanometer** (1 nm = 10^{-9} m) and the **picometer** (1 pm = 10^{-12} m). We will use picometers because the numbers are easier to express and compare (e.g., 37 pm rather than 0.037 nm).

Notice that there is a general decrease in the radii from left to right across representative element groups, but an increase in radii down a group. It is understandable that the size of atoms *increases* down a group since it is reasonable to expect that a heavier atom is also a larger atom. In fact, the radii of atoms increase down a group because the outermost electron is in a shell farther from the nucleus. On the other hand, it may seem surprising that the radii *decrease* from Li to F and

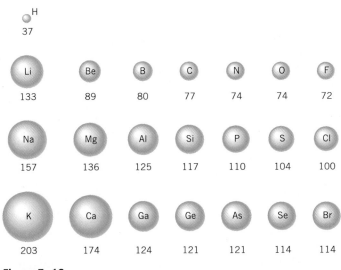

Figure 5–16
Atomic Radii Radii of these atoms are given in picometers (pm).

How can one atom be heavier yet smaller than another?

from Na to Cl, even though the atoms are heavier as we move to the right. From left to right, each successive element has one more proton in the nucleus and one more electron in the outer *s* or *p* subshell. *All* of the electrons of the outer shell, including the added electron, feel an increased electrostatic attraction for the higher nuclear charge. This results in a general shrinkage in the size of the atoms as the outer *s* and *p* subshells are filling. An analogous situation would be two people trying to read by the light of a single lamp. If one person is directly between the lamp and the other, the outer person will be shielded from some of the light. If both are at the same distance, they both share the same amount of light. Electrons between the nucleus and the outer electrons shield nuclear charge from the outer electrons. Electrons in the same shell do not shield each other from the nuclear attraction.

A similar phenomenon applies to the transition metals. The radius of Sc (which has one 3*d* electron) is 144 pm, and the radius of Zn (which has ten 3*d* electrons) is 125 pm.

One of the most important periodic properties of the elements relates to the degree of difficulty in removing an electron from an atom. It stands to reason that the smaller the atom the stronger the attraction of the nucleus for the outer electrons. If they are tightly held, obviously they are hard to remove. **Ionization energy (IE)** *is the energy required to remove an electron from a gaseous atom to form a gaseous ion.* Since the outermost electron is generally the least firmly attached, it naturally will be the first to go.

$$M(g) \longrightarrow M^+(g) + e^-$$

M(a gaseous atom) forms a gaseous cation and an electron.

The process of removing an electron from an atom always requires energy (an endothermic process) since all electrons are held by electrostatic forces of attraction to the nucleus. Since the amount of energy required relates to the size of the atom, we might expect an inverse relationship between size and ionization energy. *That is, ionization energy generally increases across a period but decreases down a group.* The ionization energies for the second and third periods are shown in Table 5-3. The energy unit abbreviated kJ/mol stands for kilojoules per mole, which is energy per a given defined quantity of atoms known as a mole. The information in Table 5-3 and Figure 5-16 is included in graphical

TABLE 5–3   IONIZATION ENERGY OF SOME ELEMENTS

Element	I.E. (kJ/mol)	Element	I.E. (kJ/mol)
Li	520	Na	496
Be	900	Mg	738
B	801	Al	578
C	1086	Si	786
N	1402	P	1102
O	1314	S	1000
F	1681	Cl	1251

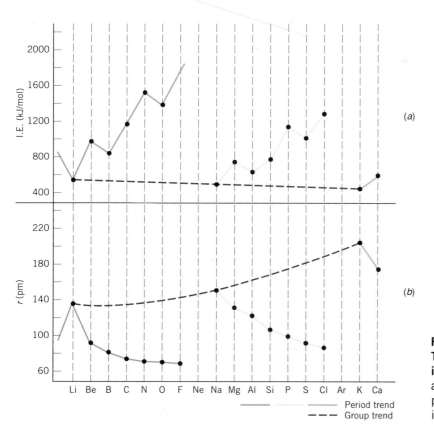

(a)

(b)

Period trend
Group trend

Figure 5–17
Trends in Atomic Radii and Ionization Energies Generally, as atomic radii decrease across a period (b), ionization energies (a) increase.

form in Figure 5-17. We can now see how size relates to ionization energy by comparing the same elements in part (a) (atomic radii) and (b) (ionization energy). While trends in size decrease, trends in ionization energy increase and vice versa.

Notice that ionization energy generally decreases to the left and down. It is no coincidence that this is the same direction as increasing metallic properties. *In fact, the most significant chemical property of metals is that they lose electrons relatively easily to form cations in compounds.* Nonmetals, on the other hand, are noted by their comparatively high ionization energies, and they do not form cations in their compounds.

The second ionization energy for an ion involves removal of an electron from a +1 ion to form a +2 ion.

$$M^+(g) \longrightarrow M^{2+}(g) + e^-$$

In a similar manner, the third ionization energy forms a +3 ion and so forth. In all cases it becomes increasingly difficult to remove each succeeding electron. When an electron is removed from an atom, the resulting ion is smaller than the parent atom (see Figure 5-18). When an electron is removed from an outer subshell, the repulsion between the remaining electrons is lessened so the ion formed is smaller. If all of the electrons in the outermost shell are removed, the resulting ion will be considerably smaller, as is the case of the two ions shown in Figure 5-18. The trends in consecutive ionization energies for Na through Al are shown in Table 5-4.

How does ionization energy relate to metallic behavior?

Figure 5–18
Ionic Radii A positive ion is always smaller than the original neutral atom.

Na
$r = 157$ pm

Na$^+$
$r = 95$ pm

Mg
$r = 136$ pm

Mg^{2+}
$r = 65$ pm

TABLE 5–4 IONIZATION ENERGIES (KJ/MOL)			
Element	First I.E.	Second I.E.	Third I.E.	Fourth I.E.
---	---	---	---	---
Na	496	4565	6912	9,540
Mg	738	1450	7732	10,550
Al	577	1816	2744	11,580

Note that the second IE for Na, the third IE for Mg, and the fourth IE for Al are all very large compared to the preceding number. For example, it takes 2188 kJ (1450 + 738) to remove the first two electrons from Mg (to form Mg^{2+}), but about three times as much energy to remove the third electron (7732 kJ to form Mg^{3+}). Why is there such a big jump? The answer lies in the electron configuration of magnesium, which is [Ne] $3s^2$. The first electron removed is from the $3s$ subshell, as is the second. To form Mg^{3+}, however, the third electron must be removed from the filled inner shell of the neon configuration (the $2p$). Because the second shell is closer to the nucleus than the third shell, this is very difficult and requires a large amount of energy. The same reasoning holds for the second electron from Na and the fourth from Al. We will refer to this observation in the next chapter. The positive charge that a representative metal forms is limited by the number of electrons beyond a noble gas configuration.

The next periodic trend concerns the opposite situation—that is, the tendency of atoms to form *negative* ions. Just as the loss of an electron from an atom results in a positive ion, the gain of an electron by a neutral atom results in a negative ion. The **electron affinity (EA)** *is the energy released when a gaseous atom adds an electron to form a gaseous ion.*

$$X(g) + e^- \longrightarrow X^-(g)$$

The representative element *nonmetals* have relatively high electron affinities, which means that the formation of negative ions is favorable. A word of caution: with reference to the ease of formation of positive and negative ions, a high electron affinity implies the opposite of a high ionization energy. Thus a high electron affinity means negative ions are easily formed, but high ionization energy means positive ions are *not* easily formed. The former is favorable (exothermic), while the latter is unfavorable (endothermic). A representative element nonmetal would be expected to have a relatively high ionization energy *and* a relatively high electron affinity. Electron affinities generally increase in the same

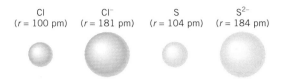

Cl
(r = 100 pm) Cl⁻
(r = 181 pm) S
(r = 104 pm) S²⁻
(r = 184 pm)

Figure 5–19
Ionic Radii A negative ion is always larger than the original neutral atom.

directions as ionization energies. To represent the removal or addition of an electron, we use actual values in kilojoules, with a positive value for the endothermic process and a negative value for the exothermic process.

Just as we find that metals can lose more than one electron to form $+2$ or even $+3$ ions, certain nonmetals may gain more than one electron to form -2 or, in the case of N and P, -3 ions. What determines the amount of charge that a nonmetal may acquire is again determined by the outer electron configuration. For example, oxygen ([He] $2s^2 2p^4$) is a nonmetal with two vacancies in its outer $2p$ subshell. It can therefore add two electrons to form a -2 anion. If oxygen were to add a third electron, that one would have to advance to the outer $3s$ subshell, and this would not be a favorable process. Likewise, nitrogen has three vacancies in its outer subshell, so an atom may add three electrons to form a -3 anion.

Just as a cation is smaller than its parent atom, an anion is considerably larger than its neutral parent atom. When an electron is added to an atom, the added repulsions between the electrons cause the radius of the ion to expand. (See Figure 5-19.)

The easiest way to arrive at an element's electron configuration is by making use of the periodic table. After all, electron configuration is the basis of the periodic table. This periodicity is apparent in groups such as the alkali metals or halogens where the outermost subshell configuration for each group can be stated by a generalized configuration. Besides electron configuration, the size, ionization energy, and electron affinity of the elements are also periodic properties. As a result, metals form cations, and representative element nonmetals form anions. If one atom loses electrons easily and another gains electrons readily, this should be the basis of a compatible marriage of atoms. This does occur, and we will follow up on this concept in the next chapter.

▲ **Looking Back**

◀ **Checking It Out**

Learning Check C

C–1. Fill in the blanks.
An element that has a general configuration beyond a noble gas of ns^2np^2 is in Group _____ while an element that has a general configuration of $ns^2(n-1)d^2$ is in Group _____. The radii of neutral atoms generally _____ down a group and _____ to the right across a period. The ionization energies generally _____ down a group and _____ to the right across a period. A metal may form a cation with a charge greater than $+1$ if it has more than one _____ beyond the previous noble gas. The elements

with the greatest tendency to gain electrons are the representative element _____ .

C–2.   Using the periodic table only, give the electron notation for the elements Zn (#30) and W (#74).

C–3.   Using the periodic table only, give the symbol of the element with the electron configuration [Ar] $4s^2 3d^{10} 4p^3$.

C–4.   Give the general electron configurations for the elements in Groups VIA and VIB.

C–5.   Given the following four elements: As, Se, Sb, and Te.
(a) Which has the smallest radius?
(b) Which has the lowest first ionization energy?
(c) Which has the highest electron affinity?
(d) Which is most likely to form a positive ion?

Additional Examples: Problems 5-31, 5-33, 5-37, 5-44, 5-58, 5-60, 5-62, and 5-70.

C H A P T E R   R E V I E W

Putting It Together ▲▼

This chapter starts us on a journey that ultimately leads to an explanation of why and how atoms of elements bond to each other. The periodic basis of chemical bonding lies in the nature of the electrons in the atoms and the various regions of space in which they exist. Our journey started with a theoretical explanation of the **discrete spectrum** of emitted **electromagnetic radiation** (light), by hot gaseous hydrogen atoms. This is quite unlike the **continuous spectrum** seen in a rainbow. The **wavelength** of light is inversely related to its energy. In 1913, Niels Bohr proposed a theory, which serves as a **model,** in which the hydrogen electron orbits the nucleus in **quantized** energy levels. He assigned a **principal quantum number** (n) to each energy level, in which the first energy level ($n = 1$) is lowest in energy and known as the **ground state.** In hydrogen, the energy levels higher than $n = 1$ are the **excited states.** Light is emitted from an atom when an electron falls from an excited state to a lower state. The difference in energy between the states becomes the energy of the light wave. If the energy of the light is within the visible part of the spectrum, we see a specific color.

Eventually, Bohr's model had to be adjusted and then mostly discarded as newer, more inclusive theories were advanced. In the **wave mechanical theory,** the electron has a probability of existing in a given region of space rather than having a specific location and velocity. This theory also tells us that electrons exist in principal energy levels known as **shells,** but that the shells contain one or more **subshells.** Each subshell is composed of one or more **orbitals.** An orbital can contain one or two electrons. The different types of orbitals (i.e., $s, p, d,$ or f) are differentiated by their shape. The s orbitals are spherical, and the others have more complex shapes.

Each shell holds $2n^2$ electrons and has n different subshells. The electrons in any atom can be assigned to a given shell and subshell. This is known as the element's **electron configuration.** Electrons fill subshells according to the **Aufbau principle,** which simply means that the lowest energy subshells fill first. As we proceed through the electron configurations of the elements, one fact makes itself apparent. Atoms of elements in vertical columns or groups have the same outermost subshell configuration but successive shells. Thus the basis of the periodic table can be established.

By using **orbital diagrams,** we can expand the representation of electron configuration to include assignment of electrons to orbitals. Two other rules are required to do this successfully. The **Pauli exclusion principle** relates to the spin of electrons in the same orbital, and **Hund's rule** relates to the electron distribution in separate orbitals of the same subshell. The subshells and orbitals in the first four shells are summarized as follows. Each orbital is represented as a box.

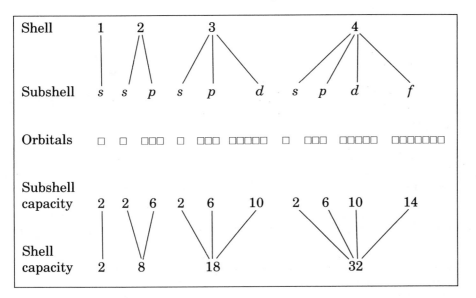

Since the periodic table and electron configuration are interrelated, we now put the periodic table to work in writing electron configurations. We see that various groups can be identified by their specific configurations as well as their properties.

The periodic table tells us even more. There are general trends in the size or **radius** of atoms. In general, atomic radii decrease up and to the right in the table. The radius of an atom is related to the shell of the outermost electrons and to the number of electrons in the outermost subshell. The higher the energy of the shell and the fewer electrons in the outermost subshell, the larger the atom. Thus atoms to the lower left are the largest atoms. The opposite trend occurs if we compare the difficulty in removing the outermost electron, which is known as the element's **ionization energy.** The same factors that affect the size of an atom affect the ionization energy. That is, in general, larger atoms have lower ionization energies than smaller atoms. Metals tend to have larger atoms than nonmetals, so they have lower ionization energies.

How many electrons can be easily removed from a metal is determined by its outer electron configuration. Atoms of representative element non-metals, on the other hand, gain electrons readily to form anions. This is measured by a quantity known as **electron affinity.** How many electrons nonmetal atoms may gain is also determined by its outer electron configuration.

E X E R C I S E S

SHELLS (ENERGY LEVELS) AND SUBSHELLS

5-1 Ultraviolet light causes sunburns, but visible light does not. Explain.

5-2 The $n = 8$ and $n = 9$ energy levels are very close in energy. Using the Bohr model, describe how the wavelengths of light compare as an electron falls from these two energy levels to the $n = 1$ energy level.

5-3 The $n = 3$ energy level is of considerably greater energy than the $n = 2$ energy level. Using the Bohr model, describe how the wavelengths of light compare as an electron falls from these two energy levels to the $n = 1$ energy level.

5-4 What is the total capacity of the fourth principal energy level or shell?

5-5 An electron in the lithium atom is in the third energy level. Is the atom in the ground or excited state? Can the atom emit light? If so, how?

5-6 Which of the following subshells do not exist: $6s$, $3p$, $2d$, $1p$, $4d$, $3f$?

5-7 What is the electron capacity of each subshell in the $n = 4$ shell?

5-8 What is the electron capacity of each subshell in the $n = 3$ and the $n = 2$ shells?

5-9 Explain how a subshell in one shell can be lower in energy than a subshell in a lower shell.

5-10 Theoretically, there should be a fifth subshell in the $n = 5$ shell. This would be called the g subshell. What is the capacity of the g subshell?

ORBITALS

5-11 How many orbitals are in the following subshells?
(a) $2p$ (d) $3s$
(b) $4d$ (e) $5f$
(c) $2d$

5-12 How many orbitals are in the $n = 3$ shell?

5-13 How many orbitals would be in the g subshell? (See Problem 5-10.)

5-14 What is meant by the "shape" of an orbital?

5-15 Describe the shape of the $4p$ orbitals.

5-16 Describe the shape of the $3s$ orbital. How does it differ from a $2s$ orbital?

ELECTRON CONFIGURATION

5-17 Which of the following subshells fills first? (Refer to Figure 5-14 or 5-15.)
(a) $6s$ or $6p$ (c) $4p$ or $4f$
(b) $5d$ or $5p$ (d) $4f$ or $4d$

5-18 Which of the following subshells is lower in energy? (Refer to Figure 5-14 or 5-15.)
(a) $6s$ or $5p$ (d) $4d$ or $5p$
(b) $6s$ or $4f$ (e) $4f$ or $6d$
(c) $5s$ or $4d$ (f) $3d$ or $4p$

5-19 Write the following subshells in order of increasing energy. (Refer to Figure 5-14 or 5-15.) $4s$, $5p$, $4p$, $5s$, $4f$, $4d$, $6s$.

5-20 Write the total electron configuration for each of the following elements. (Refer to Figure 5-14 or 5-15.)
(a) Mg   (b) Ge   (c) Pd   (d) Si

5–21 Write the electron configuration that is implied by [Ar]. (Refer to Figure 5-14 or 5-15.)

5–22 Using the noble gas shorthand, write the electron configurations for the following elements. (Refer to Figure 5-14 or 5-15.)
(a) S (b) Zn (c) Pu (d) I

5–23 Using the noble gas shorthand, write the electron configurations for the following elements. (Refer to Figure 5-14 or 5-15.)
(a) Sn (b) Ni (c) Cl (d) Au

ORBITAL DIAGRAMS

5–24 Which of the following orbital diagrams is excluded by the Aufbau principle? Which by the Pauli exclusion principle? Which by Hund's rule? Which is correct? Explain how a principle or rule is violated in the others.

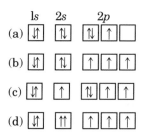

5–25 Write the orbital diagrams for electrons beyond the previous noble gas core for the following elements. (Refer to Figure 5-14 or 5-15.)
(a) S (b) V (c) Br (d) Pm

5–26 Write the orbital diagrams for electrons beyond the previous noble gas core for the following elements. (Refer to Figure 5-14 or 5-15.)
(a) As (b) Ar (c) Tc (d) Tl

5–27 How many unpaired electrons are in the atoms of elements in groups IIB, VB, VIA, VIIA; an atom with electron configuration given by $[ns^1(n-1)d^5]$; and Pm (atomic number 61)?

5–28 The atoms of the elements in three groups in the periodic table have all of their electrons paired. What are the groups?

5–29 What group or groups have two unpaired p electrons?

5–30 What group or groups have five unpaired d electrons?

ELECTRON CONFIGURATION AND THE PERIODIC TABLE

5–31 Write the symbol of the first element that has a
(a) $5p$ electron
(b) $4d$ electron
(c) $4f$ electron
(d) filled $3p$ subshell

5–32 Write the symbol of the first element that has a
(a) filled $4d$ subshell
(b) half-filled $4p$ subshell
(c) $6s$ electron
(d) filled fourth shell

5–33 Which subshell begins to fill for each of the following elements?
(a) Al (b) In (c) La (d) Rb

5–34 Which subshell begins to fill for each of the following elements?
(a) Y (b) Th (c) Cs (d) Ga

5–35 What do Groups IIIA and IIIB have in common? How are they different?

5–36 What do the elements Al and Ga have in common? How are they different?

5–37 Which elements have the following electron configurations? (Use only the periodic table.)
(a) $1s^2 2s^2 2p^5$
(b) [Ar] $4s^2 3d^{10} 4p^1$
(c) [Xe] $6s^2$
(d) [Xe] $6s^2 5d^1 4f^7$
(e) [Ar] $4s^1 3d^{10}$ (exception to rules)

5–38 Which elements have the following electron configurations? (Use only the periodic table.)
(a) [He] $2s^2 2p^3$
(b) [Kr] $5s^2 4d^2$
(c) [Ar] $4s^2 3d^{10} 4p^6$
(d) [Rn] $7s^2 6d^1$
(e) [Xe] $6s^2 5d^{10} 4f^{14} 6p^1$
(f) [Ar] $4s^2 3d^{10}$

5–39 If any of the elements in Problem 5-38 belong to a numerical group (e.g., Group IA) in the periodic table, indicate the group.

5–40 Write the number designation of the two groups that may have five electrons beyond a noble gas configuration.

5–41 Write the number designation of a group that has 2 electrons beyond a noble gas configuration. Write the number designation of a group with 12 electrons beyond a noble gas configuration.

5–42 Which two configurations belong to the same group in the periodic table?
(a) $[Kr] 5s^2$
(b) $[Kr] 5s^2 4d^{10} 5p^2$
(c) $[Xe] 6s^2 5d^1 4f^2$
(d) $[Ar] 4s^2 3d^2$
(e) $[Ne] 3s^2 3p^2$

5–43 Which two configurations belong to the same group in the periodic table?
(a) $[Kr] 5s^2 4d^{10}$
(b) $[Ar] 4s^1 3d^{10}$
(c) $[Ne] 3s^1$
(d) $[Xe] 6s^2$
(e) $[Xe] 6s^2 4f^{14} 5d^{10}$

5–44 Which group in the periodic table has the following general electron configuration? (n is the principal quantum number.)
(a) $ns^2 np^2$
(b) $ns^2 np^6$
(c) $ns^1 (n-1)d^{10}$ (e.g., $4s^1 3d^{10}$)
(d) $ns^2 (n-1)d^1 (n-2)f^2$ (two elements)

5–45 Write the general electron configuration for the following groups:
(a) IIA   (c) VIA
(b) IIB   (d) IVB

5–46 How does He differ from the other elements in Group VIII?

5–47 Which of the following elements fits the general electron configuration $ns^2 (n-1)d^{10} np^4$?
(a) Cr   (b) Te   (c) S   (d) O   (e) Si

5–48 What is the electron configuration of the noble gas at the end of the third period?

5–49 What is the electron configuration of the noble gas at the end of the fourth period?

*5–50 What would be the atomic number of an element with one electron in the $5g$ subshell?

*5–51 Element number 114 has not yet been made, but scientists believe that it will form a comparatively long-lived isotope. What would be its electron configuration and in what group would it be?

*5–52 How many elements would be in the seventh period if it were complete?

*5–53 How many elements would be in a complete eighth period? (Hint: Consider the $5g$ subshell.)

*5–54 Classify the following electron configurations into one of the four main categories of elements.

(a) $[Ar] 4s^2 3d^2$ (c) $[Ar] 4s^2 3d^{10} 4p^6$
(b) $[Kr] 5s^2 4d^{10} 5p^5$ (d) $[Rn] 7s^2 6d^1 5f^3$

*5–55 Classify the following electron configurations into one of the four main categories of elements.
(a) $[Ne] 3s^2 3p^6$ (c) $[Xe] 6s^2 4f^{14} 5d^2$
(b) $[Ne] 3s^2 3p^5$ (d) $[Xe] 6s^2 5d^1 4f^7$

*5–56 Assume that elements with atomic numbers greater than 109 will eventually be made. If so, what would be the atomic number and group number of the next nonmetal?

*5–57 What is the atomic number of the heaviest metal that would appear in the periodic table after element number 109?

PERIODIC TRENDS

5–58 In each of the following pairs of elements, which has the larger radius?
(a) As or Se   (c) Sr or Ba
(b) Ru or Rh   (d) F or I

5–59 In each of the following pairs of elements, which has the larger radius?
(a) Tl or Pb   (c) Pr or Ce
(b) Sc or Y   (d) As or P

5–60 Four elements have the following radii: 117 pm, 122 pm, 129 pm, and 134 pm. The elements, in random order, are V, Cr, Nb, and Mo. Which element has a radius of 117 pm? Which has a radius of 134 pm?

5–61 Four elements have the following radii: 180 pm, 154 pm, 144 pm, and 141 pm. The elements, in random order, are In, Sn, Tl, and Pb. Which element has a radius of 141 pm? Which has a radius of 180 pm?

5–62 Which of the following elements has the higher ionization energy?
(a) Ti or V   (d) Fe or Os
(b) P or Cl   (e) B or Br
(c) Mg or Sr

5–63 Four elements have the following first ionization energies (in kJ/mol): 762, 709, 579, and 558. The elements, in random order, are Ga, Ge, In, and Sn. Which element has an ionization energy of 558 kJ/mol? Which element has an ionization energy of 762 kJ/mol?

5–64 Four elements have the following first ionization energies (in kJ/mol): 869, 941, 1010, and

1140. The elements, in random order, are Se, Br, Te, and I. Which element has an ionization energy of 869 kJ/mol? Which element has an ionization energy of 1140 kJ/mol?

5–65 The first four ionization energies for Ga are 578.8, 1979, 2963, and 6200 kJ/mol. How much energy is required to form each of the following ions: Ga^+, Ga^{2+}, Ga^{3+}, and Ga^{4+}? Why does the formation of Ga^{4+} require a comparatively large amount of energy?

5–66 Which of the following monatomic cations is the easiest to form, and which is the hardest to form?

(a) Cs^+ (b) Rb^{2+} (c) Ne^+ (d) Sc^{3+}

5–67 Which of the following atoms would most easily form a cation?

(a) B (b) Al (c) Si (d) C

5–68 Which of the following atoms would most likely form an anion?

(a) Be (b) Al (c) Ga (d) I

5–69 Noble gases form neither anions nor cations. Why?

5–70 Arrange the following ions and atoms in order of increasing radii.

(a) Mg (c) S (e) K^+
(b) S^{2-} (d) Mg^{2+} (f) Se^{2-}

5–71 Arrange the following ions and atoms in order of increasing radii.

(a) Br (c) K (e) Br^-
(b) K^+ (d) I^- (f) Ca^{2+}

*5–72 Zirconium and hafnium are in the same group and have almost the same radius despite the general trend down a group. As a result, the two elements have almost identical chemical and physical properties. The fact that these elements have almost the same radius is due to what is referred to as the *lanthanide contraction*. With the knowledge that atoms get progressively smaller as a subshell is being filled, can you explain this phenomenon? (Hint: Follow all of the expected trends between the two elements.)

5–73 The first five ionization energies for carbon are 1086, 2353, 4620, 6223, and 37,830 kJ/mol. How much energy is required to form the following ions: C^+, C^{2+}, C^{3+}, C^{4+}, and C^{5+}? In fact, even C^+ does not form in compounds. Compare the energy required to form this ion with that needed to form some metal ions. Explain.

GENERAL PROBLEMS

5–74 Write the symbol of the element that corresponds to the following:

(a) the first element with a p electron
(b) the first element with a filled $4p$ subshell
(c) three elements with one electron in the $4s$ subshell
(d) the first element with one p electron that also has a filled d subshell
(e) the element after Xe that has two electrons in a d subshell

5–75 Write the symbol of the element that corresponds to the following:

(a) the first element with a half-filled p subshell
(b) the element with three electrons in a $4d$ subshell
(c) the first two elements with a filled $3d$ subshell
(d) the element with three electrons in a $5p$ subshell

5–76 Write the symbol of the element that corresponds to the following:

(a) a nonmetal with one electron in a p subshell
(b) an element that is a liquid at room temperature that has five p electrons
(c) the first metal to have three p electrons
(d) a metalloid with two p electrons and no d electrons

5–77 Write the symbol of the element that corresponds to the following:

(a) a transition metal in the fifth period with three unpaired electrons
(b) a representative metal with three unpaired electrons with no f electrons
(c) an element that is a liquid at room temperature with no unpaired electrons
(d) a metalloid with two unpaired electrons and a filled d subshell

5–78 A certain isotope has a mass number of 30. The element has three unpaired electrons. What is the element?

5–79 A certain isotope has a mass number of 196. It has two unpaired electrons. What is the element?

5–80 Given two elements, X and Z, identify the elements from the following information.

(a) They are both metals, but one is a representative element.

(b) The first ionization energy of Z is greater than X.

(c) They are both in the same period that does not have *f* electrons.

(d) They both have two unpaired electrons and neither is used in jewelry.

(e) A nonmetal in the same period is a diatomic solid.

5–81 Given two elements, Q and R, identify the elements from the following information.

(a) One is a gas and one is a solid.

(b) One forms a +1 ion and the other does not.

(c) Q is larger than R, but both elements are the first elements with a full shell.

(d) One is used in coins and the other in fluorescent lights.

5–82 Which of the following monatomic cations would require a particularly large amount of energy to form? If a certain ion requires a large amount of energy to form, give the reason.

(a) In^{3+} (c) Ca^{2+} (e) B^{3+}
(b) I^+ (d) K^+

5–83 Which of the following atoms would not be likely to form a +2 cation? Explain.

(a) Sr   (b) Li   (c) B   (d) Ba

5–84 Which of the following atoms would not be likely to form a −2 anion? Explain.

(a) Se   (b) Po   (c) Cl   (d) H

5–85 Which of the following monatomic anions are likely to form? If a given ion is unlikely, explain.

(a) F^{2-} (c) Be^- (e) Ar^-
(b) Se^{2-} (d) N^{3-}

5–86 Which of the following monatomic anions are likely to form? If a given ion is unlikely, explain.

(a) I^- (c) K^- (e) Sb^{3-}
(b) Bi^{2-} (d) Se^{2-}

*5–87 On the planet Zerk, the periodic table of elements is slightly different from ours. On Zerk, there are only two *p* orbitals, so a *p* subshell holds only four electrons. There are only four *d* orbitals, so a *d* subshell holds only eight electrons. Everything else is the same as on Earth, such as the order of filling (1*s*, 2*s*, etc.) and the characteristics of noble gases, metals, and nonmetals. Construct a Zerkian periodic table using numbers for elements up to element number 50. Then answer these questions.

(a) How many elements are in the second period? In the fourth period?

(b) What are the atomic numbers of the noble gases at the ends of the third and fourth periods?

(c) What is the atomic number of the first inner transition element?

(d) Which element is more likely to be a metal: element number 5 or element number 11; element number 17 or element number 27?

(e) Which element has the larger radius: element number 12 or element number 13; element number 6 or element number 12?

(f) Which element has a higher ionization energy: element number 7 or element number 13; element number 7 or element number 5; element number 7 or element number 9?

(g) Which ions are reasonable?
(1) 16^{2+}
(2) 9^{2+}
(3) 7^+
(4) 13^-
(5) 17^{4+}
(6) 15^+
(7) 1^-

SOLUTIONS TO LEARNING CHECKS

A–1   wave, inversely, longest, lowest, discrete, quantized, ground, excited, shell, subshells, *s*, *p*, *d*, *f*, orbitals, 1, 3, 5, 7, two, spherical

A–2   For the $n = 3$ shell:
(a) shell capacity: $2n^2 = 18$
(b) subshells: *s*, *p*, and *d*
(c) subshell capacities: $s = 2, p = 6, d = 10$
(d) orbitals: $s = 1, p = 3, d = 5$

B–1 five, d, fifth, Aufbau, Pauli, Hund's rule

B–2 Ni [Ar] $4s^2 3d^8$ Sr [Kr] $5s^2$ Pb [Xe] $6s^2 4f^{14} 5d^{10} 6p^2$

B–3 Nb [Kr] $5s^2 4d^3$

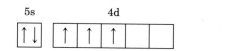

C–1 IVA, IVB, increase, decrease, decrease, increase, electron, nonmetals

C–2 Zn [Ar] $4s^2 3d^{10}$ W [Xe] $6s^1 4f^{14} 5d^5$ (an exception to rules)

C–3 As

C–4 VIA [NG] $ns^2 np^4$ VIB [NG] $ns^1 (n-1) d^5$ (an exception to rules)

C–5 (a) Se (b) Sb (c) Se (d) Sb

This is our home as seen from far-out space. Its surface and atmosphere are composed of some free elements as well as ionic and molecular compounds. We look deeper into the nature of compounds in this chapter.

THE CHEMICAL BOND

This Earth of ours is a fascinating yet complex world of chemicals. First, consider the air. Its major components are free elements: two molecular elements, nitrogen and oxygen, and smaller quantities of noble gases such as helium, neon, and argon that exist as solitary atoms. The surface of Earth is made up primarily of compounds. Lakes and oceans are composed of water containing dissolved compounds, whereas the solid earth contains compounds of living things, rocks, and minerals. Perhaps we can begin to bring some order to all of this chaos by dividing all of the substances in, on, and above Earth into roughly two categories: hard stuff and soft stuff. The rocks and minerals of the mountains are certainly hard, whereas the air, oceans, and stuff of living things (solid but still flexible) are soft. Let us focus on two prime examples that occur in nature, one hard—table salt—and one soft—water. In an earlier chapter, we emphasized two basic types of compounds: ionic and molecular. Although there are several important types of hard substances in nature that are not ionic, such as quartz, diamond, and certain metals, many of Earth's minerals and rocks are ionic compounds that resemble sodium chloride. On the other hand, water is typical of a molecular compound that makes up much of the soft stuff of nature. What is the difference between the two compounds? Sodium chloride is a binary compound formed from a metal, sodium, and a nonmetal, chlo-

◀ **Setting The Stage**

rine. Water is also a binary compound, but it is formed from two non-metals, hydrogen and oxygen. Perhaps we are on to something! Does the combination of a metal and a nonmetal result in ionic compounds, whereas the combination of two nonmetals results in molecular compounds? The answer is "yes" (generally), and we will see why in this chapter.

Formulating
Some Questions ▶

Why do noble gases rarely bond to other elements and therefore exist as solitary atoms in nature? How does this relate to why the atoms of other elements do form bonds? Why do certain elements combine to form ionic and others molecular compounds? Why is the formula of water H_2O and not H_3O or HO_2? These are some of the important questions that we will address in this chapter.

These questions all relate to the electron configurations of the elements. In the first section of this chapter, we will take a closer look at why and how representative elements combine.

6-1   BOND FORMATION AND REPRESENTATIVE ELEMENTS

Why do two atoms bond together?

Why does one atom bond with another? The answer is fairly simple. Since chemical bonds involve electrons, a bond forms if it produces a more stable electron configuration for that atom. Since the atoms of noble gases generally do not bond, they are obviously stable as solitary atoms. Let us focus more on why this is so. Noble gases (except He) have filled outer s and p subshells (i.e., ns^2np^6). Since this is a total of eight electrons, it is referred to as an **octet** of electrons. Eight electrons in the outer s and p subshells forms a particularly stable configuration. As it turns out, this observation explains the nature of many of the compounds formed by the representative elements. Bonding is correlated by the **octet rule,** which states that *the atoms of the representative elements form bonds so as to have access to eight outer electrons. The outer* s *and* p *electrons in the atoms of a representative element are referred to as the* **valence electrons.** Representative elements that border helium in the periodic table, however, follow a "duet" rule. These elements (H, Li, and Be) tend to alter their electron configuration to be like helium, which has only two electrons. The octet rule is particularly helpful in describing the bonding of many, but certainly not all, of the compounds of the representative elements. Some elements in the third period (Si to Cl) and in higher periods form compounds that are not explained by the octet rule. In this text, however, we will emphasize the compounds that do follow the octet rule.

How can an atom alter its electron configuration to obtain an octet (or duet) of electrons of a noble gas? There are three ways.

How can an atom attain more or less electrons?

1 A metal may *lose* one to three electrons to form a cation with the electron configuration of the previous noble gas.

2 A nonmetal may *gain* one to three electrons to form an anion with the electron configuration of the next noble gas.

3 Atoms (usually two nonmetals) may *share* electrons with other atoms to obtain access to the number of electrons in the next noble gas.

The first two processes complement each other in the formation of ionic compounds. Case 3 produces molecular compounds.

Since the bonding in these three cases involves the loss, gain, or sharing of valence electrons exclusively, we are free to focus on these electrons only. **Lewis dot symbols*** *of these elements represent valence electrons as dots around the symbol of the element.* Electrons are represented with up to four pairs of electrons on four sides of the element's symbol. Since the elements in each group have the same number of valence electrons (same subshells but different shells), each element in a group has the same number of dots representing electrons. The Lewis dot symbols of the first four periods of representative elements and noble gases are shown in Table 6-1. The dot symbols are usually shown first with one electron on each side of the element (Groups IA through IVA in Table 6-1) and then with paired electrons on each side (Groups VA through VIII). Note that the Roman numeral of the group number also represents the number of valence electrons (dots) for a neutral atom.

In the previous chapter, we noted that metals lose electrons comparatively easily and nonmetals gain electrons. We are now ready to observe how the octet rule determines the magnitude of the charge formed by a particular representative element.

▼**Looking Ahead**

6–2 FORMATION OF IONS

Most metals have many familiar *physical* properties such as the ability to conduct heat and electricity and the capacity to be drawn into wires and pounded into sheets. The one *chemical* property of metals that we have established is that it takes a comparatively small amount of energy to remove one or, in some cases, two or three electrons to form cations. The octet rule tells us that if a representative metal loses all of its va-

TABLE 6–1 LEWIS DOT SYMBOLS

IA	IIA	IIIA	IVA	VA	VIA	VIIA	VIIIA
Ḣ							He
Li	Be·	Ḃ·	·Ċ·	·N̈·	·Ö:	:F̈:	:N̈e:
Na	Mg·	Al·	·Si·	·P̈·	·S̈:	:C̈l:	:Är:
K̇	Ca·	Ga·	·Ge·	·Äs·	·S̈e:	:Br:	:K̇r:

*Named after the American chemist G. N. Lewis (1875–1946), who developed this theory of bonding.

lence electrons, it acquires the octet of the previous noble gas. The loss of any additional electrons is prohibitively expensive in terms of energy, so it does not occur in compound formation. We can illustrate the octet rule and cation formation using the Lewis dot symbol of sodium.

$$\dot{Na} \longrightarrow Na^+ + e^-$$
$$[Ne]3s^1 \qquad [Ne]$$

The Lewis representation of the Na^+ ion does not include any electrons (dots) because the octet of electrons of Na^+ are in an inner subshell. All of the *metals* in Group IA have the same dot symbol, so they can lose one electron to form $+1$ ions with electron configurations of the preceding noble gas.

Now consider the alkaline earth metals. For example, the loss of two electrons from magnesium produces the octet of electrons of neon.

$$\dot{Mg}\cdot \longrightarrow Mg^{2+} + 2e^-$$
$$[Ne]3s^2 \qquad [Ne]$$

All other metals in this group form $+2$ ions in the same manner.

Group IIIA metals (Al down) can lose three electrons in order to form an octet of electrons.* Boron is not a metal and does not form a $+3$ ion in its compounds. Boron bonds by electron sharing, which is discussed later in this chapter.

$$\cdot\dot{Al}\cdot \longrightarrow Al^{3+} + 3e^-$$
$$[Ne]3s^23p^1 \quad [Ne]$$

Group IVA metals such as tin and lead have four electrons in their outer subshells. Loss of all four of these electrons to produce a $+4$ ion requires a rather large amount of energy. Instead, these metals can lose two of their four outer electrons to form a $+2$ ion that does not follow the octet rule. They do form compounds where all four of their outer electrons are involved, but the bonding in these compounds is best described by electron sharing rather than ion formation. In Group VA, bismuth forms a $+3$ ion that does not follow the octet rule.

Positive ions do not exist alone in compounds. There are also enough negative ions to balance the positive charge. In the previous chapter, we found that representative nonmetals complement metals by forming negative ions. First, we will consider the negative ions formed by the VIIA nonmetals.

All of the atoms of the halogens shown in Group VIIA are one electron short of a noble gas configuration. An octet of electrons can be achieved by adding one electron. The result is an anion with a -1 charge and the electron configuration of the next noble gas. In this respect, hydrogen is also one electron short of a noble gas configuration (a duet in this case), so it can also add one electron to form an anion with a -1 charge.

$$e^- + H\cdot \longrightarrow H:^-$$
$$1s^1 \qquad 1s^2 = [He]$$

Is there a limit to the positive charge on a metal?

---

*Ions such as Tl^{3+} and Ga^{3+} have a filled d subshell in addition to a noble gas configuration. This is sometimes referred to as a pseudo-noble gas configuration. The filled d subshell does not seem to affect the stability of these ions. In this text, we do not distinguish between noble gas and pseudo-noble gas electron configurations. Transition metals also form positive ions, but for the most part, these ions do not relate to a noble gas configuration. Some of these ions were discussed in Chapter 4.

$$e^- + :\overset{\cdot\cdot}{\underset{\cdot\cdot}{Cl}}\cdot \longrightarrow :\overset{\cdot\cdot}{\underset{\cdot\cdot}{Cl}}:^-$$
$$[Ne]3s^2 3p^5 \qquad\qquad [Ne]3s^2 3p^6 = [Ar]$$

The atoms of the elements in Group VIA are two electrons short of a noble gas configuration. By gaining two electrons to form a -2 ion, they also attain an octet of electrons.

$$2e^- + :\overset{\cdot\cdot}{O}\cdot \longrightarrow :\overset{\cdot\cdot}{\underset{\cdot\cdot}{O}}:^{2-}$$
$$[He]2s^2 2p^4 \qquad\qquad [He]2s^2 2p^6 = [Ne]$$

Two nonmetals (N and P) in Group VA gain three electrons to form -3 ions. (A -3 ion is not known for As.)

$$3e^- + \cdot\overset{\cdot}{\underset{\cdot\cdot}{N}}\cdot \longrightarrow :\overset{\cdot\cdot}{\underset{\cdot\cdot}{N}}:^{3-}$$
$$[He]2s^2 2p^3 \qquad\qquad [He]2s^2 2p^6 = [Ne]$$

For the most part, Group IVA nonmetals bond by electron sharing rather than forming monatomic ions. Although there is some evidence for a C^{4-} ion with an octet of electrons, formation of such highly charged ions is an energetically unfavorable process.

We are now ready to discuss how metals and nonmetals come together to exchange electrons. As we will see, the formulas of the ionic compounds that result are determined by the charges that the metals and nonmetals attain to satisfy the octet rule.

▼**Looking Ahead**

6–3 FORMULAS OF BINARY IONIC COMPOUNDS

When a small piece of sodium metal is placed in a bottle containing chlorine gas, a chemical reaction is obvious. (See Figure 6-1.) The sodium ignites, and a white coating of sodium chloride forms on the sides of the bottle. The formation of sodium chloride involves a transfer of one electron from the sodium atom to the chlorine atom.

$$Na + :\overset{\cdot\cdot}{\underset{\cdot\cdot}{Cl}}: \longrightarrow Na^+ :\overset{\cdot\cdot}{\underset{\cdot\cdot}{Cl}}:^-$$
$$\text{Formula} = NaCl$$

As indicated in the previous section, both of the ions formed have octets of electrons. Now let us consider the reaction that occurs when lithium metal reacts with oxygen. The oxygen atom needs two electrons to achieve an octet and form an anion with a -2 charge. Since a lithium atom can lose only one electron, two lithium atoms are needed to supply the two electrons.

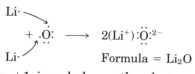

$$\text{Formula} = Li_2O$$

Notice that the two $+1$ ions balance the charge of the one -2 ion [i.e., $2(+1) - 2 = 0$]. The chemical formula of the compound lithium oxide is therefore Li_2O.

Figure 6–1
Reaction of Sodium with Chlorine Sodium reacts with chlorine to form sodium chloride.

Now consider the compound formed when calcium combines with bromine. Calcium loses two electrons to form a +2 cation, but a bromine can only add one electron to form a −1 anion. Two bromine atoms are needed to accept the two electrons lost by one calcium.

$$Ca \cdot \begin{array}{c} \nearrow \cdot \ddot{B}\ddot{r}\ddot{:} \\ \\ \searrow \cdot \ddot{B}\ddot{r}\ddot{:} \end{array} \longrightarrow \begin{array}{c} Ca^{2+}2(:\ddot{B}\ddot{r}\ddot{:}^-) \\ Formula = CaBr_2 \end{array}$$

When aluminum combines with oxygen, it is somewhat more complex to follow the transfer of three electrons from aluminum to oxygen, which can accept only two. In this case, two Al's give up six electrons, which are then accepted by three O's. The formula is thus Al_2O_3 and the charges cancel $[2(+3) + 3(-2) = 0]$.

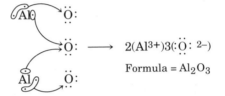

$$\longrightarrow \quad 2(Al^{3+})3(:\ddot{O}:^{2-})$$
$$Formula = Al_2O_3$$

At this point, it is well to recall the "cross-charge" method (see Chapter 4, page xxx) that we used to write formulas of binary metal-nonmetal compounds. If we know the charge on the metal and the nonmetal, the charge on one ion becomes the subscript of the other. In effect, this balances the total positive and negative charge. Note, however, that this method predicts formulas such as Ca_2O_2 and Al_3N_3. Ionic compounds are always expressed as empirical formulas (smallest whole-number ratios of atoms), so these should be expressed as CaO and AlN, respectively.

$$Na^{\textcircled{1}+}N^{\textcircled{3}-} = Na_3N$$
$$Ga^{\textcircled{3}+}S^{\textcircled{2}-} = Ga_2S_3$$
$$Ca^{\textcircled{2}+}O^{\textcircled{2}-} = Ca_2O_2 = CaO \quad \text{(write the simplest formula)}$$

EXAMPLE 6-1

The Formulas of Binary Ionic Compounds

Working It Out ●

What is the formula of the ionic compound formed between (a) aluminum and fluorine and (b) barium and sulfur?

Solution
(a) Aluminum is in Group IIIA and fluorine is in Group VIIA. Their dot symbols are

$$\cdot \dot{A}l \cdot \qquad \dot{\ddot{F}}\ddot{:}$$

To have a noble gas configuration (an octet), the Al, a metal, must lose all three outer electrons to form a +3 ion. Three fluorine atoms are needed to add one electron each to form three −1 ions. Note that each fluorine can add only one electron, which gives the F^- ion an octet. The compound formed is

$$Al^{3+}3(:\ddot{F}\ddot{:}^-) = \underline{AlF_3}$$

We could also determine the formula from the charges of the respective ions formed. Al becomes Al^{3+} and F becomes F^{1-}.

$$Al^{③+} \diagdown F^{①-} = \underline{AlF_3}$$

(b) Barium is in Group IIA and sulfur is in Group VIA, and they have the dot symbols

$$\dot{Ba}\cdot \qquad \overset{..}{\underset{..}{S}}:$$

One Ba atom gives up two electrons, and one S atom takes up two electrons, forming the compound

$$Ba^{2+}:\overset{..}{\underset{..}{S}}:^{2-} = \underline{BaS}$$

We could also determine the formula from the charges of the respective ions formed. Ba becomes Ba^{2+} and S becomes S^{2-}

$$Ba^{②+} \diagdown S^{②-} = Ba_2S_2 = \underline{\underline{BaS}}$$

Ionic compounds are solids at room temperature. They tend to have high melting points and are often brittle. As mentioned in the introduction, this type of compound comprises a large portion of the matter that we think of as "hard." If we look into the basic structure of a crystal of table salt, we can see why. Ionic compounds do not exist as discrete molecular units with one Na^+ attached to one Cl^-. As shown in Figure 6-2, each Na^+ is actually surrounded by six Cl^- ions, and each Cl^- ion is surrounded by six Na^+ ions in a *three-dimensional array of ions called a* **lattice.** Recall from Chapter 5 that cations are smaller than their parent atoms but anions are larger. Thus, in most cases, we can assume that the anion is larger than the cation. The lattice is held together strongly and rigidly by electrostatic interactions. *These electrostatic attractions are known as* **ionic bonds.** There are several other arrays of ions (lattices). For example, in CsCl both the Cs^+ and the Cl^- are surrounded by eight oppositely charged ions.

Besides the monatomic ions, polyatomic ions exist where two or more atoms are bound together by electron sharing, and the total species carries a net charge [e.g., the carbonate ion (CO_3^{2-})]. These species exist as ions because they have an imbalance of electrons compared to the total number of protons in their nuclei. We will discuss the bonding within a polyatomic ion later in this chapter, but for now we acknowledge their existence in ionic compounds.

Why are ionic compounds hard and brittle?

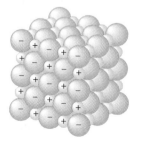 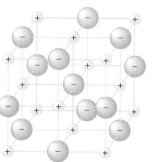

Figure 6–2
An Ionic Solid Each cation (small spheres) is surrounded by six anions. Each anion (large spheres) is surrounded by six cations.

Looking Back ▲

Metal and nonmetal atoms fit together perfectly. An exchange of electrons between the two leads to charged atoms that satisfy the octet rule. The result is the formation of ionic compounds. Rocks and most other hard substances are composed of ionic compounds. The hard, brittle nature of ionic compounds is explained by the rigid lattices formed by the oppositely charged ions.

Checking It Out ▶

Learning Check A

A-1. Fill in the blanks.

Elements that have the most stable electron configurations are in Group _____ . The outer electrons are also called the _____ electrons. Since noble gases have _____ valence electrons, other representative elements attain access to this number of electrons by _____ , _____ , or _____ electrons. By following the _____ rule, metals may form _____ ions with a maximum charge of _____ , and nonmetals may form _____ ions. The formula of a binary ionic compound is determined by the _____ on the cation and anion.

A-2. What is the Lewis dot symbol for Be and Se?

A-3. What are the charges on the ions formed by Be and Se?

A-4. What is the formula of a compound formed when Be combines with Se?

Additional Examples: Problems 6-1, 6-5, 6-9, 6-12, 6-17, and 6-20.

Looking Ahead ▼

We are now ready to turn our attention to the softer part of nature. Metals are not involved in most of these compounds, so one atom does not give up electrons to another to form ions. In these cases, the octet rule is followed by means of electron sharing.

6–4   THE COVALENT BOND

Can two nonmetals form a bond?

In compounds such as water (H_2O), the two nonmetals attain noble gas configurations by electron sharing. *A shared pair of electrons between two atoms is known as a* **covalent bond.** A molecular compound is composed of molecules having only covalent bonds and no ions. Compared to the interaction between oppositely charged ions, the discrete molecular units are only weakly attracted to each other; thus they can move past one another more freely than ions. For this reason, compounds composed of molecules tend to be gases, liquids, or soft solids with low melting points.

It is easy to appreciate how a complete exchange of electrons can satisfy the octet rule, but the concept of electron sharing and the octet rule is more subtle. A simple analogy may help. Assume that we have two people each of whom wishes to have access to $8 in savings accounts. If one has $9 and the other $7, the solution is obvious. The one with $9

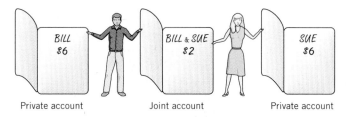

Figure 6–3

Increase Your Money by Sharing The money in a joint account is analogous to electrons in a covalent bond.

gives \$1 to the other and both are satisfied. This is analogous to a metal and a nonmetal forming an ionic bond. In a second situation, assume that both people have only \$7. There is still a solution. If each keeps \$6 in separate accounts and each contributes \$1 (for a total of \$2) to a joint account, they both can claim access to \$8. (See Figure 6-3.)

Fluorine, as well as all members of Group VIIA, exist in nature as diatomic molecules (F_2). Each fluorine has seven electrons. An octet for each can be achieved if each fluorine holds six electrons to itself and shares one electron (for a total of two shared electrons). This is illustrated as follows.

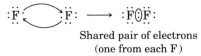

Shared pair of electrons
(one from each F)

We can now extend the concept of Lewis dot representations to covalent bonds. *A* **Lewis structure** *for a molecule shows the order and arrangement of atoms in a molecule (the structural formula) as well as all of the valence electrons for the atoms involved.* There are several variations of how Lewis structures represent molecules. A pair of electrons is sometimes shown as a pair of dots (:) or as a dash (—). In this text, we use a pair of dots to represent unshared pairs (also called *lone pairs*) of electrons on an atom, and a dash to represent a pair of electrons that are shared between atoms. In this way, shared and unshared electrons can be distinguished.

The Lewis structure of F_2* is illustrated as follows:

Total of 14 outer electrons (7 from each F)

:F̈—F̈— Three lone pairs on each F

Two shared electrons in a covalent bond

Similarly, other halogens exist as diatomic molecules, like F_2 with the same Lewis structures. Hydrogen, which forms the simplest of all molecules, also exists as a diatomic gas with one covalent bond between atoms:

$$H—H$$

*An F atom has one unpaired electron in a $2p$ orbital. Formation of a covalent bond pairs the electrons in the two F atoms so that the F_2 molecule has no unpaired electrons. Although most atoms of the representative elements have unpaired electrons, most molecules or ions formed from these elements do not have unpaired electrons.

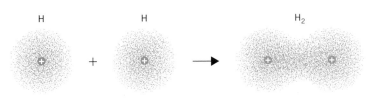

Figure 6–4
A Covalent Bond The high electron density between the two nuclei holds the two atoms together.

What holds the atoms together in a covalent bond?

Recall that hydrogen follows a duet rule in order to attain the noble gas configuration of He.

Why do two H atoms combine to form an H_2 molecule? Again the answer is "because that is a more stable arrangement." Two hydrogen atoms alone have one electron in a $1s$ orbital, which is spherically diffuse. The electron has a probability of existing in any direction from the nucleus in the separate atoms. When two hydrogen atoms come together, however, the two electrons become more localized between the two nuclei in a region where the two $1s$ orbitals overlap. Each positive hydrogen nucleus is attracted to two negative electrons between the atoms rather than just one. Although there are also forces of repulsion between the two electrons, the mutual attraction of two nuclei for two electrons predominates and holds them together. (See Figure 6-4.)

Just as we were able to justify the formulas of simple binary ionic compounds by the octet rule, we can do the same with simple binary molecular compounds. In fact, this works so well that we can predict the formulas of compounds based on the octet rule.

First, we will consider the compounds formed by hydrogen with the halogens in Group VIIA. For example, consider the compound formed from hydrogen and fluorine, which has the formula HF. A shared pair of electrons (one from each atom) gives both atoms access to the same number of electrons as a noble gas.

$$H\cdot \longrightarrow \cdot \ddot{\underset{\cdot\cdot}{F}}\colon \longrightarrow H\!-\!\ddot{\underset{\cdot\cdot}{F}}\colon$$

Now we will consider the compounds formed between hydrogen and the Group VIA elements. Our primary example, of course, is water. Since an oxygen needs access to two more electrons to have an octet, we will need two hydrogens to form two covalent bonds to one oxygen.

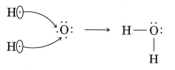

As we move across the periodic table to consider the hydrogen compounds formed between Group VA and Group IVA nonmetals, we see that the octet rule serves us well. Three hydrogens are needed by N (Group VA) and four by C (Group IVA). Recall that hydrogen is written second in binary compounds with Groups IVA and VA elements but is written first with Group VIA and VIIA elements. (See also some hydrogen compounds of third-period elements in Figure 6-5.)

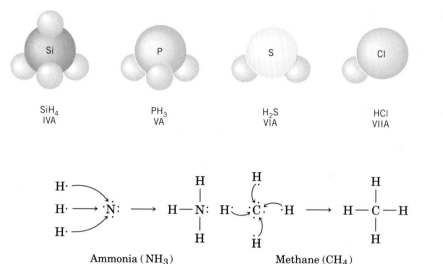

Figure 6–5
Formulas of Hydrogen Compounds The formulas of some simple hydrogen compounds can be predicted from the octet rule.

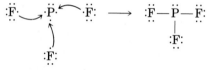

Ammonia (NH_3)　　　　Methane (CH_4)

Let's try to predict the formula of the simplest compound formed between phosphorus (Group VA) and fluorine (Group VIIA). Since the dot symbol of P indicates that it needs to add three electrons and F needs to add one, the solution points to one P sharing a pair of electrons with each of three different F's.

Simplest formula = PF_3

In addition to molecular compounds, atoms within polyatomic ions share electrons in covalent bonds. For example, consider the hypochlorite ion (ClO^-). The -1 charge on the ion tells us that there is one more electron present in this species than the valence electrons of Cl and O. The total number of electrons is calculated as follows:

$$
\begin{aligned}
\text{From a neutral Cl} &= 7 \\
\text{From a neutral O} &= 6 \\
\text{Additional electron indicated by charge} &= \underline{1} \\
\text{Total number of electrons} &= 14
\end{aligned}
$$

Two atoms bonded together with 14 electrons have a Lewis structure like F_2, which also has 14 electrons:

$$\left[:\ddot{C}l - \ddot{O}:^- \right]$$

The brackets indicate that the total ion has a -1 charge. The extra electron has not been specifically identified because all electrons are identical and belong to the ion as a whole.

In the examples illustrated so far, two atoms share one pair of electrons. There are also examples, especially among the second-period nonmetals (B through O), where two or even three pairs of electrons are shared between two atoms. *The sharing of two pairs of electrons between the same two atoms is known as a* **double bond.** *The sharing of three*

Solid carbon dioxide is known as dry ice. It is a molecular compound whose molecules contain double bonds.

How many electrons can be shared between two atoms?

pairs is known as a **triple bond.** A double bond is illustrated as ═, and a triple bond is illustrated as ≡. The use of a double bond to satisfy the octet rule is analogous to two people who wish to have access to $8 but only have a total of $12 between them. In this case, each person could have $4 in a private account while sharing $4 in a joint account. A triple bond would be analogous to the two people having only $10 between them. Each could have only $2 in a private account while sharing $6 in the joint account.

The molecules of carbon dioxide have double bonds.* In the Lewis structure shown below, notice that the octets of both carbon and oxygen are satisfied by the sharing of two pairs of electrons in each carbon-oxygen bond. Elemental nitrogen, N_2 is an example of a molecule that satisfies the octet rule with a triple bond.

$$\ddot{O}=C=\ddot{O} \qquad :N≡N:$$

Looking Ahead ▼

We have shown examples of simple binary molecules and a polyatomic ion with single covalent bonds. The existence of multiple bonds, however, can make the writing of Lewis structures somewhat complex unless we formalize some useful rules. That is the next topic.

6–5   WRITING LEWIS STRUCTURES

By following the octet rule in writing Lewis structures, we can not only justify certain formulas of compounds but predict a few as well. Actually, the Lewis structure of a compound tells us much more. For example, as we will see later in this chapter, it can be used to predict the geometry of a molecule. The molecule's geometry is very much related to both the physical and chemical properties of the compound. As we progress in the study of chemistry, we will continually refer to the Lewis structure of many compounds. Therefore, writing Lewis structures correctly is considered a fundamental skill to be acquired early in the study of chemistry.

Writing Lewis structures according to the octet rule is quite straightforward when certain guidelines or rules are systematically applied. These rules, of course, require considerable practice in their application. Starting with the formula of a compound (either ionic or molecular), the rules are as follows:

1   Check to see whether any ions are involved in the compounds. Write any ions present.
   (a) Metal-nonmetal binary compounds are mostly ionic.

---

*It would seem that O_2, with 12 valence electrons, would be an excellent example of the simplest molecule with a double bond:

$$\ddot{O}=\ddot{O}$$

This Lewis structure implies that all of the electrons in O_2 are paired. However, experiments show that O_2 has two unpaired electrons. Although writing Lewis structures works very well in explaining the bonding in most simple molecules, you should keep in mind that a Lewis structure is simply the representation of a theory. For O_2 the theory doesn't work perfectly. Thus O_2 is usually not represented by a Lewis structure. Other theories on bonding work well in explaining the bonding in this molecule, but they are not discussed here.

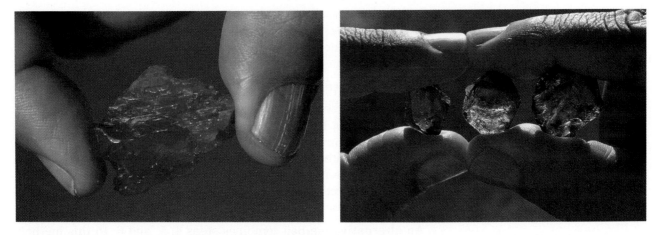

Rubies and sapphires are hard substances found in nature. They are ionic compounds.

(b) If Group IA or Group IIA metals (except Be) are part of the formula, ions are present. For example, KClO is K$^+$ClO$^-$ because K is a Group IA *element and forms only a* +1 *ion*. If K is +1, the ClO must be -1 to have a neutral compound. Likewise, Ba(NO$_3$)$_2$ contains ions because Ba is a Group IIA *element and forms only a* +2 *ion*. To maintain neutrality, each NO$_3$ ion must have a -1 charge. The ions are represented as Ba^{2+}2(NO$_3^-$).

(c) Compounds composed of nonmetals only contain covalent bonds only.

2 For a molecule, add all of the outer (valence) electrons of the neutral atoms. For an ion, add (if negative) or subtract (if positive) the number of electrons indicated by the charge.

3 Write the symbols of the atoms of the molecule or ion in a skeletal arrangement.

(a) A hydrogen atom can form only one covalent bond and therefore bonds to only one atom at a time. They are situated on the periphery of the molecule.

(b) The atoms in molecules and polyatomic ions tend to be arranged in a symmetrical pattern around a central atom. The central atom is generally a nonmetal other than oxygen or hydrogen. Oxygens usually do not bond to each other. Thus SO$_3$ has an S surrounded by three O's,

$$\begin{array}{c} \text{O} \\ \text{S} \\ \text{O} \quad \text{O} \end{array}$$

rather than such structures as

$$\text{S O O O} \qquad \text{O S O O} \qquad \begin{array}{cc} \text{S} & \text{O} \\ \text{O} & \text{O} \end{array}$$

In most cases, the first atom in a formula is the central atom, and the other atoms are bound to it.

4 Put a dash representing a shared pair of electrons between adjacent atoms that have covalent bonds (not between ions). Subtract the electrons used for this (two for each bond) from the total calculated in step 2.

5 Distribute the remaining electrons among the atoms so that no atom has more than eight electrons.

6 Check all atoms for an octet (except H). If an atom has access to fewer than eight electrons, put an electron pair from an adjacent atom into a double bond. Each double bond increases by two the number of electrons available to the atom needing electrons. Remember that you cannot satisfy an octet for an atom by adding any electrons at this point.

An alternative method combines steps 4, 5, and 6. In this method, a count of electrons is used to determine the number of multiple bonds present, and then electrons and bonds are added to the skeletal structure accordingly. This has been conveniently summarized as the "$6N + 2$ rule," where N stands for the number of atoms other than hydrogen in the formula. If the number of valence electrons in the formula equals $6N + 2$, then only single bonds are present. If the number of valence electrons is two less than $6N + 2$, then one double bond is present. (It could also mean that a ring structure is present; these are discussed briefly in Chapter 16.) If the number of valence electrons is four less than $6N + 2$, then one triple bond or two double bonds are present. Consider the following molecules:

(a) PF_3 $N = 4$ $6(4) + 2 = 26$
Valence electrons: $5[P] + (3 \times 7)[F] = 26$
Therefore, only single bonds are present.

(b) $CO_3{}^{2-}$ $N = 4$ $6(4) + 2 = 26$
Valence electrons: $4[C] + (3 \times 6)[O] + 2[\text{charge}] = 24$
$26 - 24 = 2$
Therefore, one double bond is present.

(c) C_6H_{10} $N = 6$ $6(6) + 2 = 38$
Valence electrons: $(6 \times 4)[C] + (10 \times 1)[H] = 34$ $38 - 34 = 4$
Therefore, two double bonds or one triple bond is present in this molecule.

In this text we use the "$6N + 2$ rule" as a check to confirm the Lewis structure determined by applying steps 1 through 6.

EXAMPLE 6-2

The Lewis Structure of a Molecular Compound

Working It Out ●

Write the Lewis structure for NCl_3.

Solution

1 This is a binary compound between two nonmetals. Therefore, it is not ionic.

2 The total number of electrons available for bonding is

$$
\begin{array}{llll}
\text{N} & 1 \times 5 = & 5 \\
\text{Cl} & 3 \times 7 = & \underline{21} \\
& \text{Total} = & \overline{26}
\end{array}
$$

3 The skeletal arrangement is

$$
\begin{array}{ccc}
\text{Cl} & \text{N} & \text{Cl} \\
& \text{Cl} &
\end{array}
$$

4 Use six electrons to form bonds:

$$
\begin{array}{c}
\text{Cl—N—Cl} \\
|\\
\text{Cl}
\end{array}
$$

5 Distribute the remaining 20 electrons $(26 - 6 = 20)$:

$$
\begin{array}{c}
\text{:Cl—N̈—Cl:} \\
|\\
\text{:Cl:}
\end{array}
$$

6 Check to make sure that all atoms satisfy the octet rule:

$$
\begin{array}{c}
\text{:Cl} \!\!-\!\!\text{N}\!\!-\!\! \text{Cl:} \\
\text{:Cl:}
\end{array}
$$

The $6N + 2$ rule confirms this structure.
For NCl_3 $N = 4$ $6(4) + 2 = 26$
From step 2 there are 26 valence electrons, so there are only single bonds.

EXAMPLE 6-3

The Lewis Structure of an Ion.

Write the Lewis structure of the cyanide ion (CN^-).

Solution

1 This is an ion.

2 The total number of electrons available is

$$
\begin{array}{lll}
\text{N} & 1 \times 5 = & 5 \\
\text{C} & 1 \times 4 = & 4 \\
\text{From charge} & = & \underline{1} \\
\text{Total} & = & \overline{10}
\end{array}
$$

3 The skeletal arrangement is

$$\text{C} \quad \text{N}$$

4 Use two electrons to form bonds:

$$\text{C—N}$$

5 Distribute the remaining 8 electrons $(10 - 2 = 8)$

$$\text{:C—N:}^-$$

6 Notice that both carbon and nitrogen have access to only six electrons each. Use two electrons from the carbon and two electrons from nitrogen to make a triple bond. Now the octet rule is satisfied.

$$:C \equiv N:^-$$

The $6N + 2$ rule confirms the structure.
For CN^- $N = 2$ $6(2) + 2 = 14$
$14 - 10 = 4$ (two double bonds or one triple bond)

EXAMPLE 6-4

The Lewis Structure of an Ionic Compound

Write the Lewis structure for $CaCO_3$.

Solution

1 This is an ionic compound composed of Ca^{2+} and CO_3^{2-} ions. (Since you know that Ca is in Group IIA, it must have a $+2$ charge; therefore the polyatomic anion must be -2.) A Lewis structure can be written for CO_3^{2-}.

2 For the CO_3^{2-} ion the total number of outer electrons available is

$$
\begin{array}{rl}
C & 1 \times 4 = 4 \\
O & 3 \times 6 = 18 \\
\text{From charge} = & \underline{2} \\
\text{Total} = & 24
\end{array}
$$

3, 4 The skeletal structure with bonds is

5 Add the remaining 18 electrons ($24 - 6 = 18$):

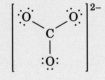

6 The C needs two more electrons, so one double bond is added using one lone pair from one oxygen:

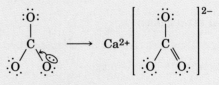

The $6N + 2$ rule confirms the structure.
For CO_3^{2-} $N = 4$ $6(4) + 2 = 26$
$26 - 24 = 2$ (one double bond)

EXAMPLE 6-5

The Lewis Structure of an Acid

Write the Lewis structure for H_2SO_4.

Solution

1 All three atoms are nonmetals, which means that all bonds are covalent.

2 The total number of outer electrons available is

$$
\begin{array}{lrcl}
\text{H} & 2 \times 1 &=& 2 \\
\text{S} & 1 \times 6 &=& 6 \\
\text{O} & 4 \times 6 &=& \underline{24} \\
& \text{Total} &=& 32
\end{array}
$$

3 In most molecules containing H and O, the H is bound to an O and the O to some other atom, which in this case is S. The skeletal structure is

$$
\begin{array}{ccccc}
& & \text{O} & & \\
\text{H} & \text{O} & \text{S} & \text{O} & \text{H} \\
& & \text{O} & &
\end{array}
$$

4 Add 12 electrons for the six bonds:

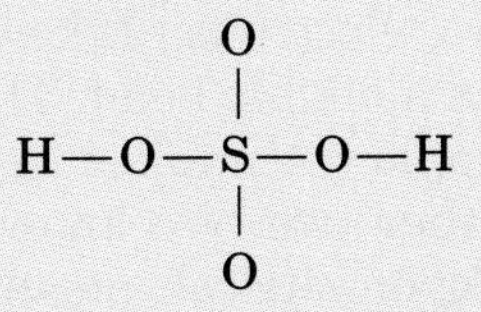

$$
\begin{array}{c}
\text{O} \\
| \\
\text{H—O—S—O—H} \\
| \\
\text{O}
\end{array}
$$

5 Add the remaining 20 electrons ($32 - 12 = 20$):

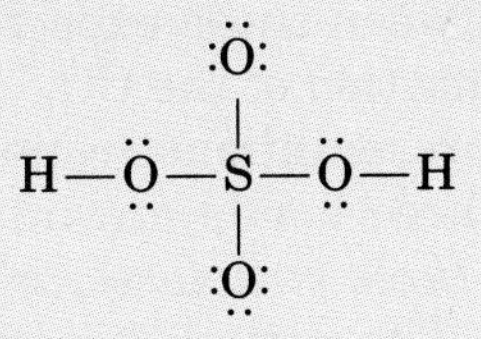

$$
\begin{array}{c}
:\!\ddot{\text{O}}\!: \\
| \\
\text{H—}\ddot{\text{O}}\text{—S—}\ddot{\text{O}}\text{—H} \\
| \\
:\!\ddot{\text{O}}\!:
\end{array}
$$

6 All octets are satisfied.
The $6N + 2$ rule is consistent with this structure.
For H_2SO_4 $N = 5$ (exclude hydrogens) $6(5) + 2 = 32$
From step 2, valence electrons $= 32$
This indicates that only single bonds are present.

The octet rule is very useful in describing the bonding in many compounds of the representative elements. A significant number of compounds, however, do not follow the octet rule. For example, in some compounds involving representative elements from the third and higher periods, the central atom has access to more than eight electrons (e.g., SF_4 and ClF_5). These cases have not been discussed in this text. In other compounds, one atom may have access to less than eight electrons (e.g., NO, which has 11 valence electrons). In still others, a Lewis structure may be written that follows the octet rule, but experiments suggest that some other structure is more likely. An example of the latter situation is illustrated by the molecule BF_3. Experiments indicate that the B—F bond has little to no double bond character. Thus its correct structure shows the boron with access to only six electrons.

Glance again at the Lewis structure of the carbonate ion shown in Example 6-4. The structure that is displayed implies that the three C—O bonds are not all identical in that one bond is double and the other two are single. Is that true? Actually, the answer is "no," but we need to explore this phenomenon in more detail.

Looking Ahead ▼

6–6   RESONANCE STRUCTURES

What if you can write more than one Lewis structure for a compound?

If we compare a double bond to a single bond between the same two elements, we find there are significant differences. The sharing of four electrons holds two atoms together more strongly, and thus more closely, than the sharing of two electrons. Likewise, a triple bond is even stronger and shorter than a double bond. The one Lewis structure of the CO_3^{2-} ion shown in Example 6-4 implies that one C—O bond is shorter and stronger than the other two. We know from experiment, however, that the ion is perfectly symmetrical, meaning that all three bonds are identical. Experiments also tell us that the lengths of the three identical bonds are somewhere between those expected for a single and a double bond. One Lewis representation of the CO_3^{2-} ion does not convey this information, but three representations (connected by double-headed arrows) indicate that all three bonds are intermediate between a single and a double bond. *The three structures as shown below are known as* **resonance structures.** The actual structure of the molecule can be viewed as a **resonance hybrid** of the three structures.

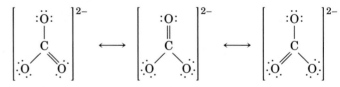

Resonance structures exist for molecules when equally correct Lewis structures can be written without changing the basic skeletal geometry or the position of any atoms.

Elemental oxygen occurs in nature in two different molecular forms, each with its own properties. *Different forms of the same element are known as* **allotropes.** The most prominent allotrope, the oxygen we breathe, has the formula O_2. The bonding in O_2 is not accurately described by the bonding theory that we have been discussing. The other form of oxygen is ozone, O_3, which is important in the stratosphere as a shield from ultraviolet light of the sun. Ozone is an example of a molecule whose bonding is described as a resonance hybrid as shown in the following example.

How is ozone related to normal oxygen?

EXAMPLE 6-6

Resonance Structures of Ozone

Write a Lewis structure and any equivalent resonance structures for ozone (O_3) where one oxygen serves as the central atom.

Solution

● Working It Out

1 No ions are involved.

2 There are 18 (3×6) outer electrons.

3 The skeletal structure is

4 Add four electrons for the two bonds:

5 Add the remaining 14 electrons:

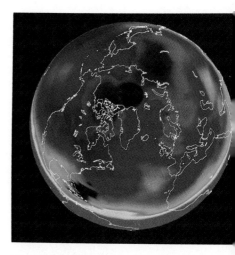

The colored area represents the ozone depletion that occurred over the Northern Hemisphere in March, 1993. This is thought to be caused to a large extent by synthetic chemicals.

6 Notice that the central oxygen does not have an octet, so make one double bond to one of the oxygens:

The $6N + 2$ rule confirms the presence of one double bond. Two resonance structures can be written, which indicates that the O—O bond is a hybrid between a single and a double bond.

The word resonance is an unfortunate choice because it is also associated with vibration or a constantly changing situation. The two O—O bonds in ozone are not changing rapidly back and forth between a single and a double bond. In fact, both bonds exist *at all times* as intermediate between a single and a double bond. It is much like a large, sweet hybrid tomato that you may grow in the garden or buy in the grocery store. This tomato is a hybrid of a large tomato and a small but sweet tomato. It isn't changing rapidly back and forth between these two but exists with properties intermediate between the two original species of tomatoes.

Looking Back ▲

Two nonmetals cannot bond by an exchange of electrons, so they bond by electron sharing. This means that nonmetal-nonmetal binary compounds, as well as molecular and atomic elements, exist in discrete units that are not as strongly attracted to each other as are ions. These are the gases, liquids, and soft solids of the world. By applying certain rules, we can easily understand the bonding and formulas of molecular compounds.

Checking It Out ▶

Learning Check B

B-1.   Fill in the blanks.
The sharing of electrons between two atoms is known as a _____ bond. Lewis structures show shared electron pairs as _____ and unshared electrons as _____ _____ _____ . The octet rule helps us understand that the simplest compound between hydrogen and chlorine is _____ , between hydrogen and sulfur is _____ , and between hydrogen and phosphorus is _____ . The Lewis structure of carbon dioxide indicates the sharing of two pairs of electrons between a carbon and oxygen, which is known as a _____ _____ . Elemental nitrogen contains a _____ _____ . Three equivalent Lewis structures can be written for the carbonate ion; they are known as _____ structures. The actual structure is a _____ of the three structures. Ozone is an _____ of oxygen.

B-2.   Write the Lewis structure for (a) SF_2 (b) NO_2Cl where N is the central atom (c) $Mg(ClO_2)_2$.

B-3.   In the NCO^- ion, the N—C bond is about midway between a double and a triple bond. Write two resonance structures that are consistent with this conclusion. (The C is the central atom in this ion.)

Additional Examples: Problems 6-22, 6-27, 6-31, and 6-37.

Looking Ahead ▼

In the formation of an ionic bond, electrons are exchanged. In the formation of a covalent bond, electrons are shared. But does this mean that the electrons in the covalent bond are shared equally? As we will see, the situation is not so simple. Atoms of different elements rarely share electrons equally in a bond between them. This unequal sharing leads to important properties of compounds that we will discuss in the next sections.

6–7 ELECTRONEGATIVITY AND POLARITY OF BONDS

You are probably aware that *sharing* a box of popcorn at a movie rarely means *equal sharing*. The hungrier, faster popcorn eater usually gets the lion's share. Likewise, in a chemical bond between two different atoms, the pair of electrons is not shared equally, and one atom gets the lion's share of the electrons. *The ability of an atom of an element to attract electrons to itself in a covalent bond is known as the element's* **electronegativity.** The value assigned for the electronegativity of each element is shown in Figure 6-6. The most electronegative element is fluorine, which is assigned an electronegativity value of 4.0. Notice that nonmetals tend to have higher values of electronegativity than metals. The values shown in Figure 6-6 were first calculated by Linus Pauling (winner of two Nobel Prizes and a proponent of vitamin C). Although more refined values are now available, the actual numbers are not as important as how the electronegativity of one element compares with that of another.

Electrons carry a negative charge. When there is a complete exchange of an electron between atoms, as in the formation of an ionic bond, one atom acquires a full negative charge. In a covalent bond between two atoms of different electronegativity, the more electronegative atom attracts the electrons in the bond partially away from the other atom and thus acquires a partial negative charge (symbolized by δ^-). This leaves the less electronegative atom with a partial positive charge (symbolized by δ^+).

Are electron pairs always shared equally?

A covalent bond that has a partial separation of charge due to the unequal sharing of electrons is known as a **polar covalent bond** (or simply, **polar bond**). A polar bond has a negative end and a positive end and is said to contain a **dipole** (two poles). A polar bond is something

Decreases ↓   Increases →

1 H 2.1																	
3 Li 1.0	4 Be 1.5											5 B 2.0	6 C 2.5	7 N 3.0	8 O 3.5	9 F 4.0	
11 Na 0.9	12 Mg 1.2											13 Al 1.5	14 Si 1.8	15 P 2.1	15 S 2.5	17 Cl 3.0	
19 K 0.8	20 Ca 1.0	21 Sc 1.3	22 Ti 1.5	23 V 1.6	24 Cr 1.6	25 Mn 1.5	26 Fe 1.8	27 Co 1.8	28 Ni 1.8	29 Cu 1.8	30 Zn 1.6	31 Ga 1.6	32 Ge 1.8	33 As 2.0	34 Se 2.4	35 Br 2.8	
37 Rb 0.8	38 Sr 1.0	39 Y 1.2	40 Zr 1.4	41 Nb 1.6	42 Mo 1.8	43 Tc 1.5	44 Ru 2.2	45 Rh 2.2	46 Pd 2.2	47 Ag 2.4	48 Cd 1.7	49 In 1.7	50 Sn 1.8	51 Sb 1.9	52 Te 2.1	53 I 2.5	
55 Cs 0.7	56 Ba 0.9	57 La 1.1	72 Hf 1.3	73 Ta 1.5	74 W 1.7	75 Re 1.9	76 Os 2.2	77 Ir 2.2	78 Pt 2.2	79 Au 2.4	80 Hg 1.9	81 Tl 1.8	82 Pb 1.8	83 Bi 1.9	84 Po 2.0	85 At 2.2	
87 Fr 0.7	88 Ra 0.9	89 Ac 1.1															

Figure 6–6
Electronegativity

like Earth itself which contains a magnetic dipole with a north and south magnetic pole. (The poles in a bond dipole are electrostatic rather than magnetic.) The dipole of a bond is represented by an arrow pointing from the positive to the negative end (↦).

$$
\begin{array}{cc}
\delta^+ & \delta^- \\
X & \!\!-\!\! Y
\end{array}
\underset{\longmapsto}{}
$$

Representation of polarity

Representation of bond dipole

The polarity of bonds has a significant effect on their chemical properties. For example, the polarity of the H—O bond in water accounts for many of its familiar properties that we take for granted. We will discuss the chemistry of water in more detail in Chapter 10.

When electrons are shared between atoms of the same element, they are obviously shared equally. *If electrons are shared equally, the bond is known as a* **nonpolar bond.** On the other hand, the greater the difference in electronegativity between two atoms, the more polar the bond. In fact, if the difference is 1.9 or greater, it may indicate that one atom has gained complete control of the pair of electrons. In other words, the bond is ionic.

In summary, when two atoms compete for a pair of electrons in a bond, there are three possibilities for the pair of electrons.

1 Both atoms share the electrons equally, forming a nonpolar bond.

2 The electron pair is not shared since one atom acquires the electrons. This is the ionic bond, in which both atoms acquire a complete charge.

3 The two atoms share electrons unequally, forming a polar bond.

These three cases are illustrated in Figure 6-7. The bond in Cl_2 is nonpolar (case 1) since both atoms are identical. To determine the charge on each Cl, we will assign electrons to the two Cl atoms. In this case, each Cl is assigned the six electrons from its three lone pairs.

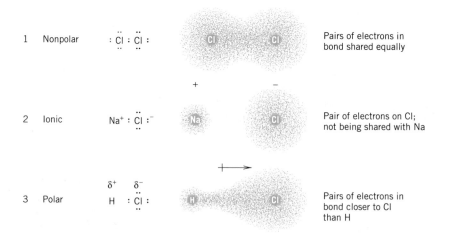

Figure 6–7
Nonpolar, Ionic, and Polar Bonds

Since the two atoms share the pair evenly, we can assign exactly one-half of the shared electrons to each Cl for a total of seven each [6 + (1/2 × 2) = 7]. Since Cl is in Group VIIA, seven valence electrons leave the Cl exactly neutral. Thus there is not a positive and a negative end.

The bond in NaCl illustrates case 2. As mentioned earlier in this chapter, there is a complete exchange of an electron between Na and Cl, leaving charged ions. Eight electrons on Cl produce an anion with a −1 charge. Notice that the difference in electronegativity between Na and Cl is rather large (2.1), indicating probable ionic nature.

Between these two extremes (equal sharing and no sharing) are the large number of polar bonds. The molecule HCl illustrates case 3. There is a difference in electronegativity between H and Cl (0.9), indicating a polar bond rather than an ionic bond. Since the Cl is more electronegative than H, it has a partial negative charge. In this case, the Cl is still assigned six electrons from its three lone pairs but more than one-half of the pair of electrons in the bond. Since it is assigned 7+ electrons, it acquires a partial negative charge. The hydrogen, with less than one-half of the electron pair, acquires an equal but opposite partial positive charge.

EXAMPLE 6-7

The Polarity of Bonds

Referring to Figure 6-6, rank the following bonds in order of increasing polarity. The positive end of the dipole is written first. On the basis of electronegativity differences, indicate if any of the bonds are predicted to be ionic.

● **Working It Out**

$$Ba—Br, C—N, Be—F, B—H, Be—Cl$$

Solution
Calculate the difference in electronegativity between the elements.

    Ba—Br $2.8 − 0.9 = 1.9$
    C—N $3.0 − 2.5 = 0.5$
    Be—F $4.0 − 1.5 = 2.5$
    B—H $2.1 − 2.0 = 0.1$
    Be—Cl $3.0 − 1.5 = 1.5$
    B—H < C—N < Be—Cl < Ba—Br < Be—F

The difference in electronegativity suggests that Ba—Br and Be—F are ionic bonds.

Since the covalent bonds in most molecules are polar, does that mean the molecule itself is polar? Surprisingly, the answer is "not necessarily." The polarity of a molecule depends on its geometry. In the next section, we will use the Lewis structure to tell us about the geometry of some simple molecules and then return to the subject of molecular polarity.

▼ **Looking Ahead**

6–8   GEOMETRY OF SIMPLE MOLECULES

Can a Lewis structure tell us the molecular geometry of a molecule?

From the complex chemistry of metabolism to the common properties of ordinary water, the arrangement of the atoms in a molecule plays an essential role. For example, if water were a linear molecule rather than bent, it would be a gas at room temperature rather than a liquid. Life, as we know it, could not exist under those conditions. When properly interpreted, however, the Lewis structure of water indicates its bent nature. The approximate geometry of the atoms around a central atom can be predicted by the **valence shell electron pair repulsion (VSEPR) theory.** *This theory tells us that electron pairs, either unshared pairs or electrons localized in a bond, repel each other to the maximum extent.* In other words, the negatively charged electron pairs get as far away from each other as possible. As a simple example, consider BeH_2 (a molecular compound that is an exception to the octet rule), which has two Be—H bonds. To be as far apart as possible, the two electron pairs in the bonds will lie on opposite sides of the beryllium atom at an angle of 180°. The geometry of the molecule is described as *linear.*

$$H \text{—} Be \text{—} H$$

180°

A similar situation exists with the CO_2 molecule. In this case, there are four electrons in each C—O bond, but the two groups of electrons in the two bonds repel each other, producing a linear molecule.

$$\ddot{O} \text{=} C \text{=} \ddot{O}$$

180°

Double and triple bonds are all considered as one group of electrons, so they are treated the same as a single bond in this theory. Thus we will consider bonded atoms as one electron group regardless of whether the bond is single, double, or triple.

Now consider the BF_3 molecule, which has three groups of electrons. In this case, the three groups can get as far away from each other as possible by assuming the geometry of an equilateral triangle with an F-B-F angle of 120°. The geometry of this molecule is described as *trigonal planar.* A similar angle is assumed by the SO_2 molecule, which also has three electron groups. The central atom is bonded to two other atoms, and it has one unshared pair of electrons. The O-S-O angle is approximately that of an equilateral triangle. *The* **molecular geometry** *of a molecule is the geometry described by the bonded atoms and does not include the unshared pairs of electrons.* Thus we describe the molecular geometry of the three atoms in SO_2 as *V-shaped.*

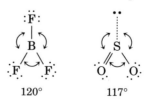

120°   117°

Now consider molecules having four groups of electrons, such as CH_4. There are two possibilities. In one, the four hydrogen atoms are located at the corners of a square, and in the other, they are located at the corners of a regular tetrahedron. Since the hydrogens are farther apart in the tetrahedron, this theory predicts such a structure with a H-C-H angle of about 109°. The geometry of CH_4 is thus described as *tetrahedral*. Ammonia (NH_3) also has four groups of electrons, but one group is an unshared pair of electrons. The H-N-H angle is found to be 107°, which is in good agreement with the angle predicted by this theory. The molecular geometry of the NH_3 molecule is described as *trigonal pyramid*. Finally, the H_2O molecule also has four groups but with two unshared pairs of electrons. The H-O-H angle is known to be 105°, which also agrees with this theory. The molecular geometry of H_2O is described as *V-shaped*. (In the V-shaped structure of SO_2, the angle of 117° is near the expected trigonal angle of 120°; in the V-shaped structure of H_2O, the angle of 105° is near the expected tetrahedral angle of 109°.)

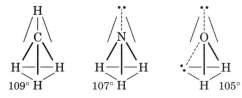

The molecular geometries for these molecules are summarized in Table 6-2.

TABLE 6–2 MOLECULAR GEOMETRIES

Number of electron groups on central atom (bonded atoms + unshared pairs)	Number of atoms bonded to central atom	Number of unshared pairs of electrons on central atom	Molecular geometry	Model
2	2	0	linear	
3	3	0	trigonal planar	
3	2	1	V-shaped (near 120°)	
4	4	0	tetrahedral	

TABLE 6–2 *Continued*

4	3	1	trigonal pyramid	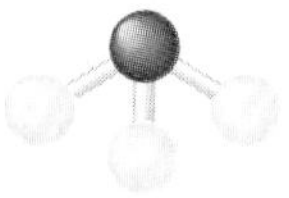
4	2	2	V-shaped (near 109°)	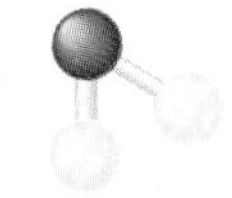

● **Working It Out**

EXAMPLE 6-8

The Geometry of Molecules

What is the molecular geometry of the following molecules?

(a) HCN   (b) H_2CO (c) HClO   (d) $SiCl_4$

In (a) and (b), the middle atom in the formula is the central atom. In (c), oxygen is the central atom.

Procedure
First write the correct Lewis structure. Then count the number of electron groups (bonded atoms [connected to the central atom] + the number of un-shared pairs of electrons).

Solution

	Lewis Structure	Number of electron groups	Molecular geometry
(a) HCN	H—C ≡ N:	2	linear (180°)
(b) H_2CO	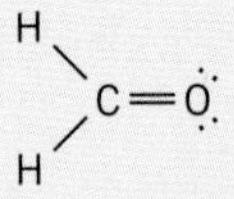 H⟍ ⟋C=O̤ / H	3	trigonal planar
(c) HClO	H—O̤ – C̤l:	4	V-shaped (109°)
(d) $SiCl_4$	:C̤l: | :C̤l—Si—C̤l: | :C̤l:	4	tetrahedral

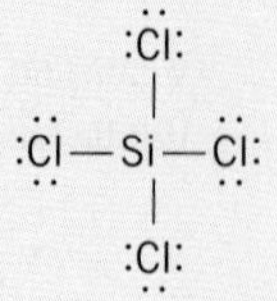

Looking Ahead ▼   *We are now ready to combine our ability to predict the geometry of some simple molecules with an understanding of the polarity of bonds.*

6–9   POLARITY OF MOLECULES

A polar bond creates a force in a molecule, much like a person pulling on a rope attached to a box. (See Figure 6-8 at bottom of page.) The force has both direction and magnitude. The magnitude of the force of the polar bond depends on the difference in electronegativity of the two atoms. The greater the difference, the greater the partial charges and the larger the magnitude of the dipole. But the force has direction as well. In Figure 6-8, if two people of exactly equal strength pull on the box in exactly opposite directions (an angle of 180°), their efforts cancel and there is no movement. Likewise, if the geometry of a molecule is such that equal dipoles cancel, there is no net force and the molecule has no net dipole. Thus, *despite having polar bonds, the molecule is nonpolar.* This is illustrated with the three perfect geometries discussed in the previous section. All have equal bond dipoles that cancel.

How does bond polarity relate to molecular polarity?

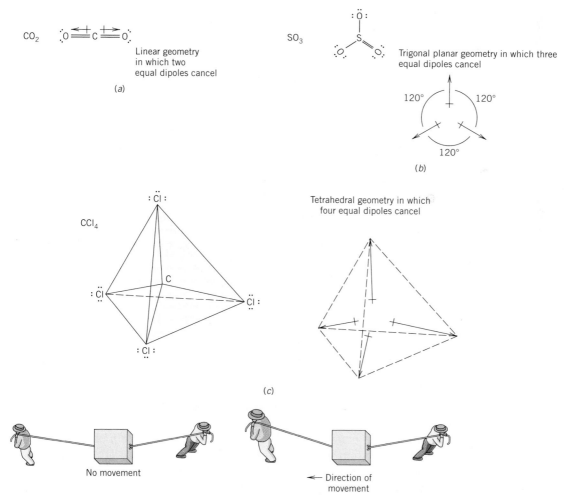

Figure 6-8
Forces at 180° If forces are equal but opposite (180°), they cancel. If the forces are not equal, there is a resultant force.

The bond dipoles do not cancel if all of the atoms are not identical or if the molecule does not have a perfect geometry. Consider, for example, the molecule COSe, which is linear like CO_2 but has bond dipoles that are not equal. There is a net or resultant dipole and the molecule is polar. *The combined effect of the bond dipoles is the* **molecular dipole.**

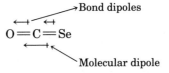

In this case, the situation would be analogous to two people of unequal strength pulling on the box in opposite directions as in Figure 6-8. Since the forces do not cancel, the box moves in the direction of the stronger person.

Now consider the case of water. The two O—H bond dipoles are equal, but they are not at an angle of 180°, so they do not cancel.

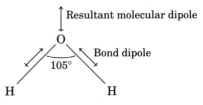

This is analogous to two people pulling on the box at an angle as shown in Figure 6-9. Both the direction and force of the movement depend on the angle between the people and their comparative strengths.

In later chapters, the importance of the polarity of molecules will become more evident. Physical state and other physical properties as well as chemical properties of compounds are determined to a large extent by the polarity of the molecules. We will refer back to this discussion.

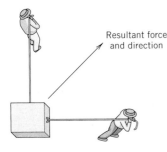

Figure 6–9
Forces at Less Than 180° Both direction and resultant force depend on the angle between the two people.

Looking Back ▲

Checking It Out ▶

Learning Check C

C-1. Fill in the blanks.
The most electronegative element is _____, and the second most electronegative is _____ . In a covalent bond between unlike atoms, the more electronegative atom has a partial _____ charge. The bond is then said to be a _____ covalent bond. If there is a large difference in electronegativity between the two atoms, the bond may be _____ . The geometry of molecules with a central atom is determined by the number of electron _____ . The three perfect geometries discussed are _____, _____, and _____ . If a molecule has one of these geometries with all the same atoms bonded to the central atom, the molecule is _____ . In other molecules, where the bonded atoms are less than 180° or the atoms bonded to the central atom are not all the same, the molecule is _____ .

C-2. If the following bonds are polar, indicate with a dipole arrow from the partially positive atom to the partially negative atom.

     (a) Al—Se   (b) As—S   (c) S—S   (d) F—Br

C-3. What is the molecular geometry of the following molecules or ions?

     (a) BF_4^- (b) SO_2 (c) SCl_2

C-4. Discuss the molecular polarity of C-3(b) and C-3(c).

Additional Examples: Problems 6-43, 6-45, 6-48, 6-51, and 6-56.

C H A P T E R   R E V I E W

The main focus of this chapter is the stability of the noble gas elements as solitary atoms. By understanding that this is related to their octet of outer electrons, we can see why and how the representative elements adjust their electron configurations to follow the **octet rule.** A metal and a nonmetal combine by an exchange of electrons, thereby achieving an octet for both. This happens because of the low ionization energy of the metal. The use of **Lewis dot symbols** of the elements aids us in focusing on the **valence electrons.** The ions formed by the electron exchange are arranged in a **lattice** held together by **ionic bonds.**

    Two nonmetals, on the other hand, follow the octet rule by electron sharing, forming a **covalent bond.** In certain cases, two atoms form **double bonds** or **triple bonds.** Covalent bonds also exist between the atoms comprising polyatomic ions. This is summarized as follows:

▲▼**Putting It Together**

Elements	Type of Bond	Comments
Metal—nonmetal	Ionic	Nonmetal can form −1, −2, or −3 ion. Metal can form +1, +2, or +3 ion.
Nonmetal—nonmetal	Covalent	Single, double, or triple bonds used to form octet.

By following certain rules, we can become proficient at writing **Lewis structures** of compounds. These rules are summarized for three compounds at the end of this review.

    Two or more equally correct Lewis structures can be written for a molecule such as ozone (O_3), an **allotrope** of oxygen. These are known as **resonance structures,** and the actual structure is a **resonance hybrid** of all of the Lewis structures.

    Atoms of two different nonmetals do not share electrons equally. **Electronegativity** is a measure of the periodic property of the atoms of an element to attract the electrons in the bond. When electrons are not shared equally, the bond is **polar** and contains a **dipole.** Atoms of the same element share electrons equally, so the bond is **nonpolar.**

Whether or not a molecule is polar depends not only on the polarity of the bonds in the molecule but on its **molecular geometry.** The molecular geometry of simple molecules can be determined from their Lewis structures and the **VSEPR theory.** If equal polar bonds are in a geometric arrangement where their dipoles cancel each other, the compound has no **molecular dipole** and it is nonpolar. If the bond dipoles are not the same in each direction or they do not cancel, the compound is polar.

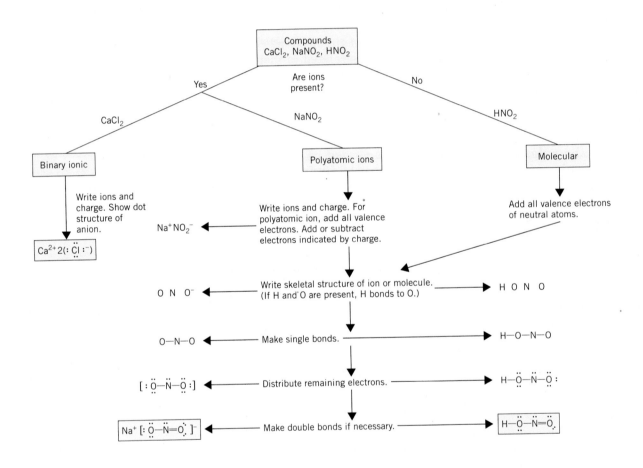

E X E R C I S E S

DOT SYMBOLS OF ELEMENTS

6–1 Write Lewis dot symbols for
(a) Ca   (c) Sn   (e) Ne
(b) Sb   (d) I   (f) Bi
(g) all group VIA elements

6–2 Identify the representative element group from the dot symbols.
(a) $\cdot\dot{M}\cdot$ (b) $\cdot\ddot{X}\cdot$ (c) $\dot{A}\cdot$

6–3 Why are only outer electrons represented in dot symbols?

6–4 Which of the following dot symbols are incorrect?
(a) $Pb\cdot$ (c) $:\ddot{He}:$ (e) $\cdot\dot{Te}\cdot$
(b) $\cdot\dot{Bi}\cdot$ (d) $Cs\cdot$ (f) $\cdot\dot{Tl}\cdot$

BINARY IONIC COMPOUNDS

6–5 From the periodic table, predict which pairs of elements can combine to form ionic bonds.
(a) H and Cl   (d) Al and F
(b) S and Sr   (e) B and Cl
(c) H and K   (f) Xe and F

6–6 From the periodic table, predict which pairs of elements can combine to form ionic bonds.
(a) Ba and I   (d) H and Se
(b) Cs and Se   (e) P and S
(c) C and O   (f) Cs and P

6–7 Which ions would not have a noble gas electron configuration?
(a) Sr^{2+} (e) In^{+}
(b) S^{-} (f) Pb^{2+}
(c) Cr^{2+} (g) Ba^{2+}
(d) Te^{2-} (h) Tl^{3+}

6–8 The ions Li^{+}, Be^{2+}, and H^{-} do not follow the octet rule. Why?

6–9 Write Lewis dot symbols for the following ions.
(a) K^{+} (d) P^{3-} (g) Sc^{3+}
(b) O^{-} (e) Ba^{+}
(c) I^{-} (f) Xe^{+}

6–10 What is the origin of the octet rule? How does the octet rule relate to s and p sublevels and to noble gases?

6–11 Which of the ions listed in Problem 6-9 do not follow the octet rule?

6–12 For the following atoms, write the charge that would give the element a noble gas configuration.
(a) Mg   (b) Ga   (c) Br   (d) S   (e) P

6–13 For the following atoms, write the charge that would give the element a noble gas configuration.
(a) Rb   (b) Ba   (c) Te   (d) N

6–14 Write six ions that have the same electron configuration as Ne.

6–15 Write five ions that have the same electron configuration as Kr.

6–16 The Tl^{3+} ion does not have the same electron configuration as Xe, even though it lost its three outermost electrons. Explain.

6–17 Complete the following table with formulas of the ionic compounds that form between the anion and cation shown.

Cation/Anion	Br^{-}	S^{2-}	N^{3-}
Cs^{+}	CsBr	_____	_____
Ba^{2+}	_____	_____	_____
In^{3+}	_____	_____	_____

6–18 Write the formula of the compound formed between the following nonmetals and the metal calcium.
(a) I   (b) O   (c) N   (d) Te   (e) F

6–19 Write the formula of the compound formed between the following metals and the nonmetal sulfur.
(a) Be   (b) Cs   (c) Ga   (d) Sr

6–20 Most transition metal ions cannot be predicted by reference to the octet rule. Determine the charge on the metal cation from the charge on the anion for the following compounds.
(a) Cr_2O_3 (c) MnS   (e) $NiBr_2$
(b) FeF_3 (d) CoO   (f) VN

6–21 Why isn't a formula unit of $BaCl_2$ referred to as a molecule?

LEWIS STRUCTURES OF COMPOUNDS

6–22 From their Lewis dot symbols, predict the formula of the simplest compound formed by the combination of the following pairs of elements.
(a) H and Se   (d) Cl and O
(b) H and Ge   (e) N and Cl
(c) Cl and F   (f) C and Br

6–23 From their Lewis dot symbols, predict the formula of the simplest compound formed by the combination of the following pairs of elements.
(a) H and I   (c) Si and Br
(b) Se and Br   (d) H and As

6–24 From a consideration of the octet rule, which of the following compounds are impossible?
(a) PH_3 (c) SCl_2 (e) H_3O
(b) Cl_3 (d) NBr_4

6–25 Determine the charge on the following polyatomic anions from the charge on the cation.
(a) K_2SO_4 (d) $Ca(H_2PO_3)_2$
(b) $Ca(IO_3)_2$ (e) BaC_2
(c) $Al_2(SeO_4)_3$

6–26 Determine the charge on the following polyatomic anions from the charge on the cation.
(a) $NaBrO_2$ (c) $AlAsO_4$
(b) $SrSeO_3$ (d) $Mg(H_2PO_4)_2$

6–27 Write Lewis structures for the following:
(a) C_2H_6 (d) SCl_2
(b) H_2O_2 (e) C_2H_6O
(c) NF_3 (There are two correct answers; both have the lone pairs on the oxygen.)

6–28 Write Lewis structures for the following:
(a) N_2H_4 (c) C_3H_8
(b) AsH_3 (d) CH_4O
                     (All unshared electrons are on the O.)

6–29 Write Lewis structures for the following:
(a) CO   (c) KCN
(b) SO_3 (d) H_2SO_3
                     (H's are on different O's.)

6–30 Use the "$6N + 2$" rule to confirm the structures in Problem 6-29.

6–31 Use the "$6N + 2$" rule to predict the number of multiple bonds (if any) in the following molecules or ions.
(a) N_2O (d) H_2S
(b) $Ca(NO_2)_2$ (e) CH_2Cl_2
(c) $AsCl_3$ (f) NH_4^+

6–32 Write Lewis structures for the compounds in Problem 6-31.

6–33 Write Lewis structures for the following.
(a) Cl_2O (c) C_2H_4 (e) BF_3
(b) SO_3^{2-} (d) H_2CO (f) NO^+

6–34 Use the "$6N + 2$" rule to predict the number of multiple bonds (if any) in the following molecules or ions.
(a) CO_2 (e) HOCN
(b) H_2NOH (f) $SiCl_4$
(c) $BaCl_2$ (g) C_2H_2
(d) NO_3^- (h) O_3

6–35 Write Lewis structures for the molecules or ions in Problem 6-34.

6–36 Write Lewis structures for the following:
(a) Cs_2Se (c) $LiClO_3$
(b) $CH_3CO_2^-$ (d) N_2O_3
(All H's are on one   (e) PBr_3
C, and both O's are
bonded to the other C.)

RESONANCE STRUCTURES

6–37 Write all equivalent resonance structures (if any) for the following:
(a) SO_3 (b) NO_2^- (c) SO_3^{2-}

6–38 Write all equivalent resonance structures (if any) for the following:
(a) NO_3^-
(b) N_2O_4 (skeletal geometry)

$$\begin{matrix} O & & & O \\ & N & N & \\ O & & & O \end{matrix}$$

6–39 Write the equivalent resonance structures for the $H_3BCO_2^{2-}$ anion. The skeletal structure for the ion is

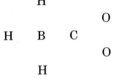

6–40 What is meant by a resonance hybrid? What is implied about the nature of the C-O bond from the resonance structures in Problem 6-39?

*6–41 A possible Lewis structure for CO_2 involves a triple bond between C and O. Write the two resonance structures involving the triple bond. What is implied about the nature of the C—O bond by these two structures? How does this relate to the common Lewis structure for CO_2 involving two double bonds?

ELECTRONEGATIVITY AND POLARITY

6–42 Rank the following elements in order of increasing electronegativity.
B, Ba, Be, C, Cl, Cs, F, O.

6–43 For bonds between the following elements, indicate the positive end of the dipole by a δ^+ and the negative end by a δ^-. Also indicate with a dipole arrow the direction of the dipole.
(a) N—H   (f) S—Se
(b) B—H   (g) C—B
(c) Li—H   (h) Cs—N
(d) F—O   (i) C—S
(e) O—Cl

6–44 Rank the bonds in Problem 6–43 in order of increasing polarity.

6–45 On the basis of difference in electronegativity, predict whether the following pairs of elements will form an ionic or a covalent bond.
(a) Sc—Br   (c) B—Br
(b) Cu—B   (d) Al—F

6–46 On the basis of difference in electronegativity, predict whether the following pairs of elements will form an ionic or a covalent bond.
(a) Al—Cl   (c) K—O
(b) Ca—I   (d) Mn—Te

6–47 Which of the following bonds is nonpolar?
(a) I—F   (d) N—Br
(b) I—I   (e) B—N
(c) C—H

MOLECULAR GEOMETRY

6–48 From the Lewis structure, determine the molecular geometry of the following molecules.
(a) SF_2 (b) CS_2 (c) CCl_2F_2 (C is the central atom.)
(d) NOCl (N is the central atom.)   (e) Cl_2O

6–49 From the Lewis structure, determine the molecular geometry of the following molecules or ions.
(a) BF_2Cl (B is the central atom.)
(b) ClO_3^-
(c) N_2O (N is the central atom.)
(d) $COCl_2$ (C is the central atom.)
(e) SO_3

MOLECULAR POLARITY

6–50 How can a molecule be nonpolar if it contains polar bonds?

6–51 Discuss the molecular polarity of the molecules in Problem 6-48.

6–52 Discuss the molecular polarity of the molecules in Problem 6-49.

6–53 Write the Lewis structure of SO_2Cl_2 where S is the central atom. Is the molecule polar?

6–54 Compare the expected molecular polarities of H_2O and H_2S. Assume that both molecules have the same angle.

6–55 Compare the expected molecular polarities of CH_4 and CH_2F_2.

6–56 Compare the expected molecular polarities of $CHCl_3$ and CHF_3.

6–57 CO_2 is a nonpolar molecule, but CO is polar. Explain.

6–58 SO_3 is a nonpolar molecule, but SO_2 is polar. Explain.

GENERAL PROBLEMS

6–59 Write the formulas of three binary, ionic compounds that contain one cation and one anion. One compound should contain Rb, one Sr, and one N.

6–60 There is a noble gas compound formed by xenon, XeO_3, that follows the octet rule. Write the Lewis structure of this compound. What is the geometry of the molecule? Is the XeO_3 molecule polar?

6–61 Which of the following compounds contains both ionic and covalent bonds?
(a) H_2SO_3 (d) H_2S
(b) K_2SO_4 (e) C_2H_6
(c) K_2S (f) $BaCl_2$

6–62 Write the Lewis structure of H_3BCO (all H's on the B). Can any resonance structures be written? What is the geometry around the B? What is the geometry around the C?

6–63 A molecule may exist that has the formula N_2O_2. The order of the bonds is O—N—N—O. Write a Lewis structure for the compound that has two N—O double bonds. Write any resonance structures with the same order of bonds that follow the octet rule.

6–64 Cyanogen has the formula C_2N_2. The order of the bonds is N—C—C—N. Write a Lewis structure for cyanogen involving a C—C single bond. What is the geometry around a C atom?

Draw dipole arrows for the bonds. Is the molecule polar? Write any other resonance structures with the same order of bonds that follow the octet rule.

*6–65 There are two compounds composed of only potassium and nitrogen. Write the formula of the compound expected between potassium and nitrogen. Another compound has the formula KN_3 and is named potassium azide. Write the Lewis structure for the azide ion plus any resonance structures. What is the geometry about the central N atom?

*6–66 A compound has the formula N_2F_2. Write a Lewis structure that contains a $N{=}N$ bond.

*6–67 A second compound of nitrogen and fluorine contains one nitrogen and the expected number of fluorines. A third has the formula N_2F_4. Write the Lewis structures of these two compounds. What is the geometry around the N in each compound? Are either or both of these compounds polar?

*6–68 Two compounds of oxygen are named oxygen difluoride and dioxygen difluoride. Write the Lewis structures of these two compounds. The latter compound contains a O—O bond. Oxygen is usually written and named second in binary compounds. Why not here? Are either or both of these compounds polar?

*6–69 A compound has a the formula $Na_2C_2O_4$. Write a Lewis structure for the compound that contains a C—C bond. Write any resonance structures present.

*6–70 Refer to Problem 5-87. Use the periodic table from the planet Zerk to answer the following:
(1) What are the simplest formulas of compounds formed between the following elements? (Example: Between 7 and 7 is 7_2.)
   (a) 1 and 7   (e) 7 and 13
   (b) 1 and 3   (f) 10 and 13
   (c) 1 and 5   (g) 6 and 7
   (d) 7 and 9   (h) 3 and 6

(2) Write Lewis structures for all of the above. Indicate which are ionic. (Remember that on Zerk there will be something different than an octet rule.)
(3) What would be the Zerkian equivalent of the $6N + 2$ rule?

SOLUTIONS TO LEARNING CHECKS

A–1 VIII, valence, eight, gaining, losing, sharing, octet, positive, +3, negative, charges

A–2 Be·   ·Se:

A–3 Be^{2+} Se^{2-}

A–4 BeSe

B–1 covalent, dashes, pairs of dots, HCl, H_2S, PH_3, double bond, triple bond, resonance, hybrid, allotrope

B–2 (a) SF_2

:S—F:
 |
:F:

(b) NO_2Cl

O
 \
  N—Cl:
 //
O

(c) $Mg(ClO_2)_2 = Mg^{2+} + 2ClO_2^{-}$

$Mg^{2+} 2[:O—Cl—O:]^{-}$

B–3 :N≡C—Ö:⁻   :N̈=C=Ö:⁻

C–1 fluorine, oxygen, negative, polar, ionic, groups, linear, trigonal planar, tetrahedral, nonpolar, polar

C–2 (a) AlSe (b) AsS (c) S—S (nonpolar) (d) FBr

C–3 (a) BF_4^- (b) SO_2 (c) SCl_2

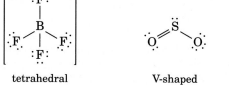

tetrahedral   V-shaped   V-shaped

C–4 (b) polar, polar bonds at an angle of about 120°
(c) polar, polar bonds at an angle of about 109°

MULTIPLE CHOICE

The following multiple-choice questions have one correct answer.

1. Which of the following elements is a halogen?
   (a) H   (b) O   (c) K   (d) Ne   (e) I

2. Which of the following elements is a metalloid?
   (a) Ga   (b) I   (c) Ge   (d) H

3. Which of the following elements is a transition metal?
   (a) tin   (c) aluminum
   (b) nickel   (d) krypton

4. Which of the following metals forms more than one charge?
   (a) Sr   (b) Tl   (c) Al   (d) Li   (e) N

5. Which of the following is the correct (Stock) name for PbO_2?
   (a) lead dioxide   (c) lead oxide
   (b) lead(IV) oxide   (d) lead(II) oxide

6. The formula of aluminum sulfide is
   (a) AlS   (b) Al_3S_2 (c) Al_2S_3 (d) AlS_2
   (e) Al_2S

7. Which of the following is the correct formula for an element?
   (a) N_4 (b) P_2 (c) C_2 (d) Br_2

8. Which of the following is the formula for aluminum hydroxide?
   (a) $Al(OH)_3$ (b) $Al(OH)_2$ (c) AlOH   (d) Al_2O_3

9. Barium superoxide has the formula $Ba(O_2)_2$. What is the charge on the superoxide ion?
   (a) +1   (b) +2   (c) −2   (d) −1   (e) −4

10. Which of the following are the subshells present in the third ($n = 3$) shell?
    (a) s, p (c) s, p, d
    (b) p, d, f (d) s, p, d, f

11. What is the electron capacity of the $4f$ subshell?
    (a) 6   (b) 7   (c) 14   (d) 8   (e) 10

12. Which of the following elements has the electron configuration $[Kr]5s^2 4d^2$?
    (a) Sn   (b) Zr   (c) Ti   (d) Pr   (e) Sr

13. What is the electron configuration of In (atomic number 49)?
    (a) $[Kr]4s^2 4p^1$ (d) $[Kr]4s^2 3d^{10} 4p^1$
    (b) $[Kr]5s^2 4d^1$ (e) $[Kr]5s^2 4d^6 5p^1$
    (c) $[Kr]5s^2 4d^{10} 5p^1$

14. Silicon has two unpaired electrons. This is a result of:
    (a) Hund's rule   (c) the Aufbau principle
    (b) the Pauli exclu-   (d) Bohr's model
        sion principle

15. How many orbitals are in a d subshell?
    (a) 5   (b) 3   (c) 10   (d) 14   (e) 2

16. Which of the following groups of elements have the general electron configuration $ns^2 np^5$?
    (a) halogens   (d) Group VA
    (b) alkaline earths   (e) transition elements
    (c) noble gases

17. Which of the following atoms has the smallest radius?
    (a) As   (b) Se   (c) Sb   (d) Te

18. Which of the following atoms has the highest first ionization energy?
    (a) N   (b) O   (c) P   (d) S

19. Which of the following ions requires the most energy to form?
    (a) K^+ (c) Al^{3+} (e) O^{2-}
    (b) Ca^{3+} (d) F^-

20. Which of the following ions violates the octet rule?
    (a) I^{2-} (c) Br^- (e) Te^{2-}
    (b) S^{2-} (d) N^{3-}

21. Which of the following pairs of elements would be expected to form an ionic bond?
    (a) Mg, Ca   (c) Mg, Br   (e) C, O
    (b) B, F   (d) S, O

22. Which of the following is the correct formula for the compound formed between beryllium and bromine?
    (a) BeBr   (c) BBr_2 (e) Be_2Br
    (b) Br_2Be (d) $BeBr_2$

23. Which of the following pairs of elements would be most likely to form a covalent bond?
    (a) Al, F   (b) K, Sn   (c) As, Br   (d) K, H

24. Which of the following is the correct formula for the simplest compound formed between As and Cl?
    (a) Cl_2As (c) $AsCl_3$ (e) $AsCl_4$
    (b) $AsCl_2$ (d) AsCl

25. Which of the following compounds contains both ionic and covalent bonds?
  (a) $BaCO_3$ (c) HNO_3 (e) CCl_4
  (b) H_2CO_3 (d) K_3N

26. Using the $6N + 2$ rule, determine the number of double bonds in the N_2O_5 molecule.
  (a) none   (b) one   (c) two   (d) three

27. Which of the following is the most electronegative element?
  (a) B   (b) Na   (c) O   (d) Cl   (e) N

28. Which of the following bonds is the most polar?
  (a) C—O   (c) Be—Cl   (e) N—N
  (b) B—Cl   (d) Be—F

PROBLEMS

1. Answer the following questions about the element aluminum (Al).
  (a) What is its group number and general classification?
  (b) What is its electron configuration?
  (c) Is it a solid, liquid, or gas?
  (d) Is it a metal or nonmetal?
  (e) What is its electrical charge in an ion?
  (f) What noble gas has the same electron configuration as its ion?
  (g) What is the simplest formula when it forms a binary compound with (1) sulfur (2) bromine (3) nitrogen?
  (h) What are the names of the three compounds in question (g)?

2. Answer the following questions about the element nitrogen (N).
  (a) What is its group number and general classification?
  (b) What is its electron configuration?
  (c) How many unpaired electrons are in an atom of nitrogen?
  (d) Is it a metal or nonmetal?
  (e) Does it exist as a solid, liquid, or gas under normal conditions?
  (f) What is the formula of the element?
  (g) What is the Lewis structure of the element?
  (h) What is its electrical charge in a monatomic ion?

  (i) What noble gas has the same electron configuration as its monatomic ion?
  (j) What is its polarity (negative or positive) when covalently bonded to (1) oxygen (2) boron?
  (k) What is the simplest formula when it forms a binary compound with (1) magnesium (2) lithium (3) fluorine?
  (l) What are the names of the three compounds in question (k)?
  (m) What is the Lewis structure of the compound with fluorine in question (k)?

3. For each of the following compounds write (1) the formula in ionic form if ions are present (2) the name of the compound (3) the Lewis structure.

Formula	Ionic Form	Name	Lewis Structure
NaClO	Na^+ClO^-	sodium hypochlorite	$Na^+[:\ddot{C}l—\ddot{O}:]^-$
(a) $MgSO_4$			
(b) HNO_3			
(c) $LiNO_3$			
(d) $Co_2(CO_3)_3$			
(e) Cl_2O			

4. For each of the following compounds write (1) the formula of the compound (2) the formula in ionic form if ions are present (3) the Lewis structure.

Name	Formula	Ionic Form	Lewis Structure
(a) dinitrogen trioxide			
(b) chromium(III) sulfite			
(c) iron(II) hydroxide			
(d) strontium oxalate			
(e) hydroiodic acid			

Modern farming techniques require the addition of nitrogen to the soil. The composition of ammonia makes it ideal for delivering a large amount of nitrogen. The composition of compounds is a topic of this chapter.

QUANTITATIVE CALCULATIONS AND THE MOLE

◀Setting The Stage

Nitrogen is an element essential to life. Primarily, it is a major component of proteins, which are part of all living systems. Although nitrogen is very abundant in elemental form in the atmosphere (80%), only limited amounts are transformed by nature into nitrogen compounds so that plant and animal life can flourish. Our complex and populous society requires that we supplement the soil with large quantities of fertilizers in the form of nitrogen compounds. Two compounds find widespread use as a source of nitrogen for the soil: ammonia (NH_3) and ammonium nitrate (NH_4NO_3). The advantage of ammonia is that 100 lb of ammonia supplies 82 lb of nitrogen to the soil. Its disadvantage is that it is a gas and must be injected into the soil. Ammonium nitrate, an ionic compound, is a solid and can be spread on the surface. The disadvantage of the latter compound, however, is that it supplies only 35 lb of nitrogen for each 100 lb of fertilizer applied. The decision about which to use depends on cost, availability, and soil conditions.

The situation just described is an example of how the composition of compounds can be important in such essential areas as agriculture. As we will see in this chapter, the mass of nitrogen in a given compound relates to its formula and to the relative masses of the elements in the compound. Thus we turn our attention to the quantitative relationships among the elements in compounds.

Our first order of business is to review some facts from Chapter 3. Recall that the smallness of the atom is quite difficult to comprehend. For example, the period at the end of this sentence contains between 10^{16} and 10^{17} atoms of carbon. Consider also the comparative masses of the isotopes of an element. The mass of a certain isotope is its mass compared to ^{12}C, which is defined as exactly 12 amu. Most elements occur in nature as mixtures of isotopes, so the atomic mass on the periodic table represents the average mass of all atoms of that element. In this chapter, we will treat the atoms of an element as if they were all identical, with the atomic mass representing the mass of an "average" atom. Also in Chapter 3, we described the mass of individual atoms in terms of the mass unit *amu* (atomic mass unit). This is valuable when comparing the masses of individual atoms, but it has no practical value in a laboratory situation. For example, the mass of one "average" carbon atom is 12.0 amu. This converts to grams as follows:

$$12.0 \text{ amu} = 2.00 \times 10^{-23} \text{ g}$$

Since even the best laboratory balance can detect no more than 10^{-5} g, it is obvious that we need many atoms at a time to register on our scales. Because we can't work with individual atoms, we must "scale up" the numbers of atoms so that the amounts are detectable with our laboratory instruments. In order to scale up our measurements, we need an appropriate counting unit for atoms.

There are many important questions that we will deal with in this chapter. How do we measure known amounts of atoms without counting? How does our counting unit relate to mass? How does the formula relate to the masses of the elements in the compound? And, finally, why is this information so important to a chemist? But first, we will describe how relative masses of elements relate to numbers.

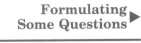

**Formulating
Some Questions** ▶

7–1   RELATIVE MASSES OF ELEMENTS

Like the chemist, the grocer uses a scale to count.

If we wanted to purchase a large number of oranges (e.g., 144) in a grocery store, it would be very tedious to actually count them one by one. It would be far easier and much quicker to weigh the oranges on a scale and know that we have the desired number. So, if one average orange has a mass of 225 g and we need 144 oranges, the following calculation tells us how much to weigh.

$$144 \text{ oranges} \times \frac{225 \text{ g}}{\text{orange}} \times \frac{1 \text{ kg}}{1000 \text{ g}} = 32.4 \text{ kg}$$

Now, suppose we want to weigh out the same number of apples. An average apple has a mass of 282 g, so we need the following mass of apples.

$$144 \text{ apples} \times \frac{282 \text{ g}}{\text{apple}} \times \frac{1 \text{ kg}}{1000 \text{ g}} = 40.6 \text{ kg}$$

Notice that the ratios of the masses of apples to oranges and oranges to apples are

$$\frac{282 \text{ g apple}}{225 \text{ g orange}} \quad \text{or} \quad \frac{225 \text{ g orange}}{282 \text{ g apple}}$$

This ratio of masses is very important. We can use it to measure any equal number of oranges and apples. *Whenever the masses of oranges and apples are in the ratio of their masses, the same number of each fruit is present.* If we wanted the same number of oranges as the number of apples present in 265 lb of apples, we do not have to count or even know what the number is. The following calculation, using the mass ratio (expressed in lb) as a conversion factor, tells us what mass we need.

How does one count by weighing?

$$265 \text{ lb apples} \times \frac{225 \text{ lb orange}}{282 \text{ lb apple}} = 211 \text{ lb oranges}$$

EXAMPLE 7–1

Counting Nails by Weighing

A builder has 250 lb of #8 common nails and wants to get an equal number of #10 common nails. Because the nails are quite uniform, an average 8 common weighs 0.197 oz and an average 10 common weighs 0.269 oz. What mass of 10 common contains the same number of nails as in the 250 lb of 8 common?

Solution
The mass ratio of nails can be used as a conversion factor between 10 common and 8 common.

$$\frac{0.269 \text{ oz (10 common)}}{0.197 \text{ oz (8 common)}} = \frac{0.269 \text{ lb (10 common)}}{0.197 \text{ lb (8 common)}}$$

Notice that the mass ratio is the same for any mass units.

$$250 \text{ lb (8 common)} \times \frac{0.269 \text{ lb (10 common)}}{0.197 \text{ lb (8 common)}} = \underline{\underline{341 \text{ lb (10 common)}}}$$

● **Working It Out**

Now, let us return to the world of atoms. In many of the calculations that follow in this chapter, we do not need to know the *actual* number of atoms involved in a certain weighed amount, but we do need to know the *relative* numbers of atoms of different elements present. However, if we know the relative masses of the individual atoms, this is no problem. For example, if one helium atom has a mass of 4.00 amu and one carbon atom has a mass of 12.0 amu, their masses are in the following ratio:

$$\frac{4.00 \text{ amu He}}{12.0 \text{ amu C}}$$

In fact, any time helium and carbon are present in a 4.00:12.0 mass ratio *regardless of the units of mass*, we can conclude that the same number of atoms of each element is present. (See Table 7-1.) We can generalize this statement to all of the elements. *When the masses of samples of any two elements are in the same ratio as that of their atomic masses, the samples have the same number of atoms.* Thus, if we wanted the same number of helium atoms as the number of atoms present in 45.0 g of carbon, we do not have to count or even know what the number is. The following calculation tells us what we want.

How do we measure equal numbers of atoms with a balance?

$$45.0 \text{ g C} \times \frac{4.00 \text{ g He}}{12.0 \text{ g C}} = 15.0 \text{ g He}$$

TABLE 7–1   MASS RELATION OF C AND HE

C	He	Number of Atoms of Each Element Present
12.0 amu	4.00 amu	1
24.0 amu	8.00 amu	2
360 amu	120 amu	30
12.0 g	4.00 g	6.02×10^{23}
24.0 g	8.00 g	1.20×10^{24}
12.0 lb	4.00 lb	2.73×10^{26}
24.0 ton	8.00 ton	1.09×10^{30}

The mass ratio of the atomic masses of any two elements can also be used to measure equivalent numbers of atoms. This is illustrated in Example 7-2.

Working It Out ●

EXAMPLE 7–2

The Relative Masses of Elements in a Compound

The formula of the compound magnesium sulfide (MgS) indicates that there is one atom of Mg for every atom of S. What mass of sulfur is combined with 46.0 lb of magnesium?

Procedure
From the atomic masses in the periodic table, note that 24.3 g of magnesium and 32.1 g of sulfur each contain the same number of atoms. Likewise, 24.3 lb of magnesium and 32.1 lb of sulfur have an equal number of atoms. This statement can be represented by two conversion factors, which we can use to change a mass of one element to an equivalent mass of the other.

$$\textbf{(1)}\ \frac{24.3\ \text{lb Mg}}{32.1\ \text{lb S}} \qquad \textbf{(2)}\ \frac{32.1\ \text{lb S}}{24.3\ \text{lb Mg}}$$

Use factor **(2)** to convert the mass of Mg to an equivalent mass of S.

Solution

$$46.0\ \text{lb Mg} \times \frac{32.1\ \text{lb S}}{24.3\ \text{lb Mg}} = \underline{60.7\ \text{lb S}}$$

Thus 60.7 lb of sulfur has the same number of atoms as 46.0 lb of magnesium.

Looking Ahead ▼

The atomic mass of an element represents the average mass of one atom of that element expressed in amu. When we express the atomic masses of the elements in grams, we can make the following conclusion: **The atomic mass expressed in grams represents the same number of atoms of each element.** *In fact, this understanding was put to use long before scientists even knew what that number was. But what is this number and just how big is it? That's next.*

7–2 THE MOLE AND THE MOLAR MASS OF ELEMENTS

We are now ready for a counting unit that represents the huge number of atoms in gram quantities of elements. *The number of atoms represented by the atomic mass of an element expressed in grams is a unit known as a* **mole.** (The SI symbol is **mol.**) This number is expressed in the following equality.

$$1.00 \text{ mol} = 6.02 \times 10^{23} \text{ objects or particles}$$

This number, 6.02×10^{23}, is referred to as **Avogadro's number** (named in honor of Amedeo Avogadro, 1776–1856, a pioneer investigator of the quantitative aspects of chemistry). Avogadro's number was determined experimentally to relate mass with number. The mass of one mole of a substance, expressed in grams, contains the same number of basic particles as there are in exactly 12 grams of ^{12}C. That number of atoms in 12 grams of ^{12}C is Avogadro's number. It is not an exact, defined number such as 12 in one dozen or 144 in one gross, but it is known to many more significant figures than the three (e.g., 6.02) that are shown and used in this text.

Many common counting units represent a number consistent with their use. For example, a baker sells a dozen doughnuts at a time because 12 is a practical number for that purpose. On the other hand, we buy a ream of typing or computer paper, which is 500 sheets. A ream of doughnuts or a dozen sheets of typing paper are not practical amounts to purchase for most purposes. Counting units of a dozen or a ream are of little use to a chemist because they don't include enough individual objects. For example, grouping 10^{20} atoms into about 10^{19} dozen atoms does us little good. The counting unit used by chemists includes an extremely large number of individual units in order to be practical. (See Figure 7-1.)

What is the chemist's version of a dozen?

Although Avogadro's number is valuable to a chemist, its size defies the ability of the human mind to comprehend. For example, if an atom were the size of a marble and one mole of marbles were spread over the

a dozen doughnuts   a ream of paper   not even one mole of sand particles

Figure 7–1
Counting Units One dozen, one ream, and one mole all have applications dependent on the amount needed.

Can we actually count one mole?

Figure 7–2
Moles of Elements One mole of iron (the paper clips), copper, liquid mercury and sulfur are shown. Each sample contains Avogadro's number of atoms but has a different mass.

surface of the earth, our planet would be covered by a 50-mile-thick layer. Or, if the marbles were laid end to end and extended into outer space, they would reach past the farthest planets almost to the center of the galaxy. It takes light moving at 186,000 miles per second over 30,000 years to travel from Earth to the center of the galaxy. A new supercomputer can count all of the people in the United States in one-quarter of a second, but it would take almost two million years for it to count one mole of people at the same rate. A glass of water, which is about 10 moles of water, contains more water molecules than there are sand grains in the Sahara desert. That's difficult to imagine.

Note that one mole of a certain element implies two things:

1 *The atomic mass expressed in grams, which is known as the **molar mass** of that element.* This is *different* for each element. (In this text, the molar mass is usually expressed to three significant figures. Thus the mass of one mole of oxygen atoms is 16.0 g, the mass of one mole of helium atoms is 4.00 g, and the mass of one mole of uranium atoms is 238 g.)

2 *Avogadro's number of atoms, which is the *same* for all elements.* (See Figure 7-2.)

At this point, we have established relationships between the mole, the molar mass of an element, and Avogadro's number. Now we need to practice conversions among these three quantities. At first, working with such a huge value as Avogadro's number seems awkward. However, working with a mole of sodium is not very different from working with a dozen oranges. Perhaps we can illustrate this similarity by first working through some conversions using a dozen oranges. After this, we will work through some directly analogous problems using the less familiar mole of sodium. In the problems that follow, note that the factor-label method of problem solving, introduced in Chapter 1, serves us well, especially with the less familiar units.

In these problems, we will assume that the oranges are all identical, with the same size, shape, and mass. The "dozen mass" (the mass of 12) of these particular oranges is 3.60 lb. The two relationships just given can be written as equalities or set up as ratios that can eventually be used as conversion factors.

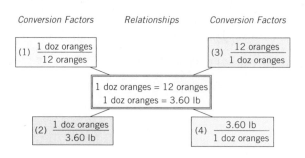

Conversion Factors	Relationships	Conversion Factors
(1) $\dfrac{1 \text{ doz oranges}}{12 \text{ oranges}}$		(3) $\dfrac{12 \text{ oranges}}{1 \text{ doz oranges}}$
	1 doz oranges = 12 oranges 1 doz oranges = 3.60 lb	
(2) $\dfrac{1 \text{ doz oranges}}{3.60 \text{ lb}}$		(4) $\dfrac{3.60 \text{ lb}}{1 \text{ doz oranges}}$

EXAMPLE 7–3A

Dozens to Mass

What is the mass of exactly 3 doz oranges?

●Working It Out

Procedure

We need a conversion factor that converts units of "dozen" to units of mass, or lb. Thus the conversion factor would have dozen in the denominator and pounds in the numerator. This is factor 4.

$$\frac{\text{lb}}{\text{doz}}$$

Solution

$$3.00 \ \cancel{\text{doz oranges}} \times \frac{3.60 \ \text{lb}}{\cancel{\text{doz oranges}}} = \underline{\underline{10.8 \ \text{lb}}}$$

EXAMPLE 7–4A

Mass to Dozens

How many dozens of oranges are there in 18.0 lb?

Procedure

This is the reverse of the previous problem, so we need a conversion factor that has pounds in the denominator and dozen in the numerator. This is factor 2.

$$\frac{\text{doz}}{\text{lb}}$$

The number of oranges, the mass, and dozens are all related.

Solution

$$18.0 \ \cancel{\text{lb}} \times \frac{1 \ \text{doz oranges}}{3.60 \ \cancel{\text{lb}}} = \underline{\underline{5.00 \ \text{doz oranges}}}$$

EXAMPLE 7–5A

Number to Mass

What is the mass of 142 oranges?

Procedure

In this conversion, a direct relationship between mass and number is not given. This requires a two-step conversion. In the first step, we convert number of oranges to dozens (factor 1); in the second step, we convert dozens to mass (factor 4).

$$\frac{\text{doz}}{12} \qquad \frac{\text{lb}}{\text{doz}}$$

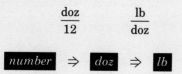

Solution

$$142 \text{ oranges} \times \frac{1 \text{ doz oranges}}{12 \text{ oranges}} \times \frac{3.60 \text{ lb}}{\text{doz oranges}} = 42.6 \text{ lb}$$

EXAMPLE 7–6A

Mass to Number

How many individual oranges are there in 73.5 lb?

Procedure
This is the reverse of the previous problem and also requires a two-step conversion. We will convert pounds to dozen using factor 2 and then dozen to number using factor 3.

$$\frac{\text{doz}}{\text{lb}} \qquad \frac{12}{\text{doz}}$$

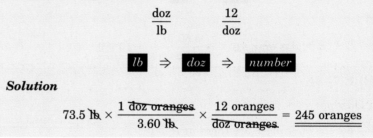

$$lb \Rightarrow doz \Rightarrow number$$

Solution

$$73.5 \text{ lb} \times \frac{1 \text{ doz oranges}}{3.60 \text{ lb}} \times \frac{12 \text{ oranges}}{\text{doz oranges}} = 245 \text{ oranges}$$

We are now ready to work some almost identical problems using moles, grams, and 6.02×10^{23} rather than dozens, pounds, and 12. If you're still a little uneasy with scientific notation, see Appendix C for a quick review. Otherwise, note that Example 7-3B is the analog of Example 7-3A, and so forth.

In these problems, we will assume that the atoms are all identical with the same size, shape, and mass. The "molar mass" (the mass of 6.02×10^{23}) of these atoms of sodium is 23.0 g. The two relationships just given can be either written as equalities or set up as ratios that can eventually be used as conversion factors.

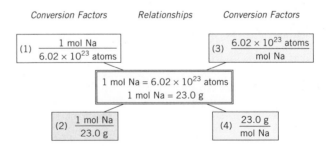

EXAMPLE 7–3B

Moles to Mass

Working It Out ●

What is the mass of exactly 3 mol of Na?

Procedure
We need a conversion factor that converts units of "moles" to units of mass, or g. Thus the conversion factor would have mole in the denominator and grams in the numerator. This is factor 4.

$$\frac{g}{mol}$$

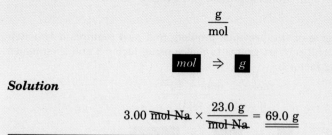

Solution

$$3.00 \ \cancel{mol \ Na} \times \frac{23.0 \ g}{\cancel{mol \ Na}} = \underline{\underline{69.0 \ g}}$$

EXAMPLE 7–4B

Mass to Moles

How many moles are there in 34.5 g of Na?

Procedure
This is the reverse of the previous problem, so we need a conversion factor that has grams in the denominator and mole in the numerator. This is factor 2.

$$\frac{mol}{g}$$

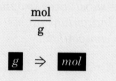

Solution

$$34.5 \ \cancel{g} \times \frac{1 \ mol \ Na}{23.0 \ \cancel{g}} = \underline{\underline{1.50 \ mol \ Na}}$$

EXAMPLE 7–5B

Number to Mass

What is the mass of 1.20×10^{24} atoms of Na?

Procedure
In this conversion, a direct relationship between mass and number is not given. This requires a two-step conversion. In the first step, we convert number of atoms to moles (factor 1); in the second step, we convert moles to mass (factor 4).

$$\frac{mol}{6.02 \times 10^{23}} \qquad \frac{g}{mol}$$

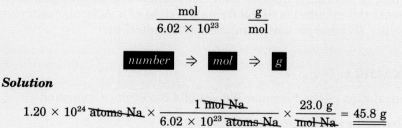

Solution

$$1.20 \times 10^{24} \ \cancel{atoms \ Na} \times \frac{1 \ \cancel{mol \ Na}}{6.02 \times 10^{23} \ \cancel{atoms \ Na}} \times \frac{23.0 \ g}{\cancel{mol \ Na}} = \underline{\underline{45.8 \ g}}$$

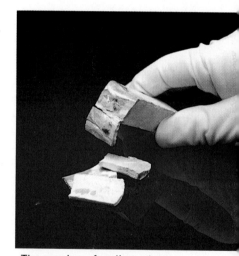

The number of sodium atoms, the mass, and moles are all related.

EXAMPLE 7–6B

Mass to Number

How many individual atoms are there in 11.5 g of Na?

Procedure
This is the reverse of the previous problem and also requires a two-step conversion. We will convert grams to moles using factor 2 and then moles to number using factor 3.

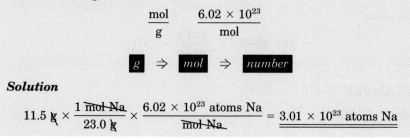

Solution

$$11.5 \not{g} \times \frac{1 \not{mol\ Na}}{23.0 \not{g}} \times \frac{6.02 \times 10^{23} \text{ atoms Na}}{\not{mol\ Na}} = \underline{\underline{3.01 \times 10^{23} \text{ atoms Na}}}$$

Looking Ahead ▼

When we express the atomic masses of the elements in grams rather than amu, we have successfully scaled up our measurements to useful laboratory quantities. Now we are using moles of atoms rather than individual atoms. Most of the substances around us, however, are composed of molecules rather than individual atoms. We are now ready to extend the concept of the mole to compounds.

7–3   THE MOLAR MASS OF COMPOUNDS

Just as the mass of an automobile is the sum of the masses of all its component parts, the mass of a molecule is the sum of the masses of its component atoms. *The **formula weight** of a compound is determined from the number of atoms and the atomic mass of each element indicated by a chemical formula.* Recall that chemical formulas represent two types of compounds: molecular and ionic. The formula of a molecular compound represents a discrete molecular unit. *The formula weight of a molecular compound is also referred to as the **molecular weight.*** The formula of an ionic compound represents the ratio of ions present, which is known as a *formula unit*. The term molecular weight applied to ionic compounds is inappropriate, but it is sometimes used for simplicity. In this text, we will use the general term formula weight for either a molecular unit or a formula unit for ionic compounds. The following examples illustrate the calculation of the formula weight of a molecular compound (Example 7-7) and an ionic compound (Example 7-8).

EXAMPLE 7–7

Calculation of the Formula Weight of a Molecular Compound

Working It Out ●

What is the formula weight of CO_2?

Solution

Atom	Number of Atoms in Molecule	Atomic Mass	Total Mass of Atom in Molecule
C	1	× 12.0 amu	= 12.0 amu
O	2	× 16.0 amu	= 32.0 amu
			44.0 amu

The formula weight of CO_2 is

$$44.0 \text{ amu}$$

EXAMPLE 7–8

Calculation of the Formula Weight of an Ionic Compound

What is the formula weight of $Fe_2(SO_4)_3$?

Solution

Atom	Number of Atoms in Formula Unit	Atomic Mass	Total Mass of Atom in Formula Unit
Fe	2	× 55.8 amu	= 112 amu
S	3	× 32.1 amu	= 96.3 amu
O	12	× 16.0 amu	= 192 amu
			400 amu

The formula weight of $Fe_2(SO_4)_3$ is

$$400 \text{ amu}$$

The formula weights that we have calculated represent one molecule or one formula unit. Once again, we need to scale up our numbers so that we have a workable amount that can be measured on a laboratory balance. Thus we extend the definition of the mole to include compounds. *The mass of one mole (6.02×10^{23} molecules or formula units) is referred to as the* **molar mass of the compound** *and is the formula weight expressed in grams.* As was the case with atoms of elements, the molar masses of various compounds differ, but the number of molecules or formula units remains the same. (See Figure 7-3.)

Figure 7–3
Moles of Compounds One mole of white sodium chloride (NaCl), blue copper sulfate hydrate ($CuSO_4 \cdot 5H_2O$), yellow sodium chromate (Na_2CrO_4), and water are shown. Each sample contains Avogadro's number of molecules or formula units.

EXAMPLE 7–9

Converting Moles to Mass And Number of Formula Units

(a) What is the mass of 0.345 mol of $Al_2(CO_3)_3$?
(b) How many individual ionic formula units does this amount represent?

Procedure
The problem is worked much like the examples in which we were dealing with moles of atoms rather than compounds. In this case, however, we

●**Working It Out**

need to find the molar mass of the compound, which is the formula weight expressed in grams. The units of molar mass are "g/mol." The formula weight equals

$$2\,Al + 3\,C + 9\,O = [(2 \times 27.0) + (3 \times 12.0) + (9 \times 16.0)] = 234\ \text{amu}$$

Thus the molar mass is 234 g/mol.

The conversions are

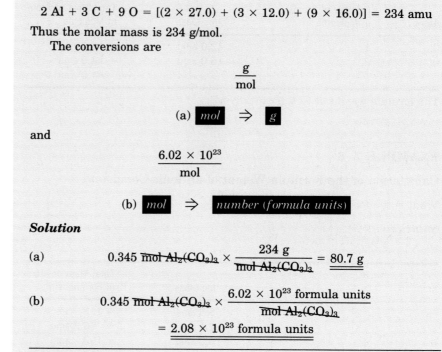

Solution

(a)

$$0.345\ \overline{\text{mol Al}_2(\text{CO}_3)_3} \times \frac{234\ \text{g}}{\overline{\text{mol Al}_2(\text{CO}_3)_3}} = \underline{\underline{80.7\ \text{g}}}$$

(b)

$$0.345\ \overline{\text{mol Al}_2(\text{CO}_3)_3} \times \frac{6.02 \times 10^{23}\ \text{formula units}}{\overline{\text{mol Al}_2(\text{CO}_3)_3}}$$

$$= \underline{\underline{2.08 \times 10^{23}\ \text{formula units}}}$$

EXAMPLE 7–10

Converting Mass to Number of Molecules

How many individual molecules are present in 25.0 g of N_2O_5?

Procedure

The formula weight of $N_2O_5 = [(2 \times 14.0) + (5 \times 16.0)] = 108$ amu. The molar mass is 108 g/mol. This is a two-step conversion as follows:

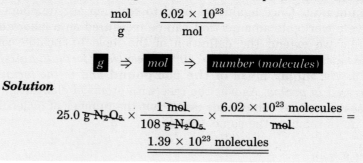

Solution

$$25.0\ \overline{\text{g N}_2\text{O}_5} \times \frac{1\ \text{mol}}{108\ \overline{\text{g N}_2\text{O}_5}} \times \frac{6.02 \times 10^{23}\ \text{molecules}}{\overline{\text{mol}}} =$$

$$\underline{\underline{1.39 \times 10^{23}\ \text{molecules}}}$$

What is the difference between a mole and a molecule?

The term "a mole of molecules" actually does sound confusing. It is somewhat unfortunate that the counting unit (mole) and the fundamental particle that is being counted (molecule) read so similarly. Try to remember that the *molecule* is the tiny, fundamental particle of a compound that is being counted. A *mole* is a whole bunch (6.02×10^{23}) of molecules. Another point of caution concerns nonmetal elements such as chlorine (Cl_2) that exist in nature as molecules rather than solitary atoms. For example, if we report the presence of "one mole of chlorine," it is not always obvious what is meant. Do we have one mole of chlorine atoms (6.02×10^{23} atoms, 35.5 g) or one mole of chlorine molecules (6.02×10^{23} molecules , 71.0 g)? Note that there are *two Cl atoms per molecule* and *two moles of Cl atoms per mole of Cl_2 molecules*. In future discussions, one mole of chlorine or any of the diatomic elements will refer to the molecules. Thus it is important to be specific in cases where we are using one mole of atoms. This situation is analogous to the difference between a dozen sneakers and dozen pairs of sneakers. Both the mass and the actual number of the dozen pair of sneakers are double the dozen sneakers.

Chemistry deals with such unbelievably small atoms that it requires an unbelievably large number of them to even register on a sensitive laboratory balance. However, the mass ratios of individual atoms can be applied to the measurement of large numbers of atoms. Our most useful mass ratios are obtained by expressing the atomic masses of the elements in grams. This is defined as the molar mass of the elements and is the mass of Avogadro's number of atoms. Then, by extending this concept to compounds, we have the molar mass of compounds, which is the mass of Avogadro's number of molecules or formula units.

▲ Looking Back

◀ Checking It Out

Learning Check A

A–1.   Fill in the blanks.
The atomic mass of an element expressed in amu represents the mass of _____ atom of the element. The atomic mass expressed in grams represents the mass of _____ atoms of the element and is known as the _____ _____ of the element. This quantity and mass refer to a unit known as the _____ . The mass of one molecule or formula unit is known as the _____ weight. This quantity expressed in grams is known as the _____ _____ of the compound and represents the mass of _____ individual molecules or formula units.

A–2.   A compound has the formula NO. What mass of oxygen is present in the compound for each 25.0 g of nitrogen?

A–3.   How many moles of vanadium atoms are represented by 4.82×10^{24} atoms? What is the mass of this number of atoms of vanadium?

A–4.   How many moles of atoms and how many individual atoms are present in 215 g of chlorine?

A–5.   How many moles of Cl_2O and how many individual molecules are present in 438 g of Cl_2O?

Additional Examples: Problems 7-1, 7-3, 7-5, 7-14, and 7-29.

When we extended the concept of the mole to compounds, we expanded the scope of the information to a considerable extent. Each compound

▼ Looking Ahead

contains two or more component parts, each present in a given mole and mass ratio. We are now ready to look deeper into the component parts of a compound.

7–4   THE COMPOSITION OF COMPOUNDS

Sometimes the math we use doesn't appear to add correctly. For example, *one plus two equals one.* But one frame plus two wheels equals one bicycle. *Likewise, one carbon atom plus two oxygen atoms equals one molecule of carbon dioxide.* Consider the compound sulfuric acid (H_2SO_4). In Table 7-2, we have illustrated the relation of one mole of compound to all its component parts. All of the relationships given in Table 7-2 could be used to construct conversion factors. To illustrate, let us again consider the 0.345-mol quantity of $Al_2(CO_3)_3$ from Example 7-9. In the following example, we will consider the component parts of the compound rather than the compound as a whole.

TABLE 7–2   THE COMPOSITION OF ONE MOLE OF H₂SO₄

Total		Components	
1.00 mol of molecules		2 mol of H atoms	2.02 g / 1.20×10^{24} H atoms
6.02×10^{23} molecules	H_2SO_4	1 mol of S atoms	32.1 g / 6.02×10^{23} S atoms
98.1 g		4 mol of O atoms	64.0 g / 2.41×10^{24} O atoms

EXAMPLE 7–11

Conversion of Moles of a Compound to its Component Elements

How many moles of each atom are present in 0.345 mol of $Al_2(CO_3)_3$? What is the total number of moles of atoms present?

Procedure
In this problem, note that there are 2 Al atoms, 3 C atoms, and 9 O atoms in each formula unit of compound. Therefore, in 1 mol of compound, there are 2 mol of Al, 3 mol of C, and 9 mol of O atoms. This can be expressed with conversion factors as

$$\frac{2 \text{ mol Al}}{\text{mol Al}_2(\text{CO}_3)_3} \quad \frac{3 \text{ mol C}}{\text{mol Al}_2(\text{CO}_3)_3} \quad \frac{9 \text{ mol O}}{\text{mol Al}_2(\text{CO}_3)_3}$$

Solution

$$\text{Al} \quad 0.345 \; \text{mol Al}_2(\text{CO}_3)_3 \times \frac{2 \; \text{mol Al}}{\text{mol Al}_2(\text{CO}_3)_3} = 0.690 \; \text{mol Al}$$

$$\text{C} \quad 0.345 \; \text{mol Al}_2(\text{CO}_3)_3 \times \frac{3 \; \text{mol C}}{\text{mol Al}_2(\text{CO}_3)_3} = 1.04 \; \text{mol C}$$

$$\text{O} \quad 0.345 \; \text{mol Al}_2(\text{CO}_3)_3 \times \frac{9 \; \text{mol O}}{\text{mol Al}_2(\text{CO}_3)_3} = 3.11 \; \text{mol O}$$

Total moles of atoms present:

$$0.690 \; \text{mol Al} + 1.04 \; \text{mol C} + 3.11 \; \text{mol O} = 4.84 \; \text{mol atoms}$$

We can now extend this concept further. Obviously, carbon dioxide is composed of one-third carbon atoms and two-thirds oxygen atoms. This, of course, is not the mass ratios of the two elements since they have different atomic masses. The mass ratios of the elements can be expressed as percent composition. **Percent composition** *expresses the mass of each element per 100 mass units of compound.* For example, if there is an 82-g quantity of nitrogen present in each 100 g of ammonia (NH_3), the percent composition is expressed as 82% nitrogen.

How does the formula relate to the mass composition?

The mass of one mole of carbon dioxide is 44.0 g, and it is composed of one mole of carbon atoms (12.0 g) and two moles of oxygen atoms (32.0 g). The percent composition is calculated by dividing the total mass of each component element by the total mass (molar mass) of the compound times 100%.

$$\frac{\text{total mass of component element}}{\text{total mass (molar mass)}} \times 100\% = \text{percent composition}$$

In CO_2, the percent composition of C is

$$\frac{12.0 \; \text{g C}}{44.0 \; \text{g CO}_2} \times 100\% = 27.3\% \; \text{C}$$

and the percent composition of O is

$$\frac{32.0 \; \text{g O}}{44.0 \; \text{g CO}_2} \times 100\% = 72.7\%$$

(It is easier to find the percent composition of oxygen by subtracting $100\% - 27.3\% = 72.7\%$. In a binary compound, the two percent compositions add to 100%. However, it is sometimes wise to calculate all percentages individually and then check your math by making sure they all add to 100%.)

EXAMPLE 7–12

The Percent Composition of Limestone ($CaCO_3$)

What is the percent composition of all of the elements in limestone, $CaCO_3$?

● **Working It Out**

Marble is also composed of calcium carbonate.

Procedure

Find the molar mass and convert the *total* mass of each element to percent of molar mass.

Solution

For $CaCO_3$e

$$Ca \quad 1 \; \cancel{mol\;Ca} \times \frac{40.1 \; g}{\cancel{mol\;Ca}} = 40.1 \; g$$

$$C \quad 1 \; \cancel{mol\;C} \times \frac{12.0 \; g}{\cancel{mol\;C}} = 12.0 \; g$$

$$O \quad 3 \; \cancel{mol\;O} \times \frac{16.0 \; g}{\cancel{mol\;O}} = 48.0 \; g$$

$$\text{molar mass} = 100.1 \; g/mol \; CaCO_3$$

$$\% \; Ca \quad \frac{40.1 \; g \; Ca}{100.1 \; g \; CaCO_3} \times 100\% = \underline{\underline{40.1\%}}$$

$$\% \; C \quad \frac{12.0 \; g \; C}{100.1 \; g \; CaCO_3} \times 100\% = \underline{\underline{12.0\%}}$$

$$\% \; O \quad \frac{48.0 \; g \; O}{100.1 \; g \; CaCO_3} \times 100\% = \underline{\underline{48.0\%}}$$

EXAMPLE 7–13

The Percent Composition of Borazine

What is the percent composition of all the elements in borazine, $B_3N_3H_6$?

Procedure

Find the molar mass and convert the *total* mass of each element to percent of molar mass.

Solution

For $B_3N_3H_6$

$$B \quad 3 \; \cancel{mol\;B} \times \frac{10.8 \; g}{\cancel{mol\;B}} = 32.4 \; g$$

$$N \quad 3 \; \cancel{mol\;N} \times \frac{14.0 \; g}{\cancel{mol\;N}} = 42.0 \; g$$

$$H \quad 6 \; \cancel{mol\;H} \times \frac{1.01 \; g}{\cancel{mol\;H}} = \underline{6.06 \; g}$$

$$\text{molar mass} = 80.5 \; g/mol \; B_3N_3H_6$$

$$\% \text{ B} \quad \frac{32.4 \text{ g B}}{80.5 \text{ g B}_3\text{N}_3\text{H}_6} \times 100\% = \underline{\underline{40.2\%}}$$

$$\% \text{ N} \quad \frac{42.0 \text{ g N}}{80.5 \text{ g B}_3\text{N}_3\text{H}_6} \times 100\% = \underline{\underline{52.2\%}}$$

$$\% \text{ H} \quad \frac{6.06 \text{ g H}}{80.5 \text{ g B}_3\text{N}_3\text{H}_6} \times 100\% = \underline{\underline{7.53\%}}$$

Given a specific mass of a compound and its molecular formula, we can find (a) the mole composition, (b) the mass composition, and (c) the percent composition. But what about the reverse situation? If we are given the mass or percent composition can we then find the molecular formula? The answer is, "not quite." We can find the simplest whole-number ratio of atoms in a compound, however. How we do that is the subject of the next section.

▼**Looking Ahead**

7–5 EMPIRICAL AND MOLECULAR FORMULAS

Two hydrocarbons (carbon-hydrogen compounds) are known as acetylene (C_2H_2) and benzene (C_6H_6). Actually, these two compounds have very little in common. Acetylene is a gas used in welding, and benzene is a liquid solvent used widely in industry. What they do have in common, however, is that they have the same percent composition. This is because they have the same empirical formula (CH). *The **empirical formula** is the simplest whole-number ratio of atoms in the compound.* All ionic compounds are represented by empirical formulas, but the formula of a molecular compound (the molecular formula) represents the actual number of atoms present in one molecule. One benzene molecule is composed of six carbons and six hydrogens, but the empirical formula is obtained by dividing the subscripts by six. The same empirical formula (CH) is obtained for acetylene by dividing the subscripts by two.

What is the difference between an empirical formula and a molecular formula?

To calculate the empirical formula of a compound from percent composition, we follow three steps.

1 Convert percent composition to an actual mass.

2 Convert mass to moles of each element.

3 Find the whole-number ratio of the moles of different elements. This procedure is best illustrated by the following examples.

EXAMPLE 7–14

The Empirical Formula of Laughing Gas

What is the empirical formula of laughing gas, which is 63.6% nitrogen and 36.4% oxygen?

●**Working It Out**

Procedure

Remember that percent means parts per 100. Therefore, if we simply assume that we have a 100-g quantity of compound, the percent converts to a specific mass as follows:

$$63.6\% \text{ N} = \frac{63.6 \text{ g N}}{100 \text{ g compound}} \quad 36.4\% \text{ O} = \frac{36.4 \text{ g O}}{100 \text{ g compound}}$$

We now convert the masses to moles of each element.

$$63.6 \text{ g N} \times \frac{1 \text{ mol N}}{14.0 \text{ g N}} = 4.54 \text{ mol N in 100 g compound}$$

$$36.4 \text{ g O} \times \frac{1 \text{ mol O}}{16.0 \text{ g O}} = 2.28 \text{ mol O in 100 g compound}$$

The ratio of N to O atoms will be the same as the ratio of N to O moles. The formula cannot remain fractional, since only whole numbers of atoms are present in a compound. To find the whole-number ratio of moles, divide through by the smaller number of moles, which in this case is 2.28 mol of O.

Solution

$$\text{N} \quad \frac{4.54}{2.28} = 2.0 \quad \text{O} \quad \frac{2.28}{2.28} = 1.0$$

The empirical formula of the compound is

$$\underline{\underline{\text{N}_2\text{O}}}$$

EXAMPLE 7–15

The Empircal Formula of Magnetite

Pure magnetite is composed of an iron-oxygen binary compound. A 3.85-g sample of magnetite is composed of 2.79 g of iron. What is the empirical formula of magnetite?

Procedure

(a) Find the mass of oxygen by subtracting the mass of the iron from the total mass.

$$3.85 \text{ g } - 2.79 \text{ g (Fe)} = 1.06 \text{ g (O)}$$

(b) Convert the masses of iron and oxygen to moles.

$$2.79 \text{ g Fe} \times \frac{1 \text{ mol Fe}}{55.8 \text{ g Fe}} = 0.0500 \text{ mol Fe}$$

$$1.06 \text{ g O} \times \frac{1 \text{ mol O}}{16.0 \text{ g O}} = 0.0663 \text{ mol O}$$

(c) Find the whole-number ratio of moles of iron and oxygen. Divide by the smaller number of moles.

$$\text{Fe} \quad \frac{0.0500}{0.0500} = 1.00 \quad \text{O} \quad \frac{0.0663}{0.0500} = 1.33$$

This time we're not quite finished, since $FeO_{1.33}$ still has a fractional number that must be cleared. (You should keep at least two decimal places in

Magnetite is a mineral composed of iron and oxygen.

these numbers, so as not to round off a number like 1.3 to 1.) This fractional number can be cleared by multiplying both subscripts by an integer that produces whole numbers. In this case, 1.33 is equivalent to $1\frac{1}{3}$ or $\frac{4}{3}$ so both subscripts (the 1 implied for Fe and the 1.33 for O) can be multiplied by 3 to clear the fraction.

Solution

$$Fe_{(1 \times 3)}O_{(1.33 \times 3)} = \underline{\underline{Fe_3O_4}}$$

A decimal value of 0.50 can be multiplied by 2, values of 0.33 and 0.67 can be multiplied by 3, values of 0.25 and 0.75 can be multiplied by 4, and values of 0.20, 0.40, 0.60, and 0.80 can be multiplied by 5. There are few examples that are more complex than these. Since the last decimal place is estimated, values such as 0.49 should be rounded off to 0.50.

To determine the molecular formula for molecular compounds, we need to know how many empirical units are present in each molecular unit. Thus it is necessary to know both the molar mass of the compound (g/mol) and the mass of one empirical unit (g/emp. unit). The ratio of these two quantities must be a whole number (represented as a below) and represents the number of empirical units in one mole of compound.

$$a = \frac{\text{molar mass}}{\text{emp. mass}} = \frac{X \text{ g/mol}}{Y \text{ g/emp. unit}} = 1, 2, 3, \text{ etc., emp. unit/mol}$$

The molecular formula is obtained by multiplying the subscripts of the empirical formula by a. For example, the empirical formula of borazine (from Example 7–13) is BNH_2. The mass of one empirical unit is $[10.8 + 14.0 + (2 \times 1.01)] = 26.8$ g/emp. unit. Its molar mass is 80.5 g/mol.

$$a = \frac{80.5 \text{ g/mol}}{26.8 \text{ g/emp. unit}} = 3 \text{ emp. units/mol}$$

The molecular formula is

$$B_{(1 \times 3)}N_{(1 \times 3)}H_{(2 \times 3)} = B_3N_3H_6$$

EXAMPLE 7–16

The Molecular Formula of a Phosphorus Oxide

A pure phosphorus-oxygen compound is 43.7% phosphorus and the remainder oxygen. The molar mass is 284 g/mol. What are the empirical and molecular formulas of this compound?

● **Working It Out**

Procedure
First, find the empirical formula. Convert percent to mass.

$$43.7\% \text{ P} = \frac{43.7 \text{ g P}}{100 \text{ g compound}}$$

$$\% \text{ O} = 100\% - 43.7\% = 56.3\%$$

$$56.3\% \text{ O} = \frac{56.3 \text{ g O}}{100 \text{ g compound}}$$

Convert mass to moles.

$$43.7 \, \cancel{g \, P} \times \frac{1 \text{ mol P}}{31.0 \, \cancel{g \, P}} = 1.41 \text{ mol P}$$

$$56.3 \, \cancel{g \, O} \times \frac{1 \text{ mol O}}{16.0 \, \cancel{g \, O}} = 3.52 \text{ mol O}$$

Divide through by the smallest number of moles.

$$\text{P} \quad \frac{1.41}{1.41} = 1.0 \quad \text{O} \quad \frac{3.52}{1.41} = 2.5$$

Multiply both numbers by 2 to remove the fraction.

$$\text{PO}_{2.5} = \text{P}_{(1 \times 2)}\text{O}_{(2.5 \times 2)} = \underline{\text{P}_2\text{O}_5}$$

To find the molecular formula, first compute the empirical mass.

$$\begin{aligned}
\text{P} \quad & 2 \times 31.0 \text{ g} = 62.0 \text{ g} \\
\text{O} \quad & 5 \times 16.0 \text{ g} = \underline{80.0 \text{ g}} \\
& \qquad\qquad\qquad 142 \text{ g/emp. unit}
\end{aligned}$$

Solution

$$a = \frac{284 \, \cancel{g}/\text{mol}}{142 \, \cancel{g}/\text{emp. unit}} = 2 \text{ emp. unit/mol}$$

The molecular formula is

$$\text{P}_{(2 \times 2)}\text{O}_{(2 \times 5)} = \underline{\text{P}_4\text{O}_{10}}$$

Looking Back ▲

A compound is the sum of its component parts. By using the atomic mass of an element along with the number of atoms of that element in a compound, we can establish the number of moles, mass, and percent composition of that element in the compound. In reverse, we can use the moles, mass, or percent composition of all of the elements in a compound to calculate its empirical formula. Ionic compounds are represented by their empirical formulas, but molecular compounds may have a molecular formula that is a multiple of the empirical formula. The molar mass is needed to establish the molecular formula.

Checking It Out ▶

Learning Check B

B–1. Fill in the blanks.
The mass of an element expressed as the mass per 100 mass units is known as the _____ _____ of the element in the compound. For this to be calculated, one must know two things about the compound: the molar mass and the _____ of the compound. The smallest whole-number ratio of atoms represents the _____ formula, and the actual number of atoms of each element represents the _____ formula. From the percent composition of elements, one can calculate the _____ formula. The _____ _____ of the compound is needed to calculate the molecular formula.

B–2. Vitamin C (ascorbic acid) is a compound with the formula $C_6H_8O_6$. The formula weight of the compound is _____ amu. In a 0.650-mole quantity of ascorbic acid, there is _____ mol of carbon, and _____ g of carbon. The compound is _____ % carbon, and the empirical formula of the compound is _____ .

B–3. A molecular compound is composed of boron and hydrogen. It is 84.4% boron by mass and has a molar mass of 76.9 g/mol. What is its empirical formula? What is its molecular formula?

Additional Examples: Problems 7-35, 7-38, 7-45, 7-56, 7-58, and 7-68.

Determination of the empirical and molecular formulas of a compound is a very important procedure to a chemist. These are essential steps in the process for identification of new compounds that may cure cancers or AIDS. How these calculations are used in research is the final topic of this chapter.

▼**Looking Ahead**

7–6 IDENTIFICATION OF A NEW COMPOUND

New substances are identified in chemistry laboratories every day. In fact, over 10 million unique chemical compounds have been reported and registered by the American Chemical Society. No matter how many there are, however, it is always a thrill for a chemical researcher, whether an undergraduate student or a professor, to be the first to see and identify a new substance. (See Figure 7-4.) Who knows whether this new material may cure a disease or have some other important application? But how does a scientist know that a new compound is truly unique before reporting it to the chemical community through the literature? In fact, it is not that difficult if the researcher carefully follows a certain procedure. Some of the steps have been discussed in this chapter. Depending on the complexity of the new substance, the following process can take from a few days to years.

1 *Determine that the new substance is pure.*
This was discussed in Chapter 2. Pure substances have definite, unchanging physical properties such as boiling points for liquids and melting points for solids. Sooner or later someone will put salt in the sugar bowl, and we will realize how closely these two white crystalline substances resemble each other. Appearances can deceive, but salt and sugar do not melt at the same temperature. If we heat a mixture of sugar and salt, we notice that the sugar melts at a relatively low temperature (186°C) and decomposes to a black gunk, but the salt crystals melt at a much higher temperature (801°C).

Having established that our potentially new compound is pure and not a mixture, we carefully measure the melting and/or boiling point because that is a very important way that this new compound can be identified by others who may wish to prepare it. We are now ready to identify the elements and ratio of the elements in the substance.

2 *Determine the empirical formula of the compound.*
Usually, we have a good idea of what elements are in the compound from the ingredients that we used to prepare it. If we do not know,

Figure 7–4
A Research Laboratory
Preparation and identification of new compounds occur frequently in chemical laboratories.

there are qualitative tests that can be performed to identify the elements present. Our next goal is to establish the percent composition of these elements. This information is then used to establish the empirical formula of the compound. Sometimes this is done in the chemist's own lab, but often a sample is sent to a commercial analytical laboratory that specializes in finding the percent composition of elements in a compound. Various methods are used, depending on the element in question. The most commonly requested elements, carbon and hydrogen, are determined by analysis of combustion products, as illustrated in Problems 7-91 and 7-92.

3 *Determine the molecular formula of the compound.*
If the new compound is molecular, the next step is to determine the molar mass, which can be used along with the empirical formula to establish the molecular formula. Often, the molar mass can be obtained commercially along with the percent composition. But it is fairly routine to measure molar mass of pure compounds in the general chemistry laboratory, and this procedure is discussed in a later chapter in the text.

When the molecular formula and the physical properties are known, it is time to research the chemical literature. The American Chemical Society indexes all known compounds in several ways, including by name and by chemical formula. Most major chemistry departments have access to these indexes, and a researcher can quickly check to see if the compound has been previously reported. It is not unusual to find that it has. If that happens, it's "back to the lab," as they say. Otherwise, we move on to the final step.

4 *Determine the structure of the molecules of the compound.*
We need to know more than just the formula of a compound. We need to know the structural formula, which is the order and arrangement of the atoms in a molecule. Recall from Chapter 3 that ethyl alcohol and dimethyl ether share the same molecular formula (C_2H_6O) but the order of the atoms is different and leads to very different properties for the two compounds. Various instruments available in most laboratories can provide information of this nature. This process may take a few minutes for simple compounds or years for extremely complex molecules. The determination of the double helix structure of DNA, a huge, complex molecule, took years and resulted in a Nobel Prize for Watson and Crick.

How does a scientist determine whether a new compound is really original?

When the structural formula of our new compound is established, we are ready to report this to the chemical community. We not only relate the results that we have just described but also give a detailed account of the process we used to prepare this new compound. Others may wish to repeat our procedures, and they will expect to get the same results. If the new compound has practical applications, however—a possible treatment for a disease, or a herbicide, for example—the new compound may be patented, giving the discoverer exclusive commercial rights for a period of time.

C H A P T E R   R E V I E W

The atomic masses of the elements can be put to greater use than just comparing the masses of individual atoms. By using the atomic mass as ratios, we can measure equivalent numbers of atoms without knowing what the actual number of atoms is. The most useful measure of a number of atoms is obtained by expressing the atomic mass of an element in grams. This is referred to as the **molar mass of the element,** and it is the mass of a number of atoms known as the **mole (mol).** The mole represents 6.02×10^{23} atoms or other individual particles, a quantity known as **Avogadro's number.**

The concept of the mole is then extended to compounds. The **molar mass of a compound** is the **formula weight** of the compound expressed in grams, and is the mass of Avogadro's number of molecules or ionic formula units. This information is summarized as follows:

▲▼**Putting It Together**

Unit	Number	Mass
1.00 mol	6.02×10^{23} atoms	Atomic mass of element in grams
	6.02×10^{23} molecules	Formula or molecular weight in grams
	6.02×10^{23} ionic formula units	Formula weight of ionic compound in grams

Perhaps the most important message of this chapter is for you to become comfortable with the interconversions among moles, mass, and numbers of atoms or molecules. The conversions are summarized as follows:

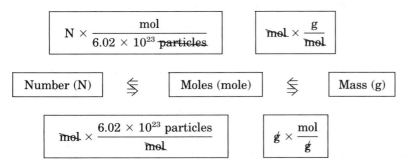

Percent composition can be obtained from the formula of a compound and the atomic masses of the elements. The **empirical formula** of a compound can be determined from its mass composition or percent composition. One needs to know the molar mass of a compound to determine the **molecular formula** from the empirical formula. The molecular formula is determined from the percent composition and the molar mass in the steps on the following page.

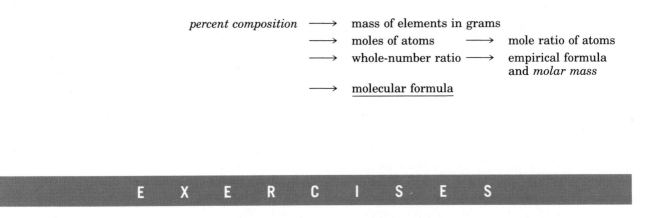

E   X   E   R   C   I   S   E   S

RELATIVE MASSES OF PARTICLES

7–1   One penny weighs 2.47 g and one nickel weighs 5.03 g. What mass of pennies has the same number of coins as there are in 1.24 lb of nickels?

7–2   A small glass bead weighs 310 mg and a small marble weighs 8.55 g. A quantity of small glass beads weighs 5.05 kg. What does the same number of marbles weigh?

7–3   A piece of pure gold has a mass of 145 g. What is the mass of the same number of silver atoms?

7–4   A large chunk of pure aluminum has a mass of 212 lb. What is the mass of the same number of carbon atoms?

7–5   A piece of copper wire has a mass of 16.0 g; the same number of atoms of a precious metal has a mass of 49.1 g. What is the metal?

7–6   In the compound CuO, what mass of copper is present for each 18.0 g of oxygen?

7–7   In the compound $NaCl$, what mass of sodium is present for each 425 g of chlorine?

7–8   In a compound containing one atom of carbon and one atom of another element, it is found that 25.0 g of carbon is combined with 33.3 g of the other element. What is the element? What is the formula of the compound?

*7–9   In the compound $MgBr_2$, what mass of bromine is present for each 46.0 g of magnesium? (Remember, there are two bromines per magnesium.)

*7–10   In the compound SO_3, what mass of sulfur is present for each 60.0 lb of oxygen?

THE MAGNITUDE OF THE MOLE

7–11   If you could count two numbers per second, how many years would it take you to count Avogadro's number? If you were helped by the whole human race of 5.5 billion people, how long would it take?

7–12   A small can of soda contains 350 mL. Planet Earth contains 326 million cubic miles (3.5×10^{20} gal) of water. How many cans of soda would it take to equal all of the water on Earth? How many moles of cans is this?

7–13   What would the number in one mole be if the atomic mass were expressed in kilograms rather than grams? In milligrams?

MOLES OF ATOMS OF ELEMENTS

7–14   Fill in the blanks. Use the factor-label method to determine the answers.

Element	Mass in Grams	Number of Moles of Atoms	Number of Atoms
S	8.00	0.250	1.50×10^{23}
(a) P	14.5	_____	_____
(b) Rb	_____	1.75	_____
(c) Al	_____	_____	6.02×10^{23}
(d) _____	363	_____	3.01×10^{24}
(e) Ti	_____	_____	1

7—15 Fill in the blanks.

Element	Mass In Grams	Number of Moles of Atoms	Number of Atoms
(a) Na	0.390	_____	_____
(b) Cr	_____	3.01×10^4	_____
(c) ____	43.2	_____	2.41×10^{24}
(d) K	4.25×10^{-6}	_____	_____
(e) Ne	_____	_____	3.66×10^{21}

7—16 What is the mass in grams of each of the following?
(a) 1.00 mol of Cu
(b) 0.50 mol of S
(c) 6.02×10^{23} atoms of Ca

7—17 What is the mass in grams of each of the following?
(a) 4.55 mol of Be
(b) 6.02×10^{24} atoms of Ca
(c) 3.40×10^5 mol of B

7—18 How many individual atoms are in each of the following?
(a) 32.1 mol of sulfur
(b) 32.1 g of sulfur
(c) 32.0 g of oxygen

7—19 How many moles of atoms are in each of the following?
(a) 281 g of Si
(b) 7.34×10^{25} atoms of phosphorus
(c) 19.0 atoms of fluorine

7—20 Which has more atoms: 50.0 g of Al or 50.0 g of Fe?

7—21 Which contains more Ni: 20.0 g, 2.85×10^{23} atoms, or 0.450 mol?

7—22 Which contains more Cr: 0.025 mol, 6.0×10^{21} atoms, or 1.5 g of Cr?

7—23 A 0.251-g sample of a certain element is the mass of 1.40×10^{21} atoms. What is the element?

7—24 A 0.250-mol sample of a certain element has a mass of 49.3 g. What is the element?

FORMULA WEIGHT

7—25 What is the formula weight of each of the following? Express your answer to three significant figures.

(a) $KClO_2$
(b) SO_3
(c) N_2O_5
(d) H_2SO_4
(e) Na_2CO_3
(f) CH_3COOH
(g) $Fe_2(CrO_4)_3$

7—26 What is the formula weight of each of the following?
(a) $CuSO_4 \cdot 6H_2O$ (Include H_2O's; they are included in one formula unit.)
(b) Cl_2O_3
(c) $Al_2(C_2O_4)_3$
(d) $Na_2BH_3CO_2$
(e) P_4O_6

7—27 A compound is composed of three sulfate ions and two chromium ions. What is the formula weight of the compound?

7—28 What is the formula weight of strontium perchlorate?

MOLES OF COMPOUNDS

7—29 Fill in the blanks. Use the factor-label method to determine answers.

Molecules	Mass in Grams	Number of Moles	Number of Molecules or Formula Units
N_2O	23.8	0.542	3.26×10^{23}
(a) H_2O	_____	10.5	_____
(b) BF_3	_____	_____	3.01×10^{21}
(c) SO_2	14.0	_____	_____
(d) K_2SO_4	_____	1.20×10^{-4}	_____
(e) SO_3	_____	_____	4.50×10^{24}
(f) $N(CH_3)_3$	0.450	_____	_____

7—30 Fill in the blanks. Use the factor-label method to determine answers.

Molecules	Mass in Grams	Number of Moles	Number of Molecules or Formula Units
(a) O_3	176	_____	_____
(b) NO_2	_____	3.75×10^{-3}	_____
(c) Cl_2O_3	_____	_____	150
(d) UF_6	_____	_____	8.50×10^{22}

7—31 Tethrahydrocarbinol (THC) is the active ingredient in marijuana. A 0.0684 mol quanitity of THC has a mass of 21.5 g. What is the molar mass of THC?

7-32 A 0.158-mol quantity of a compound has a mass of 7.28 g. What is the molar mass of the compound?

7-33 A 287-g quantity of a compound is the mass of 1.07×10^{24} molecules. What is the molar mass of the compound?

7-34 A 0.0681-g quantity of ethylene glycol (antifreeze) represents the mass of 7.88×10^{20} molecules. What is the molar mass of the compound?

THE COMPOSITION OF COMPOUNDS

7-35 How many moles of each type of atom are in 2.55 mol of grain alcohol, C_2H_6O? What is the total number of moles of atoms? What is the mass of each element? What is the total mass?

7-36 How many moles are in 28.0 g of $Ca(ClO_3)_2$? How many moles of each element are present? How many moles of atoms are present?

7-37 How many moles are in 84.0 g of $K_2Cr_2O_7$? How many moles of each element are present? How many total moles of atoms are present?

7-38 What mass of each element is in 1.50 mol of H_2SO_3?

7-39 What mass of each element is in 2.45 mol of boric acid (H_3BO_3)?

7-40 How many moles of O_2 are in 1.20×10^{22} O_2 molecules? How many moles of oxygen atoms? What is the mass of oxygen molecules? What is the mass of oxygen atoms?

7-41 How many moles of Cl_2 molecules are in 985 g of Cl_2? How many moles of Cl atoms are present? What is the mass of Cl atoms present?

7-42 What is the percent composition of a compound composed of 1.375 g of N and 3.935 g of O?

7-43 A sample of a compound has a mass of 4.86 g and is composed of silicon and oxygen. What is the percent composition if 2.27 g of the mass is silicon?

7-44 The mass of a sample of a compound is 7.44 g. Of that mass, 2.88 g is potassium, 1.03 g is nitrogen, and the remainder is oxygen. What is the percent composition of the elements?

7-45 What is the percent composition of all of the elements in the following compounds?

(a) C_2H_6O (d) Na_2SO_4
(b) C_3H_6 (e) $(NH_4)_2CO_3$
(c) C_9H_{18}

7-46 What is the percent composition of all of the elements in the following compounds?
(a) H_2CO_3 (c) $Al(NO_3)_3$
(b) Cl_2O_7 (d) $NH_4H_2PO_4$

7-47 What is the percent composition of all of the elements in borax $(Na_2B_4O_7 \cdot 10H_2O)$?

7-48 What is the percent composition of all of the elements in acetaminophen, $C_8H_9O_2N$? (Acetaminophen is an aspirin substitute.)

7-49 What is the percent composition of all of the elements in saccharin, $C_7H_5SNO_3$?

7-50 What is the percent composition of all of the elements in amphetamine, $C_9H_{13}N$? (Amphetamine is a stimulant.)

*7-51 What mass of carbon is in a 125-g quantity of sodium oxalate?

*7-52 What is the mass of phosphorus in a 25.0-lb quantity of sodium phosphate?

*7-53 What mass of chromium is in a 275-kg quantity of chromium(III) carbonate?

*7-54 Iron is recovered from iron ore, Fe_2O_3. How many pounds of iron can be recovered from each ton of iron ore? (1 ton = 2000 lb)

EMPIRICAL FORMULAS

7-55 Which of the following are not empirical formulas?
(a) N_2O_4 (d) $H_2C_2O_4$
(b) Cr_2O_3 (e) Mn_2O_7
(c) $H_2S_2O_3$

7-56 Convert the following mole ratios of elements to empirical formulas.
(a) 0.25 mol of Fe and 0.25 mol of S
(b) 1.88 mol of Sr and 3.76 mol of I
(c) 0.32 mol of K, 0.32 mol of Cl, and 0.96 mol of O
(d) 1.0 mol of I and 2.5 mol of O
(e) 2.0 mol of Fe and 2.66 mol of O
(f) 4.22 mol of C, 7.03 mol of H, and 4.22 mol of Cl

7-57 Convert the following mole ratios of elements to empirical formulas.
(a) 1.20 mol of Si and 2.40 mol of O
(b) 0.045 mol of C and 0.022 mol of S

(c) 1.00 mol of X and 1.20 mol of Y

(d) 3.11 mol of Fe, 4.66 mol of C, and 14.0 mol of O

7–58 What is the empirical formula of a compound that has the composition 63.2% oxygen and 36.8% nitrogen?

7–59 What is the empirical formula of a compound that has the composition 41.0% K, 33.7% S, and 25.3% O?

7–60 In an experiment it was found that 8.25 g of potassium combines with 6.75 g of O_2. What is the empirical formula of the compound?

7–61 Orlon is composed of very long molecules with a composition of 26.4% N, 5.66% H, and 67.9% C. What is the empirical formula of Orlon?

7–62 A compound is 21.6% Mg, 21.4% C, and 57.0% O. What is the empirical formula of the compound?

7–63 A compound is composed of 9.90 g of carbon, 1.65 g of hydrogen, and 29.3 g of chlorine. What is the empirical formula of the compound?

7–64 A compound is composed of 0.46 g of Na, 0.52 g of Cr, and 0.64 g of O. What is the empirical formula of the compound?

*7–65 A compound is composed of 24.1% nitrogen, 6.90% hydrogen, 27.6% sulfur, and the remainder oxygen. What is the empirical formula of the compound?

*7–66 Methyl salicylate is also known as oil of wintergreen. It is composed of 63.2% carbon, 31.6% oxygen, and 5.26% hydrogen. What is its empirical formula?

*7–67 Nitroglycerin is used as an explosive and a heart medicine. It is composed of 15.9% carbon, 18.5% nitrogen, 63.4% oxygen, and 2.20% hydrogen. What is its empirical formula?

MOLECULAR FORMULAS

7–68 A compound has the following composition: 20.0% C, 2.2% H, and 77.8% Cl. The molar mass of the compound is 545 g/mol. What is the molecular formula of the compound?

7–69 A compound is composed of 1.65 g of nitrogen and 3.78 g of sulfur. If its molar mass is 184 g/mol, what is its molecular formula?

7–70 A compound has a composition of 18.7% B, 20.7% C, 5.15% H, and 55.4% O. Its molar mass is about 115 g/mol. What is the molecular

7–71 A compound reported in 1970 has a composition of 34.9% K, 21.4% C, 12.5% N, 2.68% H, and 28.6% O. It has a molar mass of about 224 g/mol. What is the molecular formula of the compound?

7–72 Fructose is also known as fruit sugar. It has a molar mass of 180 g/mol and is composed of 40.0% carbon, 53.3% oxygen, and 6.7% hydrogen. What is the molecular formula of the compound?

7–73 A compound reported in 1982 has a molar mass of 834 g/mol. A 20.0-g sample of the compound contains 18.3 g of iodine and the remainder carbon. What is the molecular formula of the compound?

7–74 Quinine is a compound discovered in the bark of certain trees. It is an effective drug in the treatment of malaria. Its molar mass is 162 g/mol. It is 22.2% carbon, 22.2% hydrogen, 25.9% nitrogen, and 29.6% oxygen. What is the molecular formula of quinine?

GENERAL QUESTIONS

7–75 The U.S. national debt was about \$4.5 trillion in 1994. How many moles of pennies would it take to pay it off?

7–76 A compound has the formula MN, where M represents a certain unknown metal. A sample of the compound weighs 1.862 g, and of that, 0.442 g is nitrogen. What is the identity of the metal, M?

7–77 What would be the number of particles in one mole if the atomic mass were expressed in ounces rather than grams? (28.375 grams = 1 ounce)

7–78 A certain alloy of copper has a density of 3.75 g/mL and is 65.0% by mass copper. How many copper atoms are in 16.8 cm^3 of this alloy?

7–79 The element phosphorus exists as P_4. How many moles of molecules are in 0.344 g of phosphorus? How many phosphorus atoms are present in that amount of phosphorus?

7–80 Rank the following in order of increasing mass:
(a) 100 hydrogen atoms
(b) 100 moles of hydrogen molecules

(c) 100 grams of hydrogen
(d) 100 hydrogen molecules

7–81 Rank the following in order of increasing number of atoms.
(a) 100 lead atoms
(b) 100 moles of helium
(c) 100 grams of lead
(d) 100 grams of helium

7–82 A mineral containing iron has the formula FeS_x. A quantity of 2.84×10^{23} formula units of pyrite has a mass of 56.6 g. What is the value of x?

7–83 A compound has the formula $Na_2S_4O_6$.
(a) What ions are present in the compound? (The anion is all one species.)
(b) How many grams of sulfur are present in the compound for each 10.0 g of Na?
(c) What is the empirical formula of the compound?
(d) What is the formula weight of the compound?
(e) How many moles and formula units are present in 25.0 g of the compound?
(f) What is the percent composition of oxygen in the compound?

7–84 Glucose (blood sugar) has the formula $C_6H_{12}O_6$. Calculate how many moles of carbon, individual hydrogen atoms, and grams of oxygen are in a 10.0 g sample of glucose.

7–85 A compound has the formula N_2O_x. It is 36.8% nitrogen. What is the name of the compound?

7–86 A compound has the formula SF_x. One sample of the compound contains 0.356 mol of sulfur and 8.57×10^{23} atoms of fluorine. What is the name of the compound?

7–87 Dioxin is a compound that is known to cause cancer in certain laboratory animals. 4.55×10^{22} molecules of this compound have a mass of 24.3 g. Analysis of a sample of dioxin indicated the sample contained 0.456 mol of carbon, 0.152 mol of hydrogen, 0.152 mol of chlorine, and 0.076 mol of oxygen. What is the molecular formula of dioxin?

7–88 A certain compound has a molar mass of 166 g/mol. It is composed of 47.1% potassium, 14.4% carbon, and the remainder oxygen. What is the name of the compound?

7–89 A compound has the general formula $Cr(ClO_x)_y$ and is 14.9% Cr, 30.4% chlorine, and the remainder oxygen. What is the name of the compound?

7–90 Nicotine is a compound containing carbon, hydrogen, and nitrogen. Its molar mass is 162 g/mol. A 1.50 g sample of nicotine is found to contain 1.11 g of carbon. Analysis of another sample indicates that nicotine is 8.70% by mass hydrogen. What is the molecular formula of nicotine?

*7–91 A hydrocarbon (a compound that contains only carbon and hydrogen) was burned, and the products of the combustion were collected and weighed. All of the carbon in the original compound is now present in 1.20 g of CO_2. All of the hydrogen is present in 0.489 g of H_2O. What is the empirical formula of the compound? (Hint: Remember that all of the moles of C atoms in CO_2 and H atoms in H_2O came from the original compound.)

*7–92 A 0.500-g sample of a compound containing C, H, and O was burned, and the products were collected. The combustion produced 0.733 g of CO_2 and 0.302 g of H_2O. The molar mass of the compound is 60.0 g/mol. What is the molecular formula of the compound? (Hint: Find the mass of C and H in the original compound; the remainder of the 0.500 g is oxygen.)

SOLUTIONS TO LEARNING CHECKS

A–1 one, 6.02×10^{23}, molar mass, mole, formula, molar mass, 6.02×10^{23}

A–2 $25.0 \text{ g N} \times \dfrac{16.0 \text{ g O}}{14.0 \text{ g N}} = \underline{28.6 \text{ g O}}$

A–3 $4.82 \times 10^{24} \text{ atoms V} \times \dfrac{1 \text{ mol V}}{6.02 \times 10^{23} \text{ atoms V}} = \underline{\underline{8.01 \text{ mol V}}}$

$8.01 \text{ mol V} \times \dfrac{50.9 \text{ g V}}{\text{mol V}} = \underline{\underline{408 \text{ g V}}}$

A–4 $215 \text{ g Cl} \times \dfrac{1 \text{ mol Cl}}{35.5 \text{ g Cl}} = 6.06 \text{ mol Cl}$

$6.06 \text{ mol Cl} \times \dfrac{6.02 \times 10^{23} \text{ atoms Cl}}{\text{mol Cl}} = \underline{\underline{3.65 \times 10^{24} \text{ atoms Cl}}}$

A–5 Molar mass of $Cl_2O = (2 \times 35.5) + 16.0 = 87.0 \text{ g/mol}$

$438 \text{ g Cl}_2\text{O} \times \dfrac{1 \text{mol Cl}_2\text{O}}{87.0 \text{ g Cl}_2\text{O}} = 5.03 \text{ mol Cl}_2\text{O}$

$5.03 \text{ mol Cl}_2\text{O} \times \dfrac{6.02 \times 10^{23} \text{ molecules}}{\text{mol Cl}_2\text{O}} = \underline{\underline{3.03 \times 10^{24} \text{ molecules}}}$

B–1 percent composition, formula, empirical, molecular, empirical, molar mass

B–2 Formula weight $= (6 \times 12.0) + (8 \times 1.01) + (6 \times 16.0) = \underline{\underline{176 \text{ amu}}}$

$0.650 \text{ mol C}_6\text{H}_8\text{O}_6 \times \dfrac{6 \text{ mol C}}{\text{mol C}_6\text{H}_8\text{O}_6} = \underline{\underline{3.90 \text{ mol C}}}$

$3.90 \text{ mol C} \times \dfrac{12.0 \text{ g C}}{\text{mol C}} = \underline{\underline{46.8 \text{ g C}}}$

$\dfrac{(6 \times 12.0)}{176} \times 100\% = \underline{\underline{40.9\% \text{ C}}}$

Empirical formula $= \underline{\underline{C_3H_4O_3}}$

B–3 In 100.0 g of compound there is 84.4 g B and $100.0 \text{ g} - 84.4 \text{ g} = 15.6 \text{ g H.}$

$84.4 \text{ g B} \times \dfrac{1 \text{ mol B}}{10.8 \text{ g B}} = 7.81 \text{ mol B}$

$15.6 \text{ g H} \times \dfrac{1 \text{ mol H}}{1.01 \text{ g H}} = 15.4 \text{ mol H}$

$\dfrac{7.81}{7.81} = 1.0 \qquad \dfrac{15.4}{7.81} = 2.0$

Empirical formula $= BH_2$

$a = \dfrac{76.9 \text{ g/mol}}{[10.8 + (2 \times 1.01)] \text{ g/emp unit}} = 6 \text{ emp units/mol}$

Molecular formula $= B_{(1 \times 6)} H_{(2 \times 6)} = \underline{\underline{B_6H_{12}}}$

The miracle of life depends on a chemical reaction, powered by the sun, occurring in the green leaves of these trees. This chemical reaction is known as photosynthesis.

THE CHEMICAL REACTION: EQUATIONS AND QUANTITIES

The winter of 1991–92 was the warmest ever recorded in the United States. Was this a sure sign of the predicted "greenhouse effect"? Maybe, but maybe not. Scientists are not yet sure. The greenhouse effect is due to the increasing amount of carbon dioxide (CO_2) present in our atmosphere, which traps heat like the inside of a closed automobile on a sunny day. Carbon dioxide is removed from the atmosphere by trees and other vegetation, however. In a chemical process known as photosynthesis, green leaves use energy from the sun, carbon dioxide, and water to produce carbohydrates (carbon, hydrogen, oxygen compounds) and elemental oxygen. Does the increase in CO_2 in the atmosphere mean that vegetation will grow faster and larger? Perhaps, but experiments so far have been inconclusive. Increasing just one ingredient will not lead to more growth unless all other necessary ingredients are also increased. It's like building an automobile. We need one engine and four tires per car. Having two engines does not mean two cars unless we have four more tires as well as all the other parts. In chemical reactions such as photosynthesis, the mass ratios of reacting compounds as well as the masses of the products that are formed are extremely important. This topic is a subject of this chapter.

Photosynthesis is a complex series of chemical reactions. We eat the carbohydrates produced by vegetation. Inside our bodies, we host a type of chemical reaction known as combustion. In this reaction, the carbohy-

◀ Setting The Stage

drates react with oxygen from the air and produce carbon dioxide and water. The energy originally absorbed from the sun in producing the carbohydrates is now released in our bodies to keep us alive. We may associate combustion with an out-of-control reaction like the burning of wood. Fortunately, the combustion reactions in the body (collectively called metabolism) proceed at a controlled rate, so we don't spontaneously ignite and go up in a puff of smoke. Combustion is just one type of chemical reaction that describes a given chemical change. There are others that we will mention in this chapter.

The real star of this chapter is the *chemical equation.* It has the dual function of representing various types of chemical reactions as well as giving us information about the mass ratios of substances involved in the reaction. A chemical equation makes use of chemical symbols of elements and formulas of compounds to represent a complete chemical reaction. There are several important questions that will be addressed about the chemical equation. How do we construct the chemical equation? What does it tell us about different types of reactions? How can we use the chemical equation to tell us about the mass relationships between the substances that react and those that are produced? How can the chemical equation include information about the energy absorbed or released in a chemical reaction? Our first order of business is to examine the wealth of information that is included in a typical chemical equation.

Formulating ▶
Some Questions

8–1   THE CHEMICAL EQUATION

The main rocket thruster of the space shuttle is powered by a simple chemical reaction. Hydrogen combines with oxygen to form water and, in the process, a lot of heat. How do we represent this information? *A* **chemical equation** *is the representation of a chemical reaction using the symbols of elements and the formulas of compounds.* We will proceed to develop a chemical equation step-by-step. The information about the chemical reaction just mentioned may appear as

$$H + O \longrightarrow H_2O$$

In a chemical equation, *the original reacting species are shown to the left of the arrow and are called the* **reactants.** *The species formed as a result of the reaction are to the right of the arrow and are called the* **products.** In this format, note that the phrase "combines with" (or "reacts with") is represented by a plus sign (+). When there is more than one reactant or product, the symbols or formulas on each side of the equation are separated by a +. The word "produces," also referred to as "yields," may be represented by an arrow (→). Note in Table 8-1 that there are other representations for the yield sign, depending on the situation.

We are far from finished with the chemical equation as shown above. First, if an element exists as molecules under normal conditions, then the formula of the element is shown. Recall from Chapter 4 that hydrogen and oxygen are diatomic molecules under normal conditions. Including this information, the equation is

$$H_2 + O_2 \longrightarrow H_2O$$

TABLE 8–1 SYMBOLS IN THE CHEMICAL EQUATION

Symbol	Use
+	Between the symbols and/or formulas of reactants or products
$\longrightarrow$	Means "yields" or "produces"; separates reactants from products
=	Same as arrow
$\rightleftharpoons$	Used for reversible reactions in place of a single arrow (see Chapter 14)
(g)	Indicates a gaseous reactant or product
↑	Sometimes used to indicate a gaseous product
(s)	Indicates a solid reactant or product
↓	Sometimes used to indicate a solid product
(l)	Indicates a liquid reactant or product
(aq)	Indicates that the reactant or product is in aqueous solution (dissolved in water)
$\xrightarrow{\Delta}$	Indicates that heat must be supplied to reactants before a reaction occurs
$\xrightarrow{MnO_2}$	An element or compound written above the arrow is a *catalyst; a catalyst speeds up a reaction but is not consumed in the reaction*

An important duty of a chemical equation is to demonstrate faithfully the law of conservation of mass, which states that *mass can be neither created nor destroyed.* In Dalton's atomic theory, this law was explained for chemical reactions. He suggested that reactions were simply rearrangements of the same number of atoms. (See Figure 8-1.) A close look at the equation above shows that there are two oxygen atoms on the left but only one on the right. To conform to the law of conservation of mass, an equation must be balanced. *A* **balanced equation** *has the same number and type of atoms on both sides of the equation. An equation is balanced by introducing* **coefficients.** Coefficients are whole numbers in front of the symbols or formulas. The equation in question is balanced in two steps. If we introduce a 2 in front of the H_2O, we have equal numbers of oxygen atoms, but the number of hydrogens is now unbalanced.

What does an equation have to do with the conservation of mass?

$$H_2 + O_2 \longrightarrow 2H_2O$$

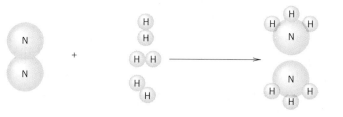

Figure 8–1
Nitrogen Plus Hydrogen Yields Ammonia. In a chemical reaction, the atoms are simply rearranged into different molecules.

This problem can be solved rather easily. Simply return to the left and place a coefficient of 2 in front of the H_2. The equation is now completely balanced.

$$2H_2 + O_2 \longrightarrow 2H_2O$$

Can we just change a subscript to balance an equation?

Note that equations cannot be balanced by changing or adjusting the subscripts of the elements or compounds. For example, the original equation could seem to be balanced in one step if the H_2O were changed to H_2O_2. However, H_2O_2 is a compound known as hydrogen peroxide, which is not the same as water.

Finally, the physical states of the reactants and products under the reaction conditions are sometimes added in parentheses after the formula for each substance. Hydrogen and oxygen are gases, and water is a liquid under the conditions of this reaction. Using the proper letters shown in Table 8-1, we have the balanced chemical equation in proper form.

$$2H_2(g) + O_2(g) \longrightarrow 2H_2O(l)$$

Note that if we describe this reaction in words, we have quite a mouthful. "Two molecules of gaseous hydrogen react with one molecule of gaseous oxygen to produce two molecules of liquid water."

Properly balanced equations are a necessity when we consider the quantitative aspects of reactants and products, as we will do later in this chapter. Before we consider some guidelines, there are three points to keep in mind concerning balanced equations.

1 The subscripts of a compound are fixed; they cannot be changed to balance an equation.

2 The coefficients used should be the smallest whole numbers possible.

3 The coefficient multiplies every number in the formula. For example, $2K_2SO_3$ indicates the presence of four atoms of K, two atoms of S, and six atoms of O.

In this chapter, equations will be balanced by *inspection*. Certainly, many complex equations are extremely tedious to balance by this method, but such equations will be reserved for later chapters in which more systematic methods can be employed. The following rules are helpful in balancing simple equations by inspection.

1 In general, it is easiest to consider balancing elements other than hydrogen or oxygen first. Look to the compound on either side of the equation that contains the most number of atoms of an element other than oxygen or hydrogen. (If polyatomic ions appear unchanged on both sides of the equation, consider them as single units.) Balance the element in question on the other side of the equation.

2 Balance all other atoms except hydrogen and oxygen.

3 Generally, it is easiest to balance hydrogen next, followed by oxygen.

4 Check to see that the atoms of all elements are balanced and that the coefficients are expressed as the smallest whole-number ratio.

EXAMPLE 8–1

Balancing a Simple Equation

Write a balanced chemical equation from the following word equation: "Nitrogen gas reacts with hydrogen gas to produce ammonia gas."

●Working It Out

Procedure

The unbalanced chemical equation using the proper formulas of the elements and compound is

$$N_2(g) + H_2(g) \longrightarrow NH_3(g)$$

First consider the N_2 molecule since it has the most atoms of an element other than hydrogen or oxygen. Balance the N's on the other side by adding a coefficient of 2 in front of the NH_3.

$$N_2(g) + H_2(g) \longrightarrow 2NH_3(g)$$

Now consider the hydrogens. We have "locked in" six hydrogens on the right, so we will need six on the left. By adding a coefficient of 3 in front of the H_2, we have completed the balancing of the equation.

Solution

$$N_2(g) + 3H_2(g) \longrightarrow 2NH_3(g)$$

EXAMPLE 8–2

Balancing an Equation

Balance the following equation:

$$B_2H_6(g) + H_2O(l) \longrightarrow H_3BO_3(aq) + H_2(g)$$

Procedure

First consider the B_2H_6 molecule since it has the most atoms of an element other than hydrogen or oxygen. Balance the B by adding a coefficient of 2 in front of the H_3BO_3 on the right.

$$B_2H_6(g) + H_2O(l) \longrightarrow 2H_3BO_3(aq) + H_2(g)$$

Now consider the hydrogens. Note that they are balanced, so next consider oxygen which is not balanced. The coefficient 6 is needed before the H_2O, but this unbalances the hydrogens. There are now 18 on the left but only 8 on the right. If the coefficient 6 is placed before the H_2, however, the equation is balanced.

Solution

$$B_2H_6(g) + 6H_2O(l) \longrightarrow 2H_3BO_3(aq) + 6H_2(g)$$

EXAMPLE 8–3

Balancing an Equation

Balance the following equation:

$$C_3H_8O(l) + O_2(g) \longrightarrow CO_2(g) + H_2O(l)$$

Procedure

First, consider the C in the C_3H_8O molecule. Add a 3 before the CO_2 on the right to balance the three carbons on the left:

$$C_3H_8O(l) + O_2(g) \longrightarrow 3CO_2(g) + H_2O(l)$$

Balance hydrogen next by adding a 4 before the H_2O to balance the eight hydrogens in C_3H_8O:

$$C_3H_8O(l) + O_2(g) \longrightarrow 3CO_2(g) + 4H_2O(l)$$

Now note that there are 10 oxygens on the right. On the left, one oxygen is in C_3H_8O so nine are needed from O_2. To get an odd number of oxygens from O_2 we need to use a fractional coefficient, in this case $\frac{9}{2}$:

$$C_3H_8O(l) + \tfrac{9}{2} O_2(g) \longrightarrow 3CO_2(g) + 4H_2O(l)$$

Since we are expressing only whole numbers in a balanced equation, all coefficients should be multiplied by 2 to clear the fraction. The final balanced equation has coefficients that are the smallest whole numbers possible.

Solution

$$2C_3H_8O(l) + 9O_2(g) \longrightarrow 6CO_2(g) + 8H_2O(l)$$

Looking Ahead ▼

We are now ready to put the representation implied by a balanced chemical equation to work. There are many ways to group chemical reactions, depending on what common criteria interest us in certain situations. A large number of reactions, however, fit into five general categories that we will discuss in the next section.

8–2 TYPES OF CHEMICAL REACTIONS

All elements and compounds can be described by their chemical properties. An important property of an element relates to its combining power with other elements or compounds. On the other hand, the chemical properties of some compounds are based on how they decompose into simpler compounds or elements. Combination and decomposition reactions are two of the five types of chemical reactions that are described by chemical equations in this section. In later chapters, other common characteristics of reactions that we will discuss at the time allow other convenient ways of classification. The five basic types discussed in this chapter are (1) combustion reactions, (2) combination reactions, (3) decomposition reactions, (4) single-replacement reactions, and (5) double-replacement reactions. A brief discussion of each type follows.

Combustion Reactions

What element is always involved in combustion?

This type of reaction refers specifically to the reaction of an element or compound with elemental oxygen. Combustion is usually accompanied by a flame and is sometimes simply referred to as "burning." In most cases, a large amount of heat energy is evolved in these reactions. When elements undergo combustion, generally only one product is formed. Examples are the combustion of carbon and aluminum shown here.

$$C(s) + O_2(g) \longrightarrow CO_2(g)$$

$$4Al(s) + 3O_2(g) \longrightarrow 2Al_2O_3(s)$$

When compounds undergo combustion, however, two or more combustion products are formed. When carbon-hydrogen compounds undergo combustion in an excess of oxygen, the combustion products are carbon dioxide and water.

$$CH_4(g) + 2O_2(g) \longrightarrow CO_2(g) + 2H_2O(l)$$

The metabolism of glucose ($C_6H_{12}O_6$, blood sugar) was described earlier as a combustion reaction that occurs at a controlled rate in our bodies to generate energy.

$$C_6H_{12}O_6(aq) + 6O_2(g) \longrightarrow 6CO_2(g) + 6H_2O(l)$$

When insufficient oxygen is present (as in the combustion of gasoline, C_8H_{18}, in an automobile engine), some carbon monoxide (CO) also forms.

$$2C_8H_{18}(l) + 17O_2(g) \longrightarrow 16CO(g) + 18H_2O(l)$$

Combination Reactions

This classification concerns the preparation of one compound from two or more elements and/or simpler compounds. For example, an important chemical property of the metal magnesium is that it reacts with elemental oxygen to form magnesium oxide. The synthesis of MgO is represented at the end of this paragraph by a balanced equation and an illustration of the magnesium and oxygen atoms in the reaction. Notice in the reaction that the Mg^{2+} cation is smaller than the parent atom, while the O^{2-} anion is larger than the parent atom. The reason for this was discussed in Chapter 5. (See also Figure 8-2.) Since one of the reactants is elemental oxygen, this reaction can also be classified as a combustion reaction.

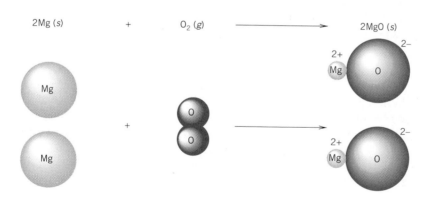

FIGURE 8–2
Combination or Combusion Reaction. When magnesium burns in air, the reaction can be classified as either a combination or a combustion reaction.

The following equations represent some other important combination reactions.

$$2Na(s) + Cl_2(g) \longrightarrow 2NaCl(s)$$

$$C(s) + O_2(g) \longrightarrow CO_2(g)$$

$$CaO(s) + CO_2(g) \longrightarrow CaCO_3(s)$$

$$SO_2(g) + H_2O(l) \longrightarrow H_2SO_3(aq)$$

Figure 8–3
Decomposition Reaction.
The fizz of carbonated water is the result of a decomposition reaction.

Decomposition Reactions

This type of reaction is simply the reverse of combination reactions. One compound is decomposed into two or more elements or simpler compounds. Many of these reactions take place only when heat is supplied, which is indicated by a Δ above the arrow. An example of this type of reaction is the decomposition of carbonic acid (H_2CO_3). This decomposition reaction causes the fizz in carbonated beverages. The reaction is illustrated here and is followed by other examples. (See also Figure 8-3.)

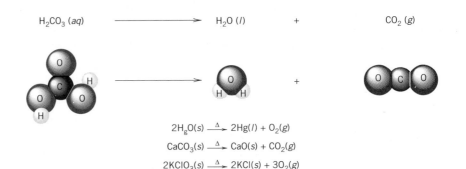

$$2H_gO(s) \xrightarrow{\Delta} 2Hg(l) + O_2(g)$$
$$CaCO_3(s) \xrightarrow{\Delta} CaO(s) + CO_2(g)$$
$$2KClO_3(s) \xrightarrow{\Delta} 2KCl(s) + 3O_2(g)$$

Single-Replacement Reactions

In these reactions one free element substitutes for another in a chemical compound. These reactions generally occur in aqueous solution. For example, when a strip of metallic zinc is immersed in a solution of $CuCl_2$, the elemental Zn is replaced by elemental copper. As a result, the zinc strip becomes coated with a layer of copper. Silver and gold will also form a coating on a strip of zinc when immersed in solutions of these compounds. The replacement of zinc ions for the copper ions and the copper metal for the zinc metal is illustrated here; other examples follow. (See also Figure 8-4.)

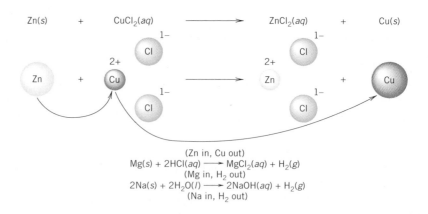

$$Mg(s) + 2HCl(aq) \longrightarrow MgCl_2(aq) + H_2(g)$$
(Mg in, H_2 out)
$$2Na(s) + 2H_2O(l) \longrightarrow 2NaOH(aq) + H_2(g)$$
(Na in, H_2 out)

In a later chapter, single-replacement reactions are discussed in more detail. Tables are available that allow you to predict which element is able to replace another. Predicting which chemical reactions

Figure 8–4
Single-Replacement Reaction. The formation of a layer of copper on a piece of zinc is a single-replacement reaction.

may or may not occur is important in chemistry, but it will not be discussed at this time.

Double-Replacement Reactions

In a single-replacement reaction, one element substitutes for another. In a double-replacement reaction, two ions exchange partners. The driving force of these reactions is the formation of a product that is insoluble in water, is a molecular compound, or both.

1 The exchanged ions may form an ionic compound that is insoluble in water.

2 The exchanged ions may lead to the formation of a molecular compound such as water or a gas.

In the first case, *mixing two water soluble ionic compounds leads to the formation of an insoluble compound known as a* **precipitate.** Thus this type of double-replacement reaction is often referred to as a **precipitation reaction.** The example that follows represents the mixing of solutions of the soluble compounds $AgNO_3$ and NaCl. Since the product AgCl is insoluble in water (indicated by the *s*), the two ions combine to form this precipitate. (See also Figure 8-5.)

The other type of double-replacement reaction involves the formation of a molecular compound from the combination of a cation and an anion. The most important type of reaction in this category involves the reaction of an **acid** *(a compound that produces H^+ ions in aqueous solution)* and a **base** *(a compound that produces OH^- ions in aqueous solution).* When acids and bases are mixed, the H^+ from the acid combines with the OH^- from the base to form water. Since the active ingredient of acids is H^+ and that of bases is OH^-, combination of the two is also re-

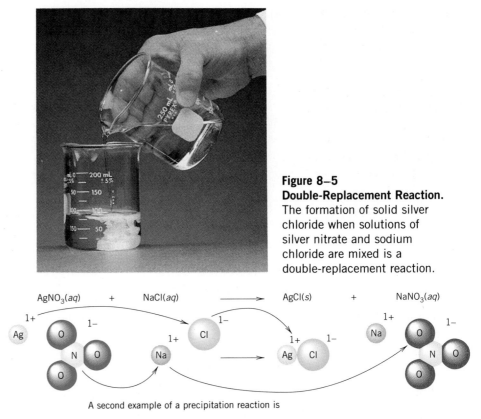

Figure 8–5
Double-Replacement Reaction.
The formation of solid silver chloride when solutions of silver nitrate and sodium chloride are mixed is a double-replacement reaction.

A second example of a precipitation reaction is

$$(NH_4)_2S(aq) + Pb(NO_3)_2(aq) \longrightarrow PbS(s) + 2NH_4NO_3(aq)$$

ferred to as a **neutralization reaction.** The reaction of the two ions is illustrated by the following equation.

$$H^+(aq) + OH^-(aq) \longrightarrow H_2O(l)$$

The other product compound is composed of the anion from the acid and the cation from the base and is an ionic compound known as a **salt.** Two examples of neutralization reactions are

$$HCl(aq) + NaOH(aq) \longrightarrow NaCl(aq) + H_2O(l)$$

$$2HNO_3(aq) + Ca(OH)_2(aq) \longrightarrow Ca(NO_3)_2(aq) + 2H_2O(l)$$

In a similar reaction, the combination of an acid with a salt produces a molecular compound that is a gas and escapes from the solution. Examples of such reactions are

$$HCl(aq) + NaCN(aq) \longrightarrow HCN(g) + NaCl(aq)$$

$$H_2SO_4(aq) + K_2S(aq) \longrightarrow H_2S(g) + K_2SO_4(aq)$$

The first reaction is very dangerous as the HCN that evolves from the solution is extremely poisonous. In the second reaction, the H_2S that evolves is also poisonous, and it has the very pungent smell of rotten eggs.

Double-replacement reactions will receive a more thorough discussion in later chapters. Precipitation reactions are discussed in Chapter 11, and neutralization reactions in Chapter 12.

The world of chemistry opens before us with the introduction of the chemical equation, which tells us "what's happening." Using the shorthand of chemical symbols and formulas of compounds, we can represent not only the changes that occur in chemical reactions but also how many molecules and formula units of compounds are involved. If the symbol of an element is the chemist's analog to the alphabet, and the formulas of compounds are equivalent to words, then the chemical equation is like a complete sentence.

▲ **Looking Back**

◀ **Checking It Out**

Learning Check A

A–1. Fill in the blanks.
A chemical reaction is represented with symbols and formulas by means of a chemical _____ . The arrow in an equation separates the _____ on the left from the _____ on the right. To conform to the law of conservation of mass, an equation must be _____ . This is accomplished by introducing _____ in front of a formula rather than changing subscripts in a formula. There are five types of chemical reactions illustrated by equations in this chapter. They are (1) _____ , (2) _____ , (3) _____ , (4) _____ , and (5) _____ reactions. Two types of reactions that fit into the last category are _____ reactions and _____ reactions.

A–2. Represent the following word equation with a balanced chemical equation: "disilane gas (Si_2H_6) reacts with oxygen to form solid silicon dioxide and water."

A–3. Balance the following equations:
(a) $Fe(s) + H_2SO_4(aq) \longrightarrow Fe_2(SO_4)_3(aq) + H_2(g)$
(b) $Na_2CO_3(aq) + Sr(NO_3)_2(aq) \longrightarrow SrCO_3(s) + NaNO_3(aq)$

A–4. Classify the reaction in Problem 2 and the two reactions in Problem 3 into one of the five types of reactions.

Additional Examples: Problems 8-2, 8-6, and 8-8.

We are now ready to look at the quantitative implications of a chemical equation. We indicated that equations must be balanced by including the correct number of reactant and product molecules or formula units. However, as in the last chapter, we need to scale up our measurements so they are appropriate for laboratory situations. To do this, we will combine our discussions of the mole in the last chapter with the chemical equation that we have introduced in this chapter.

▼ **Looking Ahead**

8–3 STOICHIOMETRY

How is an equation like a recipe?

An important goal in the chemical industry is to produce a given compound in the least expensive manner. This means that reactants must be mixed in the proper proportion, and that is determined from a balanced chemical equation. *The quantitative relationships between reactants and/or products is known as* **stoichiometry.** One of the most important industrial processes is the production of ammonia (NH_3) from hydrogen and nitrogen. We will use the balanced equation illustrating this reaction as our example. The most basic relationship of the balanced equation refers to the ratio of molecules and is shown in line 1. But it is important to note that the equation also implies any multiple of the basic molecular ratios as shown in lines 2, 3, and 4. In lines 5 and 6, the numbers have been changed to mole units and mass.

Production of Ammonia			
$N_2(g)$	$+$ $3H_2(g)$	$\longrightarrow$	$2NH_3(g)$
1 1 molecule	$+$ 3 molecules	$\longrightarrow$	2 molecules
2 12 molecules	$+$ 36 molecules	$\longrightarrow$	24 molecules
3 1 dozen molecules	$+$ 3 dozen molecules	$\longrightarrow$	2 dozen molecules
4 6.02×10^{23} molecules	$+$ 18.1×10^{23} molecules	$\longrightarrow$	12.0×10^{23} molecules
5 1 mol	$+$ 3 mol	$\longrightarrow$	2 mol
6 28 g	$+$ 6 g	$\longrightarrow$	34 g

We will focus on the mole relationships shown in line 5, as this successfully scales up the stoichiometry of the equation to laboratory situations. The relationships of moles to number (line 4) and to mass (line 6) have been discussed in Chapter 7.

First we will extract from the balanced equation and express in words the mole relationships between the reactant and product molecules.

1 mol of N_2 produces 2 mol of NH_3

This can be expressed in ratio form as (1) $\dfrac{1 \text{ mol } N_2}{2 \text{ mol } NH_3}$ and (2) $\dfrac{2 \text{ mol } NH_3}{1 \text{ mol } N_2}$.

1 mol of N_2 reacts with 3 mol of H_2

This can be expressed in ratio form as (3) $\dfrac{1 \text{ mol } N_2}{3 \text{ mol } H_2}$ and (4) $\dfrac{3 \text{ mol } H_2}{1 \text{ mol } N_2}$.

3 mol of H_2 produces 2 mol of NH_3

This can be expressed in ratio form as (5) $\dfrac{3 \text{ mol } H_2}{2 \text{ mol } NH_3}$ and (6) $\dfrac{2 \text{ mol } NH_3}{3 \text{ mol } H_2}$.

The mole relation factors generated by the balanced equation will be referred to as the "mole ratios" in this text. *Note that the coefficients in the balanced equation must be included in the mole ratios.* The six mole ratios generated by the sample balanced equation will be used in the stoichiometry problems that follow. In these examples, the central conversion is between moles of one reactant or product to moles of another reactant or product. In Example 8-5, 8-6, and 8-7, additional conversions are necessary.

The Examples below illustrate the following conversions:

mole	$\longrightarrow$	mole	(Example 8–4)
mass	$\longrightarrow$	mole	(Example 8–5)
mass	$\longrightarrow$	mass	(Example 8–8)
mass	$\longrightarrow$	number	(Example 8–7)

EXAMPLE 8–4

Mole → Mole Conversions

How many moles of NH_3 can be produced from 5.00 mol of H_2?

● **Working It Out**

Procedure
Convert moles of what's given (mol of H_2) to moles of what's requested (mol NH_3). This requires mole ratio 6, which has *mol H_2* in the denominator and *mol NH_3* in the numerator.

Scheme

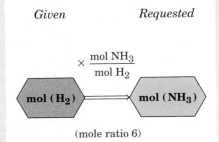

Solution

$$5.00 \text{ mol } H_2 \times \frac{2 \text{ mol } NH_3}{3 \text{ mol } H_2} = \underline{\underline{3.33 \text{ mol } NH_3}}$$

EXAMPLE 8–5

Mass → Mole Conversions

How many moles of NH_3 can be produced from 33.6 g of N_2?

Procedure
Before we can convert moles of N_2 to moles of NH_3, we must first convert the mass of N_2 to moles of N_2. This means a two-step conversion. In the first step, mass of N_2 is converted to moles using the molar mass of N_2 as

the conversion factor. In the second step, moles of N_2 is converted to moles of NH_3 using mole ratio 2.

Scheme

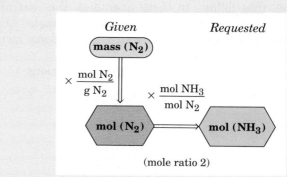

(mole ratio 2)

Solution

$$33.6 \text{ g } N_2 \times \frac{1 \text{ mol } N_2}{28.0 \text{ g } N_2} \times \frac{2 \text{ mol } NH_3}{1 \text{ mol } N_2} = \underline{\underline{2.40 \text{ mol } NH_3}}$$

EXAMPLE 8–6

Mass → Mass Conversions

What mass of H_2 is needed to produce 119 g of NH_3?

Procedure

As in the last example, we must first convert the mass of what's given (119 g NH_3) into moles using the molar mass of NH_3 (step **a** below). We then convert moles of NH_3 to moles of H_2 using mole ratio 5 (step **b** below). Finally, since the mass of H_2 is requested, we must convert moles of H_2 to mass using the molar mass of H_2 as a conversion factor (step **c** below).

Scheme

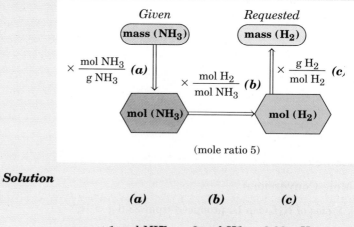

(mole ratio 5)

Solution

$$119 \text{ g } NH_3 \times \underbrace{\frac{1 \text{ mol } NH_3}{17.0 \text{ g } NH_3}}_{(a)} \times \underbrace{\frac{3 \text{ mol } H_2}{2 \text{ mol } NH_3}}_{(b)} \times \underbrace{\frac{2.02 \text{ g } H_2}{\text{mol } H_2}}_{(c)} = \underline{\underline{21.2 \text{ g } H_2}}$$

EXAMPLE 8–7

Mass → Number Conversions

How many molecules of N_2 are needed to react with 17.0 g of H_2?

Procedure
This problem reminds us that the mole relates to a number as well as a mass. In fact, this type of problem generally is not encountered because the actual numbers of molecules involved is not as important as their relative masses. In any case, this problem is much like the previous example except that in the final step moles of N_2 is converted to number of molecules (step **c** below).

Scheme

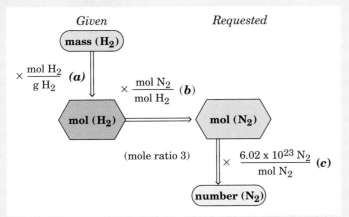

Solution

$$\text{(a)} \qquad \text{(b)} \qquad \text{(c)}$$

$$17.0 \, \cancel{\text{g H}_2} \times \frac{1 \, \cancel{\text{mol H}_2}}{2.02 \, \cancel{\text{g H}_2}} \times \frac{1 \, \cancel{\text{mol N}_2}}{3 \, \cancel{\text{mol H}_2}} \times \frac{6.02 \times 10^{23} \text{ molecules N}_2}{\cancel{\text{mol N}_2}} =$$

$$1.69 \times 10^{24} \text{ molecules N}_2$$

Before we work through more examples of stoichiometry problems, let us summarize the procedure relating moles, mass, and number of molecules, as follows:

1 Write down (a) what is given and (b) what is requested.

2 (a) If a mass is given, use the molar mass to convert mass to moles of what is given.

  (b) If a number of molecules is given, use Avogadro's number to convert to moles of what is given.

3 Using the correct mole ratio from the balanced equation, convert moles of what is given to moles of what is requested.

How can we summarize the procedures for working stoichiometry problems?

4 (a) If a mass is requested, convert moles of what is requested to mass of what is requested.
(b) If a number of molecules is requested, use Avogadro's number to convert to number of molecules of what is requested.

We represent this information in the general scheme shown in Figure 8-6.

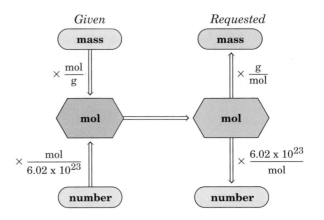

EXAMPLE 8–8

Conversion of Mass O_2 to Moles of Fe_2O_3

Some sulfur is present in coal in the form of pyrite, FeS_2 (also known as "fool's gold"). When it burns, it pollutes the air with the combustion product SO_2, as shown by the following chemical equation:

$$4FeS_2(s) + 11O_2(g) \longrightarrow 2Fe_2O_3(s) + 8SO_2(g)$$

How many moles of Fe_2O_3 are produced from 145 g of O_2?

Procedure
Given: 145 g of O_2. Requested: ? mol of Fe_2O_3.
Convert mass of O_2 to moles of $\overline{Fe_2O_3}$.

Scheme

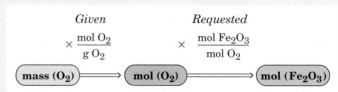

From the equation, a mole ratio relating O_2 to Fe_2O_3 is needed. The factor will have what's requested in the numerator.

$$\frac{2 \text{ mol } Fe_2O_3}{11 \text{ mol } O_2}$$

Working It Out ●

The color of pyrite caused it to be known as "fool's gold."

Solution

$$145 \ \cancel{g \ O_2} \times \frac{1 \ \cancel{mol \ O_2}}{32.0 \ \cancel{g \ O_2}} \times \frac{2 \ mol \ Fe_2O_3}{11 \ \cancel{mol \ O_2}} = \underline{\underline{0.824 \ mol \ Fe_2O_3}}$$

EXAMPLE 8–9

Conversion of Mass of FeS_2 to Mass of SO_2

From the equation used in the preceding example, determine the mass of SO_2 that is produced from the combustion of 38.8 g of FeS_2.

Procedure
Given: 38.8 g of FeS_2. Requested: __?__ g of SO_2.
Convert grams of FeS_2 to grams of SO_2.

Scheme

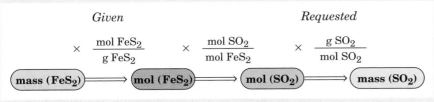

The mole ratio that relates moles of given to moles of requested (in the numerator) is

$$\frac{8 \ mol \ SO_2}{4 \ mol \ FeS_2}$$

Solution

$$38.8 \ \cancel{g \ FeS_2} \times \frac{1 \ \cancel{mol \ FeS_2}}{120 \ \cancel{g \ FeS_2}} \times \frac{8 \ \cancel{mol \ SO_2}}{4 \ \cancel{mol \ FeS_2}} \times \frac{64.1 \ g \ SO_2}{\cancel{mol \ SO_2}} = \underline{\underline{41.5 \ g \ SO_2}}$$

Perhaps the most important stoichiometry problem that we have encountered is the conversion of a given mass of reactant to an equivalent mass of product. So far, we have just assumed that the needed amounts of all other reactants are present. But how do we determine the amount of product formed if there are limited amounts of reactants present? This is the subject of the next section.

▼ **Looking Ahead**

8–4 LIMITING REACTANT

Assume that your instructor is trying to put together a three-page exam for the class. The instructor may have 85 copies of page 1, 85 copies of page 2, but only 80 copies of page 3. Obviously, only 80 copies of the test can be put together even though there is an excess of pages 1 and 2. The page that produces the least number of complete tests can be called "the limiting page." Likewise, *if specific amounts of each reactant are mixed, the reactant that produces the least amount of product is called the* **limiting reactant.** In other words, the amount of product formed is limited

How can a reactant be "limiting"?

by the reactant that is completely consumed. We can illustrate this with the simple example of the production of water from its elements, hydrogen and oxygen.

$$2H_2 + O_2 \longrightarrow 2H_2O$$

The stoichiometry of the reaction tells us that two moles (4.0 g) of hydrogen react with one mole (32 g) of oxygen to produce two moles (36 g) of water. Thus any time hydrogen and oxygen react in a 4.0:32 mass ratio, all reactants are consumed and only products appear. *When reactants are mixed in exactly the mass ratio determined from the balanced equation, the mixture is said to be* **stoichiometric.**

$$4.0 \text{ g of } H_2 + 32.0 \text{ g of } O_2 \longrightarrow 36.0 \text{ g of } H_2O \text{ (stoichiometric)}$$

What if we mix a 6.0-g quantity of H_2 with a 32.0-g quantity of O_2? Do we produce 38.0 g water? No, we still produce only 36.0 g of H_2O using only 4.0 g of the H_2. Thus H_2 is present in excess, and the amount of product is limited by the amount of O_2 present. In this case, O_2 becomes the limiting reactant.

$$6.0 \text{ g of } H_2 + 32.0 \text{ g } O_2 \longrightarrow 36.0 \text{ g } H_2O + 2.0 \text{ g of } H_2 \text{ in excess}$$
$$(H_2 \text{ in excess, } O_2 \text{ the limiting reactant})$$

If we mix a 4.0-g quantity of H_2 with a 36.0-g quantity of O_2, 36.0 g of H_2O is again produced. In this case, the H_2 is completely consumed and limits the amount of water formed. Thus H_2 is now the limiting reactant and O_2 is present in excess.

$$4.0 \text{ g of } H_2 + 36.0 \text{ g of } O_2 \longrightarrow 36.0 \text{ g of } H_2O + 4.0 \text{ g of } O_2 \text{ in excess}$$
$$(O_2 \text{ in excess, } H_2 \text{ the limiting reactant})$$

How do you know which reactant is the limiting reactant?

When quantities of two or more reactants are given, it is necessary to determine which is the limiting reactant (unless they are mixed in exactly stoichiometric amounts). This can be simplified by two procedures.

1   Determine the number of moles of the product in question that would be produced *by each reactant* using the general procedure shown in Figure 8-6.

2   The reactant producing the smallest amount of the product is the limiting reactant.

We will illustrate the calculation of the limiting reactant with two examples.

Working It Out ●

EXAMPLE 8–10

Determination of the Limiting Reactant

Silver tarnishes (turns black) in homes because of the presence of small amounts of H_2S (a gas that orginates from the decay of food and smells like rotten eggs). The reaction is

$$4Ag(s) + 2H_2S(g) + O_2(g) \longrightarrow 2Ag_2S(s) + 2H_2O(l)$$
$$\text{(black)}$$

If 0.145 mol of Ag is present with 0.0872 mol of H_2S and excess O_2:
(a) What mass of Ag_2S is produced?
(b) What mass of the other reactant remains in excess?

Procedure *(a)*
Convert moles of Ag and moles of H_2S to moles of Ag_2S produced. Then convert the number of moles of Ag_2S to mass of Ag_2S based on the limiting reactant.

Scheme *(a)*

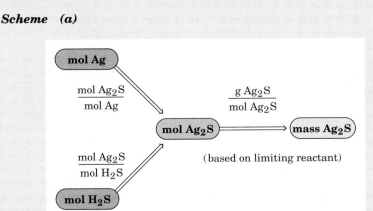

Silver tarnishes from the presence of H_2S in the atmosphere.

Solution *(a)*

$$\text{Ag: } 0.145 \text{ mol Ag} \times \frac{2 \text{ mol Ag}_2\text{S}}{4 \text{ mol Ag}} = 0.0725 \text{ mol Ag}_2\text{S}$$

$$\text{H}_2\text{S: } 0.0872 \text{ mol H}_2\text{S} \times \frac{2 \text{ mol Ag}_2\text{S}}{2 \text{ mol H}_2\text{S}} = 0.0872 \text{ mol Ag}_2\text{S}$$

Since Ag produces the smaller yield of Ag_2S, *Ag is the limiting reactant*. To find the mass of Ag_2S formed, convert moles of Ag_2S produced by the Ag to mass of Ag_2S.

$$0.0725 \text{ mol Ag}_2\text{S} \times \frac{248 \text{ g Ag}_2\text{S}}{\text{mol Ag}_2\text{S}} = \underline{\underline{18.0 \text{ g Ag}_2\text{S}}}$$

Procedure *(b)*
To find the mass of H_2S in excess, we first find the mass of H_2S that was consumed in the reaction along with the Ag. Use the mole ratio relating moles of H_2S to moles of Ag.

Scheme *(b)*

Next, find the moles of H_2S remaining or unreacted by subtracting the moles consumed from the original amount present.

mol H_2S (original) − mol H_2S (consumed) = mol H_2S (in excess)

Finally, convert moles of unreacted H_2S to mass.

$$\boxed{\text{mol } H_2S \text{ (in excess)}} \Longrightarrow \boxed{\text{mass } H_2S}$$

Solution **(b)**

$$0.145 \text{ mol Ag} \times \frac{2 \text{ mol } H_2S}{4 \text{ mol Ag}} = 0.0725 \text{ mol } H_2S \text{ consumed}$$

$$0.0872 \text{ mol} - 0.0725 \text{ mol} = 0.0147 \text{ mol } H_2S \text{ in excess.}$$

$$0.0147 \text{ mol } H_2S \times \frac{34.1 \text{ g } H_2S}{\text{mol } H_2S} = \underline{\underline{0.501 \text{ g } H_2S \text{ in excess}}}$$

EXAMPLE 8–11

Determination of the Limiting Reactant

Methanol (CH_3OH) is used as a fuel for racing cars. It burns in the engine according to the equation

$$2CH_3OH(l) + 3O_2(g) \longrightarrow 2CO_2(g) + 4H_2O(g)$$

If 40.0 g of methanol is mixed with 46.0 g of O_2, what is the mass of CO_2 produced?

Procedure
(a) Convert mass of CH_3OH to moles of CH_3OH and then to moles of CO_2, and convert mass of O_2 to moles of O_2 and then to moles of CO_2.
(b) Convert the smaller number of moles of CO_2 to mass of CO_2.

Race cars at the Indianapolis 500 burn methanol as a fuel

Scheme

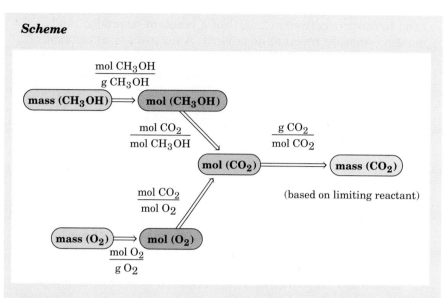

Solution

(a) CH_3OH: $40.0 \text{ g CH}_3\text{OH} \times \dfrac{1 \text{ mol CH}_3\text{OH}}{32.0 \text{ g CH}_3\text{OH}} \times \dfrac{2 \text{ mol CO}_2}{2 \text{ mol CH}_3\text{OH}}$

$$= 1.25 \text{ mol CO}_2$$

O_2: $46.0 \text{ g O}_2 \times \dfrac{1 \text{ mol O}_2}{32.0 \text{ g O}_2} \times \dfrac{2 \text{ mol CO}_2}{3 \text{ mol O}_2} = 0.958 \text{ mol CO}_2$

Therefore, O_2 is the limiting reactant.

(b) The yield is determined from the amount of product formed *from the limiting reactant.* Thus we simply convert the 0.958 mol of CO_2 produced by the O_2 to grams of CO_2.

$$0.958 \text{ mol CO}_2 \times \dfrac{44.0 \text{ g CO}_2}{\text{mol CO}_2} = \underline{\underline{42.2 \text{ g CO}_2}}$$

Most of us need a recipe in order to bake a cake from scratch. The recipe tells us the exact proportions of ingredients and how big a cake this will produce. Chemical equations serve as recipes for stoichiometry calculations. They tell us the proper mass proportions of reactants and exactly how much product we may obtain from these reactants. In cases where reactants are not mixed in exactly the stoichiometric proportions, the equation tells which reactant limits the amount of product.

▲Looking Back

Learning Check B

◀Checking It Out

B–1.   Fill in the blanks.
Balanced chemical equations can be used to construct _____ ratios relating moles of reactants and products. These ratios are

used to convert between _____ of a reactant or product and moles of another reactant or product. A mass-mass conversion requires a _____-step calculation. For example, the mass of a given reactant is converted to moles using the _____ _____ of that reactant. If a number of molecules is given, _____ number is used to convert to moles. If quantities of more than one reactant are given, the reactant that produces the _____ amount of product is the limiting reactant.

B–2. Consider the following combustion of methylamine (CH_3NH_2).

$$4CH_3NH_2(l) + 9O_2(g) \longrightarrow 4CO_2(g) + 2N_2(g) + 10H_2O(l)$$

(a) How many moles of O_2 react with 4.50 moles of methylamine?
(b) How many moles of N_2 are produced from 322 g of methylamine?
(c) What mass of H_2O is produced along with 6.45 g of CO_2?
(d) How many molecules of O_2 are required to produce 0.568 g of N_2?

B–3. Using the equation from Problem B-2, what mass of H_2O is produced when 20.0 g of CH_3NH_2 is allowed to react with 40.0 g of O_2?

Additional Examples: Problems 8-12, 8-15, 8-21, 8-29, 8-32, and 8-42.

Looking Ahead ▼

There is something else that we have been taking for granted in the problems worked so far. We have assumed that at least one reactant has been completely converted into products and that the reactants form only the products shown. This is not always the case, however. We will look into incomplete reactions and competing reactions in the next section.

8–5   PERCENT YIELD

An efficient automobile engine burns gasoline (mainly a hydrocarbon, C_8H_{18}) to form carbon dioxide and water. Untuned engines, however, do not burn gasoline efficiently. They produce carbon monoxide and may even exhaust unburned fuel. The two combustion reactions are shown below.

Complete combustion $2C_8H_{18}(g) + 25O_2(g) \longrightarrow 16CO_2(g) + 18H_2O(l)$

Incomplete combustion $2C_8H_{18}(g) + 17O_2(g) \longrightarrow 16CO(g) + 18H_2O(l)$

Note that if we were asked to calculate the mass of CO_2 produced from a given amount of C_8H_{18}, our answer would not be correct if all the hydrocarbon were not converted to CO_2. *The measured amount of product obtained in any reaction is known as the* **actual yield**. *The* **theoretical yield** *is the calculated amount of product that would be obtained if all of the reactant were converted to a given product. The* **per-**

Do reactions ever stop before the reactants are completely consumed?

cent yield *is the actual yield in grams or moles divided by the theoretical yield in grams or moles times 100%:*

$$\frac{\text{actual yield}}{\text{theoretical yield}} \times 100\% = \text{percent yield}$$

In other reactions, there is another reason for the incomplete conversion of reactants to products. An example is the reaction illustrated previously in which nitrogen and hydrogen are converted to ammonia. In this reaction, all of the hydrogen and nitrogen do not react to form ammonia. Actually, what is happening in such a reaction is that while reactants are forming products, products are reacting to reform reactants. The two reactions (forward and reverse) offset each other at some point, leaving measurable quantities of both reactants and products present. Such a reaction is *reversible* and is known as an *equilibrium reaction*; the reversibility is indicated by a double arrow rather than the yield sign. As in the case of the combustion of octane, the actual yield and the theoretical yield are not the same so that the percent yield is less than 100%. (Equilibrium reactions will be discussed in more detail in Chapters 13 and 14.) The reversible reaction involving the formation of ammonia is

$$N_2(g) + 3H_2(g) \rightleftharpoons 2NH_3(g)$$

To determine the percent yield, it is necessary to determine the theoretical yield (which is what we've been doing all along) and compare this with the actual yield.

EXAMPLE 8–12

Determination of Percent Yield

In a given experiment, a 4.70-g quantity of H_2 is allowed to react with N_2; a 12.5-g quantity of NH_3 is formed. What is the percent yield based on the H_2?

●Working It Out

Procedure
Find the mass of NH_3 that would form if all 4.70 g of H_2 were converted to NH_3 (the theoretical yield). Using the actual yield (12.5 g), find the percent yield.

Scheme

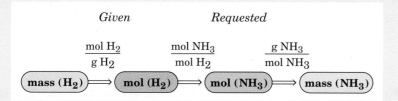

	Given	*Requested*	
	$\dfrac{\text{mol } H_2}{\text{g } H_2}$	$\dfrac{\text{mol } NH_3}{\text{mol } H_2}$	$\dfrac{\text{g } NH_3}{\text{mol } NH_3}$
mass (H_2) $\Longrightarrow$	mol (H_2) $\Longrightarrow$	mol (NH_3) $\Longrightarrow$	mass (NH_3)

The mole ratio needed is

$$\frac{2 \text{ mol } NH_3}{3 \text{ mol } H_2}$$

Solution

$$4.70 \ \cancel{g \ H_2} \times \frac{1 \ \cancel{mol \ H_2}}{2.02 \ \cancel{g \ H_2}} \times \frac{2 \ \cancel{mol \ NH_3}}{3 \ \cancel{mol \ H_2}} \times \frac{17.0 \ g \ NH_3}{\cancel{mol \ NH_3}}$$

$$= 26.4 \ g \ NH_3 \ \text{(theoretical yield)}$$

$$\frac{12.5 \ \cancel{g}}{26.4 \ \cancel{g}} \times 100\% = 47.3\% \ \text{yield}$$

EXAMPLE 8–13

Determination of Actual Yield from Percent Yield

Zinc and silver undergo a single-replacement reaction according to the following equation:

$$Zn(s) + 2AgNO_3(aq) \longrightarrow Zn(NO_3)_2(aq) + 2Ag(s)$$

When 25.0 g of zinc is added to the silver nitrate solution, the percent yield of silver is 72.3%. What mass of silver is formed?

Procedure
Find the theoretical yield of silver by converting the mass of zinc to the equivalent mass of silver. Use the percent yield to calculate the actual yield. This can be accomplished algebraically as follows:

$$\text{actual yield} = \frac{\text{percent yield}}{100\%} \times \text{theoretical yield}$$

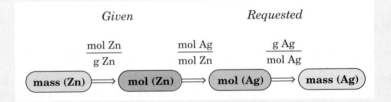

Solution

$$25.0 \ \cancel{g \ Zn} \times \frac{1 \ \cancel{mol \ Zn}}{65.4 \ \cancel{g \ Zn}} \times \frac{2 \ \cancel{mol \ Ag}}{1 \ \cancel{mol \ Zn}} \times \frac{108 \ g \ Ag}{\cancel{mol \ Ag}}$$

$$= 82.6 \ g \ Ag \ \text{(theoretical yield)}$$

$$\frac{72.3\%}{100\%} \times 82.6 \ g \ Ag = 59.7 \ g \ Ag \ \text{(actual yield)}$$

Looking Ahead ▼

Matter is not all that is involved in chemical reactions. The other component of the universe, energy, is also an intimate part of chemical reactions. Like matter, a measurable amount of energy is involved in any chemical change. In the next section, we will expand our chemical equations to include heat energy.

8–6 HEAT ENERGY IN CHEMICAL REACTIONS

We are well aware that in combustion reactions an important component of the reaction is the liberation of large amounts of heat energy. Not only is the amount of products limited by the amount of reactants present, but so is the amount of heat. *The study of heat and its relationship to chemical changes is known as* **chemical thermodynamics.**

All elements and compounds contain potential energy in the form of chemical energy. When the total potential energy of the products is less than that of the reactants, the difference in potential energy between products and reactants is evolved during the course of the reaction. Usually, this potential energy is evolved in the form of heat energy. (Light and electrical energy may also be involved in some cases.) Thus most chemical reactions move from a position of higher potential energy to a lower position, just as a car moves down a hill or a pencil drops to the floor. A reaction that evolves heat energy is said to be *exothermic*. In order for a reaction to move from reactants at a lower potential energy to products at a higher potential energy, heat energy must be supplied. A reaction that absorbs heat energy is said to be *endothermic*. The amount of heat energy (measured in kilocalories or kilojoules) involved in a reaction is a constant amount that depends on the amount of reactants present. For example, if one mole of hydrogen undergoes combustion to form liquid water, 286 kJ of heat is evolved.

A balanced equation that includes heat energy is referred to as a **thermochemical equation.** A thermochemical equation can be represented in either of two ways. In the first, the heat is shown separately from the balanced equation using the symbol Δ**H.** This is referred to as "heat of the reaction." The symbol ΔH is known technically as the *change in enthalpy.* This refers to the *change* in potential energy when reactants convert into products. By convention, *a negative sign for ΔH corresponds to an exothermic reaction, and a positive sign corresponds to an endothermic reaction.* Written in this manner, the thermochemical equation for the combustion of two moles of hydrogen is

$$2H_2(g) + O_2(g) \longrightarrow 2H_2O(l) \quad \Delta H = -572 \text{ kJ}$$

Notice that this thermochemical equation indicates that 572 kJ of heat energy is evolved in the formation of *two* moles of liquid H_2O from its elements. The heat evolved *per mole* of H_2O formed is also a useful quantity and can be adapted to any equation, as illustrtated in the example that follows this discussion. The information is written as

$$\Delta H = -286 \, \frac{\text{kJ}}{\text{mol } H_2O}$$

The second way a thermochemical equation is represented shows the heat energy as if it were a reactant or product. In exothermic reactions, heat energy is a product, and in endothermic reactions heat energy is a reactant. A positive sign is used in either case.

Can heat energy be considered a reactant or product?

$$2H_2(g) + O_2(g) \longrightarrow 2H_2O(l) + 572 \text{ kJ} \quad \text{(exothermic reaction)}$$

$$N_2(g) + O_2(g) + 181 \text{ kJ} \longrightarrow 2NO(g) \quad \text{(endothermic reaction)}$$

Heat energy from the combustion of liquid hydrogen with liquid oxygen in the main thruster lifts the space shuttle from the launch pad.

The heat energy involved in a chemical reaction may be treated quantitatively in a manner similar to the amount of a reactant or product. The following examples illustrate the calculations implied by thermochemical equations.

EXAMPLE 8–14

Mass → Heat Conversion

Oxygen is often prepared in the laboratory by heating potassium chlorate. In addition to oxygen gas, potassium chloride is produced. The process requires the addition of 45.0 kJ of heat energy per mole of potassium chlorate.
(a) Write the balanced thermochemical equation in both ways.
(b) How much heat is required to decompose 10.0 g of potassium chlorate?

Procedure
(a) The balanced equation is

$$2KClO_3(s) \longrightarrow 2KCl(s) + 3O_2(g)$$

Note that 2 mol of $KClO_3$ is in the balanced equation. Thus

$$\Delta H = 2 \text{ mol } KClO_3 \times \frac{45.0 \text{ kJ}}{\text{mol } KClO_3} = 90.0 \text{ kJ}$$

(b) Two steps are required:

$$\frac{\text{mol } KClO_3}{\text{g } KClO_3} \quad \text{and} \quad \frac{\text{kJ}}{\text{mol } KClO_3}$$

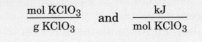

$$\boxed{\text{mass (KClO}_3\text{)}} \Longrightarrow \boxed{\text{mol (KClO}_3\text{)}} \Longrightarrow \boxed{\text{kJ}}$$

Solution

(a) The thermochemical equations are

$$2KClO_3(s) \longrightarrow 2KCl(s) + 3O_2(g) \quad \Delta H = 90.0 \text{ kJ}$$

$$2KClO_3(s) + 90.0 \text{ kJ} \longrightarrow 2KCl(s) + 3O_2(g)$$

(b)

$$10.0 \cancel{\text{ g KClO}_3} \times \frac{1 \cancel{\text{ mol KClO}_3}}{123 \cancel{\text{ g KClO}_3}} \times \frac{45.0 \text{ kJ}}{\cancel{\text{mol KClO}_3}} = \underline{\underline{3.66 \text{ kJ}}}$$

EXAMPLE 8–15

Heat → Mass Conversion

Acetylene, which is used in welding torches, undergoes combustion according to the following thermochemical equation:

$$2C_2H_2(g) + 5O_2(g) \longrightarrow 4CO_2(g) + 2H_2O(l) \quad \Delta H = -2602 \text{ kJ}$$

If 550 kJ of heat is evolved in the combustion of a sample of C_2H_2, what is the mass of CO_2 formed?

Procedure

Two steps are required as illustrated below.

$$\frac{\text{mol } CO_2}{\text{kJ}} \qquad \frac{\text{g } CO_2}{\text{mol } CO_2}$$

$$\boxed{\text{kJ}} \Longrightarrow \boxed{\text{mol } (CO_2)} \Longrightarrow \boxed{\text{mass } (CO_2)}$$

Solution

$$550 \text{ kJ} \times \frac{4 \cancel{\text{ mol CO}_2}}{2602 \text{ kJ}} \times \frac{44.0 \text{ g } CO_2}{\cancel{\text{mol CO}_2}} = \underline{\underline{37.2 \text{ g } CO_2}}$$

The combustion of acetylene produces a flame hot enough to melt iron.

Sometimes, even when we mix reactants in stoichiometric amounts, we don't obtain as much product as a calculation would indicate. In these cases we express the actual amount of product as a percent of the calculated amount. The result is the percent yield. Finally, we note that heat energy can be included as part of a balanced equation, which allows us to include heat as part of stoichiometry.

▲ **Looking Back**

Learning Check C

◀ **Checking It Out**

C–1. Fill in the blanks.

The calculated amount of a product that forms if at least one reactant is completely consumed is the _____ yield. The amount of product that is measured is the _____ yield. The ratio of the latter divided by the former times 100% is the _____ yield. When heat energy is included, a balanced equation is called a _____ equation. When the sign of ΔH is negative or the amount of heat is included as a product, the reaction is _____ .

C–2. The balanced equation describing the combustion of ethylene (C_2H_4) is

$$C_2H_4(g) + 3O_2(g) \longrightarrow 2CO_2(g) + 2H_2O(l)$$

In a given experiment, 25.0 g of CO_2 was obtained as a product, which represented a 74.0% yield. What mass of C_2H_4 was consumed in this reaction?

C–3. In the reaction in Problem C-2, $\Delta H = -556$ kJ/mol CO_2. Write a thermochemical reaction that includes heat energy as a reactant or product. What mass of C_2H_4 is required to produce 250 kJ of heat, assuming 100% yield?

Additional Examples: Problems 8-46, 8-49, 8-51, and 8-56.

CHAPTER REVIEW

Putting It Together ▲▼

A concise statement of a chemical property is relayed by the **chemical equation.** With a few symbols, formulas, and other abbreviations, a sizable amount of chemical information can be communicated. This includes the elements or compounds involved as **reactants** and **products,** their physical states, and the number of molecules of each compound involved in the reaction. Three aspects of the chemical equation were developed in this chapter: (1) the balancing of the equations, (2) the representation of five basic types of reactions, and (3) the quantitative relationships between and among reactants and products, as implied by the equations.

The first topic involved some rules and practice in balancing simple equations by inspection. **Balanced equations** have the same number of atoms of each element on both sides of the equation. This is accomplished by means of **coefficients** that multiply all of the atoms of a compound by a whole number.

After balancing equations, we looked into five different types of reactions. Each type has a general equation that characterizes that kind of reaction. The five types discussed in this chapter are illustrated below, with letters illustrating elements, compounds, or ions.

1. **Combustion reactions**

$$A + O_2 \longrightarrow AO$$

$$AB + O_2 \longrightarrow AO + BO$$

2. **Combination reactions**

$$A + B \longrightarrow C$$

3. **Decomposition reactions**

$$C \longrightarrow A + B$$

4. **Single-replacement reactions**

$$A + BC \longrightarrow AC + B$$

5. **Double-replacement reactions (precipitation** or **neutralization reactions)**

$$AB + CD \longrightarrow AD + CB$$

In the final type of reaction, if one of the products is an insoluble solid compound, it is called a **precipitate.** If AB and CD represent an **acid** and a **base,** then the products are water and a **salt.**

Next we examined how chemical equations tell us about the relationships of masses of reactants to masses of products. These are the calculations of **stoichiometry.** In these problems, the balanced equation provided the mole ratios for conversions between moles of one compound in the reaction and moles of another. All of these conversions are summarized in Figure 8-6.

When reactants are mixed in other than **stoichiometric** amounts, it is necessary to determine the **limiting reactant.** The limiting reactant is completely consumed; thus it determines the amount of product formed.

Some reactions lead to more than one set of products, and others reach a state of equilibrium before all reactants are converted to products. In either case, the **actual yield** of a product may be less than the **theoretical yield.** The actual yield can be expressed as a percent of the theoretical yield. This is called the **percent yield.**

Finally, we studied the relation of heat energy to chemical reactions, which is known as **chemical thermodynamics.** In a **thermochemical equation,** the heat evolved or absorbed is represented as either a reactant or product, or as the heat of the reaction, which is referred to as the change in enthalpy and is symbolized as $\Delta \mathbf{H}$.

E X E R C I S E S

CHEMICAL EQUATIONS

8–1 The physical state of an element is included in a chemical equation. Each of the following compounds is a gas, a solid, or a liquid under normal conditions. Indicate the proper physical state by adding either (g), (s), or (l) after the formula.
(a) Cl_2 (e) P_4 (i) S_8
(b) C   (f) H_2 (j) Na
(c) K_2SO_4 (g) Br_2 (k) Hg
(d) H_2O (h) NaBr   (l) CO_2

8–2 Balance the following equations.
(a) $CaCO_3 \xrightarrow{\Delta} CaO + CO_2$
(b) $Na + O_2 \longrightarrow Na_2O$
(c) $H_2SO_4 + NaOH \longrightarrow Na_2SO_4 + H_2O$
(d) $H_2O_2 \longrightarrow H_2O + O_2$
(e) $Al + H_3PO_4 \longrightarrow AlPO_4 + H_2$
(f) $Ca(OH)_2 + HCl \longrightarrow CaCl_2 + H_2O$
(g) $Mg + N_2 \longrightarrow Mg_3N_2$
(h) $C_2H_6 + O_2 \longrightarrow CO_2 + H_2O$
(i) $B_4H_{10} + O_2 \longrightarrow B_2O_3 + H_2O$
(j) $SF_6 + SO_3 \longrightarrow O_2SF_2$
(k) $CS_2 + O_2 \longrightarrow CO_2 + SO_2$

8–3 Balance the following equations.
(a) $NaBr + Cl_2 \longrightarrow NaCl + Br_2$
(b) $KOH + H_3AsO_4 \longrightarrow K_2HAsO_4 + H_2O$
(c) $Ti + Cl_2 \longrightarrow TiCl_4$
(d) $Al + H_2SO_4 \longrightarrow Al_2(SO_4)_3 + H_2$
(e) $Ca(CN)_2 + HBr \longrightarrow CaBr_2 + HCN$
(f) $C_3H_6 + O_2 \longrightarrow CO + H_2O$
(g) $P_4 + S_8 \longrightarrow P_4S_3$
(h) $Cr_2O_3 + Si \longrightarrow Cr + SiO_2$

8–4 Balance the following equations.
(a) $Mg_3N_2 + H_2O \longrightarrow Mg(OH)_2 + NH_3$
(b) $H_2S + O_2 \longrightarrow S + H_2O$
(c) $Si_2H_6 + H_2O \longrightarrow Si(OH)_4 + H_2$

(d) $C_2H_6 + Cl_2 \longrightarrow C_2HCl_5 + HCl$
(e) $NH_3 + Cl_2 \longrightarrow NHCl_2 + HCl$
(f) $PBr_3 + H_2O \longrightarrow HBr + H_3PO_3$

8–5 Balance the following equations.
(a) $Na_2NH + H_2O \longrightarrow NH_3 + NaOH$
(b) $CaC_2 + H_2O \longrightarrow C_2H_2 + Ca(OH)_2$
(c) $XeF_6 + H_2O \longrightarrow XeO_3 + HF$
(d) $PCl_5 + H_2O \longrightarrow H_3PO_4 + HCl$
(e) $BF_3 + NaH \longrightarrow B_2H_6 + NaF$
(f) $Bi + O_2 \longrightarrow Bi_2O_3$

8–6 Write balanced chemical equations from the following word equations. Include the physical state of each element or compound.
(a) Sodium metal plus water yields hydrogen gas and an aqueous sodium hydroxide solution.
(b) Potassium chlorate when heated yields potassium chloride plus oxygen gas. (Ionic compounds are solids.)
(c) An aqueous sodium chloride solution plus an aqueous silver nitrate solution yields a silver chloride precipitate (solid) and a sodium nitrate solution.
(d) An aqueous phosphoric acid solution plus an aqueous calcium hydroxide solution yields water and solid calcium phosphate.

8–7 Write balanced chemical equations from the following word equations. Include the physical state of each element or compound.
(a) Solid phenol (C_6H_6O) reacts with oxygen to form carbon dioxide gas and liquid water.
(b) An aqueous calcium hydroxide solution reacts with gaseous sulfur trioxide to form a precipitate of calcium sulfate and water.
(c) Lithium is the only element that combines with nitrogen at room temperature. The reaction forms lithium nitride.
(d) Magnesium dissolves in an aqueous chromium(III) nitrate solution to form chromium and a magnesium nitrate solution.

TYPES OF CHEMICAL REACTIONS

8–8 Which of the five types of reactions is represented by each equation in Problems 8-2 and 8-6?

8–9 Which of the five types of reactions is represented by each equation in Problems 8-3 and 8-7?

*8–10 Write balanced equations by predicting the products of the following reactions. Include the physical state of each element or compound.

(a) combination of potassium and chlorine
(b) single-replacement reaction of calcium metal with water to produce calcium hydroxide solution and a gas
(c) combustion of liquid benzene (C_6H_6)
(d) double-replacement reaction between an aqueous sodium sulfide solution and an aqueous copper(II) nitrate solution (The sulfide product is a solid.)
(e) decomposition of gold(III) oxide into its elements by heating

*8–11 Write balanced equations by predicting the products of the following reactions. Include the physical state of each element or compound.
(a) combustion of liquid butane (C_4H_{10})
(b) single-replacement reaction involving chlorine gas and a potassium bromide solution
(c) decomposition of a sulfurous acid solution to water and a gas
(d) double-replacement reaction involving aqueous perchloric acid and aqueous cesium hydroxide
(e) combination of sodium oxide and gaseous sulfur trioxide

STOICHIOMETRY

8–12 Given the balanced equation
$$Mg + 2HCl \longrightarrow MgCl_2 + H_2$$
provide the proper mole ratio that makes the following mole conversions.
(a) Mg to H_2 (c) HCl to H_2
(b) Mg to HCl   (d) $MgCl_2$ to HCl

8–13 Given the balanced equation
$$2C_4H_{10} + 13O_2 \longrightarrow 8CO_2 + 10H_2O$$
provide the proper mole ratio that makes the following mole conversions.
(a) CO_2 to C_4H_{10} (c) CO_2 to O_2
(b) O_2 to C_4H_{10} (d) O_2 to H_2O

8–14 Given the balanced equation
$$Cu + 4HNO_3 \longrightarrow Cu(NO_3)_2 + 2NO_2 + 2H_2O$$
provide the proper mole ratio that makes the following mole conversions.
(a) Cu to NO_2 (d) $Cu(NO_3)_2$ to HNO_3
(b) HNO_3 to Cu   (e) NO_2 to Cu
(c) H_2O to NO_2 (f) Cu to H_2O

8–15 The reaction that takes place in the reusable solid rocket booster for the space shuttle is shown by the following equation.

$$3Al(s) + 3NH_4ClO_4(s) \longrightarrow Al_2O_3(s)$$
$$+ AlCl_3(s) + 3NO(g) + 6H_2O(g)$$

(a) How many moles of each product is formed from 10.0 moles of Al?

(b) How many moles of each product is formed from 3.00 moles of NH_4ClO_4?

8-16 Phosphine (PH_3) is a poisonous gas once used as a fumigate for stored grain. It is prepared according to the following equation.

$$Ca_3P_2(s) + 6H_2O(l) \longrightarrow 3Ca(OH)_2(s)$$
$$+ 2PH_3(g)$$

(a) How many moles of phosphine are prepared from 5.00 moles of Ca_3P_2?

(b) How many moles of phosphine are prepared from 5.00 moles of H_2O?

8-17 Hydrogen cyanide is an important industrial chemical used to make a plastic, acrylonitrile. HCN is prepared according to the following equation.

$$2NH_3(g) + 3O_2(g)$$
$$+ 2CH_4(g) \longrightarrow 2HCN(g) + 6H_2O(l)$$

(a) How many moles of O_2 and CH_4 react with 10.0 moles of NH_3?

(b) How many moles of HCN and H_2O are produced from 10.0 moles of O_2?

8-18 Iron rusts according to the equation

$$4Fe(s) + 3O_2(g) \longrightarrow 2Fe_2O_3(s)$$

(a) What mass of rust (Fe_2O_3) is formed from 0.275 mole of Fe?

(b) What mass of rust is formed from 0.275 mole of O_2?

(c) What mass of O_2 reacts with 0.275 mole of Fe?

8-19 Glass (SiO_2) is etched with hydrofluoric acid according to the equation

$$SiO_2(s) + 4HF(aq) \longrightarrow SiF_4(g) + 2H_2O(l)$$

If 4.86 moles of HF reacts with SiO_2,

(a) What mass of SiF_4 forms?

(b) What mass of H_2O forms?

(c) What mass of SiO_2 reacts?

8-20 Consider the reaction

$$2H_2 + O_2 \longrightarrow 2H_2O$$

(a) How many moles of H_2 are needed to produce 0.400 mol of H_2O?

(b) How many moles of H_2O will be produced from 0.640 g of O_2?

(c) How many moles of H_2 are neeeded to react with 0.032 g of O_2?

(d) What mass of H_2O would be produced from 0.400 g of H_2?

8-21 Propane burns according to the equation

$$C_3H_8 + 5O_2 \longrightarrow 3CO_2 + 4H_2O$$

(a) How many moles of CO_2 are produced from the combustion of 0.450 mol of C_3H_8? How many moles of H_2O? How many moles of O_2 are needed?

(b) What mass of H_2O is produced if 0.200 mol of CO_2 is also produced?

(c) What mass of C_3H_8 is required to produce 1.80 g of H_2O?

(d) What mass of C_3H_8 is required to react with 160 g of O_2?

(e) What mass of CO_2 is produced by the reaction of 1.20×10^{23} molecules of O_2?

(f) How many moles of H_2O are produced if 4.50×10^{22} molecules of CO_2 are produced?

8-22 The alcohol component of gasohol burns according to the equation

$$C_2H_5OH(l) + 3O_2(g) \longrightarrow 2CO_2(g)$$
$$+ 3H_2O(g)$$

(a) What mass of alcohol is needed to produce 5.45 mol of H_2O?

(b) How many moles of CO_2 are produced along with 155 g of H_2O?

(c) What mass of CO_2 is produced from 146 g of C_2H_5OH?

(d) What mass of C_2H_5OH reacts with 0.898 g of O_2?

(e) What mass of H_2O is produced from 5.85×10^{24} molecules of O_2?

8-23 In the atmosphere, N_2 and O_2 do not react with each other. In the high temperatures of an automobile engine, however, the following reaction occurs:

$$N_2(g) + O_2(g) \longrightarrow 2NO(g)$$

When the NO reaches the atmosphere through the engine exhaust, a second reaction takes place.

$$2NO(g) + O_2(g) \longrightarrow 2NO_2(g)$$

The NO_2 is a brownish gas that contributes to the haze of smog and is irritating to the nasal passages and lungs. What mass of N_2 is required to produce 155 g of NO_2?

8-24 Elemental iron is produced in what is called the *thermite reaction* because it produces enough heat that the iron is initially in the molten state. The liquid iron can then be used to weld iron bars together.

$$2Al(s) + Fe_2O_3(s) \longrightarrow Al_2O_3(s) + 2Fe(l)$$

What mass of Al is needed to produce 750 g of Fe? How many formula units of Fe_2O_3 are used in the process?

8–25 Antacids, which contain calcium carbonate, react with stomach acid according to the equation

$$CaCO_3(s) + 2HCl(aq) \longrightarrow CaCl_2(aq)$$
$$+ CO_2(g) + H_2O(l)$$

What mass of stomach acid reacts with 1.00 g of $CaCO_3$?

8–26 Elemental copper can be recovered from the mineral chalcocite (Cu_2S). From the following equation, determine what mass of Cu is formed from 7.82×10^{22} molecules of O_2.

$$Cu_2S(s) + O_2(g) \longrightarrow 2Cu(s) + SO_2(g)$$

8–27 Fool's gold (pyrite) is so named because it looks much like gold. When it is placed in aqueous HCl, however, it dissolves and gives off the pungent gas H_2S. Gold itself does not react with aqeuous HCl. From the following equation, determine how many individual molecules of H_2S are formed from 0.520 mol of pyrite (FeS_2).

$$FeS_2(s) + 2HCl(aq) \longrightarrow FeCl_2(aq)$$
$$+ H_2S(g) + S(s)$$

8–28 Nitrogen dioxide may form so-called acid rain by reaction with water in the air according to the equation

$$3NO_2(g) + H_2O(l) \longrightarrow 2HNO_3(aq) + NO(g)$$

What mass of nitric acid is produced from 18.5 kg of NO_2?

8–29 Elemental chlorine can be generated in the laboratory according to the equation

$$MnO_2(s) + 4HCl(aq) \longrightarrow MnCl_2(aq)$$
$$+ 2H_2O(l) + Cl_2(g)$$

What mass of Cl_2 is produced from the reaction of 665 g of HCl?

8–30 The fermentation of sugar to produce ethyl alcohol is represented by the equation

$$C_6H_{12}O_6(s) \longrightarrow 2C_2H_5OH(l) + 2CO_2(g)$$

What mass of alcohol is produced from 25.0 mol of sugar?

8–31 Methane gas can be made from carbon monoxide gas according to the equation

$$2CO(g) + 2H_2(g) \longrightarrow CH_4(g) + CO_2(g)$$

What mass of CO is required to produce 8.75×10^{25} molecules of CH_4?

LIMITING REACTANT

8–32 Nitrogen gas can be prepared by passing ammonia over hot copper(II) oxide according to the equation

$$3CuO(s) + 2NH_3(g) \longrightarrow N_2(g) + 3Cu(s)$$
$$+ 3H_2O(g)$$

How many moles of N_2 are prepared from the following mixtures?
(a) 3.00 moles of CuO and 3.00 moles of NH_3
(b) 3.00 moles of CuO and 2.00 moles of NH_3
(c) 3.00 moles of CuO and 1.00 moles of NH_3
(d) 0.628 mole of CuO and 0.430 mole of NH_3
(e) 5.44 moles of CuO and 3.50 moles of NH_3

8–33 How many moles remain of the reactant in excess in Problem 8-32 (a and c).

8–34 Consider the equation illustrating the combustion of arsenic.

$$4As(s) + 5O_2(g) \longrightarrow 2As_2O_5(s)$$

How many moles of As_2O_5 are prepared from the following mixtures?
(a) 4.00 moles of As and 4.00 moles of O_2
(b) 3.00 moles of As and 4.00 moles of O_2
(c) 5.62 moles of As and 7.50 moles of O_2
(d) 3.86 moles of As and 4.75 moles of O_2

8–35 How many moles remain of the reactant in excess in Problem 8-34 (a and b)?

8–36 Consider the equation

$$2Al + 3H_2SO_4 \longrightarrow Al_2(SO_4)_3 + 3H_2$$

If 0.800 mol of Al is mixed with 1.00 mol of H_2SO_4, how many moles of H_2 are produced? How many moles of one of the reactants remain?

8–37 Consider the equation

$$2C_5H_6 + 13O_2 \longrightarrow 10CO_2 + 6H_2O$$

If 3.44 mol of C_5H_6 is mixed with 20.6 mol of O_2, what mass of CO_2 is formed?

8–38 Elemental fluorine is very difficult to prepare by ordinary chemical reactions. In 1986, however, a chemical reaction was reported that produces fluorine.

$$2K_2MnF_6 + 4SbF_5 \longrightarrow$$
$$4KSbF_6 + 2MnF_3 + F_2$$

If a 525-g quantity of K_2MnF_6 is mixed with a 900-g quantity of SbF_5, what mass of F_2 is produced?

8-39 Consider the equation

$$4NH_3(g) + 3O_2(g) \longrightarrow 2N_2(g) + 6H_2O(l)$$

If a 40.0-g sample of O_2 is mixed with 1.50 mol of NH_3, which is the limiting reactant? How many moles of N_2 form?

8-40 Consider the equation

$$2AgNO_3(aq) + CaCl_2(aq) \longrightarrow$$
$$2AgCl(s) + Ca(NO_3)_2(aq)$$

If a solution containing 20.0 g of $AgNO_3$ is mixed with a solution containing 10.0 g of $CaCl_2$, which compound is the limiting reactant? What mass of $AgCl$ forms? What mass of one of the reactants remains?

8-41 Limestone ($CaCO_3$) dissolves in hydrochloric acid as shown by the equation

$$CaCO_3(s) + 2HCl(aq) \longrightarrow$$
$$CaCl_2(aq) + CO_2(g) + H_2O(l)$$

If 20.0 g of $CaCO_3$ and 25.0 g of HCl are mixed, what mass of CO_2 is produced? What mass of one of the reactants remains?

8-42 Consider the balanced equation

$$2HNO_3(aq) + 3H_2S(aq) \longrightarrow$$
$$2NO(g) + 4H_2O(l) + 3S(s)$$

If a 10.0-g quantity of HNO_3 is mixed with 5.00 g of H_2S, what is the mass of each product and of the reactant present in excess after reaction occurs?

THEORETICAL AND PERCENT YIELD

8-43 Sulfur trioxide is prepared from SO_2 according to the equation

$$2SO_2(g) + O_2(g) \rightleftharpoons 2SO_3(g)$$

In this reaction, not all SO_2 is converted to SO_3 even with excess O_2 present. In a given experiment, 21.2 g of SO_3 was produced from 24.0 g of SO_2. What is the theoretical yield of SO_3? What is the percent yield?

8-44 The following is a reversible decomposition reaction:

$$2N_2O_5 \rightleftharpoons 4NO_2 + O_2$$

When 25.0 g of N_2O_5 decomposes, it is found that 10.0 g of NO_2 forms. What is the percent yield?

8-45 The following equation represents a reversible combination reaction.

$$P_4O_{10} + 6PCl_5 \rightleftharpoons 10POCl_3$$

If 25.0 g of PCl_5 react, there is a 45.0% yield of $POCl_3$. What is the actual yield in grams?

8-46 Octane in gasoline burns in an automobile engine according to the equation

$$2C_8H_{18}(l) + 25O_2(g) \longrightarrow$$
$$16CO_2(g) + 18H_2O(g)$$

If a 57.0-g sample of octane is burned, 152 g of CO_2 is formed. What is the percent yield of CO_2?

*8-47 In Problem 8-46, the C_8H_{18} that is *not* converted to CO_2 forms CO. What is the mass of CO formed? (CO is a poisonous pollutant that is converted to CO_2 in a car's catalytic converter.)

*8-48 Given the reaction

$$2NO_2 + 4H_2 \rightleftharpoons N_2 + 4H_2O$$

what mass of hydrogen is required to produce 250 g of N_2 if the yield is 70.0%?

*8-49 When benzene reacts with bromine, the principal reaction is

$$C_6H_6 + Br_2 \longrightarrow C_6H_5Br + HBr$$

If the yield of bromobenzene (C_6H_5Br) is 65.2%, what mass of bromobenzene is produced from 12.5 g of C_6H_6?

*8-50 A second reaction between C_6H_6 and Br_2 (see Problem 8-49) produces dibromobenzene ($C_6H_4Br_2$).
(a) Write the balanced equation illustrating this reaction.
(b) If the remainder of the benzene from Problem 6-49 reacts to form dibromobenzene, what is the mass of $C_6H_4Br_2$ produced?

HEAT IN CHEMICAL REACTIONS

8-51 When one mole of magnesium undergoes combustion to form magnesium oxide, 602 kJ of heat energy is evolved. Write the thermochemical equation in both forms.

8-52 In Problem 8-23, the reaction between N_2 and O_2 was discussed. A 90.5-kJ quantity of heat energy must be supplied per mole of NO formed. Write the balanced thermochemical equation in both forms. Is the reaction exothermic or endothermic?

8–53 To decompose one mole of $CaCO_3$ to CaO and CO_2, 176 kJ must be supplied. Write the balanced thermochemical equation in both forms.

8–54 The complete combustion of one mole of octane (C_8H_{18}) in gasoline evolves 5480 kJ of heat. The complete combustion of one mole of methane in natural gas (CH_4) evolves 890 kJ. How much heat is evolved per 1.00 g for each of these fuels?

8–55 Methyl alcohol (CH_3OH) is used as a fuel in racing cars. It burns according to the equation

$$2CH_3OH(l) + 3O_2(g) \longrightarrow$$
$$2CO_2(g) + 4H_2O(l) + 1750 \text{ kJ}$$

What amount of heat is evolved per 1.00 g of alcohol? How does this compare with the amount of heat per gram of octane in gasoline? (See Problem 8-54.)

8–56 The thermite reaction was discussed in Problem 8-24. For the balanced equation, $\Delta H = -850$ kJ. What mass of aluminum is needed to produce 35.8 kJ of heat energy?

8–57 Photosynthesis is an endothermic reaction that forms glucose ($C_6H_{12}O_6$) from carbon dioxide, water, and energy from the sun. The balanced equation is

$$6CO_2(g) + 6H_2O(l) \longrightarrow$$
$$C_6H_{12}O_6(aq) + 6O_2(g) \quad \Delta H = +2519 \text{ kJ}$$

What mass of glucose is formed from 975 kJ of energy?

8–58 When butane (C_4H_{10}) in a cigarette lighter burns, it evolves 2880 kJ per mole of butane. What is the mass of water formed if 1250 kJ of heat evolves?

GENERAL PROBLEMS

8–59 Calcium cyanamide ($CaCN_2$) is used as a fertilizer. When it reacts with water, it produces ammonia (which fertilizes the soil) and $CaCO_3$ (which counteracts acidity in the soil). Write the balanced equation illustrating the reaction, and calculate the mass of NH_3 produced from a 1.00-kg quantity of $CaCN_2$.

*8–60 Liquid iron is made from iron ore (Fe_2O_3) in a three-step process in a blast furnace as follows:
1. $3Fe_2O_3(s) + CO(g) \longrightarrow 2Fe_3O_4(s) + CO_2(g)$

2. $Fe_3O_4(s) + CO(g) \longrightarrow 3FeO(s) + CO_2(g)$
3. $FeO(s) + CO(g) \longrightarrow Fe(l) + CO_2(g)$
What mass of iron would eventually be produced from 125 g of Fe_2O_3?

8–61 A 50.0-g sample of *impure* $KClO_3$ is decomposed to KCl and O_2. If a 12.0-g quantity of O_2 is produced, what percent of the sample is $KClO_3$? (Assume that all of the $KClO_3$ present decomposes.)

8–62 In Example 8-9, SO_2 was formed from the burning of pyrite (FeS_2) in coal. If a 312-g quantity of SO_2 was collected from the burning of 6.50 kg of coal, what percent of the original sample was pyrite?

*8–63 Copper metal can be recovered from an ore ($CuCO_3$) by the decomposition reaction

$$2CuCO_3(s) \longrightarrow 2Cu(s) + 2CO_2(g) + O_2(g)$$

What is the mass of a sample of *impure* ore if it is 47.5% $CuCO_3$ and produces 350 g of Cu? (Assume complete decomposition of $CuCO_3$.)

*8–64 Consider the equation

$$4NH_3(g) + 5O_2(g) \longrightarrow 4NO(g) + 6H_2O(l)$$

When an 80.0-g quantity of NH_3 is mixed with 200 g of O_2, a 40.0-g quantity of NO is formed. Calculate the percent yield based on the limiting reactant.

*8–65 Consider the equation

$$3K_2MnO_4 + 4CO_2 + 2H_2O \longrightarrow$$
$$2KMnO_4 + 4KHCO_3 + MnO_2$$

How many moles of MnO_2 are produced if 9.50 mol of K_2MnO_4, 6.02×10^{24} molecules of CO_2, and 90.0 g of H_2O are mixed?

*8–66 Calcium chloride hydrate ($CaCl_2 \cdot 6H_2O$) is a solid used to melt ice at low temperatures. It is prepared according to the equation

$$CaCO_3(s) + 2HCl(g) + 5H_2O(l) \longrightarrow$$
$$CaCl_2 \cdot 6H_2O(s) + CO_2(g)$$

What mass of the hydrate is prepared from a mixture of 0.250 mole of H_2O, 9.50×10^{22} molecules of HCl, and 15.0 g of $CaCO_3$?

*8–67 A 2.85-g quantity of gaseous methane is mixed with 15.0 g of chlorine to produce a liquid, compound X, and gaseous hydrogen chloride. Compound X is 14.1% C, 2.35% H, and 83.5% Cl. Its molecular formula is the same as its empirical formula.
(a) From the formula of X, write a balanced equation. (Hint: Balance hydrogens before chlorines.)

(b) What is the theoretical yield of compound X?

*8–68 A 10.00-g sample of gaseous ethane (C_2H_6) reacts with chlorine to form gaseous hydrogen chloride and a liquid compound (Y) that has a molar mass of 168 g/mol. Compound Y is 14.3% carbon, 84.5% chlorine, and the remainder hydrogen. The reaction produces 12.0 g of compound Y, which is 57.0% yield.
(a) Write the balanced equation illustrating the reaction.
(b) What mass of ethane was required?

*8–69 The remainder of the ethane from Problem 8-68 reacts to form a liquid compound (Z) that is 18.0% carbon, 79.8% chlorine, and 2.25% hydrogen. The empirical and molecular formulas of this compound are the same.
(a) Write the balanced equation illustrating the reaction.
(b) What mass of compound Z is formed?

SOLUTIONS TO LEARNING CHECKS

A–1 equation, reactants, products, balanced, coefficients, (1) combustion, (2) combination, (3) decomposition, (4) single-replacement, (5) double-replacement, precipitation, neutralization.

A–2 $2Si_2H_6(g) + 7O_2(g) \longrightarrow 4SiO_2(s) + 6H_2O(l)$

A–3 (a) $2Fe(s) + 3H_2SO_4(aq) \longrightarrow Fe_2(SO_4)_3(aq) + 3H_2(g)$
(b) $Na_2CO_3(aq) + Sr(NO_3)_2(aq) \longrightarrow SrCO_3(s) + 2NaNO_3(aq)$

A–4 A-2: combustion
A-3(a): single-replacement
A-3(b): double-replacement

B–1 mole, moles, three, molar mass, Avogadro's, least

B–2 (a) $\boxed{\text{mol } CH_3NH_2} \Longrightarrow \boxed{\text{mol } O_2}$

(b)

(c) $\boxed{\text{g } CO_2} \Longrightarrow \boxed{\text{mol } CO_2} \Longrightarrow \boxed{\text{mol } H_2O} \Longrightarrow \boxed{\text{g } H_2O}$

(d)

B–3

$$20.0 \text{ g } \cancel{CH_3NH_2} \times \frac{1 \cancel{\text{ mol } CH_3NH_2}}{31.1 \cancel{\text{ g } CH_3NH_2}} \times \frac{10 \text{ mol } H_2O}{4 \cancel{\text{ mol } CH_3NH_2}} = 1.61 \text{ mol } H_2O$$

$$40.0 \text{ g } \cancel{O_2} \times \frac{1 \cancel{\text{ mol } O_2}}{32.0 \cancel{\text{ g } O_2}} \times \frac{10 \text{ mol } H_2O}{9 \cancel{\text{ mol } O_2}} = 1.39 \text{ mol } H_2O$$

Limiting reactant is O_2. $1.39 \cancel{\text{ mol } H_2O} \times \frac{18.0 \text{ g } H_2O}{\cancel{\text{mol } H_2O}} = \underline{\underline{25.0 \text{ g } H_2O}}$

C–1 theoretical, actual, percent, thermochemical, exothermic

C–2 $\dfrac{\text{actual yield}}{\text{theoretical yield}} \times 100\% = \% \text{ yield}$

theoretical yield $= \dfrac{\text{actual yield} \times 100\%}{\% \text{ yield}} \quad \dfrac{25.0 \text{ g} \times 100\%}{74.0\%} = 33.8 \text{ g } CO_2$

$$33.8 \text{ g } \cancel{CO_2} \times \frac{1 \cancel{\text{ mol } CO_2}}{44.0 \cancel{\text{ g } CO_2}} \times \frac{1 \cancel{\text{ mol } C_2H_4}}{2 \cancel{\text{ mol } CO_2}} \times \frac{28.0 \text{ g } C_2H_4}{\cancel{\text{mol } C_2H_4}} = \underline{\underline{10.8 \text{ g } C_2H_4}}$$

C–3 $C_2H_4(g) + 3O_2(g) \longrightarrow 2CO_2(g) + 2H_2O(l) + 1112 \text{ kJ}$

$$250 \cancel{\text{ kJ}} \times \frac{1 \cancel{\text{ mol } CO_2}}{556 \cancel{\text{ kJ}}} \times \frac{1 \cancel{\text{ mol } C_2H_4}}{2 \cancel{\text{ mol } CO_2}} \times \frac{28.0 \text{ g } C_2H_4}{\cancel{\text{mol } C_2H_4}} = \underline{\underline{6.29 \text{ g } C_2H_4}}$$

The following multiple-choice questions have one correct answer.

1. Which of the following elements is not a metal?
   (a) tin   (c) arsenic
   (b) tungsten   (d) uranium

2. Which of the following elements is an alkali metal?
   (a) Sc   (d) K
   (b) Al   (e) H
   (c) Cl

3. Which of the following is a representative element metal?
   (a) chromium   (c) lead
   (b) boron   (d) cerium

4. Uranium (U, at. no. 92) is in which classification of elements?
   (a) representative   (c) noble gas
   (b) transition   (d) inner transition

5. Which of the following metals forms more than one cation?
   (a) Na   (d) Al
   (b) Be   (e) Ra
   (c) Bi

6. The name of $BaBr_2$ is
   (a) barium dibromate   (c) barium (II) bromide
   (b) barium bromide   (d) barium bromate

7. The formula of a compound between a Group IIA metal (M) and a Group VIA nonmetal (X) is
   (a) M_2X_2 (d) MX
   (b) MX_2 (e) M_2X_3
   (c) M_2X

8. The formula of cobalt(III) oxide is
   (a) Co_3O_2 (d) CoO_3
   (b) Co_2O_3 (e) Co_2O
   (c) Co_3O

9. Which of the following is the formula for the permanganate ion?
   (a) MnO_4 (d) MgO_4^-
   (b) MnO_4^{2-} (e) MnO_4^-
   (c) MnO_3^{2-}

10. The formula of lithium sulfate is
    (a) $LiSO_4$ (c) Li_2S
    (b) $(Li)_2SO_3$ (d) Li_2SO_4

11. Which of the following is the commonly accepted name for SeO_3?
    (a) selenium(VI) oxide   (c) selenium dioxide
    (b) selenium oxide   (d) selenium trioxide

12. Which of the following is the name for $HClO_3$?
    (a) hydrogen chlorate   (d) chloric acid
    (b) chlorous acid   (e) hydrochloric acid
    (c) perchloric acid

13. A quantity of aluminum has a mass of 54.0 g. What is the mass of the same number of magnesium atoms?
    (a) 12.1 g   (d) 97.2 g
    (b) 24.3 g   (e) 6.0 g
    (c) 48.6 g

14. To which is the number 6.02×10^{22} equivalent?
    (a) 0.100 mol   (d) 0.500 mol
    (b) 1.00 mol   (e) no such number
    (c) 10.0 mol   exists

15. What is the mass in grams of 0.250 mol of oxygen atoms?
    (a) 16.0 g   (d) 4.00 g
    (b) 1.50×10^{23} g   (e) 32.0 g
    (c) 8.00 g

16. What is the mass of 3.01×10^{24} He atoms?
    (a) 20.0 g   (d) 12.0×10^{24} g
    (b) 4.00 g   (e) 2.00 g
    (c) 200 g

17. What is the approximate mass of one carbon atom?
    (a) 12 g   (d) 0.50×10^{-23} g
    (b) 2.0×10^{-23} g   (e) 6.0 g
    (c) 2.0×10^{23} g

18. What is the mass of one mole of $H_2C_2O_4$?
    (a) 90.0 amu   (d) 58.0 g
    (b) 46.0 g   (e) 46.0 amu
    (c) 90.0 g

19. How many moles of oxygen atoms are in 0.50 mol of $Ca(ClO_3)_2$?
    (a) 3.0   (d) 6.0
    (b) 1.0   (e) 1.50
    (c) 0.50

20. How many moles of oxygen atoms are in 3.01×10^{23} molecules of O_2?
    (a) 0.50   (d) 0.25
    (b) 6.02×10^{23} (e) 1.50×10^{23}
    (c) 1.0

21. A compound is composed of 0.24 mol of Fe, 0.36 mol of S, and 1.44 mol of O. What is its empirical formula?

(a) FeS_3O_6
(b) $Fe_4S_6O_{24}$
(c) $Fe_2S_3O_6$
(d) $Fe_2S_3O_{12}$
(e) $Fe_2S_3O_8$

22. A compound has a molar mass of 84.0 g/mol and an empirical formula of CH_2N. What is its molecular formula?

(a) $C_3H_4N_2O$
(b) $C_3H_6N_3$
(c) $C_2H_4N_2$
(d) $C_3H_6N_2$
(e) $C_4H_8N_4$

23. Which of the following elements should be represented in an equation as a diatomic molecule?

(a) C   (b) He   (c) P   (d) B   (e) Br

24. Which type of reaction does the following equation represent?

$$Ba(s) + 2H_2O(l) \longrightarrow Ba(OH)_2(aq) + H_2(g)$$

(a) combustion
(b) combination
(c) decomposition
(d) single-replacement
(e) double-replacement

25. Which type of reaction does the following equation represent?

$$SiH_4(g) + 2O_2(g) \longrightarrow SiO_2(s) + 2H_2O(l)$$

(a) combustion
(b) combination
(c) decomposition
(d) single-replacement
(e) double-replacement

26. Determine which of the answers represents the proper balanced equation for the following statement: Calcium reacts with aqueous hydrocholoric acid to produce aqueous calcium chloride and hydrogen.

(a) $Ca(g) + 2HCl(aq) \longrightarrow CaCl_2(l) + H_2(g)$
(b) $Ca(s) + 2HCl(aq) \longrightarrow CaCl_2(aq) + 2H(g)$
(c) $2Ca(s) + 2HCl(aq) \longrightarrow 2CaCl(aq) + H_2(g)$
(d) $Ca(s) + 2HCl(aq) \longrightarrow CaCl_2(aq) + H_2(g)$
(e) $2Ca(s) + H_2Cl(aq) \longrightarrow CaCl(aq) + H_2(g)$

27. Given the equation

$$2C_2H_2 + 5O_2 \longrightarrow 4CO_2 + 2H_2O$$

how many mol of O_2 are needed to produce 4 mol of CO_2?

(a) 4 mol
(b) 2 mol
(c) 3 mol
(d) 1 mol
(e) 5 mol

28. Sulfur trioxide is prepared by the following two reactions:

$$S_8(s) + 8O_2(g) \longrightarrow 8SO_2(g)$$

$$2SO_2(g) + O_2(g) \longrightarrow 2SO_3(g)$$

How many mol of SO_3 are produced from 1 mol of S_8?

(a) 1
(b) 2
(c) 4
(d) 8
(e) 16

29. In the combustion of a certain hydrocarbon 16.0 g of CO_2 is produced, which represents a 75% yield. What is the theoretical yield?

(a) 12.0 g
(b) 8.0 g
(c) 21.3 g
(d) 32.0 g
(e) 44.0 g

30. Given the combustion reaction

$$CH_4 + 2O_2 \longrightarrow CO_2 + 2H_2O$$

when 16.0 g of CH_4 is burned with 32.0 g of O_2, which statement is true?

(a) CH_4 is the limiting reactant.
(b) CO_2 is present in excess.
(c) CH_4 is present in excess.
(d) O_2 is present in excess.
(e) H_2O is the limiting reactant.

PROBLEMS

1. Fill in the blanks.

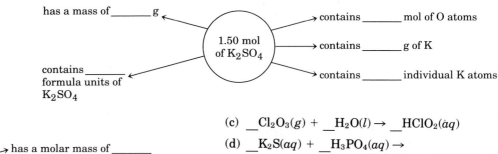

has a mass of _____ g

1.50 mol of K_2SO_4

contains _____ mol of O atoms

contains _____ g of K

contains _____ individual K atoms

contains _____ formula units of K_2SO_4

2. Fill in the blanks.

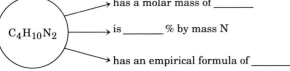

$C_4H_{10}N_2$

has a molar mass of _____

is _____ % by mass N

has an empirical formula of _____

3. A compound is composed of 1.75 g of Fe and 0.667 g of O. What is its empirical formula?

4. A compound is 9.60% H, 16.4% C, and 74% B by mass. Its molar mass is 146 g/mol. What is its molecular formula?

5. Complete the following equations. Add subscripts on formulas of elements, coefficients to balance the equations, and physical states where needed.

(a) __K(__) + $Al_2Cl_6(s)$ $\xrightarrow{\Delta}$ __KCl(__) + __Al(s)

(b) __$C_6H_6(l)$ + __O(__) →
        __CO_2(__) + __$H_2O(l)$

(c) __$Cl_2O_3(g)$ + __$H_2O(l)$ → __$HClO_2(aq)$

(d) __$K_2S(aq)$ + __$H_3PO_4(aq)$ →
        __H_2S(__) + __$K_3PO_4(aq)$

(e) __$B_4H_{10}(g)$ + __$H_2O(l)$ →
        __$B(OH)_3(aq)$ + __H(__)

6. Freon 12 (CCl_2F_2) is a gas used as a refrigerant. It is prepared according to the equation

$$3CCl_4(l) + 2SbF_3(s) \rightarrow$$
$$3CCl_2F_2(g) + 2SbCl_3(s)$$

(a) How many moles of SbF_3 are needed to produce 0.0350 mol of Freon?

(b) How many moles of $SbCl_3$ are produced if 150 g of Freon is also produced?

(c) What mass of CCl_4 is needed to react with 850 g of SbF_3?

(d) How many individual molecules of CCl_4 are needed to produce 12.0 kg of $SbCl_3$?

Our atmosphere is a protective blanket of gases. The behavior of these gases can be predicted by "gas laws" which are studied in this chapter.

THE GASEOUS STATE

Our atmosphere of gases is like the combination of a warm blanket and a strong shield. It nurtures life in many ways. Oxygen fuels the metabolism that gives energy to living things. Nitrogen dilutes the oxygen so that combustion does not occur explosively. Carbon dioxide and water vapor lock in the daytime heat so that the temperature does not become bitter cold at night. Ozone high in the stratosphere absorbs the harmful portion of the sun's radiation. The thickness of the air burns up most incoming meteorites and other debris from space. Life as we know it requires the constant support and protection of this sea of gases.

◀**Setting The Stage**

Just two hundred years ago there was little understanding of the gases that make up the air we breathe. It wasn't until the late 1700s that the experiments of Antoine Lavoisier in France and Joseph Priestly in England gave us some understanding about the gaseous mixture of the atmosphere. These scientists proved that air was not just one substance, as had been previously thought, but was mainly a mixture of two elements: nitrogen and oxygen. Their work also laid the foundation for the development of the law of conservation of mass. This was the beginning of modern quantitative chemistry.

With a few exceptions, gases are invisible. This makes the study of gases somewhat difficult. Except on a windy day, it is easy to forget that the gases of the air surround us. There is one advantage, however, when we study the nature of gases. In many ways, they all behave the same, which allows convenient generalizations, including "the gas laws." Solids and liquids are much less abstract than gases simply because we can see them. The down side of working with solids and liquids, however, is that few generalizations apply and each substance must be studied individually.

Perhaps the first question to ask is why gases are so diffuse compared to the other states of matter. How do the atoms or molecules of gases behave that explains their common properties? If we have a certain volume of a gas, how is it affected by such variables as pressure, temperature, and the number of molecules? Since gases are part of chemical reactions, how do we include the volume of a gas in stoichiometry calculations? Such questions will be discussed in this chapter. First, we will list some of the unique properties of gases and see how these are explained by an extension of the atomic theory.

Formulating Some Questions ▶

9–1   THE NATURE OF GASES AND THE KINETIC MOLECULAR THEORY

Gases are certainly the softest form of nature. It seems like we can move through the air with ease and grace, especially compared to moving through water. Of course, we have no chance of moving through a solid unless we are in a science fiction movie. By living in a sea of gases, we tend to take many common properties of the gaseous state for granted. Five of these properties are described here.

Why is the gaseous state so unique?

1 Gases Are Compressible.
   When a glass of water is full, we mean what we say—no more can be added. But when is an automobile tire full of air? Actually, never. We can always add more air (at least until the tire bursts). The nature of the gaseous state allows us to press a volume of gas into a smaller volume. Liquids and solids are not like that. They are essentially incompressible.

2 Gases Have Low Densities.
   The density of a typical liquid or solid is about 2 g/mL. The density of a typical gas is about 2 g/L. This means that the density of a solid or liquid is roughly 1000 times greater than that of a gas. This is a big difference.

3 Gases Mix Thoroughly.
   Chemistry buildings are well known around campus for their noxious odors, especially when experiments using hydrogen sulfide are assigned. If the resulting odor of rotten eggs was confined to one chemistry lab, it wouldn't be so bad, but all gases mix thoroughly and rapidly. Thus the unpleasant smell permeates the surroundings. In contrast, some liquids do not mix at all (e.g., oil and water). If they do mix (e.g., water and alcohol), the mixing process occurs quite slowly.

4 Gases Fill a Container Uniformly.

When we blow air into a round balloon, the balloon becomes spherical and uniform. The air in the balloon obviously pushes out in all directions. If we were to fill the balloon with water instead (for scientific reasons only, of course), we would notice that the balloon sags. The water accumulates in the bottom of the balloon.

5 A Gas Exerts Pressure Uniformly on All Sides of a Container.

This follows from property 4 preceding. In the context of a room full of air, it would seem that gases defy gravity. The pressure of the atmosphere is the same on the ceiling, walls, and floor.* If we were to fill the room with water, however, we would notice that the pressure increases markedly the greater the depth. Anyone diving off the high dive at a swimming pool knows the feel of the increase in pressure ten feet below the water's surface. A balloon "full" of air is spherical because the gas pushes out with the same pressure in all directions.

The balloon is spherical because a gas exerts equal pressure in all directions.

Obviously, the gaseous state of matter is quite different from the other two physical states. Gases are composed of either individual atoms (in the case of noble gases) or molecules (for all others). This fact and others can be summarized in a set of assumptions collectively known as the **kinetic molecular theory,** or simply **kinetic theory.** The kinetic theory was advanced in the late 1800s to explain the nature of gases. The major points of this theory as applied to gases are as follows:

What accounts for the behavior of gases?

1 A gas is composed of very small particles called molecules, which are widely spaced and occupy negligible volume. A gas is thus mostly empty space.

2 The molecules of a gas are in rapid, random motion, colliding with each other and the sides of the container. Pressure is a result of these collisions with the sides of the container. (See Figure 9-1.)

3 All collisions involving gas molecules are elastic. (The total energy of two colliding molecules is conserved. A ball bouncing off the pavement undergoes inelastic collisions; it does not bounce as high each time.)

4 Gas molecules have negligible attractive (or repulsive) forces between them.

5 The temperature of a gas is related to the average kinetic energy of the gas molecules. Also, at the same temperature, different gases have the same average kinetic energy (K.E.).

The kinetic theory explains the properties described above as well as the gas laws that follow. For example, since gas molecules have essentially no volume and a gas is mostly empty space, it has a low density and is easily compressed. The rapid motion explains why gases mix rapidly and thoroughly. Because gas molecules are not attracted to each other, they do not "clump together" as do the molecules in liquids and

Figure 9–1
Gases. Gas molecules are in rapid, random motion.

*In fact, gases do not really defy gravity since our atmosphere gets thinner as the altitude increases. However, the thinning of the atmosphere occurs on a scale of several kilometers. It would not be noticed on the scale of a few meters in a room.

The balloon on the left is filled with air, and the one on the right is filled with helium. One day later, the helium balloon has shrunk because the small atoms of helium have effused out of the balloon.

Do all gases have the same average velocity at the same temperature?

solids. Thus the molecules in gases spread out and fill a container uniformly, exerting pressure equally in all directions.

Another consequence of the kinetic theory concerns the relative velocities of gas molecules at the same temperature. The kinetic energy of a gas is given by the equation

$$K.E. = \frac{1}{2} mv^2$$

where

$$m = \text{mass} \quad \text{and} \quad v = \text{velocity}$$

If two different gases have the same kinetic energies at the same temperature, then we can derive the following relationship between their velocities.

$$K.E. \text{ (gas 1)} = \frac{1}{2} m_1 v_1^2 \quad K.E. \text{ (gas 2)} = \frac{1}{2} m_2 v_2^2$$

Since the two kinetic energies are equal, we have

$$\frac{1}{2} m_1 v_1^2 = \frac{1}{2} m_2 v_2^2$$

or rearranging and simplifying

$$\frac{v_1}{v_2} = \sqrt{\frac{m_2}{m_1}}$$

The average velocity of a gas molecule relates directly to two other aspects of molecular motion. These are the rates of **diffusion** *(mixing)* and **effusion** *(moving through an opening or hole)*. In fact, the relationship that we have derived from an assumption of the kinetic theory was first proposed in 1832 by Thomas Graham some time before the acceptance of kinetic theory. **Graham's law** *states that the rates of diffusion of gases are inversely proportional to the square roots of their molar masses.*

Consider the comparison of a helium atom to a nitrogen molecule at the same temperature. Since helium is much lighter than nitrogen, it has a higher velocity. (See Figure 9-2.) In Example 9-1, we have calculated the comparative velocities of the two species.

Figure 9–2
Relative Velocities. At the same temperature, a helium atom travels almost three times faster than a nitrogen molecule.

EXAMPLE 9–1

The Relative Velocities of Helium and Nitrogen

At the same temperature, how much faster does a He atom travel than a N_2 molecule?

●**Working It Out**

Solution

At the same temperature, $(K.E.)_{He} = (K.E.)_{N_2}$. Therefore,

$$m_{He} = 4.00 \text{ amu} \qquad m_{N_2} = 28.0 \text{ amu}$$

$$\frac{v_{He}}{v_{He}} = \sqrt{\frac{m_{N_2}}{m_{He}}} = \sqrt{\frac{28.00 \text{ amu}}{4.00 \text{ amu}}} = \sqrt{7.00} = 2.65$$

$$v_{He} = 2.65 \ v_{N_2}$$

On the average, He atoms travel 2.65 times faster than N_2 molecules.

Graham's law and many other quantitative laws were formulated well before the concepts of the kinetic molecular theory were accepted. Since kinetic theory makes everything seem so simple and reasonable, it is sometimes difficult to appreciate the advances made by those who were not aware of the existence of atoms, molecules, or their motions. Discussing kinetic theory before the gas laws is somewhat like reading the last chapter of a mystery novel first—we know how it's going to turn out. However, since we do know about atoms and molecules, we will note how the gas laws are obvious extensions of our modern theories. Our first order of business, however, is to take a closer look at the property known as "pressure."

▼**Looking Ahead**

9–2 THE PRESSURE OF A GAS

A newscast on TV would not be complete without the weather forecast. The weather data probably includes the atmospheric pressure as read from a **barometer.** Rising pressure usually means improving weather, and a dropping barometer often means a storm is approaching.

The barometer has been around for quite a while. It was invented by an Italian scientist named Evangelista Torricelli in 1643. Torricelli filled a long glass tube with mercury, a dense liquid metal, and then inverted the tube into a bowl of mercury so that no air would enter the tube. Torricelli found that the mercury in the tube seemed to defy gravity by staying suspended to a height of about 76 cm no matter how long or wide the tube. (See Figure 9-3.)

At the time, many scientists thought that since "nature abhors a vacuum," the vacuum created at the top of the tube suspended the mercury. Torricelli suggested instead that it was the weight of the air on the outside that pushed the mercury up. Otherwise, it was reasoned, the

Is the mercury in a barometer held up or pushed up?

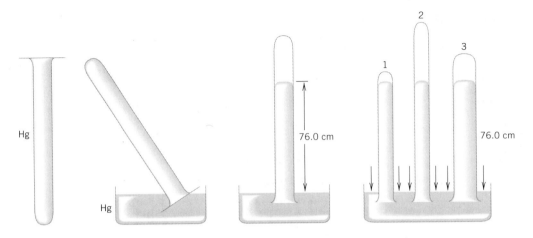

Figure 9–3
The Barometer. When a long tube is filled with mercury and inverted in a bowl of mercury, the atmosphere supports the column to a height of 76.0 cm.

The pressure of the atmosphere is measured by a barometer. The barometric pressure can be used to forecast weather.

greater vacuum present in tubes 2 and 3 in Figure 9-3 would support a higher level of mercury. Torricelli also suggested that the thinner atmosphere at higher levels in the mountains would support less mercury. He was correct. The higher the elevation, the lower the level of mercury in the barometer.

The weight of a quantity of matter pressing on a surface exerts a **force. Pressure** *is defined as the force applied per unit area.* This can be expressed mathematically as

$$P(\text{pressure}) = \frac{F(\text{force})}{A(\text{area})}$$

In a barometer, the pressure exerted by the atmosphere on the outside is balanced by the pressure exerted by the column of mercury on the inside.

If, at times, it seems like you are under a lot of "pressure," it is not necessarily from the atmosphere. You *are* under a lot of "force" from the atmosphere, however—about 15,000 lb of it. Fortunately, this force is spread out over your entire body surface (about 10,000 cm^2, with much individual variation). When we divide the force by the area, the resulting pressure is a reasonable 15 lb/in^2.

One atmosphere (1 atm) is defined as the average pressure of the atmosphere at sea level and is the standard of pressure. As we have seen, this is equivalent to the pressure exerted by a column of mercury 76.0 cm (760 mm) high.

1.00 atm = 76.0 cm Hg = 760 mm Hg = 760 torr

The unit of mm of Hg is also known as the *torr* in honor of Torricelli. In addition to the torr, there are several other units of pressure (e.g., lb/in.^{2}) that have special uses. The relationships of the units to 1 atm and their applications are listed in Table 9-1. An example of a conversion between units is shown in Example 9-2.

TABLE 9–1   ONE ATMOSPHERE

Unit of Pressure	Special Use
760 mm Hg or 760 torr	Most chemistry laboratory measurements for pressures in the neighborhood of one atmosphere
14.7 lb/in.^{2}	U.S. pressure gauges
29.9 in. Hg	U.S. weather reports
101.325 kPa (kilopascals)	The metric unit of pressure [1 N (newton)/m^2]
1.013 bars	Used in physics and astronomy mainly for very low pressures (millibars) or very high pressures (kilobars)

EXAMPLE 9–2

Conversion Between Torr and Atmospheres

What is 485 torr expressed in atmospheres?

Procedure
The conversion factors are

$$\frac{760 \text{ torr}}{\text{atm}} \quad \text{and} \quad \frac{1 \text{ atm}}{760 \text{ torr}}$$

Solution

$$485 \text{ torr} \times \frac{1 \text{ atm}}{760 \text{ torr}} = \underline{\underline{0.638 \text{ atm}}}$$

● **Working It Out**

The study of the invisible physical state of matter requires that we delve into the abstract. However, an understanding of how the tiny molecules of the gas behave makes its properties both reasonable and predictable. One thing we can directly experience about the gases of the atmosphere is that they exert pressure. The pressure of the atmosphere and any confined quantity of gas can be expressed in various units.

▲ **Looking Back**

Learning Check A

A–1.   Fill in the blanks.
Unlike solids and liquids, a gas has a _____ density and fills a container _____ . The reason for the unique behavior of gases can be explained by the _____ _____ theory. In this theory, gas molecules are in _____ , _____ motion, exerting pressure by _____ with the walls of the container. Another aspect of this theory is that the temperature relates to the average _____ en-

◀ **Checking It Out**

ergy of the gas. By relating the energies of two gases at the same _____ , we can derive a relationship known as _____ law. At the same temperature, the gas with the largest molar mass will have the _____ velocity. Atmospheric pressure is read from a device called a _____ . Pressure is defined as_____per unit _____ . One atmosphere of pressure is equivalent to _____ torr.

A–2.   At the same temperature, what is the average velocity of a SF_6 molecule compared to that of a N_2 molecule ?

A–3.   What is 0.650 atm expressed in (a) torr (b) kPa and (c) lb/in.^{2}?

Additional Examples: Problems 9-7, 9-8, 9-14, and 9-17.

Looking Ahead ▼

We know that gases are compressible. That is, when we increase the pressure on a gas, the volume is decreased. The first of the gas laws established a quantitative relationship between the volume of a gas and the pressure applied. This is our next topic.

9–3   BOYLE'S LAW

Every breath we take illustrates the interaction between the volume and the pressure of a gas. When the diaphragm under our rib cage relaxes, it moves up, squeezing our lungs and decreasing their volume. The decreased volume increases the air pressure inside the lungs relative to the outside atmosphere, and we expel air. When the diaphragm contracts, it moves down, increasing the volume of the lungs and, as a result, decreasing the pressure. Air rushes in from the atmosphere until the pressures are equal.

The relationship between the volume and pressure of a sample of gas under constant temperature conditions was first advanced in 1660 by the English scientist Sir Robert Boyle. In an apparatus similar to that shown in Figure 9-4, Boyle found that the decrease in volume can be predicted by the amount of pressure increase. This observation is now expressed as **Boyle's law,** *which states that there is an inverse relationship between the pressure exerted on a quantity of gas and its volume if the temperature is held constant.*

The inverse relationship suggested by Boyle is represented as

$$V \, \alpha \, \frac{1}{P}$$

or as an equality with k the constant of proportionality:

$$V = \frac{k}{P} \quad \text{or} \quad PV = k$$

Boyle's law can be illustrated with the apparatus shown in Figure 9-4. In experiment **1,** a certain quantity of gas ($V_1 = 10.0$ mL) is trapped

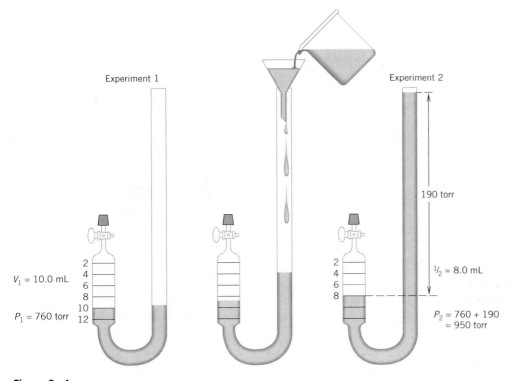

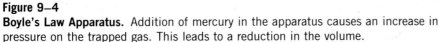

Experiment 1

$V_1 = 10.0$ mL

$P_1 = 760$ torr

Experiment 2

190 torr

$^1/_2 = 8.0$ mL

$P_2 = 760 + 190$
$= 950$ torr

Figure 9–4
Boyle's Law Apparatus. Addition of mercury in the apparatus causes an increase in pressure on the trapped gas. This leads to a reduction in the volume.

in a U-shaped tube by some mercury. Since the level of mercury is the same in both sides of the tube and the right side is open to the atmosphere, the pressure on the trapped gas is the same as the atmospheric pressure ($P_1 = 760$ torr). In this experiment,

$$P_1V_1 = 760 \text{ torr} \times 10.0 \text{ mL} = 7600 \text{ torr} \cdot \text{mL} = k$$

When mercury is added to the tube, the pressure on the trapped gas is increased to 950 torr (760 torr originally plus 190 torr from the added mercury). Note in experiment **2** that the increase in pressure has caused a decrease in volume to 8.0 mL. Applying Boyle's law to this experiment, we have

$$P_2V_2 = 950 \text{ torr} \times 8.0 \text{ mL} = 7600 \text{ torr} \cdot \text{mL} = k$$

As predicted by Boyle's law, PV equals the same value in both experiments. Therefore, for a quantity of gas under two sets of conditions at the same temperature,

$$P_1V_1 = P_2V_2 = k$$

We can use this equation to calculate how a volume of gas changes when the pressure changes. For example, if V_2 is the new volume that we are to find at a given new pressure, P_2, the equation becomes

$$V_2 = V_1 \times \frac{P_1}{P_2}$$

new volume = old volume × pressure correction factor

If a series of measurements are made of volume and pressure, the results can be graphed. The construction of such a graph illustrating Boyle's law is found in Appendix E on graphs.

Gas law problems can be worked by properly substituting values for the known variables. They can also be worked by logic since we know qualitatively how a gas should react to a change of conditions. For example, let us consider a Boyle's law problem where the pressure on a given volume of gas has been increased. We can reason that this would lead to a decrease in the volume of the gas. We therefore choose a pressure correction factor that will do the job. This requires the use of a proper fraction (a fraction less than one) with the lower pressure in the numerator of the fraction. The following example emphasizes logic in working Boyle's law problems.

Can gas law problems be worked using logic?

Working It Out ●

EXAMPLE 9–3

Volume and a Change in Pressure

Inside a certain automobile engine, the volume of a cylinder is 475 mL when the pressure is 1.05 atm. When the gas is compressed, the pressure increases to 5.65 atm at the same temperature. What is the volume of the compressed gas?

Procedure
The final volume equals the initial volume times the pressure factor. Since the pressure increases, the volume decreases. Thus the pressure factor must be less than one.

Solution

$$V_2 = V_1 \times \frac{P_1}{P_2}$$

$$V_2 = 475 \text{ mL} \times \frac{1.05 \text{ atm}}{5.65 \text{ atm}} = \underline{\underline{88.3 \text{ mL}}}$$

(Note that the units of pressure are the same. *If the initial and final pressures are given in different units, one must be converted to the other.*)

Although it certainly wasn't known in the 1660s, Boyle's law is a natural consequence of the kinetic theory of gases. In order to focus on how a change in volume affects the pressure, we will assume that there is an average molecule of a gas at a given temperature. In Figure 9-5, the path of this average molecule is traced. The pressure exerted by this molecule is a result of collisions with the sides of the container. The distance traveled in a given unit of time is represented by the length of the arrow. Since this is an average molecule, it travels the same distance per unit time in both the high-volume situation (on the left) and the low-volume situation (on the right). Note that, on the right, the lower volume leads to an increased number of collisions with the sides of the container. More frequent collisions means higher pressure.

Looking Ahead ▼

In addition to the changes in volume of gases with changes in pressure, we are probably aware that gases expand when they are heated. One hundred years after Boyle's relationship was advanced, the relation-

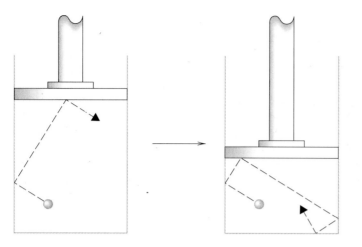

Figure 9–5
Boyle's Law And Kinetic Molecular Theory. When the volume decreases, the pressure increases because of more frequent collisions with the walls of the container.

ship between volume and temperature was studied. This and the relationship between pressure and temperature are discussed next.

9–4   CHARLES'S LAW AND GAY-LUSSAC'S LAW

Very few sights appear more tranquil than brightly colored hot-air balloons drifting across a blue summer sky. Except for the occasional "swoosh" of gas burners, they cruise by in majestic silence. Hot air expands. When it expands, it becomes less dense than the surrounding cooler air and thus rises. When the hot, expanded air is trapped, the balloon and attached gondola become "lighter than air" and lift into the sky. The quantitative effect of temperature on the volume of a sample of gas at constant pressure was first advanced by a French scientist, Jacques Charles, in 1787. Charles showed that any gas expands by a definite fraction as the temperature rises. He found that the volume increases by a fraction of 1/273 for each 1°C rise in the temperature. (See Figure 9-6.)

What keeps a hot-air balloon suspended in air?

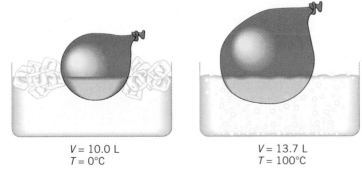

$V = 10.0$ L
$T = 0$°C

$V = 13.7$ L
$T = 100$°C

Figure 9–6
Effect of Temperature on Volume. When the temperature increases, the volume of the balloon increases.

Let us assume that we have made four measurements of the volume of a gas at four temperatures at constant pressure. The results of these experiments are listed in the table on the left in Figure 9-7. The four points are also plotted in a graph on the right. When the four points in the graph are connected, we have a straight line. The relationship between volume and temperature is known as a *linear relationship*. In the graph, the straight line has been extended to what the volume would be at temperatures lower than those in the experiment. (The procedure of extending data beyond experimental results is known as *extrapolation*.) If extended all the way to where the gas would theoretically have zero volume, the line would intersect the temperature axis at −273°C. Certainly, matter could never have zero volume, and it is impossible to cool gases indefinitely. At some point, all gases condense to become liquids or solids. However, this temperature does have significance, because it—or −273.15°C, to be more precise—is the lowest possible temperature. As noted in kinetic molecular theory, the average kinetic energy or velocity of molecules is related to the temperature. *The lowest possible temperature, known as* **absolute zero,** *is the temperature at which translational motion (motion from point to point) ceases. The* **Kelvin scale** *assigns zero as absolute zero.* Thus there are no negative values on the Kelvin scale, just as there are no negative values on any pressure scale. Since the magnitudes of Celsius and Kelvin degrees are the same, we have the following simple relationship between scales, where $T(K)$ is the temperature in kelvins and $t(C)$ represents the Celsius temperature.

$$T(K) = [t(C) + 273] \, K$$

The relationship between volume and temperature noted by Charles can now be summarized. **Charles's law** *states that the volume of a gas is directly proportional to the Kelvin temperature (T) at constant pressure.* This can be expressed mathematically as a proportion or as an equality with k as a constant of proportionality.

How much can a gas contract?

What temperature scale do we use in gas law problems?

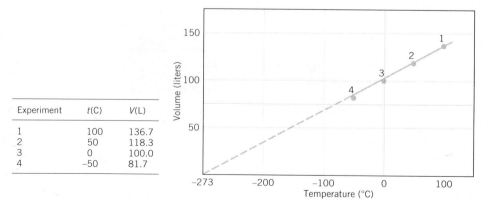

Experiment	t(C)	V(L)
1	100	136.7
2	50	118.3
3	0	100.0
4	−50	81.7

Figure 9–7
Volume and Temperature.

$$V \alpha T \quad V = kT \quad \text{or} \quad \frac{V}{T} = k$$

For a quantity of gas under two sets of conditions at the same pressure,

$$\frac{V_1}{T_1} = \frac{V_2}{T_2}$$

This equation can be used to calculate how the volume of gas changes when the temperature changes. For example, if V_2 is a new volume that we are to find at a given new temperature T_2, the equation becomes

$$V_2 = V_1 \times \frac{T_2}{T_1}$$

final volume = initial volume × temperature correction factor

In this case, if the temperature decreases, the volume must also decrease, which means that the temperature factor is less than one. On the other hand, if the temperature increases, the volume must increase, which means that the temperature factor is more than one.

EXAMPLE 9–4

Volume and Change in Temperature

A given quantity of gas in a balloon has a volume of 185 mL at a temperature of 52°C. What is the volume of the balloon if the temperature is lowered to −17°C? Assume that the pressure remains constant.

Procedure
Since the temperature decreases, the volume decreases. The temperature factor must therefore be less than one.

Solution

$$V_2 = V_1 \times \frac{T_2}{T_1} \qquad \begin{array}{l} T_2 = (-17 + 273)\text{ K} = 256\text{ K*} \\ T_1 = (52 + 273)\text{ K} = 325\text{ K} \end{array}$$

$$V_2 = 185\text{ mL} \times \frac{256\text{ K}}{325\text{ K}} = \underline{\underline{146\text{ mL}}}$$

Note that temperature must be expressed in the Kelvin scale. Also, a Celsius reading with two significant figures (e.g., −17°C) becomes a Kelvin reading with three significant figures (e.g., 256 K).]

● **Working It Out**

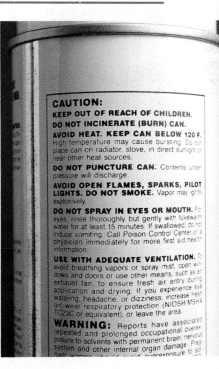

Heat increases the pressure of gas in a confined volume. If heated enough, the can will explode.

Now let's consider how a confined (constant) volume of gas is affected by temperature. A good example of this effect is provided by a warning on a pressurized can containing hair spray: DO NOT INCINERATE OR STORE NEAR HEAT. An increase in temperature causes an increase in pressure on a confined volume of gas such as that in a pressurized can. At some high temperature, the seals will fail and the can will explode. The relationship between temperature and pressure was proposed in 1802 by Gay-Lussac. **Gay-Lussac's law** *states that the pressure is directly proportional to the Kelvin temperature at constant volume.* The law can be expressed mathematically as follows.

$$P \alpha T \quad P = kT$$

For a sample of gas under two sets of conditions at the same volume,

$$\frac{P_1}{T_1} = \frac{P_2}{T_2} \qquad P_2 = P_1 \times \frac{T_2}{T_1}$$

EXAMPLE 9–5

Pressure and Change in Temperature

A quantity of gas in a steel container has a pressure of 760 torr at 25°C. What is the pressure in the container if the temperature is increased to 50°C?

Procedure
Since the temperature increases, the pressure increases. The temperature correction factor must therefore be greater than one.

Solution

$$P_2 = P_1 \times \frac{T_2}{T_1} \qquad \begin{array}{l} T_2 = (50 + 273)\ K = 323\ K \\ T_1 = (25 + 273)\ K = 298\ K \end{array}$$

$$P_2 = 760\ torr \times \frac{323\ \cancel{K}}{298\ \cancel{K}} = \underline{\underline{824\ torr}}$$

Both Charles's law and Gay-Lussac's law follow naturally from kinetic theory. In Figure 9-8, we again picture one average molecule of gas moving in a container. In the center, we have the situation before any changes are made. To the left, we have raised the temperature at constant pressure. The molecule travels farther on the average at the higher temperature, as shown by its longer path. It also collides with more force on the walls of the container. In order for the pressure to remain constant, the volume must expand correspondingly, which confirms Charles's law. To the right, we have again raised the temperature but this time at constant volume. The more frequent and more forceful collisions mean that the pressure must now rise, which confirms Gay-Lussac's law.

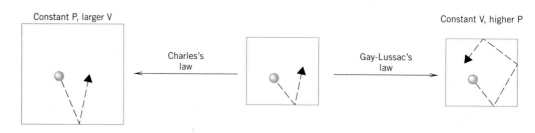

Figure 9–8
Charles's Law and Gay-lussac's Law. When temperature increases, molecules travel faster. If the pressure is constant, the volume increases in order to permit the frequency of collisions to remain the same. If the volume is constant, the pressure increases as a result of the more frequent collisions.

It is probably obvious that the volume of a gas is also dependent on the number of molecules of gas present, all other conditions being equal. However, we will discuss the relationship of volume to the number of molecules in the next section.

9–5 AVOGADRO'S LAW

The more air we blow into a balloon, the larger it gets. This very simple observation is the basis of still another gas law. **Avogadro's law** *states that equal volumes of gases at the same pressure and temperature contain equal numbers of molecules.* Another way of stating this law *is that the volume of a gas is proportional to the number of molecules (moles) of gas present at constant pressure and temperature.* Mathematically, this can be stated in three ways, where n represents the number of moles of gas.

$$V \alpha n \qquad V = kn \qquad \frac{V_1}{n_1} = \frac{V_2}{n_2}$$

In Figure 9-9, the expansion of a balloon is illustrated by adding carbon dioxide. The following example illustrates the use of Avogadro's law in a sample calculation.

EXAMPLE 9–6

Volume and Change in Moles of Gas

A balloon that is not inflated but is full of air has a volume of 275 mL and contains 0.0120 mol of air. As shown in Figure 9-9, a piece of dry ice (solid CO_2) weighing 1.00 g is placed in the balloon and the neck tied. What is the volume of the balloon after the dry ice has vaporized? (Assume constant T and P.)

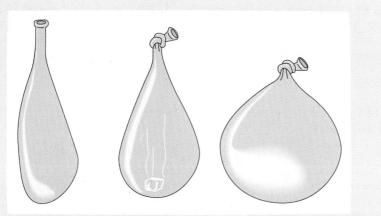

Figure 9–9
Illustration of Avogadro's Law. The addition of carbon dioxide to the balloon increases the number of gas molecules, which increases the volume.

Solution

Initial Conditions

$V_1 = 275$ mL

$n_1 = 0.0120$ mol

Final Conditions

$V_2 = ?$

$n_2 = $ mol air + mol CO_2

$$= 0.0120 + \left(1.00 \ \cancel{\text{g } CO_2} \times \frac{1 \ \text{mol}}{44.0 \ \cancel{\text{g } CO_2}}\right)$$

$$= 0.0120 + 0.0227 = 0.0347 \ \text{mol}$$

$$\frac{V_1}{n_1} = \frac{V_2}{n_2}$$

$$V_2 = V_1 \times \frac{n_2}{n_1} = 275 \ \text{mL} \times \frac{0.0347 \ \cancel{\text{mol}}}{0.0120 \ \cancel{\text{mol}}}$$

$$V_2 = \underline{\underline{795 \ \text{mL}}}$$

Looking Ahead ▼

In the next section, we will find that we can arrive at even more gas laws by putting the information in the last three sections together in the form of one equation. This very important equation is known as the ideal gas law and summarizes much of our previous discussion.

9–6   THE IDEAL GAS LAW

How many ways can you expand a balloon?

How can we make a balloon larger? In fact, we have discovered three ways. First, we could take it up a mountain to a higher elevation. The lower atmospheric pressure would allow it to expand (Boyle's law). Second, we could raise the temperature. If we placed the balloon in boiling water it would certainly expand (Charles's law). And finally, we could just put more gas in it (Avogadro's law). We can summarize these three relationships as follows:

$$V \, \alpha \, \frac{1}{P} \qquad V \, \alpha \, T \qquad V \, \alpha \, n$$

We can put all three individual relationships into one general relationship as follows:

$$V \, \alpha \, \frac{nT}{P}$$

As usual, we can change a proportionality to an equality by introducing a constant of proportionality. The constant used in this case is **R** and is called the **gas constant.** Traditionally, the constant is placed between the n and T.

Is there a gas law that puts it all together?

$$V = \frac{nRT}{P} \quad \text{or} \quad PV = nRT$$

This relationship is known as the **ideal gas law.**

In cases where we have a sample of gas (constant n) where P, V, and T can vary, we have the relationship

$$\frac{P_1 V_1}{T_1} = \frac{P_2 V_2}{T_2}$$

This is referred to as the **combined gas law** and allows calculations of a missing variable (P, V, or T) under two sets of conditions.

EXAMPLE 9–7

Volume and Change in Pressure and Temperature

A 25.8-L quantity of gas has a pressure of 690 torr and a temperature of 17°C. What is the volume if the pressure is changed to 1.85 atm and the temperature to 345 K?

● Working It Out

Solution

Initial Conditions	Final Conditions
$V_1 = 25.8$ L	$V_2 = ?$
$P_1 = 690 \text{ torr} \times \dfrac{1 \text{ atm}}{760 \text{ torr}}$	$P_2 = 1.85$ atm
$= 0.908$ atm	
$T_1 = (17 + 273)\text{ K} = 290$ K	$T_2 = 345$ K

$$\frac{P_1 V_1}{T_1} = \frac{P_2 V_2}{T_2} \qquad V_2 = V_1 \times \frac{P_1}{P_2} \times \frac{T_2}{T_1}$$

$$\text{final volume} = \text{initial volume} \times \underset{\text{factor}}{\text{correction}}^{\text{pressure}} \times \underset{\text{factor}}{\text{correction}}^{\text{temperature}}$$

In this problem note that the pressure increases, which decreases the volume. The pressure factor should therefore be less than one. On the other hand, the temperature increases, which increases the volume. The temperature factor is greater than one.

$$V_2 = 25.8 \text{ L} \times \frac{0.908 \text{ atm}}{1.85 \text{ atm}} \times \frac{345 \text{ K}}{290 \text{ K}} = \underline{\underline{15.1 \text{ L}}}$$

Obviously, a major property of a gas is that its volume is very dependent on temperature and pressure. Yet gas is sold by the local gas utility in volume units of cubic feet. Does that mean we get more or less gas in the warm summer than in the winter? Actually, we get the same amount because the volume of the gas is measured under certain universally accepted conditions known as **standard temperature and pressure (STP).**

Standard temperature: 0°C or 273 K
Standard pressure: 760 torr or 1 atm

The example on the following page illustrates the use of STP in the combined gas law.

EXAMPLE 9–8

Temperature and Change in Pressure and Volume

A 5850-ft^3 quantity of natural gas measured at **STP** was purchased from the gas company. Only 5625 ft^3 was received at the house. Assuming that all of the gas was delivered, what was the temperature at the house if the delivery pressure was 1.10 atm?

Solution

Initial Conditions	Final Conditions
$V_1 = 5850$ ft^3	$V_2 = 5625$ ft^3
$P_1 = 1.00$ atm	$P_2 = 1.10$ atm
$T_1 = 273$ K	$T_2 = ?$

$$\frac{P_1 V_1}{T_1} = \frac{P_2 V_2}{T_2} \quad T_2 = T_1 \times \frac{P_2}{P_1} \times \frac{V_2}{V_1}$$

In this case, the final temperature is corrected by a pressure and a volume correction factor. Since the final pressure is higher, the pressure factor must be greater than one to increase the temperature. The final volume is lower, so the volume correction factor must be less than one to decrease the temperature.

$$T_2 = 273 \text{ K} \times \frac{1.10 \text{ atm}}{1.00 \text{ atm}} \times \frac{5625 \text{ ft}^3}{5850 \text{ ft}^3} = 289 \text{ K}$$

$$t\,(\text{C}) = 289 \text{ K} - 273 = \underline{\underline{16°\text{C}}}$$

Now let us return our focus to the ideal gas law. By itself, this relationship is used mostly to calculate a missing variable (P, V, T, or n) when the other three are known. For these calculations, however, we need the value of the gas constant, which is

$$R = 0.0821 \frac{\text{L} \cdot \text{atm}}{\text{K} \cdot \text{mol}} = 62.4 \frac{\text{L} \cdot \text{torr}}{\text{K} \cdot \text{mol}}$$

Note that the units of R are a result of specific units of P, V, and T. The following two examples illustrate the use of the ideal gas law with a missing variable under one set of conditions.

EXAMPLE 9–9

Calculation of Pressure from V, T, and n

What is the pressure of a 1.45-mol sample of a gas if the volume is 20.0 L and the temperature is 25°C?

Solution

$$P = ? \quad V = 20.0 \text{ L} \quad n = 1.45 \text{ mol}$$

$$T = (25 + 273) \text{ K} = 298 \text{ K}$$

$$PV = nRT$$

$$P = \frac{nRT}{V} = \frac{1.45 \ \cancel{mol} \times 0.0821 \ \frac{\cancel{L} \cdot atm}{\cancel{K} \cdot \cancel{mol}} \times 298 \ \cancel{K}}{20.0 \ \cancel{L}}$$

$$= \underline{\underline{1.77 \ atm}}$$

EXAMPLE 9–10

Calculation of Volume from T, P, and n

What is the volume of 1.00 mol of gas at **STP?**

Solution

$$P = 1.00 \ atm \quad V = ? \quad T = 273 \ K \quad n = 1.00 \ mol$$

$$PV = nRT \qquad V = \frac{nRT}{P}$$

$$V = \frac{1.00 \ \cancel{mol} \times 0.0821 \ \frac{L \cdot \cancel{atm}}{\cancel{K} \cdot \cancel{mol}} \times 273 \ \cancel{K}}{1.00 \ \cancel{atm}} = \underline{\underline{22.4 \ L}}$$

This law is called "ideal" because it follows from the assumptions of the kinetic theory, which describes an ideal gas. The molecules of an ideal gas have negligible volume and no attraction for each other. The molecules of a "real" gas obviously do have a volume, and there is some interaction between molecules, especially at high pressures (i.e., the molecules are pressed close together) and low temperatures (i.e., the molecules move more slowly). Fortunately, at normal temperatures and pressures found on the surface of Earth, gases have close to ideal behavior. Therefore, the use of the ideal gas law is justified. On the other hand, the atmosphere of the planet Jupiter is composed of cold gases under very high pressures. Under these conditions, the ideal gas law would not provide acceptable values for a variable. Other, more complex equations would have to be used that take into account the volume of the molecules and the interactions between molecules.

What is "ideal" about the ideal gas law?

This diffuse form of matter that we call the gaseous state is much more sensitive to conditions than the other two states. Pressure and temperature changes have little effect on solids and liquids, especially compared to the rather large effect such changes have on gases. All of the conditions of a gas (P, V, T, and n) relate to each other by one equation called the ideal gas law. All other laws can be derived from this one relationship.

▲Looking Back

Learning Check B

B–1. Fill in the blanks.
An increase in pressure on a volume of gas causes the volume to _____, whereas an increase in temperature causes the same

◀Checking It Out

volume to _____ . Increasing the number of moles of gas present will _____ the volume if other conditions are constant. These three relationships are known, in order, as _____ law, _____ law, and _____ law. Gay-Lussac's law tells us that if the temperature of a confined volume of gas increases, the pressure _____ . The ideal gas law combines these relationships into the equation _____ . R is known as the __ __ and has the value _____ $\dfrac{L \cdot atm}{K \cdot mol}$.

B–2. The volume of 0.0227 mol of gas is 550 mL at a pressure of 1.00 atm and a temperature of 22°C.
(a) What is the volume at 1.32 atm and 22°C?
(b) What is the volume at 1.00 atm and 44°C?
(c) What is the volume if 0.0115 mol of gas is added at the same T and P?
(d) What is the pressure at 122°C if the volume remains 550 mL?
(e) What is the pressure at 102°C if the volume expands to 825 mL?
(f) What is the volume if 0.0115 mole of gas is added at 1.75 atm and 63°C?

B–3. Name the laws used in each of the calculations in Problem B-2.

Additional Examples: Problems 9-20, 9-30, 9-36, 9-43, 9-51, 9-59, 9-61, and 9-62.

Looking Ahead ▼

Avogadro's law tells us that equal volumes contain equal numbers of molecules regardless of the masses or nature of the molecules. If the volumes are independent of the nature of the gas, what about the pressure? The nature of the pressure exerted by mixtures of gases was first observed by John Dalton and is discussed next.

9–7   DALTON'S LAW OF PARTIAL PRESSURES

Our atmosphere is composed of a mixture of gases. About 78% of the atmosphere is composed of nitrogen molecules, 21% is oxygen molecules, and a little less than 1% is argon atoms. There are also traces of other gases such as carbon dioxide (0.04%) and water. According to kinetic theory, these gas molecules have negligible attraction for each other, and all have the same average kinetic energy at the same temperature. Since gas molecules all behave the same toward each other, we would expect from kinetic theory that properties such as volume and pressure depend only on the number of molecules (moles) present and not on their identities. We have already noted in Avogadro's law that the volumes of gases are independent of the identity of the gas. The independence of pressure and the nature of the gases was first advanced by John Dalton, author of modern atomic theory. Another gas law resulted.

Dalton's law *states that the total pressure of a gas in a system is the sum of the partial pressures of each component gas.* Mathematically, this can be stated as $P_T = P_1 + P_2 + P_3 +$ etc. (P_1 is the pressure due to gas 1, etc.)

EXAMPLE 9–11

Total Pressure from Partial Pressures

Three gases, Ar, N_2, and H_2, are mixed in a 5.00-L container. Ar has a pressure of 255 torr, N_2 has a pressure of 228 torr, and H_2 has a pressure of 752 torr. What is the total pressure in the container?

● **Working It Out**

Solution

$$P_{Ar} = 255 \text{ torr} \quad P_{N_2} = 228 \text{ torr} \quad P_{H_2} = 752 \text{ torr}$$

$$P_T = P_{Ar} + P_{N_2} + P_{H_2} = 255 \text{ torr} + 228 \text{ torr} + 752 \text{ torr}$$

$$= \underline{1235 \text{ torr}}$$

We can now apply Dalton's law to the composition of our atmosphere. Since 21% of the molecules of the atmosphere are oxygen, 21% of the volume and pressure of the atmosphere are due to oxygen. Therefore, the partial pressure of O_2 at sea level is

$$P(O_2) = 0.21 \times 760 \text{ torr} = 160 \text{ torr}$$

Most of us function best breathing this partial pressure of oxygen. When we live at higher elevations, the partial pressure of oxygen is lower and our bodies adjust accordingly. On top of the highest mountain, Mt. Everest, the total atmospheric pressure is 270 torr, so the partial pressure of oxygen is only 57 torr or about one-third of normal. A human cannot survive for long at such a low pressure of oxygen. At that altitude, even the most conditioned climber must use an oxygen tank and mask, which gives an increased partial pressure of oxygen to the lungs.

Why is "thin air" hard to breathe?

EXAMPLE 9–12

Partial Pressures of Atmospheric Gases

On a humid summer day, the partial pressure of water in the atmosphere is 18 torr. If the barometric pressure on this day is a high 766 torr, what is the partial pressure of nitrogen and of oxygen if 78.0% of the dry atmosphere is composed of nitrogen molecules and 21.0% is oxygen? (1% of the dry atmosphere is composed of all other gases.)

● **Working It Out**

Procedure

1 Find the pressure of the dry atmosphere by subtracting the pressure of water from the total.

2 Find the partial pressure of N_2 and of O_2 from the total pressure of the dry atmosphere and the percent.

Solution

1 $P(\text{dry atmosphere}) = P_T - P(H_2O) = 766 - 18 = 748$ torr

2 $P(N_2) = 0.780 \times 748$ torr $= \underline{\underline{583 \text{ torr}}}$

$P(O_2) = 0.210 \times 748$ torr $= \underline{\underline{157 \text{ torr}}}$

The concept of total pressure can be extended to the ideal gas law. For example, for a mixture of three gases in one container,

$$P_T = P_1 + P_2 + P_3 = \frac{n_1 RT}{V} + \frac{n_2 RT}{V} + \frac{n_3 RT}{V} = \frac{n_T RT}{V}$$

We see that, at a given temperature, the pressures or volumes of gases depend only on the total number of moles present. This is illustrated in Figure 9-10 and Example 9-13.

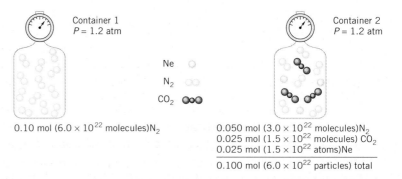

Container 1
P = 1.2 atm

Container 2
P = 1.2 atm

Ne
N_2
CO_2

0.10 mol (6.0×10^{22} molecules)N_2

0.050 mol (3.0×10^{22} molecules)N_2
0.025 mol (1.5×10^{22} molecules) CO_2
0.025 mol (1.5×10^{22} atoms)Ne
_____
0.100 mol (6.0×10^{22} particles) total

Figure 9–10
Pressures of a Pure Gas and a Mixture of Gases. Pressure depends only on the number of molecules at a certain temperature and not on their identity.

EXAMPLE 9–13

Pressure of a Mixture of Gases

Working It Out ●

What is the pressure (in atmospheres) exerted by a mixture of 12.0 g of N_2 and 12.0 g of O_2 in a 2.50-L container at 25°C?

Procedure

1 Find the number of moles of N_2 and O_2 present.

2 Use the total number of moles, temperature, and volume in the ideal gas law.

Solution

1 $n(O_2) = 12.0 \text{ g } O_2 \times \dfrac{1 \text{ mol } O_2}{32.0 \text{ g } O_2} = 0.375 \text{ mol } O_2$

$$n(\text{N}_2) = 12.0 \; \cancel{\text{g N}_2} \times \frac{1 \text{ mol N}_2}{28.0 \; \cancel{\text{g N}_2}} = 0.429 \text{ mol N}_2$$

$$n_T = 0.375 + 0.429 \text{ mol} = 0.804 \text{ mol of gas}$$

2 $\quad PV = nRT \quad P = \dfrac{nRT}{V}$

$$P = \frac{0.804 \; \cancel{\text{mol}} \times 0.0821 \; \dfrac{\cancel{\text{L}} \cdot \text{atm}}{\cancel{\text{K}} \cdot \cancel{\text{mol}}} \times (273 + 25)\cancel{\text{K}}}{2.50 \; \cancel{\text{L}}} = \underline{\underline{7.87 \text{ atm}}}$$

In previous chapters, we found that one mole of a compound represents a set number of molecules and a given mass. If the compound also happens to be a gas, we find that one mole also represents a set volume under specified conditions. We explore this next.

▼Looking Ahead

9–8 THE MOLAR VOLUME AND DENSITY OF A GAS

In Example 9-10, we calculated the volume of one mole of a gas at STP. This result represents the volume of one mole of *any* gas, or even one mole of a mixture of gases, at STP. This phenomenon is certainly not true of equal volumes of liquids and solids. *The volume of one mole of gas at STP, 22.4 L, is known as the* **molar volume.** We can now expand on the significance of the mole that was described in Chapter 7. One mole of a gas refers to three quantities: (1) the molar volume and (2) Avogadro's number, which are independent of the identity of the gas, and (3) the molar mass, which depends on the formula of the gaseous compound or element.

Does the volume of a gas relate directly to the number of moles?

Before we work two examples of problems relating to the molar volume, it may be helpful to summarize the relationships between moles of gas, volume at STP, and volume at some other T and P. In Figure 9-11, we notice that moles of gas can be converted to volume at STP using the molar volume as a conversion factor (path 1). Moles of gas can also be converted to volume at some other temperature (T) and pressure (P) using the ideal gas law (path 2). Finally, the volume of a quantity of gas at STP and the volume at some other temperature and pressure are related by the combined gas law.

EXAMPLE 9–14

Calculation of Mass from Volume at STP

What is the mass of 4.55 L of O_2 measured at STP?

●**Working It Out**

Procedure
Utilizing path 1 in Figure 9-11, convert volume to moles and then moles to mass. The two possible conversion factors for converting volume to moles

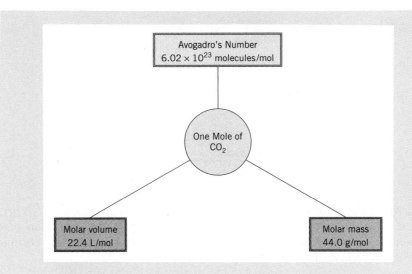

Figure 9–11
Moles and the Volume of a Gas. The number of moles relates to volume by two paths.

are

$$\frac{1 \text{ mol}}{22.4 \text{ L}} \quad \text{and} \quad \frac{22.4 \text{ L}}{\text{mol}}$$

$$\text{volume (STP)} \longrightarrow \text{moles} \longrightarrow \text{mass}$$

Solution

$$4.55 \text{ L} \times \frac{1 \text{ mol}}{22.4 \text{ L}} \times \frac{32.0 \text{ g}}{\text{mol}} = \underline{6.50 \text{ g}}$$

EXAMPLE 9–15

Calculation of the Molar Mass of a Gas

A sample of gas has a mass of 3.20 g and occupies 2.00 L at 17°C and 380 torr. What is the molar mass of the gas?

Procedure
There are two ways that one can work this problem. The first uses paths 1 and 3 in Figure 9-11 as shown above in Example 9-14.

1 Convert given volume to volume at STP.

2 Convert volume at STP to moles.

3 Convert moles and mass to molar mass as follows: from Chapter 8,

$$n \text{ (number of moles)} = \text{mass (in g)} \times \frac{1}{\text{molar mass (MM)}}$$

or

$$n = \frac{\text{mass}}{MM} \quad MM = \frac{\text{mass}}{n}$$

Solution

Initial Conditions | Final Conditions

$V_1 = 2.00$ L $V_2 = ?$

$P_1 = 380$ torr $P_2 = 760$ torr

$T_1 = (17 + 273)$ K $= 290$ K $T_2 = 273$ K

1
$$\frac{P_1 V_1}{T_1} = \frac{P_2 V_2}{T_2} \quad V_2 = V_1 \times \frac{P_1}{P_2} \times \frac{T_2}{T_1}$$

$$V_2 = 2.00 \text{ L} \times \frac{380 \text{ torr}}{760 \text{ torr}} \times \frac{273 \text{ K}}{290 \text{ K}} = \underline{\underline{0.941 \text{ L (STP)}}}$$

2
$$0.941 \text{ L} \times \frac{1 \text{ mol}}{22.4 \text{ L}} = 0.0420 \text{ mol of gas}$$

3
$$\text{molar mass} = \frac{\text{mass}}{n} = \frac{3.20 \text{ g}}{0.0420 \text{ mol}} = \underline{\underline{76.2 \text{ g/mol}}}$$

The alternative procedure is more direct, since it uses the ideal gas law as shown in path 2 in Figure 9-11. To find the molar mass using the ideal gas law, substitute the relationship for $n(n = \text{mass}/MM)$ into the gas law.

$$PV = nRT \qquad n = \frac{\text{mass}}{MM} \qquad PV = \frac{\text{mass}}{MM} RT$$

Solving for MM, we have

$$MM = \frac{(\text{mass})(RT)}{PV}$$

$$= \frac{3.20 \text{ g} \times 62.4 \dfrac{\text{L} \cdot \text{torr}}{\text{K} \cdot \text{mol}} \times 290 \text{ K}}{380 \text{ torr} \times 2.00 \text{ L}}$$

$$= \underline{\underline{76.2 \text{ g/mol}}}$$

In Chapter 2, the densities of solids and liquids were given in units of g/mL. Since gases are much less dense, units in this case are usually given in g/L (STP). The density of a gas at STP can be calculated by dividing the molar mass by the molar volume.

$$CO_2 \quad 44.0 \text{ g/mol} = 22.4 \text{ L/mol}$$

$$\frac{44.0 \text{ g/mol}}{22.4 \text{ L/mol}} = 1.96 \text{ g/L STP}$$

The densities of several gases are given in Table 9-2. The density of air (a mixture) is 1.29 g/L. Gases such as He and H_2 are less dense than air. Gases less dense than air rise in the air just as solids or liquids less dense than water float in water. Helium is used as the gas in blimps to make the whole craft "lighter than air." (See Problem 9-109.)

How does a blimp stay suspended in the air?

TABLE 9–2   DENSITIES OF SOME GASES

Gas	Density [g/L (STP)]	Gas	Density [g/L (STP)]
H_2	0.090	O_2	1.43
He	0.179	CO_2	1.96
N_2	1.25	CF_2Cl_2	5.49
Air (average)	1.29	SF_6	6.52

Looking Ahead ▼

Gases are formed or consumed in many chemical reactions. Since the volumes of gases relate directly to the number of moles, we can use volume as we did mass or number of molecules in previous stoichiometry calculations. For this discussion, we return to the topic of stoichiometry introduced in Chapter 8.

9–9   STOICHIOMETRY INVOLVING GASES

In chemical reactions, given masses or numbers of molecules react in ratios reflected by a balanced equation. When volumes of gases react or are produced as a product, these volumes are also related by means of a balanced equation. To include the volumes of gases in the stoichiometry scheme, we have rewritten the general scheme first presented in Chapter 8. (See Figure 9-12.) We have deleted number of molecules from the scheme because problems involving actual numbers of molecules are not usually encountered. The volume and number of moles of a gas are shown to be interrelated by the ideal gas law. In fact, if the volume of

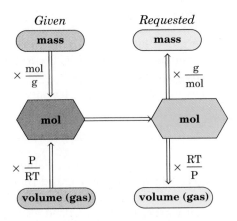

Figure 9–12
General Procedure for Stoichiometry Problems

the gas is measured at STP, the conversion can be simplified by using the molar volume relationship

$$n = \frac{V(\text{STP})}{22.4 \text{ L/mol}}$$

The following two examples illustrate the relationship of the gas laws to stoichiometry.

EXAMPLE 9–16

Mass of a Reactant from the Volume of a Product

Given the balanced equation

$$2Al(s) + 6HCl(aq) \longrightarrow 2AlCl_3(aq) + 3H_2(g)$$

what mass of Al is needed to produce 50.0 L of H_2 measured at STP?

●**Working It Out**

Procedure
The general procedure is shown below. In this case, it is easier to use the molar volume as a conversion factor between volume and moles.

Given		Requested
$\times \dfrac{\text{mol } H_2}{22.4 \text{ L}}$	$\times \dfrac{\text{mol Al}}{\text{mol } H_2}$	$\times \dfrac{\text{g Al}}{\text{mol Al}}$
volume (H_2) ⟹ mol (H_2) ⟹	mol (Al) ⟹	mass (Al)

Solution

$$50.0 \text{ L (STP)} \times \frac{1 \text{ mol } H_2}{22.4 \text{ L (STP)}} \times \frac{2 \text{ mol Al}}{3 \text{ mol } H_2} \times \frac{27.0 \text{ g Al}}{\text{mol Al}} = \underline{\underline{40.2 \text{ g Al}}}$$

EXAMPLE 9–17

Volume of a Product from Mass of a Reactant

Given the balanced equation

$$4NH_3(g) + 5O_2(g) \longrightarrow 4NO(g) + 6H_2O(l)$$

what volume of NO gas measured at 550 torr and 25°C will be produced from 19.5 g of O_2?

Procedure

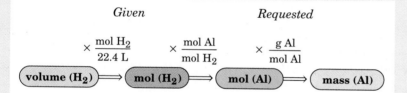

Given		Requested
$\times \dfrac{\text{mol } O_2}{\text{g } O_2}$	$\times \dfrac{\text{mol NO}}{\text{mol } O_2}$	$\times \dfrac{RT}{P}$
mass (O_2) ⟹	mol (O_2) ⟹	mol (NO) ⟹ volume (NO)

Solution

$$19.5 \text{ g O}_2 \times \frac{1 \text{ mol O}_2}{32.0 \text{ g O}_2} \times \frac{4 \text{ mol NO}}{5 \text{ mol O}_2} = 0.488 \text{ mol NO}$$

1 Using the ideal gas law

$$V = \frac{nRT}{P} = \frac{0.488 \text{ mol} \times 62.4 \frac{\text{L} \cdot \text{torr}}{\text{K} \cdot \text{mol}} \times 298 \text{ K}}{550 \text{ torr}} = 16.5 \text{ L}$$

2 Convert moles of NO to volume at STP and then volume at STP to volume at 550 torr and 25°C

$$0.488 \text{ mol} \times 22.4 \text{ L/mol} = 10.9 \text{ L (STP)}$$

$$V_2 = V_1 \times \frac{P_1}{P_2} \times \frac{T_2}{T_1}$$

Initial Conditions

$V_1 = 10.9 \text{ L}$
$P_1 = 760 \text{ torr}$
$T_1 = 273 \text{ K}$

Final Conditions

$V_2 = ?$
$P_2 = 550 \text{ torr}$
$T_2 = (25 + 273) \text{ K} = 298 \text{ K}$

$$V_2 = 10.9 \text{ L} \times \frac{760 \text{ torr}}{550 \text{ torr}} \times \frac{298 \text{ K}}{273 \text{ K}} = 16.4 \text{ L}$$

Looking Back ▲

At a given temperature, the volume and pressure of a gas depend only on the number of molecules (or moles) present and not on their identity, or even whether we have a mixture of gases. The only variable that concerns the identity of the gas is the density, which, of course, depends on the molar mass of the gas. The relationship of the volume to the number of moles of gas allows us to apply the ideal gas law (or the molar volume relationship) directly to stoichiometry problems.

Checking It Out ▶

Learning Check C

C–1. Fill in the blanks.
The relationship between the partial pressure of a gas and its proportion in a mixture was first observed by _____ _____.
One mole of any gas occupies _____ L at STP and is known as the _____ _____. The density of a gas at STP is obtained by dividing the _____ _____ by the molar volume. In stoichiometry calculations, the volume of a gas can be converted to number of moles by use of the _____ _____ if the volume is measured at STP. If other conditions are present, the volume is related to the number of moles by the _____ _____ _____.

C–2.   What is the partial pressure of O_2 if 0.0450 mole of O_2 is mixed with 0.0328 mole of N_2 and the total pressure in the container is 596 torr?

C–3.   What is the volume occupied by 142 g of SF_4 gas at STP? What is the density of SF_4 at STP?

C–4.   Humans obtain energy from the combustion of blood sugar (glucose) according to the equation

$$C_6H_{12}O_6(aq) + 6O_2(g) \longrightarrow 6CO_2(g) + 6H_2O(l)$$

(a) What volume of O_2 measured at STP is required to react with 0.122 mole of glucose?
(b) What volume of CO_2 measured at 25°C and 1.10 atm pressure is released from the combustion of 228 g of glucose?
(c) What mass of glucose is required to react with 2.48 L of oxygen measured at 22°C and 655 torr?

Additional Examples: Problems 9-68, 9-73, 9-79, 9-84, 9-86, 9-92, and 9-96.

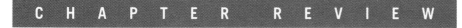

C H A P T E R   R E V I E W

Our modern theories about matter depended on some basic understanding about the most abstract state of matter—gases. Beginning in the 1600s, the quantitative laws that we know as the gas laws began to be advanced. These laws and other observations of the nature of gases led to the accepted model of behavior known as the **kinetic molecular theory.** One important result of kinetic theory is the relationship between temperature and the average kinetic energy of molecules. The velocity of the molecules, rate of **effusion,** and rate of **diffusion** all relate to the molar mass of the gas.

▲▼ Putting It Together

An understanding of the gaseous state began in the mid-1600s with Torricelli. His studies of the **barometer** led to a description of **pressure** as a **force** per unit area exerted by gases of the atmosphere. The gas laws now seem quite reasonable and predictable from kinetic theory. Except for average velocity and density, which depend on the mass of the molecules, the other gas laws are independent of the identity of the gas. Avogadro advanced the observation that equal volumes contain equal numbers of molecules under the same conditions, and Dalton observed that pressures depended only upon the amount of gas present.

The gas laws in this chapter and their applications are summarized in the following table.

Gas Law	Relationship	Meaning	Constant Conditions	Application
Graham's	$v \propto \dfrac{1}{\sqrt{MM}}$	$v \uparrow MM \downarrow$	T	Relates MM and v of two different gases at a specific T
Boyle's	$V \propto \dfrac{1}{P}$	$V \uparrow P \downarrow$	T, n	Relates V and P of a gas under two different sets of conditions
Charles's	$V \propto T$	$V \uparrow T \uparrow$	P, n	Relates V and T of a gas under two different sets of conditions
Gay-Lussac's	$P \propto T$	$P \uparrow T \uparrow$	V, n	Relates P and T of a gas under two different sets of conditions
Avogadro's	$V \propto n_T$	$V \uparrow n_T \uparrow$	P, T	Relates V and n of a gas under two different sets of conditions
Combined	$PV \propto T$	$PV \uparrow T \uparrow$	n	Relates P, V, and T of a gas under two different sets of conditions
Ideal	$PV \propto nT$	$PV \uparrow nT \uparrow$	—	Relates P, V, T, or n to the other three variables
Dalton's	$P_T = P_1 + P_2$, etc.	$P_T \uparrow n_T \uparrow$	T, V	Relates P_T to partial pressures of component gases

V = volume	n = moles	v = average velocity
T = Kelvin temperature	n_T = total moles	$\uparrow$ = quantity increases
P = pressure	MM = molar mass	$\downarrow$ = quantity decreases
P_T = total pressure		

The gas laws require the use of the **Kelvin temperature** scale, which begins at **absolute zero**. The volumes of gases are often described under **standard temperature and pressure (STP)** conditions. At STP, the volume of one mole of a gas is 22.4 L and is known as the **molar volume.** The molar volume can be used to calculate the density of a gas at STP, or it can be used directly as a conversion factor between moles and volume at STP. To convert between moles and volume under other conditions, the use of the ideal gas law is convenient. In the ideal gas law ($PV = nRT$), **R** is known as the **gas constant.**

$$\text{volume } (T, P) \xleftarrow{\text{ideal gas law}} \underline{\text{moles of gas}} \xrightarrow{\text{molar volume}} \text{volume (STP)}$$

In the final section, we related the volume of a gas to stoichiometry as summarized in Figure 9-12.

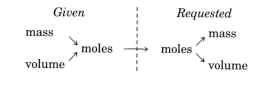

THE KINETIC THEORY OF GASES

9–1 It is harder to move your arms in water than in air. Explain on the basis of kinetic molecular theory.

9–2 A balloon filled with water is pear-shaped, but a balloon filled with air is spherical. Explain.

9–3 When a gasoline tank is filled, no more gasoline can be added. When a tire is "filled," however, more air can be added. Explain.

9–4 The pressure inside an auto tire is the same regardless of the location of the nozzle (i.e., up, down, or to the side). Explain.

9–5 A sunbeam forms when light is reflected from dust suspended in the air. Even if the air is still, the dust particles can be seen to bounce around randomly. Explain.

9–6 A bowling ball weighs 6.00 kg and a bullet weighs 1.50 g. If the bowling ball is rolled down an alley at 20.0 mi/hr, what is the velocity of a bullet having the same kinetic energy?

9–7 Arrange the following gases in order of increasing average speed (rate of effusion) at the same temperature.
(a) CO_2 (c) N_2 (e) N_2O
(b) SO_2 (d) SF_6 (f) H_2

9–8 What is the rate of effusion of N_2 molecules compared with Ar atoms at the same temperature?

9–9 Compare the rates of effusion of H_2 molecules and Kr atoms at the same temperature.

*9–10 A certain gas effuses twice as fast as SF_6 molecules. What is the molar mass of the unknown gas?

*9–11 Carbon monoxide effuses 2.13 times faster than an unknown gas. What is the molar mass of the unknown gas?

*9–12 To make enriched uranium for use in nuclear reactors or for weapons, ^{235}U must be separated from ^{238}U. Although ^{235}U is the isotope needed for fission, only 0.7% of U atoms are this isotope. Separation is a difficult and expensive process. Since UF_6 is a gas, Graham's law can be applied to separate the isotopes. How much faster does a $^{235}UF_6$ molecule travel on the average compared to a $^{238}UF_6$ molecule?

*9–13 The kinetic theory assumes that the volume of molecules and their interactions are negligible for gases. Explain why these assumptions may not be true when the pressure is very high and the temperature is very low.

UNITS OF PRESSURE

9–14 Make the following conversions:
(a) 1650 torr to atm
(b) 3.50×10^{-5} atm to torr
(c) 185 lb/in.^{2} to torr
(d) 5.65 kPa to atm
(e) 190 torr to lb/in.^{2}
(f) 85 torr to kPa

9–15 Make the following conversions:
(a) 30.2 in. of Hg to torr
(b) 25.7 kilobars to atm
(c) 57.9 kPa to lb/in.^{2}
(d) 0.025 atm to torr

9–16 Complete the following table:

torr	lb/in.^{2}	in. Hg	kPa	atm
455	——	——	——	——
——	2.45	——	——	——
——	——	117	——	——
——	——	——	783	——
——	——	——	——	0.0768

9–17 The atmospheric pressure on the planet Mars is 10.3 millibars. What is this pressure in Earth atmospheres?

9–18 The atmospheric pressure on the planet Venus is 0.0920 kilobar. What is this pressure in Earth atmospheres?

9–19 The density of water is 1.00 g/mL. If water is substituted for mercury in the barometer, how high (in feet) would a column of water be supported by 1 atm? A water well is 40 ft deep. Can suction be used to raise the water to ground level?

BOYLE'S LAW

9–20 A gas has a volume of 6.85 L at a pressure of 0.650 atm. What is the volume of the gas if the pressure is decreased to 0.435 atm?

9–21 If a gas has a volume of 1560 mL at a pressure of 81.2 kPa, what is its volume if the pressure is increased to 2.50 atm?

9–22 At sea level, a balloon has a volume of 785 mL. What is its volume if it is taken to a place in Colorado where the atmospheric pressure is 610 torr?

9–23 A gas has a volume of 125 mL at a pressure of 62.5 torr. What is the pressure if the volume is decreased to 115 mL?

9–24 How does the kinetic molecular theory explain Boyle's law?

9–25 A few miles above the surface of Earth the pressure drops to 1.00×10^{-5} atm. What would be the volume of a 1.00-L sample of gas at sea-level pressure (1.00 atm) if it were taken to that altitude? (Assume constant temperature.)

*9–26 A gas in a piston engine is compressed by a ratio of 15:1. If the pressure before compression is 0.950 atm, what pressure is required to compress the gas? (Assume constant temperature.)

*9–27 The volume of a gas is measured as the pressure is varied. The four measurements are reported as follows:

Experiment	Volume (mL)	Pressure (torr)
1	125	450
2	145	385
3	175	323
4	220	253

Make a graph using volume on the x axis and pressure on the y axis. What is the average value of the constant of proportionality, k?

CHARLES'S LAW

9–28 A balloon has a volume of 1.55 L at 25°C. What would be the volume if the balloon is heated to 100°C? (Assume constant P.)

9–29 A sample of gas has a volume of 677 mL at 63°C. What is the volume of the gas if the temperature is decreased to 46°C?

9–30 A balloon has a volume of 325 mL at 17°C. What is the temperature if the volume increases to 392 mL?

9–31 How does the kinetic molecular theory explain Charles's law?

9–32 A quantity of gas has a volume of 3.66×10^4 L. What will be the volume if the temperature is changed from 455 K to 50°C?

*9–33 The temperature of a sample of gas is 0°C. When the temperature is increased, the volume increases by a factor of 1.25 (i.e., $V_2 = 1.25 \, V_1$.) What is the final temperature in degrees Celsius?

*9–34 The volume of a gas is measured as the temperature is varied. The four measurements are reported as follows:

Experiment	Volume (L)	Temperature (°C)
1	1.54	20
2	1.65	40
3	1.95	100
4	2.07	120

Make a graph of the volume on the x axis and the Kelvin temperature on the y axis. What is the average value of the constant of proportionality, k?

GAY-LUSSAC'S LAW

9–35 A confined quantity of gas is at a pressure of 2.50 atm and a temperature of $-22°C$. What is the pressure if the temperature increases to 22°C?

9–36 A quantity of gas has a volume of 3560 mL at a temperature of 55°C and a pressure of 850 torr. What is the temperature if the volume remains unchanged but the pressure is decreased to 0.652 atm?

9–37 A metal cylinder contains a quantity of gas at a pressure of 558 torr at 25°C. At what temperature does the pressure inside the cylinder equal 1 atm pressure?

9–38 An aerosol spray can has gas under pressure of 1.25 atm at 25°C. The can explodes when the pressure reaches 2.50 atm. At what temperature will this happen? (Do not throw these cans into a fire!)

9–39 The pressure in an automobile tire is 28.0 lb/in.^{2} on a chilly morning of 17°C. After it is driven awhile, the temperature of the tire rises to 40°C. What is the pressure in the tire if the volume remains constant?

9–40 How does the kinetic molecular theory explain Gay-Lussac's law?

*9–41 The pressure of a confined volume of gas is measured as the temperature is raised. The four measurements are reported as follows:

Experiment	Pressure (torr)	Temperature (K)
1	550	295
2	685	372
3	745	400
4	822	445

Make a graph of the pressure on the x axis and the temperature on the y axis. What is the average value of the constant of proportionality, k?

AVOGADRO'S LAW

9–42 A 0.112-mol quantity of gas has a volume of 2.54 L at a certain temperature and pressure. What is the volume of 0.0750 mol of gas under the same conditions?

9–43 A balloon has a volume of 188 L and contains 8.40 mol of gas. How many moles of gas would be needed to expand the balloon to 275 L? Assume the same temperature and pressure in the balloon.

9–44 A balloon has a volume of 275 mL and contains 0.0212 mol of CO_2. What mass of N_2 must be added to expand the balloon to 400 mL?

9–45 A balloon has a volume of 75.0 mL and contains 2.50×10^{-3} mol of gas. What mass of N_2 must be added to the balloon for the volume to increase to 164 mL at the same temperature and pressure?

9–46 A 48.0-g quantity of O_2 in a balloon has a volume of 30.0 L. What is the volume if 48.0 g of SO_2 is substituted for O_2 in the same balloon?

THE COMBINED GAS LAW

9–47 Which of the following are legitimate expressions of the combined gas law?
(a) $PV = kT$
(b) $PT \propto V$
(c) $\dfrac{P_1 T_1}{V_1} = \dfrac{P_2 T_2}{V_2}$
(d) $\dfrac{P}{T} \propto \dfrac{1}{V}$
(e) $VT \propto P$

9–48 Which of the following are not STP conditions?
(a) $T = 273$ K
(b) $P = 760$ atm
(c) $T = 0$ K
(d) $P = 1$ atm
(e) $t(C) = 273°C$
(f) $P = 760$ torr
(g) $t(C) = 0°C$

9–49   In the following table, indicate whether the pressure, volume, or temperature increases or decreases.

Experiment	P	V	T
1	increases	constant	————
2	constant	————	decreases
3	————	decreases	constant
4	increases	increases	————

9–50   In the following table, indicate whether the pressure, volume, or temperature increases or decreases.

Experiment	P	V	T
1	decreases	————	constant
2	constant		T(initial) = 350K T(final) = 40°C
3	P(initial) = 1.75 atm P(final) = 2200 torr	constant	
4	————	increases	decreases

9–51   A 5.50-L volume of gas has a pressure of 0.950 atm at 0°C. What is the pressure if the volume decreases to 4.75 L and the temperature increases to 35°C?

9–52   A quantity of gas has a volume of 17.5 L at a pressure of 6.00 atm and temperature of 100°C. What is its volume at STP?

9–53   A quantity of gas has a volume of 88.7 mL at STP. What is its volume at 0.845 atm and 35°C?

9–54   A quantity of gas has a volume of 4.78×10^{-4} mL at a temperature of −50°C and a pressure of 78.0 torr. If the volume changes to 9.55×10^{-5} mL and the pressure to 155 torr, what is the temperature?

9–55   A gas has a volume of 64.2 L at STP. What is the temperature if the volume decreases to 58.5 L and the pressure increases to 834 torr?

9–56   A quantity of gas has a volume of 6.55×10^{-5} L at 7°C and 0.882 atm. What is the pressure if the volume changes to 4.90×10^{-3} L and the temperature to 273 K?

9–57   A balloon has a volume of 1.55 L at 25°C and 1.05 atm pressure. If it is cooled in the freezer, the volume shrinks to 1.38 L and the pressure drops to 1.02 atm. What is the temperature in the freezer?

9–58   A bubble from a deep-sea diver in the ocean starts with a volume of 35.0 mL at a temperature of 17°C and a pressure of 11.5 atm. What is the volume of the bubble when it reaches the surface? Assume that the pressure at the surface is 1 atm and the temperature is 22°C.

IDEAL GAS LAW

9–59   What is the temperature (in degrees Celsius) of 4.50 L of a 0.332-mol quantity of gas under a pressure of 2.25 atm?

9–60   A quantity of gas has a volume of 16.5 L at 32°C and a pressure of 850 torr. How many moles of gas are present?

9–61   What mass of NH_3 gas has a volume of 16,400 mL, a pressure of 0.955 atm, and a temperature of −23°C?

9–62   What is the pressure (in torr) exerted by 0.250 g of O_2 in a 250-mL container at 29°C?

9–63   A container of Cl_2 gas has a volume of 750 mL and is at a temperature of 19°C. If there is 7.88 g of Cl_2 in the container, what is the pressure in atmospheres?

9–64   What mass of Ne is contained in a large neon light if the volume is 3.50 L, the pressure 1.15 atm, and the temperature 23°C?

9–65   A sample of H_2 is collected in a bottle over water. The volume of the sample is 185 mL at a temperature of 25°C. The pressure of H_2 in the bottle is 736 torr. What is the mass of H_2 in the bottle?

*9–66   A blimp has a volume of about 2.5×10^7 L. What is the mass of He (in lb) in the blimp at 27°C and 780 torr? The average molar mass of air is 29.0 g/mol. What mass of air (in lb) would the blimp contain? The difference between these two values is the lifting power of the blimp. What mass could the blimp lift? If H_2 is substituted for He, what is the lifting power? Why isn't H_2 used?

*9–67 A good vacuum pump on Earth can produce a vacuum with a pressure as low as 1.00×10^{-8} torr. How many molecules are present in each milliliter at a temperature of 27.0°C?

DALTON'S LAW

9–68 Three gases are mixed in a 1.00-L container. The partial pressure of CO_2 is 250 torr, that of N_2 375 torr, and that of He 137 torr. What is the pressure of the mixture of gases?

9–69 The total pressure in a cylinder containing a mixture of two gases is 1.46 atm. If the partial pressure of one gas is 750 torr, what is the partial pressure of the other gas?

9–70 Air is about 0.90% Ar. If the barometric pressure is 756 torr, what is the partial pressure of Ar?

9–71 A sample of oxygen is collected in a bottle over water. The pressure inside the bottle is made equal to the barometric pressure, which is 752 torr. When collected over water, the gas is a mixture of oxygen and water vapor. The partial pressure of water (known as the vapor pressure) at that temperature is 24 torr. What is the pressure of the pure oxygen?

9–72 A container holds two gases, A and B. Gas A has a partial pressure of 325 torr and gas B has a partial pressure of 488 torr. What percent of the molecules in the mixture is gas A?

9–73 A volume of gas is composed of N_2, O_2, and SO_2. If the total pressure is 1050 torr, what is the partial pressure of each gas if the gas is 72.0% N_2 and 8.00% O_2?

*9–74 A mixture of two gases is composed of CO_2 and O_2. The partial pressure of O_2 is 256 torr, and it represents 35.0% of the molecules of the mixture. What is the total pressure of the mixture?

*9–75 A volume of gas has a total pressure of 2.75 atm. If the gas is composed of 0.250 mol of N_2 and 0.427 mol of CO_2, what is the partial pressure of each gas?

*9–76 The following gases are all combined into a 2.00-L container: a 2.00-L volume of N_2 at 300 torr, a 4.00-L volume of O_2 at 85 torr, and a 1.00-L volume of CO_2 at 450 torr. What is the total pressure?

*9–77 The total pressure of a mixture of two gases is 0.850 atm in a 4.00-L container. Before mixing, gas A was in a 2.50-L container and had a pressure of 0.880 atm. What is the partial pressure of gas B in the 4.00-L container?

*9–78 What is the pressure (in atm) in a 825-mL container at 33°C if it contains 6.25 g of N_2 and 12.6 g of CO_2?

MOLAR VOLUME AND DENSITY

9–79 What is the volume of 15.0 g of CO_2 measured at STP?

9–80 What is the mass (in kg) of 850 L of CO measured at STP?

9–81 What is the volume of 3.01×10^{24} molecules of N_2 measured at STP?

9–82 A 6.50-L quantity of a gas measured at STP has a mass of 39.8 g. What is the molar mass of the compound?

9–83 What is the mass of 6.78×10^{-4} L of NO_2 measured at STP?

9–84 What is the density in g/L (STP) of B_2H_6?

9–85 What is the density in g/L (STP) of BF_3?

9–86 A gas has a density of 1.52 g/L (STP). What is the molar mass of the gas?

9–87 A gas has a density of 6.14 g/L (STP). What is the molar mass of the gas?

*9–88 A gas has a density of 3.60 g/L at a temperature of 25°C and a pressure of 1.20 atm. What is its density at STP?

*9–89 What is the density (in g/L) of N_2 measured at 500 torr and 22°C?

*9–90 What is the density (in g/L) of SF_6 measured at 0.370 atm and 37°C?

STOICHIOMETRY INVOLVING GASES

9–91 Limestone is dissolved by CO_2 and water according to the equation

$$CaCO_3(s) + H_2O(l) + CO_2(g)$$
$$\longrightarrow \ Ca(HCO_3)_2(aq)$$

What volume of CO_2 measured at STP would dissolve 115 g of $CaCO_3$?

9–92 Magnesium in flashbulbs burns according to the equation

$$2Mg(s) + O_2(g) \longrightarrow 2MgO(s)$$

What mass of Mg combines with 5.80 L of O_2 measured at STP?

9–93 Oxygen gas can be prepared in the laboratory by decomposition of potassium nitrate according to the equation

$$2KNO_3(s) \xrightarrow{\Delta} 2KNO_2(s) + O_2(g)$$

What mass of KNO_2 forms along with 14.5 L of O_2 measured at 1 atm and 25°C?

9–94 Acetylene (C_2H_2) is produced from calcium carbide as shown by the reaction

$$CaC_2(s) + 2H_2O(l) \longrightarrow Ca(OH)_2(s) + C_2H_2(g)$$

What volume of acetylene measured at 25°C and 745 torr would be produced from 5.00 g of H_2O?

9–95 Nitrogen dioxide is an air pollutant. It is produced from NO (from car exhaust) as follows:

$$2NO(g) + O_2(g) \longrightarrow 2NO_2(g)$$

What volume of NO measured at STP is required to react with 5.00 L of O_2 measured at 1.25 atm and 17°C?

9–96 Butane (C_4H_{10}) burns according to the equation

$$2C_4H_{10}(g) + 13O_2(g) \longrightarrow 8CO_2(g) + 10H_2O(l)$$

(a) What volume of CO_2 measured at STP would be produced by 85.0 g of C_4H_{10}?

(b) What volume of O_2 measured at 3.25 atm and 127°C would be required to react with 85.0 g of C_4H_{10}?

(c) What volume of CO_2 measured at STP would be produced from 45.0 L of C_4H_{10} measured at 25°C and 0.750 atm?

9–97 In March 1979, a nuclear reactor overheated, producing a dangerous hydrogen bubble at the top of the reactor core. The following reaction occurring at the high temperatures (about 1500°C) accounted for the hydrogen. (Zr alloys hold the uranium pellets in long rods.)

$$Zr(s) + 2H_2O(g) \longrightarrow ZrO_2(s) + 2H_2(g)$$

If the bubble had a volume of about 28,000 L at 250°C and 70.0 atm, what mass (in kg and tons) of Zr had reacted?

9–98 Nitric acid is produced according to the equation

$$3NO_2(g) + H_2O(l) \longrightarrow 2HNO_3(aq) + NO(g)$$

What volume of NO_2 measured at −73°C and 1.56×10^{-2} atm would be needed to produce 4.55×10^{-3} mol of HNO_3?

*9–99 Natural gas (CH_4) burns according to the equation

$$CH_4(g) + 2O_2(g) \longrightarrow CO_2(g) + 2H_2O(l)$$

What volume of CO_2 measured at 27°C and 1.50 atm is produced from 27.5 L of O_2 measured at −23°C and 825 torr?

GENERAL PROBLEMS

9–100 A column of mercury (density 13.6 g/mL) is 15.0 cm high. A cross-section of the column has an area of 12.0 cm². What is the force (weight) of the mercury at the bottom of the tube? What is the pressure in grams per square centimeter and in atmospheres?

9–101 A tube containing an alcohol (density 0.890 g/mL) is 1.00 m high and has a cross-section of 15.0 cm². What is the total force at the bottom of the tube? What is the pressure? How high would be an equivalent amount of mercury assuming the same cross-section?

9–102 A 1.00-L volume of a gas weighs 8.37 g. The gas volume is measured at 1.45 atm pressure and 35°C. What is the molar mass of the gas?

9–103 A gaseous compound is 85.7% C and 14.3% H. A 6.58-g quantity of this gas occupies 4500 mL at 77.0°C and a pressure of 1.00 atm. What is the molar mass of the compound? What is its molecular formula?

9–104 What is the volume at STP of a mixture of 10.0 g each of Ar, Cl_2, and N_2? What is the partial pressure of each gas?

9–105 What is the volume occupied by a mixture of 0.265 mol of O_2, 9.88 g of N_2, and 9.65×10^{22} molecules of CO_2 at a temperature of 37°C and a pressure of 2.86 atmospheres? What is the partial pressure of each gas?

9–106 Molecular clouds in space contain 30,000 molecules per mL at a temperature of 10 K. What is the pressure in atmospheres?

9–107 Neptune is a planet that orbits the sun about 4.5 billion miles from Earth. It has a moon, Triton, with a thin atmosphere. Voyager 2 measured a surface temperature of 38 K and a pressure of 10 microbars (1 microbar = 10^{-6} bars). What is the density (in g/L) of the atmosphere at the surface? Assume that the atmosphere on Triton is nitrogen. How does this compare with the density of air at STP on Earth?

9–108 What is the molar volume at 25°C and 1.25 atm? What is the density of CO_2 under these conditions?

9–109 A hot-air balloon rises because the heated air trapped in the balloon is less dense than the surrounding air. What is the density of air (assume an average molar mass of 29.0 g/mol) at 400°C and 1 atm pressure? Compare this to the density of air at STP.

9–110 A compound is 80.0% carbon and 20.0% hydrogen. Its density at STP is 1.34 g/L. What is its molecular formula?

9–111 Given the following *unbalanced* equation

$$H_3BCO(g) + H_2O(l) \longrightarrow B(OH)_3(aq) + CO(g) + H_2(g)$$

A 425.0-mL quantity of H_3BCO measured at 565 torr and 100°C was allowed to react with excess H_2O. What volume of gas was produced measured at 25°C and 0.900 atm?

9–112 Given the following *unbalanced* equation

$$C_3H_8O(g) + O_2(g) \longrightarrow CO_2(g) + H_2O(l)$$

What mass of water forms if 6.50 L of C_3H_8O measured at STP is allowed to react with 42.0 L of O_2 measured at 27°C and 1.68 atm pressure? Assume that this is the only reaction that occurs.

9–113 Given the following *unbalanced* equation

$$Al(s) + F_2(g) \longrightarrow AlF_3(s)$$

A 8.23-L quantity of F_2 measured at 35°C and 725 torr was allowed to react with some Al. At the end of the reaction, 3.50 g of F_2 remained. What mass of AlF_3 formed?

9–114 Sulfuric acid is made from SO_3, which is obtained from the combustion of sulfur according to the following *unbalanced* equations.

$$S(s) + O_2(g) \longrightarrow SO_2(g)$$

$$SO_2(g) + O_2(g) \longrightarrow SO_3(g)$$

What volume of SO_3 measured at 2.75 atm and 400°C is prepared from 50.0 kg of sulfur?

SOLUTIONS TO LEARNING CHECKS

A–1 low, uniformly, kinetic molecular, constant, random, collisions, kinetic, temperature, Graham's, lowest, barometer, force, area, 760

A–2 Molar mass $SF_6 = [32.1 + 6(19.0)] = 146$ g/mol $N_2 = 2(14.0) = 28.0$ g/mol

$$\frac{v(SF_6)}{v(N_2)} = \sqrt{\frac{MM(N_2)}{MM(SF_6)}} = \sqrt{\frac{28 \text{ g/mol}}{146 \text{ g/mol}}} = 0.438$$

$v(SF_6) = \underline{0.438\ v(N_2)}$ (SF$_6$ moves less than half as fast as N$_2$.)

A–3 $0.650 \text{ atm} \times \dfrac{760 \text{ torr}}{\text{atm}} = \underline{\underline{494 \text{ torr}}}$

$0.650 \text{ atm} \times \dfrac{101.3 \text{ kPa}}{\text{atm}} = \underline{\underline{65.8 \text{ kPa}}}$

$0.650 \text{ atm} \times \dfrac{14.7 \text{ lb/in.}^2}{\text{atm}} = \underline{\underline{9.56 \text{ lb/in.}^2}}$

B–1   decrease, increase, increase, Boyle's, Charles's, Avogadro's, increases, $PV = nRT$, gas constant, 0.0821

B–2   (a) $550 \text{ mL} \times \dfrac{1.00 \text{ atm}}{1.32 \text{ atm}} = \underline{\underline{417 \text{ mL}}}$

(b) $550 \text{ mL} \times \dfrac{(273 + 44) \text{ K}}{(273 + 22) \text{ K}} = \underline{\underline{591 \text{ mL}}}$

(c) $550 \text{ mL} \times \dfrac{(0.0227 + 0.0115) \text{ mol}}{0.0227 \text{ mol}} = \underline{\underline{829 \text{ mL}}}$

(d) $1.00 \text{ atm} \times \dfrac{(273 + 122) \text{ K}}{(273 + 22) \text{ K}} = \underline{\underline{1.34 \text{ atm}}}$

(e) $1.00 \text{ atm} \times \dfrac{550 \text{ mL}}{825 \text{ mL}} \times \dfrac{(273 + 102) \text{ K}}{(273 + 22) \text{ K}} = \underline{\underline{0.847 \text{ atm}}}$

(f) $V = \dfrac{nRT}{P} = \dfrac{(0.0115 + 0.0227) \text{ mol} \times 0.0821 \dfrac{\text{L} \cdot \text{atm}}{\text{K} \cdot \text{mol}} \times (273 + 63) \text{K}}{1.75 \text{ atm}}$

$$= \underline{\underline{0.539 \text{ L}}} = \underline{\underline{539 \text{ mL}}}$$

B–3   (a) Boyle's law
(b) Charles's law
(c) Avogadro's law
(d) Gay-Lussac's law
(e) combined gas law
(f) ideal gas law

C–1   John Dalton, 22.4, molar volume, molar mass, molar volume, ideal gas law

C–2 $n_T = 0.0450 + 0.0328 = 0.0778 \text{ mol}$

decimal fraction of $O_2 = \dfrac{0.0450}{0.0778} = 0.578$

$0.578 \times 596 \text{ torr} = \underline{\underline{344 \text{ torr}}}$

C–3 $SF_4 [32.1 + 4(19.0)] = 108 \text{ g/mol}$

$$\boxed{\text{g } SF_4} \Longrightarrow \boxed{\text{mol } SF_4} \Longrightarrow \boxed{\text{vol } SF_4}$$

$142 \text{ g } SF_4 \times \dfrac{1 \text{ mol } SF_4}{108 \text{ g } SF_4} \times \dfrac{22.4 \text{ L}}{\text{mol } SF_4} = \underline{\underline{29.5 \text{ L (STP)}}}$

C–4   (a) $\boxed{\text{mol } C_6H_{12}O_6} \Longrightarrow \boxed{\text{mol } O_2} \Longrightarrow \boxed{\text{vol } O_2}$

$0.122 \text{ mol } C_6H_{12}O_6 \times \dfrac{6 \text{ mol } O_2}{1 \text{ mol } C_6H_{12}O_6} \times \dfrac{22.4 \text{ L}}{\text{mol } O_2} = \underline{\underline{16.4 \text{ L (STP)}}}$

(b) $\boxed{\text{g } C_6H_{12}O_6} \Longrightarrow \boxed{\text{mol } C_6H_{12}O_6} \Longrightarrow \boxed{\text{mol } CO_2} \Longrightarrow \boxed{\text{vol } CO_2 \text{ (P, T)}}$

$C_6H_{12}O_6 = [(6 \times 12.0) + (12 \times 1.01) + (6 \times 16.0)] = 180 \text{ g/mol}$

$$228 \text{ g } C_6H_{12}O_6 \times \frac{1 \text{ mol } C_6H_{12}O_6}{180 \text{ g } C_6H_{12}O_6} \times \frac{6 \text{ mol } CO_2}{1 \text{ mol } C_6H_{12}O_6} = 7.60 \text{ mol } CO_2$$

$$V = \frac{nRT}{P} = \frac{7.60 \text{ mol} \times 0.0821 \dfrac{\text{L} \cdot \text{atm}}{\text{K} \cdot \text{mol}} \times 298 \text{ K}}{1.10 \text{ atm}} = \underline{\underline{169 \text{ L}}}$$

(c) $\boxed{\text{vol } O_2} \Longrightarrow \boxed{\text{mol } O_2} \Longrightarrow \boxed{\text{mol } C_6H_{12}O_6} \Longrightarrow \boxed{\text{g } C_6H_{12}O_6}$

$$n = \frac{PV}{RT} = \frac{655 \text{ torr} \times 2.48 \text{ L}}{62.4 \dfrac{\text{L} \cdot \text{torr}}{\text{K} \cdot \text{mol}} \times 295 \text{ K}} = 0.0882 \text{ mol } O_2$$

$$0.0882 \text{ mol } O_2 \times \frac{1 \text{ mol } C_6H_{12}O_6}{6 \text{ mol } O_2} \times \frac{180 \text{ g } C_6H_{12}O_6}{\text{mol } C_6H_{12}O_6} = \underline{\underline{2.65 \text{ g } C_6H_{12}O_6}}$$

Life can flourish on this planet because all three physical states
of water can exist. The vapor condenses to form rain and snow.
The iceberg floating in the ocean represents the solid and liquid states
which are the subjects of this chapter.

THE SOLID AND LIQUID STATES

Ice cubes floating in a glass of water—what could be more familiar? What we may not appreciate in this common sight is that it represents very unusual behavior. Most solids are more dense than their liquid states, so the solid sinks to the bottom rather than floating on top. Life on this planet could not occur as we know it if water and ice behaved as most other liquids and their solid forms. Since the ice would sink as it forms, lakes would freeze solid in winter. The hot summer sun would thaw only the top layer of a lake, so very little life could survive. Heat could not be distributed from warm to cold climates by ocean currents. Nothing would be the same on this planet. We wouldn't be here if water were not a very unusual compound.

◀ Setting The Stage

Water is certainly our favorite liquid. Water and other liquids and their solid forms are, in a way, easier to study than the gaseous state. These forms of matter are more concrete—we can see them, feel them, and conveniently isolate and measure them. On the other hand, we do not have the benefit of quantitative laws that uniformly govern their behavior, as was the case with gases.

In this chapter, we will give the solid and liquid states appropriate attention, especially the most common but important compound, water. An understanding of the nature of water on the molecular level prepares us for the discussion of water as a solvent and medium for chemical reactions, which follows in the next chapter. Some of the questions that we will address now include the following: How does the kinetic molecular theory describe solids and liquids? What forces cause molecules to stick together in solids and liquids? How does temperature affect physical state? How does energy cause changes of state? What happens when a liquid evaporates? These and other questions relate to some previously discussed topics such as Lewis structures, electronegativity, molecular geometry, and molecular polarity (Chapter 6). Other topics that we will refer to are kinetic molecular theory, discussed in Chapter 9, and specific heat, discussed in Chapter 2.

Our first task is to contrast the solid and liquid states with the gaseous state and to see how kinetic theory can also be applied to these states.

10–1   PROPERTIES OF THE SOLID AND LIQUID STATES

Most of the properties of the solid and liquid states are obvious to us. Still, it is worthwhile to note these common properties in order to picture the actions and interactions of the ions or molecules that comprise these states.

1. **They Have a High Density.**
   Solids and liquids are about 1000 times denser than a typical gas.

2. **They Are Essentially Incompressible.**
   Tall buildings can be supported by bricks and other solids because they don't compress as a gas would. Likewise, a hydraulic jack uses a liquid to support weights such as that of a huge truck. Unlike the behavior of a gas, an increase in pressure on a solid or liquid does not result in a significant decrease in volume.

3. **They Undergo Little Thermal Expansion.**
   When a bridge is constructed, a small space must be left between sections for expansion on a hot day. Still, this space amounts to only a few inches for a bridge span many yards long. In addition, the degree of expansion varies for different solids and liquids. Gases, on the other hand, expand significantly as the temperature rises, and all gases expand by the same factor.

4. **They Have a Fixed Volume.**
   The volume of a gas is that of the container. Also, the volume is the same for the same number of particles under the same conditions. There is no such convenient relationship for solids and liquids. The same volumes of different solids or liquids have no relationship to the number of molecules present.

In addition to these common characteristics, solids and liquids differ with regard to shape. Solids are rigid and thus have a definite shape. Liquids flow and thus do not have a definite shape.

The characteristics of gases were adequately explained by the kinetic molecular theory. Two of the basic assumptions of the kinetic theory are also applicable to the other states. That is, solids and liquids are composed of basic particles that have kinetic energy. The average kinetic energy of the particles is related to the temperature. However, to explain the characteristics of the other two states, there are obviously some assumptions related to gases that no longer apply and must be modified. In the solid and liquid states

1 The basic particles have significant attractions for each other and so are held close together.

2 Since the basic particles are close together, the particles occupy a significant portion of the volume of the substance.

3 The basic particles are not in random motion; their motion is restricted by interactions with other particles.

The properties of the solid and liquid states are understandable on the basis of these assumptions. Because they are already close together, the basic particles cannot be pressed together easily, so the substances are incompressible and have high densities. In fact, both solids and liquids are referred to as *condensed states*. The attraction of the particles for each other holds them together, which essentially counteracts the tendency of heat to move them apart. Thus solids and liquids undergo little thermal expansion.

How are solids and liquids alike?

In Figure 10-1, we illustrate the fundamental difference in behavior of molecules in the three states of matter. We use the water molecule as

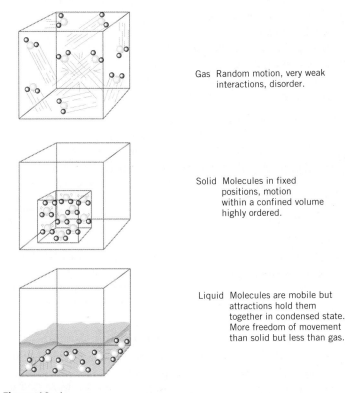

Gas Random motion, very weak
     interactions, disorder.

Solid Molecules in fixed
       positions, motion
       within a confined volume
       highly ordered.

Liquid Molecules are mobile but
        attractions hold them
        together in condensed state.
        More freedom of movement
        than solid but less than gas.

Figure 10–1
Physical States of Water. Interactions between water molecules are different in the three physical states.

an example. This simple but amazing compound exists in all three physical states on Earth: gas (vapor), liquid, and solid (ice). In fact, in a thermos of ice water, all three states exist at once, although the presence of some H_2O molecules in the gaseous state above the ice may not be obvious.

First, let's consider the solid state of water, which we know as ice. In this case, the forces of attraction between molecules hold them in fixed positions relatively close together. In the solid state, the molecules have kinetic energy, meaning that they do have motion. But the motion is restricted to various types of vibrations within a confined space. This is much like a dancer whose motions are confined to shaking and vibrating rather than moving from point to point (translational motion). In the liquid state, the water molecules are also held close together by forces of attraction, but the molecules are not held in fixed positions and thus have more freedom of motion. That is, individual molecules or groups of molecules have translational motion as well as vibrational motion. Since the molecules can move past one another, liquids can flow and take the shape of the bottom of the container. Finally, we are already familiar with the behavior of water molecules in the gaseous state. In this case, the molecules have so much translational motion that they move freely throughout the whole container unaffected by the attractions to other molecules. Their motion scatters them as far apart as possible.

Looking Ahead ▼

Obviously, water molecules have a tendency to "stick together." In fact, the molecules of any compound have at least some attraction for one another. This attraction varies a great deal, however. Before we proceed to further discussions of condensed states, we will explore the origin of the forces that make molecules (or atoms of noble gases) stick together. This should give us a better appreciation for why some compounds are solids, others are liquids, and still others are gases at the same temperature.

10–2 INTERMOLECULAR FORCES AND PHYSICAL STATE

Why does heat make ice melt?

We are all aware that higher temperatures cause some solids to melt and some liquids to vaporize. On a hot summer afternoon, ice cream melts rapidly and water evaporates quickly. As the temperature goes up, the average kinetic energy of molecules increases, which means that they move faster and faster. The increased motion of the molecules in the solid or liquid overcomes whatever forces are holding the molecules together and a phase change occurs.

Why do some solids melt at a certain temperature and others require a higher temperature to melt? The situation is analogous to what happens in an earthquake—the flimsiest buildings are the first to fall. As the earthquake intensifies, more and more buildings may be damaged or even collapse. In order to maintain a rigid structure during a severe shaking, the studs, beams, and walls of a building must be firmly attached. Heating solids is much like subjecting them to an earthquake.

Heating causes the molecules of a solid to vibrate more and more vigorously. The flimsiest solids, whose basic particles are not firmly attached to each other, are the first to collapse (i.e., melt or vaporize). Thus the temperature at which a molecular solid melts to become a liquid is dependent on the forces of attraction between the molecules. These are known as **intermolecular forces.**

Molecular compounds are held together by three types of intermolecular forces: London forces, dipole-dipole forces, and hydrogen bonding. In some cases, more than one force may be active in the same compound.

London Forces

Since atoms and molecules are surrounded by negatively charged electrons, it may seem reasonable that one molecule would repel another. In fact, there are electrostatic forces of attraction between molecules. These forces are known as **London forces.** London forces also have the rather imposing name of *instantaneous dipole-induced dipole forces.* In a molecule, positively charged nuclei exist within negatively charged electron clouds. However, the electrons in these clouds are in constant motion. Because of this constant motion, a molecule may have an imbalance of electrons on one side at a given moment. For that instant, the molecule becomes somewhat polar (i.e., forms a dipole with a negative side and an equivalent positive side). In other words, the molecule has an *instantaneous dipole.* If another molecule happens to be nearby, it is influenced by this instantaneous dipole and that molecule becomes *polarized.* Thus a dipole is *induced* in the second molecule. The attraction between these induced dipoles then spreads from molecule to molecule, thus providing an attractive force between molecules. (See Figure 10-2.)

Inert gas atoms (e.g., He, Ne) or small molecules (e.g., F_2, N_2, CH_4) have very weak interactions of this nature. The small atoms or molecules have comparatively few electrons, and these are tightly held. Such atoms or molecules are not easily polarized. As a result, these substances can exist in the solid or liquid states only at very low temperatures where molecular motions are very slow.

In larger molecules with more electrons, London forces become more and more significant. Larger molecules are generally more *polarizable* than smaller ones. *Since molar mass usually indicates a larger molecule, we can state that the heavier the molecule, the more likely we are to find it in the liquid or solid state at a given temperature.* For example, at room temperature, natural gas [CH_4(molar mass = 16 g/mol)] is, of

Which elements or compounds are likely to be gases at room temperature?

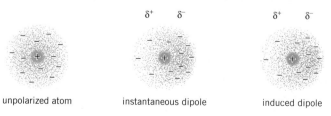

δ^+ δ^- δ^+ δ^-

unpolarized atom   instantaneous dipole   induced dipole

Figure 10–2
London Forces. An instantaneous dipole in one atom or molecule creates an induced dipole in a neighbor.

course, a gas. A major component of gasoline, called octane [$C_8 H_{18}$ (molar mass = 114 g/mol)], is a liquid, and paraffin [$C_{24} H_{50}$(molar mass = 338 g/mol)] is a solid with a low melting point. All molecules are attracted by London forces, but in nonpolar molecules, London forces act exclusively.

Dipole-Dipole Attractions

What does molecular geometry have to do with physical state?

In Chapter 6, we discussed how the polarity of a molecule is determined by its geometry. Individual bonds between unlike atoms in a molecule are most likely polar, but if all of the bonds are the same and the geometry is such that the bond dipoles cancel, the molecule is nonpolar. Highly symmetrical molecular geometries of linear, trigonal planar, and tetrahedral produce nonpolar molecules if all of the bonds are the same. Carbon dioxide is an example of a molecule that has two polar C—O bonds and a linear geometry. Since the two C—O bonds lie at a 180° angle, the individual bond dipoles cancel and the molecule is nonpolar. On the other hand, carbonyl sulfide (OCS) is also linear, like CO_2, but since the bond dipoles are unequal, they do not cancel and the molecule is polar. The SO_2 molecule is also polar. Since the geometry is bent rather than linear, the two equal bond dipoles do not cancel. The OCS and SO_2 molecules each have a permanent dipole (two poles, one negative and one positive). *Molecules with a permanent dipole can align themselves so that the negative end of one molecule is attracted to the positive end of another.* (See Figure 10-3.) *These attractions are known as* **dipole-dipole attractions.** Polar molecules thus have two forces of attraction: London forces, as described previously, and dipole-dipole attractions. When two different molecular compounds have similar molar masses, they have similar London forces. If the molecules of one of the compounds also have dipole-dipole attractions, that compound has a greater tendency to exist in a condensed state than the compound with nonpolar molecules. For example, CO_2 (44 g/mol) is nonpolar and a gas at room temperature, whereas CH_3CN (41 g/mol) is polar and a liquid at room temperature. *In general, a compound whose molecules are polar is more likely to exist as a solid or a liquid at a given temperature than is a compound of similar molar mass whose molecules are nonpolar.* We can get

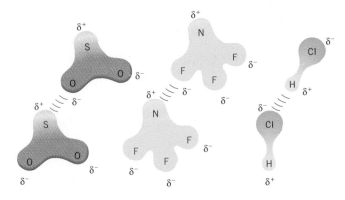

Figure 10–3
Dipole-dipole Interactions. SO_2, NF_3, and HCl are all polar molecules and have dipole-dipole interactions.

into trouble, however, if we try to compare the London forces of a compound with heavy molecules to the dipole-dipole attractions of a compound with lighter molecules.

Hydrogen Bonding

In Chapter 5, we discussed trends in the size of atoms. Atoms of the elements to the upper right in the periodic table were the smallest. In Chapter 6, we also mentioned that these same atoms (fluorine, oxygen, and nitrogen in particular) were the most electronegative elements. This means that these small, highly electronegative atoms tend to attract a significant amount of negative charge to themselves when chemically bonded to other atoms. As an example, consider molecules of water. There is a large difference in electronegativity between oxygen and hydrogen $(3.5 - 2.1 = 1.4)$. This difference is not enough to indicate an ionic bond, but it does point to a highly polar covalent bond. In Chapter 6, we discussed the geometry of water molecules. According to VSEPR theory, the mutually repulsive effect of the two electron pairs on the oxygen atom to the two bonded pairs causes the electron pairs and the hydrogens to be located at the corners of a rough tetrahedron. (See Figure 10-4a.) Since this structure implies that the molecular geometry of H_2O is V-shaped, the bond dipoles do not cancel and the water molecule is significantly polar. (See Figure 10-4b.) The partially positive charge is centered on the hydrogens at two of the corners of the tetrahedron, and the partially negative charge is centered on the electron pairs at the other two corners. As shown in Figure 10-5, in solid ice, hydrogens on two different water molecules interact with the two electron pairs. In liquid water, the structure is less orderly because the interactions between molecules are more random. Individual and groups of water molecules can slide past one another in the liquid state.

How do we know that water does not have linear molecules?

The interactions between a partially positive hydrogen atom on one molecule and the electron pair of the oxygen on another molecule is an example of what is called a hydrogen bond. *A **hydrogen bond** is an electrostatic attraction between a hydrogen bonded to an F, O, or N atom on one molecule and an unshared electron pair on an F, O, or N atom of another molecule.* Hydrogen bonding usually involves only these three

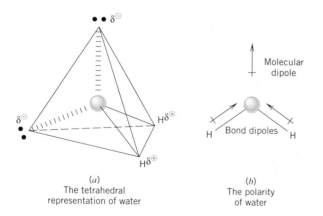

(a)
The tetrahedral
representation of water

(b)
The polarity
of water

Figure 10–4
Structure of Water. Water molecules are polar because they are V-shaped.

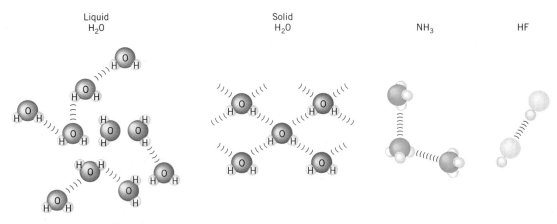

Figure 10–5
Hydrogen Bonding. H_2O molecules in both liquid and solid states, as well as NH_3 and HF molecules, are attracted by hydrogen bonds.

What makes water such a unique compound?

atoms because of their high electronegativity and small size. This interaction at first may seem like a case of an extreme dipole-dipole attraction. In fact, it is more complex than that. A hydrogen bond is not nearly as strong as a regular covalent bond, but it is considerably stronger than typical dipole-dipole attractions. Whereas dipole-dipole interactions usually do not have a large effect on the physical properties of the compound, *hydrogen bonding has a significant effect on the properties of a compound.* Consider the water molecule. It has a very small molar mass of 18 g/mol. If only London forces were present, it would be a gas at temperatures as low as −75°C. Even if regular dipole-dipole attractions were present, it would still boil at a very low temperature. The presence of hydrogen bonding between water molecules provides the "glue" that holds the molecules together so that it exists as a liquid and even a solid under normal conditions on this planet. Other compounds whose properties are altered considerably by the presence of hydrogen bonding are ammonia (NH_3) and hydrogen fluoride (HF). (See Figure 10-5.)

Hydrogen bonding is also important in the huge, complex molecules on which life is based. Consider for example, a molecule of DNA, the "messenger of life." DNA is an extremely large molecule composed of two strands of covalently bonded atoms offset from each other in a helical arrangement (somewhat like a spiral staircase). One of the miracles of life is that these two strands can separate from each other and replicate themselves from smaller molecules. The two strands can separate because they are held together by hydrogen bonds much like Velcro holds two pieces of clothing together. The hydrogen bonds are not as strong as the covalent bonds within the strands. (See Figure 10-6.)

Looking Ahead ▼

So far, we have discussed only the interactions between molecules. There is another large group of compounds, however, whose basic particles are ions. The properties of solids composed of molecules or ions, as well as other types of solids, are discussed next.

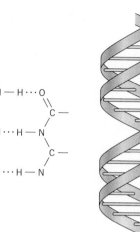

A typical
hydrogen bonding
interaction between
strands

Two strands of DNA.
The horizontal lines
represent hydrogen
bonds.

Figure 10–6
Hydrogen Bonding in DNA. The double helix structure of DNA consists of two
strands twisted about each other. The strands are connected by hydrogen bonds.

10–3 THE SOLID STATE: MELTING POINT

There are basically two types of solids: amorphous and crystalline.
Amorphous solids *are so named because they have no defined shape.*
The basic particles in amorphous solids are not located in any particular
positions. Examples of amorphous solids are glass, rubber, and many
plastics. In **crystalline solids,** *the molecules or ions are arranged in a
regular, symmetrical structure called a* **crystal lattice.** A salt crystal, a
piece of quartz, and many minerals found in the earth naturally form
crystalline solids, which is a reflection of the ordered geometry of the
molecules or ions that lie within. (See Figure 10-7.)

   *The temperature at which a crystalline solid melts (the melting point)
is a definite and constant physical property.* It is the temperature at
which the kinetic energy of the basic particles is great enough so that
they can no longer remain in fixed positions. The forces holding the
molecules or ions in fixed positions are overcome, and the basic particles
begin to move past one another. When amorphous solids melt, the melting process is a gradual softening that does not occur at a specific temperature.

   Different types of solids have different melting characteristics. We
will examine the four types of solids individually.

*Do all solids have a
definite melting point?*

Ionic Solids
Ionic solids *are crystalline solids in which ions are the basic particles
making up the crystal lattice.* (See Figure 10-8a.) Forces between oppositely charged ions are quite strong, especially when compared to forces

aquamarine   fluorite   dolomite

Figure 10–7
Crystalline Solids. The minerals shown above are crystalline solids that are ionic compounds. Their symmetrical shape reflects the ordered geometry of the ions within.

Why does table salt have such a high melting point?

between individual molecules. The strong ion-ion forces result in solid compounds that have relatively high melting points, which can be as high as 3000°C, as for ZrN. As mentioned in Chapter 6, ionic compounds are the stuff of rocks and minerals. They are always solids at room temperature. Ionic compounds are also very hard and brittle (they shatter into pieces when struck).

Molecular Solids
In **molecular solids,** *the basic particles of the crystal lattice are individual molecules, which are held together by London forces, dipole-dipole attractions, or hydrogen bonding.* (See Figure 10-8b.) These attractions are not nearly as strong as the ion-ion attractions found in ionic compounds. As a result, molecular compounds tend to be soft solids with low melting points. These melting points range from very low for small, nonpolar molecules such as N_2 and CH_4 to well above room temperature for large

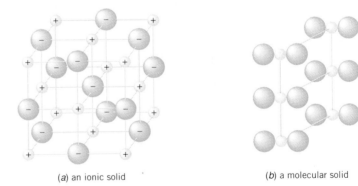

(a) an ionic solid   (b) a molecular solid

Figure 10–8
Ionic and Molecular Solids. In ionic compounds, ions occupy lattice points. In molecular solids, molecules occupy lattice points.

molecules such as table sugar, where molecules are held together by hydrogen bonds.

In the introduction, we mentioned that the solid state of water is less dense than the liquid state. In the solid state, the H_2O molecules have a relatively open structure. When ice melts, the structure becomes slightly more compact. (See Figure 10-5.) The more compact structure of liquid water accounts for its greater density.

Network Solids

In a few solids, atoms are all covalently bonded to one another throughout the entire crystal. These are known as **network solids.** An example of this is an allotrope of pure carbon known as **diamond.** In a diamond, one carbon is in the center of a regular tetrahedron and is bonded to four other carbons located at the corners of the tetrahedron. These carbons are in turn bonded to four others and so forth throughout the entire crystal. (See Figure 10-9.) To melt a diamond, a great deal of energy is needed to break a large number of covalent bonds. As a result, network solids tend to be very hard and have very high melting points. Indeed, liquid diamond has only recently been observed at a very high temperature and pressure. The other two allotropes of carbon are also shown in Figure 10-9. In **graphite,** the carbons are arranged in parallel sheets. Since these sheets can slip past one another, graphite is used as a lubricant or as the "lead" in pencils. The third allotrope of carbon has only recently been discovered and characterized and is known as **buckminsterfullerene.** It was named after the architect, Buckminster Fuller, the designer of the geodesic dome, which has a spherical shape similar to this allotrope (i.e., somewhat like a soccer ball). This allotrope of carbon is actually found in a variety of shapes and sizes. The two most familiar have the formulas C_{60} and C_{70} (see Figure 10-9) and are referred to as "bucky balls" in scientific slang. A great deal of chemical research is currently centered around this intriguing new discovery. Some very practical applications may come from this research.

The beautiful symmetry of snowflakes reflects the ordering of water molecules within the crystals.

What do a diamond and pencil "lead" have in common?

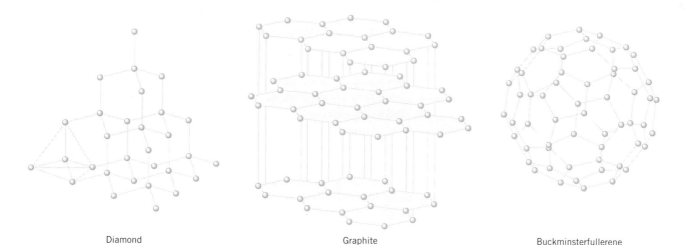

Diamond   Graphite   Buckminsterfullerene

Figure 10–9
The Allotropes of Carbon. There are three allotropes of carbon.

Quartz is a network solid with an empirical formula of SiO₂.

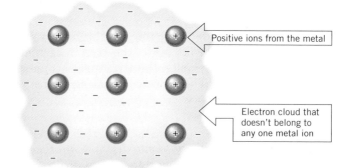

Positive ions from the metal

Electron cloud that doesn't belong to any one metal ion

Figure 10–10
Metallic Solids. In metals, positive ions occupy the lattice points.

Tungsten is used in drill bits because of its strength and high melting point.

Another example of a network solid is SiO_2 (quartz). This represents an empirical formula since the oxygens in this compound are each attached to two silicons, forming a network throughout the crystal. Quartz also has a high melting point (1610°C).

Metallic Solids

Metals are also crystalline solids. As mentioned in Chapter 5, the outer electrons of metals are loosely held. In **metallic solids,** *the positive metal ions occupy regular positions in the crystal lattice, with the valence electrons moving freely among these positive ions.* (See Figure 10-10 below.) This is the reason metals are good conductors of electricity. All metals except mercury (melting point, −39°C) are solids at room temperature. Some, like the alkali metals, are soft and melt at comparatively low temperatures. Others such as iron are hard and have high melting points. Tungsten has one of the highest melting points of any substance known (3380°C). Metals are generally ductile (can be drawn into wires) and malleable (can be pounded into sheets).

Looking Back ▲

Except for a few amorphous solids, the solid state is characterized by an orderly arrangement of ions or molecules in a crystal lattice. The temperature at which this orderly arrangement of basic particles collapses to the liquid state depends on the strength of the forces holding them in their fixed positions. The stronger the forces, the higher the temperature needed to break the interactions.

Checking It Out ▶

Learning Check A

A–1. Fill in the blanks.
Compared to gases, liquids and solids have a _____ density and occupy a _____ volume. The molecules of the condensed states have appreciable _____ attractions for each other. The three types of forces are _____ _____, and _____ _____. Of these, _____ _____ is the strongest and has a significant effect

on physical properties. Nonpolar molecules have only _____ forces. The two types of solids are _____ and _____. _____ solids have a definite melting point. Of the four types of crystalline solids, _____ solids are composed of ions and are always solids at room temperature. _____ solids have covalent bonds throughout the crystal. An allotrope of carbon that has a very high melting point is _____. Molecular solids are usually soft and melt at _____ temperatures. _____ solids have positive ions in a sea of _____.

A–2. Write the Lewis structures for BCl_3, NCl_3, and $HNCl_2$
(a) What are the intermolecular forces between molecules in each case?
(b) Which is more likely to be in a condensed state at a given temperature? Which is least likely?

A–3. Which of the following has the higher melting point and why: CaF_2 or SeF_2?

Additional Examples: Problems 10-3, 10-6, 10-8, 10-19, 10-20, and 10-25.

All elements and compounds are solids at some temperature. When these solids are heated, the solid gives way to the liquid state where the atoms, molecules, or ions can flow past one another. The nature of the liquid state is our next topic.

▼Looking Ahead

10–4 THE LIQUID STATE: SURFACE TENSION AND VISCOSITY

Liquids are the intermediate physical state. The basic particles are still attracted to each other, so liquids remain cohesive as is the case in the solid state. On the other hand, the basic particles can move past one another. In this respect, the liquid state is like the gaseous state. At a lower temperature, liquids freeze to the solid state, and at some higher temperature, liquids vaporize to the gaseous state. There are two properties of liquids that we will discuss: surface tension and viscosity.

Surface Tension
Have you ever noticed that certain insects can walk on water? Also, if one carefully sets a needle or a small metal grate on water, it remains on the surface despite the fact that the metal is much denser than water. (See Figure 10-11.) This tells us that there is some tendency for the surface of the water to stay together. **Surface tension** *is the force that causes the surface of a liquid to contract.* Because of surface tension, drops of liquid are spherical. A molecule within the body of the liquid is equally attracted in all directions by the intermolecular forces. Molecules on the surface are pulled to the side and downward but not upward. This unequal attraction means that a portion of a liquid will tend to have a minimum amount of surface area. A liquid placed on a

Water forms spherical drops because of surface tension.

(a) Molecules on the surface are pulled down and to the sides.

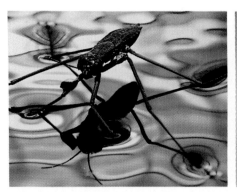

(b) An insect can walk on water.

(c) The metal grid floats on the surface.

Figure 10–11
Surface Tension. The unequal attraction for molecules on the surface accounts for surface tension.

Why don't water bugs sink?

flat surface draws itself into a "bead," or into a sphere if it is suspended in space. Although rain drops are not completely spherical because of gravity, water released in the space shuttle forms perfectly spherical drops. The force that is required to break through a surface relates to its surface tension. Water obviously has a high enough surface tension to support small insects. Other liquids, where the intermolecular forces are greater, have higher surface tensions.

Dissolved soaps reduce the surface tension of water, thus allowing the surface to expand and "wet" clothes or skin rather than form beads. An insect would do well to avoid trying to walk on soapy water. Because of the reduced surface tension, the bug would sink.

Viscosity

We all know it seems to take forever to pour ketchup on french fries when we are hungry. Ketchup is more viscous than water. *The* **viscosity** *of a liquid is a measure of its resistance to flow.* Water and gasoline flow freely because they have a low viscosity. Motor oil and syrup flow slowly because they have a high viscosity. (See Figure 10-12.) The viscosity of a liquid depends to some extent on the intermolecular forces between molecules. Strong intermolecular attractions usually mean a more viscous liquid. Water is an exception—even though its molecules interact relatively strongly, it has low viscosity. Compounds with long, complex molecules also tend to form viscous liquids because the molecules tangle together.

A breakfast of pancakes with syrup on a cold morning is hard to beat. Unfortunately, the syrup barely moves when it is cold. A little warming of the syrup solves the problem. All liquids become less viscous as the temperature increases. The higher kinetic energy of the molecules counteracts the intermolecular forces, allowing the molecules to move past one another more easily.

Figure 10–12
Viscosity. Syrup is a viscous liquid.

Perhaps the most striking property of a liquid is its tendency to change to the vapor state. We all know that water left in the open will slowly disappear (evaporate). The tendency of liquids to vaporize is the subject of the next discussion.

▼ **Looking Ahead**

10–5   VAPOR PRESSURE AND BOILING POINT

Perhaps one of the most accepted facts of life is that wet things eventually become dry. *The liquid water has changed to the gaseous state in a process known as* **vaporization.** *When this process occurs below the boiling point, it is known as* **evaporation.** In order for a molecule of a liquid to escape to the vapor state, however, it must overcome the intermolecular forces attracting it to its neighbors in the liquid. Two conditions allow a molecule in a liquid to escape the liquid state to the gaseous state. First, it must be at or near the surface of the liquid. Second, it must have at least the minimum amount of kinetic energy to overcome the intermolecular forces. So, why don't all molecules on the surface escape? To answer this, we must recall that at a given temperature, the molecules have a large variety of kinetic energies. The temperature relates to the *average* kinetic energy. The wide distribution of kinetic energies of the molecules at two temperatures (T_1 and T_2) can be represented graphically as in Figure 10-13. The unbroken vertical lines represent the average kinetic energy at each of these two temperatures. T_2 represents a higher temperature than T_1 because it has a higher av-

How do molecules in a liquid escape to the vapor state?

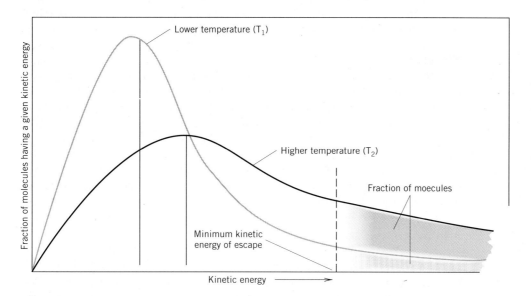

Figure 10–13
Distribution of Kinetic Energies at Two Temperatures. The average kinetic energy is higher at T_2 than at T_1.

erage. The broken vertical line represents the minimum kinetic energy that a molecule must have in order to escape from the surface. Notice that a small fraction have the minimum energy at T_1, but a larger fraction have the minimum energy at the higher temperature, T_2. As a result, a liquid evaporates faster at a higher temperature.

When water, or any other liquid, evaporates, the molecules with the highest kinetic energy escape. This lowers the average of those molecules that are left behind in the liquid state, which means that the temperature of the remaining liquid falls. Evaporation causes a liquid to cool and the gas above the liquid to warm. (The molecules with above-average energy are now in the gaseous state.)

How does one keep "cool"?

The cooling effect of evaporating water is important in maintaining health in warm climates. Perspiration covers our bodies with a layer of water when it is warm. The evaporation of this liquid cools the water on our bodies and us along with it. The cool feeling after a hot shower is not just a feeling but a reality.

Now, instead of letting the water vapor escape to the air, we will place a beaker of water in a closed glass container so that the vapor is trapped. Attached to the apparatus is an open-end mercury manometer for measuring any buildup of pressure within the container. We will assume that, initially, the air is dry, meaning that any water vapor will come from our beaker of water. Before any water evaporates, the manometer indicates the same pressure inside and outside the container. As the water begins to evaporate, the additional molecules in the gas above the beaker cause the total pressure to increase (Dalton's law). The pressure increases rather rapidly at first, but then increases more and more slowly until it does not increase further. As the number of water molecules in the vapor increases, some of the vapor molecules collide with the surface of the water and return to the liquid state. *The change of state from the gaseous to the liquid state is known as* **condensation.** As more molecules enter the gaseous state, more gaseous molecules enter the liquid state. Eventually, a point is reached where the rate of evaporation equals the rate of condensation and the pressure remains constant. The system is said to have reached a point of *equilibrium.* (See Figure 10-14.)

Equilibrium
rate of vaporization = rate of condensation

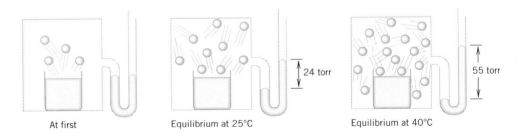

At first   Equilibrium at 25°C   Equilibrium at 40°C

24 torr   55 torr

Figure 10–14
Equilibrium Vapor Pressure. A given fraction of molecules escape to the vapor state above a liquid.

The difference in the levels of the manometer when the pressure has reached equilibrium is the pressure exerted by the water vapor. *The pressure exerted by the vapor above a liquid at a given temperature is called its* **equilibrium vapor pressure.** At a higher temperature, more molecules have the minimum energy needed to escape and, as a result, the vapor pressure is higher. (We say the liquid has become more *volatile,* meaning that it has a higher tendency to vaporize.) Solids may also have an equilibrium vapor pressure. *The vaporization of a solid is known as* **sublimation.** Dry ice (solid CO_2) has a high vapor pressure and sublimes rapidly. Ordinary ice also has a small but significant vapor pressure. Snow slowly sublimes to the vapor state even if the temperature remains below the melting point of ice.

Can ice evaporate?

On humid days, the cooling of our bodies by evaporation becomes less efficient. The relative humidity refers to the percent of saturation of the atmosphere with water vapor. For example, if the relative humidity is 80% and the equilibrium vapor pressure of water at this temperature is 30 torr, the actual vapor pressure of water in the atmosphere is about 0.80×30 torr = 24 torr. Water evaporates very slowly when the relative humidity is already 80% of saturation. "It's not the heat, it's the humidity." Actually, both factors make us uncomfortable, but a high humidity makes it difficult for our natural air conditioner (evaporation of perspiration) to work properly.

The vapor pressure of liquids varies regularly as a function of temperature. This is represented graphically in Figure 10-15. The vapor pressure curves of water, diethyl ether, and ethyl alcohol are included. Notice that the vapor pressure of water at 100°C is equal to the pressure of the atmosphere (760 torr). *When the vapor pressure of a liquid equals the restraining pressure, bubbles of vapor form in the liquid and rise to*

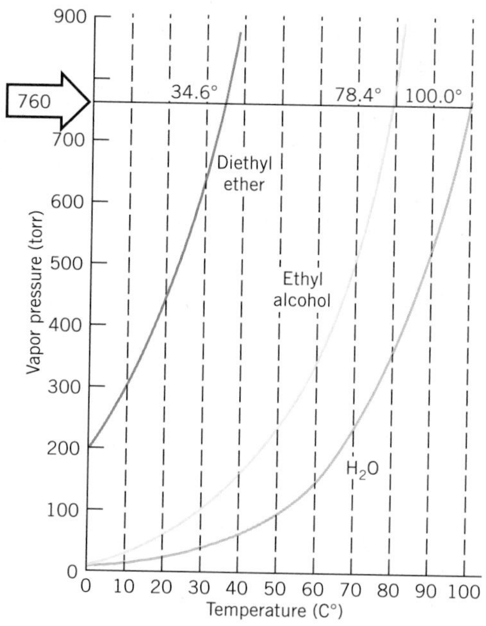

Figure 10–15
Vapor Pressure and Temperature.

On the top of this mountain (Mt. Everest) water boils at about 76°C.

the surface. This is the **boiling point** of the liquid. *The **normal boiling point** of a liquid is the temperature at which the vapor pressure is equal to 760 torr.* Diethyl ether and ethyl alcohol are more volatile than water, which means that their vapor pressures reach 1 atm at temperatures below that of water. They boil at around 34°C and 78°C, respectively. Unlike the melting point of a solid, the boiling point of a liquid is quite dependent on atmospheric conditions. For example, on Pike's Peak in Colorado, water boils at 86°C because the atmospheric pressure is only 450 torr. The highest point on Earth is the top of Mt. Everest. Water boils at 76°C at this elevation. Cooking takes much longer at high elevations and may not even kill all harmful bacteria because of the lower boiling point of water. Appliances called pressure cookers are sometimes used. They retain some of the water vapor, thus increasing the pressure inside the cooker and increasing the boiling point.

Looking Ahead ▼

Melting, boiling, and vaporization all refer to changes in physical states where forces of attraction are overcome. Energy must be supplied to effect these changes. The amount of energy needed is measurable and is discussed in the next section.

10−6 ENERGY AND CHANGES IN STATE

A solid does not melt and a liquid does not boil unless energy is supplied to overcome the forces holding the basic particles together in a crystal lattice or in the liquid state. We have determined previously that the temperature at which a substance melts depends on the forces between the basic particles. The amount of energy it takes to melt a solid and boil a liquid also depends on the magnitude of these forces. We will first consider the energy involved in transitions between the solid and the liquid states (melting and freezing), and then transitions between the liquid and the gaseous states (boiling and condensation).

Melting and Freezing

Melting a solid substance is an endothermic process. The opposite process, freezing, is exothermic. The same amount of heat energy is released when a given amount of liquid freezes as would be required if the same amount of solid were to melt at the melting point. Each compound requires a specific amount of heat energy to melt a specific mass of sample. *The* **heat of fusion** *of a substance is the amount of heat in calories or joules required to melt one gram of the substance.* Table 10-1 lists the heats of fusion and the melting points of several substances. Note that sodium chloride, which is ionic, has the strongest attractions between particles of those listed and thus has the highest melting point and the highest heat of fusion.

Nonpolar compounds of low molar mass have small heats of fusion. Water, because of hydrogen bonds, has a rather high heat of fusion for such a light molecule. The high heat of fusion has profound consequences for the climate of this planet, especially in the vicinity of the American Great Lakes. When a lake freezes in winter, it releases heat to the environment. This heat has the same effect as a giant natural furnace and helps keep the air temperature from falling as much as it otherwise would. Since the Great Lakes do not usually completely freeze over, heat is released all through the winter from the freezing lake. On the down side, spring is delayed in this region because the melting ice absorbs the heat and keeps the temperature from rising as much as it would without the lake. Siberia, in Russia, lies about as far north as

When the Great Lakes freeze in the winter, the heat released warms the surroundings. How can "freezing" water keep the surroundings warm?

TABLE 10–1 HEATS OF FUSION AND MELTING POINTS

Compound	Type of Compound	Heat of Fusion cal/g	Heat of Fusion J/g	Melting Point (°C)
NaCl	Ionic	124	519	801
H_2O	Polar covalent (hydrogen-bonding)	79.8	334	0
Ethyl alcohol	Polar covalent (hydrogen-bonding)	24.9	104	−114
Ethyl ether	Polar covalent	22.2	92.5	−116
Benzene	Nonpolar covalent	30.4	127	5.5
Carbon tetrachloride	Nonpolar covalent	4.2	17.6	− 24

Minnesota but has few large lakes. As a result, it is much cooler there in the winter and also much hotter in the summer.

Working It Out ●

EXAMPLE 10–1

The Heat Released Due to the Freezing of Water

How many kilojoules of heat are released when 185 g of water freezes?

Procedure

The heat of fusion can be used as a conversion factor relating mass in grams to joules.

$$\text{mass} \times \text{heat of fusion} = \text{heat}$$

Solution

$$185 \text{ g} \times \frac{334 \text{ J}}{\text{g}} \times \frac{1 \text{ kJ}}{10^3 \text{ J}} = \underline{\underline{61.8 \text{ kJ}}}$$

Boiling and Condensation

The temperature at which a liquid begins to boil also depends on the attractive forces between the basic particles. Ionic compounds naturally have very high boiling points, followed by large molecular compounds or compounds with hydrogen bonding. Perhaps one of the most dramatic effects on boiling points is demonstrated by compounds that have hydrogen bonding between molecules. For example, consider the boiling points of the Group VIA hydrogen compounds: H_2O, H_2S, H_2Se, and H_2Te. The molar masses of these compounds increase in the order listed. Thus London forces should increase in the same order, and, as a result, so should the intermolecular attractions. From this result alone, we may predict steadily higher boiling points for these compounds. Note in Table 10-2 that this is indeed the case for H_2S, H_2Se, and H_2Te. However, the boiling point of H_2O, by far the highest, is way out of line. The explanation for the unusually high boiling point of water is that hydrogen bonding provides significant intermolecular interactions. The other three compounds are polar, but the normal dipole-dipole attractions do not seem to have a major effect on the boiling points compared with the trend in London forces. In Table 10-2, the boiling points of Group VA and Group VIIA hydrogen compounds are also listed. Hydrogen bonding

TABLE 10–2   BOILING POINTS (°C) OF SOME BINARY HYDROGEN COMPOUNDS

Group	2nd Period		3rd Period		4th Period		5th Period	
VIIA	HF	17	HCl	−84	HBr	−70	HI	−37
VIA	H_2O	100	H_2S	−61	H_2Se	−42	H_2Te	−2.0
VA	NH_3	−33	PH_3	−88	AsH_3	−62	SbH_3	−18

explains the unusually high boiling points of NH_3 and HF compared with other hydrogen compounds in their group.

The vaporization of a liquid is also an endothermic process. The opposite process, condensation, is exothermic. As with melting and freezing, the same amount of energy is released when a given amount of vapor condenses as would be required if the same amount of liquid were to vaporize at the boiling point. Each compound also requires a specific amount of heat energy to vaporize a specific mass of the sample. *The* **heat of vaporization** *of a substance is the amount of heat in calories or joules required to vaporize one gram of the substance.* The heats of vaporization and the boiling points of several substances are given in Table 10-3.

Note again that water has an unusually high heat of vaporization compared with other molecular compounds. Once again, the strength and number of hydrogen bonds between H_2O molecules is responsible.

We can now add the information concerning the heat of fusion and heat of vaporization of water to the information discussed in Chapter 2 on specific heats of ice and water (Table 2-2). *Specific heat refers to the amount of heat required to raise one gram of a substance one degree Celsius.* We observed that water also has an unusually high specific heat compared with other compounds or elements. The heat of fusion or heat of vaporization refers to the heat energy required to change the physical state at the melting or boiling points, respectively. The specific heat refers to the heat required to change the temperature of the substance in a given phase. (The specific heat of liquid water, ice, and steam are not the same.) For example, consider the heat energy required to change one gram of ice at 0°C to steam (vapor) at 100°C. It takes 334 J to melt the ice at 0°C, 418 J to heat the water from 0°C to 100°C, and 2260 J to vaporize the water at 100°C. The total heat required is 3012 J. The fol-

Why are steam burns more serious than hot water burns?

TABLE 10–3 HEATS OF VAPORIZATION AND BOILING POINTS

Compound	Type of Compound	Heat of Vaporization		Normal Boiling Point (°C)
		cal/g	J/g	
NaCl	Ionic	3130	13,100	1465
H_2O	Polar covalent (hydrogen-bonding)	540	2,260	100
Ethyl alcohol	Polar covalent (hydrogen-bonding)	204	854	78.5
Ethyl ether	Polar covalent	89.6	375	34.6
Benzene	Nonpolar covalent	94	393	80
Carbon tetrachloride	Nonpolar covalent	46	192	76

lowing examples illustrate the use of the heats of fusion and vaporization and specific heat.

Working It Out ●

EXAMPLE 10–2

The Heat Released Due to Condensation and Cooling

Steam causes more severe burns than an equal mass of water at the same temperature. Compare the heat released in calories when 3.00 g of water at 100°C cools to 60°C with the heat released when 3.00 g of steam at 100°C condenses and then cools to 60°C.

Procedure
Use the specific heat of water as a conversion factor to convert mass and temperature change to calories.

$$\text{mass} \times \Delta t \times \text{specific heat} = \text{heat}$$

Use the heat of vaporization as a conversion factor to convert mass to calories.

$$\text{mass} \times \text{heat of vaporization} = \text{heat}$$

Solution
Cooling of water

$$3.00 \cancel{g} \times (100 - 60)°\cancel{C} \times \frac{1.00 \text{ cal}}{\cancel{g} \cdot °\cancel{C}} = \underline{\underline{120 \text{ cal}}}$$

Condensation of steam

$$3.00 \cancel{g} \times \frac{540 \text{ cal}}{\cancel{g}} = 1620 \text{ cal}$$

Total heat released from condensation and cooling = 1620 + 120 = $\underline{\underline{1740 \text{ cal}}}$

Note that almost 15 times more heat energy is released by the steam at 100°C than by the water at the same temperature.

EXAMPLE 10–3

The Heat Required to Melt, Heat, and Vaporize Water

How many kilojoules of heat are required to convert 250 g of ice at −15°C to steam at 100°C?

Procedure
The heats of fusion and vaporization as well as the specific heats of ice and water (Table 2-2) are needed.

(a) To calculate the heat required to heat the ice to 0°C, use the specific heat of ice to convert mass and temperature change to heat.

$$\text{mass} \times \Delta t \times \text{sp. heat (ice)} = \text{heat}$$

(b) To calculate the heat required to melt the ice, use the heat of fusion to convert mass to heat.

$$\text{mass} \times \text{heat of fusion} = \text{heat}$$

(c) To calculate the heat required to heat the water, use the specific heat of water to convert mass and temperature change to heat.

$$\text{mass} \times \Delta t \times \text{sp. heat (water)} = \text{heat}$$

(d) To calculate the heat required to vaporize the water, use the heat of vaporization to convert mass to heat.

$$\text{mass} \times \text{heat of vaporization} = \text{heat}$$

Solution

To heat ice from $-15°C$ to its melting point ($0°C$), or a total of $15°C$:

$$250 \text{ g} \times 15°C \times 2.06 \frac{J}{g \cdot °C} \times \frac{1 \text{ kJ}}{10^3 \text{ J}} = 7.7 \text{ kJ}$$

To melt the ice at $0°C$:

$$250 \text{ g} \times 334 \frac{J}{g} \times \frac{1 \text{ kJ}}{10^3 \text{ J}} = 83.5 \text{ kJ}$$

To heat the water from $0°C$ to its boiling point ($100°C$), or a total of $100°C$:

$$250 \text{ g} \times 100°C \times 4.18 \frac{J}{g \cdot °C} \times \frac{1 \text{ kJ}}{10^3 \text{ J}} = 104 \text{ kJ}$$

To vaporize the water at $100°C$:

$$250 \text{ g} \times 2260 \frac{J}{g} \times \frac{1 \text{ kJ}}{10^3 \text{ J}} = 565 \text{ kJ}$$

$$\text{total} = 7.7 + 83.5 + 104 + 565 = \underline{760 \text{ kJ}}$$

In the final section, we will put much of this information together by following what happens when we heat a sample of cold ice until it becomes hot steam.

▼Looking Ahead

10–7   HEATING CURVE OF WATER

Assume that we take an ice cube cooled to $-10°C$ from the freezer compartment of the refrigerator. Imagine that the ice crystal could be magnified so that the motions and positions of the individual molecules could be seen. We would notice that all the molecules are in fixed positions and that ice is a crystalline molecular solid. The molecules are certainly in motion, but the motion consists mainly of vibrations about their fixed locations. We are now going to supply heat at a constant rate to the ice cube and observe the changes that occur. *The graphical representation of the temperature as a solid is heated through the two phase changes plotted as a function of the time of heating is called the* **heating curve.** There are five regions of interest in the curve.

(a) Heating Ice From $-10°C$ to $0°C$, the added heat causes an increase in the kinetic energy or motion about the fixed positions of the water

molecules. In other words, the added heat causes the temperature to rise. How fast the temperature rises in a solid is dependent on the specific heat of that solid. The larger the specific heat, the slower the temperature rises. This and the subsequent changes that occur are represented in the heating curve shown in Figure 10-16. Part (a) in the graph illustrates the temperature change as the ice is heated.

(b) Melting Ice At 0°C (and 1 atm pressure), the situation begins to change. The vibrations of the molecules become so great that some hydrogen bonds break and the molecules begin to move out of their fixed positions in the crystal lattice. We notice that the solid ice begins to melt. Not all of the hydrogen bonds break, however, so groups of molecules still stick together in the liquid state. The added heat is now causing molecules to move apart rather than move faster. When particles that attract each other move apart, there is an increase in the potential energy. Since changes in temperature are related to changes in kinetic energy only, the temperature remains constant while the ice is melting. A constant temperature as a solid melts is a property of pure matter. (See Figure 10-16b.) How long it takes a given solid to melt depends on the heat of fusion.

Why is the temperature constant in an ice and water mixture?

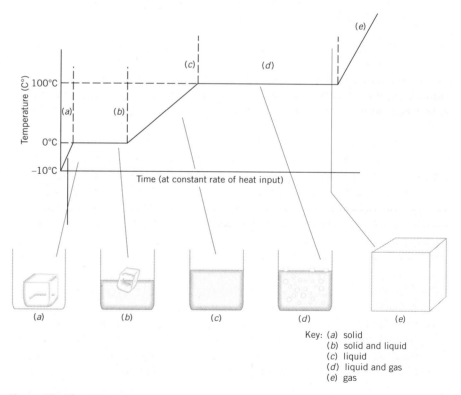

Figure 10–16
Heating Curve of Water.

(c) Heating Water Between 0°C and 100°C, the temperature of the liquid water rises. Once again, the added heat causes the kinetic energy of the water molecules to increase. Unlike the solid state, the motion of water molecules in the liquid state includes translational motion (movement from point to point). The rate of rise of temperature of a liquid depends on the specific heat of the liquid. Notice that the temperature does not appear to rise as fast in the liquid as in the solid. This is because the specific heat of water (1.00 cal/g · °C) is about twice that of ice (0.490 cal/g · °C). (See Figure 10-16c.)

(d) Vaporizing Water At 100°C (and 1 atm pressure), the situation changes again. As the temperature increases, the vapor pressure of the liquid has also been rising. At this temperature, the vapor pressure of water is equal to the atmospheric pressure, and the water begins to boil. The kinetic energy of the water molecules has now become great enough to break the remaining hydrogen bonds holding them in the liquid state. Just as in the melting process, the added heat is now causing molecules to move apart rather than increasing the velocity of the molecules. As with melting, the added heat causes an increase in potential energy, and the temperature remains constant. How long it takes to vaporize a given liquid depends on the heat of vaporization of that liquid. Notice that it takes much longer to vaporize the water than to melt it. This is because the heat of vaporization of water is more than six times that of the heat of fusion. (See Figure 10-16d.)

(e) Heating Vapor Above 100°C, the added heat causes the temperature of the gas to increase. In the gaseous state, the kinetic energy of the gas molecules is great enough to overcome the interactions between molecules. As a result, water vapor acts as any other gas and is subject to the gas laws discussed in the previous chapter. (See Figure 10-16e.)

The properties of the liquid state reflect the cohesive forces between the molecules. A molecule may escape to the vapor state only if it has enough kinetic energy to overcome these forces. The formation of a liquid from a solid (melting), the formation of a gas from a liquid (vaporization), as well as the heating of a solid, liquid, or gas are all endothermic processes, each requiring a specific amount of heat energy. This is sometimes represented as a heating curve.

▲Looking Back

Learning Check B

B–1. Fill in the blanks.
In the liquid state, _____ _____ causes water to form spherical drops. _____ is a property relating to the rate of flow. When evaporation of a liquid occurs, the remaining liquid becomes _____. The equilibrium _____ _____ is the pressure exerted by the vapor at a given _____. The vaporization of a solid is known as _____. The normal boiling point is the temperature at

◀Checking It Out

which the _____ _____ is equal to 760 torr. The energy required to melt one gram of a solid is known as the _____ _____, and the energy required to vaporize one gram of liquid is known as the _____ _____. A heating curve represents the time of _____ on one axis and the _____ on the other. When a pure solid is melting, the _____ does not change.

B–2.   What is the boiling point of water at an altitude where the atmospheric pressure is 550 torr? (See Figure 10-15.) What is the boiling point of diethyl ether at this altitude?

B–3.   What mass of ethyl ether can be melted with 1.00 kJ of heat energy? What mass of ethyl ether can be vaporized by 1.00 kJ? (Refer to Tables 10-1 and 10-3.)

B–4.   How many kJ of heat energy are released when 10.0 g of ethyl alcohol vapor is condensed at its boiling point, allowed to cool, and then frozen at its melting point? The specific heat of ethyl alcohol is 2.47 J/g · °C.

Additional Examples: Problems 10-38, 10-51, 10-58, 10-63, 10-64, and 10-68.

CHAPTER REVIEW

Putting It Together ▲▼

This chapter gives the liquid and solid state their proper consideration. The most significant difference between these "condensed" states and the gaseous state is that cohesive forces hold the basic particles together and the basic particles occupy a significant volume. Motions are confined to vibrations about fixed positions in the solid state or vibrational and translational motion in the liquid state. For this reason, solids and liquids are essentially incompressible, undergo little thermal expansion, and have definite volumes and high densities.

The temperature at which a molecular substance undergoes a phase transition (melting or boiling) depends on the **intermolecular forces** between molecules. **London forces** occur between all molecules but act exclusively for nonpolar molecules. These forces depend to a large extent on the size of the molecules. They increase as the size (or molar mass) increases. Compounds composed of heavy molecules generally have higher melting points than lighter ones. If a molecule has a permanent dipole, it has **dipole-dipole attractions** between molecules in addition to London forces. Compounds composed of polar molecules usually have higher melting and boiling points than nonpolar compounds of similar molar mass. **Hydrogen bonding** is a particular type of electrostatic interaction that has a significant effect on physical properties. It involves hydrogen bonded to an F, O, or N interacting with an F, O, or N on another molecule. Hydrogen bonding is responsible for many of the unusual properties of water.

We first examined the solid state of matter. Solids may be either **amorphous** or **crystalline.** Most solids are crystalline, which means they have atoms, ions, or molecules occupying fixed positions in a **crystal lattice.** They display sharp and definite melting points. There are four types of crystalline solids. **Ionic solids** are composed of ions as the basic particles in the crystal lattice. The ion-ion attractions are strong, so ionic compounds are solids at room temperature. **Molecular solids** are composed of molecules in the crystal lattice. The melting points of such solids depend on the intermolecular forces between molecules. Generally, they are soft solids melting at lower temperatures than ionic solids. **Network solids** are somewhat rare but generally melt at high temperatures because covalent bonds must be broken to cause melting. All of the three allotropes of carbon—**diamond, graphite,** and **buckminsterfullerene**—melt at high temperatures, but diamond has the highest melting point. **Metallic solids** have metal cations at the crystal lattice positions surrounded by a sea of electrons. These solids have melting points that range from low to extremely high temperatures.

Type of Solid	Interaction	Melting points	Examples
Molecular	London only	low	O_2, Ne, CH_4
Molecular	dipole-dipole	low	SO_2, CH_3F
Molecular	hydrogen bonding	intermediate	H_2O, CH_3OH
Network	covalent	very high	diamond, quartz
Metallic	metal ion-electron	generally high	Na, Ca, Cr, Fe

The liquid state is intermediate between the solid and gaseous state. The cohesive forces between the basic particles in the liquid state are evident because of the liquid's **surface tension** and **viscosity.** Some molecules in the liquid state may overcome these cohesive forces and escape to the gaseous state. When this process of **vaporization** is exactly counteracted by the process of **condensation,** an **equilibrium vapor pressure** above the liquid is established. The **boiling point** of the liquid relates to its vapor pressure. **Evaporation** occurs when the liquid vaporizes below the boiling point. A solid may also vaporize in a process known as **sublimation.** These phase transitions are summarized as follows:

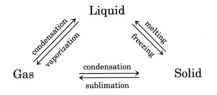

The kinetic molecular theory can be used to explain the behavior of basic particles in these states. This is illustrated with the common example of water.

Gas	Liquid	Solid
There is negligible attraction between individual H_2O molecules, and molecular motion is random.	Molecules are mobile, but significant hydrogen bonding holds the molecules in a condensed state.	Water molecules are held in fixed positions by additional hydrogen bonding; motion is restricted to vibrations within a fixed volume.

Melting and boiling each take place at a specific temperature for most substances. It also takes a specific amount of energy to change one gram of solid to liquid, which is known as the **heat of fusion,** and a specific amount of energy to change one gram of liquid to vapor, which is known as the **heat of vaporization.** The magnitude of these quantities depends on the forces between the basic particles of the solid or liquid. A substance with a high melting point usually has a high heat of fusion and a high heat of vaporization. Much of the information on phase transitions can be shown in a **heating curve.** The rate of heating of a single phase is dependent upon the specific heat of that phase. The time it takes for a solid to melt and a liquid to boil at a given pressure depends on the magnitude of the heat of fusion and vaporization, respectively. The heating curve of a pure substance shows that, when two phases are present in a phase change, the temperature remains constant. When melting or boiling is occurring, the applied heat energy is increasing the potential energy of the system rather than the kinetic energy.

E X E R C I S E S

THE NATURE OF THE SOLID AND LIQUID STATES

10–1 Liquids mix more slowly than gases. Why?

10–2 Why is a gas compressible whereas a liquid is not compressible?

10–3 Describe why a drop of food coloring in a glass of water slowly becomes evenly distributed without the need for stirring.

10–4 What properties do liquids have in common with solids? With gases?

10–5 Review the densities of the solids listed in Table 2-1. In general, which have higher densities, solids or liquids? Why is this so?

INTERMOLECULAR FORCES

10–6 Arrange the following compounds in order of increasing magnitude of the London forces between their molecules.
(a) CH_4 (b) CCl_4 (c) $GeCl_4$

10–7 Arrange the following compounds in order of increasing magnitude of the London forces between their molecules or atoms.
(a) Ne
(b) Xe
(c) N_2
(d) SF_6

10–8 At room temperature, Cl_2 is a gas, Br_2 is a liquid, and I_2 is a solid. Explain the trend.

10–9 At room temperature, CO_2 is a gas and CS_2 is a liquid. Why is this reasonable?

10–10 If H_2O were a linear molecule, could it have hydrogen bonding interactions?

10–11 The H_2S molecule is also V-shaped, similar to H_2O, but it has a very small molecular dipole. How does one H_2S molecule interact with other H_2S molecules?

10–12 Write the Lewis structure of NH_3, with the electron pair and the hydrogens in a tetrahedral arrangement. (See Figure 10-4a.) How does one NH_3 molecule interact with other NH_3 molecules?

10–13 The PH_3 molecule can be represented in a tetrahedral arrangement similar to NH_3. How does one PH_3 molecule interact with other PH_3 molecules?

10–14 Which of the following molecules have dipole-dipole attractions in a condensed state?
(a) HBr
(b) SO_2 (nonlinear)
(c) CO_2 (linear)
(d) BF_3 (trigonal planar)
(e) CO

10–15 Which of the following molecules have dipole-dipole interactions in a condensed state?
(a) SCl_2 (nonlinear)
(b) PH_3
(c) CCl_4 (tetrahedral)
(d) BF_3 (trigonal planar)
(e) CO

10–16 What is the difference between a covalent bond and hydrogen bonding?

10–17 Which of the following molecules may have hydrogen bonding in the liquid state?
(a) HF
(b) NCl_3
(c) H_2NCl
(d) H_2O
(f) $H-C{\overset{\displaystyle O}{\underset{\displaystyle O-H}{}}}$

(e) CH_4 (tetrahedral)
(g) CH_3Cl (tetrahedral)

10–18 Which should have stronger hydrogen bonding, NH_3 or H_2O?

THE SOLID STATE

10–19 Identify the forces that must be overcome to cause melting in the following solids.
(a) KF
(b) HF
(c) HCl
(d) F_2

10–20 Rank the compounds in Problem 10-19 in order of expected melting points (lowest one first).

10–21 Identify the forces that must be overcome to cause melting in the following solids.
(a) diamond   (b) CF_4 (c) CrF_2 (d) SCl_2

10–22 Rank the compounds in Problem 10-21 in order of expected melting points (lowest one first).

10–23 The two Group IVA oxides, CO_2 and SiO_2, have similar formulas but very different melting points. Why?

10–24 The two Group IIIA fluorides, BF_3 and AlF_3, have similar formulas but very different melting points. Why?

10–25 Lead forms two compounds with chlorine, $PbCl_2$ and $PbCl_4$. The melting point of $PbCl_2$ is 501°C and that of $PbCl_4$ is −15°C. Interpret in terms of types of solids.

THE LIQUID STATE

10–26 Why is motor oil more viscous than water? Does motor oil have a greater surface tension than water?

10–27 Explain how molecules of a liquid may go into the vapor state if the temperature is below the boiling point.

10–28 Why does a summer rainstorm lower the temperature?

10–29   Why does rubbing alcohol feel cool on the skin even if the alcohol is initially at room temperature when first applied?

10–30   Ethyl chloride boils at 12°C. When it is sprayed on the skin, it freezes a small part of the skin and thus serves as a local anesthetic. Explain how it cools the skin.

10–31   Given a sample of water at 90°C and a sample of water at 30°C. In which liquid does the temperature change at a faster rate when both are allowed to evaporate?

10–32   A beaker of a liquid with a vapor pressure of 350 torr at 25°C is set alongside a beaker of water, and both are allowed to evaporate. In which liquid does the temperature change at a faster rate? Why?

10–33   What is implied by the word "equilibrium" in *equilibrium vapor pressure?*

10–34   What is the difference between boiling point and normal boiling point?

10–35   A liquid has a vapor pressure of 850 torr at 75°C. Is the substance a gas or a liquid at 75°C and 1 atm pressure?

10–36   If the atmospheric pressure is 500 torr, what are the approximate boiling points of water, ethyl alcohol, and ethyl ether? (Refer to Figure 10-16.)

10–37   How can the boiling point of a pure liquid be raised?

10–38   On top of Mt. Everest, the atmospheric pressure is about 260 torr. What is the boiling point of ethyl alcohol at that pressure? If the temperature is 10°C, is ethyl ether a gas or a liquid under conditions on Mt. Everest? (Refer to Figure 10-15.)

10–39   On the planet Mars the temperature can reach as high as a comfortable 50°F (10°C) at the equator. The atmospheric pressure is about 8 torr on Mars, however. Can liquid water exist on Mars under these conditions? What would the atmospheric pressure have to be before liquid water could exist at this temperature? What would happen to a glass of water set out on the surface of Mars under these conditions? (Refer to Figure 10-15.)

ENERGY AND CHANGES OF STATE

10–40   At a given temperature, one liquid has a vapor pressure of 240 torr and another measures 420 torr. Which liquid probably has the lower boiling point? Which probably has the lower heat of vaporization?

10–41   Ethyl ether ($C_2H_5OC_2H_5$) and ethyl alcohol (C_2H_5OH) are both polar covalent molecules. What accounts for the considerably higher heat of vaporization for alcohol?

10–42   The heats of vaporization of liquid O_2, liquid Ne, and liquid CH_3OH are in the order Ne < O_2 < CH_3OH. Why?

10–43   A given compound has a heat of fusion of about 600 cal/g. Is it likely to have a comparatively high or low melting point?

10–44   A given compound has a boiling point of −75°C. Is it likely to have a comparatively high or low heat of vaporization?

10–45   A given compound has a boiling point of 845°C. Is it likely to have a comparatively high or low melting point?

10–46   Graph the data in Table 10-2 for the Group VIA hydrogen compounds. Plot the boiling points on the y axis and the molar masses of the compounds on the x axis. What would be the expected boiling point of H_2O if only London forces were important, as is the case with the other compounds in this series? (Determine from the graph.)

10–47   Graph the data in Table 10-2 for the Group VA hydrogen compounds. Plot the boiling points on the y axis and the molar masses of the compounds on the x axis. What would be the expected boiling point of NH_3 if only London forces were important, as is the case with the other compounds in this series?

10–48   How many kilojoules are required to vaporize 3.50 kg of H_2O at its boiling point? (Refer to Table 10-3.)

10–49   How many joules of heat are released when an 18.0-g sample of benzene condenses at its boiling point? (Refer to Table 10-3.)

10–50   Molten ionic compounds are used as a method to store heat. How many kilojoules of heat are released when 8.37 kg of NaCl solidifies at its melting point? (Refer to Table 10-1.)

10–51 If 850 J of heat is added to solid H_2O, NaCl, and benzene at their respective melting points, what mass of each is changed to a liquid? (Refer to Table 10-1.)

10–52 What mass of carbon tetrachloride can be vaporized by addition of 1.00 kJ of heat energy to the liquid at its boiling point? What mass of water would be vaporized by the same amount of heat? (Refer to Table 10-3.)

10–53 When a 25.0-g quantity of ethyl ether freezes, how many calories are liberated? When 25.0 g of water freezes, how many calories are liberated? In a large lake, which liquid would be more effective in modifying climate? How many joules of heat are required to melt 125 g of ethyl alcohol at its melting point? (Refer to Table 10-1.)

10–54 Refrigerators cool because a liquid extracts heat when it is vaporized. Before the synthesis of Freon (CF_2Cl_2), ammonia was used. Freon is nontoxic, and NH_3 is a pungent and toxic gas. The heat of vaporization of NH_3 is 1.36 kJ/g and that of Freon is 161 J/g. How many joules can be extracted by the vaporization of 450 g of each of these compounds? Which is the better refrigerant on a mass basis?

10–55 How many joules of heat are released when 275 g of steam at 100.0°C is condensed and cooled to room temperature (25.0°C)? (Refer to Table 10-3.)

10–56 How many kilocalories of heat are required to melt 0.135 kg of ice and then heat the liquid water to 75.0°C? (Refer to Table 10-1.)

10–57 How many calories of heat are needed to heat 120 g of ethyl alcohol from 25.5°C to its boiling point and then vaporize the alcohol? (The specific heat of alcohol is 0.590 cal/g · °C. Refer to Table 10-3.)

10–58 How many calories of heat are required to change 132 g of ice at −20.0°C to steam at 100.0°C? (The specific heat of ice is 0.492 cal/g · °C. Refer to Tables 10-1 and 10-3.)

10–59 How many kilojoules of heat are released when 2.66 kg of steam at 100.0°C is condensed, cooled, frozen and then cooled to −25.0°C? (The specific heat of ice is 2.06 J/g · °C. Refer to Tables 10-1 and 10-3.)

*10–60 A sample of steam is condensed at 100.0°C and then cooled to 75.0°C. If 28.4 kJ of heat is released, what is the mass of the sample? (Refer to Table 10-3.)

*10–61 What mass of ice at 0°C can be changed into steam at 100°C by 2.00 kJ of heat? (Refer to Tables 10-1 and 10-3.)

*10–62 A 10.0-g sample of benzene is condensed from the vapor at its boiling point, and the liquid is allowed to cool. If 5000 J is released, what is the final temperature of the liquid benzene? (The specific heat of benzene is 1.72 J/g · °C. Refer to Table 10-3.)

THE HEATING CURVE

10–63 Which of the following processes are endothermic?
(a) freezing   (c) boiling
(b) melting   (d) condensation

10–64 Which has the higher kinetic energy: H_2O molecules in the form of ice at 0°C, or in the form of liquid water at 0°C?

10–65 Which has the higher potential energy: H_2O molecules in the form of ice at 0°C, or in the form of liquid water at 0°C?

10–66 Which has the higher potential energy: H_2O molecules in the form of steam at 100°C, or in the form of liquid water at 100°C?

10–67 If water is boiling and the flame supplying the heat is turned up, does the water become hotter? What happens?

*10–68 Construct a heating curve for ethyl alcohol. Refer to Tables 10-1 and 10-3. How should the time of constant temperature for melting compare with that for boiling? (Assume that the specific heats of the three phases of water are about twice those of ethyl alcohol.)

*10–69 Construct a heating curve for ethyl ether. Refer to Tables 10-1 and 10-3. How should the time of constant temperature for melting compare with that for boiling? (Assume that the specific heats of the three phases of water are about four times those of ethyl ether.)

GENERAL PROBLEMS

10-70 The following three compounds have similar molar masses: $C_2H_5NH_2$, CH_3OCH_3, and CO_2. The temperatures at which these compounds boil are $-78°C$, $-25°C$, and $17°C$. Match the boiling point with the compound and give the respective intermolecular forces that account for this order.

10-71 At room temperature, SF_6 is a gas and SnO is a solid. Both have similar formula weights. What accounts for the difference in physical states of the two compounds?

10-72 At room temperature, CH_3OH is a liquid and H_2CO is a gas. Both are polar and have similar molar masses. What accounts for the difference in physical state of the two compounds?

10-73 CH_3F and CH_3OH have almost the same molar mass, and both are polar compounds. Yet CH_3OH boils at $65°C$ and CH_3F at $-78°C$. What accounts for the large difference?

10-74 SiH_4, PH_3, and H_2S melt at $-185°C$, $-133°C$, and $-85°C$, respectively. Since all have about the same molar mass, what accounts for the order?

10-75 The boiling point of F_2 is $-188°C$ and that of Cl_2 is $-34°C$, yet the boiling point of HF is much higher than HCl. Explain.

10-76 Nitrogen gas and carbon monoxide have the same molar masses. Carbon monoxide boils at a slightly higher temperature, however ($-191°C$ versus $-196°C$ for N_2). Account for the difference.

*10-78 On a hot, humid day the relative humidity is 70% of saturation. If the temperature is $34°C$, the vapor pressure of water is 39.0 torr. (This would be 100% of saturation.) What mass of water is in each 100 L of air under these conditions?

10-77 The heat of fusion of iron is 266 J/g. Iron is formed in industrial processes in the molten state and is solidified with water. The water vaporizes when it comes in contact with the liquid iron. What mass of water is needed to solidify or freeze 1.00 ton (2000 lb) of iron? Assume that the water is originally at $25.0°C$ and that the steam remains at $100.0°C$.

SOLUTIONS TO LEARNING CHECKS

A-1 high, definite, intermolecular, London forces, dipole-dipole attractions, hydrogen bonding, hydrogen bonding, London, amorphous, crystalline, crystalline, ionic, network, diamond, low, metallic, electrons.

A-2

(a) BCl_3: London
   NCl_3: dipole-dipole
   $HNCl_2$: hydrogen bonding
(b) $HNCl_2$ is most likely to be in the condensed state and BCl_3 is the least likely.

A-3 CaF_2 has the higher melting point because it is an ionic solid. SeF_2 is a molecular solid.

B–1 surface tension, viscosity, cooler, vapor pressure, temperature, sublimation, vapor pressure, heat of fusion, heat of vaporization, heating, temperature, temperature.

B–2 Water boils at about 90°C at this altitude, and diethyl ether at about 25°C.

B–3 melting: $1.00 \text{ kJ} \times \dfrac{10^3 \text{ J}}{\text{kJ}} \times \dfrac{1 \text{ g}}{92.5 \text{ J}} = \underline{\underline{10.8 \text{ g}}}$

vaporizing: $1.00 \text{ kJ} \times \dfrac{10^3 \text{ J}}{\text{kJ}} \times \dfrac{1 \text{ g}}{375 \text{ J}} = \underline{\underline{2.67 \text{ g}}}$

B–4 condensation at 78.5°C:

$10.0 \text{ g} \times 854 \dfrac{\text{J}}{\text{g}} = \underline{\underline{8540 \text{ J}}}$

cooling from 78.5°C to −114°C ($\Delta T = 193 \text{ °C}$):

$10.0 \text{ g} \times 193°\text{C} \times 2.47 \dfrac{\text{J}}{\text{g} \cdot °\text{C}} = \underline{\underline{4770 \text{ J}}}$

freezing at −114°C:

$10.0 \text{ g} \times 104 \dfrac{\text{J}}{\text{g}} = \underline{\underline{1040 \text{ J}}}$

$$\text{TOTAL} = 14{,}350 \text{ J} = \underline{\underline{14.35 \text{ kJ}}}$$

The water in the beautiful Morning Glory Spring in Yellowstone National Park is actually a concentrated solution of several compounds. The deposits on the sides of the spring formed from this solution.

AQUEOUS SOLUTIONS

One of the most common and essential compounds on Earth—water—is actually quite unique. For example, we found in the previous chapter that water has a surprisingly high melting point and boiling point for a compound with such small molecules. The reason for this behavior is *hydrogen bonding*. This comparatively strong electrostatic interaction makes the small water molecules "stick together" so that water has properties similar to compounds with much larger molecules. Because of these properties, water exists as a liquid under most weather conditions on this planet. In addition, liquid water is a unique solvent. It is that property of water that we will explore in this chapter.

◀ **Setting The Stage**

The oceans of the world are examples of water's ability to hold compounds in solution. There are so many other substances in ocean water that it is not drinkable or useable for irrigation of crops. The blood coursing in our veins is yet another example of an aqueous solution. This red waterway dissolves oxygen and nutrients for life processes and carries away dissolved waste products for removal by the kidneys. In fact, water is indispensable for any life form as we know it.

This chapter builds on many topics that have come before. We will note first that water tends to dissolve ionic compounds, so we will need to recall how ionic compounds are identified and the ions of which they are composed (Chapters 4 and 6). Water is a powerful solvent for polar compounds because water molecules are also polar. Thus we will need to be aware of the polarity of molecules (Chapter 6). The hydrogen bonding of water covered in the previous chapter is also important in this discussion. Finally, since many chemical reactions take place in aqueous solution, we will include stoichiometry in these discussions. Stoichiometry was discussed in Chapters 8 and 9.

In this chapter, we will address many questions about water as a solvent. How does water dissolve a substance? What happens when a chemical reaction takes place between dissolved substances? How do we indicate how much of a dissolved compound is present? And finally, how does the presence of a dissolved substance affect the physical properties of water? Our first order of business is to explore how and why water dissolves so many other compounds.

Formulating Some Questions ▶

11–1   THE NATURE OF AQUEOUS SOLUTIONS

A **solution** *is a homogeneous mixture of a* **solute,** *the substance that dissolves, in a* **solvent,** *the medium that disperses the solute. When two liquids form a solution, we say they are* **miscible.** *Two such liquids are alcohol and water. If two liquids do not mix to form a solution, we say they are* **immiscible.** Oil and water, for example, are immiscible and thus form a heterogeneous mixture. A typical solution is formed when a solid, ionic solute, table salt (NaCl), dissolves in a liquid solvent, water. In this case, the solution assumes the same physical state as the solvent. (See Figure 11-1.)

NaCl is an ionic compound composed of Na^+ and Cl^- ions. When placed in water, a crystal of NaCl immediately attracts the polar water molecules. Even though the water molecules are attracted to one another by hydrogen bonding, an even stronger attraction develops between the ions on the surface of the NaCl crystal and the dipoles of the water molecules. The positive end of the dipoles, the H atoms, will be attracted to the Cl^- ions, while the negative ends of the dipoles, the unshared electron pairs, will be attracted to the Na^+ ions. (See Figure 11-2.) Strong forces hold the solid crystal together (ion-ion), but other forces lift the ions on the surface of the crystal into the aqueous medium. *These forces between the ions and dipoles of solvent molecules are referred to as* **ion-dipole forces.** An individual ion-dipole force is not as strong as one ion-ion force, but there are many ion-dipole forces at work. As a result, a tug-of-war develops between the ion-ion forces, holding the crystal together, and the ion-dipole forces, pulling the ions into solution. In the case of NaCl, the solution forces are strong enough to lift the ions, one by one, into the solvent. Since the ions in solution are now surrounded by water molecules, we say that the ions are *hydrated.* In solution, the positive and negative ions are no longer associated with one another. In fact, the charges on the ions are diminished or insulated by the "escort" of hydrating water molecules around each ion. The solution of NaCl in water can be represented by the equation

$$NaCl(s) \xrightarrow{x\,H_2O} Na^+(aq) + Cl^-(aq)$$

The x before H_2O (above the reaction arrow) represents a large, undetermined number of water molecules needed to hydrate the ions. The (aq)

Figure 11–1
A Solution A solution (the red liquid) is a homogeneous mixture of solute (the solid chromium compound) and solvent (water) that is in the same physical state as the solvent.

How does water break up an ionic compound?

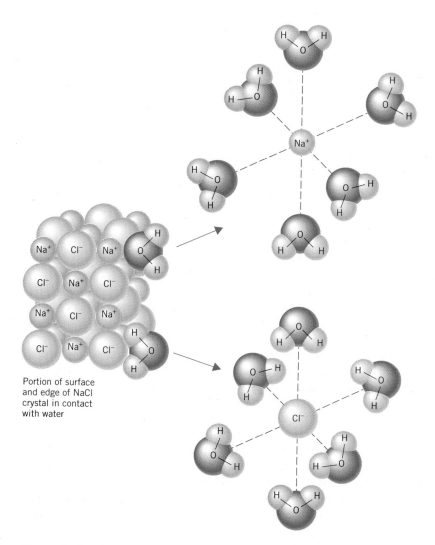

Figure 11–2
Interaction of Water and Ionic Compounds. There is an electrostatic interaction between the polar water molecules and the ions. This is a dipole-ion force.

after the symbol of the ions indicates that they are hydrated in the solution. As we will see shortly, however, not all ionic compounds dissolve to any appreciable extent in water. In these cases, the ion-ion forces are considerably stronger than the ion-dipole attractions, so the crystal remains intact when placed in water. Example 11-1 reviews the ions in solution when selected ionic compounds dissolve.

EXAMPLE 11–1

Ions In Solution When Ionic Compounds Dissolve

Write equations illustrating the solution of the following ionic compounds in water: (a) Na_2CO_3, (b) $CaCl_2$, (c) 2 mol of K_3PO_4, and (d) 3 mol of $(NH_4)_2SO_4$.

● **Working It Out**

Solution

Review the charges and the formulas of the ions discussed in Chapters 4 and 6. The ions in solution are the same as those in the solid ionic compound.

(a) $Na_2CO_3(s) \xrightarrow{x\,H_2O} 2Na^+(aq) + CO_3{}^{2-}(aq)$

(b) $CaCl_2(s) \xrightarrow{x\,H_2O} Ca^{2+}(aq) + 2Cl^-(aq)$

(c) $2K_3PO_4(s) \xrightarrow{x\,H_2O} 6K^+(aq) + 2PO_4{}^{3-}(aq)$

(d) $3(NH_4)_2SO_4(s) \xrightarrow{x\,H_2O} 6NH_4{}^+(aq) + 3SO_4{}^{2-}(aq)$

The forces leading to the solution of an ionic compound in water are much like the forces leading to mixing of the sexes. Groups of men seem to be happy together (strong man-man forces), as do groups of women (strong woman-woman forces). When the two groups come together, however, another force becomes apparent. Solution (homogeneous mixing) of the two sexes occurs if the mixing force (man-woman) predominates. It often does. In life, as in the solution process, the stronger force usually wins.

The formation of aqueous solutions is not the exclusive domain of ionic compounds. Many molecular compounds also dissolve in water if the molecules of the compound are polar. In the solution process, a dipole-dipole attraction forms between the dipoles of the solute molecules and the dipoles of the water molecules. Methyl alcohol is a polar molecular compound that dissolves in water. (See Figure 11-3.)

Some molecular compounds dissolve in water with the formation of ions in a process known as **ionization.** An example is the HCl molecule, which in the pure gaseous state is held together by a single covalent bond. In water, the dipole-dipole interactions between water and solute are strong enough to cause the H—Cl bond to break, leading to the for-

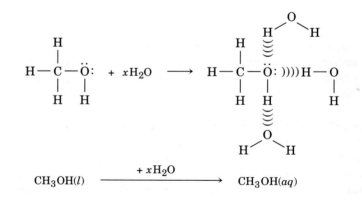

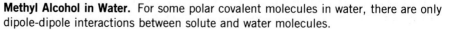

Figure 11–3

Methyl Alcohol in Water. For some polar covalent molecules in water, there are only dipole-dipole interactions between solute and water molecules.

mation of hydrated H^+ and Cl^- ions in aqueous solution. This is represented by the equation

$$HCl(g) \xrightarrow{x\ H_2O} H^+(aq) + Cl^-(aq)$$

If a molecule is nonpolar, there are no centers of positive and negative charge to attract the dipoles of water. Since the water molecules are moderately attracted to one another by hydrogen bonding, there is limited incentive for the water molecules to interact with a solute molecule. *As a result, most nonpolar compounds do not dissolve in polar solvents such as water.* On the other hand, nonpolar compounds do tend to dissolve in nonpolar solvents such as benzene or carbon tetrachloride. This happens because neither solvent nor solute molecules are held together by particularly strong forces, so they are free to mix. In summary, *like solvents dissolve like solutes (e.g., polar solvents dissolve ionic or polar molecular compounds, whereas nonpolar solvents dissolve nonpolar compounds).*

Our next step is to define more of the terms used in connection with solutions and to systematize which common ionic compounds are or are not soluble in water.

What if a potential solute in water is nonpolar?

▼Looking Ahead

11–2 SOLUBILITY OF COMPOUNDS IN WATER

The terms used to define the relative amount of a substance that dissolves almost constitute a separate vocabulary. *When a compound dissolves in the solvent to an appreciable amount, we say that the compound is **soluble**. If very little or none of the substance dissolves, we say that the compound is **insoluble** in that solvent.* Various amounts of a compound can dissolve in a solvent. *The maximum amount of a solute that dissolves in a specific amount of solvent at a certain temperature is known as that compound's **solubility**.*

Now let's turn our attention from the solute to the solution. *When a specific amount of solvent contains the maximum amount of dissolved solute, the solution is said to be **saturated**. If less than the maximum amount is present, the solution is **unsaturated**. In certain unusual situations, an unstable condition may exist in which there is actually more solute present in solution than its solubility would indicate. Such a solution is said to be **supersaturated**.* Supersaturated solutions often shed the excess solute if a tiny "seed" crystal of solute is added, or if the solution is shaken. The excess solute solidifies and falls to the bottom of the container. *A solid that comes out of a solution is known as a **precipitate**.*

Given that certain compounds are soluble in water, how do we express the **concentration**, *which is the amount of solute present in a given amount of solvent?* Several units are available, and the one we choose depends on what needs to be emphasized. Although three additional units are discussed later in the chapter, we will introduce one unit at this time. This concentration unit expresses the amount of solute present in a saturated solution as mass of solute present in 100 g of solvent (g solute/100 g H_2O in the case of aqueous solutions). The solubili-

Can a solute ever exceed its solubility in a solvent?

TABLE 11–1 SOLUBILITIES OF COMPOUNDS (AT 20°C)

Compound	Solubility (g solute/100 g H_2O)
Sucrose (table sugar)	205
HCl	63
NaCl	38
KNO_3	34
$PbSO_4$	0.04
$Mg(OH)_2$	0.01
AgCl	1.9×10^{-4}

ties of several compounds in water are listed in Table 11-1. Although there is no definite dividing point, note that $PbSO_4$, $Mg(OH)_2$, and AgCl have very low solubilities and thus are considered insoluble. The other compounds listed are definitely considered soluble, although to various extents. Fortunately, it is possible to predict which of the common ionic compounds are soluble and which are comparatively insoluble. In Table 11-2, information concerning compounds of the more common anions has been summarized. Note that the first group of anions produces compounds that are *soluble,* with certain exceptions, while the second group produces compounds that are *insoluble,* with exceptions.

TABLE 11–2 SOLUBILITY RULES FOR SOME IONIC COMPOUNDS

Anion	Solubility Rule
Cl^-, Br^-, I^-	All cations form *soluble* compounds except Ag^+, Hg_2^{2+}, and Pb^{2+}. ($PbCl_2$ and $PbBr_2$ are slightly soluble.)
NO_3^-, ClO_3^-, ClO_4^-, $C_2H_3O_2^-$	All cations form *soluble* compounds. ($KClO_4$ and $AgC_2H_3O_2$ are slightly soluble.)
SO_4^{2-}	All cations form *soluble* compounds except Pb^{2+}, Ba^{2+}, and Sr^{2+}. (Ca^{2+} and Ag^+ form slightly soluble compounds.)
CO_3^{2-}, SO_3^{2-}, PO_4^{3-}	All cations form *insoluble* compounds except Group IA metals and NH_4^+.
S^{2-}	All cations form *insoluble* compounds except Groups IA and IIA metals, and NH_4^+.
OH^-	All cations form *insoluble* compounds except Group IA metals, Ba^{2+}, Sr^{2+}, and NH_4^+. ($Ca(OH)_2$ is slightly soluble.)
O^{2-}	All cations form *insoluble* compounds except Group IA metals, Ca^{2+}, Sr^{2+}, and Ba^{2+}.

The following example illustrates the use of Table 11-2 to determine whether an ionic compound is soluble or insoluble in water.

EXAMPLE 11–2

Predicting Whether an Ionic Compound Is Soluble

Use Table 11-2 to predict whether the following compounds are soluble or insoluble: (a) NaI, (b) CdS, (c) Ba(NO$_3$)$_2$, and (d) SrSO$_4$.

Working It Out

Solution

(a) According to Table 11-2, all alkali metal (Group IA) compounds of the ions listed are *soluble*. Therefore, <u>NaI is soluble.</u>

(b) All S^{2-} compounds are *insoluble* except those formed with Groups IA and IIA metals and NH$_4^+$. Since Cd is in Group IIB, <u>CdS is insoluble.</u>

(c) All NO$_3^-$ compounds are *soluble*. Therefore, <u>Ba(NO$_3$)$_2$ is soluble.</u>

(d) The Sr^{2+} ion forms an *insoluble* compound with SO$_4^{2-}$. Therefore, <u>SrSO$_4$ is insoluble.</u>

It is apparent that the nature of a solute affects its solubility in water. Temperature is also an important factor. From practical experience, most of us know that more sugar or salt dissolves in hot water than in cold. This is generally true. *Most solid and liquids are more soluble in water at higher temperatures*. In Figure 11-4, the solubility of several ionic compounds is graphed as a function of temperature. Note that all except Li$_2$SO$_4$ are more soluble as the temperature increases. The information shown in Figure 11-4 has important laboratory implications. An impure solid can be purified by a process called **recrystallization.** In this procedure, a solution is saturated with the solute to be purified at a

Why can hot water be made sweeter than cold water?

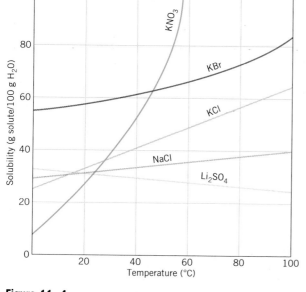

Figure 11–4
Solubility and Temperature. The solubility of most solids increases as their temperature increases.

The hot solution of KNO_3 on the left is unsaturated. A precipitate has formed in the cold solution on the right because of its lower solubility.

The fish maintain life by extracting dissolved oxygen from water.

high temperature such as the boiling point. Insoluble impurities are then filtered from the hot solution. As the solution is allowed to cool, the solvent can hold less and less solute. The excess solute precipitates from the solution as it cools. This solid, now more pure, can then be filtered from the cold solution. Most of the soluble impurities are left behind in the solution.

Despite their nonpolar nature, many gases also dissolve in water to a small extent (e.g., O_2, N_2, and CO_2). Indeed, the presence of dissolved oxygen in water provides the means of life for fish and other aquatic animals. *Unlike solids, gases become less soluble as the temperature increases.* This can be witnessed by observing water being heated. As the temperature increases, the water tends to fizz somewhat as the dissolved gases are expelled. High temperatures in lakes can be a danger to aquatic animals and may cause fish kills. The lower solubility of oxygen at the higher temperatures can lead to an oxygen-depleted lake.

Looking Ahead ▼

Certain combinations of cations and anions form insoluble compounds. Such a compound does not dissolve in water. What happens when a cation in one solution is mixed with a solution containing an anion with which it forms an insoluble compound? They come together and precipitate. This is the basis of a chemical reaction and is discussed next.

11–3   PRECIPITATION REACTIONS AND IONIC EQUATIONS

In the solid phase of an ionic compound, the cations and anions are strongly attracted to each other. In aqueous solution, however, the surrounding water molecules insulate the charge, and the two oppositely charged ions behave independently. What happens when we mix solutions of ionic compounds? It depends. If we mix a solution of $CaCl_2$ and

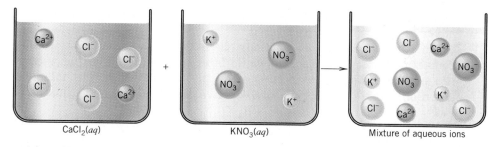

Figure 11–5
A Mixture of CaCl₂ and KNO₃ Solutions. No reaction occurs when these solutions
are mixed.

KNO_3, we simply have a solution of the four ions as illustrated in Figure
11-5. No cation is associated with a particular anion.

Now let us consider a case of two solutions, one containing the soluble compound $NaCl$, and the other containing the soluble compound $AgNO_3$. When we mix these two clear solutions, something obviously occurs. The mixture immediately becomes cloudy, and eventually a solid settles to the bottom of the container. The solid is a precipitate of an insoluble compound, $AgCl$. Whenever Ag^+ and Cl^- are mixed into the same solution, they come together to form solid $AgCl$. This leaves the Na^+ and the NO_3^- ions in solution since $NaNO_3$ is soluble. (See Figure 11-6.)

The formation of a precipitate by mixing solutions of two soluble compounds is known as a **precipitation reaction.** This type of reaction was introduced in Chapter 8 as one of two classes of *double-replacement reactions.* There are some very practical applications of precipitation reactions in industry as well as in the laboratory. Our example is of particular value. Silver actually finds more use in development of photographs and film than in jewelry and coins. It is obviously worthwhile to recover this precious metal from film negatives whenever possible. Silver metal in film can be dissolved in aqueous nitric acid to form the water soluble compound $AgNO_3$. Although the solution contains many other dissolved substances, addition of a soluble compound containing the Cl^- ion (e.g., $NaCl$) leads to the formation of solid $AgCl$ as shown in Figure 11-6. As you notice from Table 11-2, very few other cations form precipitates with Cl^-, so this is a reaction more or less specific to remove Ag^+ from

Can silver be recycled?

Figure 11–6
A Mixture of NaCl and AgNO₃ Solutions. When these solutions are mixed, a precipitate forms.

aqueous solution. The AgCl can then be filtered from the solution and silver metal eventually recovered.

In other precipitation reactions, we may wish to recover the soluble compound. In such a case, we would filter the solution and then recover the soluble compound by boiling away the solvent water. The following equation shows the reaction that we have described and have illustrated in Figure 11-6.

$$NaCl(aq) + AgNO_3(aq) \longrightarrow AgCl(s) + NaNO_3(aq)$$

This is known as the molecular form of the equation. *In a* **molecular equation,** *all reactants and products are shown as neutral compounds.* The word molecular is used even though the formulas may represent ionic compounds in addition to actual molecular compounds.

When all cations and anions in solution are shown separately, the resulting equation is known as a **total ionic equation.**

$$Na^+(aq) + Cl^-(aq) + Ag^+(aq) + NO_3^-(aq) \longrightarrow AgCl(s) + Na^+(aq) + NO_3^-(aq)$$

Are all ions involved in a precipitation reaction?

In the preceding total ionic equation, the $Na^+(aq)$ and the $NO_3^-(aq)$ ions appear the same on both sides of the equation. *Ions that are in an identical state on both sides of an equation are not directly involved in the reaction and are called* **spectator ions.** A chemical equation is much like an algebra equation, which means that identical quantities can be subtracted from both sides of the equation. *When spectator ions are subtracted from both sides of the equation, the result is known as a* **net ionic equation.** A net ionic equation focuses on the "main action," or what causes the reaction to occur. It represents what has actually changed. In our example, the net ionic equation is

$$Ag^+(aq) + Cl^-(aq) \longrightarrow AgCl(s)$$

By careful use of Table 11-2, we can predict the occurrence of many precipitation reactions. To accomplish this, we write the formulas of the compounds formed by "switching partners" and examine the table to determine if one of these compounds is insoluble. If so, a precipitation reaction occurs, and we can write the equation illustrating the reaction. The following examples illustrate the use of Table 11-2 to predict and write precipitation reactions. In the following three Examples, 11-3, 11-4, and 11-5, write the balanced molecular equation, the total ionic equation, and the net ionic equation for any reaction that occurs when solutions of the two compounds are mixed.

EXAMPLE 11–3

A Possible Precipitation Reaction

Working It Out ●

A solution of Na_2CO_3 is mixed with a solution of $CaCl_2$.

Solution
To begin, write out the formulas of the possible products resulting from a precipitation reaction. In this case, the possible products from the exchange of ions are NaCl and $CaCO_3$.

If both of these compounds are soluble, no reaction occurs. In this case, however, Table 11-2 tells us that $CaCO_3$ is insoluble. Thus a reaction occurs, as illustrated by the following balanced equation written in molecular form:

$$Na_2CO_3(aq) + CaCl_2(aq) \longrightarrow CaCO_3(s) + 2NaCl(aq)$$

The equation written in total ionic form is

$$2Na^+(aq) + CO_3^{2-}(aq) + Ca^{2+}(aq) + 2Cl^-(aq) \longrightarrow$$
$$CaCO_3(s) + 2Na^+(aq) + 2Cl^-(aq)$$

Note that the Na^+ and the Cl^- ions are spectator ions. Elimination of the spectator ions on both sides of the equation leaves the net ionic equation.

$$Ca^{2+}(aq) + CO_3^{2-}(aq) \longrightarrow CaCO_3(s)$$

EXAMPLE 11–4

A Possible Precipitation Reaction

A solution of KOH is mixed with a solution of MgI_2.

Solution

The possible precipitation reaction products are KI and $Mg(OH)_2$. The information in Table 11-2 indicates that $Mg(OH)_2$ is insoluble. The balanced molecular equation for this reaction is

$$2KOH(aq) + MgI_2(aq) \longrightarrow Mg(OH)_2(s) + 2KI(aq)$$

The total ionic equation is

$$2K^+(aq) + 2OH^-(aq) + Mg^{2+}(aq) + 2I^-(aq) \longrightarrow$$
$$Mg(OH)_2(s) + 2K^+(aq) + 2I^-(aq)$$

Elimination of spectator ions gives the net ionic equation.

$$Mg^{2+}(aq) + 2OH^-(aq) \longrightarrow Mg(OH)_2(s)$$

EXAMPLE 11–5

A Possible Precipitation Reaction

A solution of KNO_3 is mixed with a solution of $CaBr_2$.

Solution

The possible reaction products are KBr and $Ca(NO_3)_2$. Since these compounds are soluble, no precipitate forms and a reaction does not occur.

▲Looking Back

Water has an important role in the support of life because it is such a remarkable solvent. It can dissolve many ionic compounds because of the strong solvent-solute interactions. Many polar covalent compounds also dissolve, as well as some nonpolar gases. Ionic compounds with comparatively strong ion-ion forces, however, are insoluble or have very low solubilities in water. A chemical reaction occurs when solutions containing cations and anions that form insoluble compounds are mixed.

◀Checking It Out

Learning Check A

A–1. Fill in the blanks.
A solution is composed of a _____ homogeneously dispersed in a _____. Two liquids that form a solution are _____. If they do

not, they are _____. When an ionic compound dissolves in water, the _____-_____ forces between solvent and solute overcome the _____-_____ forces in the crystal. Polar covalent compounds may also dissolve in water due to the _____-_____ forces between solute and solvent. If a soluble compound dissolves to the limit of its solubility at a given temperature, the solution is _____; if the solute is not dissolved to the limit, the solution is _____. If more solute is present than indicated by its solubility, the solution is _____ and the excess solute may eventually form a _____. Solids generally become _____ soluble at higher temperatures, whereas gases become _____ soluble. When a solution containing a cation is mixed with a solution containing an anion that together form an insoluble compound, a _____ forms. In this case, balanced molecular, total _____, and _____ _____ equations can be written.

A–2.   Write the ions that form when the following ionic compounds dissolve in water: (a) $BaBr_2$, (b) $(NH_4)_2CrO_4$, and (c) K_2SO_3.

A–3.   By referring to Table 11-2, consider whether a precipitation reaction occurs when the following solutions are mixed. If a reaction occurs, write the balanced molecular, total ionic, and net ionic equations.
(a) $Sr(OH)_2$ and Na_2SO_3
(b) $NaC_2H_3O_2$ and $FeCl_3$
(c) $Ni(NO_3)_2$ and $(NH_4)_3PO_4$

Additional Examples: 11-1, 11-4, 11-6, 11-13, 11-16, and 11-18.

Looking Ahead ▼

A concentrated solution means that a large amount of a given solute is present in a given amount of solvent. A dilute solution means that comparatively little of the same solute is present. Obviously, we need more quantitative methods of expressing concentrations for laboratory situations. Some of these means of expressing the solute content are discussed next.

11–4   CONCENTRATION: PERCENT BY MASS

Basically, there are two types of concentration units. One type relates the amount of solute to the amount of *solvent*. We put such a unit to work in the previous discussion on solubility when we related the grams of solute to 100 grams of solvent. The second type of concentration unit relates the amount of solute to the amount of *solution*. Two such units are *percent by mass,* which expresses the mass of solute per 100 grams of solution, and *molarity*. We will discuss percent by mass in this section and molarity in the next.

Percent by mass, a rather straightforward method of expressing concentration, simply relates the mass of solute as a percent of the total mass of the solution. Therefore, in 100 g of a solution that is 25% by mass HCl, there are 25 g of HCl and 75 g of H_2O. The formula for percent by mass is

$$\% \text{ by mass (solute)} = \frac{\text{mass of solute}}{\text{mass of solution}} \times 100\%$$

EXAMPLE 11–6

Calculation of Mass Percent

What is the percent by mass of NaCl if 1.75 g of NaCl is dissolved in 5.85 g of H_2O?

● **Working It Out**

Procedure
Find the total mass of the solution and then the percent of NaCl.

Solution
The total mass is

$$
\begin{array}{l}
1.75 \text{ g NaCl (solute)} \\
\underline{5.85 \text{ g } H_2O \text{ (solvent)}} \\
7.60 \text{ g solution}
\end{array}
$$

$$\frac{1.75 \text{ g NaCl}}{7.60 \text{ g solution}} \times 100\% = \underline{\underline{23.0\% \text{ by mass NaCl}}}$$

EXAMPLE 11–7

Calculation of Solute Amount from Mass Percent

A solution is 14.0% by mass H_2SO_4. What quantity (in moles) of H_2SO_4 is in 155 g of solution?

Procedure

1 Find the mass of H_2SO_4 in the solution by multiplying the mass of compound by the percent in fraction form.

2 Convert mass to moles using the molar mass of H_2SO_4.

Solution

1 $$155 \text{ g solution} \times \frac{14.0 \text{ g } H_2SO_4}{100 \text{ g solution}} = 21.7 \text{ g } H_2SO_4$$

2 The molar mass is
$$2.0 \text{ g(H)} + 32.1 \text{ g(S)} + 64.0 \text{ g (O)} = 98.1 \text{ g/mol}$$

$$21.7 \text{ g } H_2SO_4 \times \frac{1 \text{ mol } H_2SO_4}{98.1 \text{ g } H_2SO_4} = \underline{\underline{0.221 \text{ mol } H_2SO_4}}$$

Another way of expressing percent by mass is simply parts per hundred. When concentrations are extremely low, however, two closely related units become more convenient. These units are **parts per million (ppm)** and **parts per billion (ppb),** and they are particularly useful for expressing concentrations of trace amounts of a substance relative to the total amount. For example, one hears of dangerous dioxin levels in the soil in ranges of ppm and even ppb. Parts per million is obtained by multiplying the ratio of the mass of solute to mass of solution by 10^6 ppm rather than 100%. Parts per billion is obtained by multiplying the same ratio by 10^9 ppb. For example, if a solution has a mass of 1.00 kg and contains only 3.0 mg of a solute, it has the following concentration in percent by mass, ppm, and ppb.

$$\frac{3.0 \times 10^{-3}\ \cancel{g}\ \text{(solute)}}{1.0 \times 10^3\ \cancel{g}\ \text{(solution)}} \times 100\% = \underline{\underline{3.0 \times 10^{-4}\%}}$$

$$\frac{3.0 \times 10^{-3}\ \cancel{g}}{1.0 \times 10^3\ \cancel{g}} \times 10^6\ \text{ppm} = \underline{\underline{3.0\ \text{ppm}}}$$

$$\frac{3.0 \times 10^{-3}\ \cancel{g}}{1.0 \times 10^3\ \cancel{g}} \times 10^9\ \text{ppb} = \underline{\underline{3.0 \times 10^3\ \text{ppb}}}$$

In this case, the most convenient expression of concentration is in units of ppm.

Looking Ahead ▼

Mass percent relates mass of solute to <u>mass</u> of the solution. A more useful concentration unit for laboratory measurements relates the amount of solute in moles to the <u>volume</u> of the solution. This is the next concentration unit that we will discuss.

11–5   CONCENTRATION: MOLARITY

Why is molarity so useful?

The most useful unit of concentration expresses an exact amount of solute in a measured volume of the solution. This concentration unit is known as molarity. **Molarity (M)** *is defined as the number of moles of solute (n) per volume in liters (V) of solution.*

$$M = \frac{n\ \text{(moles of solute)}}{V\ \text{(liters of solution)}}$$

The following examples illustrate the calculation of molarity and its use in determining the amount of solute in a specific solution.

Working It Out ●

EXAMPLE 11–8

Calculation of Molarity

What is the molarity of H_2SO_4 in a solution made by dissolving 49.0 g of pure H_2SO_4 in enough water to make 250 mL of solution?

Procedure
Write down the formula for molarity and what you have been given, and then solve for what's requested.

Solution

$$M = \frac{n}{V}$$

$$(n)\ 49.0\ \text{g H}_2\text{SO}_4 \times \frac{1\ \text{mol}}{98.1\ \text{g H}_2\text{SO}_4} = 0.499\ \text{mol}$$

$$(V)\ 250\ \text{mL} \times \frac{10^{-3}\ \text{L}}{\text{mL}} = 0.250\ \text{L}$$

$$\frac{n}{V} = \frac{0.499\ \text{mol}}{0.250\ \text{L}} = \underline{\underline{2.00\ M}}$$

EXAMPLE 11–9

Calculation of Amount of Solute Present

What mass of HCl is present in 155 mL of a 0.540 M solution?

Solution

$$M = \frac{n}{V} \quad n = M \times V$$

$$M = 0.540\ \text{mol/L} \quad V = 155\ \text{mL} = 0.155\ \text{L}$$

$$n = 0.540\ \text{mol/L} \times 0.155\ \text{L} = 0.0837\ \text{mol HCl}$$

$$0.0837\ \text{mol HCl} \times \frac{36.5\ \text{g}}{\text{mol HCl}} = \underline{\underline{3.06\ \text{g HCl}}}$$

EXAMPLE 11–10

Calculation of Molarity from Percent Composition

Concentrated laboratory acid is 35.0% by mass HCl and has a density of 1.18 g/mL. What is its molarity?

Procedure

Since a volume was not given, you can start with any volume you wish. The molarity will be the *same* for 1 mL as for 25 L. To make the problem as simple as possible, assume that you have exactly 1 L of solution ($V = 1.00$ L) and go from there. The number of moles of HCl (n) in 1 L can be obtained as follows:

1 Find the mass of 1 L from the density.

2 Find the mass of HCl in 1 L using the percent by mass and the mass of 1 L.

3 Convert the mass of HCl to moles of HCl.

Solution

Assume that $V = 1.00$ L

1 The mass of 1.00 L (10^3 mL) is

$$10^3\ \text{mL} \times 1.18\ \text{g/mL} = 1180\ \text{g solution}$$

2 The mass of HCl in 1.00 L is

$$1180\ \text{g solution} \times \frac{35.0\ \text{g HCl}}{100\ \text{g solution}} = 413\ \text{g HCl}$$

3 The number of moles of HCl in 1.00 L is

$$413 \not{g} \times \frac{1 \text{ mol}}{36.6 \not{g}} = 11.3 \text{ mol HCl}$$

$$\frac{n}{V} = \frac{11.3 \text{ mol}}{1.00 \text{ L}} = \underline{\underline{11.3 \; M}}$$

Dilute solutions are prepared
from concentrated solutions.

Looking Ahead ▼

A common laboratory exercise is to prepare a dilute solution of a specific concentration from a more concentrated solution. This procedure is not complex once we set up a simple relationship between the dilute and concentrated solutions.

11–6 DILUTION OF CONCENTRATED SOLUTIONS

What if an experiment calls for 200 mL of a 0.20 *M* solution but the only solution of this compound available is 0.60 *M*? In fact, it is quite easy to prepare the dilute solution (0.20 *M*) from the concentrated solution (0.60 *M*). This is called a *dilution problem* and involves measuring the needed amount of concentrated solution and adding it to the appropriate amount of water. (See Figure 11-7.) In the following discussion, the number of moles in the dilute solution is designated n_d, and the volume and molarity of that solution are designated V_d and M_d, respectively. The number of moles of solute in the dilute solution is

$$M_d \times V_d = n_d$$

The moles of solute taken from the concentrated solution is designated

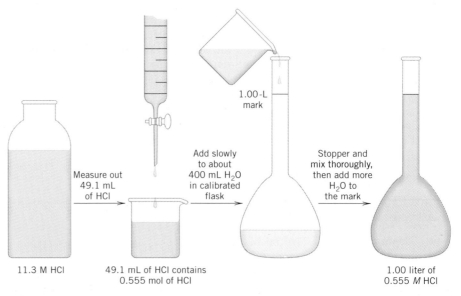

Figure 11–7
Dilution of Concentrated HCl (*Note:* Never add water directly to concentrated acid, because it may splatter and cause severe burns.)

n_c. The moles of solute in the concentrated solution relates to the volume and molarity of the concentrated solution:

$$M_c \times V_c = n_c$$

Since the moles of solute used in the dilute solution is the same as that taken from the concentrated solution, moles of solute remains constant:

$$\text{moles solute} = n_d = n_c$$

Because, in algebra, quantities equal to the same quantity are equal to each other, we have the simple relationship between the dilute solution and the volume and molarity of the concentrated solution that is used:

$$M_c \times V_c = M_d \times V_d$$

EXAMPLE 11–11

Calculation of Volume of Concentrated Solution

What volume of 11.3 M HCl must be mixed with water to make 1.00 L of 0.555 M HCl? (See also Figure 11-7.)

● **Working It Out**

Procedure

$$M_c \times V_c = M_d \times V_d \qquad M_d = \frac{M_c \times V_c}{V_d}$$

Solution

$$M_c = 11.3\ M \qquad\qquad M_d = 0.555\ M$$
$$V_c = ? \qquad\qquad\qquad V_d = 1.00\ L$$

$$V_c = \frac{0.555\ \text{mol/L} \times 1.00\ L}{11.3\ \text{mol/L}} = 0.0491\ L = \underline{\underline{49.1\ mL}}$$

EXAMPLE 11–12

Calculation of Molarity of a Dilute Solution

What is the molarity of a solution of KCl that is prepared by dilution of 855 mL of a 0.475 M solution to a volume of 1.25 L?

Procedure

$$M_c \times V_c = M_d \times V_d \qquad M_d = \frac{M_c \times V_c}{V_d}$$

Solution

$$M_c = 0.475\ M \qquad\qquad M_d = ?$$
$$V_c = 855\ mL = 0.855\ L \qquad V_d = 1.25\ L$$

$$M_d = \frac{0.475\ M \times 0.855\ \cancel{L}}{1.25\ \cancel{L}} = \underline{\underline{0.325\ M}}$$

Stoichiometry problems basically involve the conversion of moles of one reactant or product to moles of another reactant or product. In Chapter 8, we related the masses of reactants and products to moles. In Chap-

▼**Looking Ahead**

ter 9, we related the volume of gases to moles. In this chapter, we extend the mole concept further to include the volume of solutions of known molarity.

11–7 STOICHIOMETRY INVOLVING SOLUTIONS

How do solutions and molarity fit into stoichiometry?

The mole concept is amazingly extensive. For a given compound, the number of molecules, its mass, its volume if a gas, or its volume if in solution can each be converted into the number of moles of that compound. Since many chemical reactions take place in aqueous solution, the latter relationship is our primary concern in this section. In Figure 11-8, the general scheme for working stoichiometry problems has been extended to include volumes of solutions. The following examples also illustrate the inclusion of solutions in stoichiometry problems.

Working It Out ●

EXAMPLE 11–13

Calculation of the Volume of a Reactant

Given the balanced equation

$$3NaOH(aq) + H_3PO_4(aq) \longrightarrow Na_3PO_4(aq) + 3H_2O(l)$$

what volume of 0.250 M NaOH is required to react completely with 4.90 g of H_3PO_4?

Procedure

$$\times \frac{\text{mol } H_3PO_4}{\text{g } H_3PO_4} \quad \times \frac{\text{mol NaOH}}{\text{mol } H_3PO_4} \quad \times \frac{1}{M \,(\text{mol NaOH/L NaOH})}$$

$$\boxed{\text{mass } H_3PO_4} \Longrightarrow \boxed{\text{mol } H_3PO_4} \Longrightarrow \boxed{\text{mol NaOH}} \Longrightarrow \boxed{\text{volume NaOH}}$$

Solution

$$4.90 \text{ g } H_3PO_4 \times \frac{1 \text{ mol } H_3PO_4}{98.0 \text{ g } H_3PO_4} \times \frac{3 \text{ mol NaOH}}{1 \text{ mol } H_3PO_4} = 0.150 \text{ mol NaOH}$$

$$\text{vol NaOH} = \frac{\text{mol NaOH}}{M} = \frac{0.150 \text{ mol}}{0.250 \text{ mol/L}} = \underline{\underline{0.600 \text{ L}}}$$

EXAMPLE 11–14

Calculation of the Mass of a Product

Given the following balanced equation

$$Cd(NO_3)_2(aq) + Na_2S(aq) \longrightarrow CdS(s) + 2NaNO_3(aq)$$

what mass of CdS is produced from 158 mL of a 0.122 M Na_2S solution with excess $Cd(NO_3)_2$ present?

Procedure

$$\times M\left(\frac{\text{mol } Na_2S}{\text{L } Na_2S}\right) \quad \times \frac{\text{mol CdS}}{\text{mol } Na_2S} \quad \times \frac{\text{g CdS}}{\text{mol CdS}}$$

$$\boxed{\text{volume } Na_2S} \Longrightarrow \boxed{\text{mol } Na_2S} \Longrightarrow \boxed{\text{mol CdS}} \Longrightarrow \boxed{\text{mass CdS}}$$

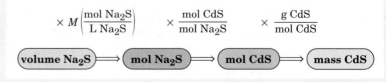

When a $Cd(NO_3)_2$ solution is added to a Na_2S solution, a yellow precipitate of CdS forms.

Solution

$$\text{mol Na}_2\text{S} = V \times M$$
$$= 0.158 \; \cancel{L} \times 0.122 \; \text{mol/}\cancel{L} = 0.0193 \; \text{mol Na}_2\text{S}$$

$$0.0193 \; \cancel{\text{mol Na}_2\text{S}} \times \frac{1 \; \cancel{\text{mol CdS}}}{1 \; \cancel{\text{mol Na}_2\text{S}}} \times \frac{145 \; \text{g CdS}}{\cancel{\text{mol CdS}}} = \underline{\underline{2.80 \; \text{g CdS}}}$$

EXAMPLE 11–15

Calculation of the Volume of Gas

Given the balanced equation

$$2\text{HCl}(aq) + \text{K}_2\text{S}(aq) \longrightarrow \text{H}_2\text{S}(g) + 2\text{KCl}(aq)$$

what volume of H_2S measured at STP would be evolved from 1.65 L of a 0.552 M HCl solution with excess K_2S present?

Procedure

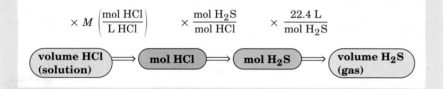

Solution

$$V(\text{solution}) \times M(\text{HCl}) = n(\text{HCl})$$

$$1.65 \; \cancel{L} \times 0.552 \; \text{mol/}\cancel{L} = 0.911 \; \text{mol HCl}$$

Since the volume of the gas is at STP, the molar volume relationship can be used rather than the ideal gas law.

$$0.911 \; \cancel{\text{mol HCl}} \times \frac{1 \; \cancel{\text{mol H}_2\text{S}}}{2 \; \cancel{\text{mol HCl}}} \times \frac{22.4 \; \text{L (STP)}}{\cancel{\text{mol H}_2\text{S}}} = \underline{\underline{10.2 \; \text{L (STP)}}}$$

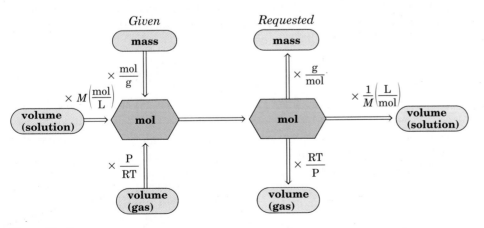

Figure 11–8
General Procedure for Stoichiometry Problems

Of the two units of concentration, mass percent and molarity, the latter is the more useful in laboratory situations. From a concentrated solution of a given compound, a quick calculation and an easy procedure lead to a dilute solution of a specific molarity. Of more importance, however, is the fact that, with molarity, we can conveniently use the volume of a solution as a means of finding the amount of compound present.

Learning Check B

B–1.   Fill in the blanks.
When the mass of solute is expressed as parts per 100 of the mass of solution, the concentration unit is known as _____ _____ . Molarity is defined as _____ per _____ of solution. In stoichiometry problems involving solutions of known molarity, the number of moles of a reactant or product is obtained by multiplying the _____ times the _____ .

B–2.   A 5.00-g quantity of $MgCl_2$ is dissolved in 200 mL of water. What are the mass percent and the molarity of the solution? Assume that the volume of the solution is the same as the original solvent.

B–3.   What mass of NaOH is in 15.0 mL of a 0.375 *M* solution?

B–4.   What is the molarity of the solution in B-3 if it is diluted to 40.0 mL?

B–5.   Given the following balanced equation,

$$3HCl(aq) + K_3PO_4(aq) \longrightarrow H_3PO_4(aq) + 3KCl(aq)$$

what mass of K_3PO_4 is needed to react with 325 mL of 0.250 *M* HCl?

Additional Examples: 11-21, 11-22, 11-30, 11-39, 11-48, and 11-50.

A solution is a mixture. Like all mixtures, the properties of an aqueous solution are distinct from pure water and vary according to the amount of solute. How the presence of solutes affects the properties of water is the final subject of this chapter.

11–8   PHYSICAL PROPERTIES OF SOLUTIONS

Perhaps one of the first lessons we learned in safety is not to have electrical appliances around a bathtub. Ordinary tap water is a good conductor of electricity, so a hair dryer in the bathroom can be very dangerous. Tap water is an aqueous solution. The solutes in the water are what makes water a conductor. In addition, solutes change the vapor pressure, the boiling point, the melting point, and the osmotic pressure of water. We will explore these effects in the order mentioned.

Conductivity

It has long been known that the presence of a solute in water may affect its ability to conduct electricity. *Electricity is simply a flow of electrons through a substance called a* **conductor.** Metals are the most familiar conductors and, as such, find use in electrical wires. Because the outer electrons of metals are loosely held, they can be made to flow through a continuous length of wire. *Other substances resist the flow of electricity and are known as* **nonconductors** *or* **insulators.** Glass and wood are examples of nonconductors of electricity. Pure water is also an example of a nonconductor. When wires are attached to a charged battery and then to a light bulb, the light shines brightly. If the wire is cut, the light goes out because the circuit is broken. If the two ends of the cut wire are now immersed in pure water, the light stays out, indicating that water does not conduct electricity. Now let us dissolve certain solutes in water and examine what happens. When compounds such as NaCl and HCl are dissolved in water, the effect is obvious. The light immediately begins to shine, indicating that the solution is a good conductor of electricity. *Compounds whose aqueous solutions conduct electricity are known as* **electrolytes.** (Some ionic compounds are not soluble in water, but if their molten state conducts electricity, they are also classified as electrolytes.)

We now understand that it is the presence of ions in the aqueous solution that allows the solution to conduct electricity. Almost all soluble ionic compounds form ions in solution and some polar covalent compounds also dissolve to form ions. For example, both NaCl (ionic) and HCl (polar covalent) are classified as electrolytes because they form ions in aqueous solution.

Does water actually conduct electricity?

Other compounds such as sucrose (table sugar) and alcohol dissolve in water, but their solutions do not conduct electricity. *Compounds whose aqueous solutions do not conduct electricity are known as* **nonelectrolytes.** Nonelectrolytes are molecular compounds that dissolve in water without formation of ions. Methyl alcohol is an example of a nonelectrolyte. (See Figure 11-9a.)

There are two classes of electrolytes: strong electrolytes and weak electrolytes. Solutions of **strong electrolytes** (e.g., NaCl and HCl) *are good conductors of electricity. Solutions of* **weak electrolytes** *allow a limited amount of conduction.* When electrodes are immersed in solutions of weak electrolytes, the light bulb glows, but very faintly. (See Figure 11-9b and c.) Even adding more of the solute does not help. An example of a weak electrolyte is HF. This gas is soluble in water but, unlike HCl, only a small percent of the HF molecules ionize at any one time. Because the number of ions is small compared with the total number of HF molecules dissolved, the solution conducts only a limited amount of electricity. The ionization of HF is an example of a reversible reaction that reaches a point of equilibrium. This type of reaction will be discussed in more detail in the next chapter. The solution and ionization of HF can be represented by the following equations.

$$HF(g) \xrightarrow{\ x\,H_2O\ } HF(aq)$$

$$HF(aq) \rightleftharpoons H^+(aq) + F^-(aq)$$

(a)

(b)

(c)

Figure 11-9
Nonelectrolytes, Strong Electrolytes, and Weak Electrolytes (a) A sugar solution is a nonelectrolyte and does not conduct electricity. (b) A $CuSO_4$ solution is a strong electrolyte and conducts electricity. (c) An ammonia solution is a weak electrolyte and conducts a limited amount of electricity.

The presence of a solute in water may or may not affect its conductivity, depending upon whether the solute is an electrolyte or a nonelectrolyte. There are other properties of water, however, that are always affected to some extent by the presence of a solute. Consider again what was mentioned in Chapter 2 as characteristic of a pure substance. Recall that a pure substance has a distinct and unvarying melting point and boiling point. Mixtures, such as aqueous solutions, freeze and boil over a range of temperatures that are lower (for freezing) and higher (for boiling) than those of the pure solvent. The more solute, the more the melting and boiling points are affected. *A property that depends only on the relative amounts of solute and solvent is known as a* **colligative property.** The remaining physical properties discussed in this section are all colligative properties.

Vapor Pressure
The presence of a nonvolatile solute* in a solvent lowers the equilibrium vapor pressure from that of the pure solvent. (See Section 10-5.) A simple explanation of why this happens is illustrated in Figure 11-10. In the solution, a certain proportion of the solute particles are near the surface. They do not escape to the vapor and also block access of some of the solvent molecules to the surface. Thus fewer of the solvent molecules can escape to the vapor, which means that the solution has a lower equilibrium vapor pressure than the pure solvent. As we might predict from this model, the more solute particles present, the greater the effect of vapor pressure lowering.

---

*A nonvolatile solute is one that has essentially no vapor pressure at the relevant temperatures.

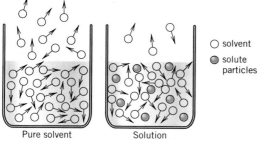

Figure 11–10
Vapor Pressure Lowering. A nonvolatile solute reduces the number of solvent molecules escaping to the vapor.

An example of the effect of vapor pressure lowering is provided by the Dead Sea in Israel. This large body of water has no outlet to the ocean, so dissolved substances have accumulated, forming a very concentrated solution. Even though it exists in an arid region with high summer temperatures, it evaporates very slowly compared with a freshwater lake or even the ocean. It evaporates slowly because of its reduced vapor pressure. If this water evaporated at the same rate as fresh water, the Dead Sea would nearly dry up in a matter of a few years.

Why doesn't a salty sea dry up?

Boiling Point

A direct effect of the lowered vapor pressure of a solution is a higher boiling point compared with the solvent. Since a liquid boils at the temperature at which its vapor pressure equals the atmospheric pressure, a higher temperature is necessary to cause a solution to reach the same vapor pressure as the solvent.

The amount of boiling point elevation is given by the equation

$$\Delta T_b = K_b m$$

where ΔT_b = the number of Celsius degrees that the boiling point is raised

K_b = a constant characteristic of the solvent (for water K_b = 0.512 °C · kg/mol. Other values of K_b are given for particular solvents in the exercises.)

m = a concentration unit called molality

Molality emphasizes a relationship between moles of solute and mass of solvent rather than the volume, as does molarity. The definition of molality is

$$\text{molality } (m) = \frac{\text{moles of solute}}{\text{kg of solvent}}$$

EXAMPLE 11–16

Calculation of Molality

What is the molality of methyl alcohol in a solution made by dissolving 18.5 g of methyl alcohol (CH_3OH) in 850 g of water?

Procedure

$$\text{molality} = \frac{\text{mol solute}}{\text{kg solvent}}$$

●**Working It Out**

Solution

$$\text{mol solute} = \frac{18.5 \text{ g}}{32.0 \text{ g/mol}} = 0.578 \text{ mol}$$

$$\text{kg solvent} = \frac{850 \text{ g}}{10^3 \text{ g/kg}} = 0.850 \text{ kg}$$

$$\text{molality} = \frac{0.578 \text{ mol}}{0.850 \text{ kg}} = \underline{\underline{0.680 \text{ m}}}$$

Freezing Point

Just as the boiling point of a solution is higher than that of the pure solvent, the freezing point is lower. This effect has many applications. When water freezes, it expands, and it will likely break its container. This would occur in an automobile radiator and engine block. Antifreeze (ethylene glycol) is used to lower the freezing point of water in the radiator and to prevent freezing and the damage it would cause. (It also raises the boiling point, preventing the fluid from boiling away in the summer.) The amount of freezing point lowering is given by the equation

$$\Delta T_f = K_f m$$

where ΔT_f = the number of Celsius degrees that the freezing point is lowered

         K_f = a constant characteristic of the solvent (for water K_f = 1.86 °C · kg/mol.)

         m = molality of the solution

Examples of these calculations follow.

Antifreeze lowers the freezing point and raises the boiling point of water.

Working It Out ●

EXAMPLE 11–17

The Boiling Point of a Solution

What is the boiling point of an aqueous solution containing 468 g of sucrose ($C_{12}H_{22}O_{11}$) in 350 g of water?

Procedure

$$\Delta T_b = K_b m = 0.512 \text{ °C} \cdot \text{kg/mol} \times \frac{\text{mol solute}}{\text{kg solvent}}$$

Solution

$$\text{mol solute} = \frac{468 \text{ g}}{342 \text{ g/mol}} = 1.37 \text{ mol}$$

$$\text{kg solvent} = \frac{350 \text{ g}}{10^3 \text{ g/kg}} = 0.350 \text{ kg}$$

$$\Delta T_b = 0.512 \text{ °C} \cdot \text{kg/mol} \times \frac{1.37 \text{ mol}}{0.350 \text{ kg}} = 2.0\text{°C}$$

Normal boiling point of water = 100.0°C

Boiling point of the solution $= 100.0°C + \Delta T_b = 100.0 + 2.0$

$$= \underline{\underline{102.0°C}}$$

EXAMPLE 11–18

The Freezing Point of a Solution

What is the freezing point of the solution in Example 11–17?

Procedure

$$\Delta T_f = K_f m = 1.86 \text{ °C} \cdot \text{kg/mol} \times \frac{\text{mol solute}}{\text{kg solvent}}$$

Solution

$$\Delta T_f = 1.86 \text{ °C} \cdot \text{kg/mol} \times \frac{1.37 \text{ mol}}{0.350 \text{ kg}} = 7.3°C$$

Freezing point of water $= 0.0°C$

Freezing point of the solution $= 0.0°C - \Delta T_f = 0.0°C - 7.3°C$

$$= \underline{\underline{-7.3°C}}$$

Note that the liquid range was extended by over 9°C for the solution.

Osmotic Pressure

An important phenomenon in the life process is osmosis. **Osmosis** *is the tendency for a solvent to move through a thin porous membrane from a dilute solution to a more concentrated solution.* The membrane is said to be **semipermeable,** which means that small solvent molecules can pass through but solute species cannot. Figure 11-11 illustrates osmosis. On the right is a pure solvent, and on the left, a solution. The two are separated by a semipermeable membrane. Solvent molecules can pass through the membrane in both directions, but the rate at which they diffuse to the right is lower because solute particles block some of the pores

How does water get into trees and plants?

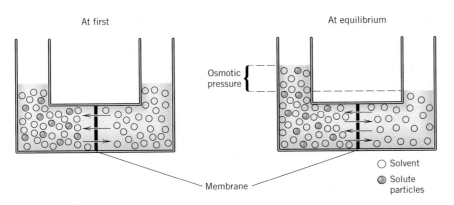

Figure 11–11
Osmotic Pressure. Osmosis causes dilution of the more concentrated solution.

By the process of reverse osmosis, a hand operated device desalinates sea water for emergency use.

Sodium chloride is used to melt ice from the sidewalks, streets, and highways.

of the membrane. As a result, the water level rises on the left and drops on the right. This creates increased pressure on the left, which eventually counteracts the osmosis, and equilibrium is established. *The extra pressure required to establish this equilibrium is known as the* **osmotic pressure.** Like other colligative properties, it depends on the concentration of the solute. In Figure 11-11, the more concentrated the solution on the left (less solvent), the higher the osmotic pressure would be.

We see an example of the osmosis process whenever we leave our hands in a soapy water or saltwater solution. The movement of water molecules from the cells of our skin to the more concentrated solution causes them to be wrinkled. Pickles are wrinkled because the cells of the cucumber have been dehydrated by the salty brine solution. In fact, brine solutions preserve many foods because the concentrated solution of salt removes water from the cells of bacteria, thus killing the bacteria. Trees and plants obtain water by absorbing water through the semipermeable membranes in their roots into the more concentrated solution inside the root cells. Osmosis has many important applications in addition to life processes. In Figure 11-11, if pressure greater than the osmotic pressure is applied on the left, reverse osmosis takes place and solvent molecules move from the solution to the pure solvent. This process is used in desalination plants that convert sea water (a solution) to drinkable water. This is important in areas of the world such as the Middle East, where there is a shortage of fresh water.

Electrolytes have a more pronounced effect on colligative properties than do nonelectrolytes. For example, one mole of NaCl dissolves in water to produce two moles of particles, one mole of Na^+, and one mole of Cl^-.

$$NaCl(s) \xrightarrow{H_2O} Na^+(aq) + Cl^-(aq)$$

Thus one mole of NaCl lowers the freezing point approximately twice as much as one mole of a nonelectrolyte. This effect is put to good use in the U.S. Snow Belt, where sodium chloride is spread on snow and ice to cause melting even though the temperature is below freezing. Even more effective in melting ice is calcium chloride ($CaCl_2$). This compound produces three moles of ions ($Ca^{2+} + 2Cl^-$) per mole of solute and therefore is three times as effective per mole as a nonelectrolyte in lowering the freezing point. Calcium chloride is occasionally used on roads when the temperature is too low for sodium chloride to be effective. Electrolytes would be very effective as an antifreeze for the water in a radiator, but, unfortunately, their presence in water causes severe corrosion of most metals.

Why does salt melt ice?

An aqueous solution may have the same appearance as pure water, but its properties can be very different. A solution may (or may not) conduct electricity, but it will always have a lower vapor pressure, a higher boiling point, and a lower melting point than pure water. A solution also has a higher osmotic pressure than pure water.

▲ **Looking Back**

◄ **Checking It Out**

Learning Check C

C–1. Fill in the blanks.
If the presence of a solute in water causes the solution to become a conductor of electricity, the solute is known as an _____ . If it allows only a limited flow of electricity, it is known as a _____ _____ . If the water remains a nonconductor, the solute is a _____ . Properties that depend only on the amount of solute present are known as _____ properties. These are _____ _____ lowering, _____ _____ elevation, and _____ _____ lowering. The movement of solvent through a semipermeable membrane is known as _____ . A solute causes the _____ pressure to be _____ .

C–2. Sodium chloride is a strong electrolyte. What are the boiling point and the freezing point of a solution made by dissolving 10.0 g of NaCl in 100 g of H_2O?

Additional Examples: 11–57, 11–59, 11–63, 11–71, and 11–72.

C H A P T E R   R E V I E W

Water, the most common and inexpensive chemical available, serves as a medium for many chemical reactions. Water acts as an effective **solvent**, dispersing **solutes** into a homogeneous mixture known as a **solution.** When two liquids are **miscible,** they form a solution, but if they are **immiscible,** they remain a heterogeneous mixture. Some ionic compounds may dissolve in water because the water-ion forces (**ion-dipole**) overcome the forces holding the crystal together. Polar covalent molecular compounds may dissolve in water as discrete molecules, or

▲▼ **Putting It Together**

they may undergo **ionization.** Such compounds are said to be **soluble.** In certain ionic compounds, the crystals remain together and the compounds are comparatively **insoluble.** The solubility of a compound is indicated by some convenient unit of **concentration.** How much of a compound can dissolve at a certain temperature to make a **saturated** solution varies from compound to compound. **Unsaturated** solutions contain less than the maximum amount of a compound so that more of the compound may dissolve. **Supersaturated** solutions are unstable solutions containing more of a compound than the solubility would indicate. A **precipitate** often forms in such a solution. Solid compounds are generally more soluble at higher temperatures, whereas gaseous compounds are less soluble at higher temperatures. This can be used to purify solids in a process called **recrystallization.**

Certain ionic compounds are insoluble when placed in water. In fact, any time the cation and anion of a particular ionic compound are combined in the same solution, a **precipitation reaction** results. Such a reaction can be written as balanced **molecular, total ionic,** and **net ionic equations.**

$$\text{Molecular} \qquad AB(aq) + CD(aq) \longrightarrow AD(s) + CB(aq)$$

$$\text{Total ionic} \qquad A^+(aq) + B^-(aq) + C^+(aq) + D^-(aq) \longrightarrow$$
$$AD(s) + C^+(aq) + B^-(aq)$$

$$\text{Net ionic} \qquad A^+(aq) + D^-(aq) \longrightarrow AD(s)$$

Besides mass of solute per 100 g of solvent that was used to illustrate comparative solubilities, other units of concentration are **percent by mass, molarity,** and **molality.**

Percent by Mass	Molarity	Molality
$\dfrac{\text{mass of solute}}{\text{mass of solvent}} \times 100\%$	$M = \dfrac{\text{mol solute}}{\text{L solution}}$	$m = \dfrac{\text{mol solute}}{\text{kg solvent}}$

Since molarity relates volume of a solution to moles of solute, it can be incorporated into the general scheme for stoichiometry problems along with the mass of a compound (Chapter 8) and the volume of a gas (Chapter 9) as follows:

$$\begin{array}{ccccccc}
\textit{Given} & & & \text{mole} & & \textit{Requested} & \\
\text{volume} & \longrightarrow & \text{moles} & \xrightarrow{\text{ratio}} & \text{moles} & \longrightarrow & \text{volume} \\
\text{(of solution)} & & & & & & \text{(of solution)}
\end{array}$$

In the final section, we studied the physical properties of solutions. In the first property mentioned, we found that certain solutes act as **nonelectrolytes** or as either **weak** or **strong electrolytes.** Electrolytes change water from a **nonconductor (insuiator)** to a **conductor** of electricity. The information is summarized on the next page.

Type of Solute	Property in Water	Reason	Examples
Nonelectrolyte	Solution is a nonconductor of electricity.	Ions are not formed in solution.	$C_{12}H_{22}O_{11}$ (sugar) CH_3OH (methyl alcohol)
Strong electrolyte	Solution is a good conductor of electricity.	Ions are formed in solution.	NaCl K_2SO_4
Weak electrolyte	Solution is a weak conductor of electricity.	Limited amounts of ions are formed in solution.	HF HNO_2

There are also four **colligative properties** of solutions. These are **vapor pressure lowering, boiling point elevation, freezing point lowering,** and **osmotic pressure elevation.** This is summarized as follows:

Property	Effect	Result
Vapor pressure	Lowered	Solutions evaporate slower than pure solvents.
Boiling point	Raised	Solutions boil at higher temperatures than pure solvents.
Freezing point	Lowered	Solutions freeze at lower temperatures than pure solvents.
Osmotic pressure	Raised	Solvent from dilute solutions diffuses through a semipermeable membrane into concentrated solutions.

E X E R C I S E S

AQUEOUS SOLUTIONS

11-1 Write equations illustrating the solution of each of the following compounds in water.
(a) Na_2S
(b) Li_2SO_4
(c) $K_2Cr_2O_7$
(d) CaS
(e) 2 mol of $(NH_4)_2S$
(f) 4 mol of $Ba(OH)_2$

11-2 Write equations illustrating the solution of each of the following compounds in water.
(a) $Ca(ClO_3)_2$
(b) CsBr
(c) $AlCl_3$
(d) 2 mol of Cs_2SO_3

11-3 What ions and how many moles of each are present when the following dissolve in water?
(a) 1 mol of $Al_2(SO_4)_3$
(b) 2 mol of $Mg_3(PO_4)_2$

11-4 Formaldehyde (HCHO) dissolves in water without formation of ions. Write the Lewis structure of formaldehyde and show what type of interactions between solute and solvent are involved.

11-5 Nitric acid is a covalent compound that dissolves in water to form ions, as does HCl. Write the equation illustrating its solution in water.

SOLUBILITY OF IONIC COMPOUNDS

11-6 Referring to Table 11-2, determine which of the following compounds are insoluble in water.

(a) Na_2S　　　　(d) Ag_2O
(b) $PbSO_4$　　　　(e) $(NH_4)_2S$
(c) $MgSO_3$　　　　(f) HgI_2

11–7 Referring to Table 11-2, determine which of the following compounds are insoluble in water.
(a) NiS　　　　(d) Rb_2SO_4
(b) Hg_2Br_2　　　　(e) CaS
(c) $Al(OH)_3$　　　　(f) $BaCO_3$

TEMPERATURE AND SOLUBILITY

11–8 Referring to Figure 11-4, determine which of the following compounds is most soluble at 10°C: $NaCl$, KCl, or Li_2SO_4. Which is most soluble at 70°C?

11–9 Referring to Figure 11-4, determine what mass of each of the following dissolves in 250 g of H_2O at 60°C: KBr, KCl, and Li_2SO_4.

11–10 Referring to Figure 11-4, determine whether each of the following solutions is saturated, unsaturated, or supersaturated. (All are in 100 g of H_2O.)
(a) 40 g of KNO_3 at 40°C
(b) 40 g of KNO_3 at 20°C
(c) 75 g of KBr at 80°C
(d) 20 g of $NaCl$ at 40°C

11–11 A 200-g sample of water is saturated with KNO_3 at 50°C. What mass of KNO_3 forms as a precipitate if the solution is cooled to the freezing point of water? (Refer to Figure 11-4.)

11–12 A 500-mL portion of water is saturated with Li_2SO_4 at 0°C. What happens if the solution is heated to 100°C? (Refer to Figure 11-4.)

PRECIPITATION REACTIONS AND IONIC EQUATIONS

11–13 Write the balanced molecular equation for any reaction that occurs when the following solutions are mixed. (Refer to Table 11-2.)
(a) KI and $Pb(C_2H_3O_2)_2$
(b) $AgClO_3$ and KNO_3
(c) $Sr(ClO_3)_2$ and $Ba(OH)_2$
(d) BaS and $Hg_2(NO_3)_2$
(e) $FeCl_3$ and KOH

11–14 Write the balanced molecular equation for any reaction that occurs when the following solutions are mixed. (Refer to Table 11-2.)
(a) $Ba(C_2H_3O_2)_2$ and Na_2SO_4

(b) $NaClO_4$ and $Pb(NO_3)_2$
(c) $Mg(NO_3)_2$ and Na_3PO_4
(d) SrS and NiI_2

11–15 Write the total ionic equation and the net ionic equation for each of the following reactions.
(a) $K_2S(aq) + Pb(NO_3)_2(aq) \longrightarrow$
$$PbS(s) + 2KNO_3(aq)$$
(b) $(NH_4)_2CO_3(aq) + CaCl_2(aq) \longrightarrow$
$$CaCO_3(s) + 2NH_4Cl(aq)$$
(c) $2AgClO_4(aq) + Na_2CrO_4(aq) \longrightarrow$
$$Ag_2CrO_4(s) + 2NaClO_4(aq)$$

11–16 Write the total ionic equation for any reaction that occurred in Problem 11-13.

11–17 Write the total ionic equation for any reaction that occurred in Problem 11-14.

11–18 Write the net ionic equation for any reaction that occurred in Problem 11-13.

11–19 Write the net ionic equation for any reaction that occurred in Problem 11-14.

*11–20 Write balanced molecular equations indicating how the following ionic compounds can be prepared by a precipitation reaction using any other ionic compounds. In some cases, the equation should reflect the fact that the desired compound is soluble and must be recovered by vaporizing the solvent water after removal of a precipitate.
(a) $CuCO_3$　　　　(d) NH_4NO_3
(b) $PbSO_3$　　　　(e) $KC_2H_3O_2$
(c) Hg_2I_2

PERCENT BY MASS

11–21 What is the percent by mass of solute in a solution made by dissolving 9.85 g of $Ca(NO_3)_2$ in 650 g of water?

11–22 What is the percent by mass of solute if 14.15 g of NaI is mixed with 75.55 g of water?

11–23 A solution is 10.0% by mass $NaOH$. How many moles of $NaOH$ are dissolved in 150 g of solution?

11–24 A solution contains 15.0 g of NH_4Cl in water and is 8.50% NH_4Cl. What is the mass of water present?

11–25 A solution is 23.2% by mass KNO_3. What mass of KNO_3 is present in each 100 g of H_2O?

11–26 A solution contains 1 mol of NaOH dissolved in 9 mol of ethyl alcohol (C_2H_5OH). What is the percent by mass NaOH?

11–27 Blood contains 10 mg of calcium ions in 100 g of blood serum (solution). What is this concentration in ppm?

11–28 A high concentration of mercury in fish is 0.5 ppm. What mass of mercury is present in each kilogram of fish? What is this concentration in ppb?

MOLARITY

11–29 What is the molarity of a solution made by dissolving 2.44 mol of NaCl in enough water to make 4.50 L of solution?

11–30 Fill in the blanks.

Solute	M	Amount of Solute	Volume of Solution
(a) KI		2.40 mol	2.75 L
(b) C_2H_5OH		26.5 g	410 mL
(c) $NaC_2H_3O_2$	0.255	3.15 mol	L
(d) $LiNO_2$	0.625	g	1.25 L
(e) $BaCl_2$		0.250 mol	850 mL
(f) Na_2SO_3	0.054	mol	0.45 L
(g) K_2CO_3	0.345	14.7 g	mL
(h) LiOH	1.24	g	1650 mL
(i) H_2SO_4	0.905	0.178 g	mL

11–31 What is the molarity of a solution of 345 g of Epsom salts ($MgSO_4 \cdot 7H_2O$) in 7.50 L of solution?

11–32 What mass of $CaCl_2$ is in 2.58 L of a solution with a concentration of 0.0784 M?

11–33 What volume in liters of a 0.250 M solution contains 37.5 g of KOH?

11–34 What is the molarity of a solution made by dissolving 2.50×10^{-4} g of baking soda ($NaHCO_3$) in enough water to make 2.54 mL of solution?

11–35 What is the molarity of the hydroxide ion and the barium ion if 13.5 g of $Ba(OH)_2$ is dissolved in enough water to make 475 mL of solution?

*11–36 A solution is 25.0% by mass calcium nitrate and has a density of 1.21 g/mL. What is its molarity?

*11–37 A solution of concentrated NaOH is 16.4 M. If the density of the solution is 1.43 g/mL, what is the percent by mass NaOH?

*11–38 Concentrated nitric acid is 70.0% HNO_3 and 14.7 M. What is the density of the solution?

DILUTION

11–39 What volume of 4.50 M H_2SO_4 should be diluted with water to form 2.50 L of 1.50 M acid?

11–40 If 450 mL of a certain solution is diluted to 950 mL with water to form a 0.600 M solution, what was the molarity of the original solution?

11–41 One liter of a 0.250 M solution of NaOH is needed. The only available solution of NaOH is a 0.800 M solution. Describe how to make the desired solution.

11–42 What is the volume in liters of a 0.440 M solution if it was made by dilution of 250 mL of a 1.25 M solution?

11–43 What is the molarity of a solution made by diluting 3.50 L of a 0.200 M solution to a volume of 5.00 L?

11–44 What volume of water in milliliters should be *added* to 1.25 L of 0.860 M HCl so that its molarity will be 0.545?

11–45 What volume of water in milliliters should be *added* to 400 mL of a solution containing 35.0 g of KBr to make a 0.100 M KBr solution?

*11–46 What volume in milliliters of *pure* acetic acid should be used to make 250 mL of 0.200 M $HC_2H_3O_2$? (The density of the pure acid is 1.05 g/mL.)

*11–47 What would be the molarity of a solution made by mixing 150 mL of 0.250 M HCl with 450 mL of 0.375 M HCl?

STOICHIOMETRY INVOLVING SOLUTIONS

11–48 Given the reaction

$$3KOH(aq) + CrCl_3(aq) \longrightarrow$$
$$Cr(OH)_3(s) + 3KCl(aq)$$

what mass of $Cr(OH)_3$ would be produced if 500 mL of 0.250 M KOH were added to a solution containing excess $CrCl_3$?

11–49 Given the reaction

$$2KCl(aq) + Pb(NO_3)_2(aq) \longrightarrow$$
$$PbCl_2(s) + 2KNO_3(aq)$$

what mass of $Pb(NO_3)_2$ is required to react with 1.25 L of 0.550 M KCl?

11–50 Given the reaction

$$Al_2(SO_4)_3(aq) + 3BaCl_2(aq) \longrightarrow$$
$$3BaSO_4(s) + 2AlCl_3(aq)$$

what mass of $BaSO_4$ is produced from 650 mL of 0.320 M $Al_2(SO_4)_3$?

11–51 Given the reaction

$$3Ba(OH)_2(aq) + 2Al(NO_3)_3(aq) \longrightarrow$$
$$2Al(OH)_3(s) + 3Ba(NO_3)_2(aq)$$

what volume of 1.25 M $Ba(OH)_2$ is required to produce 265 g of $Al(OH)_3$?

11–52 Given the reaction

$$2AgClO_4(aq) + Na_2CrO_4(aq) \longrightarrow$$
$$Ag_2CrO_4(s) + 2NaClO_4(aq)$$

what volume of a 0.600 M solution of $AgClO_4$ is needed to produce 160 g of Ag_2CrO_4?

11–53 Given the reaction

$$3Ca(ClO_3)_2(aq) + 2Na_3PO_4(aq) \longrightarrow$$
$$Ca_3(PO_4)_2(s) + 6NaClO_3(aq)$$

what volume of a 2.22 M solution of Na_3PO_4 is needed to react with 580 mL of a 3.75 M solution of $Ca(ClO_3)_2$?

11–54 Consider the reaction

$$2HNO_3(aq) + 3H_2S(aq) \longrightarrow$$
$$2NO(g) + 3S(s) + 4H_2O\ (l)$$

(a) What volume of 0.350 M HNO_3 will completely react with 275 mL of 0.100 M H_2S?

(b) What volume of NO gas measured at 27°C and 720 torr will be produced from 650 mL of 0.100 M H_2S solution?

*11–55 Given the reaction

$$2NaOH(aq) + MgCl_2(aq) \longrightarrow$$
$$Mg(OH)_2(s) + 2NaCl(aq)$$

what mass of $Mg(OH)_2$ would be produced by mixing 250 mL of 0.240 M NaOH with 400 mL of 0.100 M $MgCl_2$?

*11–56 Given the reaction

$$CO_2(g) + Ca(OH)_2(aq) \longrightarrow$$
$$CaCO_3(s) + H_2O(l)$$

what is the molarity of a 1.00-L solution of $Ca(OH)_2$ that would completely react with 10.0 L of CO_2 measured at 25°C and 0.950 atm?

PROPERTIES OF SOLUTIONS

11–57 Three hypothetical binary compounds dissolve in water. AB is a strong electrolyte, AC is a weak electrolyte, and AD is a nonelectrolyte. Describe the extent to which each of these solutions conducts electricity and how each compound exists in solution.

11–58 Chlorous acid ($HClO_2$) is a weak electrolyte, and perchloric acid ($HClO_4$) is a strong electrolyte. Write equations illustrating the difference in behavior of these two polar covalent molecules in water.

11–59 What is the molality of a solution made by dissolving 25.0 g of NaOH in (a) 250 g of water and (b) 250 g of alcohol (C_2H_5OH)?

11–60 What is the molality of a solution made by dissolving 1.50 kg of KCl in 2.85 kg of water?

11–61 What mass of NaOH is in 550 g of water if the concentration is 0.720 m?

11–62 What mass of water is in a 0.430 m solution containing 2.58 g of CH_3OH?

11–63 What is the freezing point of a 0.20 m aqueous solution of a nonelectrolyte?

11–64 What is the boiling point of a 0.45 m aqueous solution of a nonelectrolyte?

11–65 When immersed in salty ocean water for an extended period, a person gets very thirsty. Explain.

11–66 Dehydrated fruit is wrinkled and shriveled up. When put in water, the fruit expands and becomes smooth again. Explain.

11–67 Explain how pure water can be obtained from a solution without boiling.

11–68 In industrial processes, it is often necessary to concentrate a dilute solution (much more difficult than diluting a concentrated solution). Explain how the principle of reverse osmosis can be applied.

*11–69 What is the molality of an aqueous solution that is 10.0% by mass $CaCl_2$?

*11–70 A 1.00 m KBr solution has a mass of 1.00 kg. What is the mass of the water?

*11–71 Ethylene glycol ($C_2H_6O_2$) is used as an antifreeze. What mass of ethylene glycol should be added to 5.00 kg of water to lower the freezing point to $-5.0°C$? (Ethylene glycol is a nonelectrolyte.)

*11–72 What is the boiling point of the solution in Problem 11-71?

*11–73 Methyl alcohol can also be used as an antifreeze. What mass of methyl alcohol (CH_3OH) must be added to 5.00 kg of water to lower its freezing point to $-5.0°C$?

11–74 What is the molality of an aqueous solution that boils at $101.5°C$?

11–75 What is the boiling point of a 0.15 m solution of a solute in liquid benzene? (For benzene, $K_b = 2.53$, and the boiling point of pure benzene is $80.1°C$.)

11–76 What is the boiling point of a solution of 75.0 g of naphthalene ($C_{10}H_8$) in 250 g of benzene? (See Problem 11-75.)

11–77 What is the freezing point of a solution of 100 g of CH_3OH in 800 g of benzene? (For benzene, $K_f = 5.12$, and the freezing point of pure benzene is $5.5°C$.)

11–78 What is the freezing point of a 10.0% by mass solution of CH_3OH in benzene? (See Problem 11-77.)

*11–79 Give the freezing point of each of the following in 100 g of water.
(a) 10.0 g of CH_3OH
(b) 10.0 g of NaCl
(c) 10.0 g of $CaCl_2$

GENERAL PROBLEMS

11–80 A mixture is composed of 10 g of KNO_3 and 50 g of KCl. What is the approximate amount of KCl that can be separated using the difference in solubility shown in Figure 11-4?

KBr and KNO_3 have equal solubility at about $42°C$. What is the composition of the precipitate if 100 g of H_2O saturated with these two salts at $42°C$ is then cooled to $0°C$? (Refer to Figure 11-4.)

11–82 Given the net ionic equation

$$Fe^{2+}(aq) + S^{2-}(aq) \longrightarrow FeS(s)$$

write an appropriate total ionic equation showing suitable spectator ions. Also write a balanced molecular equation.

11–83 What is the percent composition by mass of a solution made by dissolving 10.0 g of sugar and 5.0 g of table salt in 150 mL of water?

11–84 What is the molarity of each ion in a solution that is 0.15 M $CaCl_2$, 0.22 M $Ca(ClO_4)_2$, and 0.18 M NaCl?

11–85 500 mL of 0.20 M $AgNO_3$ is mixed with 500 mL of 0.30 M NaCl. What is the concentration of Cl^- ion in the solution? The net ionic equation of the reaction that occurs is

$$Ag^+(aq) + Cl^-(aq) \longrightarrow AgCl(s)$$

11–86 400 mL of 0.15 M $Ca(NO_3)_2$ is mixed with 500 mL of 0.20 M Na_2SO_4. Write the net ionic equation of the precipitation reaction that occurs. Of the two anions involved, which remains in solution after precipitation? What is its concentration?

*11–87 A certain metal (M) reacts with HCl according to the equation

$$M(s) + 2HCl(aq) \longrightarrow MCl_2(aq) + H_2(g)$$

1.44 g of the metal reacts with 225 mL of 0.196 M HCl. What is the metal?

*11–88 Another metal (Z) also reacts with HCl according to the equation

$$2\,Z(s) + 6HCl(aq) \longrightarrow 2\,ZCl_3(aq) + 3H_2(g)$$

24.0 g of Z reacts with 0.545 L of 2.54 M HCl. What is the metal? What volume of H_2 measured at STP is produced?

*11–89 A certain compound dissolves in a solvent known as nitrobenzene. For nitrobenzene, $K_f = 8.10$. A solution with 3.07 g of the compound dissolved in 120 g of nitrobenzene freezes at $2.22°C$. The freezing point of pure nitrobenzene is $5.67°C$. Analysis of the compound shows it to be 40.0% C, 13.3% H, and 46.7% N. What is the formula of the compound?

*11–90 An electrolyte dissolves in water to form three ions. A 9.21-g quantity of this compound is dissolved in 175 g of water. The freezing point of this solution is $-1.77°C$. The compound is 47.1% K, 14.5% C, and 38.6% O. What is the formula of this compound?

SOLUTIONS TO LEARNING CHECKS

A–1 solute, solvent, miscible, immiscible, ion-dipole, ion-ion, dipole-dipole, saturated, unsaturated, supersaturated, precipitate, more, less, precipitate, ionic, net ionic

A–2 (a) $BaBr_2 \longrightarrow Ba^{2+}(aq) + 2Br^-(aq)$
(b) $(NH_4)_2CrO_4 \longrightarrow 2NH_4^+(aq) + CrO_4^{2-}(aq)$
(c) $K_2SO_3 \longrightarrow 2K^+(aq) + SO_3^{2-}(aq)$

A–3 (a) $Sr(OH)_2(aq) + Na_2SO_3(aq) \longrightarrow SrSO_3(s) + 2NaOH(aq)$
$Sr^{2+}(aq) + 2OH^-(aq) + 2Na^+(aq) + SO_3^{2-}(aq) \longrightarrow SrSO_3(s)$
$+ 2Na^+(aq) + 2OH^-(aq)$

$Sr^{2+}(aq) + SO_3^{2-}(aq) \longrightarrow SrSO_3(s)$
(b) No reaction
(c) $3Ni(NO_3)_2(aq) + 2(NH_4)_3PO_4(aq) \longrightarrow Ni_3(PO_4)_2(s) + 6NH_4NO_3(aq)$
$3Ni^{2+}(aq) + 6NO_3^-(aq) + 6NH_4^+(aq) + 2PO_4^{3-}(aq) \longrightarrow Ni_3(PO_4)_2(s)$
$+ 6NH_4^+(aq) + 6NO_3^-(aq)$

$3Ni^{2+}(aq) + 2PO_4^{3-}(aq) \longrightarrow Ni_3(PO_4)_2(s)$

B–1 mass percent, moles, liter, molarity, volume

B–2 $200 \text{ mL} \times 1.00 \text{ g/mL} = 200 \text{ g } H_2O$ $\dfrac{5.00 \text{ g solute,}}{205 \text{ g solution}} \times 100\% = \underline{\underline{2.44\%}}$

mass of solution $= 200 + 5.00 = 205$ g
$MgCl_2 = [24.3 + 2(35.5)] = 95.3$ g/mol

$5.00 \text{ g} \times \dfrac{1 \text{ mol}}{95.3 \text{ g}} = 0.0525 \text{ mol } MgCl_2$ $\dfrac{n}{V} = \dfrac{0.0525 \text{ mol}}{0.200 \text{ L}} = \underline{\underline{0.263 \text{ M}}}$

B–3 $n = M \times V = 0.375 \text{ mol/L} \times 0.015 \text{ L} = 5.63 \times 10^{-3} \text{ mol NaOH}$

$5.63 \times 10^{-3} \text{ mol} \times \dfrac{40.0 \text{ g NaOH}}{\text{mol}} = \underline{\underline{0.225 \text{ g NaOH}}}$

B–4 $M_d = \dfrac{M_c \times V_c}{V_d} = \dfrac{0.375 \text{ M} \times 15.0 \text{ mL}}{40.0 \text{ mL}} = \underline{\underline{0.141 \text{ M}}}$

B–5 volume HCl $\Longrightarrow$ mol HCl $\Longrightarrow$ mol K_3PO_4 $\Longrightarrow$ mass K_3PO_4

$K_3PO_4 = [(3 \times 39.1) + 31.0 + (4 \times 16.0)] = 212$ g/mol

$0.325 \text{ L} \times \dfrac{0.250 \text{ mol HCl}}{\text{L}} \times \dfrac{1 \text{ mol } K_3PO_4}{3 \text{ mol HCl}} \times \dfrac{212 \text{ g } K_3PO_4}{\text{mol } K_3PO_4} = \underline{\underline{5.74 \text{ g } K_3PO_4}}$

C–1 electrolyte, weak electrolyte, nonelectrolyte, colligative, vapor pressure, boiling point, freezing point, osmosis, osmotic, higher

C–2 $n(NaCl) = 10.0 \text{ g NaCl} \times \dfrac{1 \text{ mol NaCl}}{58.5 \text{ g NaCl}} = 0.171 \text{ mol NaCl}$

$m = \dfrac{\text{mol NaCl}}{\text{kg } H_2O} = \dfrac{0.171 \text{ mol}}{0.100 \text{ kg}} = 1.71 \text{ m}$

$\Delta T_b = K_b \times m = 0.512 \times 1.71 = 0.876°C$

Since NaCl is a strong electrolyte producing two ions per mole of NaCl,

$\Delta T_b = m \times 2 = 1.8°C$ Boiling point $= 100.0°C + 1.8 = \underline{\underline{101.8°C}}$

$\Delta T_f = K_f \times m \times 2 = 1.86 \times 1.71 \times 2 = 6.36°C$

Freezing point $= 0.0°C - 6.4 = \underline{\underline{-6.4°C}}$

The following multiple-choice questions have one correct answer.

1. Which of the following is the SI unit of pressure?
   (a) atm
   (b) torr
   (c) Pa
   (d) in. of Hg
   (e) lb/in.^{2}

2. Which of the following is *not* an assumption of the kinetic molecular theory applied to gases?
   (a) Molecules have negligible volume.
   (b) Molecules of all gases have the same average velocity at a given temperature.
   (c) Gas molecules have negligible interactions.
   (d) Temperature is related to the average kinetic energy of the system.
   (e) Molecules are in rapid, random motion.

3. Which of the following is a representation of Boyle's law?
   (a) $P \propto \dfrac{1}{V}$
   (b) $V \propto P$
   (c) $V \propto T$
   (d) $P \propto \dfrac{1}{T}$
   (e) $V \propto \dfrac{1}{T}$

4. The temperature of a volume of gas is increased from 20 to 40°C at constant pressure. Its volume
   (a) doubles.
   (b) decreases by half.
   (c) increases by a factor of $\frac{313}{293}$.
   (d) decreases by a factor of $\frac{20}{273}$.
   (e) decreases by a factor of $\frac{293}{313}$.

5. Which of the following is a representation of Gay-Lussac's law?
   (a) $P_1 T_1 = P_2 T_2$
   (b) $P_1 V_1 = P_2 V_2$
   (c) $V_1 T_2 = V_2 T_1$
   (d) $P_1 V_2 = P_2 V_1$
   (e) $P_1 T_2 = P_2 T_1$

6. Which of the following is the set of conditions known as standard temperature and pressure (STP)?
   (a) 0 K and 1 atm
   (b) 0°F and 760 torr
   (c) 0°C and 760 atm
   (d) 273°C and 1 atm
   (e) 273 K and 760 torr

7. Which of the following gases has the highest average velocity at a given temperature?
   (a) oxygen
   (b) carbon monoxide
   (c) neon
   (d) sulfur dioxide
   (e) hydrogen chloride

8. A mixture of gases has a total pressure of 2.00 atm. If one gas has a partial pressure of 0.50 atm, what part of the mixture is this gas?
   (a) 50%
   (b) 75%
   (c) 25%
   (d) 1.00 atm
   (e) 1.50 atm

9. A 22.4-L quantity of O_2 at STP
   (a) contains 1 mol of oxygen atoms.
   (b) has a mass of 16.0 g.
   (c) contain 1.20×10^{24} oxygen atoms.
   (d) contains 2 mol of O_2 molecules.
   (e) has a mass of 48.0 g.

10. A gas has a density of 2.68 g/L (STP). What is the gas?
    (a) CO_2
    (b) SO_2
    (c) NO_2
    (d) COS
    (e) He

11. Which of the following is a value for the gas constant R?
    (a) $62.4 \dfrac{L \cdot atm}{mol \cdot K}$
    (b) $62.4 \dfrac{L \cdot torr}{mol \cdot K}$
    (c) $82.1 \dfrac{L \cdot atm}{mol \cdot K}$
    (d) $0.0821 \dfrac{L \cdot atm}{mol \cdot °C}$
    (e) $0.0821 \dfrac{L \cdot torr}{mol \cdot K}$

12. Given the equation
    $$C(s) + H_2O(l) \longrightarrow CO(g) + H_2(g)$$
    what volume of gas measured at STP would be produced from 24.0 g of carbon?
    (a) 22.4 L   (c) 44.8 L   (e) 4.0 L
    (b) 89.6 L   (d) 11.2 L

13. Which of the following is not a property of the liquid state?
    (a) surface tension   (c) viscosity
    (b) melting point   (d) boiling point

14. Which of the following is a property of an ionic compound?
    (a) high melting point   (c) amorphous
    (b) network bonding   (d) soft

15. The basic particles (atoms, molecules, or ions) in a liquid
    (a) have completely random motion.
    (b) move about fixed points.
    (c) have translational motion.
    (d) are not attracted to each other.

16. In which of the following compounds would there be dipole-dipole interactions in the liquid state?
    (a) NaCl   (d) COS
    (b) H_2 (e) CO_2
    (c) CCl_4

17. Which of the following has hydrogen bonding in the liquid state?
    (a) H_2S (d) CH_3OH
    (b) NaH   (e) NCl_3
    (c) CH_4

18. Which of the following nonpolar molecules should have the highest boiling point?
    (a) H_2 (d) CH_4
    (b) SF_6 (e) N_2
    (c) CO_2

19. Based on interactions in the liquid state, which of the following has the highest heat of vaporization?
    (a) H_2O (d) H_2
    (b) H_2S (e) NH_3
    (c) CH_4

20. What happens when heat is applied to a liquid at its boiling point?
    (a) The heat energy is converted into potential energy.
    (b) The heat energy is converted into kinetic energy.
    (c) The heat energy is converted into both potential and kinetic energy.
    (d) It makes the molecules of the vapor move faster than those in the liquid.

21. The normal boiling point of a liquid is defined as the temperature at which
    (a) bubbles form in the liquid.
    (b) the vapor pressure equals the atmospheric pressure.

(c) the vapor pressure equals one atmosphere.
(d) the vapor and the liquid exist in equilibrium.

22. A compound has a melting point of 950°C. Which of the following statements is most likely true?
    (a) Its molecules are nonpolar.
    (b) The compound has a low heat of fusion.
    (c) Its molecules are polar covalent.
    (d) The compound has a low boiling point.
    (e) The compound has a high heat of vaporization.

23. When a liquid in an insulated container is allowed to evaporate
    (a) the temperature of the liquid rises.
    (b) the temperature of the liquid does not change.
    (c) no liquid evaporates unless heat is supplied from the outside.
    (d) the vapor that escapes is warmer than the liquid.

24. Which of the following equations illustrates the solution of Na_2SO_4 in water?

$$Na_2SO_4(s) \xrightarrow{H_2O}$$

    (a) $Na_2^{2+}(aq) + SO_4^{2-}(aq)$
    (b) $2Na^+(aq) + SO_4^{2-}(aq)$
    (c) $Na^{2+}(aq) + SO_4^-(aq)$
    (d) $2Na^{2+}(aq) + SO_4^-(aq)$

25. Which of the following equations represents a precipitation reaction?

    (a) $CaCO_3 \rightarrow CaO + CO_2$
    (b) $HCl + KOH \rightarrow H_2O + KCl$
    (c) $Na_2S + Ni(NO_3)_2 \rightarrow NiS + 2NaNO_3$
    (d) $Zn + CuCl_2 \rightarrow ZnCl_2 + Cu$
    (e) $CH_2 + 2O_2 \rightarrow CO_2 + 2H_2O$

26. Which of the following is a net ionic equation?

    (a) $CaCl_2 + Na_2CO_3 \rightarrow CaCO_3 + 2NaCl$
    (b) $Ca^{2+} + 2Cl^- + 2Na^+ + CO_3^{2-} \rightarrow CaCO_3 + 2Cl^- + 2Na^+$
    (c) $2Cl^- + 2Na^+ \rightarrow 2Na^+ + 2Cl^-$
    (d) $Ca^{2+} + CO_3^{2-} \rightarrow CaCO_3$

27. A 2.0-L quantity of 0.10 M HCl contains
    (a) 1.0 mol of HCl.   (d) 0.05 mol of H_2O.
    (b) 0.20 mol of H_2O.   (e) 0.20 mol of HCl.
    (c) 20 mol of HCl.

28. Which of the following is not a colligative property?
  (a) the boiling point of a solvent
  (b) freezing point depression
  (c) vapor pressure lowering
  (d) osmotic pressure

29. What is the freezing point of a 0.100 m aqueous solution of a nonelectrolyte? For water $K_f = 1.86$.

  (a) $-1.86°C$ (d) $18.6°C$
  (b) $0.186°C$ (e) $-18.6°C$

30. If solutions of each of the following have the same concentration, which has the lowest freezing point?
  (a) CH_3OH (a nonelectrolyte)
  (b) NaCl
  (c) Na_2SO_4
  (d) CaS
  (e) all are the same

PROBLEMS

1. Fill in the blanks.

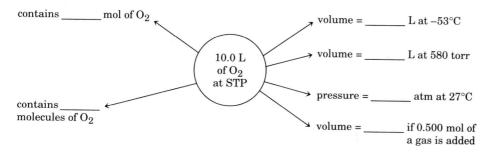

contains _____ mol of O_2

volume = _____ L at $-53°C$

volume = _____ L at 580 torr

10.0 L of O_2 at STP

pressure = _____ atm at $27°C$

contains _____ molecules of O_2

volume = _____ if 0.500 mol of a gas is added

2. A compound can be vaporized at 100°C. At that temperature, 1.19 g of the compound occupies 250 mL at 756 torr. Analysis of the compound shows that it is 49.0% C, 48.3% Cl, and 2.72% H. What is the molecular formula of the compound?

3. The following compounds have nearly the same formula weights: $(CH_3)_2O$ (polar covalent), C_2H_5OH (polar covalent), C_3H_8 (nonpolar covalent), and BeF_2 (ionic). The melting points of these compounds are 800°C, $-117°C$, $-138°C$ and $-189°C$. Match the melting point with the compound and explain what forces must be overcome in the solid to allow melting.

4. How many calories are required to change 25.0 g of ice at $-20.0°C$ to steam at 100°C? (The specific heat of ice is 0.492 cal/g · °C, the heat of fusion is 79.8 cal/g, and the heat of vaporization is 540 cal/g.)

5. An aqueous solution is made by dissolving 40.0 g of H_2SO_4 in enough water to make 850 mL of solution.
  (a) What is the molarity of the H_2SO_4 solution?
  (b) If 100 mL of this solution is diluted to 500 mL, what is the resulting molarity?
  (c) What volume of the original solution is needed to form 2.00 L of a 0.150 M solution?

6. When aqueous solutions of lead(II) nitrate and sodium iodide are mixed, a precipitate of lead(II) iodide forms.
  (a) Write the balanced equation representing this reaction.
  (b) What volume of 0.40 M lead nitrate is needed to react with 2.50 L of 0.55 M sodium iodide?

The lemon is known for its sour taste which is caused by citric acid. Acids are a unique class of compounds that are discussed in this chapter.

ACIDS, BASES, AND SALTS

It can make your mouth pucker, your body shudder, and your eyes water. That's the reaction one gets from taking a bite into a fresh lemon. A taste of vinegar has the same effect. Even carbonated beverages produce a subtle sour taste that peps up the drink. All of these substances have a similar effect because of the presence of a compound that produces a sour taste (but to different degrees). These compounds, known as acids, were characterized over five hundred years ago in the Middle Ages in the chemical laboratories of alchemists. Substances were classified as acids because of their common properties (such as sourness) rather than a certain chemical composition. Acids are well known to the general population. Some are in foods or drugs, such as citric acid in lemons, acetic acid in vinegar, lactic acid in sour milk, or acetylsalicylic acid in aspirin. Some must be handled with caution, such as sulfuric acid in a car battery or hydrochloric acid used to clean concrete. We also relate the word acid with the serious environmental hazard, "acid rain." As we will discuss in this chapter, acids undergo many characteristic chemical reactions. When rain is acidic, these reactions lead to the degradation of stone used in buildings and the liberation of poisonous metal ions locked in soil. The control of acid rain has become an issue of international conferences and negotiations.

◀ **Setting The Stage**

Bases are compounds with properties that in some ways complement those of acids. Acids and bases react with one another to form a third class of compounds known as salts. We will be concerned with the identity and reactions of these three types of compounds.

There are several important questions that will be addressed in this chapter. What active ingredients in acids and bases produce their specific properties? Why are some acids, like citric acid in fruits, comparatively tame and others, like battery acid, so harsh? How do we measure different degrees of acidity? How is acidity controlled? And finally, how does burning coal lead to acid formation in rain? First, we will examine some of the properties of acids and bases.

12–1   PROPERTIES OF ACIDS AND BASES

The sour taste of acids accounts for the origin of the word itself. The word *acid* originates from the Latin *acidus*, meaning "sour," or the closely related Latin *acetum*, meaning "vinegar." This ancient class of compounds has several characteristic chemical properties. Acids are compounds that

1   Taste sour (of course, one never tastes laboratory chemicals).

2   React with certain metals (e.g., Zn and Fe), with the liberation of hydrogen gas. (See Figure 12-1.)

Figure 12–1
Zinc and Limestone in Acid. Zinc (*left*) reacts with acid to liberate hydrogen; limestone ($CaCO_3$, *right*) reacts with acid to liberate carbon dioxide.

3 Cause certain organic dyes to change color (e.g., litmus turns from blue to red in acids).

4 React with limestone ($CaCO_3$), with the liberation of carbon dioxide gas. (See Figure 12-1.)

5 React with bases to form salts and water.

The counterparts to acids are bases. Bases are compounds that

1 Taste bitter.

2 Feel slippery or soapy.

3 React with oils and grease.

4 Cause certain organic dyes to change color (e.g., litmus turns from red to blue in bases).

5 React with acids to form salts and water.

Notice that these properties relate to what acids and bases do and not to their composition. Thus simple laboratory procedures could be used to classify appropriate compounds into one of these two classes. A little over one hundred years ago, however, the secret of acid or base behavior was discovered. At that time, the role of ions in solutions was recognized. *In 1884, Swedish chemist Svante Arrhenius suggested that* **acids** *are substances that produce H⁺ ions and* **bases** *are substances that produce OH⁻ ions in aqueous solution.* This model is still quite important today even though its use is limited to aqueous solutions. However, we can use Arrhenius's definition until we need to employ a more extensive and broader model.

How is a compound classified as either an acid or a base?

First, let us consider the action of acids. Four common acids that most of us recognize are hydrochloric acid (also known as muriatic acid and available in most hardware stores), sulfuric acid (oil of vitriol, or battery acid), acetic acid (the source of the sour taste of vinegar), and carbonic acid (carbonated water). The following equations illustrate how these four acids could produce the common ingredient, the H⁺ ion, as Arrhenius envisioned. Each acid would produce a different anion.

Muriatic or hydrochloric acid	$HCl \rightarrow H^+ + Cl^-$
Oil of vitriol or sulfuric acid	$H_2SO_4 \rightarrow H^+ + HSO_4^-$
Acetic acid in vinegar	$HC_2H_3O_2 \rightleftharpoons H^+ + C_2H_3O_2^-$
Carbonated water or carbonic acid	$H_2CO_3 \rightleftharpoons H^+ + HCO_3^-$

These acids are not ionic compounds when pure. They form ions only when mixed with water. In fact, a chemical reaction occurs between the acid molecules and the H_2O molecules that results in removal of a H⁺ ion from the rest of the molecule. As discussed in Chapters 6 and 11, acids (such as HCl) and H_2O are polar covalent compounds. When HCl dissolves in water, there is a dipole-dipole electrostatic attraction between the negative dipole of one molecule and the positive dipole of the other, as illustrated in Figure 12-2. (Actually, there are many more H_2O molecules involved with each HCl than those shown in the figure.)

Are acids ionic before they dissolve in water?

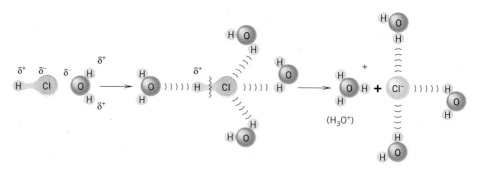

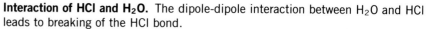

Figure 12–2
Interaction of HCl and H₂O. The dipole-dipole interaction between H_2O and HCl leads to breaking of the HCl bond.

A tug-of-war develops. On one side, the H_2O's pull on the HCl molecule, attempting to split the molecule into a H^+ ion and a Cl^- ion. On the other side, the HCl covalent bond tries mightily to hold the molecule together. For HCl, the H_2O molecules are clearly the winners, since the interaction is strong enough to break the bonds of all dissolved HCl molecules. Therefore, HCl is an acid since a H^+ is produced in the solution as a result of the dissociation (breaking apart) of HCl. (Since dissociation produces ions in this case, the process is called ionization.)

To illustrate the importance of water in the ionization process, the reaction of an acid with water can be represented as

$$HCl + H_2O \longrightarrow H_3O^+(aq) + Cl^-(aq)$$

Instead of H^+, the acid species is often represented as H_3O^+, which is known as the **hydronium ion.** The hydronium ion is simply a representation of the H^+ ion in a hydrated form. The acid species is represented as H_3O^+ rather than H^+ because it is somewhat closer to what is believed to be the actual species. In fact, the nature of H^+ in aqueous solution is even more complex than H_3O^+ (i.e., $H_5O_2^+$, $H_7O_3^+$, etc.). In any case, the acid species is represented as H^+, $H^+(aq)$, or $H_3O^+(aq)$ depending on the convenience of the particular situation. *Just remember that all refer to the same species in aqueous solution.* If $H^+(aq)$ is used, it should be understood that it is not just a bare proton in aqueous solution but is associated with water molecules. (It is hydrated.)

The common property of acids is that they produce $H^+(aq)$ ions in aqueous solution. It is this ion that is responsible for the common chemical reactions associated with acids. The following equations focus on the role of the H^+ ion in the reactions of acids with metals, limestone, and bases. The anion of the acid is not included in the first two reactions.

1   Acids react with zinc and give off hydrogen gas:

$$Zn(s) + \underline{2H^+(aq)} \longrightarrow Zn^{2+}(aq) + H_2(g)$$

2   Acids react with limestone to give off carbon dioxide gas:

$$CaCO_3(s) + \underline{2H^+(aq)} \longrightarrow Ca^{2+}(aq) + H_2O(l) + CO_2(g)$$

3   Acids react with bases:

$$ACID \quad + BASE \quad \longrightarrow \quad SALT \quad + WATER$$

$$HCl(aq) + NaOH(aq) \longrightarrow NaCl(aq) + H_2O(l)$$

Some common acids.   Some common bases.

The last reaction is of prime importance and is discussed in more detail in the next section.

Other molecules containing hydrogen may not be acidic in H_2O because the hydrogens are (1) too strongly attached to the rest of the molecule to be ionized, or (2) not polar enough to create a strong interaction with H_2O. Sometimes both reasons are important. The hydrogens in CH_4 and NH_3 and the three hydrogens on the carbon in acetic acid are examples of hydrogens that do not ionize in aqueous solution to form H^+ ions.

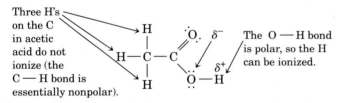

Three H's on the C in acetic acid do not ionize (the C — H bond is essentially nonpolar).

The O — H bond is polar, so the H can be ionized.

Now let us turn our attention to bases. Bases are compounds that produce OH^- ions in water, forming what are known as basic solutions, sometimes referred to as *alkaline* or *caustic* solutions. Some of the commonly known bases are sodium hydroxide (also known as caustic soda, or lye), potassium hydroxide (caustic potash), calcium hydroxide (slaked lime), and ammonia. Except for ammonia, these compounds are all solid ionic compounds. Solution in water simply releases the OH^- ion into the aqueous medium:

$$NaOH(s) \xrightarrow{x\,H_2O} Na^+(aq) + OH^-(aq)$$

$$Ba(OH)_2(s) \xrightarrow{x\,H_2O} Ba^{2+}(aq) + 2OH^-(aq)$$

The action of ammonia (NH_3) as a base is somewhat different from that of the ionic hydroxides and is discussed in Section 12-3.

Acids and bases react with one another. When this happens, the characteristic properties of both the acid and the base are destroyed, or neutralized. The products of the reaction are a salt (a third class of com-

▼ **Looking Ahead**

pound) and water. We will look at the interactions of acids with bases next.

12–2   NEUTRALIZATION AND SALTS

A neutralization reaction is one of two types of double-replacement reactions first introduced in Chapter 8. The familiar neutralization of hydrochloric acid with sodium hydroxide is an example. The equations illustrating this reaction are written in molecular, total ionic, and net ionic form as follows:

$$\text{ACID} + \text{BASE} \longrightarrow \text{SALT} + \text{WATER}$$

molecular: $HCl(aq) + NaOH(aq) \longrightarrow NaCl(aq) + H_2O(l)$

total ionic: $H^+(aq) + Cl^-(aq) + Na^+(aq) + OH^-(aq) \longrightarrow$
$$Na^+(aq) + Cl^-(aq) + H_2O(l)$$

net ionic: $H^+(aq) + OH^-(aq) \longrightarrow H_2O(l)$

The key to what drives neutralization reactions is found in the net ionic equation. The active ingredient from the acid [$H^+(aq)$] reacts with the active ingredient from the base [$OH^-(aq)$] to form the molecular compound water. A salt is what is left over—usually present as spectator ions. *Thus a* **salt** *is formed from the anion of the acid (e.g., Cl^-) and the cation from the base (e.g., Na^+).* By themselves, solutions of hydrochloric acid and sodium hydroxide are corrosive compounds that must be handled with caution. Mixing the two in stoichiometric amounts produces a harmless saltwater solution.*

How do acids neutralize bases?

When most people use the word salt, they are usually referring to just one substance, sodium chloride, formed in the previous reaction. Actually, a salt can result from many different combinations of anions and cations from a variety of neutralizations. The following neutralization reactions illustrate the formation of some other salts.

Acid	+	Base	$\longrightarrow$	Salt	+	Water
1 $HNO_3(aq)$	+	$KOH(aq)$	$\longrightarrow$	$KNO_3(aq)$	+	H_2O
2 $H_2SO_4(aq)$	+	$2NaOH(aq)$	$\longrightarrow$	$Na_2SO_4(aq)$	+	$2H_2O$
3 $2HClO_4(aq)$	+	$Ca(OH)_2(aq)$	$\longrightarrow$	$Ca(ClO_4)_2(aq)$	+	$2H_2O$
4 $2H_3PO_4(aq)$	+	$3Ba(OH)_2(aq)$	$\longrightarrow$	$Ba_3(PO_4)_2(s)$	+	$6H_2O$

Sodium chloride is just one of many salts.

Hydrochloric acid (HCl) and nitric acid (HNO_3) are examples of acids that produce *one mole of H^+ ions in solution per mole of acid.* Such acids are known as **monoprotic acids.** On the other hand, note in example 2

*Don't try this without supervision. Much heat evolution, with boiling and splattering, can occur.

above that one mole of sulfuric acid (H_2SO_4) requires two moles of sodium hydroxide for complete neutralization. Sulfuric acid is an example of a **diprotic acid,** *which means that it can produce two moles of H^+ per mole of acid:*

$$H_2SO_4 \longrightarrow 2H^+(aq) + SO_4{}^{2-}(aq)$$

There are several examples of **triprotic acids** *(three potential H^+ ions per molecule),* but the most important one is phosphoric acid (H_3PO_4).

The alkaline earth hydroxides are analogous to the diprotic acids except that they produce two OH^- ions per formula unit dissolved. An example is the solution of calcium hydroxide in water:

$$Ca(OH)_2(s) \xrightarrow{\ x\,H_2O\ } Ca^{2+}(aq) + 2OH^-(aq)$$

Notice in example 3 above that two moles of a monoprotic acid such as $HClO_4$ are required to neutralize one mole of calcium hydroxide.

The neutralizations of sulfuric acid (example 2) and calcium hydroxide (example 3) are illustrated further with the total ionic equations and the net ionic equations.

$$2H^+(aq) + \cancel{SO_4{}^{2-}(aq)} + \cancel{2Na^+(aq)} + 2OH^-(aq)$$
$$\longrightarrow \cancel{2Na^+(aq)} + \cancel{SO_4{}^{2-}(aq)} + 2H_2O(l)$$

$$H^+(aq) + OH^-(aq) \longrightarrow H_2O(l)$$

$$2H^+(aq) + \cancel{2ClO_4{}^-(aq)} + \cancel{Ca^{2+}(aq)} + 2OH^-(aq)$$
$$\longrightarrow \cancel{Ca^{2+}(aq)} + \cancel{2ClO_4{}^-(aq)} + 2H_2O(l)$$

$$H^+(aq) + OH^-(aq) \longrightarrow H_2O(l)$$

Polyprotic acids *(acids with more than one acidic hydrogen)* can be partially neutralized. The ionization of H_2SO_4 can be represented in two steps.

$$H_2SO_4 \longrightarrow H^+(aq) + HSO_4{}^-(aq) \text{ (first ionization of } H_2SO_4)$$

$$HSO_4{}^- \rightleftharpoons H^+(aq) + SO_4{}^{2-}(aq) \text{ (second ionization of } H_2SO_4)$$

The second ionization is represented by a double arrow, which indicates that this ionization is not complete and reaches a point of equilibrium. This phenomenon will be discussed in more detail in the next section.

If *one mole* of NaOH is added to *one mole* of H_2SO_4, only the first ionization is relevant. The balanced equations representing this reaction are as follows:

$$H_2SO_4(aq) + NaOH(aq) \longrightarrow NaHSO_4(aq) + H_2O(l)$$

$$\underline{H^+(aq)} + \cancel{HSO_4{}^-(aq)} + \cancel{Na^+(aq)} + \underline{OH^-(aq)} \longrightarrow$$
$$\cancel{Na^+(aq)} + \cancel{HSO_4{}^-(aq)} + H_2O(l)$$

$$H^+(aq) + OH^-(aq) \longrightarrow H_2O(l)$$

Note that the salt formed in the partial neutralization ($NaHSO_4$, sodium hydrogen sulfate, or sodium bisulfate) contains a hydrogen on the anion. This is known as an **acid salt,** *an ionic compound containing an anion with one or more hydrogens that can be neutralized with a base.* An acid salt results from the partial neutralization of a diprotic or triprotic acid. When additional base is added to a solution of an acid salt, further neutralization takes place. For example, the reaction of a solution of $NaHSO_4$ with $NaOH$ is illustrated as follows:

$$NaHSO_4(aq) + NaOH(aq) \longrightarrow Na_2SO_4(aq) + H_2O(l)$$

Neutralization of the triprotic acid H_3PO_4 with KOH can occur in three steps. Consider the reaction when one mole of KOH is added to one mole of H_3PO_4.

1

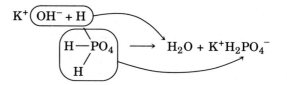

If one additional mole of KOH is then added to the resulting KH_2PO_4 solution, reaction **2** occurs.

2

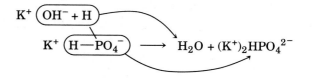

Finally, by adding one mole of KOH to the resulting K_2HPO_4 solution, reaction **3** occurs.

3

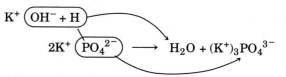

Note that each additional OH^- removes one H^+ from the acid or acid salt. The total reaction, which is the algebraic sum of all three reactions, is

$$H_3PO_4(aq) + 3KOH(aq) \longrightarrow K_3PO_4(aq) + 3H_2O(l)$$

This is the reaction that occurs regardless of whether the KOH is added one mole at a time or all three moles at once. In these reactions, the H_3PO_4 is identified as an acid, KH_2PO_4 and K_2HPO_4 as acid salts, and K_3PO_4 as a normal salt. (See Figure 12-3.)

Figure 12–3
Neutralization of H_3PO_4.
The hydrogens of H_3PO_4
can be neutralized one
at a time.

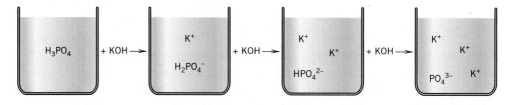

EXAMPLE 12–1

Identification of Acids, Bases, and Salts

Identify each of the following as an acid, a base, a normal salt, or an acid salt:
(a) $KClO_4$ (b) H_2S (c) $Ba(HSO_3)_2$ (d) $Al(OH)_3$

●Working It Out

Solution
(a) normal salt
(b) acid
(c) acid salt
(d) base

EXAMPLE 12–2

Writing a Complete Neutralization Reaction

Write the balanced equation in molecular form illustrating the complete neutralization of $Al(OH)_3$ with H_2SO_4.

Procedure
Each mole of base produces *3 mol* of OH^-. Each mole of acid produces *2 mol* of H^+. Therefore, 2 mol of base (producing 6 mol of OH^-) reacts with 3 mol of acid (producing 6 mol of H^+) for complete neutralization.

Solution

$$2Al(OH)_3 + 3H_2SO_4 \longrightarrow Al_2(SO_4)_3 + 6H_2O$$

EXAMPLE 12–3

Writing a Partial Neutralization Reaction

Write the balanced molecular equation illustrating the reaction of 1 mol of H_3PO_4 with 1 mol of $Ca(OH)_2$.

Procedure
One mole of base produces 2 mol of OH^-, which reacts with 2 mol of H^+. The removal of 2 mol of H^+ from 1 mol of H_3PO_4 leaves the HPO_4^{2-} ion in solution.

Solution

$$H_3PO_4 + Ca(OH)_2 \longrightarrow CaHPO_4 + 2H_2O$$

Battery (sulfuric) acid requires careful handling. It is extremely corrosive, dissolving many metals and causing severe burns to skin. On the other hand, acetic acid is mild enough to use on a salad. Obviously, there is a wide range of acidic behavior. The difference in these two acids lies in their acid strengths. That is our next subject.

▼Looking Ahead

12–3   STRENGTHS OF ACIDS AND BASES

Why isn't it a good idea to use battery acid on a salad?

Hydrochloric (muriatic) acid is a strong acid used to clean concrete.

The corrosive power of acids and bases has been known for centuries. Sulfuric and hydrochloric acid rapidly dissolve iron and zinc. Sodium hydroxide is used to unclog drains filled with hair and grease. This base must be handled and stored with caution. Yet there are many acids and bases that are comparatively easy to handle and even to ingest. The large difference in acid or base behavior relates to the concentration of the active ingredient (H^+ or OH^-) produced by the acid or base in water. This depends on the strength of the acid or base. First, we will consider the strength of acids.

Nitric and hydrochloric acids are known as strong acids. **Strong acids** *are essentially 100% ionized in aqueous solutions.* Thus strong acids are molecular compounds that are completely ionized when dissolved in water. They are also known as strong electrolytes. The ionization of HCl is illustrated in the following equation:

$$HCl(g) + H_2O(l) \longrightarrow H_3O^+(aq) + Cl^-(aq)$$

Besides HNO_3 and HCl, other common strong acids are $HClO_4$, HBr, HI, and H_2SO_4. (Only the first ionization of H_2SO_4 is complete, however.) Most other common acids behave as weak acids in water. A **weak acid** *is partially ionized (usually less than 5% at typical molar concentrations).* Because of the small concentration of ions, weak acids behave as weak electrolytes in water. (See Section 11-8.) The ionization of a weak acid is incomplete because it is a reversible reaction. A reaction in which the reverse reaction also occurs to an appreciable extent is illustrated by double arrows ($\rightleftharpoons$) rather than the single arrow that implies an essentially complete reaction such as was previously shown for HCl. Ionization of the weak acid HF is illustrated as follows:

$$HF(aq) + H_2O(l) \rightleftharpoons H_3O^+(aq) + F^-(aq)$$

The partial ionization of a weak acid is one example of a chemical reaction that reaches a state of equilibrium. *In a reaction at* **equilibrium,** *two reactions are occurring simultaneously.* In the ionization of HF, for example, a forward reaction occurs to the right, producing ions (H_3O^+ and F^-), and a reverse reaction occurs to the left, producing covalent compounds (HF and H_2O.)

$$\text{Forward} \quad HF(aq) + H_2O(l) \longrightarrow H_3O^+(aq) + F^-(aq)$$

$$\text{Reverse} \quad H_3O^+(aq) + F^-(aq) \longrightarrow HF(aq) + H_2O(l)$$

At equilibrium, the forward and reverse reactions occur at the same rate. For weak acids, the point of equilibrium lies far to the left side of the original ionization equation. This means that most of the fluorine is present in the form of covalent HF molecules rather than fluoride ions. (See Figure 12-4.)

When a system is at equilibrium, the concentrations of all species (reactants and products) remain the same, but the identities of the individual molecules change. The reaction thus *appears* to have gone so far to the right and then stopped. In fact, at equilibrium, a *dynamic (constantly changing)* situation exists in which two reactions going in opposite directions at the same rate keep the concentrations of all species constant.

HCl, a strong acid   HF, a weak acid

Figure 12–4
Strong Acids and Weak Acids. Strong acids are completely ionized in water; weak acids are only partially ionized.

The following examples illustrate the difference in acidity (the difference in H_3O^+ concentration) between a strong acid and a weak acid. In these examples, *the appearance of a species in brackets (e.g., $[H_3O^+]$) represents the numerical value of the concentration of that species in moles per liter (M)*.

EXAMPLE 12–4

The H_3O^+ Concentration in a Strong Acid Solution

What is $[H_3O^+]$ in a 0.100 M HNO_3 solution?

Solution
HNO_3 is a strong acid, so the following reaction goes 100% to the right.

$$HNO_3(aq) + H_2O(l) \longrightarrow H_3O^+(aq) + NO_3^-(aq)$$

The initial concentration of HNO_3 is 0.100 M, and all of the HNO_3 ionizes to form H_3O^+. The reaction stoichiometry tells us that 1 mol of H_3O^+ is formed for every mole of HNO_3 initially present. Therefore, 0.100 mol of HNO_3 forms 0.100 mol of H_3O^+ in 1 L. Thus

$$[H_3O^+] = \underline{\underline{0.100\ M}}$$

●**Working It Out**

EXAMPLE 12–5

The H_3O^+ Concentration in a Weak Acid Solution

What is $[H_3O^+]$ in a 0.100 M $HC_2H_3O_2$ solution that is 1.34% ionized?

Solution
Since $HC_2H_3O_2$ is a weak acid, the following ionization reaches equilibrium when 1.34% of the initial $HC_2H_3O_2$ ionizes.

$$HC_2H_3O_2(aq) + H_2O(l) \rightleftharpoons H_3O^+(aq) + C_2H_3O_2^-(aq)$$

The $[H_3O^+]$ is calculated by multiplying the original concentration of acid by the percent *expressed in fraction form*.

$$[H_3O^+] = [\text{original conc. of acid}] \times \frac{\%\ \text{ionization}}{100\%}$$

In this case,

$$[H_3O^+] = [0.100] \times \frac{1.34\%}{100\%} = \underline{\underline{1.34 \times 10^{-3}\ M}}$$

Notice in these two examples that the concentration of H_3O^+ is about one hundred times greater in the nitric acid solution than in the acetic

Milk of Magnesia is a base taken for acid indigestion.

acid solution. Yet both acids were at the same original concentration. The large difference in acid strengths lies in the proportion of the original acid molecules that ionize. We should take note that the concentration of an acid in water does not necessarily correspond to the concentration of H_3O^+ ions present in solution.

Bases also have a wide range of strengths. Lye is effective in unclogging drains but is far too strong for use in removing grease stains from a linoleum floor. It would dissolve the floor and burn the hand that applied it. A more practical base for the job is an ammonia solution. Lye (sodium hydroxide) is a strong base whereas ammonia is a weak base.

Strong bases *are those compounds that completely ionize in water to produce OH^- ions.* All alkali metal hydroxides are strong bases and are quite soluble in water. The solution of the ionic compound NaOH is illustrated as follows:

$$NaOH(s) \xrightarrow{\; x\,H_2O \;} Na^+(aq) + OH^-(aq)$$

All of the alkaline earth hydroxides [except $Be(OH)_2$] also completely dissociate into ions in solution. However, $Mg(OH)_2$ has a very low solubility in water and so produces a very small concentration of aqueous OH^-. Because of its low solubility, it can be taken internally to combat excess stomach acid (milk of magnesia).

As mentioned, ammonia exhibits much more moderate basic properties than caustic soda. This is because ammonia is a **weak base** in water:

$$NH_3(aq) + H_2O(l) \rightleftharpoons NH_4^+(aq) + OH^-(aq)$$

Only a small portion of the dissolved ammonia molecules react to form ions, so the concentration of OH^- remains low. *In this equilibrium, like that of weak acids, the position of equilibrium lies far to the left.* Ammonia is the most familiar weak base, and this reaction is encountered often. Many other neutral nitrogen compounds such as methylamine (CH_3NH_2) and pyridine (C_5H_5N) also react with water in a manner similar to that of ammonia to produce weakly basic solutions.

Looking Back ▲

Acids and bases are two ancient classes of compounds that are classified as such because of (a) some common properties and (b) the particular ions that they produce in aqueous solution. The degree to which they produce these ions determines their acid or base strength. When acids or bases are mixed together, they form a third class of compounds called salts (or in some cases, acid salts) and water. The formation of water is the driving force for the neutralization reaction between acids and bases.

Checking It Out ▶

Learning Check A

A–1. Fill in the blanks.
Acids are compounds that produce _____ ions in solution, which is also written as the hydronium ion (_____). Bases are compounds that produce an ion with the formula _____ and the name _____ ion. The reaction between acids and bases is known as a _____

reaction. The net ionic equation of this reaction always has _____ as a product. The spectator ions of the reaction form a _____. The partial neutralization of polyprotic acids produces an _____ _____ and water. Strong acids are essentially _____ % ionized in aqueous solution, whereas weak acids are _____ ionized. Sodium hydroxide is a strong base, but ammonia is a _____ _____.

A–2. Write the formula and name of the acid or base formed from the following ions:
(a) ClO_4^- (b) Fe^{2+} (c) S^{2-} (d) Li^+

A–3. Write the balanced molecular, total ionic, and net ionic equations illustrating the neutralization of HNO_3 with $Sr(OH)_2$.

A–4. Write the balanced molecular equation illustrating the reaction of one mole of $H_2C_2O_4$ (oxalic acid) and one mole of CsOH.

A–5. A certain 0.10 M solution of an acid is 10.0% ionized. Is this a strong or a weak acid? What is the $[H_3O^+]$ in this solution?

Additional Examples: 12-1, 12-2, 12-6, 12-7, 12-9, 12-17, 12-20, and 12-21.

We are now ready to expand the concept of how we express the strengths of acids and bases. As it turns out, there is more going on in pure water than our earlier discussion may have indicated. We take a closer look at water next.

▼ Looking Ahead

12–4 EQUILIBRIUM OF WATER

In the previous chapter, pure water was classified as a nonconductor of electricity. This implied that no ions were present. This isn't entirely true, however. With more sensitive instruments, we find that there actually is a small concentration of ions in pure water. Since there is no solute, the ions must come from the water itself. The presence of ions in water is due to a process known as autoionization. **Autoionization** *produces positive and negative ions from the dissociation of the molecules of a liquid.* For water this is represented as follows. The double arrow again indicates that this is a reversible reaction reaching a point of equilibrium.

From where do the ions in pure water originate?

$$H_2O + H_2O \rightleftharpoons H_3O^+ + OH^-$$

Although this equilibrium lies very far to the left, there is a small but important amount of H_3O^+ ions and OH^- ions coexisting in pure water. It is this small concentration that we will focus on as a means of expressing acid or base behavior and their relative strengths.

The concentration of each ion at 25°C has been found by experiment to be 1.0×10^{-7} mol/L. This means that only about one out of every 500 million water molecules is actually ionized at any one time. Other experimental results tell us that the product of the ion concentrations is a constant. This phenomenon will be explained in more detail in Chapter 14, but for now we accept it as fact. Therefore, at 25°C

$$[H_3O^+][OH^-] = K_w \quad \text{(a constant)}$$

Substituting the actual concentrations of the ions, we can now find the numerical value of the constant.

$$[1.0 \times 10^{-7}][1.0 \times 10^{-7}] = 1.0 \times 10^{-14}$$

K_w *(1.0 × 10^{-14}) is known as the* **ion product** *of water* at 25°C. The importance of this constant is that it tells us the concentrations of H_3O^+ and OH^- not only in pure water but also in acidic and basic solutions. The following example illustrates this relationship.

Working It Out ●

EXAMPLE 12–6

Calculation of [OH$^-$] from [H$_3$O$^+$]

In a certain solution, $[H_3O^+] = 1.5 \times 10^{-2}$ *M*. What is [OH$^-$] in this solution?

Procedure
Use the relationship for K_w, $[H_3O][OH^-] = 1.0 \times 10^{-14}$, and solve for [OH$^-$].

Solution

$$[H_3O^+][OH^-] = 1.0 \times 10^{-14}$$

$$[OH^-] = \frac{1.0 \times 10^{-14}}{[H_3O^+]} = \frac{1.0 \times 10^{-14}}{1.5 \times 10^{-2}}$$

$$= \underline{\underline{6.7 \times 10^{-13}\ M}}$$

How can OH$^-$ ions be present in an acidic solution?

As you can see, there is an inverse relationship between [H$_3$O$^+$] and [OH$^-$]. Although inverse relationships have been discussed and illustrated several times previously in this text, the relationship between

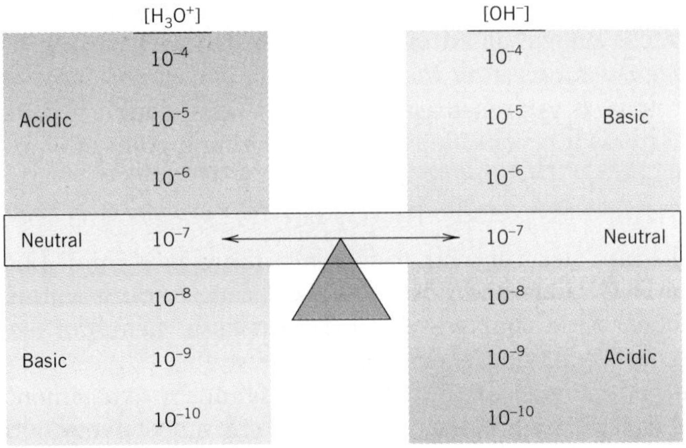

Figure 12–5
Pure Water. In pure water or a neutral solution, [H$_3$O$^+$] and [OH$^-$] are both equal to 10^{-7} *M*.

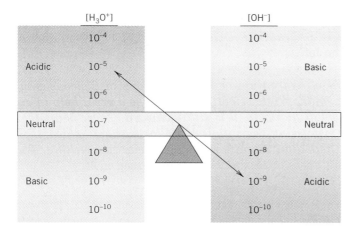

Figure 12–6
An Acid Solution. There is an inverse relationship between $[H_3O^+]$ and $[OH^-]$.

$[H_3O^+]$ and $[OH^-]$ is illustrated in Figures 12-5 and 12-6. In these figures, concentrations of $[H_3O^+]$ and $[OH^-]$ between 10^{-4} and 10^{-10} M are listed. (Higher and lower concentrations exist, however.) Concentrations of both ($[H_3O^+]$ on the left and $[OH^-]$ on the right) are shown to decrease from top to bottom. In between, there is a rigid pointer with an arrow at each end. Figure 12-5 illustrates the situation in pure water, which is neutral. In pure water or a neutral solution, the pointer is exactly balanced, indicating that the concentrations of both ions are equal to 10^{-7} M.

We will now add acid to pure water, which means that we are adding H_3O^+. This causes the pointer to go up on the left and down on the right. The pointer on the right indicates a correspondingly smaller $[OH^-]$. If, instead, a base had been added, the situation would have been just the opposite. Figure 12-6 illustrates the relationship in an acid solution. A calculation similar to the one we made in Example 12-6 would confirm that when $[H_3O^+]$ is *increased* from 10^{-7} to 10^{-5} M, $[OH^-]$ is *decreased* from 10^{-7} to 10^{-9} M.

In summary, an acidic, basic, or neutral solution can now be defined in terms of concentrations of ions:

$$\text{Neutral} \quad [H_3O^+] = [OH^-] = 1.0 \times 10^{-7}$$

$$\text{Acidic} \quad [H_3O^+] > 1.0 \times 10^{-7} \quad \text{and} \quad [OH^-] < 1.0 \times 10^{-7}$$

$$\text{Basic} \quad [H_3O^+] < 1.0 \times 10^{-7} \quad \text{and} \quad [OH^-] > 1.0 \times 10^{-7}$$

We should now incorporate this information into our understanding of acids and bases. A modern version of the Arrhenius definition of an acid is a substance that produces H_3O^+ ions in aqueous solution. But now we see that H_3O^+ is present in neutral and even basic solutions as well. A slight modification of the definitions solves this problem. *An* **acid** *is any substance that increases $[H_3O^+]$ in water, and a* **base** *is any substance that increases $[OH^-]$ in water.* With our new understanding of the equilibrium, note that a substance can be an acid by directly donating $H^+(aq)$ to the solution (e.g., HCl, H_2S), or a substance can be an acid by reacting with OH^- ions, thus removing them from the solution. As illustrated in Figure 12-6, lowering $[OH^-]$ on the right has the effect of raising $[H_3O^+]$ on the left.

Looking Ahead ▼

The concentration of hydronium or hydroxide ions in aqueous solution is usually quite small. While scientific notation is a great help in expressing these very small numbers, it is still awkward. There is another way. This involves expressing the numbers as logarithms, which then gives us a three- or four-digit number that tells us the same thing. The expression of these numbers in this manner is discussed next.

12–5   pH SCALE

What does pH have to do with acidity?

The producers of commercial television advertising assume that the general population is not only aware of the importance of acidity but how it is scientifically expressed. We often hear references to controlled pH in hair shampoo commercials. (See Figure 12-7.) In fact, pH is an important and convenient method for expressing $[H_3O^+]$ in aqueous solution. For example, in Figure 12-6, we described an acidic solution as one where $[H_3O^+]$ is equal to 10^{-5} M. Scientific notation is certainly better than using a string of nonsignificant zeros (i.e., 0.00001 M), but expressed as pH, the number is simply 5.0. The pH scale represents the negative exponent of 10 as a positive number. The exponent of 10 in a number is the number's common logarithm or, simply, log. **pH** is a logarithmic expression of $[H_3O^+]$:

$$pH = -\log[H_3O^+]$$

Therefore, a solution of pH = 1.00 has $[H_3O^+]$ equal to 1.0×10^{-1} M and pure water has pH = 7.00 ($[H_3O^+] = 1.0 \times 10^{-7}$ M). In expressing pH, the number to the right of the decimal place should have the same number of significant figures as the original number. That is,

if $[H_3O^+] = \underset{\uparrow}{1.00} \times 10^{-4}$ M pH = $4.\underset{\uparrow}{000}$

3 significant figures   =   3 places to the right
of the decimal

A much less popular but valid way of expressing $[OH^-]$ is pOH:

$$pOH = -\log[OH^-]$$

A simple relationship between pH and pOH can be derived from the ion product of water:

$$[H_3O^+][OH^-] = 1.0 \times 10^{-14}$$

If we now take $-\log$ of both sides of the equation, we have

$$-\log[H_3O^+][OH^-] = -\log(1.0 \times 10^{-14})$$

Since $\log(A \times B) = \log A + \log B$, the equation can be written as

$$-\log[H_3O^+] - \log[OH^-] = -\log 1.0 - \log 10^{-14}$$

Since $\log 10.0 = 0$ and $\log 10^{-14} = -14$, the equation is

$$pH + pOH = 14.00$$

Figure 12–7
pH and Commercial Products.
This shampoo is supposedly desirable because it is "pH balanced."

Generally, pOH is not used extensively since pH relates to the OH⁻ concentration as well as the H_3O^+ concentration. However, the relationships among $[H_3O^+]$, $[OH^-]$, pH, and pOH can be summarized as follows:

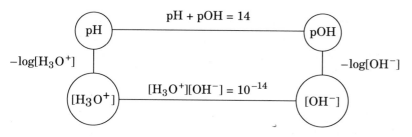

Although most of us are tempted to go straight to our calculators to change from scientific notation to logarithms, it is helpful to review the *meaning* of common logs and some of the rules of their use. You are encouraged to read Appendix D, which contains a brief discussion of common logarithms, and then proceed to Appendix F for help in using your calculator to convert $[H_3O^+]$ to pH and back again.

EXAMPLE 12–7

Expressing the pH and pOH of a Solution

What is the pH of a solution with $[H_3O^+] = 1.0 \times 10^{-11}$ *M*? What is the pOH?

● Working It Out

Solution

$$[H_3O^+] = 1.0 \times 10^{-11} \, M$$

$$pH = \underline{11.00}$$

(Since there are two significant figures, there should be two places to the right of the decimal.)

$$pH + pOH = 14.00$$

$$pOH = 14.00 - pH$$

$$pOH = 14.00 - 11.00 = \underline{3.00}$$

EXAMPLE 12–8

Expressing the pH and pOH of a Strong Acid Solution

What is the pH of a 0.015 *M* solution of $HClO_4$? What is the pOH of this solution?

Procedure

$HClO_4$ is a strong acid, which means that it is 100% ionized in solution. Therefore, $[H_3O^+]$ is equal to the original $HClO_4$ concentration. In this case, it is necessary to use a calculator or a log table to determine the pH since the number is not an integral power of ten. If you use a calculator, enter the number and press the log function. The number that is shown on the display is a negative number, so you must change the sign.

Solution

$$[H_3O^+] = 0.015\ M = 1.5 \times 10^{-2}\ M$$

$$-\log[1.5 \times 10^{-2}] = \underline{\underline{1.82}}$$

If you use a log table, the number is divided into two parts.

$$-\log[1.5 \times 10^{-2}] = -\log 1.5 - \log 10^{-2} = -0.18 + 2.00 = 1.82$$

$$pOH = 14.00 - 1.82 = \underline{\underline{12.18}}$$

EXAMPLE 12–9

Conversion of pH to [H₃O⁺]

In a given weakly basic solution, pH = 9.62, what is $[H_3O^+]$?

Procedure

In this case, it is necessary to take the antilog (or the inverse log on a calculator). Enter the 9.62 on the calculator, change the sign (+/− key), and press the inv or shift key, then the log key. The number should be rounded off to two significant figures since there are two numbers to the right of the decimal place in the original number.

Solution

$$pH = -9.62 = -\log[H_3O^+]$$

$$[H_3O^+] = \underline{\underline{2.4 \times 10^{-10}\ M}}$$

(If you use a log table, the number −9.62 is changed to 0.38 − 10, and the antilog is taken of each of these two numbers. The antilog of 0.38 is 2.4, and the antilog of −10 is 10^{-10}. $[H_3O^+] = 2.4 \times 10^{-10}\ M$. Refer to Appendix D-4.)

Acidic, basic, and neutral solutions were previously defined in terms of concentrations. We can now do the same thing in terms of pH and pOH.

Neutral   pH = pOH = 7.00

Acidic   pH < 7.00   and   pOH > 7.00

Basic   pH > 7.00   and   pOH < 7.00

Remember, a high pH (greater than 7) means basic and a low pH (less than 7) means acidic.

You've probably heard of the Richter scale for measuring earthquakes, especially if you're from California. The Richter scale measures the amplitude of the seismic waves set off by an earthquake. A reading of 7.0 indicates that the waves are ten times larger than one of magnitude 6.0. The Richter scale, like the pH scale, is a logarithmic scale. *In such a scale, a difference of one integer actually represents a tenfold change in what is being measured.* For example, a solution of pH = 4 is ten times more acidic than a solution of pH = 5. A pH = 2 solution is one thousand times more acidic than a pH = 5 solution, although both are labeled acidic. The pH values of some common substances are given in Table 12-1.

TABLE 12–1 THE pH SCALE.

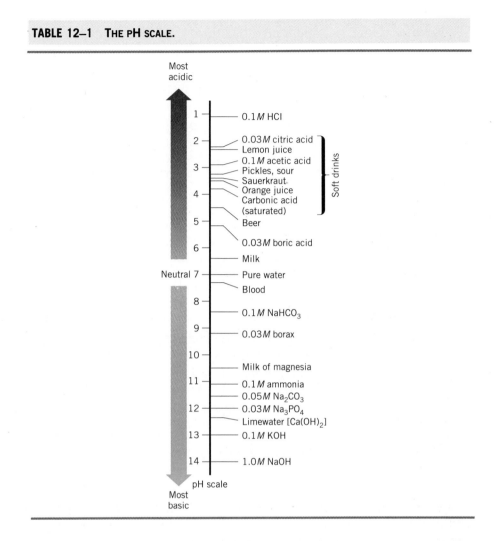

The situation in water is somewhat more complex then we originally implied. Since water autoionizes, there is an equilibrium concentration of both the active ingredient of acids [H_3O^+] and that of bases [OH^-] in the same solution. Whether a solution is acidic, neutral, or basic depends on which ion predominates. One convenient thing about this, however, is that since both ions are present, we can focus on just one to express either acidity or basicity. This fact is even more apparent in the expression of both acidity and basicity by use of one unit, pH.

▲Looking Back

Learning Check B

◀Checking It Out

B–1. Fill in the blanks.
Pure water contains a small but important concentration of _____ and _____ ions. The product of the concentrations of these two ions is known as the _____ _____ of water, is symbolized by _____, and has a value of _____ at 25°C. The pH of a solution is defined as _____. An acidic solution is one that has a [H_3O^+]

greater than _____ and a pH _____ than 7. A basic solution is one that has a $[OH^-]$ greater than _____ and a pH _____ than 7.

B–2. What is the $[H_3O^+]$ in a solution that has $[OH^-] = 7.2 \times 10^{-5}$? Is the solution acidic or basic? Is it strongly or weakly acidic or basic?

B–3. What is the pH of the solution in Problem B-2? What is the pOH?

Additional Examples: 12-26, 12-28, 12-32, 12-34, 12-36, and 12-45.

Looking Ahead ▼

So far, we have focused on electrically neutral compounds such as HCl and NaOH that act as acids or bases by formation of H_3O^+ or OH^- ions in water. In fact, acid-base behavior in water may not directly involve these two ions if a broader definition of acids and bases is employed. We do this in the following discussion.

12–6 BRØNSTED-LOWRY ACIDS AND BASES

A somewhat straightforward approach to acid-base behavior simply focuses on the role of the H^+ ion in solution. (As you will recall from Chapter 3, since the hydrogen atom is composed of a single proton and one electron, the H^+ ion is simply the representation of a proton.) *In the* **Brønsted-Lowry** *definition, an* **acid** *is a proton (H^+) donor and a* **base** *is a proton acceptor.* To illustrate this definition, we again look at the equilibrium involving the common weak base ammonia.

$$NH_3(aq) + H_2O(l) \rightleftharpoons NH_4^+(aq) + OH^-(aq)$$

Ammonia is a base by the Arrhenius definition because, by removing a proton from an H_2O molecule, *it produces OH^-.* It is also a base by the Brønsted-Lowry definition because *it accepts a H^+ from H_2O.*

Although NH_3 is a base by both definitions, the H_2O molecule plays the role of an acid in the Brønsted-Lowry sense because it donates the H^+ to NH_3.

As mentioned previously, this reaction is a reversible reaction that reaches a point of equilibrium. This means that the reverse reaction, in which a NH_4^+ ion donates a H^+ to a OH^- to form the original reactants, is also occurring. *Thus in the reverse reaction, NH_4^+ acts as an acid and OH^- as a base.*

In the Brønsted-Lowry sense, the reaction can be viewed as simply an exchange of a H^+. When the base (NH_3) reacts, it adds H^+ to form an acid (NH_4^+). When the acid (H_2O) reacts, it loses a H^+ to form a base

(OH$^-$). The NH_3, $NH_4{}^+$ and the H_2O, OH$^-$ pairs are known as **conjugate acid-base pairs.** Thus a reaction of an acid with a base is said to produce a conjugate base and acid as illustrated in the following equation (A = Brønsted-Lowry acid, B = Brønsted-Lowry base).

$$NH_3 + H_2O \rightleftharpoons NH_4{}^+ + OH^-$$
$$B_1 \quad A_2 \qquad\quad A_1 \quad B_2$$

The conjugate base of a compound or ion results from removal of an H^+. *The conjugate acid of a compound or ion results from the addition of an* H^+.

$$\text{conjugate acid} \underset{+H^+}{\overset{-H^+}{\rightleftharpoons}} \text{conjugate base}$$

For example,

$$\begin{array}{ccc} \text{Acid} & & \text{Base} \\ H_3PO_4 & \underset{+H^+}{\overset{-H^+}{\rightleftharpoons}} & H_2PO_4{}^- \end{array}$$

We have seen that bases by the Arrhenius definition, such as NH_3, are also bases in the Brønsted-Lowry sense. Arrhenius acids such as HCl also maintain their status as acids when they are considered as H^+ donors. Consider the following equation, which illustrates the acid behavior of HCl.

$$HCl + H_2O \longrightarrow H_3O^+ + Cl^-$$
$$A_1 \quad B_2 \qquad\quad A_2 \quad B_1$$

In this reaction, H_2O is a base, since it accepts a H^+ to form H_3O^+. Recall that H_2O acted as *an acid* when NH_3 is present. *A compound or ion that can either donate or accept* H^+ *ions is called* **amphiprotic.** Water is amphiprotic since it can accept H^+ ions when an acid is present or donate H^+ when a base is present. An amphiprotic substance has both a conjugate acid and a conjugate base. Examples of other amphiprotic substances include HS$^-$ and $H_2PO_4{}^-$.

Does water act as an acid or a base?

EXAMPLE 12–10

Determining Conjugate Bases of Compounds

What are the conjugate bases of **(a)** H_2SO_3 and **(b)** $H_2PO_4{}^-$?

● **Working It Out**

Procedure

$$\text{acid} - H^+ = \text{conjugate base}$$

Solution

(a) $H_2SO_3 - H^+ = \underline{HSO_3{}^-}$

(b) $H_2PO_4{}^- - H^+ = \underline{\underline{HPO_4{}^{2-}}}$

EXAMPLE 12–11

Determining Conjugate Acids of Compounds

What are the conjugate acids of (**a**) CN^- and (**b**) $H_2PO_4^-$?

Procedure

$$base + H^+ = conjugate\ acid$$

Solution

(a) $CN^- + H^+ = \underline{\underline{HCN}}$

(b) $H_2PO_4^- + H^+ = \underline{\underline{H_3PO_4}}$

EXAMPLE 12–12

Writing Acid-Base Reactions

Write the equations illustrating the following Brønsted-Lowry acid-base reactions.

(a) H_2S as an acid with H_2O
(b) $H_2PO_4^-$ as an acid with OH^-
(c) $H_2PO_4^-$ as a base with H_3O^+
(d) CN^- as a base with H_2O

Procedure

A Brønsted-Lowry acid-base reaction produces a conjugate acid and base.

Solution

	Acid		Base	$\longrightarrow$	Acid		Base
(a)	H_2S	+	H_2O	$\longrightarrow$	H_3O^+	+	HS^-
(b)	$H_2PO_4^-$	+	OH^-	$\longrightarrow$	H_2O	+	HPO_4^{2-}
(c)	H_3O^+	+	$H_2PO_4^-$	$\longrightarrow$	H_3PO_4	+	H_2O
(d)	H_2O	+	CN^-	$\longrightarrow$	HCN	+	OH^-

Note that equations (b) and (c) indicate that the $H_2PO_4^-$ is amphiprotic.

Looking Ahead ▼

In the Brønsted-Lowry sense, an acid-base reaction can be viewed as a competition for a proton between two conjugate acid-base pairs. By using some common laboratory instruments, we can quickly establish a pecking order for the acids and bases. In the next section, we will find that, by ranking acids and bases in order of their strengths, we can produce a useful predictive tool.

12–7 PREDICTING ACID AND BASE REACTIONS IN WATER

The strength of an acid is judged by its ability to donate a H^+ to a base. By choosing a reference base (H_2O) and a specific concentration of acid, we can then determine the strength of the acid by the concentration of H_3O^+ formed.

The strength of an acid is illustrated as follows:

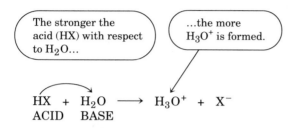

The stronger the acid (HX) with respect to H_2O...

...the more H_3O^+ is formed.

$$HX \; + \; H_2O \; \longrightarrow \; H_3O^+ \; + \; X^-$$
ACID　　BASE

The concentration of H_3O^+ (the pH) is determined directly with a common laboratory instrument called a pH meter. (See Figure 12-8.)

Table 12-2 relates the strengths of some acids and bases. The strongest acid is at the top of the table, and the weakest acid is at the bottom.

Now let us focus on the strength of the conjugate bases on the right in Table 12-2. They are also listed in order of their strength as bases with H_2O now serving as the reference acid.

The strength of a base is illustrated as follows:

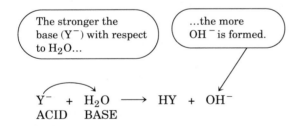

The stronger the base (Y^-) with respect to H_2O...

...the more OH^- is formed.

$$Y^- \; + \; H_2O \; \longrightarrow \; HY \; + \; OH^-$$
ACID　　BASE

In the case of the bases, however, the strongest base is at the bottom of the table and the weakest base is at the top. This is just the opposite of the acids. In other words, *the stronger the acid, the weaker is its conjugate base.* This is an inverse relationship, much like the relationship between $[H_3O^+]$ and $[OH^-]$.

The table has been divided into three groups. Group **1** at the top includes the very strong acids. *They are stronger than H_3O^+, which means that the donation of H^+ to H_2O is essentially 100% complete.* Since 100% ionization is as strong as possible, these acids all have the same strength in H_2O. (The actual rank of the very strong acids relative to each other is determined by comparison to a reference base other than H_2O.) The reaction of an acid in this group is as follows:

$$HNO_3 + H_2O \; \longrightarrow \; H_3O^+ \; + \; NO_3^- \quad \text{(100\% to the right)}$$

Figure 12–8
pH is read directly with this common laboratory instrument.

TABLE 12–2   RELATIVE STRENGTHS OF SOME ACIDS AND BASES*

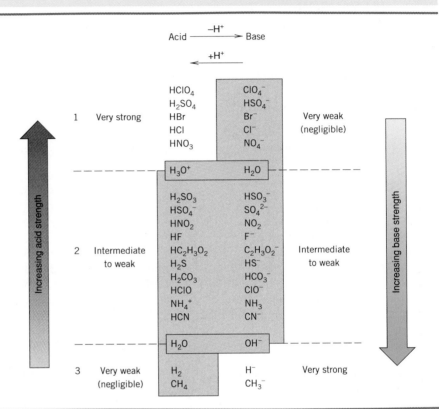

The species in the shaded region are all capable of existence in appreciable concentrations in aqueous solutions.

Notice that the stronger acid (HNO_3) reacts with the stronger base (H_2O) to produce a weaker acid (H_3O^+) and a weaker base (NO_3^-). We can now extend this observation into a generalization regarding the use of Table 12-2 in predicting the extent of a large number of acid-base reactions. If, for example, we mix a given acid and base, three things may occur.

1 The reactants may not react at all, leaving essentially all of the original reactants in solution *(negligible reaction)*.

2 The reactants may react to a slight extent, leaving mostly reactants but a small concentration of products *(limited reaction)*.

3 The reactants may react to a large extent, forming essentially all products in solution *(favorable reaction)*.

Can we predict favorable acid-base reactions?

A *favorable reaction* occurs between a reactant acid (on the left) with a reactant base (on the right) that is lower in the table than the acid. A favorable reaction implies that, when an acid and a base are mixed, essentially all reactants will be converted into products. As illustrated in Figure 12-9, we notice that a favorable reaction takes place in a clockwise direction. We will apply this observation as we look at other reactions.

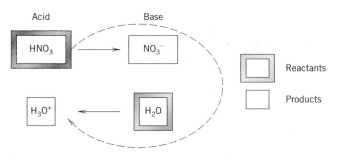

Figure 12–9
A Favorable Reaction. The stronger acid (HNO_3) reacts with the stronger base (H_2O) to form H_3O^+ and NO_3^-.

Now let's consider those acids listed in group **2**. Notice that these acids are weaker than H_3O^+. This suggests that the reaction of these acids with H_2O is not favorable. Even though this is true, a reaction does occur to a slight extent. The ionization of these acids in H_2O consti- tutes *limited reactions producing a small but significant concentration of* H_3O^+. These are the intermediate to weak acids, and the ionization is represented by equilibrium arrows. For these acids, however, the equi- librium lies quite far to the left (usually more than 95%). The reaction of HNO_2, which is in group **2**, with water is

Does an unfavorable reaction mean no reaction?

$$HNO_2 + H_2O \rightleftharpoons H_3O^+ + NO_2^- \quad \text{(Equilibrium favors the left)}$$

The acids in group **3** rank not only below the acid H_3O^+ but also be- low the acid H_2O. The ionization of these acids in water is a *negligible reaction, so their presence in water has no measurable effect on the pH.* In effect, they are acids in name only when H_2O is the reference base. The acids in group **3** are just the opposite of those in group **1**. In group **1**, the molecular form of the acid cannot exist in water since its molecules are completely ionized. In group **3**, only the molecular form can exist in water since there is no ionization. The reactions of these three groups are summarized as follows:

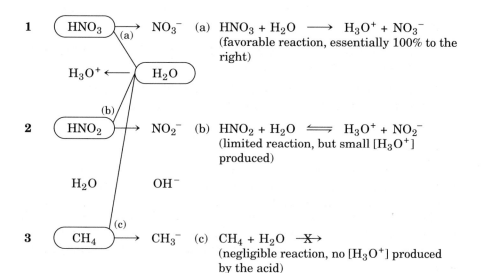

	ACID	BASE	

1 HNO₃ →(a) NO₃⁻ (a) $HNO_3 + H_2O \longrightarrow H_3O^+ + NO_3^-$
  (favorable reaction, essentially 100% to the right)

   H₃O⁺ ← H₂O

2 HNO₂ →(b) NO₂⁻ (b) $HNO_2 + H_2O \rightleftharpoons H_3O^+ + NO_2^-$
  (limited reaction, but small $[H_3O^+]$ produced)

   H₂O OH⁻

3 CH₄ →(c) CH₃⁻ (c) $CH_4 + H_2O \xrightarrow{\;\;\times\;\;}$
  (negligible reaction, no $[H_3O^+]$ produced by the acid)

Now we will consider the reaction of the bases (on the right) with H_2O as the reference acid (on the left). Once again, a favorable reaction occurs between a reactant acid that is higher in the table than the reactant base. First consider the bases in group **3**. These are very strong bases since they are the conjugate bases of very weak acids. Notice that H_2O is a stronger acid than the acids in this group, and OH^- is a weaker base than the group **3** bases. *Thus the reaction of these bases with H_2O to form OH^- is favorable and essentially 100% complete.*

$$CH_3^- + H_2O \longrightarrow CH_4 + OH^- \quad \text{(100\% to the right)}$$

The bases in group **2**, however, undergo *a limited reaction with H_2O as an acid, producing a small but significant concentration of OH^-.*

$$NO_2^- + H_2O \rightleftharpoons HNO_2 + OH^- \quad \text{(Equilibrium favors the left)}$$

Do all conjugate bases actually act as bases in water?

The bases in group **1** are very weak conjugate bases of very strong acids. *The reaction of the bases in this group with water is negligible and will not affect the pH of the solution.* The HSO_4^- ion is an exception to this, however, since this is the anion of an acid salt and is a weak acid, as you will notice by its position in group **2**. This group of bases is the counterpart of the acids in group **3**. Neither shows any acid or base activity with H_2O as the appropriate reference. These three reactions are summarized as follows:

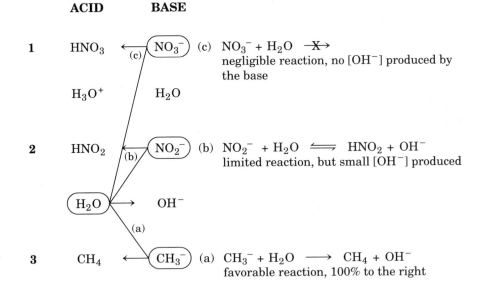

EXAMPLE 12–13

Writing Acid-Base Reactions from the Table

Referring to Table 12-2, write an equation illustrating the reaction of each of the following as either an acid or a base in water. If a partial reaction occurs, indicate with double arrows. If the species does not exhibit acid or base nature in water, write "no reaction."

(a) HBr   (b) CH_3^- (c) ClO^- (d) ClO_4^- (e) H_2 (f) HCN

Working It Out ●

Solution
(a) HBr is a very strong acid: $HBr + H_2O \rightarrow H_3O^+ + Br^-$.
(b) CH_3^- is a very strong base: $CH_3^- + H_2O \rightarrow CH_4 + OH^-$.
(c) ClO^- is a weak base: $ClO^- + H_2O \rightleftharpoons HClO + OH^-$.
(d) ClO_4^- is a very weak base: no reaction with water.
(e) H_2 is a very weak acid: no reaction with water.
(f) HCN is a weak acid: $HCN + H_2O \rightleftharpoons H_3O^+ + CN^-$.

Notice in our discussion that we have written reactions where an ion acts as a base. The reaction of ions with water is called a hydrolysis reaction. *A* **hydrolysis reaction** *is the reaction of an anion with water to produce OH^-, or the reaction of a cation with water to produce H_3O^+.*

When we dissolve a salt in water, its solution may be neutral, basic, or acidic, depending on the extent to which none, both, or one of the ions undergo hydrolysis reactions. We will consider four cases.

1 Neutral solutions of salts When neither the cation nor the anion undergoes hydrolysis reactions, the pH of the water is not affected and the solution remains neutral. These are salts that are formed from the neutralization of a strong acid and a strong base. The anions are the conjugate bases in group **1** (except HSO_4^-), and the cations are those of the alkali and alkaline earth metals (except Be^{2+}). Examples of neutral salts are NaCl, $Ba(ClO_4)_2$, and $CsNO_3$.

2 Basic solutions of salts When the salt forms from the cation of a strong base and the anion of a weak acid in group **2**, the cation does not affect the pH but the anion does. In the anion hydrolysis reaction, a small equilibrium concentration of OH^- makes the solution basic. How basic depends on where the anion ranks as a conjugate base in Table 12-2. Examples of basic salts are KCN, $Ca(NO_2)_2$, and NaF.

How can a salt solution be acidic or basic?

3 Acidic solutions of salts When a salt forms from the cation of a weak base and the anion of a strong acid, the anion does not affect the pH, but the cation does. The only example of a potentially acidic cation in group **2** that undergoes hydrolysis is NH_4^+. The hydrolysis of the ammonium ion is shown as

$NH_4^+ + H_2O \rightleftharpoons NH_3 + H_3O^+$ (Limited reaction, but produces some H_3O^+)

Since the anions do not undergo hydrolysis, solutions of salts such as NH_4Br and NH_4NO_3 are acidic.

4 Complex cases There are other salt solutions that are not easy to predict without quantitative data. For example, in a solution of the salt NH_4NO_2 both anion and cation undergo hydrolysis, but not to the same extent. In other cases, it is not immediately obvious whether solutions of acid salts are acidic or basic. For instance, the HCO_3^- ion is potentially an acid since it has a hydrogen that can react with a base. However, it is also the conjugate base of the weak acid H_2CO_3, so it can act as a base by undergoing hydrolysis. Sodium bicarbonate solutions are actually slightly basic and for that reason are sometimes used as an "antacid" to treat upset stomachs. On the other hand, solutions of the acid salts $NaHSO_3$ and $NaHSO_4$ are acidic. Quantitative information, which is discussed in Chapter 14, would be needed, however, for us to have predicted these facts.

EXAMPLE 12–14

Predicting the Acidity of Salt Solutions

Working It Out ●

Indicate whether the following solutions are acidic, basic, or neutral. If the solution is acidic or basic, write the equation illustrating the appropriate reaction.
(a) KCN
(b) $Ca(NO_3)_2$
(c) $H_2N(CH_3)_2{}^+Br^-$ [$HN(CH_3)_2$ is a weak base like NH_3.]

Solution
(a) KCN
K^+ is the cation of the strong base KOH and does not hydrolyze. The CN^- ion, however, is the conjugate base of the weak acid HCN and hydrolyzes as follows:

$$CN^- + H_2O \rightleftharpoons HCN + OH^-$$

Since OH^- is formed in this solution, the solution is <u>basic</u>.

(b) $Ca(NO_3)_2$
Ca^{2+} is the cation of the strong base $Ca(OH)_2$ and does not hydrolyze. $NO_3{}^-$ is the conjugate base of the strong acid HNO_3 and also does not hydrolyze. Since neither ion hydrolyzes, the solution is <u>neutral</u>.

(c) $H_2N(CH_3)_2{}^+Br^-$
$H_2N(CH_3)_2{}^+$ is the conjugate acid of the weak base $HN(CH_3)_2$. It undergoes hydrolysis according to the equation

$$H_2N(CH_3)_2{}^+ + H_2O \rightleftharpoons HN(CH_3)_2 + H_3O^+$$

The Br^- ion is the conjugate base of the strong acid HBr and does not hydrolyze. Since only the cation undergoes hydrolysis, the solution is <u>acidic</u>.

Do acid-base reactions always involve H_2O?

In all of the reactions discussed so far, either the H_3O^+ or the OH^- ion is produced. As such, the Brønsted-Lowry definition is simply a helpful extension of the Arrhenius definition. But there are many other H^+ exchange reactions that do not directly form the H_3O^+ ion or the OH^- ion. These reactions are more specific to the Brønsted-Lowry model. In fact, Table 12-2 can be used to predict many other H^+ exchange reactions that do not directly involve H_2O. For example, consider the following two potential reactions between the H_2SO_3, $HSO_3{}^-$ and the HF, F^- conjugate acid-base pairs. Two H^+ exchange reactions are possible.

$$HF(aq) + HSO_3{}^-(aq) \longrightarrow H_2SO_3(aq) + F^-(aq)$$

and the reverse reaction

$$H_2SO_3(aq) + F^-(aq) \longrightarrow HF(aq) + HSO_3{}^-(aq)$$

Notice from Table 12-2 that H_2SO_3 is a stronger acid than HF and that F^- is a stronger base than $HSO_3{}^-$. Therefore, we can predict that the second reaction is favorable.

EXAMPLE 12–15

Predicting Favorable Reactions Using Table 12-2

Use Table 12-2 to write equations for favorable reactions between the following conjugate acid-base pairs.
(a) HNO_2, NO_2^- and $HClO$, ClO^-
(b) HBr, Br^- and H_3O^+, H_2O

Procedure
The stronger acid (e.g., HNO_2) and the stronger conjugate base (e.g., ClO^-) are shown as reactants on the left.

Solution
(a) $HNO_2 + ClO^- \rightarrow NO_2^- + HClO$
(b) $HBr + H_2O \rightarrow H_3O^+ + Br^-$

EXAMPLE 12–16

Constructing Favorable Reactions from Conjugate Acid-Base Pairs

Use Table 12-2 to predict which of the following reactions are favorable.
(a) $HCl + HS^- \rightarrow H_2S + Cl^-$
(b) $NH_4^+ + HS^- \rightarrow H_2S + NH_3$
(c) $CH_4 + OH^- \rightarrow CH_3^- + H_2O$
(d) $HNO_2 + OH^- \rightarrow NO_2^- + H_2O$

Procedure
Consider the two acids (e.g., HCl and H_2S). If the stronger acid is on the left side of the equation, the reaction is favorable.

Solution
Reaction **(a)** and **(d)** are favorable.

● Working It Out

To any of us, the most important chemical system involves the blood coursing through our veins. Our blood has a pH of about 7.4, but a variation of as little as 0.2 pH units causes coma or even death. How is the pH of the blood so rigidly controlled despite all of the acidic and basic substances we ingest? The control of pH is the job of buffers, which are discussed next.

▼ Looking Ahead

12–8   BUFFER SOLUTIONS

The pH level is important not only in our bodies but, according to TV commercials, also in our hair shampoo. In addition, ads stress pH in counteracting the acidity produced by aspirin (which is a weak acid) and its effect on the stomach. The control of pH is accomplished by means of a buffer. *A **buffer solution** resists changes in pH caused by the addition of limited amounts of a strong acid or a strong base.*

   One type of buffer solution is a weak acid solution that also contains a salt of its conjugate base. An example would be a solution containing HCN and the salt NaCN. Notice that HCN is a weak acid listed in group **2** in Table 12-2. This means that HCN exists mostly as un-ionized

The buffer in this product controls the pH.

molecules in solution. Thus this buffer solution contains a significant concentration of HCN molecules and CN^- ions, as well as Na^+ ions (from the NaCN), and a small concentration of H_3O^+. *The Na^+ ions (the cation of a strong base) are strictly spectator ions and thus can be ignored in this discussion.* The H_3O^+ concentration is due to the very limited ionization of HCN. The CN^- concentration is mostly from the NaCN, with a very small concentration provided by the ionization of HCN. The equilibrium is illustrated as follows:

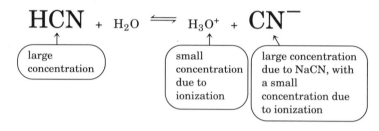

$$HCN + H_2O \rightleftharpoons H_3O^+ + CN^-$$

large concentration

small concentration due to ionization

large concentration due to NaCN, with a small concentration due to ionization

If a strong acid (H_3O^+) or a strong base (OH^-) is added, an additional reaction takes place. The HCN reacts with the added OH^- and the CN^- reacts with the added H_3O^+, as illustrated by the following equations:

(added base) $HCN + OH^- \longrightarrow H_2O + CN^-$

(added acid) $CN^- + H_3O^+ \longrightarrow HCN + H_2O$

Notice from Table 12-2 that both of these reactions are favorable (essentially complete) since the reactant acid is higher in the table than the reactant base. All buffer solutions control pH by containing one substance that reacts with H_3O^+ (e.g., CN^-) and another that reacts with OH^- (e.g., HCN). Thus addition of limited amounts of a strong acid or base is counteracted by the species present in the buffer solution, and the pH changes very little. (See Figure 12-10.) A strong acid cannot act as a buffer because it is not present in solution in molecular form. Therefore, there is no reservoir of molecules that can react with added OH^- ions.

Why must an acid be weak to be a buffer?

Consider the following analogy. One person has $20 in his pocket with no savings in the bank; another person has $20 in her pocket and $100 in the bank. A $10 expense will change the first person's pocket money drastically. The second person is "buffered" from this expense and can cover it with bank savings; this person can then maintain the same amount of pocket money. The un-ionized HCN is like "money in the bank." It is available to react with added OH^-, leaving the concentration of H_3O^+ essentially unchanged. The person who keeps all of his money in his pocket is analogous to a strong acid solution in which there is no un-ionized acid in reserve. In this case, since all of the acid ionizes to H_3O^+, any added OH^- decreases the H_3O^+ concentration and thus increases the pH.

There is a limit to how much a buffer system can resist change. If the added amount of OH^- exceeds the reserve of HCN (referred to as the *buffer capacity*), then the pH will rise. This is analogous to a $110 expense for the person with the savings. It is more than she can cover with bank savings, so the amount in her pocket decreases.

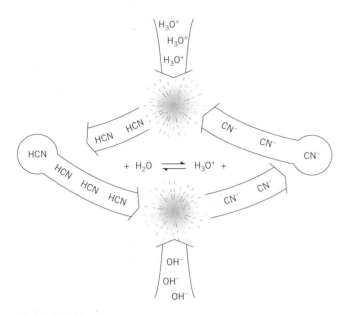

Figure 12–10
Action of a Buffer. Added OH^- is removed by the reservoir of HCN, and added H_3O^+ is removed by the reservoir of CN^-.

Solutions of weak bases and salts containing their conjugate acids (e.g., NH_3 and NH_4Cl) also serve as buffers. The two relevant reactions in this case are

$$\text{(added acid)} \qquad NH_3 + H_3O^+ \longrightarrow NH_4^+ + H_2O$$

$$\text{(added base)} \qquad NH_4^+ + OH^- \longrightarrow H_2O + NH_3$$

The pH of the buffer solution depends on the strength of the acid or base chosen. In Chapter 14, actual pH values of buffer solutions are calculated from quantitative information related to acid strength.

We began this section by mentioning the importance of the buffer system of our blood. Most of the buffering of blood is carried out by a carbonic acid-bicarbonate buffer. This equilibrium is shown as

$$\text{(1)} \quad H_2CO_3 + H_2O \rightleftharpoons HCO_3^- + H_3O^+$$

Any additional H_3O^+ that forms in the blood is removed by reaction with the bicarbonate ion.

$$\text{(added acid)} \quad \text{(2)} \quad HCO_3^- + H_3O^+ \longrightarrow H_2CO_3 + H_2O$$

As we will again mention in the next section, carbonic acid also breaks down into water and gaseous carbon dioxide.

$$\text{(3)} \quad H_2CO_3(aq) \rightleftharpoons H_2O(l) + CO_2(g)$$

The extra carbon dioxide is released in the lungs and exhaled. Strenuous exercise releases lactic acid ($HC_3H_5O_3$) from the muscles as a product of metabolism. This temporarily increases the acidity and thus the H_2CO_3 concentration (equation 2). The body reacts to this stress by increased respiration to expel the CO_2 from the lungs (equation 3). If the pH of the blood drops below 7.2, a potentially fatal condition called *acidosis* results. Acidosis may also be caused by chronic medical conditions such as diabetes.

When extra OH^- finds its way into the blood system, it reacts with the H_2CO_3.

$$\text{(added base)} \quad (4) \quad H_2CO_3 + OH^- \longrightarrow HCO_3^- + H_2O$$

How do you cure hiccups?

The extra bicarbonate ion formed is eventually expelled by the kidneys. Sometimes when we breathe too fast (hyperventilate) because of stress or panic, we inadvertently exhale too much CO_2. This reduces the concentration of H_2CO_3 (equation 3), and the reaction in equation 4 occurs to a lesser extent. Thus the blood becomes too basic. Hiccups are a symptom of high blood pH. The body responds by slower respiration for a short while in order to increase the concentration of CO_2 in the blood and increase the acidity (reverse of equation 3). If the pH of the blood increases to over 7.5, another serious condition known as *alkalosis* results.

Another buffer system is at work in the cells of our bodies. This is the $H_2PO_4^-$-HPO_4^{2-} buffer system. These two ions can react with acid or base produced in the cells as follows:

$$\text{(added acid)} \quad HPO_4^{2-} + H_3O^+ \longrightarrow H_2PO_4^- + H_2O$$

$$\text{(added base)} \quad H_2PO_4^- + OH^- \longrightarrow HPO_4^{2-} + H_2O$$

Looking Ahead ▼

We have one more important question concerning acids. How do they get into the atmosphere so as to make rain acidic? This, of course, is the question of acid rain that is a matter of international consequence. It is our final topic in this chapter.

12–9   OXIDES AS ACIDS AND BASES

One of the more important consequences of the human race's progress is acid rain. Such rain originates from exhaust gases that may be produced in one country but can be deposited in another. Probably no other issue is currently more touchy between the United States and Canada and between countries of Northern Europe than control of acid rain. Its effect on lakes and forests can be devastating, and there is little doubt that the problem must be faced and solved.

Most acid rain originates from the combustion of coal or other fossil fuels that contain sulfur as an impurity. Combustion of sulfur or sulfur compounds produces sulfur dioxide (SO_2). In the atmosphere, sulfur dioxide reacts with oxygen to form sulfur trioxide (SO_3).

$$2SO_2(g) + O_2(g) \longrightarrow 2SO_3(g)$$

The sulfur trioxide reacts with water in the atmosphere according to the equation

$$SO_3(g) + H_2O(l) \longrightarrow H_2SO_4(aq)$$

The product is a sulfuric acid solution, which is one of the strong acids—corrosive and destructive. As mentioned earlier, acids react with metals and limestone, which are both used externally in buildings. (See Figure 12-11.) In the above reaction, sulfur trioxide can be considered as simply the dehydrated form of sulfuric acid. It is thus known as an **acid anhydride,** *which means acid without water.* Many nonmetal oxides are acid anhydrides. When dissolved in water, the acid is formed. Three other reactions of nonmetal oxides to form acids follow.

$$CO_2(g) + H_2O(l) \rightleftharpoons H_2CO_3(aq) \quad \text{(carbonic acid)}$$

$$SO_2(g) + H_2O(l) \rightleftharpoons H_2SO_3(aq) \quad \text{(sulfurous acid)}$$

$$N_2O_5(l) + H_2O(l) \longrightarrow 2HNO_3(aq) \quad \text{(nitric acid)}$$

We have previously mentioned that carbon dioxide, the acid anhydride of carbonic acid, is responsible for the fizz and the tangy taste in carbonated soft drinks and beer. When all of the carbon dioxide escapes, the beverage goes flat. Carbon dioxide is also present in the atmosphere and dissolves in rain water to make rain naturally acidic. Carbon dioxide by itself lowers the pH to about 5.7. The presence of oxides of sulfur and nitrogen, however, lowers the pH to 4.0 or even lower.* The oxides of nitrogen originate mainly from engines in automobiles. The high temperature in the engine causes the two elements of air, nitrogen and oxygen, to combine to form nitrogen oxides. In eastern North America and western Europe, the acidity is about two-thirds due to sulfuric acid and the remainder due to nitric acid. There is also a small amount of hydrochloric acid in acid rain.

Ionic metal oxides dissolve in water to form bases and thus are known as base **anhydrides.** Some examples of these reactions are

$$Na_2O(s) + H_2O(l) \longrightarrow 2NaOH(aq)$$

$$CaO(s) + H_2O(l) \longrightarrow Ca(OH)_2(aq)$$

Salt is formed by the reaction between an acid anhydride and a base anhydride. For example, the following reaction produces the same salt formed in the neutralization of H_2SO_3 with $Ca(OH)_2$ in aqueous solution.

$$SO_2(g) + CaO(s) \longrightarrow CaSO_3(s)$$

$$H_2SO_3(aq) + Ca(OH)_2(aq) \longrightarrow CaSO_3(s) + 2H_2O(l)$$

The first reaction is typical of reactions that are being studied as a possible way of removing SO_2 from the combustion products of an industrial plant so that some of our abundant high-sulfur coal can be used without harming the environment.

There is a lot more to acids than sourness. Actually, any substance that donates a H^+ ion can be considered an acid. This broader definition allows us to categorize as acid-base interactions many more reactions in water besides just a neutralization reaction. In fact, some H^+ exchange reactions do not even directly involve the H_2O molecule. With a table of acids, we can predict many of these reactions and decide which solutions may act as buffers. Buffers and the resulting controlled pH allow life processes to occur.

Figure 12–11
Effect of Acid Rain The deterioration of this ancient statue is blamed on acid rain and air pollution.

Can anything be done about acid rain?

▲Looking Back

*On April 10, 1974, a rain fell on Pilochry, Scotland, that had a pH of 2.4, which is about the same as that of vinegar. This is the most acidic rain ever recorded. Nitrogen oxides formed in automobile engines also contribute to acid rain by forming nitric acid.

Learning Check C

C–1. Fill in the blanks.

In the Brønsted-Lowry definition, acids are _____ donors and bases are _____ acceptors. A conjugate acid of a compound or ion results from the _____ of a _____ ion. A substance that has both a conjugate acid and a conjugate base is said to be _____. In the table of acids and bases, the acids at the top (group **1**) are _____% ionized in water, whereas those at the bottom (group **3**) are _____% ionized. Those in group **2** are _____ ionized. The bases in group **3** react with water to the extent of _____%. The reaction of an ion with water is known as a _____ reaction. Salts composed of cations of strong bases and anions of weak acids are _____ in water. A buffer solution _____ change in pH. It is usually made up of a weak acid and a salt containing its _____ _____. Most non-metal oxides react with water to form _____ solutions, and ionic metal oxides react with water to form _____ solutions.

C–2. Write the conjugate acid and the conjugate base of the $HC_2O_4^-$ ion.

C–3. Write the reaction of the $HC_2O_4^-$ ion with (a) water acting as a base and (b) water acting as an acid.

C–4. Write the favorable reaction of the $H_2C_2O_4$-$HC_2O_4^-$ conjugate acid-base pair with the $HClO$-ClO^- pair. Oxalic acid ($H_2C_2O_4$) is slightly stronger than sulfurous acid but weaker than H_3O^+. Write the reaction of $H_2C_2O_4$ with water.

C–5. A mixture of $H_2C_2O_4$ and $NaHC_2O_4$ can act as a buffer. (a) Write the reaction that occurs when H_3O^+ is added to the solution. (b) Write the reaction when OH^- is added to the solution.

Additional Examples: 12-51, 12-54, 12-57, 12-59, 12-63, 12-67, 12-71, and 12-78.

C H A P T E R   R E V I E W

Compounds have been classified as **acids** or **bases** for hundreds of years on the basis of common sets of chemical characteristics. In this century, however, acid character has been attributed to formation of $H^+(aq)$ [also represented as the **hydronium** ion (H_3O^+)] in aqueous solution. Base character is due to the formation of $OH^-(aq)$ in solution. When acid and base solutions are mixed, the two ions combine in a **neutralization reaction** to form water. Complete neutralization of a **monoprotic acid** results in a **salt** and water (reaction **1** below). Incomplete neutralization of a **polyprotic acid** (either **diprotic** or **triprotic**) produces an **acid salt** and water (reaction **2** below).

	Acid	+ Base	$\longrightarrow$	Salt	+ Water
1	$HX(aq)$	$+ M^+OH^-(aq)$	$\longrightarrow$	$M^+X^-(aq)$	$+ H_2O(l)$
2	$H_2Y(aq)$	$+ M^+OH^-(aq)$	$\longrightarrow$	$M^+HY^-(aq)$	$+ H_2O(l)$

Acids and bases can also be classified as to strength. **Strong acids** and **strong bases** are 100% ionized in water, whereas **weak acids** and **weak bases** are only partially ionized. Partial ionization results when a reaction reaches a point of equilibrium in which both molecules (on the left of the equation) and ions (on the right) are present. For weak acids and bases, the point of equilibrium favors the left, or molecular, side of the equation. Therefore, the H_3O^+ concentration in a weak acid solution is considerably lower than in a strong acid solution at the same initial concentration of acid.

Even in pure water, there is a very small equilibrium concentration of H_3O^+ and OH^- ($1.0 \times 10^{-7} M$) due to the **autoionization** of water. The product of these concentrations is a constant known as the **ion product** of water. The ion product can be used to calculate the concentration of one ion from that of the other in any aqueous solution.

A convenient method to express the H_3O^+ or OH^- concentrations of solutions involves the use of the logarithmic definition, **pH** and **pOH**. The following are examples of solutions of various acidities in terms of $[H_3O^+]$ and pH.

Solution	$[H_3O^+]$	$[OH^-]$	pH	pOH
Strongly acidic	$>10^{-1}$	$<10^{-13}$	< 1.0	>13.0
Weakly acidic	10^{-4}	10^{-10}	4.0	10.0
Neutral	10^{-7}	10^{-7}	7.0	7.0
Weakly basic	10^{-10}	10^{-4}	10.0	4.0
Strongly basic	$<10^{-13}$	$>10^{-1}$	>13.0	< 1.0

Our understanding of acid-base behavior can be broadened somewhat by use of the **Brønsted-Lowry** definition, which defines **acids** as proton (H^+) donors and **bases** as proton acceptors. **Amphiprotic** substances can either donate or accept a proton. An acid-base reaction constitutes an H^+ exchange between an acid and a base to form a **conjugate base** and **conjugate acid,** respectively. When potential acids and bases are mixed, the stronger acid reacts with the stronger base to form a weaker acid and base. Acids can be ranked with respect to their strength as proton donors, as in Table 12-2. Because of the inverse relationship between the strengths of an acid and its conjugate base, the bases are ranked in reverse order with respect to strength as proton acceptors. Reactions that proceed essentially 100% to the right are considered favorable reactions and can be predicted from Table 12-2 and summarized on the following page.

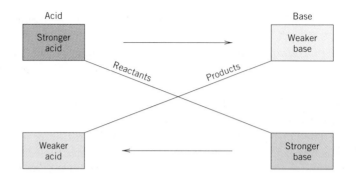

Acids and their conjugate bases are divided into three groups in Table 12-2. A reaction of a group **2** acid with H_2O is not predicted to be favorable. However, such a reaction does occur to a small extent, producing a measurable concentration of H_3O^+. Likewise, anions in this group react with water to produce a small but measurable concentration of OH^-. This latter reaction is known as **hydrolysis.**

The extent of ionization of the acids in each group and the extent of hydrolysis of their anions are summarized as follows:

Group	Acid in Water	Conjugate Base in Water	Comments
1	$HCl + H_2O \xrightarrow{100\%}$ $Cl^- + H_3O^+$	$Cl^- + H_2O \rightarrow$ no reaction	100% ionization of acid No hydrolysis of anion
2	$HF + H_2O \rightleftharpoons$ $F^- + H_3O^+$	$F^- + H_2O \rightleftharpoons$ $HF + OH^-$	Partial ionization of acid Partial hydrolysis of anion
3	$H_2 + H_2O \rightarrow$ no reaction	$H^- + H_2O \xrightarrow{100\%}$ $H_2 + OH^-$	No ionization of acid 100% hydrolysis of anion

The possibility of hydrolysis reactions means that solutions of salts may not be neutral. To predict the acidity or basicity of a solution of a salt, possible hydrolysis reactions of both cation and anion must be examined. Five possible combinations follow.

Salt	Solution	Comments
1 $NaClO_4$	Neutral	Neither cation nor anion hydrolyzes.
2 K_2S	Basic	Anion (S^{2-}) undergoes hydrolysis. $S^{2-} + H_2O \rightleftharpoons HS^- + OH^-$
3 NH_4Br	Acidic	Cation (NH_4^+) undergoes hydrolysis. $NH_4^+ + H_2O \rightleftharpoons NH_3 + H_3O^+$
4 $(NH_4)_2S$	Unable to predict	Both ions undergo hydrolysis, but more information is needed as to the extent of hydrolysis of each ion.
5 $NaHSO_3$	Unable to predict	This is an acid salt. The anion can ionize further (be acidic) or hydrolyze (be basic). More information is needed concerning the extent of these two reactions.

When a solution of a weak acid is mixed with a solution of a salt providing its conjugate base, a **buffer solution** is formed. Buffer solutions resist changes in pH from addition of limited amounts of a strong acid or base. In a buffer, the reservoir of un-ionized acid (e.g., HCN) reacts with added OH^-, while the reservoir of the conjugate base (e.g., CN^-) reacts with added H_3O^+. Weak bases and salts providing their conjugate acids also act as buffers (e.g, NH_3 and $NH_4^+Cl^-$).

Finally, the list of acids and bases was expanded to include oxides, which can be classified as **acid anhydrides** or **base anhydrides.**

The acids and bases studied in this chapter are summarized as follows:

Type	Example	Reaction
ACIDS		
1 Molecular hydrogen compounds (H^+ + ion)	$HClO_4$	$HClO_4 + H_2O \rightarrow H_3O^+ + ClO_4^-$
2 Cations (conjugate acids of weak bases)	NH_4^+	$NH_4^+ + H_2O \rightleftharpoons H_3O^+ + NH_3$
3 Nonmetal oxides	SO_3	$SO_3 + H_2O \rightarrow H_2SO_4$ $H_2SO_4 + H_2O \rightarrow H_3O^+ + HSO_4^-$
BASES		
1 Ionic hydroxides	$Ca(OH)_2$	$Ca(OH)_2 \xrightarrow{H_2O} Ca^{2+} + 2OH^-$
2 Molecular nitrogen compounds	$HN(CH_3)_2$	$HN(CH_3)_2 + H_2O \rightleftharpoons H_2N(CH_3)_2^+ + OH^-$
3 Anions (conjugate bases of weak acids)	CN^-	$CN^- + H_2O \rightleftharpoons HCN + OH^-$
4 Metal oxides	K_2O	$K_2O + H_2O \rightarrow 2KOH$ $KOH \xrightarrow{H_2O} K^+ + OH^-$

E X E R C I S E S

ACIDS AND BASES

12–1 Give the formulas and names of the acid compounds derived from the following anions.
(a) NO_3^- (c) ClO_3^-
(b) NO_2^- (d) SO_3^{2-}

12–2 Give the formulas and names of the base compounds derived from the following cations.
(a) Cs^+ (c) Al^{3+}
(b) Sr^{2+} (d) Mn^{3+}

12–3 Give the formulas and names of the acid or base compounds derived from the following ions.
(a) Ba^{2+} (d) ClO^-
(b) Se^{2-} (e) $H_2PO_4^-$
(c) Pb^{2+} (f) Fe^{3+}

12–4 Write equations illustrating the reactions with water of the acids formed in Problem 12-1.

12–5 Write equations illustrating the reactions with water of the acids and bases formed in Problem 12-3.

NEUTRALIZATION AND SALTS

12–6 Identify each of the following as an acid, base, normal salt, or acid salt.
(a) H_2S (d) $Ba(HSO_4)_2$
(b) $BaCl_2$ (e) K_2SO_4
(c) H_3AsO_4 (f) $LiOH$

12–7 Write the balanced equation showing complete neutralization of the following:
(a) KOH by $HC_2H_3O_2$
(b) $Ca(OH)_2$ by HI
(c) H_2SO_4 by $Ca(OH)_2$

12–8 Write the balanced equation showing the complete neutralization of the following:
(a) HNO_2 by $NaOH$
(b) H_2S by $CsOH$
(c) $H_2C_2O_4$ by $Ba(OH)_2$

12–9 Write the total ionic equations and the net ionic equations for reactions (b) and (c) in Problem 12-7.

12–10 Write balanced acid-base neutralization reactions that would lead to formation of the following salts or acid salts.
(a) $CaBr_2$ (c) $Ba(HS)_2$
(b) $Sr(ClO_2)_2$ (d) Li_2S

12–11 Write balanced acid-base neutralization reactions that would lead to formation of the following salts or acid salts.
(a) Na_2SO_3 (c) $Mg_3(PO_4)_2$
(b) AlI_3 (d) $NaHCO_3$

12–12 Write two equations illustrating the stepwise neutralization of H_2S with $LiOH$.

12–13 Write three equations illustrating the stepwise neutralization of H_3AsO_4. Write the total reaction.

12–14 Write the equation illustrating the reaction of 1 mol of H_2S with 1 mol of $NaOH$.

*12–15 Write the equation illustrating the reaction between 1 mol of $Ca(OH)_2$ and 2 mol of H_3PO_4.

STRENGTHS OF ACIDS AND BASES

12–16 Describe the difference between a strong acid and a weak acid.

12–17 Dimethylamine $[HN(CH_3)_2]$ is a weak base that reacts in water like ammonia (NH_3). Write the equilibrium illustrating this reaction.

12–18 The concentration of a weak monoprotic acid (HX) in water is 0.10 M. The concentration of H_3O^+ ion in this solution is 0.010 M. Is HX a weak or a strong acid? What percent of the acid is ionized?

12–19 A 0.50-mol quantity of an acid is dissolved in 2.0 L of water. In the solution, $[H_3O^+] = 0.25$ M. Is this a strong or a weak acid? Explain.

12–20 What is $[H_3O^+]$ in a 0.55 M $HClO_4$ solution?

12–21 What is $[H_3O^+]$ in a 0.55 M solution of a weak acid, HX, that is 3.0% ionized?

12–22 What is $[OH^-]$ in a 1.45 M solution of NH_3 if the NH_3 is 0.95% ionized?

*12–23 What is $[H_3O^+]$ in a 0.354 M solution of H_2SO_4? Assume that the first ionization is complete but that the second is only 25% complete.

EQUILIBRIUM OF WATER AND K_W

12–24 If some ions are present in pure water, why isn't pure water considered to be an electrolyte?

12–25 Why can't $[H_3O^+] = [OH^-] = 1.00 \times 10^{-2}$ M in water? What would happen if we tried to make such a solution by mixing 10^{-2} mol/L of KOH with 10^{-2} mol/L of HCl?

12–26 (a) What is $[H_3O^+]$ when $[OH^-] = 10^{-12}$ M?
(b) What is $[H_3O^+]$ when $[OH^-] = 10$ M?
(c) What is $[OH^-]$ when $[H_3O^+] = 2.0 \times 10^{-5}$ M?

12–27 (a) What is $[OH^-]$ when $[H_3O^+] = 1.50 \times 10^{-3}$ M?
(b) What is $[H_3O^+]$ when $[OH^-] = 2.58 \times 10^{-7}$ M?
(c) What is $[H_3O^+]$ when $[OH^-] = 56.9 \times 10^{-9}$ M?

12–28 When 0.250 mol of the strong acid $HClO_4$ is dissolved in 10.0 L of water, what is $[H_3O^+]$? What is $[OH^-]$?

12–29 Lye is a very strong base. What is $[H_3O^+]$ in a 2.55 M solution of $NaOH$? In the weakly basic household ammonia, $[OH^-] = 4.0 \times 10^{-3}$ M. What is $[H_3O^+]$?

12–30 Identify the solutions in Problem 12-26 as acidic, basic, or neutral.

12–31 Identify the solutions in Problem 12-27 as acidic, basic, or neutral.

12–32 Identify each of the following as an acidic, basic, or neutral solution.
(a) $[H_3O^+] = 6.5 \times 10^{-3} M$
(b) $[H_3O^+] = 5.5 \times 10^{-10} M$
(c) $[OH^-] = 4.5 \times 10^{-8} M$
(d) $[OH^-] = 50 \times 10^{-8} M$

12–33 Identify each of the following as an acidic, basic, or neutral solution.
(a) $[OH^-] = 8.1 \times 10^{-8} M$
(b) $[H_3O^+] = 10.0 \times 10^{-8} M$
(c) $[H_3O^+] = 4.0 \times 10^{-3} M$
(d) $[OH^-] = 55 \times 10^{-8} M$

pH AND pOH

12–34 What is the pH of the following solutions?
(a) $[H_3O^+] = 1.0 \times 10^{-6} M$
(b) $[H_3O^+] = 1.0 \times 10^{-9} M$
(c) $[OH^-] = 1.0 \times 10^{-2} M$
(d) $[OH^-] = 2.5 \times 10^{-5} M$
(e) $[H_3O^+] = 6.5 \times 10^{-11} M$

12–35 What is the pH of the following solutions?
(a) $[H_3O^+] = 1.0 \times 10^{-2} M$
(b) $[OH^-] = 1.0 \times 10^{-4} M$
(c) $[H_3O^+] = 1.0 M$
(d) $[OH^-] = 3.6 \times 10^{-9} M$
(e) $[OH^-] = 7.8 \times 10^{-4} M$
(f) $[H_3O^+] = 42.2 \times 10^{-5} M$

12–36 What is $[H_3O^+]$ when
(a) pH = 3.00 (d) pOH = 6.38
(b) pH = 3.54 (e) pH = 12.70
(c) pOH = 8.00

12–37 What is $[H_3O^+]$ when
(a) pH = 9.0 (c) pH = 2.30
(b) pOH = 9.0 (d) pH = 8.90

12–38 Identify each of the solutions in Problems 12-34 and 12-36 as acidic, basic, or neutral.

12–39 Identify each of the solutions in Problems 12-35 and 12-37 as acidic, basic, or neutral.

12–40 What is the pH of a 0.075 M solution of the strong acid HNO_3?

12–41 What is the pH of a 0.0034 M solution of the strong base KOH?

12–42 What is the pH of a 0.018 M solution of the strong base $Ca(OH)_2$?

12–43 A weak monoprotic acid is 10.0% ionized in solution. What is the pH of a 0.10 M solution of this acid?

12–44 A weak base is 5.0% ionized in solution. What is the pH of a 0.25 M solution of this base? (Assume one OH^- per formula unit.)

12–45 Identify each of the following solutions as strongly basic, weakly basic, neutral, weakly acidic, or strongly acidic.
(a) pH = 1.5 (e) pOH = 7.0
(b) pOH = 13.0 (f) pH = 8.5
(c) pH = 5.8 (g) pOH = 7.5
(d) pH = 13.0 (h) pH = −1.0

12–46 Arrange the following substances in order of increasing acidity.
(a) household ammonia, pH = 11.4
(b) vinegar, $[H_3O^+] = 2.5 \times 10^{-3} M$
(c) grape juice, $[OH^-] = 1.0 \times 10^{-10} M$
(d) sulfuric acid, pOH = 13.6
(e) eggs, pH = 7.8
(f) rain water, $[H_3O^+] = 2.0 \times 10^{-6} M$

12–47 Arrange the following substances in order of increasing acidity.
(a) lime juice, $[H_3O^+] = 6.0 \times 10^{-2} M$
(b) antacid tablet in water, $[OH^-] = 2.5 \times 10^{-6} M$
(c) coffee, pOH = 8.50
(d) stomach acid, pH = 1.8
(e) saliva, $[H_3O^+] = 2.2 \times 10^{-7} M$
(f) a soap solution, pH = 8.3
(g) a solution of lye, pOH = 1.2
(h) a banana, $[OH^-] = 4.0 \times 10^{-10} M$

*12–48 What is the pH of a 0.0010 M solution of H_2SO_4? (Assume that the first ionization is complete but the second is only 25% complete.)

BRØNSTED-LOWRY ACIDS AND BASES

12–49 What is the conjugate base of each of the following?
(a) HNO_3 (d) CH_4
(b) H_2SO_4 (e) H_2O
(c) $HPO_4{}^{2-}$ (f) NH_3

12–50 What is the conjugate acid of each of the following?
(a) NH_2CH_3 (d) O^{2-}
(b) $HPO_4{}^{2-}$ (e) HCN
(c) $NO_3{}^-$ (f) H_2O

12–51 Identify conjugate acid-base pairs in the following reactions.
(a) $HClO_4 + OH^- \rightarrow H_2O + ClO_4^-$
(b) $HSO_4^- + ClO^- \rightarrow HClO + SO_4^{2-}$
(c) $H_2O + NH_2^- \rightarrow NH_3 + OH^-$
(d) $NH_4^+ + H_2O \rightarrow NH_3 + H_3O^+$

12–52 Identify conjugate acid-base pairs in the following reactions.
(a) $HCN + H_2O \rightarrow H_3O^+ + CN^-$
(b) $HClO_4 + NO_3^- \rightarrow HNO_3 + ClO_4^-$
(c) $H_2S + NH_3 \rightarrow NH_4^+ + HS^-$
(d) $H_3O^+ + HCO_3^- \rightarrow H_2CO_3 + H_2O$

12–53 Write reactions indicating Brønsted-Lowry acid behavior with H_2O for the following. Indicate conjugate acid-base pairs.
(a) H_2SO_3 (c) HBr (e) H_2S
(b) $HClO$ (d) HSO_3^- (f) NH_4^+

12–54 Write reactions indicating Brønsted-Lowry base behavior with H_2O for the following. Indicate conjugate acid-base pairs.
(a) NH_3 (c) HS^- (e) F^-
(b) N_2H_4 (d) H^-

12–55 Most acid salts are amphiprotic, as illustrated in Example 12-12. Write equations showing how HS^- can act as a Brønsted-Lowry base with H_3O^+ and as a Brønsted-Lowry acid with OH^-.

12–56 Calcium carbonate is the principal ingredient in Tums and other products used to neutralize the H_3O^+ from excess stomach acid (HCl). Write an equation illustrating how the CO_3^{2-} ion acts as a Brønsted-Lowry base with H_3O^+. Bicarbonate of soda ($NaHCO_3$) also acts as an antacid (base) in water. Write an equation illustrating how the HCO_3^- ion reacts with H_3O^+.

PREDICTING ACID-BASE REACTIONS

12–57 Complete the following equations in which water acts as a base. If the reaction is favorable, use a single arrow. If it occurs to a limited extent, use double arrows. If it does not occur, write "no reaction."
(a) $H_2S + H_2O$ (c) $HBr + H_2O$
(b) $H_2 + H_2O$ (d) $NH_4^+ + H_2O$

12–58 Complete the following equations in which water acts as an acid. If the reaction is favorable, use a single arrow. If it occurs to a limited extent, use double arrows. If it does not occur, write "no reaction."

(a) $ClO_4^- + H_2O$
(b) $C_2H_3O_2^- + H_2O$
(c) $NH_3 + H_2O$
(d) $H^- + H_2O$

12–59 Which of the following equations represents a reaction that does not occur to a measurable extent?
(a) $HClO + OH^- \rightarrow ClO^- + H_2O$
(b) $HNO_3 + OH^- \rightarrow NO_3^- + H_2O$
(c) $H_2 + OH^- \rightarrow H^- + H_2O$
(d) $NH_4^+ + OH^- \rightarrow NH_3 + H_2O$

12–60 Which of the following equations represents a reaction that does not occur to a measurable extent?
(a) $ClO_4^- + H_3O^+ \rightarrow HClO_4 + H_2O$
(b) $HS^- + H_3O^+ \rightarrow H_2S + H_2O$
(c) $H^- + H_3O^+ \rightarrow CH_4 + H_2O$

12–61 Complete the following hydrolysis equilibria.
(a) $S^{2-} + H_2O \rightleftharpoons$ _____ $+ OH^-$
(b) $N_2H_5^+ + H_2O \rightleftharpoons N_2H_4 +$ _____
(c) $HPO_4^{2-} + H_2O \rightleftharpoons H_2PO_4^- +$ _____
(d) $H_2N(CH_3)_2^+ + H_2O \rightleftharpoons$ _____ $+ H_3O^+$

12–62 Complete the following hydrolysis equilibria.
(a) $CN^- + H_2O \rightleftharpoons HCN +$ _____
(b) $NH_4^+ + H_2O \rightleftharpoons$ _____ $+ H_3O^+$
(c) $B(OH)_4^- + H_2O \rightleftharpoons H_3BO_3 +$ _____
(d) $Al(H_2O)_6^{3+} + H_2O \rightleftharpoons Al(H_2O)_5(OH)^{2+} +$

_____

12–63 Write the hydrolysis equilibria (if any) for the following ions.
(a) F^- (d) HPO_4^{2-} (f) Li^+
(b) SO_3^{2-} (e) CN^-
(c) $H_2N(CH_3)_2^+$ [$HN(CH_3)_2$ is a weak base like ammonia.]

12–64 Write the hydrolysis equilibria (if any) for the following ions.
(a) Br^- (c) ClO_4^- (e) Ca^{2+}
(b) HS^- (d) H^-

12–65 Sodium hypochlorite ($NaClO$) is the principal ingredient in commercial bleaches. When dissolved in water, it makes a slightly basic solution. Write the equation illustrating this basic behavior.

12–66 Aqueous NaF solutions are slightly basic, whereas aqueous NaCl solutions are neutral. Write the appropriate equation that illustrates this. Why aren't NaCl solutions also basic?

12–67 Predict whether aqueous solutions of the following salts are acidic, neutral, or basic.

(a) $Ba(ClO_4)_2$
(b) $N_2H_5^+NO_3^-$ (N_2H_4 is a weak base.)
(c) $LiC_2H_3O_2$
(d) KBr
(e) NH_4Cl
(f) BaF_2

12–68 Predict whether aqueous solutions of the following salts are acidic, neutral, or basic.
(a) Na_2CO_3 (c) NH_4ClO_4
(b) K_3PO_4 (d) SrI_2

12–69 Refer to Table 12-2 to predict which of the following equations represent favorable reactions.
(a) $HClO + Cl^- \rightarrow HCl + ClO^-$
(b) $H_2S + CN^- \rightarrow HCN + HS^-$
(c) $CH_4 + H^- \rightarrow CH_3^- + H_2$
(d) $HCl + H_2O \rightarrow H_3O^+ + Cl^-$

12–70 Refer to Table 12-2 to predict which of the following equations represent favorable reactions.
(a) $H_2S + F^- \rightarrow HS^- + HF$
(b) $HNO_3 + NO_2^- \rightarrow HNO_2 + NO_3^-$
(c) $NH_4^+ + OH^- \rightarrow NH_3 + H_2O$
(d) $HC_2H_3O_2 + HCO_3^- \rightarrow H_2CO_3 + C_2H_3O_2^-$

12–71 Use Table 12-2 to write equations for favorable reactions between the following conjugate acid-base pairs.
(a) H_3O^+, H_2O and H_2S, HS^-
(b) $HClO$, ClO^- and HCN, CN^-
(c) H_2O, OH^- and H_2SO_3, HSO_3^-
(d) HBr, Br^- and CH_4, CH_3^-

12–72 Use Table 12-2 to write equations for favorable reactions between the following conjugate acid-base pairs.
(a) HCN, CN^- and H_2, H^-
(b) NH_4^+, NH_3 and H_2CO_3, HCO_3^-
(c) $HC_2H_3O_2$, $C_2H_3O_2^-$ and HCN, CN^-
(d) H_2O, OH^- and HNO_3, NO_3^-

*12–73 Both C_2^{2-} and its conjugate acid HC_2^- hydrolyze 100% in water. From this information, complete the following equation.
$CaC_2(s) + 2H_2O(l) \longrightarrow$ _____ (g)
              $+ Ca^{2+}(aq) + 2$ _____ (aq)
(The gas formed—acetylene—can be burned as it is produced. This reaction was once important for this purpose as a source of light in old miners' lamps.)

*12–74 Aqueous solutions of NH_4CN are basic. Write the two hydrolysis reactions and indicate which takes place to the greater extent.

*12–75 Aqueous solutions of $NaHSO_3$ are acidic. Write the two equations (one hydrolysis and one ionization) and indicate which takes place to the greater extent.

BUFFERS

12–76 Identify which of the following form buffer solutions when 0.50 mol of each compound is dissolved in 1 L of water.
(a) HNO_2 and KNO_2
(b) NH_4Cl and NH_3
(c) HNO_3 and KNO_2
(d) HNO_3 and KNO_3
(e) $HClO$ and $Ca(ClO)_2$
(f) HCN and $KClO$
(g) NH_3 and $BaBr_2$
(h) H_2S and $LiHS$
(i) KH_2PO_4 and K_2HPO_4

12–77 A certain solution contains dissolved HCl and NaCl. Why can't this solution act as a buffer?

12–78 Write the equilibrium involved in the N_2H_4, N_2H_5Cl buffer system (N_2H_4 is a weak base.). Write equations illustrating how this system reacts with added H_3O^+ and added OH^-.

12–79 Write the equilibrium involved in the HCO_3^-, CO_3^{2-} buffer system. Write equations illustrating how this system reacts with added H_3O^+ and added OH^-.

12–80 Write the equilibrium involved in the HPO_4^{2-}, PO_4^{3-} buffer system. Write equations illustrating how this system reacts with added H_3O^+ and added OH^-.

*12–81 If 0.5 mol of KOH is added to a solution containing 1.0 mol of $HC_2H_3O_2$, the resulting solution is a buffer. Explain.

*12–82 A solution contains 0.50 mol each of HClO and NaClO. If 0.60 mol of KOH is added, will the buffer prevent a significant change in pH? Explain.

OXIDES AS ACIDS AND BASES

12–83 Write the formula of the acid or base formed when each of the following anhydrides is dissolved in water.
(a) SrO (c) P_4O_{10} (e) N_2O_3
(b) SeO_3 (d) Cs_2O (f) Cl_2O_5

12–84   Write the formula of the acid or base formed when each of the following anhydrides is dissolved in water.
(a) BaO (c) Cl_2O (e) K_2O
(b) SeO_2 (d) Br_2O

12–85   Carbon dioxide is removed from manned space capsules by bubbling the air through a LiOH solution. Show the reaction and the product formed.

*12–86   Complete the following equation.

$$Li_2O(s) + N_2O_5(g) \longrightarrow \underline{\hspace{1cm}} (s)$$

GENERAL PROBLEMS

12–87   There are other acid-base systems based on solvents other than H_2O. One is ammonia (NH_3), which is also amphiprotic. Write equations illustrating each of the following:
(a) the reaction of HCN with NH_3 acting as a base
(b) the reaction of H^- with NH_3 acting as an acid
(c) the reaction of HCO_3^- with NH_3 acting as a base
(d) the reaction between NH_4Cl and $NaNH_2$ in ammonia

12–88   The conjugate base of methyl alcohol (CH_3OH) is CH_3O^-. Its conjugate acid is $CH_3OH_2^+$. Write equations illustrating each of the following:
(a) the reaction of HCl with methyl alcohol acting as a base
(b) the reaction of NH_2^- with methyl alcohol acting as an acid

12–89   Sulfite ion (SO_3^{2-}) and sulfur trioxide (SO_3) look similar at first glance, but one forms a strongly acidic solution whereas the other is weakly basic. Write equations illustrating this behavior.

12–90   Tell whether each of the following compounds forms acidic, basic, or neutral solutions when added to pure water. Write the equation illustrating the acidic or basic behavior where appropriate.
(a) H_2S (f) $Ba(OH)_2$
(b) KClO   (g) $Sr(NO_3)_2$
(c) NaI   (h) $LiNO_2$
(d) NH_3 (i) H_2SO_3
(e) $N_2H_5^+Br^-$ (j) Cl_2O_3

12–91   Tell whether each of the following compounds forms acidic, basic, or neutral solutions when added to pure water. Write the equation illustrating the acidic or basic behavior where appropriate.
(a) HBrO   (e) SO_2
(b) CaO   (f) $Ba(C_2H_3O_2)_2$
(c) NH_4ClO_4 (g) RbBr
(d) N_2H_4

12–92   In a lab there are five different solutions with pH's of 1.0, 5.2, 7.0, 10.2, and 13.0. The solutions are LiOH, $SrBr_2$, KClO, NH_4Cl, and HI, all at the same concentration. Which pH corresponds to which compound? What must be the concentration of all of these solutions?

12–93   When one mixes a solution of baking soda ($NaHCO_3$) with vinegar ($HC_2H_3O_2$), bubbles of gas appear. Write equations for two reactions that indicate the identity of the gas.

*12–94   High-sulfur coal contains 5.0% iron pyrite (FeS_2). When the coal is burned, the iron pyrite also burns according to the equation:

$$4FeS_2(s) + 11O_2(g) \longrightarrow 2Fe_2O_3(s) + 8SO_2(g)$$

$$2SO_2(g) + O_2(g) \longrightarrow 2SO_3(g)$$

$$SO_3(g) + H_2O(l) \longrightarrow H_2SO_4(aq)$$

What mass of sulfuric acid can eventually form from the combustion of 100 kg of coal?

*12–95   A solution is prepared by mixing 10.0 g of HCl with 10.0 g of NaOH. What is the pH of the solution if the volume is 1.00 L?

*12–96   A solution is prepared by mixing 25.0 g of H_2SO_4 with 50.0 g of KOH. What is the pH of the solution if the volume is 500 mL?

*12–97   A solution is prepared by mixing 500 mL of $0.10\ M$ HNO_3 with 500 mL of $0.10\ M$ $Ca(OH)_2$. What is the pH of the solution after mixing?

*12–98   A solution is prepared by mixing 250 mL of $0.250\ M$ $HClO_4$ with 500 mL of $0.150\ M$ KOH. What is the pH of the solution after mixing?

SOLUTIONS TO LEARNING CHECKS

A–1 H^+, H_3O^+, OH^-, hydroxide, neutralization, water, salt, acid salt, 100, partially, weak base

A–2 (a) $HClO_4$—perchloric acid (b) $Fe(OH)_2$—iron(II) hydroxide
(c) H_2S—hydrosulfuric acid (d) $LiOH$—lithium hydroxide

A–3 molecular $2HNO_3(aq) + Sr(OH)_2(aq) \longrightarrow Sr(NO_3)_2(aq) + 2H_2O(l)$

total ionic $2H^+(aq) + 2NO_3^-(aq) + Sr^{2+}(aq) + 2OH^-(aq) \longrightarrow$
$Sr^{2+}(aq) + 2NO_3^-(aq) + 2H_2O(l)$

net ionic $H^+(aq) + OH^-(aq) \longrightarrow H_2O(l)$

A–4 $H_2C_2O_4(aq) + CsOH(aq) \longrightarrow CsHC_2O_4(aq) + H_2O(l)$

A–5 This is a weak acid solution.

$$0.10 \times 0.10\ M = 0.010\ M = [H_3O^+]$$

B–1 H_3O^+, OH^-, ion product, K_W, 1.0×10^{-14}, $-\log[H_3O^+]$, 10^{-7}, less, 10^{-7}, greater

B–2 $[H_3O^+] = K_W/[OH^-] = 1.0 \times 10^{-14}/7.2 \times 10^{-5} = \underline{\underline{1.4 \times 10^{-10}}}$
The solution is weakly basic.

B–3 $pH = -\log[1.4 \times 10^{-10}] = 9.85$ $pOH = 14.00 - 9.85 = \underline{\underline{4.15}}$

C–1 H^+, H^+, addition, H^+, amphiprotic, 100, 0, partially, 100, hydrolysis, basic, resists, conjugate base, acidic, basic

C–2 conjugate acid conjugate base
$H_2C_2O_4 \longleftarrow HC_2O_4^- \longrightarrow C_2O_4^{2-}$

C–3 (a) Base
$HC_2O_4^- + H_2O \rightleftharpoons C_2O_4^{2-} + H_3O^+$
(b) Acid
$HC_2O_4^- + H_2O \rightleftharpoons H_2C_2O_4 + OH^-$

C–4 $H_2C_2O_4 + ClO^- \longrightarrow HClO + HC_2O_4^-$
$H_2C_2O_4 + H_2O \rightleftharpoons H_3O^+ + HC_2O_4^-$

C–5 (a) $HC_2O_4^- + H_3O^+ \longrightarrow H_2C_2O_4 + H_2O$
(b) $H_2C_2O_4 + OH^- \longrightarrow HC_2O_4^- + H_2O$

Lightning is a spectacular flow of electricity. On a more modest scale, we can use redox reactions to generate electricity.

OXIDATION-REDUCTION REACTIONS

◀**Setting The Stage**

A distant rumble signals the ominous gathering of thunderstorms. We may cast a cautious eye towards the sky and think of shelter. The roll of thunder warns us about one force of nature for which we have great respect, so we try to get out of its way. That, of course, is lightning. Lightning has no doubt caused fear as well as amazement in the human race since people first looked to the sky and wondered about its nature. But this force was not harnessed until modern times. The use of electricity (the same force as lightning) is so common to us now that it is taken for granted. Huge generating plants dot the rural landscape with towering smokestacks discharging smoke and steam. Not many decades ago, however, electricity was mainly a laboratory curiosity, until the experiments of inventors such as James Watt, Alexander Graham Bell, and Thomas Edison tapped its limitless potential. The electricity used by these early investigators originated from chemical reactions. Even now, when we turn on a calculator or a flashlight or start a car, a specific chemical reaction in a battery causes a flow of electrons that is the source of electricity in the battery. Because one reactant has a greater affinity for electrons than the other, the exchange of electrons takes place spontaneously. This type of reaction is much like the acid-base reactions discussed in the preceding chapter that are favorable because one reactant has a greater affinity for a H^+ ion than the other. Electron exchanges have not previously been defined as a specific type of reaction, but many of the classifications discussed in Chapter 8 fit into this category. Most combination, all combustion, and all single-replacement reactions can also be categorized as electron exchange reactions.

Formulating Some Questions ▶

There are many practical questions that will be discussed in this chapter. First of all, how do we keep track of electrons in these special reactions? What are the terms we use? How does a battery produce electricity? Can we predict favorable electron exchange reactions like we did for acid-base reactions? And finally, can we make unfavorable reactions occur? Our first topic in this chapter concerns how we do "electron bookkeeping."

13–1   OXIDATION STATES

How do you tell which element loses electrons?

In Chapter 6, we discussed a simple but important combination reaction involving the formation of sodium chloride from elemental sodium and chlorine. We traced the loss of one electron from a sodium atom and the gain of that electron by a chlorine atom. The gain and loss of an electron was predicted by the octet rule. As we will soon notice, the exchange of electrons in other reactions is not as easy to see. To aid in following this exchange, it is helpful to have a method of electron bookkeeping. Such a system is known as the oxidation state of an element in a compound. *The* **oxidation state (or oxidation number)** *of an atom in a compound is the charge that atom would have if all atoms were present as monatomic ions.* In other words, all of the electrons in a molecule or ion are assigned to specific atoms. The assignment of oxidation states is not complicated if we apply a few rules and practice on a few examples. The rules for assigning oxidation states are as follows:

1   The oxidation state of an element in its free state is zero [e.g., Cu (s), B (s)]. This includes polyatomic elements [e.g., Cl_2 (g), P_4 (s)].

2   The oxidation state of a monatomic ion is the same as the charge on that ion (e.g., $Na^+ = +1$ oxidation state, $O^{2-} = -2$, $Al^{3+} = +3$).
(a) Alkali metal ions are always $+1$.
(b) Alkaline earth metal ions are always $+2$.

3   The halogens are in a -1 oxidation state in binary (two-element) compounds, whether ionic or covalent, *when bound to a less electronegative element.*

4   Oxygen in a compound is usually -2. Certain compounds (which are rare) called peroxides or superoxides contain O in a lower negative oxidation state. Oxygen is positive when bound to F.

5   Hydrogen in a compound is usually $+1$. When combined with a less electronegative element, usually a metal, H has a -1 oxidation state (e.g., LiH).

6   The sum of the oxidation states of all of the atoms in a neutral compound is zero. For polyatomic ions, the sum of the oxidation states equals the charge on the ion.

The following examples illustrate the assignment of oxidation states.

EXAMPLE 13–1

Calculation of Oxidation States

What is the oxidation state of the following?
(a) Fe in FeO   (c) S in H_2SO_3
(b) N in N_2O_5 (d) As in AsO_4^{3-}

●Working It Out

Procedure
An algebra equation can be constructed from rule 6. For example, assume that we have a hypothetical compound M_2A_3. Then from rule 6

$$[2 \times (\text{ox. state M})] + [3 \times (\text{ox. state A})] = 0, \text{ or simply}$$

$$2M + 3A = 0$$

If the formula represents a polyatomic ion, the quantity on the left is equal to the charge.

Solution
(a) FeO
The oxidation states of the two elements add to zero (rule 6).

$$(\text{ox. state Fe}) + (\text{ox. state O}) = 0$$

Since O is -2 (rule 4),

$$Fe + (-2) = 0$$
$$Fe = +2 \quad \text{(This is an ionic compound.)}$$

(b) N_2O_5
The oxidation states add to zero (rule 6), as shown by the equation

$$2(\text{ox. state N}) + 5(\text{ox. state O}) = 0$$

Since O is -2 (rule 4),

$$2N + 5(-2) = 0$$
$$2N = +10$$
$$N = +5 \quad \text{(This is a molecular compound.)}$$

(c) H_2SO_3
The oxidation states add to zero (rule 6).

$$2(\text{ox. state H}) + (\text{ox. state S}) + 3(\text{ox. state O}) = 0$$

H is usually $+1$ and O is usually -2 (rules 4 and 5).

$$2(+1) + S + 3(-2) = 0$$
$$S = +4$$

(d) AsO_4^{3-}
The sum of the oxidation states of the atoms equals the charge on the ion (rule 6).

$$(\text{ox. state As}) + 4(\text{ox. state O}) = -3$$

Since O is -2 (rule 4),

$$As + 4(-2) = -3$$
$$As = +5$$

We now have an important tool. By noting the change of oxidation state of the same atom in going from a reactant to a product, we can trace the exchange of electrons in the reaction. We will apply oxidation

▼Looking Ahead

state and define the terms used in electron exchange reactions in the next section.

13-2 THE NATURE OF OXIDATION AND REDUCTION

A dramatic chemical demonstration involves the electron exchange reaction that leads to the formation of ordinary table salt from its elements. When a small chunk of sodium metal is inserted into a closed jar containing chlorine gas, spectacular flames leap from the sodium. Almost immediately, a white coating of sodium chloride begins to deposit on the walls. (See Figure 13-1.) Two dangerous elements—one a reactive, solid metal; the other a toxic, gaseous nonmetal—combine to form a compound that is necessary for our good health. This reaction is represented by the equation

$$2Na(s) + Cl_2(g) \longrightarrow 2NaCl(s)$$

In the precipitation and acid-base reactions discussed so far, the atoms in the reactants kept their quota of electrons in changing to products. Such is not the case in the reaction shown above, however. This is best illustrated by taking the reaction apart and examining the change that each reactant undergoes as the product is formed. An electron exchange reaction can be viewed as the sum of two half-reactions. A **half-reaction** *illustrates either the loss of electrons or the gain of electrons as a separate balanced equation.* Thus the half-reaction involving only sodium is

$$Na \longrightarrow Na^+ + e^-$$

Recall that the oxidation state of an element is zero and that of a monatomic ion is the charge of the ion. In the half-reaction, note that sodium has changed from zero to +1 and, in so doing, has lost one electron. *A substance that loses electrons in a chemical reaction (as indicated by an increase in oxidation state) is said to be* **oxidized.** Originally, the term "oxidation" referred only to a substance that added oxygen atoms in a reaction. But since this meant that the substance adding oxygen underwent an increase in oxidation state, the process of oxidation now refers to any substance containing an element undergoing an increase in oxidation state. This is so regardless of whether oxygen is even involved in the reaction.

Now let's consider what happens to the chlorine in going from reactant to product.

$$2e^- + Cl_2 \longrightarrow 2Cl^-$$

In this half-reaction, note that the oxidation state of the chlorine has decreased from zero to -1 for each chloride ion formed. Two electrons are gained by the elemental chlorine in the reaction. *A substance that gains electrons in a chemical reaction (as indicated by a decrease in oxidation state) is said to be* **reduced.** Originally, the term "reduction" referred to a process in which oxygens were removed from a substance, leaving its mass "reduced." Now, any substance containing an element undergoing

Figure 13-1
Formation of NaCl.
An active metal reacts with a poisonous gas to form ordinary table salt (NaCl).

a decrease in oxidation state is said to be reduced or to undergo reduction.

Obviously, the two processes (oxidation and reduction) complement each other, giving us the basis of this classification of chemical reactions. *Reactions involving an exchange of electrons are known as* **oxidation–reduction** *or, simply,* **redox** *reactions.*

Can oxidation take place without reduction?

Note that the reactant that is oxidized (Na) provides the electrons for the reactant reduced. *The reactant oxidized is therefore referred to as the* **reducing agent.** Conversely, the reactant reduced (Cl_2) accepts the electrons from the reactant oxidized. *The reactant reduced is referred to as the* **oxidizing agent.**

The reaction of Na and Cl_2 is summarized as follows:

Reactant	Change	Agent	Product
Na	Oxidation	Reducing	Na^+
Cl_2	Reduction	Oxidizing	Cl^-

Let's now see how the two half-reactions add together to make a complete, balanced equation. *An important principle of a redox reaction is that the electrons gained in the reduction process must equal the electrons lost in the oxidation process.* Note in our sample reaction that the reduction process involving Cl_2 requires two electrons. Therefore, the oxidation process must involve two Na's to provide these two electrons. In adding the two half-reactions, the electrons on both sides of the equation must be equal so that they can be eliminated by subtraction.

How are electrons "balanced"?

Oxidation half-reaction: $2Na \longrightarrow 2Na^+ + 2e^-$

Reduction half-reaction: $2e^- + Cl_2 \longrightarrow 2Cl^-$

Total reaction: $2Na + Cl_2 \longrightarrow 2Na^+ + 2Cl^-$ (or 2NaCl)

In many reactions, the substance undergoing a change is not as obvious as in the sample reaction we discussed above. In most compounds, usually only one of the elements undergoes a change. Thus it is often necessary to calculate the oxidation states of all of the elements in all compounds so that we can see which ones have undergone the change. With experience, however, the oxidized and reduced species are more easily recognized. In the following examples, we will find all the oxidation states so that we can identify the changes and label them appropriately.

EXAMPLE 13–2

Identification of Oxidized and Reduced Species

In the following unbalanced equations, indicate the reactant oxidized, the reactant reduced, the oxidizing agent, and the reducing agent. Indicate the products that contain the elements that were oxidized or reduced.
(a) $Al + HCl \rightarrow AlCl_3 + H_2$
(b) $CH_4 + O_2 \rightarrow CO_2 + H_2O$

●**Working It Out**

(c) $MnO_2 + HCl \rightarrow MnCl_2 + Cl_2 + H_2O$

(d) $K_2Cr_2O_7 + SnCl_2 + HCl \rightarrow CrCl_3 + SnCl_4 + KCl + H_2O$

Procedure

In the equations, we wish to identify the species that contain atoms of an element undergoing a change in oxidation state. At first, it may be necessary to calculate the oxidation state of every atom in the equation until you can recognize the changes by inspection. You will notice that any substance present as a free element is involved in either the oxidation or the reduction process, and that hydrogen and oxygen are generally not oxidized or reduced unless they are present as free elements.

Solution

(a) Oxidation state of element:

$$\overset{0}{Al} + \overset{+1-1}{HCl} \longrightarrow \overset{+3-1}{AlCl_3} + \overset{0}{H_2}$$

Reactant	Change	Agent	Product
Al	Oxidation	Reducing	$AlCl_3$
HCl	Reduction	Oxidizing	H_2

(b) Oxidation state of element:

$$\overset{-4+1}{CH_4} + \overset{0}{O_2} \longrightarrow \overset{+4-2}{CO_2} + \overset{+1-2}{H_2O}$$

Reactant	Change	Agent	Product
CH_4	Oxidation	Reducing	CO_2
O_2	Reduction	Oxidizing	CO_2, H_2O

(c) Oxidation state of element:

$$\overset{+4-2}{MnO_2} + \overset{+1-1}{HCl} \longrightarrow \overset{+2-1}{MnCl_2} + \overset{0}{Cl_2} + \overset{+1-2}{H_2O}$$

Reactant	Change	Agent	Product
HCl	Oxidation	Reducing	Cl_2
MnO_2	Reduction	Oxidizing	$MnCl_2$

(d) Oxidation state of element:

$$\overset{+1+6-2}{K_2Cr_2O_7} + \overset{+2-1}{SnCl_2} + \overset{+1-1}{HCl} \longrightarrow \overset{+3-1}{CrCl_3} + \overset{+4-1}{SnCl_4} + \overset{+1-2}{H_2O}$$

Reactant	Change	Agent	Product
$SnCl_2$	Oxidation	Reducing	$SnCl_4$
$K_2Cr_2O_7$	Reduction	Oxidizing	$CrCl_3$

An important principle of redox reactions is that "electrons lost equal electrons gained." In the next two sections, we will put this concept to use as the key to balancing some rather complex equations. As you will notice, balancing these equations by inspection, as discussed in Chapter 6, would be some chore indeed. There are two widely used procedures for balancing redox equations. The oxidation state method, discussed in the next section, treats all atoms in the compounds as if they were ions. The second method, described in a later section, focuses on the whole molecule or ion undergoing a change.

▼ **Looking Ahead**

13–3 BALANCING REDOX EQUATIONS: OXIDATION STATE METHOD

*The **oxidation state** or **bridge method** focuses on the atoms of the elements undergoing a change in oxidation state.* This method can serve as a helpful introduction to the concepts involved and is effective in balancing uncomplicated redox equations.

The following reaction will be used to illustrate the procedures for balancing equations by the oxidation state method:

$$HNO_3(aq) + H_2S(aq) \longrightarrow NO(g) + S(s) + H_2O(l)$$

1 Identify the atoms whose oxidation states have changed.

$$\overset{+5}{HNO_3} + \overset{-2}{H_2S} \longrightarrow \overset{+2}{NO} + \overset{0}{S} + H_2O$$

2 Draw a bridge between the same atoms whose oxidation states have changed, indicating the electrons gained or lost. This is the change in oxidation state.

$$\begin{array}{c} +3e^- \\ \overset{+5}{HNO_3} + H_2S \longrightarrow \overset{+2}{NO} + S + H_2O \\ -2 \qquad\qquad 0 \\ -2e^- \end{array}$$

3 Multiply the two numbers ($+3$ and -2) by whole numbers that produce a common number. For 3 and 2 the common number is 6. (For example, $+3 \times \underline{2} = +6$; $-2 \times \underline{3} = -6$.) Use these multipliers as coefficients of the respective compounds or elements.

$$\begin{array}{c} +3e^- \times \textcircled{2} = +6e^- \\ \underline{2}HNO_3 + \underline{3}H_2S \longrightarrow \underline{2}NO + \underline{3}S + H_2O \\ -2e^- \times \textcircled{3} = -6e^- \end{array}$$

Note that six electrons are lost (bottom) and six are gained (top).

4 Balance the rest of the equation by inspection. Note that there are eight H's on the left, so *four* H_2O's are needed on the right. If the equation has been balanced correctly, the O's should balance. Note that they do.

$$2HNO_3 + 3H_2S \longrightarrow 2NO + 3S + 4H_2O$$

Working It Out ●

Silver metal forms a coating on zinc when it is immersed in a silver nitrate solution.

When a copper penny reacts with nitric acid, nitrogen dioxide gas (brown) is formed.

EXAMPLE 13–3

Balancing Equations by the Oxidation State Method

Balance the following equations by the oxidation state method.
(a) $Zn + AgNO_3 \rightarrow Zn(NO_3)_2 + Ag$

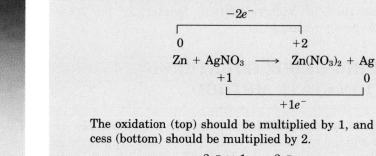

$$\overset{\displaystyle -2e^-}{\overbrace{}}$$

$$\underset{+1}{\overset{0}{Zn}} + AgNO_3 \longrightarrow \underset{}{\overset{+2}{Zn(NO_3)_2}} + \underset{0}{Ag}$$

$$\underset{+1e^-}{\underbrace{}}$$

The oxidation (top) should be multiplied by 1, and the reduction process (bottom) should be multiplied by 2.

$$\overset{-2e^- \times 1 = -2e^-}{\overbrace{}}$$

$$Zn + 2AgNO_3 \longrightarrow Zn(NO_3)_2 + 2Ag$$

$$\underset{+1e^- \times 2 = +2e^-}{\underbrace{}}$$

The final balanced equation is

$$Zn + 2AgNO_3 \longrightarrow Zn(NO_3)_2 + 2Ag$$

(b) $Cu + HNO_3 \rightarrow Cu(NO_3)_2 + H_2O + NO_2$

$$\overset{-2e^- \times 1 = -2e^-}{\overbrace{}}$$

$$\underset{+5}{\overset{0}{Cu}} + HNO_3 \longrightarrow \underset{}{\overset{+2}{Cu(NO_3)_2}} + H_2O + \underset{+4}{NO_2}$$

$$\underset{+1e^- \times 2 = +2e^-}{\underbrace{}}$$

The equation, so far, is

$$Cu + 2HNO_3 \longrightarrow Cu(NO_3)_2 + H_2O + 2NO_2$$

Note, however, that four N's are present on the right, but only two are on the left. The addition of two more HNO_3's balances the N's, and the equation is completely balanced with two H_2O's on the right.

$$Cu + 4HNO_3 \longrightarrow Cu(NO_3)_2 + 2H_2O + 2NO_2$$

(In this aqueous reaction, HNO_3 serves two functions. Two HNO_3's are reduced to two NO_2's, and the other two HNO_3's provide anions for the Cu^{2+} ion. These latter NO_3^- ions are present in the solution as spectator ions. Spectator ions are not oxidized, reduced, or otherwise changed during the reaction.)

(c) $O_2 + HI \rightarrow H_2O + I_2$

The elements undergoing a change in oxidation state are oxygen and iodine. *If an atom that has changed is in a compound where it has a subscript other than one, first balance these atoms by adding a temporary coefficient.*
(e.g., $2HI \rightarrow I_2$ and $O_2 \rightarrow 2H_2O$)

$$2(-1e^-) \times 2 = -4e^-$$

$$\overset{-1}{O_2 + 2HI} \longrightarrow \overset{}{2H_2O} + \overset{0}{I_2}$$
$$\overset{0}{} \qquad\qquad \overset{-2}{}$$

$$2(+2e^-) \times 1 = +4e^-$$

$$O_2 + 4HI \longrightarrow 2H_2O + 2I_2$$

Some of the species in this section do not actually exist as ions in the oxidation states shown (e.g., N^{5+}). Another method for balancing equations is more realistic because it looks at the changes involved in a whole molecule or ion. This is the ion-electron method, which is discussed next.

▼**Looking Ahead**

13–4   BALANCING REDOX EQUATIONS: ION-ELECTRON METHOD

In the **ion-electron method** *(also known as* **the half-reaction method**)*, the total reaction is separated into half-reactions, which are balanced individually and then added.* The ion-electron method recognizes the complete change of an ion or molecule as it changes from reactant to products. Balanced half-reactions are also represented in this manner in discussions of batteries later in this chapter.

The rules for balancing equations are somewhat different in acidic solution [containing $H^+(aq)$ ion] than in basic solution [containing $OH^-(aq)$ ion]. The two solutions are considered separately, with acid solution reactions discussed first. To simplify the equations, only the net ionic equations are balanced.

The balancing of an equation in aqueous acid solution is illustrated by the following unbalanced equation.

What is the ion that always shows up in an acidic solution? In a basic solution?

$$Cr_2O_7^{2-}(aq) + Cl^-(aq) + H^+(aq) \longrightarrow Cr^{3+}(aq + Cl_2(g) + H_2O$$

1 Separate the molecule or ion that contains atoms of an element that has been oxidized or reduced and the product containing atoms of that element. If necessary, calculate the oxidation states of individual atoms until you are able to recognize the species that changes. In this method, it is actually not necessary to know the oxidation state. The reduction process is

$$Cr_2O_7^{2-} \longrightarrow Cr^{3+}$$

2 If a subscript of the atoms of the element undergoing a change in oxidation state is more than one, balance those atoms with a temporary coefficient. In this case, it is the Cr.

$$Cr_2O_7^{2-} \longrightarrow 2Cr^{3+}$$

3 Balance the oxygens by adding H_2O on the side needing the oxygens (one H_2O for each O needed).

$$Cr_2O_7^{2-} \longrightarrow 2Cr^{3+} + 7H_2O$$

4 Balance the hydrogens by adding H^+ on the other side of the equation from the H_2Os ($2H^+$ for each H_2O added). Note that the H and O have not undergone a change in oxidation state.

$$14H^+ + Cr_2O_7^{2-} \longrightarrow 2Cr^{3+} + 7H_2O$$

5 The atoms in the half-reaction are now balanced. Check to make sure. The charge on both sides of the reaction must now be balanced. To do this, add the appropriate number of electrons to the *more positive* side. The total charge on the left is $(14 \times +1) + (-2) = +12$. The total charge on the right is $(2 \times +3) = +6$. By adding $6e^-$ on the left, the charges balance on both sides, and the half-reaction is balanced.

$$6e^- + 14H^+ + Cr_2O_7^{2-} \longrightarrow 2Cr^{3+} + 7H_2O$$

6 Repeat the same procedure for the other half-reaction.

$$Cl^- \longrightarrow Cl_2$$
$$2Cl^- \longrightarrow Cl_2$$
$$2Cl^- \longrightarrow Cl_2 + 2e^-$$

7 Before the two half-reactions are added, we must make sure that electrons gained equal electrons lost. Sometimes, the half-reactions must be multiplied by factors that give the same number of electrons. In this case, if the oxidation process is multiplied by 3 (and the reduction process by 1), there will be an exchange of $6e^-$. When these two half-reactions are added, the $6e^-$ can be subtracted from both sides of the equation.

$$3[2Cl^- \longrightarrow Cl_2 + 2e^-]$$
$$6Cl^- \longrightarrow 3Cl_2 + 6e^-$$

Addition produces the balanced net ionic equation.

$$6e^- + 14H^+ + Cr_2O_7^{2-} \longrightarrow 2Cr^{3+} + 7H_2O$$
$$6Cl^- \longrightarrow 3Cl_2 + 6e^-$$

$$14H^+(aq) + 6Cl^-(aq + Cr_2O_7^{2-}(aq) \longrightarrow 2Cr^{3+}(aq) + 3Cl_2(g) + 7H_2O(l)$$

EXAMPLE 13–4

Balancing Redox Equations in Acid Solution

Balance the following equations for reactions occurring in acid solution by the ion-electron method.
(a) $MnO_4^-(aq) + SO_2(g) + H_2O(l) \rightarrow Mn^{2+}(aq) + SO_4^{2-}(aq) + H^+(aq)$

Reduction:	$MnO_4^- \longrightarrow Mn^{2+}$
H_2O:	$MnO_4^- \longrightarrow Mn^{2+} + 4H_2O$
H^+:	$8H^+ + MnO_4^- \longrightarrow Mn^{2+} + 4H_2O$
e^-:	$5e^- + 8H^+ + MnO_4^- \longrightarrow Mn^{2+} + 4H_2O$
Oxidation:	$SO_2 \longrightarrow SO_4^{2-}$
H_2O:	$2H_2O + SO_2 \longrightarrow SO_4^{2-}$
H^+:	$2H_2O + SO_2 \longrightarrow SO_4^{2-} + 4H^+$
e^-:	$2H_2O + SO_2 \longrightarrow SO_4^{2-} + 4H^+ + 2e^-$

The reduction reaction is multiplied by 2 and the oxidation by 5 to produce 10 electrons for each process as shown below.

$$2(5e^- + 8H^+ + MnO_4^- \longrightarrow Mn^{2+} + 4H_2O)$$

$$5(2H_2O + SO_2 \longrightarrow SO_4^{2-} + 4H^+ + 2e^-)$$

$$\cancel{10e^-} + 16H^+ + 2MnO_4^- \longrightarrow 2Mn^{2+} + 8H_2O$$

$$10H_2O + 5SO_2 \longrightarrow 5SO_4^{2-} + 20 H^+ + \cancel{10e^-}$$

$$10H_2O + 16H^+ + 5SO_2 + 2MnO_4^- \longrightarrow$$
$$5SO_4^{2-} + 2Mn^{2+} + 8H_2O + 20 H^+$$

Note that H_2O and H^+ are present on both sides of the equation. Therefore, $8H_2O$ and $16H^+$ can be subtracted from *both sides,* leaving the final balanced net ionic equation:

$$2MnO_4^-(aq) + 5SO_2(g) + 2H_2O(l) \longrightarrow$$
$$2Mn^{2+}(aq) + 5SO_4^{2-}(aq) + 4H^+(aq)$$

(b) $Cu(s) + NO_3^-(aq) \rightarrow Cu^{2+}(aq) + H_2O + NO(g)$

Reduction:	$NO_3^- \longrightarrow NO$
H_2O:	$NO_3^- \longrightarrow NO + 2H_2O$
H^+:	$4H^+ + NO_3^- \longrightarrow NO + 2H_2O$
e^-:	$3e^- + 4H^+ + NO_3^- \longrightarrow NO + 2H_2O$
Oxidation:	$Cu \longrightarrow Cu^{2+}$
e^-:	$Cu \longrightarrow Cu^{2+} + 2e^-$

Multiply the reduction half-reaction by 2 and the oxidation half-reaction by 3, and then add the two half-reactions:

$$\cancel{6e^-} + 8H^+ + 2NO_3^- \longrightarrow 2NO + 4H_2O$$

$$3Cu \longrightarrow 3Cu^{2+} + \cancel{6e^-}$$

$$8H^+(aq) + 2NO_3^-(aq) + 3Cu(s) \longrightarrow 3Cu^{2+}(aq) + 2NO(g) + 4H_2O(l)$$

In a basic solution, OH^- ion is predominant rather than H^+. Therefore, the procedure is adjusted to allow the half-reactions to be balanced with OH^- ions and H_2O molecules in the basic solution.

The balancing of an equation in aqueous base solution is illustrated by the following unbalanced equation.

$$MnO_4^-(aq) + C_2O_4^{2-}(aq) + OH^-(aq) \longrightarrow MnO_2(s) + CO_3^{2-}(aq) + H_2O(l)$$

1 Separate the molecule or ion that contains an atom that has been oxidized or reduced and the product containing that atom. The reduction process is

$$MnO_4^- \longrightarrow MnO_2$$

2 For every oxygen needed on the oxygen-deficient side, add one OH^- ion.

$$MnO_4^- \longrightarrow MnO_2 + 2\ OH^-$$

3 To balance H, add one OH^- for each H in either OH^- or in any other compound, and then add one H_2O to the other side for each additional OH^- added in this step.

$$\underline{2H_2O} + MnO_4^- \longrightarrow MnO_2 + 2OH^- + \underline{2\ OH^-}$$

$$2H_2O + MnO_4^- \longrightarrow MnO_2 + 4OH^-$$

An example of a half-reaction where hydrogens are present in a compound is

$$NO_3^- \longrightarrow NH_3$$

(balance O) $\quad NO_3^- \longrightarrow NH_3 + 3OH^-$

(balance H) $\quad \underline{6H_2O} + NO_3^- \longrightarrow NH_3 + 3OH^- + \underline{6OH^-}$

4 Balance the charge by adding electrons to the more positive side as before.

$$3e^- + 2H_2O + MnO_4^- \longrightarrow MnO_2 + 4OH^-$$

5 Repeat the same procedure for the other half-reaction.

$$C_2O_4^{2-} \longrightarrow CO_3^{2-}$$

(balance C) $\quad C_2O_4^{2-} \longrightarrow 2CO_3^{2-}$

(balance O and H) $\quad 4OH^- + C_2O_4^{2-} \longrightarrow 2CO_3^{2-} + 2H_2O$

(balance charge) $\quad 4OH^- + C_2O_4^{2-} \longrightarrow 2CO_3^{2-} + 2H_2O + 2e^-$

6 Multiply the half-reactions so that electrons gained equal electrons lost. In this example, the reduction half-reaction is multiplied by 2, and the oxidation half-reaction is multiplied by 3.

$$\cancel{6e^-} + 4H_2O + 2MnO_4^- \longrightarrow 2MnO_2 + 8OH^-$$

$$12OH^- + 3C_2O_4^{2-} \longrightarrow 6CO_3^{2-} + 6H_2O + \cancel{6e^-}$$

$$\overset{4}{\cancel{4H_2O}} + 2MnO_4^- + \overset{4}{\cancel{12}}OH^- + 3C_2O_4^{2-} \longrightarrow$$
$$6CO_3^{2-} + \overset{2}{\cancel{6}}H_2O + 2MnO_2 + \cancel{8OH^-}$$

$$2MnO_4^-(aq) + 3C_2O_4^{2-}(aq) + 4OH^-(aq) \longrightarrow$$
$$6CO_3^{2-}(aq) + 2MnO_2(s) + 2H_2O(l)$$

EXAMPLE 13–5

Balancing Redox Equations in Basic Solution

Working It Out ●

Balance the following equation in basic solution by the ion-electron method.

$$Bi_2O_3(s) + ClO^-(aq) + OH^-(aq) \longrightarrow BiO_3^-(aq) + Cl^-(aq) + H_2O(l)$$

Reduction: $ClO^- \longrightarrow Cl^-$

OH^-: $ClO^- \longrightarrow Cl^- + 2OH^-$

H_2O: $H_2O + ClO^- \longrightarrow Cl^- + 2OH^-$

Balance e^-: $2e^- + H_2O + ClO^- \longrightarrow Cl^- + 2OH^-$

Oxidation: $Bi_2O_3 \longrightarrow BiO_3^-$

Bi: $Bi_2O_3 \longrightarrow 2BiO_3^-$

OH^-: $6OH^- + Bi_2O_3 \longrightarrow 2BiO_3^-$

H_2O: $6OH^- + Bi_2O_3 \longrightarrow 2BiO_3^- + 3H_2O$

e^-: $6OH^- + Bi_2O_3 \longrightarrow 2BiO_3^- + 3H_2O + 4e^-$

Multiply the reduction half-reaction by 2 and add to the oxidation half-reaction:

$$\cancel{4e^-} + 2H_2O + 2ClO^- \longrightarrow 2Cl^- + 4OH^-$$

$$6OH^- + Bi_2O_3 \longrightarrow 2BiO_3^- + 3H_2O + \cancel{4e^-}$$

$2H_2O + 6OH^- + Bi_2O_3 + 2ClO^- \longrightarrow 2Cl^- + 4OH^- + 2BiO_3^- + 3H_2O$

By eliminating H_2O and OH^- duplications, we have the balanced net ionic equation.

$Bi_2O_3(s) + 2ClO^-(aq) + 2OH^-(aq) \longrightarrow 2BiO_3^-(aq) + 2Cl^-(aq) + H_2O(l)$

Just as acid-base reactions involve an exchange of a H^+ ion between reactants, redox reactions involve an exchange of electrons. One substance is oxidized and one is reduced. An important point is that the same number of electrons are gained by the substance reduced as are lost by the substance oxidized. Thus a redox reaction can be balanced by emphasizing the equal exchange of electrons.

▲ **Looking Back**

Learning Check A

◀ **Checking It Out**

A–1. Fill in the blanks.
If all of the atoms in a compound were ions, the charge on the ion would be the same as its _____ _____. For oxygen in compounds, this is usually _____. Hydrogen is usually _____ in compounds. A substance oxidized undergoes a _____ of electrons and an _____ in oxidation state. This substance is also known as a _____ agent. The principle used to aid us in balancing equations is that electrons _____ equal electrons _____.

A–2. What is the oxidation state of
(a) B in H_3BO_3
(b) S in $S_2O_3^{2-}$

A–3. In the following reaction, indicate the substance oxidized, the substance reduced, the oxidizing agent, and the reducing agent.

$$ClO_2^- + H_2O_2 \longrightarrow O_2 + Cl^-$$

A–4. Balance the following equations by the ion-electron method.
(a) $NO_3^- + H_2SO_3 \rightarrow SO_4^{2-} + NO$ (acid solution)
(b) The reaction in A–3, which occurs in basic solution.

Additional Examples: 13-1, 13-6, 13-10, 13-12, 13-16, 13-18, 13-20, 13-22, and 13-24.

Looking Ahead ▼

In the previous chapter, we found that, when we mixed a strong conjugate acid with a strong conjugate base, the exchange of a H^+ ion was favorable. We say that such a reaction is spontaneous. Electron exchange reactions may also be favorable and spontaneous. That is how the flow of electrons can be used to obtain electricity from a battery. We will look into spontaneous redox reactions and their application next.

13–5 VOLTAIC CELLS

Why is a chemical reaction spontaneous in only one direction?

Releasing the brake on an automobile parked on the side of a hill causes no surprise. The automobile spontaneously rolls down the hill because the bottom of the hill represents a lower potential energy than the top. Chemical reactions occur spontaneously for the same reason. The products represent a position of lower potential energy than the reactants. In the case of the car, the difference in potential energy is due to position and the attraction of gravity; in the case of a chemical reaction, the difference in potential energy is due to the composition of the reactants and products. (See Figure 13-2.) When the car rolls down the hill, the difference in energy is transformed into the kinetic energy of the moving car and heat from friction. When a chemical reaction proceeds spontaneously from reactants to products, the difference is given off as heat, light, or electrical energy.

Two experiments confirm that a redox reaction proceeds spontaneously in only one direction. When a strip of copper metal is placed in a

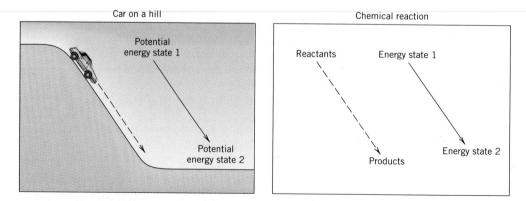

Figure 13–2
Energy States. A chemical reaction proceeds in a certain direction for the same reason that a car rolls down a hill.

Figure 13–3
The Direction of a Spontaneous Reaction. Copper does not react with Zn^{2+} (a) but zinc does react with a Cu^{2+} solution (b, c, and d).

solution containing Zn^{2+}[$ZnSO_4(aq)$], no reaction occurs. (See Figure 13-3a.) However, when we immerse a zinc strip in a Cu^{2+}[$CuSO_4(aq)$] solution, a coating of Cu forms on the zinc strip. (See Figure 13-3b, c, and d.) *It is important to distinguish between the two separate species in a half-reaction; these are the metal ion (e.g., Cu^{2+}, which is dissolved in water) and the neutral metal (e.g., Cu, which is present as a solid).* These simple experiments indicate that the following is the spontaneous reaction that occurs:

$$Zn(s) + CuSO_4(aq) \longrightarrow ZnSO_4(aq) + Cu(s)$$

A **voltaic cell** *(also called a* **galvanic cell***) uses a favorable or spontaneous redox reaction to generate electrical energy through an external circuit.* In the early days of the use of electricity, the reaction between Zn and Cu^{2+} discussed above was used to generate current for the new telegraph and doorbells in homes. This voltaic cell is known as the Daniell cell.

To generate electricity, however, the oxidation reaction must be separated from the reduction reaction so that the electrons exchanged can flow in an external wire where they can be put to use. The Daniell cell is illustrated in Figure 13-4. A zinc strip is immersed in a Zn^{2+} solution, and, in a separate compartment, a copper strip is immersed in a Cu^{2+} solution. A wire connects the two metal strips. The two metal strips are called electrodes. *The* **electrodes** *are the surfaces in a cell at which the reactions take place. The electrode at which oxidation takes place is called the* **anode.** *Reduction takes place at the* **cathode.**

In the compartment on the left, the strip of Zn serves as the anode, since the following reaction occurs when the circuit is connected.

$$Zn \longrightarrow Zn^{2+} + 2e^-$$

When the circuit is complete, the two electrons travel in the external wire to the Cu electrode, which serves as the cathode. The reduction reaction occurs at the cathode.

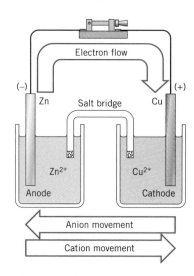

Figure 13–4
The Daniell Cell. This chemical reaction produced electricity for the first telegraphs.

$$2e^- + Cu^{2+} \longrightarrow Cu$$

To maintain neutrality in the solution, some means must be provided for the movement of a SO_4^{2-} ion (or some other negative ion) from the right compartment where a Cu^{2+} has been removed to the left compartment where a Zn^{2+} has been produced. The *salt bridge* is an aqueous gel that allows ions to migrate between compartments but does not allow the mixing of solutions. (If Cu^{2+} ions wandered into the left compartment, they would form a coating of Cu on the Zn electrode, thus short-circuiting the cell.)

As the cell discharges (generates electrical energy), the Zn electrode becomes smaller but the Cu electrode becomes larger. (See the solid lines on the electrodes in Figure 13-4.) An important feature of this cell is that the reaction can be stopped by interrupting the external circuit with a switch. If the circuit is open, the electrons can no longer flow, so no further reaction occurs until the switch is again closed.

Two of the most common voltaic cells in use today are the dry cell (flashlight battery) and the lead-acid cell (car battery). (*The word* **battery** *means a collection of one or more separate cells joined together in one unit.*)

Where does the power come from when you start a car?

One cell of a lead-acid storage battery is illustrated in Figure 13-5. Each cell is composed of two grids separated by an inert spacer. One grid of a fully charged battery contains metallic lead. The other contains PbO_2, which is insoluble in H_2O. Both grids are immersed in a sulfuric acid solution (battery acid). When the battery is discharged by connecting the electrodes, the following half-reactions take place spontaneously.

Anode: $Pb(s) + H_2SO_4(aq) \longrightarrow PbSO_4(s) + 2H^+(aq) + 2e^-$

Cathode: $2e^- + 2H^+(aq) + PbO_2(s) + H_2SO_4(aq) \longrightarrow PbSO_4(s) + 2H_2O$

Total
reaction: $Pb(s) + PbO_2(s) + 2H_2SO_4(aq) \longrightarrow 2PbSO_4(s) + 2H_2O$

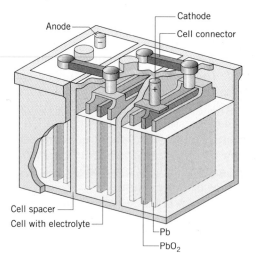

Figure 13-5
Lead Storage Battery. The lead storage battery is rechargeable.

The electrons released at the Pb anode travel through the external circuit to run lights, starters, radios, or whatever is needed. The electrons return to the PbO_2 cathode to complete the circuit. As the reaction proceeds, both electrodes are converted to $PbSO_4$, and the H_2SO_4 is depleted. Since $PbSO_4$ is also insoluble, it remains attached to the grids as it forms. The degree of discharge of a battery can be determined by the density of the battery acid. Since the density of a fully discharged battery is 1.05 g/mL, the difference in density between this value and the density of a fully charged battery (1.35 g/mL) gives the amount of charge remaining in the battery. The hydrometer discussed in Chapter 2 is used to determine the density of the acid. As the electrodes convert to $PbSO_4$, the battery loses power and eventually becomes "dead."

The convenience of a car battery is that it can be recharged. After the engine starts, an alternator or generator is engaged to push electrons back into the cell in the opposite direction from which they came during discharge. This forces the reverse, nonspontaneous reaction to proceed.

$$2PbSO_4(s) + 2H_2O \longrightarrow Pb(s) + PbO_2(s) + 2H_2SO_4(aq)$$

When the battery is fully recharged, the alternator shuts off, the circuit is open, and the battery is ready for the next start.

The dry cell (invented by Leclanché in 1866) is not rechargeable to any extent but is comparatively inexpensive and easily portable. (In contrast, the lead-acid battery is heavy and expensive and must be kept upright.) The dry cell illustrated in Figure 13-6 consists of a zinc anode, which is the outer rim, and an inert graphite electrode. (An inert electrode provides a reaction surface but does not itself react.) In between is an aqueous paste containing NH_4Cl, MnO_2, and carbon. The reactions are as follows:

Anode: $$Zn(s) \longrightarrow Zn^{2+}(aq) + 2e^-$$

Cathode: $$2NH_4^+(aq) + 2MnO_2(s) + 2e^- \longrightarrow Mn_2O_3(s) + 2NH_3(aq) + H_2O$$

Figure 13–6
Dry Cell. The dry cell is comparatively inexpensive, light, and portable.

Space travel requires a tremendous source of electrical energy. The requirements are that the source be continuous (no recharging necessary), light-weight, and dependable. Solar energy directly converts rays from the sun into electricity but is not practical for shorter runs such as the space shuttle. Although expensive, a source of power that fills the bill nicely is the fuel cell. A **fuel cell** *uses the direct reaction of hydrogen and oxygen to produce electrical energy.* Figure 13-7 is a picture of a fuel cell. Hydrogen and oxygen gases are fed into the cell where they form water. As long as the gases enter the cell, power is generated. The water that is formed can be removed and used for other purposes in the spacecraft. Since reactants are supplied from external sources, the electrodes are not consumed, and the cell does not have to be shut down to be regenerated like a car battery. Fuel cells have had some large-scale application in power generation for commercial purposes, but they are quite expensive. The best inert electrode surfaces at which the gases react are made of the extremely expensive metal platinum.

The reactions that take place in the fuel cell are as follows:

Anode: $\quad H_2(g) + 2OH^-(aq) \longrightarrow 2H_2O + 2e^-$

Cathode: $\quad O_2(g) + 2H_2O + 4e^- \longrightarrow 4OH^-(aq)$

Overall: $\quad 2H_2(g) + O_2(g) \longrightarrow 2H_2O$

Many other batteries have their special advantages. Nickel-cadmium batteries are popular as replacements for dry cells. Although initially more expensive, they are rechargeable and so last longer. A tiny battery made of silver oxide and zinc can supply a small amount of current for a long time. These have applications in digital watches and calculators. More efficient and durable batteries are a subject of much research at the current time. For example, an efficient and relatively inexpensive

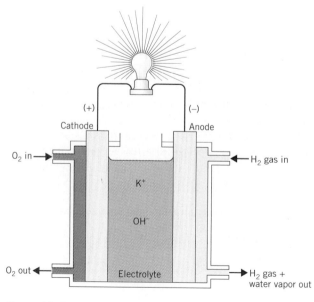

Figure 13–7
Fuel Cell. The fuel cell can generate power without interruption for recharging.

electrical automobile is the object of intensive efforts. A small electric car using lead-acid batteries requires at least 18 batteries. These have to be replaced every year or so, depending on use. Also, much of the power of these heavy batteries must be used just to move the batteries around, not including the car and passengers. There have been some encouraging possibilities for lighter, more durable batteries, but large-scale production of such an automobile is not planned for the immediate future.

The reactions in the Daniell cell and other batteries proceed spontaneously in one direction only. Is there a way to establish a pecking order of oxidizing and reducing agents as we did for acids and bases? Since each agent has a different tendency to give up or accept electrons, we can indeed construct a table that is a powerful predictive tool for spontaneous reactions. This is our next topic.

▼Looking Ahead

13–6 PREDICTING SPONTANEOUS REDOX REACTIONS

Let's return to the net ionic reaction that occurs in the Daniell cell.

$$Zn(s) + Cu^{2+}(aq) \longrightarrow Zn^{2+}(aq) + Cu(s)$$

In Figure 13-4, the cell is arranged so that the reaction could proceed in either direction. That is, Cu^{2+} could oxidize Zn, or Zn^{2+} could oxidize Cu. The direction of the spontaneous reaction indicates that *the Cu^{2+} ion is a better oxidizing agent than the Zn^{2+} ion.* In other words, of the two possible half-reactions,

How can we decide which is the more powerful oxidizing agent?

$$Cu^{2+} + 2e^- \longrightarrow Cu$$

or

$$Zn^{2+} + 2e^- \longrightarrow Zn$$

the first has a stronger potential to occur.

The competition between the two ions as oxidizing agents is analogous to a tug-of-war between two competitors. In Figure 13-8, two people of unequal strength are pulling on a rope looped around a post. Obviously, they both can't move in the same direction since only the stronger of the two prevails. The weaker of the two is thus forced to move in the opposite direction. Likewise, since Cu^{2+} has a greater potential to be reduced (is a stronger oxidizing agent), it is reduced to form copper metal.

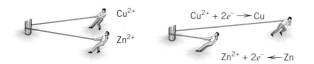

Figure 13–8
A Tug-of-War for Electrons. The stronger of the two moves in the desired direction. The other moves in the opposite direction.

The other half-reaction is forced to go in the opposite direction, which means that zinc metal is oxidized to Zn^{2+} ion.

$$Zn \longrightarrow Zn^{2+} + 2e^-$$

Note that, when an ion is reduced, it forms a species (a metal in this case) that is a potential reducing agent. *The strength of the reduced form as a reducing agent is inversely related to the strength of the original ion as an oxidizing agent.* In other words, if a *metal ion* is a strong oxidizing agent, the *metal* is a weak reducing agent and vice versa. This is the same inverse relationship we noticed in the previous chapter with regard to an acid and its conjugate base. Thus for the reaction in the Daniell cell, we also could have focused on the competition between the metals as reducing agents. In the reaction, Zn is a stronger reducing agent than Cu. We can now make another useful generalization. *A spontaneous reaction occurs when the stronger oxidizing agent reacts with the stronger reducing agent to form a weaker oxidizing and a weaker reducing agent.* Again, notice the similarity to the favorable acid-base reaction between the stronger acid and the stronger base to form a weaker acid and base.

What always happens in a spontaneous redox reaction?

Now let's conduct experiments similar to those illustrated in Figure 13-3 in order to compare the strengths of Cu^{2+} and Ni^{2+} as oxidizing agents. We find that a coating of Cu forms on a strip of nickel metal immersed in a Cu^{2+} solution. If we immerse a strip of Cu in a Ni^{2+} solution, no reaction occurs. The following equation illustrates the spontaneous reaction that these observations indicate.

$$Ni(s) + Cu^{2+}(aq) \longrightarrow Ni^{2+}(aq) + Cu(s)$$

Once again Cu^{2+} prevails. *This indicates that the Cu^{2+} ion is a stronger oxidizing agent than the Ni^{2+} ion.*

Now let's compare Ni^{2+} to Zn^{2+} as oxidizing agents. When a strip of nickel is placed in a Zn^{2+} solution, no coating of Zn appears. However, if a strip of Zn is placed in a Ni^{2+} solution, a coating of nickel forms on the zinc. These experiments indicate that the following reaction is spontaneous.

$$Zn(s) + Ni^{2+}(aq) \longrightarrow Zn^{2+}(aq) + Ni(s)$$

The results indicate that Ni^{2+} ion is a stronger oxidizing agent than Zn^{2+} ion.

All of these experimental observations can be summarized as follows.

Metal Ions	Metal Formed	Metal Ion Reduced	Oxidizing Strength
Cu^{2+} and Zn^{2+}	Cu	Cu^{2+}	$Cu^{2+} > Zn^{2+}$
Cu^{2+} and Ni^{2+}	Cu	Cu^{2+}	$Cu^{2+} > Ni^{2+}$
Ni^{2+} and Zn^{2+}	Ni	Ni^{2+}	$Ni^{2+} > Zn^{2+}$

As a result of our observations, we can rank these three ions as oxidizing agents in order of decreasing strength:

1 Cu^{2+} **2** Ni^{2+} **3** Zn^{2+}

Because of the inverse relationship, the ranking of the three metals as reducing agents is in the reverse order.

1 Zn   **2** Ni   **3** Cu

More experiments can provide additional ions and molecular species to our ranking. Eventually, we can construct a table of oxidizing agents ordered according to strength. Such a ranking is given in Table 13-1. In some cases, instruments are required to give quantitative values, known as *reduction potentials,* which indicate the comparative strength of a given oxidizing agent to a defined standard. The strengths of the oxidizing agents are compared in these measurements at the same concentration for all ions involved $(1.00\ M)$ and at the same partial pressure of all gases involved (1.00 atm).

The strongest oxidizing agent is at the top of Table 13-1, on the left. On the other hand, the reducing agents, shown on the right, become stronger down the table. A redox reaction takes place between an oxidizing agent on the left and a reducing agent on the right. *A favorable or spontaneous reaction occurs between an oxidizing agent and any reducing agent in the table that is lower than the oxidizing agent.* The reaction can be visualized as taking place in a clockwise direction, with the oxidizing agent reacting to the right and the reducing agent reacting in the opposite direction to the left. (See Figure 13-9.)

Can we predict favorable redox reactions?

Table 13-1 can be used to predict spontaneous redox reactions in the same manner that Table 12-2 was used to predict favorable acid-base reactions. Of particular interest is the use of Table 13-1 to predict the reactions of certain elements with water. (All of the reactions shown in the

---

TABLE 13–1   OXIDIZING AGENTS AND REDUCING AGENTS

Strongest oxidizing agent	$F_2 + 2e^- \rightleftarrows 2F^-$	Weakest reducing agent
	$Cl_2 + 2e^- \rightleftarrows 2Cl^-$	
	$\boxed{O_2 + 4H^+ + 4e^- \rightleftarrows 2H_2O}$	
	$Br_2 + 2e^- \rightleftarrows 2Br^-$	
	$Ag^+ + e^- \rightleftarrows Ag$	
	$Cu^{2+} + 2e^- \rightleftarrows Cu$	
	$\boxed{2H^+ + 2e^- \rightleftarrows H_2}$	
Increasing strength of oxidizing agent	$Pb^{2+} + 2e^- \rightleftarrows Pb$	Increasing strength of reducing agent
	$Sn^{2+} + 2e^- \rightleftarrows Sn$	
	$Ni^{2+} + 2e^- \rightleftarrows Ni$	
	$Fe^{2+} + 2e^- \rightleftarrows Fe$	
	$Cr^{3+} + 3e^- \rightleftarrows Cr$	
	$Zn^{2+} + 2e^- \rightleftarrows Zn$	
	$\boxed{2H_2O + 2e^- \rightleftarrows H_2 + 2OH^-}$	
Weakest oxidizing agent	$Al^{3+} + 3e^- \rightleftarrows Al$	Strongest reducing agent
	$Mg^{2+} + 2e^- \rightleftarrows Mg$	
	$Na^+ + e^- \rightleftarrows Na$	

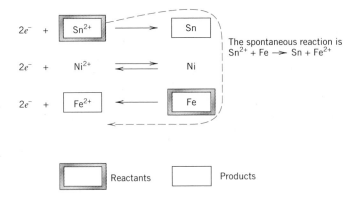

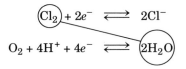

Figure 13–9
Spontaneous Reaction. The stronger oxidizing agent reacts with the stronger reducing agent.

table are assumed to occur in aqueous solution.) The highlighted reaction near the top of the table represents the oxidation of water when read from right to left. Note that the gaseous elements F_2 and Cl_2 spontaneously oxidize water to produce oxygen gas and an acid. (The reaction of Cl_2 with water is quite slow, however.)

$$Cl_2 + 2e^- \rightleftharpoons 2Cl^-$$
$$O_2 + 4H^+ + 4e^- \rightleftharpoons 2H_2O$$

The spontaneous reaction is

$$2Cl_2 + 2H_2O \longrightarrow O_2 + 4H^+ + 4Cl^-$$

The highlighted reaction near the bottom of the table represents the reduction of water when read from left to right. Note that the metals Al, Mg, and Na spontaneously reduce water to produce hydrogen gas and a base. *(We can say that Na reduces water, or, conversely, we can say that water oxidizes Na.)*

$$2H_2O + 2e^- \rightleftharpoons H_2 + 2OH^-$$
$$Na^+ + e^- \rightleftharpoons Na$$

The spontaneous reaction is

$$2H_2O + 2Na \longrightarrow H_2 + 2Na^+ + 2OH^-$$

These metals are known as *active* metals because of their chemical reactivity with water.

The third highlighted reaction near the middle of the table represents the reduction of aqueous acid solutions (1.00 M H^+) to form hydrogen gas. Note that metals such as Ni and Fe are not oxidized by water but are oxidized by strong acid solutions.

$$2H^+ + 2e^- \rightleftharpoons H_2$$
$$Fe^{2+} + 2e^- \rightleftharpoons Fe$$

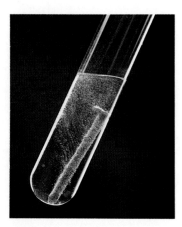

Iron is oxidized by a strong acid solution to form iron(II) ions and hydrogen gas.

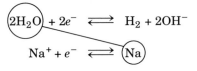

The spontaneous reaction is

$$2H^+ + Fe \longrightarrow Fe^{2+} + H_2$$

Acid rain contains a considerably higher H^+ concentration than ordinary rain. From this discussion, we can understand why metals such as iron and nickel are more likely to be corroded by acid rain.

Two of the metals shown in the table—Cu and Ag—are not oxidized by either water or acid solutions. Thus these metals are relatively unreactive and find use in jewelry and coins.

EXAMPLE 13–6

The Direction of a Spontaneous Reaction

In which direction will the following reaction be spontaneous?

$$Pb^{2+}(aq) + 2Cl^-(aq) \longleftrightarrow Pb(s) + Cl_2(g)$$

●**Working It Out**

Procedure
In Table 13-1, note that Cl_2 is a stronger oxidizing agent than Pb^{2+} and that Pb is a stronger reducing agent than Cl^-. Therefore, the reaction is spontaneous to the left. Metallic lead will dissolve in an aqueous solution of Cl_2.

Solution

$$Pb(s) + Cl_2(g) \longrightarrow PbCl_2(aq)$$

EXAMPLE 13–7

A Spontaneous Single-Replacement Reaction

A strip of tin metal is placed in a $AgNO_3$ solution. If a reaction takes place, write the equation illustrating the spontaneous reaction.

Procedure
In Table 13-1, note that the oxidizing agent Ag^+ is above the reducing agent Sn. Therefore, a spontaneous reaction does occur.

Solution

$$2Ag^+(aq) + Sn(s) \longrightarrow Sn^{2+}(aq) + 2Ag(s)$$

EXAMPLE 13–8

The Spontaneous Reaction of a Metal with Water

A length of aluminum wire is placed in water. Does the aluminum react with water?

Procedure
In Table 13-1, note that aluminum is an active metal and should react with water (as an oxidizing agent).

Solution

$$6H_2O(l) + 2Al(s) \longrightarrow 2Al^{3+}(aq) + 6OH^-(aq) + 3H_2(g)$$

Since $Al(OH)_3$ is insoluble in water, however, the equation should be

$$6H_2O(l) + 2Al(s) \longrightarrow 2Al(OH)_3(s) + 3H_2(g)$$

Theoretically, aluminum should dissolve in water. Metallic aluminum is actually coated with Al_2O_3, which protects the metal from coming into contact with water. Thus it is a useful metal even for the hulls of boats, despite its high chemical reactivity.

Looking Ahead ▼

Spontaneous reactions are like cars rolling down a hill. They go to lower energy states. Can a car or a reaction go up a hill? Of course, but in order for this to happen, energy in the form of a push for a car or energy for a chemical reaction must be supplied. Nonspontaneous redox reactions can occur if sufficient electrical energy is supplied from an outside source.

13–7 ELECTROLYTIC CELLS

Voltaic cells convert the chemical energy in reactants into electrical energy. The opposite process can be made to occur. *Cells that convert electrical energy into chemical energy are called* **electrolytic cells.**

Can a nonspontaneous reaction occur?

An example of an electrolytic cell is shown in Figure 13-10. When sufficient electrical energy is supplied to the electrodes from an outside source, the following nonspontaneous reaction occurs:

$$2H_2O(l) \longrightarrow 2H_2(g) + O_2(g)$$

For this electrolysis to occur, an electrolyte such as K_2SO_4 must be present in solution. Pure water alone does not have a sufficient concentration of ions to allow conduction of electricity.

Another example of an electrolytic cell is the recharge cycle of the lead-acid battery described in an earlier section. When energy from the engine activates the alternator, electrical energy is supplied to the battery, and the nonspontaneous reaction occurs as an electrolysis reaction. This reaction reforms the original reactants.

How is silver plate made?

Electrolysis has many useful applications. For example, silver or gold can be electroplated onto cheaper metals. In Figure 13-11, the metal spoon is the cathode and the silver bar serves as the anode. When electricity is supplied, the Ag anode produces Ag^+ ions, and the spoon

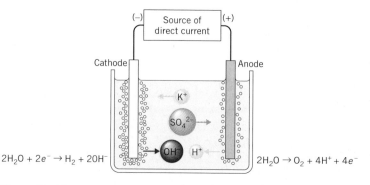

Figure 13–10
An Electrolytic Cell. Electrolysis of a solution of potassium sulfate gives hydrogen gas and oxygen gas as products.

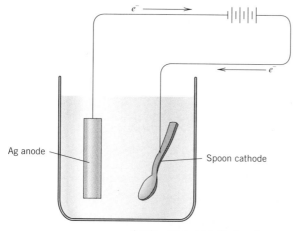

Anode: $Ag(s) \longrightarrow Ag^+(aq) + e^-$
Cathode: $Ag^+(aq) + e^- \longrightarrow Ag(s)$

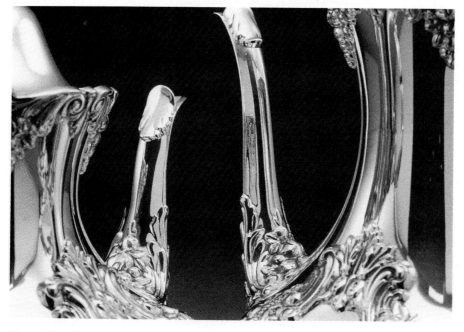

Figure 13–11
Electroplating. With an input of energy, a spoon can be coated with silver. The service has been electroplated with silver.

cathode reduces Ag^+ ions to form a layer of Ag. The silver-plated spoon can be polished and made to look as good as sterling silver.

Electrolytic cells are used to free elements from their compounds. Such cells are especially useful where metals are held in their compounds by strong chemical bonds. Examples are the metals aluminum, sodium, and magnesium. All aluminum is produced by the electrolysis of molten aluminum salts. Commercial quantities of sodium and chlorine are also produced by electrolysis of molten sodium chloride. An apparatus used for the electrolysis of molten sodium chloride is illustrated in Figure 13-12. At the high temperature required to keep the NaCl in the

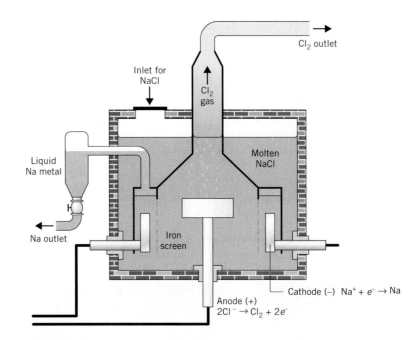

Figure 13–12
Electrolysis of NaCl. Cross section of the Downs cell used for the electrolysis of molten sodium chloride. The cathode is a circular ring that surrounds the anode. The electrodes are separated from each other by an iron screen. During the operation of the cell, molten sodium collects at the top of the cathode compartment, from which it is periodically drained. The chlorine gas bubbles out of the anode compartment and is collected.

liquid state, sodium forms as a liquid and is drained from the top of the cell.

Looking Back ▲

Redox reactions proceed spontaneously in only one direction. If the direction of the reaction is toward lower energy, the energy difference can be harnessed in voltaic cells. All common batteries use a given spontaneous chemical reaction as a source of electricity. By matching oxidizing agents against each other and noting the direction of the reaction, we can construct a table of oxidizing and reducing agents. The table is a powerful predictive tool. Reactions that proceed to a higher energy state can occur with an input of energy. These reactions have important applications.

Checking It Out ▶

Learning Check B

B–1. Fill in the blanks.
A spontaneous chemical reaction is used in a _____ cell. An example is the Daniell cell where _____ is oxidized at the _____ and _____ is reduced at the _____. An electrolytic balance is maintained by means of a _____ bridge. In the lead-acid battery, _____ is _____ at the anode and _____ is reduced at the _____.

Spontaneous redox reactions occur between an oxidizing agent
_____ and a _____ agent lower in the table. Nonspontaneous
reactions occur in _____ cells.

B–2. In which of the following would a spontaneous reaction occur? Write the balanced equation for that reaction.
(a) a strip of Pb is placed in a Cr^{3+} solution, or (b) a strip of Cr is placed in a Pb^{2+} solution.

B–3. A cell is constructed of a Br_2, Br^- half-cell connected to a Fe^{2+}, Fe half-cell. Write a balanced equation representing the spontaneous reaction that occurs.

B–4. Write the reaction that occurs when molten $MgCl_2$ is electrolyzed.

Additional Examples: 13-28, 13-30, 13-32, 13-34, 13-40, and 13-43.

C H A P T E R   R E V I E W

A common characteristic of a large number of chemical reactions is an exchange of electrons between reactants. To keep track of the electron exchange, we can follow the change in **oxidation states** of the elements in the compounds. These reactions are known as **oxidation-reduction** reactions, or simply **redox** reactions. In such reactions, the reactant that gives up or loses the electrons is **oxidized,** and the reactant that gains the electrons is **reduced.** The reactants are also classified as **oxidizing agents** and **reducing agents** as follows:

▲▼**Putting It Together**

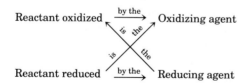

A redox reaction can be divided into two **half-reactions:** an oxidation and a reduction. These two processes take place so that all electrons lost in the oxidation process are gained in the reduction process. This fact is useful in balancing oxidation-reduction reactions. Most of these reactions would, at best, be very difficult to balance by inspection methods as described in Chapter 8.

Balancing equations by the **oxidation state method** requires that the atoms undergoing a change in oxidation state be identified. By equalizing electron gain and electron loss with proper coefficients, equations can be balanced. The more useful method is the **ion-electron** or **half-reaction method.** In this method, the entire molecule or ion containing the atom undergoing a change is balanced in two half-reactions.

The rules for balancing equations by this method are summarized as follows:

1 Identify the ion or molecule containing the atom that has a change in oxidation state and the product species containing the same atom.

2 Balance the atom undergoing the change.

3 Balance oxygens. $\left[\begin{array}{l}\text{Acidic: add one } H_2O \text{ for each O needed.} \\ \text{Basic: add two } OH^- \text{ ions for each O needed.}\end{array}\right.$

4 Balance H. $\left[\begin{array}{l}\text{Acidic: add } H^+ \text{ for each H needed.} \\ \text{Basic: add one } H_2O \text{ for each two H's needed.}\end{array}\right.$

5 Add electrons on the more positive side to balance the charge.

6 Multiply each half-reaction by a whole number so that electrons cancel when the half-reactions are added.

7 Subtract any species common to both sides of the equation.

Spontaneous chemical reactions occur because reactants are higher in chemical (potential) energy than products. We are familiar with many exothermic reactions where the difference in energy is released as heat. The energy can also be released as electrical energy in a **voltaic cell.** In a voltaic cell, the two half-reactions are physically separated so that electrons travel in an external circuit or wire between **electrodes.** The **anode** is the electrode at which oxidation takes place, and the **cathode** is the electrode at which reduction takes place. The Daniell cell, the car **battery,** the dry cell, and the fuel cell all involve spontaneous chemical reactions in which the chemical energy is converted directly into electrical energy.

Each substance has its own inherent strength as either an oxidizing agent or a reducing agent. A table can be constructed in which oxidizing and reducing agents are ranked by strength as determined by observation or measurements with electrical instruments. From this table, a great many other spontaneous reactions can be predicted. In Table 13-1, stronger oxidizing agents are ranked higher on the left and stronger reducing agents are ranked lower on the right. Reactions can thus be predicted as shown in Figure 13-9.

The reactions (or lack of reactions) of certain elements with water can be predicted from Table 13-1. Four groups of elements can be classified.

1 Nonmetals (e.g., F_2 and Cl_2) that are very strong oxidizing agents. These elements *oxidize* water to produce oxygen gas.

2 Metals (e.g., Na and Mg) that are very strong reducing agents. These elements *reduce* water to produce hydrogen gas.

3 Metals (e.g., Cu and Ag) that are very weak reducing agents. These elements are not oxidized by water or a strong acid (1.00 M) solution.

4 Metals (e.g., Ni and Fe) that are moderate reducing agents. These elements are not oxidized by water but are oxidized by aqueous acid.

Many reactions that are predicted to be unfavorable or nonspontaneous can be made to occur if electrical energy is supplied from an outside source. These are referred to as **electrolytic cells** and are useful in the commercial production of certain elements and in electroplating.

E X E R C I S E S

OXIDATION STATES

13-1 Give the oxidation states of the elements in the following compounds.
(a) PbO_2 (d) N_2H_4 (g) Rb_2Se
(b) P_4O_{10} (e) LiH (h) Bi_2S_3
(c) C_2H_2 (f) BCl_3

13-2 Give the oxidation states of the elements in the following compounds.
(a) ClO_2 (c) CO (e) Mn_2O_3
(b) XeF_2 (d) O_2F_2 (f) Bi_2O_5

13-3 Which of the following elements form *only* the +1 oxidation state in compounds?
(a) Li   (c) Ca   (e) K   (g) Rb
(b) H   (d) Cl   (f) Al

13-4 Which of the following elements form *only* the +2 oxidation state in compounds?
(a) O   (c) Be   (e) Sc   (g) Ca
(b) B   (d) Sr   (f) Hg

13-5 What is the only oxidation state of Al in its compounds?

13-6 What is the oxidation state of each of the following?
(a) P in H_3PO_4 (e) S in SF_6
(b) C in $H_2C_2O_4$ (f) N in $CsNO_3$
(c) Cl in ClO_4^- (g) Mn in $KMnO_4$
(d) Cr in $CaCr_2O_7$

13-7 What is the oxidation state of the specified atom?
(a) S in SO_3 (d) N in HNO_3
(b) Co in Co_2O_3 (e) Cr in K_2CrO_4
(c) U in UF_6 (f) Mn in $CaMnO_4$

13-8 What is the oxidation state of each of the following?
(a) Se in SeO_3^{2-} (d) Cl in $HClO_2$
(b) I in H_5IO_6 (e) N in $(NH_4)_2S$
(c) S in $Al_2(SO_3)_3$

OXIDATION-REDUCTION

13-9 Which of the following reactions are oxidation-reduction reactions?
(a) $2H_2 + O_2 \rightarrow 2H_2O$
(b) $CaCO_3 \rightarrow CaO + CO_2$
(c) $2Na + 2H_2O \rightarrow 2NaOH + H_2$
(d) $2HNO_3 + Ca(OH)_2 \rightarrow Ca(NO_3)_2 + 2H_2O$
(e) $AgNO_3 + KCl \rightarrow AgCl + KNO_3$
(f) $Zn + CuCl_2 \rightarrow ZnCl_2 + Cu$

13-10 Identify each of the following half-reactions as either oxidation or reduction.
(a) $Na \rightarrow Na^+ + e^-$
(b) $Zn^{2+} + 2e^- \rightarrow Zn$
(c) $Fe^{2+} \rightarrow Fe^{3+} + e^-$
(d) $O_2 + 4H^+ + 4e^- \rightarrow 2H_2O$
(e) $S_2O_8^{2-} + 2e^- \rightarrow 2SO_4^{2-}$

13-11 Identify each of the following changes as either oxidation or reduction.
(a) $P_4 \rightarrow H_3PO_4$ (d) $Al \rightarrow Al(OH)_4^-$
(b) $NO_3^- \rightarrow NH_4^+$ (e) $S^{2-} \rightarrow SO_4^{2-}$
(c) $Fe_2O_3 \rightarrow Fe^{2+}$

13-12 For each of the following unbalanced equations, complete the table below.
(a) $MnO_2 + H^+ + Br^- \rightarrow$
$$Mn^{2+} + Br_2 + H_2O$$
(b) $CH_4 + O_2 \rightarrow CO_2 + H_2O$
(c) $Fe^{2+} + MnO_4^- + H^+ \rightarrow$
$$Fe^{3+} + Mn^{2+} + H_2O$$

Reaction	Reactant Oxidized*	Product of Oxidation	Reactant Reduced	Product of Reduction	Oxidizing Agent	Reducing Agent
(a)						
(b)						
(c)						

*Element, molecule, or ion.

13–13 For the following two unbalanced equations, construct a table such as that in Problem 13-12.
(a) $Al + H_2O \rightarrow AlO_2^- + H_2 + H^+$
(b) $Mn^{2+} + Cr_2O_7^{2-} + H^+ \rightarrow$
$$MnO_4^- + Cr^{3+} + H_2O$$

13–14 For the following equations, identify the reactant oxidized, the reactant reduced, the oxidizing agent, and the reducing agent.
(a) $Sn + HNO_3 \rightarrow SnO_2 + NO_2 + H_2O$
(b) $IO_3^- + SO_2 + H_2O \rightarrow I_2 + SO_4^{2-} + H^+$
(c) $CrI_3 + OH^- + Cl_2 \rightarrow$
$$CrO_4^{2-} + IO_4^- + Cl^- + H_2O$$
(d) $I^- + H_2O_2 \rightarrow I_2 + H_2O + OH^-$

13–15 Identify the product or products containing the elements oxidized and reduced in Problem 13-14.

BALANCING EQUATIONS BY THE OXIDATION STATE METHOD

13–16 Balance each of the following equations by the oxidation state method.
(a) $NH_3 + O_2 \rightarrow NO + H_2O$
(b) $Sn + HNO_3 \rightarrow SnO_2 + NO_2 + H_2O$
(c) $Cr_2O_3 + Na_2CO_3 + KNO_3 \rightarrow$
$$CO_2 + Na_2CrO_4 + KNO_2$$
(d) $Se + BrO_3^- + H_2O \rightarrow H_2SeO_3 + Br^-$

13–17 Balance the following equations by the oxidation state method.
(a) $I_2O_5 + CO \rightarrow I_2 + CO_2$
(b) $Al + H_2O \rightarrow AlO_2^- + H_2 + H^+$
(c) $HNO_3 + HCl \rightarrow NO + Cl_2 + H_2O$
(d) $I_2 + Cl_2 + H_2O \rightarrow HIO_3 + HCl$

BALANCING EQUATIONS BY THE ION-ELECTRON METHOD

13–18 Balance the following half-reactions in acidic solution.
(a) $Sn^{2+} \rightarrow SnO_2$ (d) $I_2 \rightarrow IO_3^-$
(b) $CH_4 \rightarrow CO_2$ (e) $NO_3^- \rightarrow NO_2$
(c) $Fe^{3+} \rightarrow Fe^{2+}$

13–19 Balance the following half-reactions in acidic solution.
(a) $P_4 \rightarrow H_3PO_4$ (d) $NO_3^- \rightarrow NH_4^+$
(b) $ClO_3^- \rightarrow Cl^-$ (e) $H_2O_2 \rightarrow H_2O$
(c) $S_2O_3^{2-} \rightarrow SO_4^{2-}$

13–20 Balance each of the following by the ion-electron method. All are in acidic solution.
(a) $S^{2-} + NO_3^- + H^+ \rightarrow S + NO + H_2O$
(b) $I_2 + S_2O_3^{2-} \rightarrow S_4O_6^{2-} + I^-$
(c) $SO_3^{2-} + ClO_3^- \rightarrow Cl^- + SO_4^{2-}$

(d) $Fe^{2+} + H_2O_2 + H^+ \rightarrow Fe^{3+} + H_2O$
(e) $AsO_4^{3-} + I^- + H^+ \rightarrow$
$$I_2 + AsO_3^{3-} + H_2O$$
(f) $Zn + H^+ + NO_3^- \rightarrow$
$$Zn^{2+} + NH_4^+ + H_2O$$

13–21 Balance each of the following by the ion-electron method. All are in acidic solution.
(a) $Mn^{2+} + BiO_3^- + H^+ \rightarrow$
$$MnO_4^- + Bi^{3+} + H_2O$$
(b) $IO_3^- + SO_2 + H_2O \rightarrow I_2 + SO_4^{2-} + H^+$
(c) $Se + BrO_3^- + H_2O \rightarrow H_2SeO_3 + Br^-$
(d) $P_4 + HClO + H_2O \rightarrow$
$$H_3PO_4 + Cl^- + H^+$$
(e) $Al + Cr_2O_7^{2-} + H^+ \rightarrow$
$$Al^{3+} + Cr^{3+} + H_2O$$
(f) $ClO_3^- + I^- + H^+ \rightarrow Cl^- + I_2 + H_2O$
(g) $As_2O_3 + NO_3^- + H_2O \rightarrow$
$$AsO_4^{3-} + NO + H^+$$

13–22 Balance the following half-reactions in basic solution.
(a) $SnO_2^{2-} \rightarrow SnO_3^{2-}$ (c) $Si \rightarrow SiO_3^{2-}$
(b) $ClO_2^- \rightarrow Cl_2$ (d) $NO_3^- \rightarrow NH_3$

13–23 Balance the following half-reactions in basic solution.
(a) $Al \rightarrow Al(OH)_4^-$
(b) $S^{2-} \rightarrow SO_4^{2-}$
(c) $N_2H_4 \rightarrow NO_3^-$

13–24 Balance each of the following by the ion-electron method. All are in basic solution.
(a) $S^{2-} + OH^- + I_2 \rightarrow SO_4^{2-} + I^- + H_2O$
(b) $MnO_4^- + OH^- + I^- \rightarrow$
$$MnO_4^{2-} + IO_4^- + H_2O$$
(c) $BiO_3^- + SnO_2^{2-} + H_2O \rightarrow$
$$SnO_3^{2-} + OH^- + Bi(OH)_3$$
(d) $CrI_3 + OH^- + Cl_2 \rightarrow$
$$CrO_4^{2-} + IO_4^- + Cl^- + H_2O$$
Hint: In (d), two ions are oxidized; include both in one half-reaction.

13–25 Balance each of the following by the ion-electron method. All are in basic solution.
(a) $ClO_2 + OH^- \rightarrow ClO_2^- + ClO_3^- + H_2O$
(b) $OH^- + Cr_2O_3 + NO_3^- \rightarrow$
$$CrO_4^{2-} + NO_2^- + H_2O$$
(c) $Cr(OH)_4^- + BrO^- + OH^- \rightarrow$
$$Br^- + CrO_4^{2-} + H_2O$$
(d) $Mn^{2+} + H_2O_2 + OH^- \rightarrow H_2O + MnO_2$
(e) $Ag_2O + Zn + H_2O \rightarrow Zn(OH)_2 + Ag$

13–26 Balance the following two equations by the ion-electron method, first in acidic solution and then in basic solution.
(a) $H_2 + O_2 \rightarrow H_2O$
(b) $H_2O_2 \rightarrow O_2 + H_2O$

VOLTAIC CELLS

13–27 What is the function of the salt bridge in the voltaic cell?

13–28 In an alkaline battery, the following two half-reactions occur:

$$Zn(s) + 2OH^-(aq) \longrightarrow Zn(OH)_2(s) + 2e^-$$

$$2MnO_2(s) + 2H_2O(l) + 2e^- \longrightarrow$$
$$2MnO(OH)(s) + 2OH^-(aq)$$

Which reaction takes place at the anode and which at the cathode? What is the total reaction?

13–29 The following overall reaction takes place in a silver oxide battery.

$$Ag_2O(s) + H_2O(l) + Zn(s) \longrightarrow$$
$$Zn(OH)_2(s) + 2Ag(s)$$

The reaction takes place in basic solution. Write the half-reaction that takes place at the anode, and the half-reaction that takes place at the cathode.

13–30 The nickel-cadmium (nicad) battery is used as a replacement for a dry cell because it is rechargeable. The overall reaction that takes place is

$$NiO_2(s) + Cd(s) + 2H_2O(l) \longrightarrow$$
$$Ni(OH)_2(s) + Cd(OH)_2(s)$$

Write the half-reactions that take place at the anode and the cathode.

13–31 Sketch a galvanic cell in which the following overall reaction occurs.

$$Ni^{2+}(aq) + Fe(s) \longrightarrow Fe^{2+}(aq) + Ni(s)$$

(a) What reactions take place at the anode and the cathode?
(b) In what direction do the electrons flow in the wire?
(c) In what direction do the anions flow in the salt bridge?

PREDICTING REDOX REACTIONS

13–32 Using Table 13-1, predict whether the following reactions occur in aqueous solution. If not, write N.R. (no reaction).
(a) $2Na + 2H_2O \rightarrow H_2 + 2NaOH$
(b) $Pb + Zn^{2+} \rightarrow Pb^{2+} + Zn$
(c) $Fe + 2H^+ \rightarrow Fe^{2+} + H_2$
(d) $Fe + 2H_2O \rightarrow Fe^{2+} + 2OH^- + H_2$
(e) $Cu + 2Ag^+ \rightarrow 2Ag + Cu^{2+}$
(f) $2Cl_2 + 2H_2O \rightarrow 4Cl^- + O_2 + 4H^+$
(g) $3Zn^{2+} + 2Cr \rightarrow 2Cr^{3+} + 3Zn$

13–33 Using Table 13-1, predict whether the following reactions occur in aqueous solution. If not, write N.R.
(a) $Sn^{2+} + Pb \rightarrow Pb^{2+} + Sn$
(b) $Ni^{2+} + H_2 \rightarrow 2H^+ + Ni$
(c) $Cu + F_2 \rightarrow CuF_2$
(d) $Ni^{2+} + 2Br^- \rightarrow Ni + Br_2$
(e) $3Ni^{2+} + 2Cr \rightarrow 2Cr^{3+} + 3Ni$
(f) $2Br_2 + 2H_2O \rightarrow 4Br^- + O_2 + 4H^+$

13–34 Tell whether a reaction occurs in each of the following situations. Write a balanced equation indicating the spontaneous reaction if one occurs.
(a) Some iron nails are placed in a $CuCl_2$ solution.
(b) Silver coins are dropped into an acid solution [$H^+(aq)$].
(c) A copper penny is placed in a $Pb(NO_3)_2$ solution.
(d) Ni metal is placed in a $Pb(NO_3)_2$ solution.
(e) Sodium metal is placed in water.
(f) Zinc metal is placed in water.
(g) Bromine is dissolved in water.
(h) Zinc strips are placed in a $Cr(NO_3)_3$ solution.

13–35 Tell whether a reaction occurs in each of the following situations. Write a balanced equation illustrating the spontaneous reaction if one occurs.
(a) Magnesium metal makes contact with water.
(b) Iron nails are placed in a $ZnBr_2$ solution.
(c) H_2 gas is bubbled into an acid solution of Cu^{2+}.
(d) Copper metal is placed in an acid [$H^+(aq)$] solution.
(e) Tin metal is placed in an acid solution.
(f) A piece of chromium is placed in a $SnCl_2$ solution.
(g) Chlorine is dissolved in water.

13–36 Which of the following elements react with water: (a) Pb (b) Ag (c) F_2 (d) Br_2 (e) Mg? Write the balanced equation illustrating any reaction of these elements with water.

13–37 Which of the following species will be reduced by hydrogen gas in aqueous solution: (a) Br_2 (b) Cr (c) Ag^+ (d) Ni^{2+}? Write the balanced equation illustrating any reaction of these species with hydrogen.

13–38 In Chapter 12, we mentioned the corrosiveness of acid rain. Why does rain containing a higher $H^+(aq)$ concentration cause more damage to iron exposed in bridges and buildings than pure H_2O? Write the reaction between Fe and $H^+(aq)$.

13-39 Br_2 can be prepared from the reaction of Cl_2 with NaBr dissolved in seawater. Explain. Write the reaction. Can Cl_2 be used to prepare F_2 from NaF solutions?

13-40 Describe how a voltaic cell could be constructed from a strip of iron, a strip of lead, a $Fe(NO_3)_2$ solution, and a $Pb(NO_3)_2$ solution. Write the anode reaction, the cathode reaction, and the total reaction.

13-41 Judging from the relative difference in the strengths of the oxidizing agents (Fe^{2+} vs. Pb^{2+}) and (Zn^{2+} vs. Cu^{2+}), which do you think would be the more powerful cell, the one in Problem 13-40 or the Daniell cell? Why?

*13-42 The power of a cell depends on the strength of both the oxidizing and the reducing agents. Write the equation illustrating the most powerful redox reaction possible between an oxidizing agent and a reducing agent *in aqueous solution*. Consider only the species shown in Table 13-1.

ELECTROLYTIC CELLS

13-43 Chrome plating is an electrolytic process. Write the reaction that occurs when an iron bumper is electroplated using a $CrCl_3$ solution. Are there any metals shown in Table 13-1 on which a chromium layer would spontaneously form?

13-44 Why can't elemental sodium be formed in the electrolysis of an aqueous NaCl solution? Write the reaction that does occur at the cathode. How is elemental sodium produced by electrolysis?

13-45 Why can't elemental fluorine be formed by electrolysis of an aqueous NaF solution? Write the reaction that does occur at the anode. How is elemental fluorine produced?

13-46 A "tin can" is made by forming a layer of tin on a sheet of iron. Is electrolysis necessary for such a process or does it occur spontaneously? Write the equation illustrating this reaction. Is electrolysis necessary to form a layer of tin on a sheet of lead? Write the relevant equation.

13-47 Certain metals can be purified by electrolysis. For example, a mixture of Ag, Zn, and Fe can be dissolved so that their metal ions are present in aqueous solution. If a solution containing these ions is electrolyzed, which metal ion would be reduced to the metal first?

GENERAL QUESTIONS

13-48 Nitrogen exists in nine oxidation states. Arrange the following compounds in order of increasing oxidation state of N: K_3N, N_2O_4, N_2, NH_2OH, N_2O, $Ca(NO_3)_2$, N_2H_4, N_2O_3, NO.

13-49 Given the following information concerning metal strips immersed in certain solutions, write the net ionic equations representing the reactions that occur.

Metal Strip	Solution	Reaction
Cd	$NiCl_2$	Ni coating formed
Cd	$FeCl_2$	No reaction
Zn	$CdCl_2$	Cd coating formed
Fe	$CdCl_2$	No reaction

Where does Cd^{2+} rank as an oxidizing agent in Table 13-1?

13-50 A hypothetical metal (M) forms a coating of Sn when placed in a $SnCl_2$ solution. However, when a strip of Ni is placed in an MCl_2 solution, a coating of the metal M forms on the nickel. Write the net ionic equations representing the reactions that occur. Where does M^{2+} rank as an oxidizing agent in Table 13-1?

13-51 A solution of gold ions (Au^{3+}) reacts spontaneously with water to form metallic gold. Metallic gold does not react with chlorine but does react with fluorine. Write the equations illustrating the two spontaneous reactions, and locate Au^{3+} in Table 13-1 as an oxidizing agent.

13-52 Given the following *unbalanced* equation,

$$H^+(aq) + Zn(s) + NO_3^-(aq) \longrightarrow Zn^{2+}(aq) + N_2(g) + H_2O(l)$$

what mass of Zn is required to produce 0.658 g of N_2?

13–53 Given the following *unbalanced* equation,

$$MnO_2(s) + HBr(aq) \longrightarrow$$
$$MnBr_2(aq) + Br_2(l) + H_2O(l)$$

what mass of MnO_2 reacts with 228 mL of 0.560 M HBr?

13–54 Given the following *unbalanced* equation,

$$H^+(aq) + NO_3^-(aq) + Cu_2O(s) \longrightarrow$$
$$Cu^{2+}(aq) + NO(g) + H_2O(l)$$

what volume of NO gas measured at STP is produced by the complete reaction of 10.0 g of Cu_2O?

*13–55 Given the following *unbalanced* equation in acid solution:

$$H_2O(l) + HClO_3(aq) + As(s) \longrightarrow$$
$$H_3AsO_3(aq) + HClO(aq)$$

If 200 g of As reacts with 200 g of $HClO_3$, what mass of H_3AsO_3 is produced? Hint: Calculate the limiting reactant.

*13–56 Given the following *unbalanced* equation in basic solution:

$$Zn(s) + NO_3^-(aq) \longrightarrow$$
$$NH_3(g) + Zn(OH)_4^{2-}(aq)$$

what volume of NH_3 is produced by 6.54 g of Zn? The NH_3 is measured at 27.0°C and 1.25 atm pressure.

SOLUTIONS TO LEARNING CHECKS

A–1 oxidation state, -2, $+1$, loss, increase, reducing, gained, lost

A–2 (a) H_3BO_3 $3H + B + 3O = 0$ (b) $S_2O_3^{2-}$ $2S + 3O = -2$

$$3(+1) + B + 3(-2) = 0 \qquad\qquad 2S + 3(-2) = -2$$
$$\underline{\underline{B = +3}} \qquad\qquad\qquad\qquad \underline{\underline{S = +2}}$$

A–3 $ClO_2^- \rightarrow Cl^-$ ClO_2^- is reduced and is the oxidizing agent.
       $H_2O_2 \rightarrow O_2$ H_2O_2 is oxidized and is the reducing agent.

A–4 (a) $3e^- + 4H^+ + NO_3^- \longrightarrow NO + 2H_2O$ | $\times 2$

         $H_2O + H_2SO_3 \longrightarrow SO_4^{2-} + 4H^+ + 2e^-$ | $\times 3$

$$\cancel{6e^-} + \cancel{8H^+} + 2NO_3^- + \cancel{3H_2O} + 3H_2SO_3 \longrightarrow 2NO + \overset{1}{\cancel{4}}H_2O + 3SO_4^{2-} + \overset{4}{\cancel{12}}H^+ + \cancel{6e^-}$$

$$\underline{\underline{2NO_3^- + 3H_2SO_3 \longrightarrow 2NO + 3SO_4^{2-} + H_2O + 4H^+}}$$

   (b) $\cancel{4}e^- + 2H_2O + ClO_2^- \longrightarrow Cl^- + 4OH^-$ | $\times 1$

         $2OH^- + H_2O_2 \longrightarrow O_2 + 2H_2O + 2e^-$ | $\times 2$

$$\cancel{4e^-} + \cancel{2H_2O} + ClO_2^- + 4OH^- + 2H_2O_2 \longrightarrow Cl^- + \cancel{4OH^-} + 2O_2 + \overset{2}{\cancel{4}}H_2O + \cancel{4e^-}$$

$$\underline{\underline{ClO_2^- + 2H_2O_2 \longrightarrow Cl^- + 2O_2 + 2H_2O}}$$

B–1 voltaic, Zn, anode, Cu^{2+}, cathode, salt, Pb, oxidized, PbO_2, cathode, higher, reducing, electrolytic

B–2 A spontaneous reaction would occur in (b).

$$3Pb^{2+}(aq) + 2Cr(s) \longrightarrow 3Pb(s) + 2Cr^{3+}(aq)$$

B–3 $Br_2(l) + Fe(s) \longrightarrow 2Br^-(aq) + Fe^{2+}(aq)$

B–4 $MgCl_2(l) \longrightarrow Mg(l) + Cl_2(g)$

Because water enters this small lake at the same rate that it leaves, the level of the lake stays the same. This is similar to the equilibrium that occurs in certain chemical reactions.

REACTION RATES AND EQUILIBRIUM

Who among us has not discovered what was first thought to be a new and possibly alien life-form in the refrigerator? Usually it is in the back, perhaps on a long forgotten slice of bologna. A ghastly, green mold of some sort slowly devours the meat and even spreads to neighboring foods. What we are witnessing is a chemical reaction connected with the decay of foods and the growth of molds. But the situation could be much worse. Without the cool temperature in the refrigerator, the strange life-form that took weeks to develop would have evolved in a matter of a few days. At room temperature, meat and dairy products immediately begin to deteriorate as a result of chemical reactions. Decay, growth of alien life-forms, and all other chemical reactions slow down as the temperature decreases. How chemical reactions occur and why temperature affects the rate at which they occur are topics of this chapter.

◀ **Setting The Stage**

A related topic concerns reactions that attain a "balance." Many of us strive for a healthy balance between the demands of school or job and the need to relax and have fun. Or perhaps we seek a balance between having money coming in as fast as it goes out. In a chemical sense, balance means an equilibrium between two competing reactions, one forming products and the other, in the reverse direction, reforming the original reactants.

Some of the questions that we will be addressing in this chapter include: How do reactants become transformed into products? What factors besides temperature affect how fast a reaction occurs? Why don't some reactions go entirely to the right? What factors affect the distribu-

◀ **Formulating
Some Questions**

tion of reactants and products if the reaction does not go entirely to the right? How can we predict the amounts of reactants and products present at equilibrium?

Our first topic concerns how reactant molecules become transformed into product molecules.

14–1 HOW REACTIONS TAKE PLACE

How do reactions get from "here to there"?

Dalton's atomic theory tells us that there is a reshuffling of atoms in a chemical reaction. The theory makes no mention of how this reshuffling occurs, however. Our modern view of how chemical reactions occur is actually quite simple. For example, we know that because molecules have kinetic energy, they are all moving in one way or another. In the gaseous and liquid states, this motion leads to frequent collisions. These collisions are responsible for chemical reactions. *The assumption that chemical reactions are due to the collisions of molecules is known as the* **collision theory.** In some cases, collisions are such that chemical bonds are broken in the colliding molecules. If the molecular fragments from the collision reform in a different arrangement, a chemical reaction results. We will illustrate the collision theory by means of a hypothetical chemical reaction that can be "custom designed" to clearly illustrate the basic principles involved. Real-life reactions are generally more complicated, and explanations are required for each complication or exception.

Our reaction involves the combination of two hypothetical diatomic gaseous elements, A_2 and B_2, to form two molecules of a gaseous product, AB. The reaction is illustrated by the equation

$$A_2(g) + B_2(g) \longrightarrow 2AB(g)$$

Do all collisions cause a reaction?

We will assume that products form from the collision of an A_2 molecule with a B_2 molecule. If all collisions led to products, however, this and all reactions would be essentially instantaneous because of the large number of collisions per second. There are two conditions for the collision to lead to formation of products. First, *the collision between two reactant molecules must take place in the right geometric orientation.* As illustrated in Figure 14-1, for new bonds to form, the two reactant molecules must meet in a side-to-side manner rather than end-to-end or side-to-end. This condition, by itself, severely limits the number of collisions that can lead to a reaction. A second condition is that *the collision must occur with enough energy to break the bonds in the reactants so that new bonds can form in the products. This minimum kinetic energy needed for the reaction is known as the* **activation energy** *and is discussed in more detail in the next section. The second condition even more severely limits the number of collisions that lead to product formation. If the two conditions are not met, the colliding molecules simply recoil from each other unchanged. (See Figure 14-1.)

Looking Ahead ▼

The reaction in a stick of dynamite is obviously instantaneous, whereas it takes months for a nail to rust. Why the big difference? Whether a reaction takes place in an instant or a month depends on several factors that relate to the rate of the reaction. This is our next subject.

The chemical reaction in the explosion of dynamite occurs almost instantaneously.

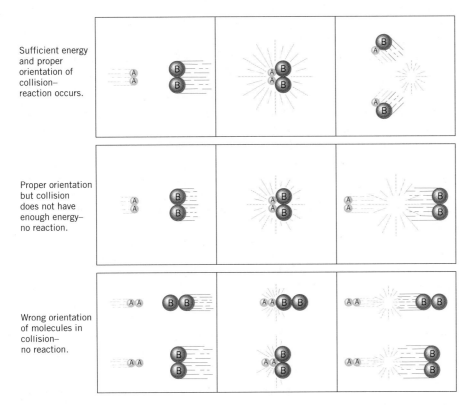

Figure 14–1
Reaction of A_2 and B_2. Reactions occur only when colliding molecules have the minimum amount of energy and the right orientation.

14–2 RATES OF CHEMICAL REACTIONS

As any chemical reaction occurs, reactants disappear and products appear. *The **rate of a reaction** measures the increase in concentration of a product or the decrease in concentration of a reactant per unit time.* There are several factors that affect the rate of a reaction. These include the magnitude of the activation energy, the temperature at which the reaction takes place, the concentration of reactants, the size of the particles of solid reactants, and the presence of a catalyst. We will discuss these factors individually.

Activation Energy
The total energy of a molecule is the sum of its kinetic energy (energy of motion) and its potential energy (energy of position and composition). Let us consider the energy and changes of our two hypothetical molecules, A_2 and B_2, as they move from positions that are far apart on a course that will eventually lead to a collision and the formation of products. As they approach each other, but before they make contact, both molecules have a given amount of kinetic energy of motion and a given amount of potential energy in the form of chemical energy stored in their bonds. This is represented by the energy of A_2 and B_2 molecules on the left side of the graph in Figure 14-2.

By moving to the right along the horizontal axis in the graph, we can follow the change in potential energy of the reactant molecules as they approach each other, collide, and eventually form products. This horizontal axis is referred to as the *reaction coordinate* and can be viewed as a time axis. The potential energy of the molecules does not change until they make contact and begin to compress together. As they collide, the molecules lose velocity, which means they lose kinetic energy. This kinetic energy changes to potential energy of the compressing molecules, so their potential energy rises. At maximum impact, the motion stops for an instant and the potential energy is at a maximum. This is analogous to hitting a tennis ball with a racquet. As the ball makes contact with the racquet, it slows down and compresses. For just an instant, the ball has no motion but has maximum potential energy from compression. As the ball recoils, the potential energy of the ball reconverts to kinetic energy of motion. Likewise, *at maximum impact, both*

How does the kinetic energy of a molecule change to potential energy?

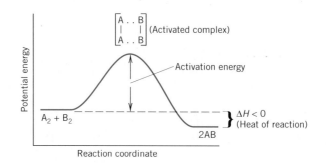

Figure 14–2
Activation Energy. The activation energy is the potential energy difference between the activated complex and the reactants.

molecules are compressed together, forming what is known as an **activated complex.** Since we are assuming reaction takes place, the activated complex represents the state that is intermediate between reactants and products. That is, old bonds are partially broken and new ones are partially formed. The activation energy for the forward reaction is represented by the difference in potential energy of the activated complex (at the peak of the curve) and the reactants. As product molecules recoil, potential energy decreases as it is converted into the kinetic energy of the recoiling molecules. Notice in Figure 14-2 that the product molecules (AB) eventually have lower potential energy than the original reactants. This difference between the potential energy of the products and that of the reactants represents the *heat of the reaction* and is given the symbol ΔH. *In an exothermic process, the potential energy of the products is lower than the potential energy of the reactants.* This means that the recoiling molecules eventually have more kinetic energy than the reacting molecules. In other words, the product molecules are "hotter." When the potential energy of the reactants is lower than that of the products, the reaction is endothermic and the product molecules are "cooler." An actual endothermic reaction between N_2 and O_2 to form NO is represented in Figure 14-3.

For a reaction to occur, the colliding molecules must have enough kinetic energy to form the activated complex (equal to the activation energy). If they don't have this amount of energy, the activated complex doesn't form and the colliding molecules simply recoil as unchanged reactants. It is much like trying to roll a bowling ball over a small hill. When we hurl the ball, we impart kinetic energy to the ball. As it goes up the hill, it slows as kinetic energy is converted into potential energy. If the ball has kinetic energy that is at least equal to the potential energy of the top of the hill, then the ball goes over and down the other side. If it does not, the ball rolls back.

Perhaps one of the most unique characteristics of any chemical reaction is its activation energy. Consider, for example, the reactions of two nonmetals with oxygen. An allotrope of phosphorus known as white phosphorus reacts almost instantly with the oxygen in the air, even at

At the instant the tennis ball is not moving, all of the energy is in the form of potential energy.

What if colliding molecules do not have enough kinetic energy to overcome the activation energy?

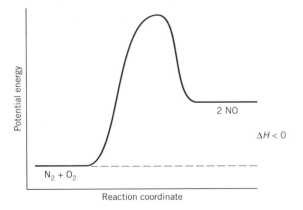

Figure 14–3
An Endothermic Reaction. In an endothermic reaction, the potential energy of the reactants is less than that of the products.

Phosphorus ignites instantly in air because the reaction has a low activation energy.

How does a higher temperature affect the rate of a reaction?

room temperature, producing an extremely hot flame. As a result, this form of phosphorus must be stored under water to prevent exposure to the air. Hydrogen also reacts with oxygen to form water. This is also a highly exothermic reaction that is used in space rockets. A mixture of hydrogen and oxygen, however, does not react at room temperature. In fact, no appreciable reaction occurs unless the temperature is raised to at least 400°C or the mixture is ignited with a spark. The difference in the rates of combustion of these two elements at room temperature lies in the activation energies of the two reactions. Figure 14-4 illustrates the activation energies for the combustion reactions of hydrogen and phosphorus. Note that the phosphorus reaction has a much lower activation energy than the hydrogen reaction. This explains why collisions between phosphorus molecules and oxygen have enough energy for reaction at room temperature, whereas collisions of the same energy between hydrogen and oxygen do not have the required energy to overcome the activation energy barrier.

Temperature

Previously, we established that the rate of a reaction depends on the kinetic energy of the reactant molecules:

$$r \text{ (rate)} \propto \text{energy of colliding molecules}$$

In Chapter 10 we discussed the distribution of kinetic energies of molecules at a given temperature. This is again illustrated in Figure 14-5. Note that, since temperature is related to average kinetic energy, the higher the temperature, the higher the average kinetic energy of the molecules. In the figure, the dashed line represents the activation energy (E_a). As mentioned, this is the minimum amount of kinetic energy that colliding molecules must have for products to form. Note that at the higher temperature (T_2) the curve intersects the dashed line at a higher point than T_1, meaning that a greater fraction of molecules can overcome the activation energy in a collision. As a result, the rates of chemical reactions increase as the temperature increases.

There is another reason why an increase in temperature increases the rate of a reaction. In addition to the energy of the collisions, the rate

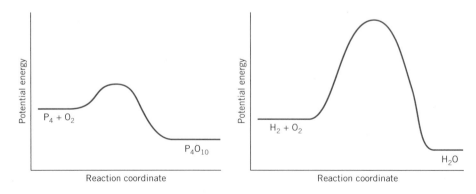

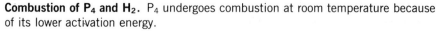

Figure 14–4
Combustion of P_4 and H_2. P_4 undergoes combustion at room temperature because of its lower activation energy.

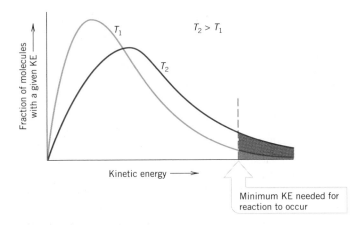

Figure 14–5
Kinetic Energy Distributions for a Reaction Mixture at Two Different Temperatures.
The sizes of the shaded areas under the curves are proportional to the total fractions
of the molecules that possess the minimum activation energy.

of a reaction depends on the frequency of collisions (the number of collisions per second):

$$r \text{ (rate)} \propto \text{frequency of collisions}$$

Recall that the kinetic energy (K.E.) of a moving object is given by
the relationship

$$\text{K.E.} = \tfrac{1}{2}mv^2 \quad (m = \text{mass}, v = \text{velocity})$$

This indicates that the more kinetic energy a moving object has, the
faster it is moving.

As the temperature increases, the average velocity of the molecules
increases. This means that the frequency of collisions also increases. As
an analogy, imagine a box containing red and blue Ping-Pong balls as
shown in Figure 14-6. If we jiggle the box, the balls move around and
collide. If we jiggle the box faster (analogous to a higher temperature),
the balls move around faster and there are more frequent collisions. The
increased noise we hear tells us that, indeed, collisions are not only

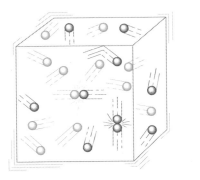

Slow jiggling

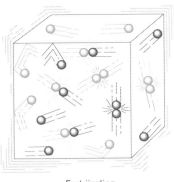

Fast jiggling

Figure 14–6
Effect of Velocity on Collisions. Collisions occur more frequently and with more
force as the velocity of the balls increases.

more energetic but more frequent. Most of the increase in the rate of a chemical reaction is due to the increased energy of the collisions, with a lesser contribution from the increased rate of collisions. The rates of many chemical reactions approximately double for each 10°C rise in temperature.

Concentration of Reactants

Increasing the rate at which we shake a box of Ping-Pong balls obviously increases the rate of collisions between the balls. Another way of increasing the rate of collisions is to increase the number of balls in the box. In Figure 14-7, we have two situations. On the left, there are four red and four blue balls. On the right, there are four red balls, but the number of blue balls has been increased to eight. The concentration (number of balls per unit volume) has been increased. Increasing the number of balls of either color increases the number of blue-red collisions, which is signaled by the more intense noise even though the box is jiggled at the same rate (analogous to the same temperature). In the reaction of A_2 and B_2, the same phenomenon applies. The greater the concentration of reactants, the more frequent the collisions and the greater the rate of the reaction.

Particle Size

If we wish to burn an old dead tree, there is a slow way and a quick way. Trying to burn it as one big log could take days. If we cut the tree up into smaller logs, we could burn it in hours. In fact, if we changed the tree into a pile of sawdust and spread it around, we might burn it in a matter of minutes. Related to this is the inherent danger in the storage of grain in large silos. Normally, a pile of grain burns slowly. However, small particles of grain dust can become suspended in air, forming a dangerous mixture that can actually detonate. Such explosions have occurred. In the reaction between a solid and a gas, a solid and a liquid, or a liquid and a gas, the surface area of the solid or liquid obviously affects the rate of the reaction. The more area that is exposed, the faster the rate of the reaction.

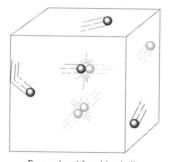

Four red and four blue balls

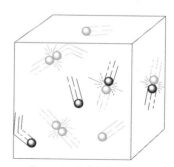

Four red and eight blue balls

Figure 14–7
Effect of Concentration on Collisions. Collisions occur more frequently when there are more balls in the box.

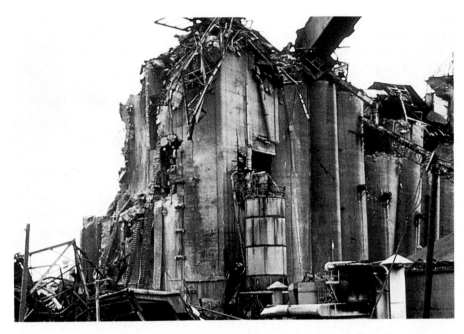

Grain dust and air make an explosive mixture because of the small particle size of the dust. This grain elevator was destroyed by such an explosion.

Catalysts

A **catalyst** *is a substance that is not consumed in a reaction but whose presence increases the rate of the reaction.* In effect, catalysts provide an alternate reaction pathway that has a lower activation energy. (See Figure 14-8.) The presence of a catalyst is analogous to having access to a tunnel through a mountain rather than having to go over a pass. It provides an easier and quicker route. Likewise, a lower activation energy for a reaction means a faster reaction rate.

An excellent example of how catalysts work is provided by the catalytic destruction of ozone (O_3) in the stratosphere, which is a topic of serious environmental concern. The following reaction is very slow in the absence of catalysts.

$$O_3(g) + O(g) \longrightarrow 2O_2(g)$$

A tunnel provides an easier, faster route than a high mountain pass.

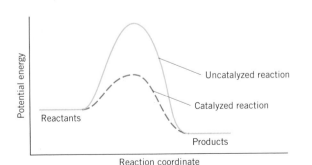

Figure 14–8
Activation Energy and Catalysis. A catalyst speeds a reaction by lowering the activation energy.

How do chlorine atoms destroy ozone?

The atoms of oxygen (O) originate from the dissociation of normal oxygen molecules (O_2) by solar radiation. Chlorofluorohydrocarbons (e.g., CF_2Cl_2), used as refrigerants and propellant gases in compressed spray cans, diffuse into the stratosphere and also are dissociated by solar radiation into Cl atoms (not Cl_2 molecules). When an atom of chlorine is present, two reactions take place that lead to the rapid conversion of ozone to normal oxygen (O_2).

$$Cl + O_3 \longrightarrow ClO + O_2$$

$$ClO + O \longrightarrow Cl + O_2$$

$$\text{Sum of the two reactions: } Cl + ClO + O_3 + O \longrightarrow Cl + ClO + 2O_2$$

$$\text{Overall reaction: } O_3 + O \longrightarrow 2O_2$$

There are two significant facts about these two reactions. First, the presence of chlorine atoms drastically increases the rate of conversion of ozone to oxygen. Of more importance is the fact that a chlorine atom is a reactant in the first reaction but is regenerated as a product in the second. This is how it acts as a catalyst. Because it is not consumed in the reaction, one chlorine atom can destroy billions of ozone molecules. In fact, the series of reactions leading to the destruction of ozone in the stratosphere appears to be more complex than that shown above. Intense studies are currently under way to help us understand exactly how chlorine atoms interact with ozone. In the meantime, the use of chlorofluorohydrocarbons is being phased out by international agreement.

In the reactions just described, the catalyst is intimately mixed with reactants and is actually involved in the reaction. Catalysts of this nature are referred to as *homogeneous catalysts*. In other cases, a catalyst may simply provide a surface on which reactions take place (*heterogeneous catalysts*).

Perhaps the most familiar application of heterogeneous catalysts is in the catalytic converter of an automobile (see Figure 14-9). The exhaust from the engine contains poisonous carbon monoxide and unburned fuel (mainly C_8H_{18}). Both contribute to air pollution, which is still a serious environmental concern in many localities. The catalytic converter, which is attached to the exhaust pipe, contains finely divided platinum and/or palladium. These metals provide a surface for the following reactions, which occur only at a very high temperature in the absence of a catalyst.

Figure 14–9
The Catalytic Converter. This device on the automobile helps to reduce air pollution.

$$2CO(g) + O_2(g) \longrightarrow 2CO_2(g)$$

$$2C_8H_{18}(g) + 25O_2(g) \longrightarrow 16CO_2(g) + 18H_2O(g)$$

$$2NO(g) \longrightarrow N_2(g) + O_2(g)$$

All three reactions are exothermic, which explains why the catalytic converter becomes quite hot when the engine is running.

Many reactions do not proceed to completion. It seems that some reactions proceed to the right just so far. These reactions have reached a point of equilibrium because the reaction is reversible. We consider reversible reactions that reach equilibrium next.

▼Looking Ahead

14–3 REACTIONS AT EQUILIBRIUM

Let's return to the hypothetical reaction discussed in Section 14-1 and illustrated in Figure 14-1. We will assume that this reaction is reversible. A **reversible reaction** *is one where both a forward reaction (forming products) and a reverse reaction (reforming reactants) can occur.* Reversible reactions where both reactions occur simultaneously reach a point of equilibrium. *The* **point of equilibrium** *in a reversible process is when both the forward and reverse processes proceed at the same rate so that the concentrations of reactants and products remain constant.*

Does a reaction stop once it reaches equilibrium?

In our hypothetical reaction, the reaction mixture initially contains only A_2 and B_2. As the reaction proceeds, the concentrations of these two molecules begin to decrease as the concentration of the product, AB, increases. The rate of buildup of AB begins to decrease until, eventually, the concentration of AB no longer changes despite the presence of excess reactant molecules. To understand this, we turn our attention to the product molecule AB. If we had started with pure AB, we would find that AB slowly decomposes to form A_2 and B_2, just the reverse of the original reaction. This reaction occurs when two AB molecules collide with the proper orientation and sufficient energy to form A_2 and B_2. (See Figure 14-10.)

Now we put these two reactions together. If we again start with pure A_2 and B_2, at first the only reaction leads to the formation of AB. As the concentration of AB increases, however, the reverse reaction becomes important. Eventually, the rate of formation of products (the forward reaction) is exactly offset by the rate of formation of reactants (the reverse reaction). At this point the concentrations of all species (A_2, B_2, and AB) remain constant. This phenomenon is referred to as a **dynamic equilibrium,** *which emphasizes the changing identities of reactants and products despite the fact that the total amounts of each do not change.* In Figure 14-11, the reaction and the point of equilibrium are illustrated graphically.

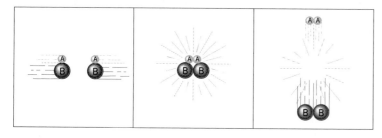

Figure 14–10
Reaction of 2AB. Collisions between AB molecules having the minimum energy and correct orientation lead to the formation of reactants.

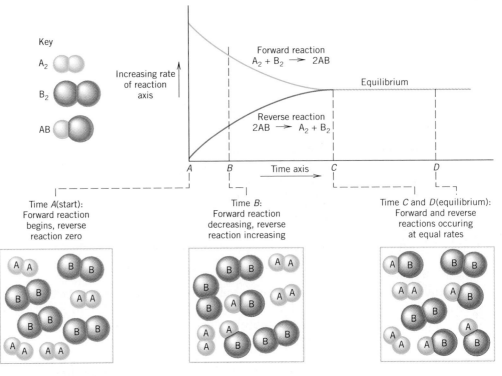

Figure 14–11

A_2, B_2, AB Equilibrium. Equilibrium is achieved when the rates of the forward and reverse reactions are equal.

If we could shrink ourselves to the molecular level, we would immediately become aware of this dynamic (changing) situation. If one A atom in an A_2 molecule was marked so that it could be traced, we would notice that at one moment it is present in an A_2 molecule, later it is part of an AB molecule, and still later it is part of another A_2 molecule.

A reversible reaction that reaches a point of equilibrium is indicated by the use of double arrows instead of a single arrow that implies a complete reaction. Thus our hypothetical reaction is represented as

$$A_2(g) + B_2(g) \rightleftharpoons 2AB(g)$$

In earlier parts of this text, we have mentioned chemical and physical systems that reach a point of equilibrium. Our first mention of this phenomenon was in Chapter 8, where we discussed the formation of ammonia from its elements. This is a classic example of a system at equilibrium since ammonia decomposes to its elements, which is the reverse reaction of its formation.

$$N_2(g) + 3H_2(g) \rightleftharpoons 2NH_3(g)$$

We encountered a system in Chapter 10 involving equilibrium between two physical states of water. When water is placed in a closed container at a certain temperature, it establishes an equilibrium vapor

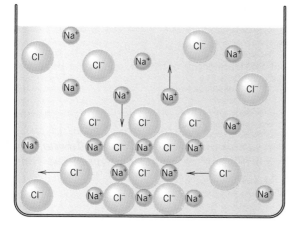

Figure 14–12
Solubility and Equilibrium. At equilibrium, the rate at which ions are dissolving equals the rate at which ions are forming the solid.

pressure. At the point of equilibrium, the vapor pressure is constant, which means that the rate of condensation equals the rate of evaporation. The equation for this equilibrium was

$$H_2O(l) \rightleftharpoons H_2O(g)$$

Our next experience with equilibrium was discussed in Chapter 11. When a solvent is saturated with a solute, such as water saturated with table salt (NaCl), it appears that no additional solute dissolves. In fact, dissolving is still occurring, but formation of the solid is occurring at the same rate, so it appears that no additional solute dissolves. (See Figure 14-12.) This is illustrated by the following equation indicating a dynamic equilibrium condition between the dissolved ions and the solid salt.

$$NaCl(s) \rightleftharpoons Na^+(aq) + Cl^-(aq)$$

A common but interesting chemical phenomenon that is often displayed at science fairs is the science (and art) of crystal growing. Crystal growing demonstrates the principle of equilibrium. If a comparatively large crystal of a water-soluble compound such as $CuSO_4$ is suspended in a saturated solution of that compound, the small crystals at the bottom get smaller and the large one grows larger. The concentrations of Cu^{2+} and SO_4^{2-} in solution remain constant as does the total mass of solid crystals. The changing size of the crystals, however, indicates that an equilibrium exists in which the identities of ions in the solid and dissolved states are changing. (See Figure 14-13.) This phenomenon occurs because small crystals, due to their greater surface area, dissolve faster than large crystals. Thus although the total mass of the crystals remains constant, the unequal rate of solution favors solution of small crystals and precipitation on the large crystal.

The equilibrium reactions discussed most recently were in Chapter 12 and involved the weak acids and bases. Although the ionization of a weak acid in water is unfavorable, it does occur to a small but measurable extent. At equilibrium, most molecules are present in water in the

Figure 14–13
Crystal Growth. The size of the crystals changes, but the total mass does not.

molecular rather than the ionized state, as represented by the following equilibrium equation:

$$HF(aq) + H_2O(l) \rightleftharpoons H_3O^+(aq) + F^-(aq)$$

The quantitative aspects of the reactions of weak acids and bases in water are discussed later in this chapter.

Looking Ahead ▼

A system at equilibrium has a remarkable ability to adapt to external changes. For example, if the temperature of a mixture of gases at equilibrium is changed, the system eventually reestablishes equilibrium by adjusting the equilibrium proportions of reactants and products. How and why this happens is our next topic.

14–4   EQUILIBRIUM AND LE CHÂTELIER'S PRINCIPLE

The point of equilibrium for a specific reaction may lie to the right, which means that products are favored over reactants or to the left where reactants are favored over products. In any case, there are also several external changes of conditions that may affect the point of equilibrium.

Can chemical systems suffer stress?

How a system at equilibrium reacts to a change in conditions is summarized by **Le Châtelier's principle,** which states: *When stress is applied to a system at equilibrium, the system reacts in such a way to counteract the stress.* The changes in a chemical system that affect the point of equilibrium include:

1   a change in concentration of a reactant or product

2   a change in the pressure on a gaseous system

3   a change in the temperature

We will discuss these three changes individually. To illustrate how these conditions affect the point of equilibrium, we will use a very important industrial process involving the following reaction.

$$SO_2(g) + O_2(g) \rightleftharpoons 2SO_3(g)$$

The SO_2 is formed by the combustion of sulfur or sulfur compounds. The SO_3 produced by the reaction is then allowed to come in contact with water to form sulfuric acid.

$$SO_3(g) + H_2O(l) \longrightarrow H_2SO_4(aq)$$

In 1992, 43 million tons of H_2SO_4, with a value of about \$4 billion, were produced by this process in the United States alone. (See Figure 14-14.)

In such an important industrial process, it is important to convert the maximum amount of SO_2 to SO_3. In other words, the goal is to shift the equilibrium to the right as much as possible in favor of the product. There are several conditions that can be adjusted in order to do this.

1   If the system is at equilibrium and then an additional portion of a reactant is introduced, the system shifts to produce more products. This

Figure 14–14
A Sulfuric Acid Plant. More sulfuric acid is manufactured than any other chemical.

happens because the rate of the forward reaction increases due to the increased number of collisions that can occur to form products. The concentration of products will increase, which then increases the rate of the reverse reaction. Eventually, equilibrium is reestablished. In our system, if additional O_2 is added, it will lead to formation of additional SO_3. The additional O_2 can be considered as a stress on a system at equilibrium. Formation of more products removes some of the additional O_2 and thus relieves the stress on the system. We can generalize this observation. *An increase in the concentration of a reactant or product ultimately leads to an increase in the concentration of the species on the other side of the equation.* A decrease in the concentration of a reactant or product ultimately leads to formation of more of the reactant or product that decreased. In reactions where a precipitate forms or a gas is evolved, that product is, in effect, removed from the system. The system shifts to replace this loss, but if the loss is complete, the reaction in the direction of that loss is complete. This is known as an *irreversible reaction.*

2 In a gaseous system, a change in the volume of the reaction container or the pressure also affects the point of equilibrium. Boyle's law tells us that a *decrease in volume corresponds to an increase in pressure.* Increasing the pressure on a gaseous mixture of reactants and products at equilibrium (or decreasing the volume of the container) is another example of stress on a system. The system can counteract the stress of increased pressure by shifting to a lower volume. In our sample system, there are three moles of gaseous reactants ($2SO_2 + O_2$) but only two moles of the gaseous product ($2SO_3$). Therefore, an increase in pressure on this system shifts the point of equilibrium to the right, increasing the concentration of SO_3 until equilibrium is eventually reestablished. *In general, an increase in pressure on a gaseous system shifts the point of equilibrium in the direction of the fewest number of moles of gas.* When gases are not involved in a reaction at equilibrium, such as one in aqueous solution, pressure changes have negligible effect since solids and liquids are virtually incompressible.

3 The reaction of SO_2 with O_2 is an exothermic reaction, which means that heat energy is produced along with SO_3. The equation showing heat as a product can be written as follows:

$$SO_2(g) + O_2(g) \rightleftharpoons 2SO_3(g) + \text{heat}$$

Since heat can be considered a component of the reaction, it can be treated in the same way as any other component, as discussed in **1** above. That is, if we add heat (increase the temperature), the reaction shifts in a direction to remove that heat. If we remove heat by cooling the reaction mixture, the equilibrium shifts in a direction to replace that heat loss, which increases the concentration of species on the other side. *For an exothermic reaction, cooling the reaction mixture increases the concentration of products when equilibrium is reestablished. For an endothermic reaction, where heat can be considered as a reactant, heating the reaction mixture increases the concentration of products at equilibrium.* In our system, the formation of SO_3 is favored at low temperatures. Cooling the reaction mixture removes heat. The system attempts to compensate for this loss by formation of more SO_3 and thus more heat.

The formation of products in the endothermic reaction illustrated in Figure 14-3 is favored at high temperatures. As a result, this reaction between the two major components of our atmosphere does not take place under normal atmospheric conditions but does occur to some extent in the hot combustion chambers of automobile engines.

In our example system, we have a potential problem as we lower the temperature to increase the concentration of SO_3. The rate of the reaction decreases as the temperature decreases. Since time is also important in industry, we must consider a compromise. The reaction must be run at a high enough temperature so that it proceeds to equilibrium in a reasonable amount of time. The discovery of a catalyst for the reaction allows the reaction to be run at a low enough temperature so that a reasonable yield of SO_3 is possible. That temperature is around 400°C.

How do reaction rates and yields of products sometimes conflict?

Catalysts play a very significant role in many chemical reactions with industrial significance. *A catalyst in a system that reaches a point of equilibrium increases the rate of both the forward and reverse reaction but does not change the point of equilibrium.* The function of a catalyst is simply to reach the point of equilibrium in less time. As a consequence, the use of a catalyst allows the reaction to be run at a considerably lower temperature than would otherwise be possible. In the reaction discussed, heterogeneous catalysts such as V_2O_5 are effective. In other industrial processes, the exact formulation of the catalyst for a specific reaction may be a closely guarded secret.

In summary, the production of SO_3 from SO_2 is affected by the following:

	Effect at Equilibrium		
Change	SO_2	O_2	SO_3
Add O_2	decrease	increase	increase
Increase P	decrease	decrease	increase
Decrease V	decrease	decrease	increase
Lower T	decrease	decrease	increase
Add catalyst	no effect	no effect	no effect

Looking Back ▲

Chemical reactions don't just happen. The molecules that react must make contact in order for old bonds to break and new bonds to form. Even then, a reaction does not occur every time two reactant molecules collide. There are important conditions involving the way the molecules interact with each other that make formation of products possible. How fast these reactions occur (the reaction rate) depends on several conditions such as the temperature and the concentration of reactants. Some reactions, however, are reversible and do not proceed completely to the right. Still, we can force the point of equilibrium to the right or left by manipulating conditions appropriately.

Learning Check A

A–1. Fill in the blanks.

Chemical reactions occur because of _____ between reactant molecules. In order for a reaction to take place, molecules must _____ in the proper orientation and with a _____ amount of energy known as the _____ energy. The rate of appearance of a product as a function of _____ is a measure of the _____ of the reaction. An _____ _____ forms at the point of maximum potential energy. The difference in potential energy between the activated complex and the reactants is the _____ _____. At _____ temperatures, more molecules can overcome this energy barrier so that the rate of the reaction _____. Increasing the concentration of colliding molecules _____ the rate of the reaction. The presence of a _____ does not change the point of equilibrium, but it _____ the rate of the reaction. Reactions that are reversible may reach a _____ _____ _____ where the concentrations of reactants and products remain _____. The point of equilibrium may be affected by a change of _____ of reactants, a change of _____, and a change of _____ (in a gaseous system).

A–2. Given the following equilibrium,

$$2H_2O(g) \rightleftharpoons 2H_2(g) + O_2(g) \qquad \Delta H = +484 \text{ kJ}$$

will the concentration of H_2 at equilibrium be increased, decreased, or not affected by the following?

   (a) an increase in concentration of H_2O
   (b) an increase in concentration of O_2
   (c) an increase in temperature
   (d) an increase in pressure
   (e) addition of a catalyst

A–3. Sketch a reaction profile for the reaction in A-2. Sketch a reaction profile when a catalyst is present.

Additional Examples: 14-1, 14-3, 14-6, 14-8, 14-10, and 14-12.

At the point of equilibrium, there are definite amounts of reactants and products present as long as conditions are held constant. In fact, if we start with any amounts of reactants and/or products, we can predict the distribution of all species at equilibrium. Equilibrium distributions are determined by a law just as the conditions of gases are determined by gas laws. The quantitative aspects of the point of equilibrium are our next subject.

14–5   POINT OF EQUILIBRIUM AND LAW OF MASS ACTION

Long before chemists knew anything about collision theory, reaction rates, or activation energy, they realized that the concentrations of reactants and products were distributed in a predictable manner in a reaction at equilibrium. This, of course, is for a specific reaction at a given temperature. To illustrate how the concentrations of reactants and products relate, we will consider a hypothetical reversible reaction involving a moles of reactant A combining with b moles of B, etc.

$$a\text{A} + b\text{B} \rightleftharpoons c\text{C} + d\text{D}$$

For a reaction at equilibrium, the distribution of reactants and products is given by the following relationship, which is known as the **law of mass action.**

$$K_{eq} = \frac{[\text{C}]^c[\text{D}]^d}{[\text{A}]^a[\text{B}]^b}$$

Note that the coefficients (a, b, c, and d) of the compounds (A, B, C, and D) become the exponents of the molar concentrations (mol/L) of the compounds. The products are written in the numerator, and the reactants are written in the denominator. The law of mass action for a particular reaction consists of two parts. K_{eq} *is called the* **equilibrium constant** *and has a definite numerical value at a given temperature. The ratio to the right of the equality is called the* **equilibrium constant expression.**

How does the law of mass action work?

The significance of the law of mass action is that, even with a variety of initial concentrations of reactants and products, the distribution of reactants and products is predictable when the reaction reaches the point of equilibrium. This is illustrated in Table 14-1. In a series of three experiments, we start with an initial concentration of reactants (expt. 1), an initial concentration of products (expt. 2), and an initial concentration of reactants and products (expt. 3). In all cases, the eventual distribution follows the law of mass action. That is, by substituting the actual equilibrium concentrations of all species in the equilibrium constant expression, a consistent value for the equilibrium constant is obtained.

In our first example, we will write the laws of mass action for three reversible reactions in the gaseous phase.

EXAMPLE 14–1

The Law of Mass Action

Working It Out ●

Write the law of mass action for each of the following reactions.
(a) $N_2(g) + 3H_2(g) \rightleftharpoons 2NH_3(g)$
(b) $4NH_3(g) + 5O_2(g) \rightleftharpoons 4NO(g) + 6H_2O(g)$
(c) $PCl_3(g) + Cl_2(g) \rightleftharpoons PCl_5(g)$

Solutions

(a)
$$K_{eq} = \frac{[NH_3]^2}{[N_2][H_2]^3}$$

(b)
$$K_{eq} = \frac{[NO]^4[H_2O]^6}{[NH_3]^4[O_2]^5}$$

(c)
$$K_{eq} = \frac{[PCl_5]}{[PCl_3][Cl_2]}$$

The numerical value of K_{eq} is a measure of the extent of the reaction, so it tells us about the *position* of equilibrium. A large value for K_{eq} indi-

| TABLE 14–1 | THE VALUE OF K_{eq} |

For the reaction $H_2(g) + I_2(g) \rightleftharpoons 2HI(g)$ at 450°C,

$$K_{eq} = \frac{[HI]^2}{[H_2][I_2]} \quad [X] = \text{concentration in mol/L}$$

Expt.	Initial Concentration			Equilibrium Concentration			
	$[H_2]$	$[I_2]$	$[HI]$	$[H_2]$	$[I_2]$	$[HI]$	K_{eq}
1	2.000	2.000	0	0.428	0.428	3.144	53.9
2	0	0	2.000	0.214	0.214	1.572	54.0
3	1.000	1.000	1.000	0.321	0.321	2.358	54.0

cates that the numerator is much larger than the denominator in the equilibrium constant expression. This means that the concentrations of products are larger than those of reactants. For example, consider the simple equilibrium

$$A \rightleftharpoons B \quad K_{eq} = 10^2$$

$$K_{eq} = \frac{[B]}{[A]} = 10^2 \quad \text{thus } [B] = 10^2[A] = 100[A]$$

The large value of the equilibrium constant signifies that, for this reaction at the point of equilibrium, the concentration of the product [B] is 100 times the concentration of the reactant [A]. That is, the equilibrium lies far to the right. On the other hand, if the value of K_{eq} were small (e.g., 10^{-2}), then the concentration of the reactant [A] would be 100 times the concentration of the product [B]. In this case, the equilibrium lies far to the left.

What does the value of the constant tell us about the point of equilibrium?

   The equilibrium constant is calculated from the experimental determination of the distribution of reactants and products at equilibrium. The following two examples illustrate such calculations.

EXAMPLE 14–2

Calculation of K_{eq}

What is the value of K_{eq} for the following system at equilibrium?

$$2NO(g) + O_2(g) \rightleftharpoons 2NO_2(g)$$

● **Working It Out**

At a given temperature, the equilibrium concentrations of the gases are [NO] = 0.890, $[O_2]$ = 0.250, and $[NO_2]$ = 0.0320.

Solution
For this reaction,

$$K_{eq} = \frac{[NO_2]^2}{[NO]^2[O_2]}$$

$$= \frac{[0.0320]^2}{[0.890]^2[0.250]} = \frac{1.024 \times 10^{-3}}{0.198}$$

$$= \underline{\underline{5.17 \times 10^{-3}}}$$

EXAMPLE 14-3

Calculation of Equilibrium Concentrations and K_{eq}

For the equilibrium

$$N_2(g) + 3H_2(g) \rightleftharpoons 2NH_3(g)$$

complete the table and compute the value of the equilibrium constant.

Initial Concentration			Equilibrium Concentration		
[H$_2$]	[N$_2$]	[NH$_3$]	[H$_2$]	[N$_2$]	[NH$_3$]
0.200	0.200	0	?	?	0.0450

Procedure

1 As in other stoichiometry problems, find the [H$_2$] and [N$_2$] that reacted to form the 0.0450 mol/L of NH$_3$ (Section 8-3).

2 Find the [H$_2$] and [N$_2$] remaining at equilibrium by subtracting the concentration that reacted from the initial concentration; that is,

$$[N_2]_{eq} = [N_2]_{initial} - [N_2]_{reacted}$$

3 Substitute the concentrations of all compounds present at equilibrium into the equilibrium constant expression and solve to find the value of K_{eq}.

Solution

1 If 0.0450 mol of NH$_3$ is formed, calculate the number of moles of N$_2$ that reacted (per liter).

$$0.0450 \ \text{mol NH}_3 \times \frac{1 \ \text{mol N}_2}{2 \ \text{mol NH}_3} = 0.0225 \ \text{mol N}_2 \ \text{reacted (per liter)}$$

The number of moles of H$_2$ that reacted is

$$0.0450 \ \text{mol NH}_3 \times \frac{3 \ \text{mol H}_2}{2 \ \text{mol NH}_3} = 0.0675 \ \text{mol H}_2 \ \text{reacted (per liter)}$$

2 At equilibrium,

$$[N_2]_{eq} = 0.200 - 0.0225 = 0.178$$

$$[H_2]_{eq} = 0.200 - 0.0675 = 0.132$$

3
$$K_{eq} = \frac{[NH_3]^2}{[N_2][H_2]^3} = \frac{(0.0450)^2}{(0.178)(0.132)^3} = \underline{\underline{4.95}}$$

EXAMPLE 14-4

Calculation of Equilibrium Concentrations from K_{eq}

In the preceding equilibrium, what is the concentration of NH$_3$ at equilibrium if the equilibrium concentrations of N$_2$ and H$_2$ are 0.22 and 0.14 mol/L, respectively?

Procedure

In this problem, use the value of K_{eq} found in the previous problem. The concentration of NH_3 can be found by substituting the concentrations of the species given and solving for the one unknown.

Solution

$$K_{eq} = \frac{[NH_3]^2}{[N_2][H_2]^3} \quad \begin{array}{l} [N_2] = 0.22 \\ [H_2] = 0.14 \end{array} \quad K_{eq} = 4.95$$

$$= \frac{[NH_3]^2}{[0.22][0.14]^3} = 4.95$$

$$[NH_3]^2 = 2.99 \times 10^{-3}$$

$$[NH_3] = 5.5 \times 10^{-2} = 0.055$$

Thus the concentration of NH_3 = <u>0.055 mol/L.</u>

We now return to weak acids, weak bases, and buffer solutions that we first discussed in Chapter 12. As you recall, weak acids and weak bases produce only a small concentration of H_3O^+ or OH^-, respectively. Even though these equilibria lie far to the left, the small concentrations of these ions are significant in the concept of acids and bases. We will apply the law of mass action to these systems in the final section of this chapter.

▼**Looking Ahead**

14–6   EQUILIBRIA OF WEAK ACIDS AND WEAK BASES IN WATER

Strong acids such as HNO_3 are completely ionized in water. Since no appreciable equilibrium is involved, the ionization is represented by a single arrow.

$$HNO_3(aq) + H_2O(l) \longrightarrow H_3O^+(aq) + NO_3^-(aq)$$

A weak acid such as hypochlorous acid (HClO), however, produces only a small concentration of ions because the ionization is a reversible reaction in which the equilibrium lies far to the left. The ionization of HClO is represented by the following equation.

$$HClO(aq) + H_2O(l) \rightleftharpoons H_3O^+(aq) + ClO^-(aq)$$

The law of mass action can be written as

$$K_{eq} = \frac{[H_3O^+][ClO^-]}{[H_2O][HClO]}$$

In this equilibrium, the concentrations of H_3O^+, ClO^-, and HClO can all be varied. H_2O is the solvent, however, and is present in a very large excess compared with the other species. Since the amount of H_2O actually reacting is very small compared with the total amount present, the concentration of H_2O is essentially a constant. The $[H_2O]$ can therefore be included with the other constant, K_{eq}, to produce another constant labeled $\boldsymbol{K_a}$, *which is known as an* **acid ionization constant.**

$$K_{eq}[H_2O] = K_a = \frac{[H_3O^+][ClO^-]}{[HClO]}$$

Ionization of the weak base NH_3 is illustrated as

$$NH_3(aq) + H_2O(l) \rightleftharpoons NH_4^+(aq) + OH^-(aq)$$

For this reaction, the *equilibrium constant is labeled* $\mathbf{K_b}$, *which is the* **base ionization constant.**

$$K_b = \frac{[NH_4^+][OH^-]}{[NH_3]}$$

The values of the equilibrium constants for weak acids and weak bases can be determined experimentally as illustrated in the following examples.

EXAMPLE 14–5

Calculation of K_a from Equilibrium Concentrations

Working It Out ●

In a 0.20 M solution of HNO_2, it is found that 0.009 mol/L of the HNO_2 dissociates. What are the concentrations of H_3O^+, NO_2^-, and HNO_2 at equilibrium, and what is the value of K_a?

Procedure

1   Write the equilibrium equation.

2   Calculate the concentration of undissociated HNO_2 present at equilibrium. Remember that the initial concentration given (0.20 M) represents the concentration of undissociated HNO_2 present *plus* the concentration that dissociates. Therefore, $[HNO_2]$ at equilibrium is the initial concentration minus the concentration that dissociates.

$$[HNO_2]_{eq} = [HNO_2]_{initial} - [HNO_2]_{dissociated}$$

3   Note that one H_3O^+ and one NO_2^- are formed for each HNO_2 that dissociates (from the equation stoichiometry).

$$[H_3O^+]_{eq} = [NO_2^-]_{eq} = [HNO_2]_{dissociated}$$

4   Calculate K_a.

Solution

1   $$HNO_2(aq) + H_2O(l) \rightleftharpoons H_3O^+(aq) + NO_2^-(aq)$$

2   $$[HNO_2]_{initial} = 0.20 \qquad [HNO_2]_{dissociated} = 0.009$$

$$[HNO_2]_{eq} = 0.20 - 0.009 = \underline{\underline{0.19}}$$

3   $$\underline{\underline{[H_3O^+] = [NO_2^-] = 0.009}}$$

4   $$K_a = \frac{[H_3O^+][NO_2^-]}{[HNO_2]} = \frac{[0.009][0.009]}{[0.19]}$$

$$= \underline{\underline{4 \times 10^{-4}}}$$

EXAMPLE 14–6

Calculation of K_a from pH

A 0.25 M solution of HCN has a pH of 5.00. What is K_a?

Procedure

1 Write the equilibrium reaction.

2 Convert pH to $[H_3O^+]$.

3 Note that $[CN^-] = [H_3O^+]$ at equilibrium.

4 Calculate [HCN] at equilibrium, which is

$$[HCN]_{eq} = [HCN]_{initial} - [HCN]_{dissociated}$$

$[HCN]_{dissociated} = [H_3O^+]_{eq}$, since one H_3O^+ is produced at equilibrium for every HCN that dissociates.

5 Use these values to calculate K_a.

Solution

$$HCN(aq) + H_2O(l) \rightleftharpoons H_3O^+(aq) + CN^-(aq)$$

$$[H_3O^+] = \text{antilog} \, (-5.00)$$

$$[H_3O^+] = 1.0 \times 10^{-5} = [CN^-]$$

$$[HCN]_{eq} = 0.25 - (1.0 \times 10^{-5}) \approx 0.25$$

($\approx$ means approximately equal)

Note that the amount of HCN that dissociates (10^{-5} M) is negligible compared with the initial concentration of HCN (0.25 M.* Therefore, $[HCN]_{eq} = [HCN]_{initial}$.

$$K_a = \frac{[H_3O^+][CN^-]}{[HCN]}$$

$$= \frac{(1.0 \times 10^{-5})(1.0 \times 10^{-5})}{(0.25)}$$

$$= \underline{\underline{4.0 \times 10^{-10}}}$$

EXAMPLE 14–7

Calculation of K_b from Percent Ionization

A 0.10 M solution of NH_3 is 1.34% ionized. What is the value of K_b (the base ionization constant)?

Procedure

Find the concentrations of all species in the mass action expression at equilibrium. Substitute these values into the expression and solve for K_b.

Solution

$$NH_3(aq) + H_2O(l) \rightleftharpoons NH_4^+(aq) + OH^-(aq)$$

$$K_b = \frac{[NH_4^+][OH^-]}{[NH_3]}$$

At equilibrium, 1.34% of the NH_3 is ionized, or

$$0.0134 \times 0.10 = 1.34 \times 10^{-3} \text{ mol/L}$$

---

*For the purposes of these calculations (two significant figures), a number is considered negligible compared with another if it is less than 10% of the larger number.

According to the equation, for every NH_3 ionized, one NH_4^+ and one OH^- form. Therefore, if 1.34×10^{-3} mol/L ionizes, at equilibrium

$$[NH_4^+] = [OH^-] = 1.34 \times 10^{-3}$$

The concentration of NH_3 at equilibrium is the initial concentration (0.10) minus the concentration that ionizes.

$$[NH_3] = 0.10 - (1.34 \times 10^{-3}) \approx 0.10$$

Substituting these values into the expression,

$$K_b = \frac{(1.34 \times 10^{-3})(1.34 \times 10^{-3})}{(0.10)}$$

$$= \underline{\underline{1.8 \times 10^{-5}}}$$

How can one tell which is a stronger acid?

In Chapter 12, we listed several acids in order of increasing acid strength in Table 12-1. We now have a numerical value (K_a) that is the quantitative measure of this acid strength. The larger the value for K_a, the stronger the acid. The acids in Table 12-1 were listed in order of decreasing values of K_a. The K_a values for some weak acids are shown in Table 14-2, and the K_b values of some weak bases are listed in Table 14-3.

TABLE 14–2 K_a FOR SOME WEAK ACIDS

$K_a = \dfrac{[H_3O^+][X^-]}{[HX]}$ (HX symbolizes a weak acid, X^- its conjugate base)

Acid	Formula	K_a	
Hydrofluoric	HF	6.7×10^{-4}	
Nitrous	HNO_2	4.5×10^{-4}	
Formic	$HCHO_2$	1.8×10^{-4}	Decreasing
Acetic	$HC_2H_3O_2$	1.8×10^{-5}	acid
Hypochlorous	HClO	3.2×10^{-8}	strength
Hypobromous	HBrO	2.1×10^{-9}	
Hydrocyanic	HCN	4.0×10^{-10}	

TABLE 14–3 K_b FOR SOME WEAK BASES

$K_b = \dfrac{[HB^+][OH^-]}{[B]}$ (B symbolizes a weak base, HB^+ its conjugate acid)

Base	Formula	K_b	
Dimethylamine	$HN(CH_3)_2$	7.4×10^{-4}	Decreasing
Ammonia	NH_3	1.8×10^{-5}	base
Hydrazine	N_2H_4	9.8×10^{-7}	strength

We are now ready to use the values for K_a or K_b to calculate the pH of a solution of a weak acid or weak base of known concentration. These calculations are illustrated in the following examples.

EXAMPLE 14-8

Calculation of the pH of a Solution of a Weak Acid

What is the pH of a 0.155 M solution of HClO?

Procedure

●Working It Out

1 Write the equilibrium involved.

2 Write the appropriate equilibrium constant expression.

3 Let $x = [H_3O^+]$ at equilibrium; since $[H_3O^+] = [ClO^-]$, $x = [ClO^-]$.

4 At equilibrium, $[HClO]_{eq} = [HClO]_{initial} - [HClO]_{dissociated}$.

5 Using the value for K_a in Table 14-2, solve for x.

6 Convert x to pH.

In summary:

	HClO	$[H_3O^+]$	$[ClO^-]$
Initial	0.155	0	0
Equilibrium	0.155 − x	x	x

Solution

1
$$HClO(aq) + H_2O(l) \rightleftharpoons H_3O^+(aq) + ClO^-(aq)$$

2
$$K_a = \frac{[H_3O^+][ClO^-]}{[HClO]} = 3.2 \times 10^{-8}$$

3 At equilibrium, $[H_3O^+] = [ClO^-] = x$.

4 At equilibrium, $[HClO] = 0.155 - x$.

5
$$\frac{[x][x]}{[0.155 - x]} = 3.2 \times 10^{-8}$$

The solution of this equation appears to require the quadratic equation. However, a simplification of this calculation is possible. Note that K_a is a small number, indicating that the degree of dissociation is small (the equilibrium lies far to the left). This means that x is a very small number. Since very small numbers make little or no difference when added to or subtracted from large numbers, they can be ignored with little or no error. (Refer to the example of the large crowd in Section 1-1.)

$$0.155 - x \approx 0.155$$

Therefore, the expression can now be simplified:

$$\frac{[x][x]}{[0.155 - x]} = \frac{x^2}{0.155} = 3.2 \times 10^{-8}$$

$$x^2 = 5.0 \times 10^{-9}$$

To solve for x, take the square root of both sides of the equation:

$$\sqrt{x^2} = \sqrt{5.0 \times 10^{-9}} = \sqrt{50 \times 10^{-10}}$$

$$x = [H_3O^+] = 7.1 \times 10^{-5}$$

(Note that the x is indeed much smaller than 0.155.)

6          $$pH = -\log[H_3O^+] = -\log(7.1 \times 10^{-5}) = \underline{\underline{4.15}}$$

EXAMPLE 14–9

Calculation of the pH of a Solution of a Weak Base

What is the pH of a 0.245 M solution of N_2H_4?

Solution

1          $$N_2H_4(aq) + H_2O(l) \rightleftharpoons HN_2H_4^+(aq) + OH^-(aq)$$

2          $$K_b = \frac{[HN_2H_4^+][OH^-]}{[N_2H_4]} = 9.8 \times 10^{-7} \text{ (from Table 14-3)}$$

3   Let $x = [OH^-] = [HN_2H_4^+]$ (at equilibrium).

4          $[N_2H_4] = 0.245 - x$ (at equilibrium). Since K_b is very small, x is very small. Therefore,

$$[0.245 - x] \approx [0.245]$$

5          $$\frac{[x][x]}{[0.245]} = 9.8 \times 10^{-7}$$

$$x^2 = 2.4 \times 10^{-7}$$

$$x = 4.9 \times 10^{-4} = [OH^-]$$

6          $$pOH = -\log[OH^-] = -\log(4.9 \times 10^{-4}) = 3.31$$

$$pH = 14.00 - pOH = 14 - 3.31 = \underline{\underline{10.69}}$$

EXAMPLE 14–10

Calculation of the pH of a Buffer Solution

What is the pH of a buffer solution that is made by dissolving 0.50 mol of $HC_2H_3O_2$ and 0.50 mol of $NaC_2H_3O_2$ in enough water to make 1.00 L of solution?

Procedure

1, 2   As in previous examples.

   3   Let $x = [H_3O^+]$ at equilibrium. In this case, $[C_2H_3O_2^-] = 0.50 + x$ (the concentration from the dissolved salt plus that from the dissociation of the acid).

   4 $$[HC_2H_3O_3] = 0.50 - x$$

5, 6   As in previous examples.

Solution

1    $HC_2H_3O_2(aq) + H_2O(l) \rightleftharpoons H_3O^+(aq) + C_2H_3O_2^-(aq)$

2    $$K_a = \frac{[H_3O^+][C_2H_3O_2^-]}{[HC_2H_3O_2]}$$

3, 4   Let $x = [H_3O^+]$, $[C_2H_3O_2^-] = 0.50 + x$, and $[HC_2H_3O_2] = 0.50 - x$.

Since K_a is small, x is small. Therefore,

$$[C_2H_3O_2^-] = 0.50 + x \approx 0.50$$

$$[HC_2H_3O_2] = 0.50 - x \approx 0.50$$

5    $$K_a = \frac{[x][\cancel{0.50}]}{[\cancel{0.50}]} = 1.8 \times 10^{-5} \text{ (from Table 14-2)}$$

$$x = [H_3O^+] = 1.8 \times 10^{-5}$$

6    $$pH = -\log(1.8 \times 10^{-5}) = \underline{\underline{4.74}}$$

Note that for buffers made up of equal molar concentrations of the acid and the salt, the pH of the solution is simply the negative log of the constant K_a. This is defined as the pK_a.

When $[HX] = [X^-]$, $pH = pK_a$, and $pK_a = -\log K_a$. pK_a values are useful in determining the pH values of various buffer solutions.

Perhaps one of the most important tools for dealing with reactions that reach a point of equilibrium is provided by the law of mass action. This law permits many calculations that can tell us about the actual distribution of reactants and products present at equilibrium. Once we know the value of the equilibrium constant, we can determine how any mixture of reactants and products will redistribute at equilibrium. A practical application of these calculations allows us to calculate the pH of weak acid and weak base solutions.

▲Looking Back

Learning Check B

◀Checking It Out

B–1. Fill in the blanks.
The law of mass action is composed of a numerical quantity known as an _____ _____ and a ratio known as the _____ _____ _____. A large value for K_{eq} tells us that the equilibrium lies in favor of _____. In the calculation of K_{eq}, the coefficient of a compound in the balanced equation becomes an _____ of the _____ of the compound in the ratio. For weak acids, the smaller the value of _____, the _____ the acid in water. One can calculate the pH of a weak base solution from a knowledge of the original _____ of the weak base and the value of _____.

B–2. Write the law of mass action for the following equilibria.
(a) $CO(g) + 3H_2(g) \rightleftharpoons CH_4(g) + H_2O(g)$
(b) $HCHO_2(aq) + H_2O(l) \rightleftharpoons H_3O^+(aq) + CHO_2^-(aq)$

B–3.   What is the value of K_{eq} in B-2(a) if there are 2.0×10^{-3} mol of CO, 1.0×10^{-2} mol of H_2, 3.5×10^{-4} mol of CH_4, and 5.0×10^{-5} mol of H_2O present in a 10.0-L container at equilibrium?

B–4.   A 0.25 M solution of a hypothetical weak base (NX_3) is 1.0% ionized. What is K_b?

B–5.   What is the pH of a 0.10 M solution of $HCHO_2$? (Refer to Table 14-2.)

Additional Examples: 14-15, 14-17, 14-20, 14-25, 14-30, 14-35, 14-39, 14-43, 14-45, 14-49, and 14-51.

C H A P T E R   R E V I E W

Putting It Together ▲▼

This chapter concerns several important chemistry topics such as how reactions take place, the rates at which they take place, and the point of equilibrium in reversible reactions. **Collision theory** tells us that reactions take place through collisions of molecules. The **rate of a reaction** is determined by the frequency of these collisions and the **activation energy** of the reaction. There are several things that influence these two factors and thus affect the rate of the reaction.

1	Activation energy	Each set of potential reactants requires a different amount of energy to form the **activated complex** that leads to the formation of products. If collisions do not have the necessary kinetic energy, no products form.
2	Temperature	Higher temperatures increase both the energy of colliding molecules and the frequency of collisions. Thus the rate increases.
3	Concentration of Reactants	The concentration of reactants relates directly to the frequency of collisions. The higher the concentration of colliding molecules, the faster the reaction.
4	Particle size	Finely divided solids allow more frequent collisions because of increased surface area. Thus reaction rates are faster.
5	Catalyst	The presence of a catalyst lowers the activation energy so the rate of the reaction increases.

Reversible reactions often reach a **point of equilibrium** where significant concentrations of reactants and products coexist. In fact, a **dynamic equilibrium** exists where both the forward and reverse reactions are occurring at the same rate. The point of equilibrium can be influenced by reaction conditions. However, changes in these reaction conditions have a predictable effect on the point of equilibrium, which is correlated by **Le Châtelier's principle.** The factors that can shift the point of equilibrium are summarized as follows:

1 Concentration

An increase in concentrations on one side of the equation results in an increase in concentrations on the other side. A decrease on one side leads to a shift to the same side.

2 Pressure

For gas phase reactions, compression of the mixture leads to a lower volume and a higher pressure. The equilibrium shifts to the side with the smaller total number of molecules of gas, thus reducing the pressure.

3 Temperature

For exothermic reactions, an increase in temperature leads to a decrease in the proportion of products. Just the opposite occurs for an endothermic reaction.

Many important industrial reactions do not take place readily, or even at all, without a catalyst present. A catalyst brings a reaction mixture to the point of equilibrium faster but does not change the actual point of equilibrium.

The distribution of concentrations of reactants and products can be predicted from the **law of mass action.** This law is composed of a numerical value (K_{eq}) called the **equilibrium constant,** which is equal to a ratio of reactants and products constructed from the balanced equation and known as the **equilibrium constant expression.** Examples of the two types of equilibria discussed are

Gaseous

$$A_2(g) + B_2(g) \rightleftharpoons 2AB(g)$$

$$K_{eq} = \frac{[AB]^2}{[A_2][B_2]}$$

Ionic

$$HX(aq) + H_2O(l) \rightleftharpoons H_3O^+(aq) + X^-(aq)$$

$$K_a = \frac{[H_3O^+][X^-]}{[HX]}$$

The value of the equilibrium constant at a given temperature is obtained by experimental measurements of the distribution of reactants and products at equilibrium. The constant and the initial concentrations can then be used to calculate the concentration of a given gas or ion at equilibrium. The **acid ionization constants** for weak acids (K_a) and the **base ionization constants** for weak bases (K_b) are useful for calculation of the pH of these solutions.

COLLISION THEORY AND REACTION RATES

14–1 Explain why all collisions between reactant molecules do not lead to products.

14–2 The equation

$$NO_2(g) + CO(g) \rightleftharpoons NO(g) + CO_2(g)$$

represents a reaction in which the mechanism of the forward reaction involves a collision between the two reactant molecules. Describe the probable orientation of the molecules in this collision.

14–3   Explain the following facts from your knowledge of collision theory and the factors that affect the rates of reactions.
(a) The rates of chemical reactions approximately double for each 10°C rise in temperature.
(b) Eggs cook slower in boiling water at higher altitudes, where water boils at temperatures less than 100°C.
(c) H_2 and O_2 do not start to react to form H_2O except at very high temperatures (unless initiated by a spark).
(d) Wood burns explosively in pure O_2 but slowly in air, which is about 20% O_2.
(e) Coal dust burns faster than a single lump of coal.
(f) Milk sours if left out for a day or two but will last two weeks in the refrigerator.
(g) H_2 and O_2 react smoothly at room temperature in the presence of finely divided platinum.

14–4   Explain the following facts from your knowledge of collision theory and the factors that affect the rates of reactions.
(a) A wasp is lethargic at temperatures below 65°C.
(b) Charcoal burns faster if you blow on the coals.
(c) A 0.10 M boric acid solution ($[H_3O^+]$ = 7.8×10^{-6}) can be used for eyewash, but a 0.10 M hydrochloric acid solution ($[H_3O^+]$ = 0.10) would cause severe damage.
(d) To keep apple juice from fermenting to apple cider, the apple juice must be kept cold.
(e) Coal does not spontaneously ignite at room temperature, but phosphorus does.
(f) Potassium chlorate decomposes at a much lower temperature when mixed with MnO_2.

EQUILIBRIUM

14–5   Explain the point of equilibrium in terms of reaction rates.

14–6   In the hypothetical reaction between A_2 and B_2, when was the rate of the forward reaction at a maximum? In the same reaction,

when was the rate of the reverse reaction at a maximum? (Refer to Figure 14-10.)

14–7   Give a reason, besides reaching a point of equilibrium, why certain reactions do not go to completion (100% to the right).

14–8   Compare the activation energy for the reverse reaction with that for the forward reaction in Figure 14-2. Which are easier to form, reactants from products or products from reactants? Which system would come to equilibrium faster, a reaction starting with pure products or one starting with pure reactants?

14–9   Consider an endothermic reaction that comes to a point of equilibrium such as shown in Figure 14-3. Which is greater, the activation energy for the forward or for the reverse reaction? Which system would come to equilibrium faster, a reaction starting with pure products or one starting with pure reactants?

LE CHÂTELIER'S PRINCIPLE AND EQUILIBRIUM

14–10   The following equilibrium represents an important industrial process, called the Haber process, used to convert N_2 to NH_3. The ammonia is used mainly for fertilizer. It requires the use of a catalyst.

$$N_2(g) + 3H_2(g) \xrightleftharpoons{500°C} 2NH_3(g) + heat$$

Determine the direction in which the equilibrium will be shifted by the following changes.
(a) increasing the concentration of N_2
(b) increasing the concentration of NH_3
(c) decreasing the concentration of H_2
(d) decreasing the concentration of NH_3
(e) compressing the reaction mixture
(f) removing the catalyst
(g) How will the yield of NH_3 be affected by raising the temperature to 750°C?
(h) How will the yield of NH_3 be affected by lowering the temperature to 0°C? How will this affect the rate of formation of NH_3?

14–11   Consider the following equilibrium:

$$4NH_3(g) + 5O_2(g) \rightleftharpoons 4NO(g) + 6H_2O(g) + heat\ energy$$

How will this system at equilibrium be affected by the following changes?
(a) increasing the concentration of O_2
(b) removing all of the H_2O as it is formed
(c) increasing the concentration of NO
(d) compressing the reaction mixture
(e) increasing the volume of the reaction container
(f) increasing the temperature
(g) decreasing the concentration of O_2
(h) adding a catalyst

14-12 The following equilibrium takes place at high temperatures.

$$N_2(g) + 2H_2O(g) + \text{heat energy} \rightleftharpoons 2NO(g) + 2H_2(g)$$

How will the concentration of NO at equilibrium be affected by these changes?
(a) increasing $[N_2]$
(b) decreasing $[H_2]$
(c) compressing the reaction mixture
(d) decreasing the volume of the reaction container
(e) decreasing the temperature
(f) adding a catalyst

14-13 Consider the following equilibrium:

$$2SO_3(g) + CO_2(g) + \text{heat energy} \rightleftharpoons CS_2(g) + 4O_2(g)$$

How will the concentration of CS_2 at equilibrium be affected by these changes?
(a) decreasing the volume of the reaction vessel
(b) adding a catalyst
(c) increasing the temperature
(d) increasing the original concentration of CO_2
(e) removing some O_2 as it forms

14-14 Fortunately for us, the major components of air, N_2 and O_2, do not react under ordinary conditions. The reaction shown is endothermic.

$$N_2(g) + O_2(g) \rightleftharpoons 2NO(g)$$

Would the formation of NO in an automobile be affected by these changes?
(a) a lower pressure
(b) a lower temperature
(c) a lower concentration of O_2

LAW OF MASS ACTION

14-15 Write the law of mass action for each of the following equilibria.

(a) $CO(g) + Cl_2(g) \rightleftharpoons COCl_2(g)$
(b) $CH_4(g) + 2H_2O(g) \rightleftharpoons CO_2(g) + 4H_2(g)$
(c) $4HCl(g) + O_2(g) \rightleftharpoons 2Cl_2(g) + 2H_2O(g)$
(d) $CH_4(g) + Cl_2(g) \rightleftharpoons CH_3Cl(g) + HCl(g)$

14-16 Write the law of mass action for each of the following equilibria.
(a) $3O_2(g) \rightleftharpoons 2O_3(g)$
(b) $N_2(g) + 2O_2(g) \rightleftharpoons 2NO_2(g)$
(c) $C_2H_2(g) + 2H_2(g) \rightleftharpoons C_2H_6(g)$
(d) $4H_2(g) + CS_2(g) \rightleftharpoons CH_4(g) + 2H_2S(g)$

VALUE OF K_{eq}

14-17 For the hypothetical reaction $A_2 + B_2 \rightleftharpoons 2AB$, $K_{eq} = 1.0 \times 10^8$, are reactants or products favored at equilibrium?

14-18 For the hypothetical reaction $2C + 3B \rightleftharpoons 2D + F$, $K_{eq} = 5 \times 10^{-7}$, are reactants or products favored at equilibrium?

14-19 For the reaction $H_2(g) + I_2(g) \rightleftharpoons 2HI(g)$, $K_{eq} = 45$ at a given temperature, are reactants or products favored at equilibrium?

14-20 Given the system

$$3O_2(g) \rightleftharpoons 2O_3(g)$$

at equilibrium, $[O_2] = 0.35$ and $[O_3] = 0.12$. What is K_{eq} for the reaction at this temperature?

14-21 Given the system

$$N_2(g) + 2O_2(g) \rightleftharpoons 2NO_2(g)$$

at a given temperature, there are 1.25×10^{-3} mol of N_2, 2.50×10^{-3} mol of O_2, and 6.20×10^{-4} mol of NO_2 in a 1.00-L container. What is K_{eq} for this reaction at this temperature?

14-22 Given the system

$$2SO_3(g) + CO_2(g) \rightleftharpoons CS_2(g) + 4O_2(g)$$

what is K_{eq} if, at equilibrium, $[SO_3] = 2.0 \times 10^{-2}$ mol/L, $[CO_2] = 4.5 \times 10^{-3}$ mol/L, $[CS_2] = 6.2 \times 10^{-4}$ mol/L, and $[O_2] = 1.0 \times 10^{-4}$ mol/L?

14-23 Given the system

$$CH_4(g) + 2H_2O(g) \rightleftharpoons CO_2(g) + 4H_2(g)$$

at equilibrium, we find 2.20 mol of CO_2, 4.00 mol of H_2, 6.20 mol of CH_4, and 3.00 mol of H_2O in a 30.0-L container. What is K_{eq} for the reaction? (Hint: Convert amount and volume to concentration.)

14–24 Given the system

$$C_2H_2(g) + 2H_2(g) \rightleftharpoons C_2H_6(g)$$

at equilibrium, we find 296 g of C_2H_6 present along with 3.50 g of H_2 and 21.0 g of C_2H_2 in a 400-mL container. What is K_{eq}? (Hint: Convert amount and volume to concentration.)

*14–25 Consider the following system:

$$2HI(g) \rightleftharpoons H_2(g) + I_2(g)$$

(a) If we start with [HI] = 0.60, what would be [H_2] and [I_2] if all of the HI reacts?
(b) If we start with [HI] = 0.60 and [I_2] = 0.20, what would be [H_2] and [I_2] if all of the HI reacts?
(c) If we start with only [HI] = 0.60 and 0.20 mol/L of HI reacts, what are [HI], [I_2], and [H_2] at equilibrium?
(d) From the information in (c), calculate K_{eq}.
(e) What is the K_{eq} for the reverse reaction? How does this value differ from the value given in Table 14-1?

*14–26 In the reaction

$$N_2(g) + 3H_2(g) \rightleftharpoons 2NH_3(g)$$

initially 1.00 mol of N_2 and 1.00 mol of H_2 are mixed in a 1.00-L container. At equilibrium, it is found that [NH_3] = 0.20.
(a) What is the concentration of N_2 and H_2 at equilibrium?
(b) What is the K_{eq} for the system at this temperature?

*14–27 At a given temperature, N_2, H_2, and NH_3 are mixed so that the initial concentration of each is 0.50 mol/L. At equilibrium, the concentration of N_2 is 0.40 mol/L.
(a) Calculate the concentration of H_2 and NH_3 at equilibrium.
(b) What is the K_{eq} at this temperature?

*14–28 At the start of the reaction

$$4NH_3(g) + 5O_2(g) \rightleftharpoons 4NO(g) + 6H_2O(g)$$

[NH_3] = [O_2] = 1.00. At equilibrium, it is found that 0.25 mol/L of NH_3 has reacted.
(a) What concentration of O_2 reacts?
(b) What are the concentrations of all species at equilibrium?
(c) Write the equilibrium constant expression and substitute the proper values for the concentrations of reactants and products.

*14–29 At the start of the reaction

$$CO(g) + Cl_2(g) \rightleftharpoons COCl_2(g)$$

[CO] = 0.650 and [Cl_2] = 0.435. At equilibrium, it is found that 10.0% of the CO has reacted. What is [CO], [Cl_2], and [$COCl_2$] at equilibrium? What is K_{eq}?

CALCULATIONS INVOLVING K_{eq}

14–30 For the reaction

$$PCl_3(g) + Cl_2(g) \rightleftharpoons PCl_5(g)$$

K_{eq} = 0.95 at a given temperature. If [PCl_3] = 0.75 and [Cl_2] = 0.40 at equilibrium, what is the concentration of PCl_5 at equilibrium?

14–31 At a given temperature, K_{eq} = 46.0 for the reaction

$$4HCl(g) + O_2(g) \rightleftharpoons 2Cl_2(g) + 2H_2O(g)$$

At equilibrium, [HCl] = 0.100, [O_2] = 0.455, and [H_2O] = 0.675. What is [Cl_2] at equilibrium?

14–32 Using the value for K_{eq} calculated in Problem 14-23, what is the concentration of H_2O at equilibrium if, at equilibrium, [CH_4] = 0.50, [CO_2] = 0.24, and [H_2] = 0.20?

*14–33 For the following equilibrium, K_{eq} = 56 at a given temperature.

$$CH_4(g) + Cl_2(g) \rightleftharpoons CH_3Cl(g) + HCl(g)$$

At equilibrium, [CH_4] = 0.20 and [Cl_2] = 0.40. What are the equilibrium concentrations of CH_3Cl and HCl if they are equal?

*14–34 Using the equilibrium in Problem 14-20, calculate the equilibrium concentration of O_3 if the equilibrium concentration of O_2 is 0.80 mol/L.

EQUILIBRIA OF WEAK ACIDS AND WEAK BASES

14–35 Write the expressions for K_a or K_b for each of the following equilibria. Where necessary, complete the equilibrium.
(a) $HBrO + H_2O \rightleftharpoons H_3O^+ + BrO^-$
(b) $NH_3 + H_2O \rightleftharpoons NH_4^+ + OH^-$
(c) $H_2SO_3 + H_2O \rightleftharpoons H_3O^+ + HSO_3^-$
(d) $HSO_3^- + H_2O \rightleftharpoons \underline{\quad} + SO_3^{2-}$
(e) $H_3PO_4 + H_2O \rightleftharpoons H_3O^+ + \underline{\quad}$
(f) $HN(CH_3)_2 + H_2O \rightleftharpoons H_2N(CH_3)_2^+ + \underline{\quad}$

14–36 Write the expressions for K_a or K_b for each of the following equilibria. Where necessary, complete the equilibrium.
(a) $N_2H_4 + H_2O \rightleftharpoons HN_2H_4^+ + OH^-$
(b) $HCN + H_2O \rightleftharpoons H_3O^+ + CN^-$
(c) $H_2C_2O_4 + H_2O \rightleftharpoons H_3O^+ + \_\_$
(d) $H_2PO_4^- + H_2O \rightleftharpoons \_\_ + HPO_4^{2-}$

14–37 A 0.10 M solution of a weak acid HX has a pH of 5.0. A 0.10 M solution of another weak acid HB has a pH of 5.8. Which is the weaker acid? Which has the larger value for K_a?

14–38 A hypothetical weak acid HZ has a K_a of 4.5×10^{-4}. Rank the following 0.10 M solutions in order of increasing pH: HZ, $HC_2H_3O_2$, and HClO.

14–39 In a 0.20 M solution of cyanic acid, HOCN, $[H_3O^+] = [OCN^-] = 6.2 \times 10^{-3}$.
(a) What is [HOCN] at equilibrium?
(b) What is K_a?
(c) What is the pH?

14–40 A 0.58 M solution of a weak acid (HX) is 10.0% dissociated.
(a) Write the equilibrium equation and the equilibrium constant expression.
(b) What are $[H_3O^+]$, $[X^-]$, and [HX] at equilibrium?
(c) What is K_a?
(d) What is the pH?

14–41 Nicotine (Nc) is a nitrogen base in water (similar to ammonia). Write the equilibrium equation illustrating this base behavior. In a 0.44 M solution of nicotine, $[OH^-] = [NcH^+] = 5.5 \times 10^{-4}$. What is K_b for nicotine? What is the pH of the solution?

14–42 In a 0.085 M solution of carbolic acid, HC_6H_5O, the pH = 5.48. What is K_a?

14–43 Novocaine (Nv) is a nitrogen base in water (similar to ammonia). Write the equilibrium equation. In a 1.25 M solution of novocaine, pH = 11.46. What is K_b?

14–44 In a 0.300 M solution of chloroacetic acid, $HC_2Cl_3O_2$, $[HC_2Cl_3O_2] = 0.277 \, M$ at equilibrium. What is the pH? What is K_a?

14–45 What is the pH of a 0.65 M solution of HBrO?

14–46 What is the pH of a 1.50 M solution of HNO_2?

14–47 What is $[OH^-]$ in a 0.55 M solution of NH_3?

14–48 What is $[H_3O^+]$ in a 0.25 M solution of $HC_2H_3O_2$?

14–49 What is the pH of a 1.00 M solution of $HN(CH_3)_2$?

14–50 What is the pH of a 0.567 M solution of N_2H_4?

14–51 What is the pH of a buffer made by mixing 0.45 mol of NaCN and 0.45 mol of HCN in 2.50 L of solution?

14–52 What is the pH of a buffer made by dissolving 1.20 mol of NH_3 and 1.20 mol of NH_4ClO_4 in 13.5 L of solution?

*14–53 What is the pH of a buffer made by dissolving 0.20 mol of KBrO and 0.60 mol of HBrO in 850 mL of solution?

*14–54 A buffer of pH = 7.50 is desired. Which two of the following compounds should be dissolved in water in equimolar amounts to provide this buffer solution: HNO_2, HClO, HCN, $NaNO_2$, NH_3, KCN, KClO, NH_4Cl?

*14–55 A buffer of pH = 9.25 is desired. Which two of the following compounds should be dissolved in water in equimolar amounts to provide this buffer solution: $HC_2H_3O_2$, HClO, NH_3, N_2H_4, $KC_2H_3O_2$, NH_4Cl, NaClO, N_2H_5Br?

GENERAL QUESTIONS

14–56 In a given gaseous equilibrium, chlorine reacts with ammonia to produce nitrogen trichloride and hydrogen chloride.
(a) Write the balanced equation illustrating this equilibrium.
(b) Write the expression for K_{eq}.
(c) How is the equilibrium concentration of ammonia affected by an increase in pressure?
(d) How is the equilibrium concentration of ammonia affected by the addition of some chlorine gas?
(e) The value of K_{eq} for this reaction is 2.4×10^{-9}. Are there more products or reactants present at equilibrium?
(f) If the equilibrium concentration of chlorine = 0.10 M, nitrogen trichloride = 2.0×10^{-4} M, and hydrogen chloride = 1.0×10^{-3} M, what is the equilibrium concentration of ammonia?

14–57 In a given gaseous equilibrium, acetylene (C_2H_2) reacts with HCl to form $C_2H_4Cl_2$.
  (a) Write the balanced equation illustrating this equilibrium.
  (b) Write the expression for K_{eq}.
  (c) How does an increase in pressure affect the equilibrium concentration of $C_2H_4Cl_2$?
  (d) The reaction is exothermic. How does an increase in temperature affect the value of K_{eq}? How does it affect the equilibrium concentration of HCl?
  (e) At a given temperature at equilibrium, $[C_2H_2] = 0.030$, $[HCl] = 0.010$, and $[C_2H_4Cl_2] = 0.60$. What is the value of K_{eq}?
  (f) What does the value of K_{eq} tell us about the point of equilibrium at this temperature?

14–58 Dinitrogen oxide (nitrous oxide) decomposes to its elements in a gaseous equilibrium reaction.
  (a) Write the balanced equation illustrating this equilibrium.
  (b) Write the expression for K_{eq}.
  (c) How does an increase in pressure affect the equilibrium concentration of nitrogen?
  (d) If the reaction is exothermic, how does a decrease in temperature affect the equilibrium concentration of dinitrogen oxide?
  (e) At a given temperature, 0.10 mol of dinitrogen oxide is placed in a 1.00-L container. At equilibrium, it is found that 1.5% of the dinitrogen oxide has decomposed. What is the value of K_{eq}?

14–59 Nitrogen dioxide is in a gaseous equilibrium with dinitrogen tetroxide.
  (a) Write the balanced equation illustrating this reaction. Show nitrogen dioxide as the reactant and dinitrogen tetroxide as the product.
  (b) Write the expression for K_{eq}.
  (c) How does an increase in pressure affect the concentration of nitrogen dioxide at equilibrium?
  (d) At equilibrium at a given temperature, it is found that there are 10.0 g of nitrogen dioxide and 12.0 g of dinitrogen teroxide in a 2.50-L container. What is the value of K_{eq}?

14–60 The bicarbonate ion (HCO_3^-) can act as either an acid or a base in water. For the acid reaction

$$HCO_3^- + H_2O \rightleftharpoons H_3O^+ + CO_3^{2-}$$
$$K_a = 4.7 \times 10^{-11}$$

For the base reaction

$$HCO_3^- + H_2O \rightleftharpoons OH^- + H_2CO_3$$
$$K_b = 2.2 \times 10^{-8}$$

  (a) Based on the values of K_a and K_b, is a solution of $NaHCO_3$ acidic or basic?
  (b) How is bicarbonate of soda used medically?
  (c) Write the reaction that occurs when sodium bicarbonate reacts with stomach acid (HCl).
  (d) Can sodium bisulfate act in the same way toward acids as sodium bicarbonate does? (Refer to Table 12-2.)

*14–61 Given 0.10 M solutions of the compounds KOH, NaCl, HNO_2, NH_3, HNO_3, $HCHO_2$, and N_2H_4, rank them in order of increasing pH.

*14–62 What is the pH of a buffer that contains 150 g of HNO_2 and 150 g of $LiNO_2$?

*14–63 What is the pH of a buffer made by adding 0.20 mol of NaOH to 1.00 L of a 1.00 M solution of $HC_2H_3O_2$? Assume no volume change. (Hint: First consider the partial neutralization of $HC_2H_3O_2$ by NaOH.)

SOLUTIONS TO LEARNING CHECKS

A–1 collisions, collide, minimum, activation, time, rate, activated complex, activation energy, higher, increases, increases, catalyst, increases, point of equilibrium, constant, concentration, temperature, pressure

A–2 (a) increased   (b) decreased   (c) increased   (d) decreased
  (e) not affected

A–3

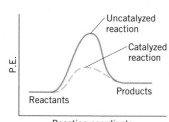

B–1 equilibrium constant, equilibrium constant expression, products, exponent, concentration, K_a, weaker, concentration, K_b.

B–2 (a) $K_{eq} = \dfrac{[CH_4][H_2O]}{[CO][H_2]^3}$ (b) $K_a = \dfrac{[H_3O^+][CHO_2^-]}{[HCHO_2]}$

B–3 $K_{eq} = \dfrac{\left[\dfrac{3.5 \times 10^{-4}\ mol}{10.0\ L}\right]\left[\dfrac{5.0 \times 10^{-5}\ mol}{10.0\ L}\right]}{\left[\dfrac{2.0 \times 10^{-3}\ mol}{10.0\ L}\right]\left[\dfrac{1.0 \times 10^{-2}\ mol}{10.0\ L}\right]^3} = \dfrac{17.5 \times 10^{-11}}{2.0 \times 10^{-13}} = \underline{\underline{880}}$

B–4 $NX_3(aq) + H_2O(l) \rightleftharpoons HNX_3^+(aq) + OH^-(aq)$

$[HNX_3^+] = [OH^-] = 0.010 \times 0.25\ M = 2.5 \times 10^{-3}\ M$ (at equilibrium)

$[NX_3] = 0.25 - (2.5 \times 10^{-3}) = 0.25\ M$

$K_b = \dfrac{(2.5 \times 10^{-3})^2}{0.25} = \dfrac{6.25 \times 10^{-6}}{0.25} = \underline{\underline{2.5 \times 10^{-5}}}$

B–5 From Table 14-2 for $HCHO_2$, $K_a = 1.8 \times 10^{-4}$.

$$HCHO_2(aq) + H_2O(l) \rightleftharpoons H_3O^+(aq) + CHO_2^-(aq)$$

Let $x = [H_3O^+] = [CHO_2^-]$ at equilibrium. Then $[HCHO_2] = 0.10 - x$ 0.10 at equilibrium.

$\dfrac{x^2}{0.10} = 1.8 \times 10^{-4}$ $x^2 = 1.8 \times 10^{-5} = 18 \times 10^{-6}$ $x = 4.2 \times 10^{-3}$

$pH = -\log(4.2 \times 10^{-3}) = \underline{\underline{2.38}}$

The following multiple-choice questions have one correct answer.

1. Which of the following is the formula for perchloric acid?
   (a) HCl
   (b) HClO
   (c) HClO$_2$
   (d) HClO$_3$
   (e) HClO$_4$

2. Which of the following is a balanced neutralization reaction that leads to the formation of an acid salt?
   (a) HCl + NaOH → NaCl + H$_2$O
   (b) H$_3$PO$_4$ + KOH → KHPO$_4$ + H$_2$O
   (c) H$_2$SO$_4$ + 2KOH → K$_2$SO$_4$ + H$_2$O
   (d) NaOH + H$_2$SO$_3$ → NaHSO$_3$ + H$_2$O

3. Which of the following is a strong acid?
   (a) HNO$_2$
   (b) HNO$_3$
   (c) H$_2$SO$_3$
   (d) HClO
   (e) KOH

4. The concentration of an aqueous HCl solution is 0.01 M. The pH is
   (a) 1   (b) 2   (c) 3   (d) 12   (e) 7

5. A solution has a pH of 8.5. The solution is
   (a) strongly basic
   (b) weakly basic
   (c) neutral
   (d) weakly acidic
   (e) strongly acidic

6. Which of the following is the conjugate acid of the HS$^-$ ion?
   (a) H$_2$S$^-$
   (b) S^{2-}
   (c) S$^-$
   (d) H$_3$O$^+$
   (e) H$_2$S

7. Which of the following conjugate bases does not exhibit basic nature in water?
   (a) S^{2-}
   (b) NO$_2^-$
   (c) ClO$_4^-$
   (d) PO$_4^{3-}$
   (e) HS$^-$

8. A 0.10M solution of a compound has a pH of 5.4. The compound is
   (a) HCl
   (b) KOH
   (c) NH$_4$Cl
   (d) K$_2$S
   (e) NaCl

9. Which of the following pairs of compounds does not form a buffer when aqueous solutions of the two are mixed?
   (a) NaC$_2$H$_3$O$_2$ and HC$_2$H$_3$O$_2$
   (b) KBr and HBr
   (c) LiNO$_2$ and HNO$_2$
   (d) NH$_3$ and NH$_4$Cl

10. Which of the following acids form when Cl$_2$O dissolves in water?
    (a) HCl
    (b) HClO$_4$
    (c) HClO
    (d) HClO$_2$
    (e) HClO$_3$

11. Which of the following is an oxidation process?
    (a) H$_2$CO$_3$ → H$_2$O + CO$_2$
    (b) 2Br$^-$ → Br$_2$
    (c) Ca^{2+} + SO$_4^{2-}$ → CaSO$_4$
    (d) H$^+$ + OH$^-$ → H$_2$O
    (e) SO$_4^{2-}$ → S^{2-}

12. What is the oxidation state of the Br in Ca(Br O$_3$)$_2$?
    (a) +10
    (b) +6
    (c) +5
    (d) +12
    (e) −1

13. In the Daniell cell, what allows for the migration of anions between compartments?
    (a) a salt bridge
    (b) the electrodes
    (c) the external wire
    (d) the electrons

14. In the following electrolytic cell, _____ is _____ at the anode.

    $$2Cl^-(aq) + Fe^{2+}(aq) \longrightarrow Fe(s) + Cl_2(g)$$

    (a) Fe^{2+}, reduced
    (b) Cl$^-$, reduced
    (c) Cl$_2$, reduced
    (d) Cl$^-$, oxidized
    (e) Fe, oxidized

15. The following equation represents a spontaneous reaction:

    $$Br_2(aq) + Ni(s) \longrightarrow Ni^{2+}(aq) + 2Br^-(aq)$$

    Which of the following statements is correct?
    (a) Br$_2$ is a stronger reducing agent than Ni^{2+}.
    (b) Br$^-$ is a stronger reducing agent than Ni.
    (c) Ni is an oxidizing agent.
    (d) Br$_2$ is oxidized.
    (e) Br$_2$ is a stronger oxidizing agent than Ni^{2+}.

16. Which of the following does not affect the rate of a reaction?
    (a) the volume of the reaction vessel
    (b) the temperature in the reaction vessel
    (c) the concentration of a reactant
    (d) the shape of the reaction vessel

17. Which of the following is not true about a catalyst?
    (a) can be homogeneous or heterogeneous
    (b) allows more products to form
    (c) is not consumed in the reaction
    (d) lowers the activation energy

18. Given the following equilibrium:

$$2CO(g) + O_2(g) \rightleftharpoons CO_2(g) + heat$$

The concentration of CO_2 is increased at equilibrium by
(a) increasing the temperature
(b) decreasing the concentration of CO
(c) increasing the pressure
(d) adding a catalyst

19. Given the following equilibrium:

$$A + B \rightleftharpoons C + D, K_{eq} = 10^{-6}$$

Which of the following is correct?
(a) Products are favored at equilibrium.
(b) Reactants are favored at equilibrium.
(c) There are significant amounts of both products and reactants present at equilibrium.
(d) No conclusions can be made about the point of equilibrium.

20. A 0.10 M solution of a weak acid is 1.0% ionized. Its K_a is
(a) 10^{-3}
(b) 10^{-4}
(c) 10^{-5}
(d) 10^{-6}
(e) 10^{-7}

PROBLEMS

1. (a) Write the balanced equation representing the complete neutralization of sulfuric acid with lithium hydroxide.
(b) What volume (in mL) of 0.25 M sulfuric acid is needed to completely react with 1.80 g of lithium hydroxide?

2. Complete the following equations:
(a) $HClO + H_2O \rightleftharpoons$
(b) $NH_3 + H_2O \rightleftharpoons$
(c) $KOH + HNO_2 \rightarrow$
(d) $H_3AsO_3 + 2NaOH \rightarrow$

3. Write equations representing reactions for the following. If no reaction occurs, write N.R.
(a) $K^+ + H_2O$ (d) $CO_2 + H_2O$
(b) $CaO + H_2O$ (e) $ClO^- + H_2O$
(c) $NH_4^+ + H_2O$ (f) $Br^- + H_2O$

4. In a 0.25 M solution of a weak base, $[OH^-] = 6.5 \times 10^{-4}$. What is the value of
(a) $[H_3O^+]$ (b) pH   (c) pOH

5. (a) When elemental tin is placed in a nitric acid solution, a spontaneous redox reaction occurs producing SnO_2 and NO_2. Write the balanced equation representing this reaction.
(b) What mass of NO_2 is produced from the complete reaction of 350 mL of 0.20 M nitric acid?

6. From the equation in Problem 5, answer the following:
(a) What is the reactant oxidized?
(b) What is the reactant reduced?
(c) What is the product of oxidation?
(d) What is the product of reduction?
(e) What is the oxidizing agent?
(f) What is the reducing agent?

7. Given the following table of oxidizing and reducing agents, answer the questions below.

Strongest ⟶ $I_2 + 2e^- \rightleftharpoons 2I^-$
oxidizing $Cr^{3+} + 3e^- \rightleftharpoons Cr$
  agent $Mn^{2+} + 2e^- \rightleftharpoons Mn$ Strongest
             $Ca^{2+} + 2e^- \rightleftharpoons Ca$ ⟵ reducing
                                             agent

(a) Is the following reaction spontaneous or nonspontaneous?

$$Mn + Ca^{2+} \longrightarrow Ca + Mn^{2+}$$

(b) Write a balanced equation representing a spontaneous reaction involving the Cr, Cr^{3+} and the Mn, Mn^{2+} half-reactions.
(c) If an aqueous solution of I_2 (an antiseptic) is spilled on a chromium-coated bumper of an automobile, will a reaction occur? If so, write the balanced equation for the reaction.

8. Given the equilibrium

$$2NO(g) + Br_2(g) \rightleftharpoons 2NOBr(g)$$

(a) Write the law of mass action.
(b) What is the value of K_{eq} if we start with 4.00 moles of NOBr only in a 20.0-L container, and, at equilibrium, it is found that [NOBr] = 0.10?
(c) What is the concentration of NOBr if [NO] = 0.45 and [Br_2] = 0.22 at equilibrium?

9. What is the value of K_b for the weak base in Problem 4?

10. What is the pH of a 0.64 M solution of boric acid? Boric acid (H_3BO_3) is a weak monoprotic acid used as an eyewash. K_a for boric acid is 6.0×10^{-10}.

The sun is the source of life. Deep in its interior nuclear reactions are occurring that liberate life-giving energy.

NUCLEAR CHEMISTRY

In July 1945, the most powerful explosion yet produced by the human race shook the New Mexico desert. The first nuclear bomb had been detonated, and our world would never be the same—the nuclear age was now upon us. But the dawn of this age actually began quite innocently forty-nine years earlier. At that time, there was certainly no intent to launch the human race into a radical new direction. In 1896, a French scientist named Henri Becquerel was investigating how sunlight interacted with a uranium mineral called pitchblende. He suspected that sunlight caused pitchblende to give off the mysterious "X rays" that had been discovered the previous year by Wilhelm Roentgen. X rays, a high-energy form of light, were known to penetrate the covering of photographic film, thus exposing the film. One cloudy day, Becquerel was unable to do an experiment, but he left the uranium on top of unexposed film in a photographic plate. Several days later, he developed the film anyway and discovered that the part underneath the sample of pitchblende was exposed. Obviously, the uranium spontaneously emitted some form of radiation. It was this discovery that marked the true beginning of the nuclear age.

Since we first discussed the atom in Chapter 3, we have addressed the results of interactions of the electrons that reside outside a tiny but comparatively massive nucleus. Interactions of electrons account for the way atoms of an element combine chemically with other atoms. Now we turn our attention to the nucleus, which is composed of two types of particles, neutrons and protons, collectively called nucleons. Also recall that elements are composed of isotopes. Isotopes of an element are those

◄Setting The Stage

atoms that have the same number of protons but different numbers of neutrons. The changes that can occur in the nucleus of a particular isotope account for the phenomenon that Becquerel first observed. We will discover the awesome implications, starting from that first observation, that range from the despair of nuclear bombs to the hope of nuclear medicine.

**Formulating
Some Questions** ▶

Questions that we will address in this chapter include: What is the nature of the radiation observed by Becquerel and other early investigators? What does this radiation imply for the nucleus? How does radiation cause damage, and how can it be detected? What good can it do? How can we change one element into another? What goes on in the atomic and hydrogen bombs, and how can we tame these devices for energy? Our first task is to examine the changes in the nucleus that give rise to natural radioactivity.

15–1   RADIOACTIVITY

*What is different about a
radioactive isotope?*

The radiation process discovered by Becquerel was more complex than originally thought. The term *radioactive* was first suggested by Marie Curie to describe elements that spontaneously emit **radiation** *(particles or energy)*. Marie Curie and her husband, Pierre Curie, pursued the observation of Becquerel and discovered several more naturally **radioactive isotopes** of elements besides uranium (thorium, polonium, and radium). Lord Rutherford later demonstrated three types of natural radiation. We will look at these three processes individually, as well as two additional processes since discovered.

Alpha (α) Particles

The first nuclear change that was identified involved the emission of a helium nucleus from the nucleus of a heavy isotope. *The emitted helium nucleus is referred to as an* **alpha (α) particle**. This process is conveniently illustrated by a nuclear equation. *A* **nuclear equation** *shows the initial nucleus to the left of the reaction arrow and the product nuclei or particles to the right of the arrow.* Just as chemical equations are balanced, so are nuclear equations. *Nuclear equations are balanced by having the same totals of positive charges and mass numbers on both sides of the equation.* The emission of an alpha particle by ^{238}U is illustrated in Figure 15-1 and represented by the following nuclear equation.

$$^{238}_{92}\text{U} \longrightarrow {}^{234}_{90}\text{Th} + {}^{4}_{2}\text{He}$$

Note that the loss of four nucleons from the original (parent) nucleus leaves the remaining (daughter) nucleus with 234 nucleons ($238 - 4 = 234$); the loss of two protons leaves 90 protons in the daughter nucleus ($92 - 2 = 90$). Because a nucleus with 90 protons is an isotope of thorium, one element has indeed changed into another.

Beta (β) Particles

*How does beta emission
change an isotope?*

Rutherford discovered a second form of radiation, which involves the emission of an electron from the nucleus. *An electron emitted from the nucleus is known as a* **beta (β) particle.** *In effect, beta particle emission*

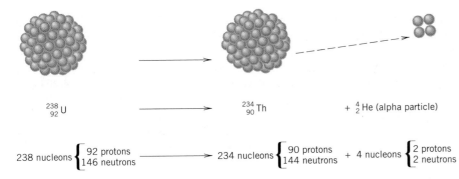

Figure 15–1
Alpha Radiation. An alpha particle is a helium nucleus emitted from a larger nucleus.

changes a neutron in a nucleus into a proton. This causes the atomic number of the isotope to increase by one while the mass number remains constant. This type of radiation is illustrated in Figure 15-2 and represented by the following nuclear equation.

$$^{131}_{53}\text{I} \longrightarrow {}^{131}_{54}\text{Xe} + {}^{0}_{-1}e$$

Gamma (γ) Rays
A third type of radioactive decay involves the emission of a high-energy form of light called a **gamma (γ) ray.** Like all light, gamma rays travel at 3.0×10^{10} cm/sec (186,000 miles/sec). This type of radiation may occur alone or in combination with alpha or beta radiation as discussed above. The following equation illustrates gamma ray emission.

$$^{60m}_{27}\text{Co} \longrightarrow {}^{60}_{27}\text{Co} + \gamma$$

Gamma radiation alone does not result in a change in either the mass number or the atomic number of the isotope. In the equation shown above, the *m* stands for *metastable,* which means that the isotope is in a high-energy state. A nucleus in a metastable, or high-energy state, emits energy in the form of gamma radiation. The product nucleus has the same identity but is then in a lower energy state.

Positron Particles
This type of radiation is not among the three types of naturally occurring radiation but is found in a few artificially produced isotopes. *A*

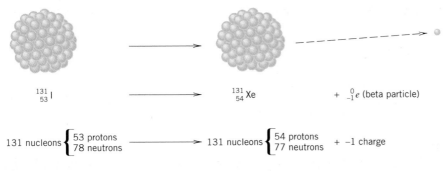

Figure 15–2
Beta Radiation. A beta particle is an electron emitted from a nucleus.

Is there such a thing as antimatter?

positron ($_{+1}^{0}e$) particle *is an electron with a positive charge rather than a negative charge. In effect, positron emission changes a proton into a neutron.* As a result, the product nucleus has the same mass number but a lower atomic number. A positron is known as a form of *antimatter* because when it comes in contact with normal matter (e.g., a negatively charged electron) the two particles annihilate each other, producing two gamma rays that travel in exactly opposite directions.

$$_{7}^{13}N \longrightarrow _{6}^{13}C + _{+1}^{0}e$$

$$_{+1}^{0}e + _{-1}^{0}e \longrightarrow 2\gamma$$

Electron Capture

Another nuclear process, known as **electron capture,** involves *the capture of an inner-shell electron by the nucleus.* This process is very rare in nature but is more common in artificially produced elements. When an isotope captures an orbital electron, the effect is the same as positron emission. *That is, a proton changes into a neutron.* Electron capture is detected by the X rays that are emitted as outer electrons fall into the inner shell where there is a vacancy due to the captured electron. In most cases, gamma rays are also emitted by the product nuclei. Electron capture is illustrated by the following nuclear equation.

$$_{23}^{50}V + _{-1}^{0}e \longrightarrow _{22}^{50}Ti + X \text{ rays} + \gamma$$

Although why one isotope is stable and another is not is a somewhat involved topic, we can make some observations. Notice that the atomic mass of some of the lighter elements in the periodic table is about double the atomic number. This is because the most common isotope of that element has an equal number of protons and neutrons, or nearly so (e.g., ^{4}He, ^{16}O, ^{40}Ca, ^{32}S). Heavier elements, however, tend to have more neutrons than protons in their most common isotopes. Isotopes that have excess neutrons (e.g., $_{6}^{14}C$) tend to decay by beta particle emission. The result of this radiation is a change of a neutron into a proton and a more stable neutron-proton ratio (e.g., $_{7}^{14}N + _{-1}^{0}e$). Isotopes that have excess protons (e.g., $_{6}^{10}C$) tend to decay by positron emission (or electron capture). The result of this radiation is a change of a proton into a neutron and a more stable nucleus (e.g., $_{5}^{10}B + _{+1}^{0}e$). Very heavy isotopes decay in a way that they lose both mass and charge. In this case, alpha particle emission carries away both mass and charge.

The five types of radiation that have been discussed are summarized in Table 15-1.

TABLE 15–1   TYPES OF RADIATION

Radiation	Mass Number	Charge	Identity
Alpha (α)	4	+2	Helium nucleus
Beta (β)	0	−1	Electron (out of nucleus)
Gamma (γ)	0	0	Light
Positron	0	+1	Positive electron
Electron capture	0	−1	Electron (into nucleus)

EXAMPLE 15–1

Nuclear Equations and Radioactivity

Complete the following nuclear equations:
(a) $^{218}_{84}\text{Po} \rightarrow$ ___ $+ \, ^4_2\text{He}$
(b) $^{210}_{81}\text{Tl} \rightarrow \, ^{210}_{82}\text{Pb} +$ ___
(c) ___ $\rightarrow \, ^8_4\text{Be} + \, ^0_{+1}e$

Procedure
For the missing isotope or particle:

1 Find the total number of nucleons; it is the same on both sides of the equation.

2 Find the total charge or atomic number; it is also the same on both sides of the equation.

3 If what's missing is the isotope of an element, find the symbol of the element that matches the atomic number from the list of elements inside the front cover.

Solution
(a) Nucleons: $218 = x + 4$, so $x = 214$.
   Atomic number: $84 = y + 2$, so $y = 82$.

From the inside cover, the element having an atomic number of 82 is Pb. The isotope is

$$^{214}_{82}\text{Pb}$$

(b) Nucleons: $210 = 210 + x$, so $x = 0$.
   Atomic number: $81 = 82 + y$, so $y = -1$.

An electron or beta particle has negligible mass (compared with a nucleon) and a charge of -1.

$$^{0}_{-1}e$$

(c) Nucleons: $x = 8 + 0$, so $x = 8$.
   Atomic number: $y = 4 + 1$, so $y = 5$.

From the inside cover, the element with an atomic number of 5 is B. The isotope is

$$^8_5\text{B}$$

Uranium is a naturally occurring element. Its most abundant isotope, however, decays to another element by alpha emission. Why hasn't it all decayed by now? The answer is that it decays at a very slow rate. The rate of decay is the next topic.

15–2 RATES OF DECAY OF RADIOACTIVE ISOTOPES

The heaviest element in the periodic table with at least one stable isotope is bismuth (#83). All elements heavier than bismuth have no stable isotopes (i.e., all isotopes are radioactive). Some heavy elements, such as uranium, still exist in the earth in significant amounts, however, because they have very long "half-lives." *The **half-life** ($t_{1/2}$) of a radioactive isotope is the time required for one-half of a given sample to decay.*

TABLE 15–2 HALF-LIVES

Isotope	$t_{1/2}$	Mode of Decay	Product
$^{238}_{92}U$	4.5×10^9 years	α, γ	$^{234}_{90}Th$
$^{234}_{90}Th$	24.1 days	β	$^{234}_{91}Pa$
$^{226}_{88}Ra$	1620 years	α	$^{222}_{86}Rn$
$^{14}_{6}C$	5760 years	β	$^{14}_{7}N$
$^{131}_{53}I$	8.0 days	β, γ	$^{131}_{54}Xe$
$^{218}_{85}At$	1.3 sec	α	$^{214}_{83}Bi$

Can the half-life of an isotope be changed?

Each radioactive isotope has a specific and constant half-life. Unlike the rate of chemical reactions, the rate of decay of an isotope is independent of conditions such as temperature, pressure, and whether the element is in the free state or part of a compound. For example, $^{238}_{92}U$ has a half-life of 4.5×10^9 years. Since that is roughly the age of Earth, about one-half of the $^{238}_{92}U$ originally present when Earth formed from hot gases and dust is still present. Half of what is now present will be gone in another 4.5 billion years. The half-lives and modes of decay of some radioactive isotopes are listed in Table 15-2.

Note in Table 15-2 that when $^{238}_{92}U$ decays, it forms an isotope ($^{234}_{90}Th$) that also decays (very rapidly compared with $^{238}_{92}U$). The $^{234}_{91}Pa$ formed from the decay of $^{234}_{90}Th$ also decays and so forth, until finally the stable isotope $^{206}_{82}Pb$ is formed. This is known as the $^{238}_{92}U$ radioactive decay series. *A* **radioactive decay series** *starts with a naturally occurring radioactive isotope with a half-life near the age of Earth.* (If it was very much shorter, there wouldn't be any left.) *The series ends with a stable isotope.* There are at least two other naturally occurring decay series: the $^{235}_{92}U$ series, which ends with $^{207}_{82}Pb$, and the $^{232}_{90}Th$ series, which ends with $^{208}_{82}Pb$. As a result of the $^{238}_{92}U$ decay series, where uranium is found in rocks, we also find other radioactive isotopes as well as lead. In fact, by examining the ratio of $^{238}_{92}U/^{206}_{82}Pb$ in a rock, its age can be determined. For example, a rock from the moon showed that about half of the original $^{238}_{92}U$ had decayed to $^{206}_{82}Pb$. This meant that the rock was about 4.5×10^9 years old.

Why is lead almost always found along with uranium?

Working It Out ●

EXAMPLE 15–2

Half-Life

If we started with 4.0 mg of $^{14}_{6}C$, how long would it take until only 0.50 mg remained?

Solution

After 5760 years, $\frac{1}{2} \times 4.0$ mg = 2.0 mg remaining

After another 5760 years, $\frac{1}{2} \times 2.0$ mg = 1.0 mg remaining

After another 5760 years, $\frac{1}{2} \times 1.0$ mg = 0.50 mg remaining

Total time = 17,280 years

Therefore, in 17,280 years, 0.50 mg remains.

Radiation is scary. We fear ingesting radioactive isotopes because they may cause cancer. Radon gas may need to be monitored in our schools and even our homes. Yet this same radiation can save lives when used in medical diagnosis or cancer treatment. How radiation interacts with matter is our next subject.

▼ Looking Ahead

15–3 EFFECTS OF RADIATION

Most chemical reactions that proceed spontaneously, like combustion of wood, give off energy. This happens because the products of the reaction have less potential energy than the reactants. If we are not careful, the heat energy from fires may cause injury or even death. Nuclear decay is a process in which a nucleus spontaneously gives off energy in the form of a high-energy particle (alpha or beta) or high-energy light (gamma). The collisions of fast-moving particles with surrounding molecules eventually result in increased velocity for all molecules and thus a higher temperature. Gamma rays can also cause heating. In fact, much of the heat generated in the interior of Earth and other planets is a direct result of natural radiation from decaying isotopes. The presence of radioactive isotopes in wastes from nuclear power plants makes these materials extremely hot; in most cases they must be stored in water continuously so that they do not melt.

Radiation also affects matter when the high-energy particles or rays penetrate. *The radiation changes neutral molecular compounds into ions in a process known as ionization.* For example, if high-energy particles collide with an H_2O molecule, an electron may be removed from the molecule, leaving an H_2O^+ ion behind. The properties of the ion are distinctly different from those of the neutral molecule. If the molecule in question is a large complex molecule that is part of a cell of a living system, the ionization causes damage or even eventual destruction of the cell. As shown in Figure 15-3, an alpha particle causes the most ionization and is the most destructive along its path. However, these particles do not penetrate matter to any extent and can be stopped, even by a piece of paper. The danger of alpha emitters (such as uranium and plutonium) is that they can be ingested through food or inhaled into the lungs. Then these heavy elements tend to accumulate in bones. There, in intimate contact with the blood-producing cells of the bone marrow, they slowly do damage. Ultimately, the radiation can cause certain cells to change into abnormal cells that reproduce rapidly. These are the dreaded cancer cells such as those found in leukemia.

Beta radiation is less ionizing than alpha radiation but is more penetrating. Still, a sheet of metal such as aluminum will stop most beta radiation. This type of radiation can cause damage to surface tissue such as skin and eyes but does not reach internal organs unless ingested. Although much less ionizing than the others, gamma radiation has tremendous penetrating power. Several feet of concrete or thick blocks of lead are needed for protection from gamma radiation. Without such protection, gamma radiation can cause damage from far away.

How does radiation make the surroundings "hot"?

These lead blocks are used to shield people from gamma radiation.

What type of radiation is the most destructive overall?

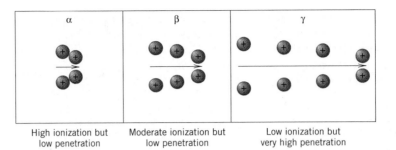

α	β	γ
High ionization but low penetration	Moderate ionization but low penetration	Low ionization but very high penetration

Figure 15–3
Ionization and Penetration. Gamma radiation has the most penetrating power of the three types of radiation.

The damage done to living cells by gamma radiation is cumulative. That means small doses over a long period of time can be as harmful as one large dose. For this reason, the dosage of radiation to which employees are exposed in radioactive environments such as a nuclear research laboratory or nuclear power plant must be carefully and continually monitored. If a worker receives too much radiation over a specified period, that person must be removed from additional exposure for a length of time until the body has had time to heal some of the damage.

Looking Ahead ▼

How do we know how much radiation we are absorbing? Since we can't see it or feel it, we must have some method of measuring radiation. The detection and measurement of radiation is obviously an important endeavor.

15–4   DETECTING RADIATION

Becquerel first discovered radiation by noticing that it exposed photographic film. This is still an important method of both detecting and measuring radiation. Film badges (see Figure 15-4) are worn by workers in the nuclear industry or any one else who works near a source of radiation. When the film is developed, the degree of darkening is proportional to the amount of radiation absorbed. Different filters are used so that the amount of each type of radiation received (alpha, beta, or gamma) can be measured.

Why should we care about radon?

Radon-222 is a radioactive isotope formed as part of the natural decay series starting with ^{238}U. Because radon is a noble gas, it can escape from the ground into the air. In certain areas of the United States, radon may accumulate in basements or other enclosed areas through foundation cracks. When inhaled, radon is usually exhaled immediately, causing no harm. Its half-life, however, is only 3.82 days, which means that some actually undergoes radioactive decay when it is inhaled. If so, it decays to ^{218}Po, which stays in the lungs because polonium is a solid. This is also a highly radioactive isotope that then decays in the lungs and may cause damage. It is estimated that between 10,000 and 20,000

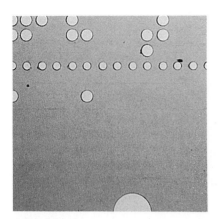
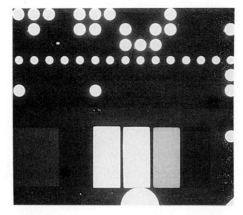

Figure 15–4
Film Badge and Radon Detector. These devices measure radiation by the amount that the film is exposed.

lung cancer deaths are caused annually by radon in the United States alone. Radon detectors are now sold commercially. When left in a basement, the detector measures the number of radon disintegrations. After a specified time, the detector is sent to a laboratory where the film is developed and the amount of radiation from radon is determined. (See Figure 15-4.)

As mentioned in the previous section, radiation causes ionization. Ionization is also used to measure radiation in the *Geiger counter*. The Geiger counter is composed of a long metal tube with a thin film that allows all types of radiation to penetrate at one end. The metal tube is filled with a gas such as argon at low pressure. A metal bar in the center of the tube is positively charged. When a radioactive particle enters the tube, it causes ionization. The electrons from this ionization move quickly to the positive electrode, causing a pulse of electricity to flow. This pulse is amplified and causes a light to flash, an audible "beep," or registers on a meter in some unit such as "counts per minute." (See Figure 15-5 on the next page.)

A third method used to detect radiation is a *scintillation counter*. Certain compounds such as zinc sulfide are known as *phosphors*. When radiation strikes a phosphor, a tiny flash of light is emitted. The flash of light is converted into a pulse of electricity that can be counted in the same manner as in the Geiger counter.

The discovery of natural radioactivity was a key that unlocked many of the mysteries of the atom. Except for gamma rays, the emission of particles from a radioactive nucleus of one element leads to the formation of the nucleus of another element. Each radioactive isotope has a specific rate of decay that is indicated by its half-life. Since radiation causes destruction of living cells, it must be monitored in many locations or situations. Radiation is detected by exposure of film or from the resulting ionization.

▲ **Looking Back**

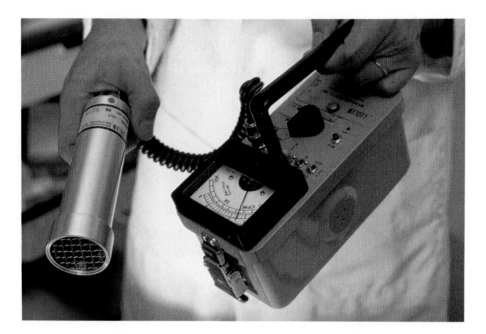

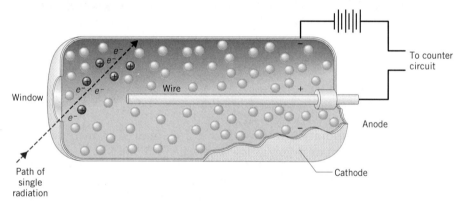

Figure 15–5
Gieger Counter. This device measures radiation by the ionization it causes.

Checking It Out ▶

Learning Check A

A–1.   Fill in the blanks.
There are five types of _____.
1. An alpha particle is a _____ nucleus.
2. A beta particle is an _____.
3. A gamma ray is a high-energy form of _____.
4. A positron is a _____ _____.
5. Electron capture has the same effect on a nucleus as _____ emission.

   The rate of decay of an isotope is indicated by its _____. The most ionizing types of radiation are _____ _____, whereas the most penetrating are _____ _____. Radiation is detected by _____ badges, and by _____ and _____ counters.

A–2. Write the radioactive isotope that leads to the indicated nucleus and particle or ray.

(a) _____ $\rightarrow$ $^{87}_{38}$Sr + beta particle

(b) _____ $\rightarrow$ $^{223}_{87}$Fr + alpha particle

(c) _____ $\rightarrow$ $^{54}_{26}$Fe + positron

(d) _____ + electron capture $\rightarrow$ $^{123}_{51}$Sb

(e) _____ $\rightarrow$ $^{137}_{55}$Cs + gamma ray

A–3. After 36.0 minutes, it was found that 4.0 μg remain of a sample of a radioactive isotope that originally weighed 64 μg. What is the half-life of the isotope?

Additional Examples: Problems 15-1, 15-16, 15-18, 15-20, 15-22, and 15-27.

In the Middle Ages, alchemists tried in vain to change one element into another. Some elements exist in nature, however, that spontaneously change into another element. The first instance of a person actually changing one element into another had to await the nuclear age and the advent of nuclear reactions.

▼Looking Ahead

15–5 NUCLEAR REACTIONS

Transmutation *of an element is the conversion of that element into another.* We have discussed how this occurs naturally by the spontaneous emission of an alpha or beta particle from a nucleus. Transmutation can also occur by artificial or synthetic means. The first example of transmutation was discovered by Ernest Rutherford in 1919; earlier, he had proposed the nuclear model of the atom. Rutherford found that $^{14}_{7}$N atoms could be bombarded with alpha particles, causing a **nuclear reaction** that produced $^{17}_{8}$O and a proton. This nuclear reaction is illustrated by the nuclear equation

$$^{14}_{7}\text{N} + {}^{4}_{2}\text{He} \longrightarrow {}^{17}_{8}\text{O} + {}^{1}_{1}\text{H}$$

Note that the total number of nucleons is conserved by the reaction (14 + 4 on the left = 17 + 1 on the right). The total charge is also conserved during the reaction (7 + 2 on the left = 8 + 1 on the right.) Both the number of nucleons and the total charge must be balanced (the same on both sides of the equation) in a nuclear reaction.

Since the time of Rutherford, thousands of isotopes have been prepared by nuclear reactions. Two other examples are

$$^{209}_{83}\text{Bi} + {}^{2}_{1}\text{H} \longrightarrow {}^{210}_{84}\text{Po} + {}^{1}_{0}n$$

$$^{27}_{13}\text{Al} + {}^{4}_{2}\text{He} \longrightarrow {}^{30}_{15}\text{P} + {}^{1}_{0}n$$

($^{2}_{1}$H is a deuterium nucleus or a deuteron; $^{1}_{0}n$ is a neutron.)

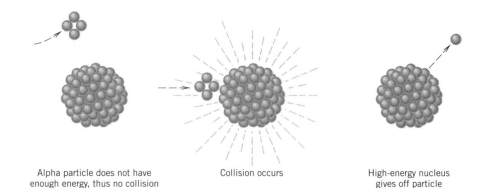

Alpha particle does not have
enough energy, thus no collision

Collision occurs

High-energy nucleus
gives off particle

Figure 15–6
Collision of an Alpha Particle with a Nucleus. Alpha particles must have sufficient
energy to overcome the repulsion of the target nucleus for a collision to take place.

Figure 15–7
**The Superconducting Super
Collider.** Construction of this
particle accelerator was suspended in late 1993 because
of its high cost.

*How did the periodic table
get lengthened?*

Because the nucleus has a positive charge and many of the bombarding nuclei (except neutrons) also have a positive charge (e.g., $^4_2$He, $^1_1$H,
$^2_1$H), the particles must have a high energy (velocity) to overcome the
natural repulsion between these two like charges. From the collision of
the nucleus and the particle, a high-energy nucleus is formed. As a result, another particle (usually a neutron) is then emitted by the nucleus
to carry away excess energy from the collision. This is analogous to a
head-on collision of two cars, where fenders, doors, and bumpers come
flying out after impact. The nucleus-particle collision is illustrated in
Figure 15-6.

The invention of particle accelerators, which speed nuclei and subatomic particles to high velocities and energies, has opened the door to
vast possibilities for artificial nuclear reactions. The larger accelerators
also allow studies of collisions so powerful that the colliding nuclei are
blasted apart. From these experiments, scientists have glimpsed the
most basic composition of matter. These accelerators are large and expensive. Completed in the early 1970s, the Fermi National Accelerator
in Batavia, Illinois, has a circumference of about six miles and cost $245
million. A huge accelerator now under construction in Texas is known
as the *Superconducting Super Collider* (SSC) and will have a circumference of 52 miles. (See Figure 15-7.) The latest estimate of cost is
about $10 billion. At this writing, however, its continued funding by
the U.S. government is in doubt because of the high cost and the budget
deficit.

One interesting development from particle accelerators has been the
preparation of new, heavy elements. In fact, all elements between plutonium (atomic number 94) and the last element made (atomic number
109, which has not yet been named) are synthetic elements. The following nuclear reactions illustrate the formation of heavy elements:

$$^{238}_{92}\text{U} + ^{12}_{6}\text{C} \longrightarrow ^{244}_{98}\text{Cf} + 6^1_0n$$

$$^{238}_{92}\text{U} + ^{16}_{8}\text{O} \longrightarrow ^{250}_{100}\text{Fm} + 4^1_0n$$

EXAMPLE 15–3

Nuclear Reactions and Equations

Complete the following nuclear equations:
(a) $^9_4\text{Be} + ^2_1\text{H} \rightarrow$ _____ $+ ^1_0n$
(b) $^{252}_{98}\text{Cf} + ^{10}_5\text{B} \rightarrow$ _____ $+ 5^1_0n$

● Working It Out

Procedure
Same as Example 15-1.
(a) Nucleons: $9 + 2 = x + 1$, so $x = 10$.
   Atomic number: $4 + 1 = y + 0$, so $y = 5$.

From the list of elements, we find that the element is boron (atomic number 5). The isotope is

$$^{10}_5\text{B}$$

(b) Nucleons: $252 + 10 = x + (5 \times 1)$, so $x = 257$.
   Atomic number: $98 + 5 = y + (5 \times 0)$, so $y = 103$.

From the list of elements, the isotope is

$$^{257}_{103}\text{Lr}$$

Many of us grew up in great fear of a holocaust caused by a nuclear war. But the many beneficial applications of radioactive isotopes are also part of our reality. A few of these applications are discussed next.

▼ Looking Ahead

15–6   USES OF RADIOACTIVITY

Carbon Dating

Perhaps one of the most useful applications of radioactive isotopes is the dating of wood or other carbon-containing substances that were once alive. Carbon-14 dating is effective in dating artifacts that are from about two thousand to about fifty thousand years old. Carbon-14 is a radioactive isotope with a half-life of 5760 years. It is produced in the stratosphere by a nuclear reaction involving *cosmic rays* (a variety of radiation and particles) from the sun.

$$^{14}_7\text{N} + ^1_0n \longrightarrow ^{14}_6\text{C} + ^1_1\text{H}$$

The ^{14}C then mixes with the normal and stable isotopes of carbon in the form of carbon dioxide. Carbon dioxide is taken up by living systems and, through photosynthesis, becomes part of the carbon structure of the organism. As long as the carbon-based system is alive, the ratio of ^{14}C to normal carbon is the same as in the atmosphere. When the system dies, the amount of ^{14}C in the organism begins to decrease. By comparing the amount of radiation from ^{14}C in the artifact with that in living systems, the age can be determined. (See Example 15-2.) Besides ^{14}C dating, the decay of other isotopes (e.g., ^{40}K) can be studied to determine the age of rocks that were formed from molten material millions of years ago.

How can radioactivity indicate the age of an ancient campsite?

Neutron Activation Analysis

Most naturally occurring isotopes are not radioactive. When these isotopes are bombarded by neutrons from a nuclear reactor, however, they often absorb a neutron, which makes them radioactive. An example is the production of ^{60}Co, which is used in cancer therapy, from a stable cobalt isotope.

$$^{59}_{27}\text{Co} + ^{1}_{0}n \longrightarrow ^{60m}_{27}\text{Co}$$

Another application of neutron activation concerns the analysis of arsenic in human hair. Arsenic compounds can be used as a slow poison, but arsenic accumulates in human hair in minute amounts. By subjecting the human hair to neutron bombardment, the stable isotope of arsenic is changed to a metastable nucleus.

$$^{75}_{33}\text{As} + ^{1}_{0}n \longrightarrow ^{76m}_{33}\text{As} \longrightarrow ^{76}_{33}\text{As} + \gamma$$

The amount of gamma radiation from the metastable arsenic can be measured and is proportional to the amount of arsenic present. The method is very sensitive to even trace amounts of arsenic.

Food Preservation

In a very simple procedure, many types of food can be irradiated with gamma radiation, which kills bacteria and other microorganisms that cause food spoilage, without changing the taste or appearance of the food. This increases the shelf life of food before decay sets in. For example, mold begins to form on strawberries, even when refrigerated, in just a few days. After irradiation, the same strawberries can be stored for two weeks without decay. (See Figure 15-8.) There has been some buyer resistance to irradiated foods. The public worries that irradiation somehow may make the food radioactive. The procedure is perfectly safe, however, and as public apprehension declines, we will probably see more and more irradiated foods at the grocery store.

Does irradiated food become radioactive?

Figure 15–8
Irradiated Strawberries. The strawberries on the left are moldy after a few days. Those on the right have been irradiated and are still fresh after two weeks.

Medical Therapy

Cobalt-60 gamma radiation can be focused into a narrow beam. Although gamma radiation destroys healthy as well as malignant cells, healthy cells recover faster. By focusing the beam at different angles on tumors located within the body, the gamma rays can be concentrated on the tumor. (See Figure 15-9.) This treatment has many unpleasant side effects because of the destruction of normal, healthy cells. Although there are several other radioactive isotopes used in therapies, the greatest application of these isotopes is in diagnosis.

Medical Diagnosis

Today, hardly any large hospital could be without its nuclear medicine division. We have come to rely on the use of radioactive isotopes to help diagnose many diseases and conditions. Various radioactive isotopes can be injected into the body, and their movement through the body or where they accumulate can be detected outside the body with radiation detectors. Iron-59 is used to measure the rate of formation and lifetime of red blood cells. Technetium-99m is used to image the brain, heart, and other organs. Iodine-131 is used to detect thyroid malfunction and can also be used for the treatment of thyroid tumors. Sodium-24 is used to study the circulatory system.

Figure 15–9
^{60}CO Radiation Treatment.
A beam of gamma rays can be used to destroy cancerous tissue.

One of the more useful (and expensive) tests is known as a *PET (positron emission tomography)* scan, which produces an image of a two-dimensional slice through a portion of the body. The body is injected with a compound (e.g., glucose) containing a radioactive isotope such as ^{11}C. The glucose containing this isotope of carbon is metabolized along with glucose produced by the body containing the stable isotope ^{12}C. The parts of the brain that are particularly active in metabolism of glucose will display increased radioactivity. Abnormal glucose metabolism in the brain can then be detected, which may indicate a tumor or Alzheimer's disease. This diagnostic procedure is considered noninvasive compared to surgery.

The ^{11}C decays by positron emission.

$$^{11}_{6}\text{C} \longrightarrow \ ^{11}_{5}\text{B} + \ ^{0}_{+1}e$$

$$^{0}_{+1}e + \ ^{0}_{-1}e \longrightarrow 2\gamma$$

Almost immediate annihilation of the positron by an electron leads to two gamma rays that exit the body in exactly opposite directions. Scintillation counters are positioned around the body so as to detect these two gamma rays and ignore others from background radiation. Computers are then used to translate the density and location of the gamma rays into two-dimensional images. (See Figure 15-10 at the top of the next page.)

In 1938, a discovery was made that at first promised unlimited destruction and, later, on the positive side, unlimited energy. This, of course, was the invention of the atomic bomb and then control of its processes in nuclear power plants. The nuclear process that produces this power is the next topic.

How does radioactivity "image" internal body parts?

▼ Looking Ahead

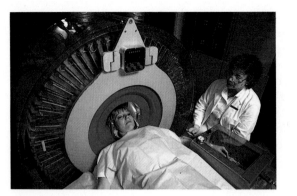

Figure 15-10
PET Scan. The PET scanner can detect abnormal
functioning of organs without the need for surgery.

15–7   NUCLEAR FISSION

In 1934, two Italian physicists, Enrico Fermi and Emilio Segre, attempted to expand the periodic table by bombarding ^{238}U with neutrons to produce isotopes that would decay by beta particle emission to form elements with atomic numbers greater than 92 (the last element on the periodic table at the time).

$$^{238}_{92}U + {}^{1}_{0}n \longrightarrow {}^{239}_{92}U \longrightarrow {}^{239}_{93}X + {}^{0}_{-1}e$$

The experiment seemed to work, but they were perplexed by the presence of several radioactive isotopes produced in addition to the presumed element number 93. In 1938, these experiments were repeated by two German scientists, Otto Hahn and Fritz Strassman. They were able to identify the excess radioactivity as coming from isotopes such as barium, lanthanum, and cerium, which had about half the mass of the uranium isotope. Two other German physicists, Lise Meitner and Otto Frisch, were able to show that the rare isotope of uranium (^{235}U, 0.7% of naturally occurring uranium) was undergoing fission into roughly two equal parts after absorbing a neutron. **Fission** *is the splitting of a large nucleus into two smaller nuclei of similar size.* It should be noted that there are a variety of products from the fission of ^{235}U in addition to those represented by the following equation. (See Figure 15-11.)

$$^{235}_{92}U + {}^{1}_{0}n \longrightarrow {}^{139}_{56}Ba + {}^{94}_{36}Kr + 3{}^{1}_{0}n$$

Two points about this reaction had monumental consequences for the world, and scientists in Europe and America were quick to grasp their meaning in a world about to go to war.

How is fission different from natural radioactivity?

1   Comparison of the masses of the product nuclei with that of the original nucleus indicated that a significant amount of mass was lost in the reaction. According to Einstein's equation (see Section 2-8), the mass lost must be converted to a tremendous amount of energy. Fission of a few kilograms of ^{235}U could produce energy equivalent to tens of thousands of tons of the conventional explosive TNT.

How did scientists conclude that an atomic bomb was possible?

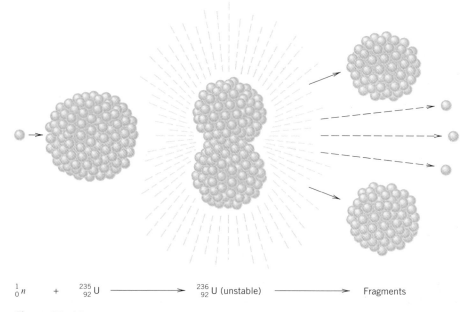

$$\begin{array}{ccccccc} {}^{1}_{0}n & + & {}^{235}_{92}U & \longrightarrow & {}^{236}_{92}U \ (\text{unstable}) & \longrightarrow & \text{Fragments} \end{array}$$

Figure 15–11
Fission. In fission, one neutron produces two fragments of the original atom and an average of three neutrons.

2 What made the rapid fission of a large sample of ^{235}U feasible was the potential for a chain reaction. *A **nuclear chain reaction** is a reaction that is self-sustaining.* The reaction generates the means to trigger additional reactions. Note in Figure 15-11 that the reaction of one original neutron caused three to be released. These three neutrons cause the release of nine neutrons and so forth, as illustrated in Figure 15-12. If a given densely packed "critical mass" of ^{235}U is present, the whole mass of uranium can undergo fission in an instant with a quick release of energy in the form of radiation.

Unfortunately, the world thus entered the nuclear age in pursuit of a bomb. After a massive secret effort, the first nuclear bomb was exploded over Alamogordo Flats in New Mexico on July 16, 1945. This bomb and the bomb exploded over Nagasaki, Japan, were made of ^{239}Pu, which is a synthetic fissionable isotope. The bomb exploded over Hiroshima, Japan, was made of ^{235}U.

This enormously destructive device, however, can be tamed. The chain reaction can be controlled by absorbing excess neutrons with cadmium bars. A typical nuclear reactor is illustrated in Figure 15-13. In a reactor core, uranium in the form of pellets is encased in long rods called fuel elements. Cadmium bars are raised and lowered among the fuel elements to control the rate of fission by absorbing neutrons. If the cadmium bars are lowered all the way, the fission process can be halted altogether. In normal operation, the bars are raised just enough so that the fission reaction occurs at the desired level. Energy released by the fission and the decay of radioactive products formed from the fission is used to heat water, which circulates among the fuel elements. This wa-

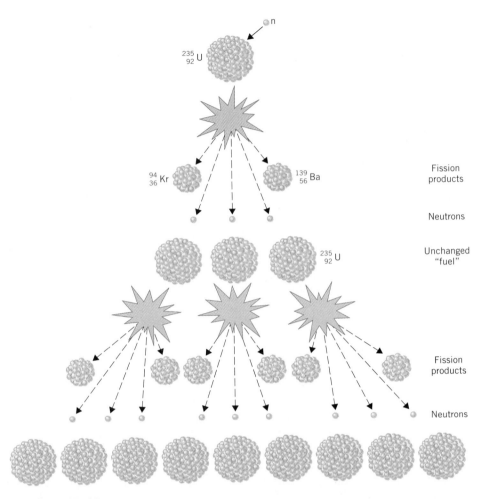

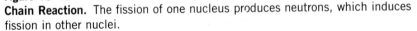

Figure 15–12
Chain Reaction. The fission of one nucleus produces neutrons, which induces fission in other nuclei.

ter, which is called the primary coolant, becomes very hot (about 300°C) without boiling because it is under high pressure. The water heats a second source of water, changing it to steam. The steam from the secondary coolant is cycled outside of the containment building, where it is used to turn a turbine that generates electricity.

Perhaps the greatest advantage of nuclear energy is that it does not pollute the air with compounds (called oxides) of sulfur and carbon as conventional power plants do. Sulfur oxides have been implicated as a major cause of acid rain. It is also feared that large amounts of carbon dioxide will eventually lead to significant changes in the weather as a result of the greenhouse effect. Another advantage is that nuclear power does not deplete the limited supply of fossil fuels (oil, coal, and natural gas). Until the tragedy in Chernobyl, Russia, in April 1986, there had been no loss of human life from the use of nuclear energy to generate electricity. In 1991, 26 countries produced electricity through nuclear power. The United States generated 21.7% and France 72.7% of their electricity from nuclear power in that year. Proponents of nuclear power

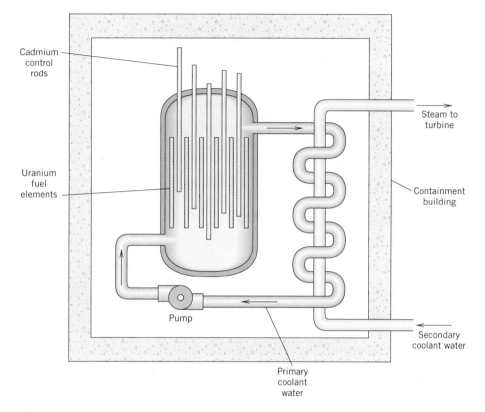

Cadmium control rods

Uranium fuel elements

Pump

Steam to turbine

Containment building

Secondary coolant water

Primary coolant water

Figure 15–13
A Nuclear Reactor. Nuclear energy is converted to heat energy, the heat energy to mechanical energy, and the mechanical energy to electrical energy.

feel that if the proper systems are used with adequate safeguards and backups, catastrophic accidents would be avoided. In fact, new designs of nuclear power plants solve many of the legitimate concerns of citizens. A safe reactor would be one that shuts itself down in case of an emergency and will not overheat or explode. American and Russian experts are also currently planning a relatively safe reactor that will use as a fuel the ^{239}Pu salvaged from dismantled nuclear bombs of the former USSR. This reactor would be highly efficient, converting about 50% of the heat generated into electricity, compared to an efficiency of about 33% for a typical commercial nuclear power plant.

The disadvantages have been made painfully obvious. The catastrophic explosion and subsequent fire of reactor 4 in Chernobyl, Ukraine, in 1986 dispersed more dangerous radiation into the environment than all of the atmospheric bomb tests by all nations put together. The surroundings for at least 30 miles around the plant, and even more downwind, are still uninhabitable. The reactor itself has been entombed in concrete and must remain so for hundreds of years. The extent of the damage is still being assessed. Several thousand people lost their lives and thousands more will have their lives shortened by radiation sickness. Many brave people died as a result of the radiation received while fighting the fire or attempting to contain and secure the building. In the United States, an accident at Three Mile Island, Pennsylvania, in March

The explosion of this reactor spread radioactive debris for miles around.

1979 took more than fourteen years to clean up at a cost greater than $1 billion. Fortunately, there was very little release of radioactivity and no loss of life in this accident.

In both of these accidents, a core "meltdown" did not occur. A meltdown would occur if the temperature of the reactor core exceeded 1130°C, the melting point of uranium. Theoretically, the mass of molten uranium, together with all of the highly radioactive decay products, could accumulate on the floor of the reactor, melt through the many feet of protective concrete, and eventually reach groundwater. If this happened, vast amounts of deadly radioactive wastes could be spread through the groundwater into streams and lakes.

What did we learn at Chernobyl?

The type of accident that occurred in Ukraine is unlikely in most of the world's commercial reactors. Unlike reactors in the former Soviet Union, these reactors are surrounded by an extremely strong structure, called a containment building, made of reinforced concrete. (See Figure 15-14.) These buildings should contain an explosion such as occurred at

Figure 15–14
A Nuclear Power Plant. In these plants, the nuclear energy of uranium is converted into electrical energy.

Chernobyl. Also, most reactors do not contain graphite, which is highly combustible and led to the prolonged fire in Chernobyl that spread so much radioactivity into the atmosphere. There is one reactor similar to the Ukrainian reactors, in Hanford, Washington, but it has been shut down indefinitely.

A growing problem in the use of nuclear energy involves used, or spent, fuel. Originally, used fuel was to be processed at designated centers where the unused fuel could be separated and reused and the radioactive wastes disposed. However, problems in transportation of this radioactive material and the danger of theft of the ^{239}Pu that is produced in a reactor have hindered solution of the problem. (^{239}Pu is also fissionable and could be used in a nuclear bomb.) Currently, spent fuel elements, which must be replaced every four years, are being stored at the reactor sites. These fuel rods are not only highly radioactive but remain extremely hot and must be continually cooled. It is estimated that they will remain dangerously radioactive for at least one thousand years. This is a problem that must soon be resolved.

Because of the accidents, new regulations regarding safeguards have made the cost of a new nuclear power plant considerably higher than that of a conventional power plant using fossil fuels. As a result, no new nuclear plants are planned in the United States in the immediate future. (See Figure 15-14.)

In the late 1930s, yet another source of nuclear energy was proposed by Hans Bethe. This involved fusing small nuclei rather than splitting large ones. It was even suggested that this was a major source of energy in the universe because it was the energy source of the sun. We now have a good (but not complete) understanding of the nuclear reactions that eventually made life on Earth possible. This is the subject of the final section of this chapter.

▼Looking Ahead

15–8 NUCLEAR FUSION

Soon after the process of nuclear fission was demonstrated, an even more powerful type of nuclear reaction was shown to be theoretically possible. *This reaction involved the* **fusion** *or bringing together of two small nuclei.* An example of a fusion reaction is

$$^{3}_{1}\text{H} + ^{2}_{1}\text{H} \longrightarrow ^{4}_{2}\text{He} + ^{1}_{0}n$$

($^{3}_{1}$H is called tritium; it is a radioactive isotope of hydrogen.)

As in the fission process, a significant amount of mass is converted into energy. Fusion energy is the origin of almost all of our energy, since it powers the sun. Millions of tons of matter are converted to energy in the sun every second. Because of its large mass, however, the sun contains enough hydrogen to "burn" for additional billions of years.

The principle of fusion was first demonstrated on this planet with a tremendously destructive device called the hydrogen bomb. This bomb can be more than one thousand times more powerful than the atomic bomb, which uses the fission process.

This small reactor is being used to study the feasibility of the generation of fusion power.

Why is controlled fusion so difficult?

Fission is controlled in nuclear power plants. What about fusion? Controlled fusion is technically extremely difficult. Temperatures on the order of 50 million degrees Celsius are needed so that the colliding nuclei have enough kinetic energy to overcome their mutual repulsions and cause fusion. The necessary temperature is hotter than the interior of the sun. No known materials can withstand these temperatures, so alternate containment procedures must be used. Producing electricity from a fusion reactor is perhaps the greatest technical challenge yet attempted by the human race. However, the effort is well under way. The design and construction of a large fusion reactor (International Thermonuclear Experimental Reactor, or ITER) is being financed and staffed by an international consortium of nations including the United States, Japan, the European Economic Community, and the Commonwealth of Independent States (former USSR). The cost is estimated at about $8 billion over a thirteen-year period. It is hoped that eventually this reactor will exceed the *break-even point* where the same amount of energy is produced by the fusion as is needed to bring about the fusion. In about twenty years, if things go well, the first fusion reactor designed to produce electricity could be built. With the amount of fossil fuel on the decline, there are not many alternatives for energy later in the next century.

The advantages of controlled fusion power are impressive.

1 It should be relatively clean. Few radioactive products are formed.

2 Fuel is inexhaustible. The oceans of the world contain enough deuterium, one of the reactants, to provide the world's energy needs for a trillion years. On the other hand, there is a very limited supply of fossil fuels and uranium.

3 There is no possibility of the reaction going out of control and causing a meltdown. Fusion will occur in power plants in short bursts of energy that can be stopped easily in case of mechanical problems.

Nuclear reactions occur when stable nuclei are bombarded with various charged particles or light nuclei. These reactions have led to an expansion of the periodic table from element 93, the heaviest naturally occurring element, to element number 109. These reactions have also produced radioactive isotopes that have broad application in the diagnosis and treatment of diseases. A nuclear reaction involving a neutron and ^{235}U led to the discovery of nuclear fission and the production of vast amounts of energy. Nuclear fusion is also a nuclear reaction that may someday provide a major source of energy.

◀ **Looking Back**

Learning Check B

◀ **Checking It Out**

B–1. Fill in the blanks.

Nuclear reactions may cause one element to be transformed into another by a process known as _____. The energy of the bombarding particle is increased by use of particle _____. Both naturally occurring and synthetic radioactive isotopes find many uses. The decay of carbon-__ can be used to date many carbon-containing fossils. Medically useful radioactive isotopes can be prepared by _____ activation. Food can be preserved by radiation with _____ _____. PET scans make use of isotopes that emit _____. A nuclear reaction known as fission can occur as a nuclear _____ reaction. The process whereby small nuclei combine to form a larger nucleus and a small particle is known as _____.

B–2. Complete the following nuclear equations.

(a) $^{51}_{23}V + {}^{2}_{1}H \rightarrow$ _____ $+ {}^{1}_{1}H$

(b) $^{235}_{92}U + {}^{1}_{0}n \rightarrow {}^{90}_{38}Sr +$ _____ $+ 2{}^{1}_{0}n$

(c) $^{246}_{96}Cm +$ __ $\rightarrow {}^{254}_{102}No + 4{}^{1}_{0}n$

(d) ${}^{2}_{1}H + {}^{2}_{1}H \rightarrow {}^{3}_{1}H +$ __

B–3. Which of the nuclear reactions in B-2 represents fission? Which represents fusion?

Additional Examples: Problems 15-30, 15-32, and 15-36.

CHAPTER REVIEW

The nuclei of **radioactive isotopes** are unstable and emit **radiation.** Although radioactive isotopes exist for each element, all isotopes with an atomic number greater than 83 are unstable. Originally, three types of radiation were discovered, **alpha particles, beta particles,** and **gamma rays.** Since then, two other types of radiation have been characterized that occur rarely in nature but more commonly in artificially

▲▼ **Putting It Together**

produced radioactive isotopes. These are **positron particles** and **electron capture**. These five modes of decay are illustrated by the following **nuclear equations:**

$$\text{Alpha particle} \qquad ^{240}_{96}\text{Cm} \longrightarrow ^{236}_{94}\text{Pu} + ^{4}_{2}\text{He}$$

$$\text{Beta particle} \qquad ^{71}_{30}\text{Zn} \longrightarrow ^{71}_{31}\text{Ga} + ^{0}_{-1}e$$

$$\text{Gamma radiation} \qquad ^{99m}_{43}\text{Tc} \longrightarrow ^{99}_{43}\text{Tc} + \gamma$$

$$\text{Positron particle} \qquad ^{19}_{10}\text{Ne} \longrightarrow ^{19}_{9}\text{F} + ^{0}_{+1}e$$

$$\text{Electron capture} \quad ^{7}_{4}\text{Be} + ^{0}_{-1}e \longrightarrow ^{7}_{3}\text{Li}$$

Radioactive isotopes decay at widely different rates. A measure of the rate of decay is called the **half-life** of the isotope. Half-lives vary from billions of years to fractions of a second. A long-lived isotope is ^{238}U, which begins a **radioactive decay series** and ends with a stable isotope of lead after a series of alpha and beta decays.

Radiation affects the molecules of surrounding matter by causing ionization. In living matter, this ionization may lead to destruction of the cells. Alpha and beta emitters are most dangerous when ingested since they cause a high degree of ionization in close proximity to cells. Gamma rays are less ionizing but very penetrating; thus they are also very dangerous. Radiation is detected and measured by means of film badges, Geiger counters, and scintillation counters.

Nuclear reactions occur when nuclei are bombarded by particles or other nuclei. This leads to the artificial **transmutation** of one element into another. **Fission** and **fusion** are also nuclear reactions. Fission is the splitting of one heavy nucleus into two more-or-less equal fragments and neutrons. Fusion is the joining of two light nuclei to form a heavier nucleus and a particle. Examples of these three processes are illustrated with the following nuclear equations:

$$\text{Nuclear reaction} \qquad ^{253}_{99}\text{Es} + ^{4}_{2}\text{He} \longrightarrow ^{256}_{101}\text{Md} + ^{1}_{0}n$$

$$\text{Fission} \qquad ^{235}_{92}\text{U} + ^{1}_{0}n \longrightarrow ^{94}_{38}\text{Sr} + ^{139}_{54}\text{Xe} + 3^{1}_{0}n$$

$$\text{Fusion} \qquad ^{2}_{1}\text{H} + ^{2}_{1}\text{H} \longrightarrow ^{3}_{2}\text{He} + ^{1}_{0}n$$

E X E R C I S E S

NUCLEAR RADIATION

15–1 Write isotope symbols in the form $^{A}_{Z}\text{M}$ for isotopes of the following composition:
(a) 84 protons and 126 neutrons
(b) 46 neutrons with a mass number of 84
(c) 100 protons with a mass number of 257
(d) lead-206
(e) uranium-233

15–2 Write isotope symbols in the form $^{A}_{Z}\text{M}$ for isotopes of the following composition:
(a) 86 protons and 134 neutrons
(b) 6 protons with a mass number of 13
(c) 22 neutrons with a mass number of 40
(d) potassium-41
(e) americium-243

15-3 Give the symbols, including mass number and charge, for
(a) an alpha particle
(b) a beta particle
(c) a neutron

15-4 A deuteron has a mass number of two and a positive charge of one. Write its isotope symbol.

15-5 A triton has a mass number of three and a positive charge of one. Write its isotope symbol.

15-6 Write the isotope symbol, including mass number, for the isotope that results when each of the following emits an alpha particle.
(a) $^{210}_{84}Po$ (c) fermium-252
(b) ^{152}Gd (d) Une-266

15-7 Write the isotope symbol, including mass number, for the isotope that results when each of the following emits an alpha particle.
(a) $^{234}_{92}U$ (c) ^{210}Bi
(b) $^{222}_{88}Ra$ (d) thorium-229

15-8 Write the isotope symbol, including mass number, for the isotope that results when each of the following emits a beta particle.
(a) $^{3}_{1}H$ (c) iron-59
(b) ^{153}Gd (d) sodium-24

15-9 Write the isotope symbol, including mass number, for the isotope that results when each of the following emits a beta particle.
(a) $^{131}_{53}I$ (c) lead-210
(b) ^{234}Pa (d) nitrogen-16

15-10 Give the symbol of a positron. What effect does the emission of a positron have on an isotope? How are positrons detected?

15-11 Manganese-51 undergoes positron emission. Write the nuclear equation illustrating this reaction.

15-12 A certain isotope undergoes positron emission to form ^{23}Na. Write the nuclear equation illustrating this reaction.

15-13 What happens to an isotope that undergoes electron capture. How is electron capture detected?

15-14 Germanium-68 undergoes electron capture. Write the nuclear equation illustrating this reaction.

15-15 A certain isotope undergoes electron capture to form ^{55}Mn. Write the nuclear equation illustrating this reaction.

15-16 Complete the following nuclear equations:
(a) $^{214}_{83}Bi \rightarrow {}^{214}_{84}Po + \underline{\quad}$
(b) $^{90}_{37}Rb \rightarrow \underline{\quad} + {}^{0}_{-1}e$
(c) $^{26}_{14}Si \rightarrow \underline{\quad} + {}^{0}_{+1}e$
(d) $^{235}_{92}U \rightarrow \underline{\quad} + {}^{4}_{2}He$
(e) $^{179}_{73}Ta + {}^{0}_{-1}e \rightarrow \underline{\quad} {}^{0}_{-1}e$
(f) $\underline{\quad} \rightarrow {}^{41}_{21}Sc + {}^{0}_{-1}e$
(g) $\underline{\quad} \rightarrow {}^{210}_{82}Pb + {}^{0}_{-1}e$

15-17 Complete the following nuclear equations:
(a) $^{239}_{93}Np \rightarrow \underline{\quad} + {}^{0}_{-1}e$
(b) $\underline{\quad} \rightarrow {}^{93}_{44}Ru + {}^{0}_{+1}e$
(c) $^{226}_{88}Ra \rightarrow {}^{222}_{86}Rn + \underline{\quad}$
(d) $\underline{\quad} \rightarrow {}^{235}_{92}U + {}^{4}_{2}He$
(e) $^{80}_{37}Rb + \underline{\quad} \rightarrow {}^{80}_{36}Kr$
(f) $^{32}_{15}P \rightarrow {}^{32}_{16}S + \underline{\quad}$

15-18 From the following information, write nuclear equations that include all isotopes and particles.
(a) $^{230}_{90}Th$ decays to $^{226}_{88}Ra$.
(b) $^{214}_{84}Po$ emits an alpha particle.
(c) $^{210}_{84}Po$ emits a beta particle.
(d) An isotope emits an alpha particle and forms lead-214.
(e) $^{14}_{6}C$ decays to form $^{14}_{7}N$.
(f) Chromium-50 is formed by positron emission.
(g) Argon-37 captures an electron.

*15-19 The decay series of $^{238}_{92}U$ to $^{206}_{82}Pb$ involves alpha and beta emissions in the following sequence: α, β, β, α, α, α, α, α, β, α, β, β, β, α. Identify all isotopes formed in the series.

NUCLEAR DECAY AND HALF-LIFE

15-20 What fraction of a radioactive isotope remains after four half-lives?

15-21 What percent of a radioactive isotope remains after five half-lives?

15-22 If one starts with 20 mg of a radioactive isotope with a half-life of 2.0 days, how much remains after
(a) 4 days   (b) 8 days
(c) four half-lives

15-23 The half-life of a given isotope is 10 years. If we start with a 10.0-g sample of the isotope, how much is left after 20 years?

15-24 Start with 12.0 g of a given radioactive isotope. After 11 years, only 3.0 g is left. What is the half-life of the isotope?

15-25 The isotope $^{14}_{6}C$ is used to date fossils of formerly living systems such as prehistoric animal bones. If the radioactivity of the carbon in a sample of bone from a mammoth is one-fourth of the radioactivity of the current level, how old is the fossil? ($t_{1/2} = 5760$ years)

THE EFFECTS OF RADIATION AND ITS DETECTION

15–26   Of the three types of radiation, which is the most ionizing? Which is the most penetrating? How does each type of radiation cause damage to cells?

15–27   How does radiation cause ionization? How does ionization cause damage to living tissues?

15–28   How does a film badge work? How can a film badge tell which type of radiation is being absorbed?

15–29   How does a Geiger counter work? How does a scintillation counter work?

NUCLEAR REACTIONS

15–30   Complete the following nuclear equations:
(a) $^{35}_{17}Cl + ^{1}_{0}n \rightarrow$ _____ $+ ^{1}_{1}H$
(b) $^{27}_{13}Al +$ _____ $\rightarrow ^{25}_{12}Mg + ^{4}_{2}He$
(c) $^{27}_{13}Al + ^{4}_{2}He \rightarrow$ _____ $+ ^{1}_{0}n$
(d) $^{238}_{92}U + 15^{1}_{0}n \rightarrow ^{253}_{100}Es +$ _____
(e) $^{244}_{96}Cm + ^{12}_{6}C \rightarrow$ _____ $+ 2^{1}_{0}n$
(f) _____ $+ ^{2}_{1}H \rightarrow ^{238}_{93}Np + ^{1}_{0}n$
(g) $^{242}_{94}Pu + ^{22}_{10}Ne \rightarrow ^{260}_{104}-\ +$ _____

15–31   Complete the following nuclear equations:
(a) $^{249}_{96}Cm +$ _____ $\rightarrow ^{260}_{103}Lr + 4\,^{1}_{0}n$
(b) $^{15}_{7}N + ^{1}_{1}H \rightarrow$ _____ $+ ^{4}_{2}He$
(c) $^{10}_{5}B + ^{4}_{2}He \rightarrow$ _____ $+ ^{1}_{1}H$
(d) $^{249}_{98}Cf +$ _____ $\rightarrow ^{263}_{106}-\ + 4\,^{1}_{0}n$

15–32   The last element on the periodic table was prepared by the following nuclear reaction:

$$^{209}_{83}Bi + ^{58}_{26}Fe \longrightarrow ^{266}_{109}Une$$

How many neutrons were emitted in the reaction?

15–33   Unp-260 plus four neutrons is prepared by bombarding Cf-249 with what isotope?

15–34   Uno-265 plus one neutron is prepared by bombarding an isotope of lead with iron-58. What is the isotope of lead?

15–35   Bismuth-209 can be bombarded with chromium-54, producing one neutron and what heavy element isotope?

FISSION AND FUSION

15–36   When $^{235}_{92}U$ and $^{239}_{94}Pu$ undergo fission, a variety of reactions take place. Complete the following:
(a) $^{235}_{92}U + ^{1}_{0}n \rightarrow$ _____ $+ ^{146}_{58}Ce + 3^{1}_{0}n$
(b) $^{239}_{94}Pu + ^{1}_{0}n \rightarrow ^{141}_{56}Ba +$ _____ $+ 2^{1}_{0}n$

15–37   What is the difference between fission and fusion? What is the source of energy for these two processes?

*15–38   In 1989, some scientists reported that they had achieved "cold fusion." This was supposedly accomplished by electrolyzing heavy water (deuterium oxide) with palladium electrodes. It was suggested that the deuterium was absorbed into the electrodes and somehow two deuterium nuclei could overcome the strong repulsions at a low temperature and fuse. Evidence was presented to indicate that more energy came out of the reaction than could be accounted for by a chemical process. There has been little support for these initial experiments, but, for awhile, it generated nightly reports on the national news. The fusion of two deuterium nuclei should produce helium-4, which, according to theory, would have too much energy to be stable and so would decompose to either helium-3 or tritium. What other two particles would be produced by this decomposition? Write the appropriate nuclear equations.

GENERAL QUESTIONS

15–39   Write equations illustrating each of the following nuclear processes:
(a) Palladium-106 absorbs an alpha particle to produce an isotope and a proton.
(b) Une-266 emits an alpha particle to form an isotope that also emits an alpha particle.
(c) Bismuth-212 emits a beta particle.
(d) An isotope emits a positron to form copper-60.
(e) Plutonium-239 absorbs a neutron to produce cesium-140, another isotope, and three neutrons.

(f) Lead-206 is bombarded by an isotope to produce Unh-257 and three neutrons.

(g) An isotope captures an electron to form niobium-93.

15–40 Write equations illustrating each of the following nuclear processes:

(a) An isotope of a heavy element is bombarded with boron-11 to form lawrencium-257 and four neutrons.

(b) An isotope emits an alpha particle to form actinium-231.

(c) Oxygen-14 emits a particle to form nitrogen-14.

(d) Sulfur-31 captures an electron to form an isotope.

(e) Uranium-235 absorbs a neutron to form rubidium-90, cesium-142, and several neutrons.

(f) Two helium nuclei fuse (in the sun) to form lithium-7 and a particle.

(g) An isotope emits a beta particle to form lead-208.

(h) Plutonium-239 is bombarded with an isotope to form curium-242 and a neutron.

15–41 The radioactive isotope $^{90}_{38}$Sr can accumulate in bones, where it replaces calcium. It emits a high-energy beta particle, which eventually can cause cancer.

(a) What is the product of the decay of $^{90}_{38}$Sr?

(b) How long would it take for a 0.10-mg sample of $^{90}_{38}$Sr to decay to where only 2.5×10^{-2} mg was left? (The half-life of $^{90}_{38}$Sr is 25 years.)

15–42 The radioactive isotope $^{131}_{53}$I accumulates in the thyroid gland. On the one hand, this can be useful in detecting diseases of the thyroid and even in treating cancer at that location. On the other hand, exposure to excessive amounts of this isotope, such as from a nuclear power plant, can cause cancer of the thyroid. $^{131}_{53}$I emits a beta particle with a half-life of 8.0 days. What is the product of the decay of $^{131}_{53}$I? If one started with 8.0×10^{-6} g of $^{131}_{53}$I, how much would be left after 32 days?

15–43 The fissionable isotope $^{239}_{94}$Pu is made from the abundant isotope of uranium, $^{238}_{92}$U, in nuclear reactors. When $^{238}_{92}$U absorbs a neutron from the fission process, $^{239}_{94}$Pu eventually forms. This is the principle of the breeder reactor, although $^{239}_{94}$Pu is formed in all reactors. Complete the following reaction:

$$^{238}_{92}U + {}^{1}_{0}n \longrightarrow \underline{\qquad}$$
$$+ {}^{0}_{-1}e \longrightarrow {}^{239}_{94}Pu + \underline{\qquad}$$

SOLUTIONS TO LEARNING CHECKS

A–1 radiation, helium, electron, light, positive electron, positron, half-life, alpha particles, gamma rays, film, Geiger, scintillation

A–2 (a) $^{87}_{37}$Rb (b) $^{227}_{89}$Ac (c) $^{54}_{27}$Co (d) $^{123}_{52}$Te (e) $^{137m}_{55}$Cs

A–3 After four half-lives, 4.0 μg of the original sample is left. $t_{1/2} = 36/4 = 9.0$ min (i.e., after 9.0 min, 32 μg remains; after 18 min, 16 μg; after 27 min, 8.0 μg; and after 36 min, 4.0 μg remains.)

B–1 transmutation, accelerators, 14, neutron, gamma rays, positrons, chain, fusion

B–2 (a) $^{52}_{23}$V (b) $^{144}_{54}$Xe (c) $^{12}_{6}$C (d) $^{1}_{1}$H

B–3 Reaction (b) is fission and reaction (d) is fusion.

The trees, deer, and all other life-forms in this picture are based on the element carbon. The compounds of carbon are the subjects of this chapter.

ORGANIC CHEMISTRY

Perfectly positioned about 90 million miles from a medium-sized star and protected by a thick atmosphere, our planet glides majestically in space. Under these ideal conditions, the formation and stable existence of chemical bonds that compose living systems are possible. These living systems grow and reproduce on the Earth, move gracefully in the oceans, and float effortlessly in the air. All of these creatures and plants are composed of compounds that include the element carbon as the central character. We, of course, consider ourselves to be the very ultimate carbon-based life-form. Could there be life-forms based on other elements? Science fiction stories would have us think so, but it is actually quite unlikely, if not impossible. In this chapter, we will see how this unique element forms the types of molecules that are part of living systems.

◀ Setting The Stage

Traditionally, all chemical compounds have been relegated to two categories: organic and inorganic. **Organic compounds** *include most of the compounds of carbon, especially those containing carbon-hydrogen bonds.* However, much of the chemistry that we have discussed so far concerns *compounds derived from minerals and other noncarbon compounds, and these are known as* **inorganic compounds.** A few compounds of carbon such as $CaCO_3$ and LiCN are considered to be inorganic compounds since they are derived from minerals and do not contain C—H bonds. Originally, the division of compounds into the two categories was determined by whether they were derived from (or part of) a living system (organic) or from a mineral (inorganic). In fact, before 1828 it was thought that organic compounds and their decay products could be synthesized from inorganic compounds (e.g., H_2O, CO_2, and NH_3) only by living matter (organisms). Only life had that magic ingredient of nature called the "vital force" that allowed the miracle of organic synthesis. In 1828, however, a German scientist named Friedrich Wohler synthesized urea (H_2NCONH_2), which had been classified as organic, from the inorganic compound ammonium cyanate (NH_4CNO). Urea is a waste product of the metabolism of proteins and, until 1828, was thought to result only from this source. Although it was just one compound, the concept of the "vital force" was doomed. Since that time, millions of organic compounds have been synthesized from basic minerals in the laboratory.

Organic compounds are certainly central to our lives. Hydrocarbons (compounds composed only of carbon and hydrogen) are used as fuel to power our cars and to heat our homes. Our bodies are fueled with organic compounds obtained from the food we eat in the form of sugars (carbohydrates), fats, and proteins. This food is made more palatable by organic flavorings, is wrapped in organic plastic, and is kept from spoiling with organic preservatives. Our clothes are made of organic compounds, whether these compounds come from plant and animal sources (cotton and wool) or are synthetic (nylon and Dacron®.) These fabrics are made colorful with organic dyes. When we are ill, we take drugs that may also be organic: aspirin relieves headaches, codeine suppresses coughs, and diazepam (Valium®) calms nerves. These are only a few examples of how we use organic chemicals daily.

The introduction of such a large topic as organic chemistry in one chapter is not an easy task. The questions that we will address for each type of organic compound are: What is unique about its molecular structure? How is it named? What are its particular properties? How is it made or otherwise obtained? What use is made of the specific type of compound? First, however, we will lay the groundwork by reviewing the bonding and general properties of the central and most important atom in organic compounds—carbon. In this endeavor, the discussions of Lewis structures, polarity of bonds, geometry, and polarity of molecules are important. These topics were all covered in Chapter 6.

Formulating ▶
Some Questions

16–1   BONDING IN ORGANIC COMPOUNDS

Of the 109 elements on the periodic table, only one has properties that make it possible for the existence of large complex molecules on which living systems are based. This, of course, is carbon. There are two reasons for carbon's unique properties.

1 Carbon-carbon bonds are strong. This makes long chains of carbon atoms possible.

2 Carbon and hydrogen have similar electronegativities. This means that the C—H bond is nearly nonpolar. As a result, the C—H bond is not chemically reactive towards many compounds such as water.

A science fiction writer may look at Group IVA in the periodic table and notice that silicon is immediately under carbon. Could the writer propose life based on silicon? In fact, such life-forms are not likely. The Si—Si bond is much weaker than the C—C bond, so long chains of silicon atoms in a molecule do not occur. Secondly, the Si—H bond is polar, meaning that it is chemically reactive with other polar compounds. For example, Si—H compounds react spontaneously with water (a polar compound) at room temperature. It is hard to imagine a life-form based on small molecules that react with water. If such life existed on some distant planet, we would hope that it wouldn't get caught in the rain.

What does life have to do with the nature of the C—C and the C—H bond?

Most organic compounds are typical molecular compounds. That is, they are usually gases or liquids at room temperature, or solids with low melting points. Since the C—H bond is nearly nonpolar, most organic molecules are nonpolar or, at most, they have low polarity. This means that, in general, they have little solubility in water. Inorganic compounds, on the other hand, are likely to be high-melting-point solids and are more likely to be soluble in water.

Carbon (in Group IVA) has four valence electrons. In its compounds, carbon follows the octet rule, which means that it must form *four* bonds to have access to eight valence electrons. It may form (a) four single bonds to four different atoms, (b) a double bond to one atom and two single bonds to two others, or (c) a triple bond to one atom and one single bond to one other atom. Examples of each of these three cases are shown, written as Lewis structures.

(a) Single Bonds

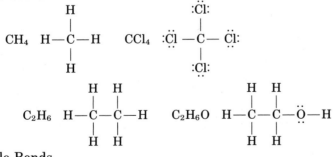

(b) Double Bonds

(c) Triple Bond

$$C_2H_2 \quad H—C \equiv C—H$$

As we will see, organic compounds are classified according to the types of bonds in their molecules. Thus writing correct Lewis structures is an important endeavor and is reviewed in the example on the next page.

Working It Out ●

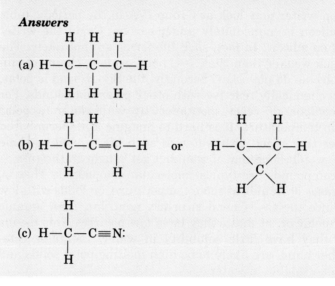

EXAMPLE 16–1

Writing Lewis Structures of Some Organic Compounds

Give the Lewis structures for the following compounds:
(a) C_3H_8 (b) C_3H_6 (c) H_3C_2N (all H's on one C)

Answers

(a) Structure of propane

(b) Two structures

(c) Structure

Many times we find that there is more than one Lewis structure for a given formula. Such is the case with butane, C_4H_{10}, since two correct structures can be drawn. *Compounds with different structures but the same molecular formula are called* **isomers.**

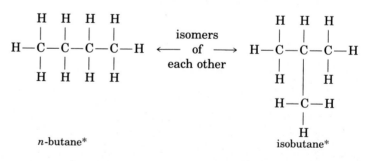

n-butane*   isobutane*

Are organic compounds "flat"?

These Lewis structures represent the molecules as flat and two-dimensional. In fact, when carbon is bonded to four other atoms, it lies in the center of a tetrahedron. In the actual three-dimensional struc-

---

*The *n* is an abbreviation of normal and refers to the isomer in which all the carbons are bound consecutively in a continuous chain. The *iso*-prefix refers to the isomer in which there is a branch involving three carbons attached to a carbon chain or another atom. For example,

$$CH_3$$
$$H \longrightarrow C—$$
$$CH_3$$

ture, the H—C—H bond angle is about 109°. A more realistic representation of the structure of *n*-butane is sometimes shown as

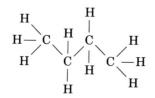

This representation is somewhat awkward, however, so long-chain structures are usually shown in a straight line even though they actually have a three-dimensional geometry. A true representation of the structure requires the use of molecular models. In Figure 16-1, the CH_4, C_2H_6, and C_3H_8 molecules are represented by "ball-and-stick" models. In a three-dimensional representation, we can see that all the hydrogens in the C_2H_6 molecule are identical.

Many of the compounds that we will mention in the following sections have several repeating units. Drawing these structures can become quite tedious if all hydrogens and carbons are written out. A **condensed formula** in which separate bonds are not written is helpful. For example,

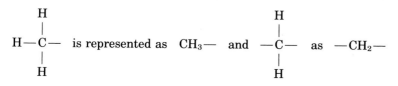

Figure 16–1
Geometry of CH_4, C_2H_6, and C_3H_8 In alkanes, the carbons are at the center of a tetrahedron.

Is there an easy way to write the formulas of long organic compounds?

Depending on what we are trying to show, the structure may be partially or fully condensed.

	Partially Condensed	Fully Condensed
n-butane	$CH_3-CH_2-CH_2-CH_3$	$CH_3(CH_2)_2CH_3$

isobutane

$$CH_3-\overset{\overset{\displaystyle H}{|}}{\underset{\underset{\displaystyle CH_3}{|}}{C}}-CH_3 \qquad (CH_3)_3CH$$

In a fully condensed structure, it is understood that the CH_2s in parentheses are in a continuous chain and the CH_3s in parentheses are attached to the same atom.

A few other compounds and their isomers are shown in Table 16-1. As you can see, the number of isomers increases as the number of carbons increases, and addition of a **hetero atom** (*any atom other than carbon or hydrogen*) also increases the number of isomers.

Looking Ahead ▼

In the next four sections, we will discuss organic compounds composed of only two elements (carbon and hydrogen). These compounds are classified into four categories based on the type of bonding in the molecules. In the first of these sections, we will investigate carbon-hydrogen compounds that contain only single covalent bonds.

TABLE 16–1 ISOMERS

Formula	Isomer, Name		
C_5H_{12}	$CH_3CH_2CH_2CH_2CH_3$ *n*-pentane	$CH_3CH_2\overset{\overset{\displaystyle CH_3}{\|}}{CH}-CH_3$ isopentane	$CH_3-\overset{\overset{\displaystyle CH_3}{\|}}{\underset{\underset{\displaystyle CH_3}{\|}}{C}}-CH_3$ neopentane
C_3H_6	$H_2C{=}CHCH_3$ propene	$H_2C\overset{\diagdown}{}\underset{\underset{\displaystyle CH_2}{}}{}\overset{\diagup}{}CH_3$ cyclopropane	(Note that the carbons can also be arranged in a ring or *cyclic* structure.)
C_2H_6O	CH_3CH_2OH ethanol	$H_3C-O-CH_3$ dimethyl ether	
C_3H_6O	$CH_3CH_2\overset{\overset{\displaystyle O}{\|\|}}{CH}$ propanal	$H_3C-\overset{\overset{\displaystyle O}{\|\|}}{C}-CH_3$ propanone	$H_2C{=}CHCH_2OH$ ally alcohol and others

16–2 ALKANES

Natural gas, gasoline, and candle wax are all composed of hydrocarbons. **Hydrocarbons** *are compounds that contain only carbon and hydrogen.* These particular hydrocarbons are also known as alkanes. **Alkanes** *are hydrocarbons that contain only single covalent bonds.* Since the carbons in these molecules bond to the maximum number of hydrogens, alkanes are known as **saturated** hydrocarbons. Alkanes can be described by the general formula C_nH_{2n+2}. This refers only to *open-chain alkanes,* meaning that the carbons do not form a ring. The simplest alkane (*n* = 1) is methane (CH_4), in which the carbon shares *four pairs of electrons* with four different hydrogen atoms. The next alkane (*n* = 2) is ethane (C_2H_6), and the third is propane (C_3H_8). These alkanes are members of a homologous series. *In a* **homologous series,** *the next member differs from the previous one by a constant structure unit, which is one carbon and two hydrogens (CH₂).*

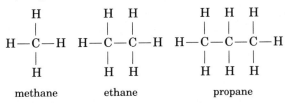

methane   ethane   propane

The names in Table 16–2 are the basis for the names of all organic compounds. By altering them slightly, we can name other classes of organic compounds that are discussed later. Two systems of nomenclature are used in organic chemistry. The most systematic is the one devised by the International Union of Pure and Applied Chemistry (the IUPAC system). Although the rules for naming complex molecules can be extensive, we will be concerned with just the basic concepts. Compounds are also known by *common* or *trivial* names. Sometimes these names follow a pattern; sometimes they do not. They have been used for so many years that it is hard to break the habit of using them. When a chemical that is frequently known by its common name is encountered, that name is given in parentheses.

Is there a system to naming organic compounds?

The name of a simple organic compound has two parts; the *prefix* gives the number of carbons in the longest carbon chain, and the *suffix* tells what kind of a compound it is. The underlined portions of the

TABLE 16–2 **ALKANES**

Formula	Name	Formula	Name
CH_4	methane	C_6H_{14}	hexane
C_2H_6	ethane	C_7H_{16}	heptane
C_3H_8	propane	C_8H_{18}	octane
C_4H_{10}	butane	C_9H_{20}	nonane
C_5H_{12}	pentane	$C_{10}H_{22}$	decane

names in Table 16-2 are the prefixes used for compounds containing 1 through 10 carbons in the longest chain; *meth-* stands for one carbon, *eth-* for two carbons, and so on. The ending used for alkanes is *-ane*. Therefore, the one-carbon alkane is methane, the two-carbon alkane is ethane, and so on.

Organic compounds can exist as unbranched compounds (all carbons bound to each other in a continuous chain) or as branched compounds. Previously, we indicated that *n*-butane is an unbranched alkane and that isobutane is a branched alkane. The IUPAC system bases its names on the longest carbon chain in the molecule, whereas the common names frequently include all of the carbons in the name (e.g., isobutane). The longest carbon chain in isobutane is three carbons long and is therefore considered a propane in the IUPAC system. Note that isobutane has a CH_3— group attached to a propane chain. In this system of nomenclature, the branches are named separately. *Since the branches can be considered as groups of atoms substituted for a hydrogen, the branches are called* **substituents.** *Substituents that contain one less hydrogen than an alkane are called* **alkyl groups.** Alkyl groups are not compounds by themselves; they must always be attached to some other group or atom. They are named by taking the alkane name, dropping the *-ane* ending, and substituting *-yl*. The most common alkyl groups are given in Table 16-3.

Thus the alkyl substituent in isobutane is a methyl group. The IUPAC name for isobutane is *methylpropane.*

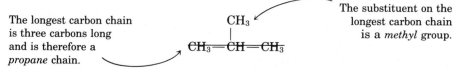

The longest carbon chain is three carbons long and is therefore a *propane* chain.

The substituent on the longest carbon chain is a *methyl* group.

If the same group is substituted on either the same carbon or another carbon, the prefix *di-* is used to indicate two groups. If the same group is substituted on three carbons, the prefix *tri-* is used to indicate three groups. For example, the IUPAC name for neopentane (see Table 16-1) is *dimethylpropane.*

The location of the substituent on the longest chain is designated by a number. The carbons are numbered from the end that is closest to the substituent. Thus the compound at the top of the next page is named 3-ethylheptane rather than 5-ethylheptane.

TABLE 16–3   ALKYL GROUPS (ONE TO FOUR CARBONS)

Alkyl Group*	Name	Alkyl Group	Name
CH_3—	methyl	$CH_3CH_2CH_2CH_2$—	*n*-butyl
CH_3CH_2—	ethyl	$(CH_3)_2CHCH_2$—	isobutyl
$CH_3CH_2CH_2$—	*n*-propyl	$CH_3CH_2CH(CH_3)$—	*sec*-butyl
$(CH_3)_2CH$—	isopropyl	$(CH_3)_3C$—	*tert*-butyl (or *t*-butyl)

* The dash (—) shows where the alkyl group is attached to a carbon chain or a hetero atom.

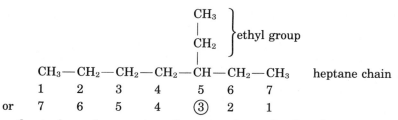

$$CH_3-CH_2-CH_2-CH_2-CH-CH_2-CH_3 \quad \text{heptane chain}$$

| 1 | 2 | 3 | 4 | 5 | 6 | 7 |

or ⑦ 6 5 4 ③ 2 1

When the carbons form a ring, the name is prefixed with *cyclo-*. Therefore, the three-carbon ring compound

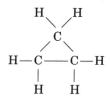

is called cyclopropane. Cycloalkanes have the general formula C_nH_{2n}.

EXAMPLE 16–2

Identification and the Name of the Longest Carbon Chain

Draw a line through the longest carbon chain in the following compounds and circle the substituents. Name the longest chain.

● **Working It Out**

(a)
$$\begin{array}{ccc} CH_3 & CH_3 & CH_3 \\ | & | & | \\ CH_2-CH-CH & & \\ & & | \\ & & CH_2-CH_3 \end{array}$$

(b)
$$\begin{array}{c} CH_2-CH_3 \\ | \\ CH_2-CH-CH_2-CH_3 \\ | \\ CH_3 \end{array}$$

(c)
$$\begin{array}{c} CH_2-CH_2-CH_3 \\ | \\ CH_3CH_2-CH-CH_2-CH_2-CH_2-CH_3 \end{array}$$

(d)
$$\begin{array}{c} CH_3 \\ | \\ H-C-CH_2-CH_3 \\ | \\ CH_3 \end{array}$$

Answers

(a)
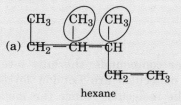

hexane

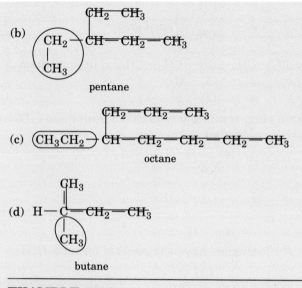

(b) pentane

(c) octane

(d) butane

EXAMPLE 16–3

Naming Compounds by the IUPAC Method

Name the compounds in Example 16–2 by the IUPAC method.

Answers
(a) 3,4-dimethylhexane
(b) 3-ethylpentane
(c) 4-ethyloctane
(d) methylbutane (no number necessary)

EXAMPLE 16–4

Writing Condensed Structures from a Name

Write the condensed structures of the following:
(a) 2-methylpentane
(b) 3-ethyloctane
(c) 3-isopropyl-2-methylhexane (Substituent groups are listed in alphabetical order.)

Answers

(a) $CH_3CHCH_2CH_2CH_3$
      |
      CH_3

(b) $CH_3CH_2CHCH_2CH_2CH_2CH_2CH_3$
          |
          CH_2CH_3

(c) $\overset{\displaystyle CH(CH_3)_2}{\overset{|}{CH_3CHCHCH_2CH_2CH_3}}$
                      |
                      CH_3

Why are hydrocarbons with small molecules gases?

Since alkanes are essentially nonpolar compounds, the only intermolecular forces of attraction between molecules are London forces. These forces depend on the volume of the molecules. Since the molar mass is roughly proportional to the volume, we can state that the lighter

Natural gas (methane) is stored in the gaseous state.

the molecules, the less are the London forces. (See Section 10-2.) As a result, the boiling points of alkanes are related to their molar masses. The lighter alkanes are the more volatile. **Volatility** *refers to the tendency of a liquid to vaporize to the gaseous state and is related to its boiling point.* Alkanes with four or fewer carbon atoms are all gases at room temperature. Those with five to eighteen carbons are liquids, whereas those with more than eighteen carbons are low-melting-point solids (resembling candle wax). All alkanes are odorless and colorless. They are also extremely flammable.

There are two major sources of alkanes, natural gas and crude oil (petroleum). Natural gas is mainly methane with smaller amounts of ethane, propane, and butanes. Unlike natural gas, petroleum contains hundreds of compounds, the majority of which are open-chain and cyclic alkanes. Before we can make use of petroleum, it must be separated into groups of compounds with similar properties. Further separation may or may not be carried out depending on the final use of the hydrocarbons.

Crude oil is separated into groups of compounds according to boiling points by distillation in a refinery. (See Figure 16-2.) In such a distillation, the liquid is boiled, and the gases move up a large column that becomes cooler and cooler toward the top. Compounds condense (become liquid) at different places in the column, depending on their boiling points. As the liquids condense, they are drawn off, providing a rough separation of the crude oil. Some of the material has too high a boiling point to vaporize and remains in the bottom of the column. A drawing showing this process and the various fractions obtained is shown in Figure 16-3. Note that the fewer the carbon atoms in the alkane, the lower the boiling point (the more volatile it is).

The composition of crude oil itself varies somewhat depending on

Figure 16–2
An Oil Refinery The crude oil is separated in various fractions in large refineries.

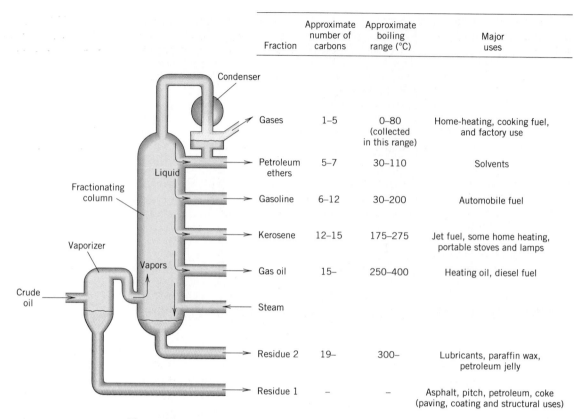

Fraction	Approximate number of carbons	Approximate boiling range (°C)	Major uses
Gases	1–5	0–80 (collected in this range)	Home-heating, cooking fuel, and factory use
Petroleum ethers	5–7	30–110	Solvents
Gasoline	6–12	30–200	Automobile fuel
Kerosene	12–15	175–275	Jet fuel, some home heating, portable stoves and lamps
Gas oil	15–	250–400	Heating oil, diesel fuel
Steam			
Residue 2	19–	300–	Lubricants, paraffin wax, petroleum jelly
Residue 1	–	–	Asphalt, pitch, petroleum, coke (paving, coating and structural uses)

Figure 16–3
Refining of Petroleum Oil is separated into fractions according to boiling points.

How is a hydrocarbon molecule changed or rearranged?

where it is found. Certain crude oil, such as that found in Nigeria and Libya, is called light oil because it is especially rich in the hydrocarbons that are present in gasoline. Otherwise, one fraction can be converted into another by three processes: cracking, reforming, and alkylation. **Cracking** *changes large molecules into small molecules.* **Reforming** *removes hydrogens from the carbons and/or changes unbranched hydrocarbons into branched hydrocarbons.* (Branched hydrocarbons perform better in gasoline; that is, they have a higher octane rating.) **Alkylation** *takes small molecules and puts them together to make larger molecules.* In all of these processes, catalysts are used, but there is a different catalyst for each process.

About 96% of all oil and gas is burned as fuel, whereas only 4% is used to make other organic chemicals. As a fuel, hydrocarbons burn to give carbon dioxide, water, and a great deal of heat energy.

$$CH_4 + 2O_2 \xrightarrow{\text{spark}} CO_2 + 2H_2O + \text{heat}$$

$$2C_8H_{18} + 25O_2 \xrightarrow{\text{spark}} 16CO_2 + 18H_2O + \text{heat}$$

Industrially, most synthetic organic chemicals have their ultimate origin in *the alkanes obtained from crude oil.* These **petrochemicals** are put to a wide range of uses in the manufacture of fibers, plastics, coatings, adhesives, synthetic rubber, some flavorings, perfumes, and pharmaceuticals.

The second class of hydrocarbons we will discuss are those that contain a double bond somewhere in the carbon chain. These are known as alkenes and are discussed next. Alkenes are used to make many plastics.

▼Looking Ahead

16–3 ALKENES

Alkenes *are hydrocarbons that contain a double bond. Organic compounds with multiple bonds are said to be* **unsaturated.** The general formula for an open-chain alkene with one double bond is C_nH_{2n}. The simplest alkene is ethene (common name ethylene), which has the structure

Alkenes are named by dropping the *-ane* ending of the corresponding alkane and substituting *-ene*. The location of the double bond is indicated by a number in a manner similar to the naming of branched alkanes by the IUPAC method. The double bond is located by numbering the C—C bonds starting from the end closest to the double bond in the main hydrocarbon chain. The smallest number for a double bond (or a triple bond) takes precedence over the numbering of any substituents on the carbon chain.

$$CH_3-CH_2-CH_2-CH=CH-CH_3$$
$$5 \quad\; 4 \quad\;\; 3 \quad\;\; ② \quad 1 \qquad \text{2-hexene}$$

Only small amounts of alkenes are found naturally in crude oil; the majority are made from alkanes by the reforming process during the refining of crude oil. When alkanes are heated over a catalyst, hydrogen is lost from the molecule and alkenes together with hydrogen are formed.

$$C_nH_{2n+2} \xrightarrow[\text{heat}]{\text{catalyst}} C_nH_{2n} + H_2$$

$$C_2H_6 \xrightarrow[\text{heat}]{\text{catalyst}} CH_2{=}CH_2 + H_2$$

A large amount of alkenes are produced industrially to make polymers. When certain compounds called initiators are added to an alkene or a mixture of alkenes, the double bond is broken and the alkenes become joined to each other by single bonds. This produces a high-molar-mass molecule called a **polymer,** *which has repeating units of the original alkene (called the monomer).*

$$CH_2{=}CH_2 \xrightarrow{\text{initiator}}$$
monomer (ethylene)

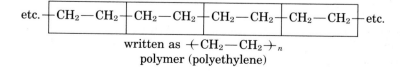

How can one locate the double bond in an alkene?

Teflon is a polymer used to coat frying pans which makes the surface "non-stick."

Why are plastics sometimes called "polymers"?

Polymers are named by adding *poly-* to the name of the alkene used to form the polymer. In the example shown above, the polymer was made from ethylene (usually common names are used for polymers), and so the polymer is called polyethylene. If the name of the polymer contains two words, the alkene name is enclosed in parenthesis [e.g., poly (methyl methacrylate)]. This rule is not always followed in some well-known polymers such as polyvinyl chloride. When writing the condensed structure of a polymer, we abbreviate the structure by giving the repeating unit in parentheses along with a subscript n to indicate that the monomer is repeated many times. Groups attached to the double bond affect the properties of the polymer, and by varying the group, we can vary the uses for which the polymer is suited. (See Figure 16-4.) Some commonly used polymers and their uses are given in Table 16-4.

Chemicals can add to the double bond to form new single bonds and therefore new compounds. A test based on such a reaction, the addition of bromine to an alkene, is used to show the presence of alkenes. It is a very simple test to perform. The disappearance of the red color of bromine when it is added to a liquid means that multiple bonds are present (the dibromide formed is colorless). We will talk more about this test when we discuss aromatic compounds.

$$CH_2{=}CH_2 + Br_2 \longrightarrow \begin{array}{cc} CH_2{-}CH_2 \\ | \quad\ \ | \\ Br \quad Br \end{array}$$

red   colorless

Looking Ahead ▼

The third category of hydrocarbons has a triple bond somewhere in the carbon chain. This class of hydrocarbons is important in the production of alkenes, which are then used to make polymers.

Figure 16–4
Polymers The plastic bag is made of polyethylene. Polypropylene is used to make artificial turf, and polystyrene (Styrofoam) is used for insulation.

TABLE 16–4 POLYMERS

Monomer Name, Structure	Polymer Name, Structure	Some Common Trade Names	Uses
ethylene $CH_2{=}CH_2$	polyethylene $+CH_2{-}CH_2{+}_n$	Polyfilm[a] Marlex[b]	Electrical insulation, packaging (plastic bags), floor covering, plastic bottles, pipes, tubing
propylene $CH_2{=}CH$ $\quad\vert$ $\quad CH_3$	polypropylene $\left(CH_2{-}CH\atop \quad\vert\atop \quad CH_3\right)_n$	Herculon[c]	Pipes, carpeting, artificial turf, molded auto parts, fibers
vinyl chloride $CH_2{=}CH$ $\quad\vert$ $\quad Cl$	polyvinyl chloride (PVC) $\left(CH_2{-}CH\atop \quad\vert\atop \quad Cl\right)_n$	Tygon[d]	Wire and cable coverings, pipes, rainwear, shower curtains, tennis court playing surfaces
styrene $CH_2{=}CH$	polystyrene $\left(CH_2{-}CH\right)_n$	Styrofoam[a] Styron[a]	Molded objects (combs, toys, brush and pot handles), refrigerator parts, insulating material, phonograph records, clock and radio cabinets
tetrafluoroethylene $CF_2{=}CF_2$	polytetrafluoroethylene $+CF_2{-}CF_2{+}_n$	Teflon[e] Halon[f]	Gaskets, valves, tubing, coatings for cookware
methyl methacrylate $CH_2{=}C\underset{CO_2CH_3}{\overset{CH_3}{\diagup}}$	poly(methyl methacrylate) $\left(CH_2{-}C\atop \quad\vert\atop \quad CO_2CH_3\right)_n^{CH_3}$	Plexiglas[g] Lucite[e]	Glass substitute, lenses, aircraft glass, dental fillings, artificial eyes, braces
acrylonitrile $CH_2{=}CH$ $\quad\vert$ $\quad CN$	polyacrylonitrile $\left(CH_2{-}CH\atop \quad\vert\atop \quad CN\right)_n$	Orlon[e] Acrilan[h]	Fibers for clothing, carpeting

[a]Dow Chemical Co.
[b]Phillips Petroleum Co.
[c]Hercules, Inc.
[d]U.S. Stoneware Co.

[e]E. I. du Pont de Nemours & Co.
[f]Allied Chemical Corp.
[g]Rohm & Haas Co.
[h]Monsanto Industrial Chemicals, Inc.

16–4 ALKYNES

Alkynes *are hydrocarbons that contain a triple bond.* The general formula for an open-chain alkyne with one triple bond is C_nH_{2n-2}. The simplest alkyne is ethyne (acetylene), which has the structure

$$H—C≡C—H$$

Alkynes are named just like alkenes except that *-yne* is substituted for the alkane ending instead of *-ene*. Thus, $CH_3CH_2CH_2CH_2C≡CCH_2CH_3$ is named 3-octyne.

Alkynes are not found in nature, but they can be prepared synthetically. Acetylene can be made from coal by first reacting the coal with calcium oxide at high temperature and then treating the calcium carbide formed with water.

$$\underset{\substack{\text{coal}}}{C} + \underset{\substack{\text{calcium} \\ \text{oxide}}}{CaO} \xrightarrow{\Delta} \underset{\substack{\text{calcium} \\ \text{carbide}}}{CaC_2} \xrightarrow{H_2O} \underset{\substack{\text{acetylene}}}{H—C≡C—H} + \underset{\substack{\text{calcium} \\ \text{hydroxide}}}{Ca(OH)_2}$$

A more common method of making acetylene is the reforming process. Methane is heated in the presence of a catalyst, forming acetylene and hydrogen.

$$2CH_4 \xrightarrow[\Delta]{\text{catalyst}} H—C≡C—H + 3H_2$$

Although acetylene is used in oxyacetylene torches, its most important application is in the synthesis of specific alkenes for polymers. For example, acetylene reacts with hydrogen chloride to form vinyl chloride, or with hydrogen cyanide to form acrylonitrile. These alkenes are then used to manufacture polyvinyl chloride and polyacrylonitrile, respectively.

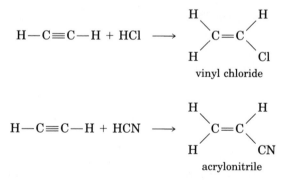

Looking Ahead ▼

The final category of hydrocarbons involves cyclic compounds that are particularly stable. The most important member of this group involves a six-membered ring. These compounds are discussed next.

16–5 AROMATIC COMPOUNDS

Benzene (C_6H_6) is a six-membered ring compound. Its Lewis structure can be written with alternating single and double bonds between adjacent carbons. In fact, two identical resonance structures can be written for benzene, which indicates that all six bonds are equivalent. The two resonance Lewis representations of benzene are shown below in (a). The Lewis representation in (b) is simplified by omitting the carbons. In (c), the structure is simplified further by omitting the hydrogens.

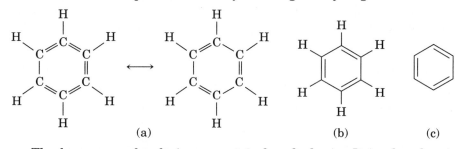

(a)   (b)   (c)

The benzene molecule is symmetrical and planar. It is also chemically unreactive compared to alkenes. For example, benzene does not "decolorize" a solution of bromine as do straight chain alkenes (see page 544). It is this special stability that allows us to classify benzene among a unique class of ring hydrocarbons known as **aromatic compounds.**

There are other ring hydrocarbons that are also aromatic and can be written with alternating single and double bonds, but we will confine our discussion to benzene and derivatives of benzene. *A **derivative** of a compound is produced by the substitution of a group or hetero atom for a hydrogen on the molecules of the original compound.* Compounds that contain the benzene ring can usually be recognized by name because they are named as derivatives of benzene. The following are the simplified structures and the names of some common derivatives of benzene.

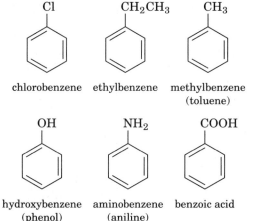

chlorobenzene   ethylbenzene   methylbenzene
(toluene)

hydroxybenzene   aminobenzene   benzoic acid
(phenol)   (aniline)

Although some benzene and toluene are present in crude oil, additional quantities can be obtained by the reforming process. When cyclohexane and/or hexane are heated with a catalyst, benzene is formed. Coal is another source of benzene. When coal is heated to high temperatures in the absence of air, some benzene is formed.

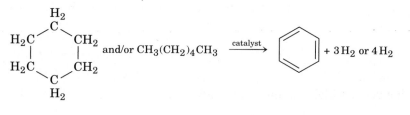

Benzene and toluene find application mainly as solvents and as starting materials to make other aromatic compounds. Benzene must be handled with care, however, because it has been found to be a potent carcinogen (it can cause cancer). Phenol and its derivatives are used as disinfectants and preservatives, and in the manufacture of dyes, explosives, drugs, and plastics.

Looking Back ▲

Organic compounds form the basis of life processes. The central atom in the molecules of these compounds is the carbon atom, which can form bonds to other carbons in long chains, or to atoms of other elements. Carbon-hydrogen compounds (hydrocarbons) are classified as alkanes, alkenes, alkynes, or aromatics, based on the types of bonds or their molecular structure.

Checking It Out ▶

Learning Check A

A–1.   Fill in the blanks.
Lewis structures of organic compounds illustrate that carbon forms a total of _____ bonds to other atoms. Two different compounds that share the same formula are known as _____. A hydrocarbon that has only single bonds in its molecules is classified as an _____. The principal sources of this class of compounds are _____ gas and _____. The latter liquid is separated into component compounds in a _____. The separation is accomplished because light alkanes are more _____ than the heavier compounds. Alkenes are known as _____ hydrocarbons because they contain a _____ bond. An important use of these compounds is in the manufacture of _____. Alkynes contain a _____ bond. Benzene is the most important example of an _____ hydrocarbon. A _____ of benzene is formed by the replacement of a hydrogen with a hydrocarbon or hetero atom group.

A–2.   Represent all of the isomers of pentane. (Show as partially condensed structures.)

A–3.   Give the IUPAC names for each of the isomers in the previous problem.

A–4.   Show the partially condensed Lewis structures of
(a) 2-pentene
(b) 4,4,dimethyl-1-pentyne
(c) isopropylbenzene

Additional Examples: 16-3, 16-8, 16-17, 16-18, 16-24, 16-27, and 16-29.

Looking Ahead ▼

Organic chemistry is not limited to hydrocarbons. Many other elements, especially oxygen and nitrogen, can also be attached to a hydrocarbon group. When this happens, the presence of one or more hetero atoms in a molecule may drastically change the properties of the compound. We begin our discussion of this area of organic chemistry with an overview of the effect of the presence of hetero atoms.

16–6   ORGANIC FUNCTIONAL GROUPS

Methane (CH_4) boils at a very low −164°C and thus is a gas at room temperature. If one chlorine is substituted for a hydrogen on the carbon, the resulting compound, CH_3Cl (chloromethane), has properties very different from those of methane. Chloromethane is a polar compound with a higher molar mass than methane, so it boils at a much higher −24°C. The presence of the chlorine hetero atom has a significant effect on the properties of the original compound. In most compounds, the nature and bonding of the hetero atom (or atoms) control the chemistry or the function of the molecule. *In an organic compound, the atom or group of atoms that determines the chemical nature of the molecule is known as the* **functional group.** Although no hetero atom is involved, the functional group of alkenes is the double bond and that of alkynes is the triple bond since those bonds account for several types of reactions that these compounds undergo. Each functional group has a strong influence on the chemistry of the compounds that contain it and thus establishes a specific class of compounds. In the following sections, we examine some of these classes of compounds that are determined by the presence of a particular functional group.

Is there a particular part of an organic molecule that is the center of chemical activity?

In the next two sections, we will examine two classes of compounds that contain oxygen as a hetero atom where the oxygen forms bonds to two other atoms. The first is a familiar class of compounds known as alcohols.

▼ **Looking Ahead**

16–7   ALCOHOLS (R—OH)*

Alcohol is most familiar to us as the active ingredient in alcoholic beverages. That is just one alcohol. Others are known to us as rubbing alcohol and wood alcohol. **Alcohols** *are a class of organic compounds that contain the OH functional group (known as a* **hydroxyl group***) in place of a hydrogen on a carbon chain.*

Alcohols are named by taking the alkane name, dropping the *-e*, and substituting *-ol*. Common names are obtained by just naming the alkyl group attached to the —OH followed by *alcohol*. Some familiar alcohols (see Figure 16-5) have more than one hydroxyl group. (In naming alcohols with more than one hydroxyl group, the *-e* in the alkane is not dropped.) Examples of four alcohols are:

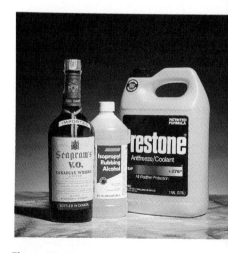

CH_3CH_2OH	$CH_3CH_2CH_2OH$	$\begin{array}{c} CH_2-CH_2 \\ \mid \qquad \mid \\ OH \quad\ OH \end{array}$	$\begin{array}{c} CH_2-CH-CH_2 \\ \mid \qquad \mid \qquad \mid \\ OH \quad OH \quad OH \end{array}$
ethanol (ethyl alcohol)	propanol (*n*-propyl alcohol)	1,2-ethane*diol* (ethylene glycol) (*diol* means two hydroxyl groups)	1,2,3-propane*triol* (glycerol) (*triol* means three hydroxyl groups)

*The R or R′ represents a group such as an alkyl or an aromatic group.

Figure 16–5
Alcohols. A major ingredient in each of these products is an alcohol.

Methanol and ethanol can both be obtained from natural sources. Methanol can be prepared by heating wood in the absence of oxygen to about 400°C; at such high temperatures, methanol, together with other organic compounds, is given off as a gas. Since methanol was once made exclusively by this process, it is often called wood alcohol. Currently, methanol is prepared from synthesis gas, a mixture of CO and H_2. When CO and H_2 are passed over a catalyst at the right temperature and pressure, methanol is formed.

$$CO + 2H_2 \xrightarrow[\Delta]{catalyst} CH_3OH$$

Ethanol (often known simply as alcohol) is formed in the fermentation of various grains. (Therefore, it is also known as grain alcohol.) In fermentation, the sugars and other carbohydrates in grains are converted to ethanol and carbon dioxide by the enzymes in yeast.

$$\underset{\text{glucose (a sugar)}}{C_6H_{12}O_6} \xrightarrow[\text{enzymes}]{\text{yeast}} 2CH_3CH_2OH + 2CO_2$$

Industrially, ethanol is prepared by reacting ethylene with water in the presence of an acid catalyst.

$$CH_2{=}CH_2 + H_2O \xrightarrow{H^+} CH_3CH_2OH$$

2-propanol (isopropyl alcohol or rubbing alcohol) can be prepared from propene (propylene) in the same way.

$$CH_3CH{=}CH_2 + H_2O \xrightarrow{H^+} CH_3-\overset{\displaystyle CH_3}{\underset{\displaystyle OH}{\overset{|}{\underset{|}{C}}}}-H$$

Methanol has been used as a solvent for shellac, as a denaturant for ethanol (it makes the ethanol undrinkable), and as an antifreeze for automobile radiators. It is very toxic when ingested. In small doses it causes blindness, and in large doses it can cause death.

Which alcohol is present in an "alcoholic" beverage?

Ethanol is present in alcoholic beverages such as beer, wine, and liquor. The "proof" of an alcoholic beverage is two times the percent by volume of alcohol. If a certain brand of bourbon is 100 proof, it contains 50% ethanol. Ethanol can also be mixed with gasoline to form a mixture called "gasohol," which is used as a fuel. Ethanol is an excellent solvent and has been used as such in perfumes, medicines, and flavorings. It has also been used as an antiseptic and as a rubbing compound to cleanse the skin and lower a feverish person's temperature. While ethanol is not as toxic as methanol, it can cause coma or death when ingested in large quantities.

Ethylene glycol is the major component of antifreeze and coolant used in automobiles. It is also used to make polymers, the most common of which is Dacron®, a polyester.

Glycerol, which can be obtained from fats, is used in many applications where a lubricant and/or softener is needed. It has been used in pharmaceuticals, cosmetics, foodstuffs, and some liqueurs. When glycerol reacts with nitric acid, it produces nitroglycerin. Nitroglycerin is a powerful explosive,

$$CH_2-CH-CH_2 + 3HNO_3 \longrightarrow CH_2-CH-CH_2 + 3H_2O$$

$$\begin{array}{ccc} | & | & | \\ OH & OH & OH \end{array} \qquad \begin{array}{ccc} | & | & | \\ ONO_2 & ONO_2 & ONO_2 \end{array}$$

glycerol
nitroglycerin

but it is also a strong smooth-muscle relaxant and vasodilator and has been used to lower blood pressure and to treat angina pectoris.

The next class of compounds also has an oxygen bonded to two other atoms. In ethers, however, the oxygen is bonded to two hydrocarbon groups. ▼ Looking Ahead

16–8 ETHERS (R—O—R)

An **ether** *contains an oxygen bonded to two hydrocarbon groups* (rather than one hydrocarbon group and one hydrogen as in alcohols). The simplest ether, dimethyl ether, is an isomer of ethanol but has very different properties.

The common names of ethers are obtained by giving the alkyl groups on either side of the oxygen and adding *ether*.

$H_3C-O-CH_3$ $CH_3OCH_2CH_3$
dimethyl ether   ethyl methyl ether

$CH_3CH_2OCH_2CH_3$ $CH_3CH_2OCH_2CH_2CH_3$
diethyl ether   ethyl propyl ether

Diethyl ether is made industrially by reacting ethanol with sulfuric acid. In this reaction, two ethanol molecules are joined together with the loss of an H_2O molecule.

$$2CH_3CH_2OH \xrightarrow{H_2SO_4} CH_3CH_2-O-CH_2CH_3 + H_2O$$

The most commonly known ether, diethyl ether (or ethyl ether, or just ether), has, in the past, been used extensively as an anesthetic. It has the advantage of being an excellent muscle relaxant that doesn't affect blood pressure, pulse rate, or rate of respiration greatly. On the other hand, ether has an irritating effect on the respiratory passages and often causes nausea. Its flammability is also a drawback because of the danger of fire and explosions. Diethyl ether is rarely used now as an anesthetic. Other anesthetics that do not have its disadvantages have taken its place.

The next two classes of compounds also have oxygen as the lone hetero atom. In these two classes, however, the oxygen is bonded to only one carbon by a double bond. The carbon is then bonded to two other atoms to form either aldehydes or ketones. ▼ Looking Ahead

16–9   ALDEHYDES $\overset{\overset{\text{O}}{\|}}{(\text{R}-\text{C}-\text{H})}$ AND KETONES $\overset{\overset{\text{O}}{\|}}{(\text{R}-\text{C}-\text{R}')}$

The functional group of aldehydes and ketones is a carbonyl group. A **carbonyl group** *is a carbon with a double bond to an oxygen.*

$$\overset{\overset{\text{O}}{\|}}{-\text{C}-}$$

In **aldehydes** *the carbonyl group is bound to at least one hydrogen, whereas in* **ketones** *the carbonyl group is bound to two hydrocarbon groups.*

Aldehydes are named by dropping the -*e* of the corresponding alkane name and substituting -*al*. Therefore, the two-carbon aldehyde is ethanal. Ketones are named by taking the alkane name, dropping the -*e*, and substituting -*one*. The common names of ketones are obtained by naming the alkyl groups on either side of the C=O and adding *ketone*. Thus the four-carbon ketone can be called butanone (IUPAC) or methyl ethyl ketone (common).

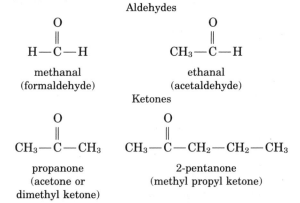

Aldehydes

methanal
(formaldehyde)

ethanal
(acetaldehyde)

Ketones

propanone
(acetone or
dimethyl ketone)

2-pentanone
(methyl propyl ketone)

Many aldehydes and ketones are prepared by the oxidation of alcohols. Industrially, the alcohol is oxidized by heating it in the presence of oxygen and a catalyst. About half of the methanol produced industrially is used to make formaldehyde by oxidation. Acetone can be prepared in the same fashion by oxidation of 2-propanol. In the following reactions, the [O] represents an oxidizing agent that removes hydrogen (to form H_2O) from the alcohol.

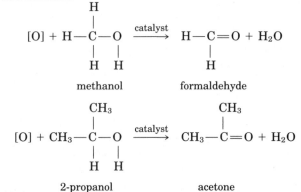

methanol

formaldehyde

2-propanol

acetone

Aldehydes and ketones have uses as solvents, in the preparation of polymers, as flavorings, and in perfumes. The simplest aldehyde, formaldehyde, has been used as a disinfectant, antiseptic, germicide, fungicide, and embalming fluid (as a 37% by mass water solution). It has also been used in the preparation of polymers such as Bakelite® (the first commercial plastic) and Melmac® (used to make dishes). Formaldehyde polymers have also been used as coatings on fabrics to give "permanent press" characteristics.

The simplest ketone, acetone, has been used mainly as a solvent. It is soluble in water and dissolves relatively polar and nonpolar molecules. It is an excellent solvent for paints and coatings.

More complex aldhydes and ketones, such as the following two examples, are used in flavorings and perfumes.

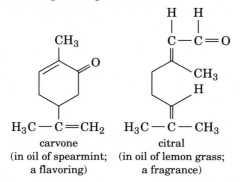

carvone
(in oil of spearmint; a flavoring)

citral
(in oil of lemon grass; a fragrance)

In the next section, we will introduce a class of compounds that has a nitrogen atom as the hetero atom. This class of organic compounds is known as amines and includes many familiar medications.

▼ Looking Ahead

16–10   AMINES (R—NH₂)

An **amine** *contains a nitrogen with single bonds to a hydrocarbon group and two other hydrocarbon groups or hydrogens.* The nitrogen in amines has one pair of unshared electrons similar to ammonia, NH_3. As we learned in Chapter 12, NH_3 utilizes the unshared pair of electrons to form weakly basic solutions in water. In a similar manner, amines are characterized by their ability to act as bases.

$$\ddot{N}H_3 + H_2O \rightleftharpoons NH_4^+ + OH^-$$

$$CH_3\ddot{N}H_2 + H_2O \rightleftharpoons CH_3NH_3^+ + OH^-$$

basic solutions

The common names of amines are obtained by listing the alkyl groups (in alphabetical order) attached to the nitrogen and adding *-amine*.

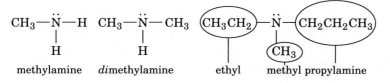

methylamine   *di*methylamine   ethyl   methyl propylamine

Simple amines are prepared by the reaction of ammonia with alkyl halides (e.g., CH_3Cl). In the reaction, the CH_3 group in the chloromethane adds on to the electron pair of the nitrogen forming a methylammonium cation. The amine can then be prepared by the reaction with a base such as NaOH.

$$H-\overset{\overset{\displaystyle H}{|}}{\underset{\underset{\displaystyle H}{|}}{N}}: \;+\; \overset{\frown}{(CH_3)}-Cl \;\longrightarrow\; \left[H-\overset{\overset{\displaystyle H}{|}}{\underset{\underset{\displaystyle H}{|}}{N}}-CH_3\right]^{+} \; Cl^{-}$$

$$[NH_3(CH_3)]^{+}Cl^{-} + NaOH \longrightarrow NH_2(CH_3) + NaCl + H_2O$$

Amines are used in the manufacture of dyes, drugs, disinfectants, and insecticides. They also occur naturally in biological systems and are important in many biological processes.

Amine groups are present in many synthetic and naturally occurring drugs. (See Figure 16-6.) They may be useful as antidepressants, antihistamines, antibiotics, antiobesity preparations, antinauseants, analgesics, antitussives, diuretics, and tranquilizers, among others. Frequently, a drug may have more than one use. Codeine, for example, is both an analgesic (pain reliever) and an antitussive agent (cough depressant). Often, a drug may be obtained from plants or animals. One such class of compounds is the alkaloids, which are amines found in plants. Two common drugs are:

What do many medications have in common?

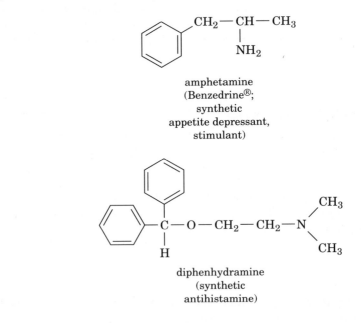

amphetamine
(Benzedrine®;
synthetic
appetite depressant,
stimulant)

diphenhydramine
(synthetic
antihistamine)

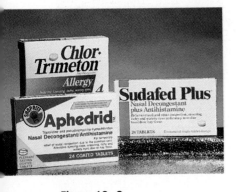

Figure 16–6
Amines. The active ingredient in an antihistamine is an organic amine.

Looking Ahead ▼

Another class of compounds that we will discuss in this chapter is known as the acids. This class has two oxygen hetero atoms; one is like an alcohol and the other is a carbonyl group. The acids and two classes that are derivatives of the acids are our final topics in this introduction to organic chemistry.

16–11 CARBOXYLIC ACIDS (R—C̈—O—H), ESTERS (R—C̈—O—R′), AND AMIDES (R—C̈—NH₂)

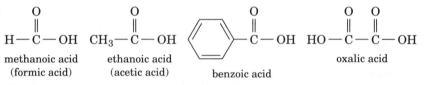

Carboxylic Acids

Acetic acid ($HC_2H_3O_2$), the sour component of vinegar, is an example of a carboxylic acid. **Carboxylic acids** *contain the functional group*

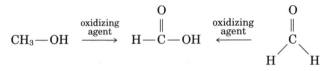

known as a **carboxyl group.** When acetic acid is written as a carboxylic acid, it is shown as CH_3COOH. Carboxylic acids are named by dropping the *-e* from the alkane name and substituting *-oic acid*. Examples of some common carboxylic acids are:

The sour taste of vinegar is due to acetic acid.

Carboxylic acids are made by the oxidation of either alcohols or aldehydes. Formic acid, which was first isolated by distilling red ants, can be made by oxidizing either methanol or formaldehyde.

Acetic acid can be made from the oxidation of ethanol. In fact, this is what happens when wine becomes "sour." Wine vinegar is produced by air oxidation of alcohol in ordinary wine.

Esters

An **ester** *is a derivative of a carboxylic acid, where a hydrocarbon group is substituted for the hydrogen in the carboxyl group.* Esters are named by giving the name of the alkyl group attached to the oxygen, followed by the acid name minus the *-ic* ending and substituting *-ate*. Some examples of esters are:

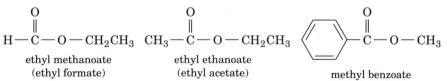

Esters are made from the reaction of alcohols and carboxylic acids. In the reaction, H_2O is split off from the two molecules (OH from the acid and H from the alcohol), leading to the union of the remnants of the two molecules.

$$H-\overset{\overset{\displaystyle O}{\|}}{C}\boxed{-OH\, +\, H}-O-CH_3 \xrightarrow{H^+} H-\overset{\overset{\displaystyle O}{\|}}{C}-O-CH_3 + H_2O$$

We all distinguish the various fruits by their unique odors. These pleasant odors are all caused by esters. Four of these esters are shown with their familiar odors.

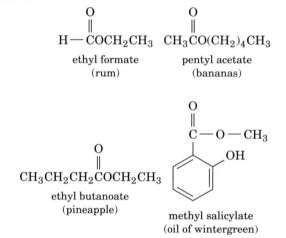

$$H-\overset{\overset{\displaystyle O}{\|}}{C}OCH_2CH_3 \qquad CH_3\overset{\overset{\displaystyle O}{\|}}{C}O(CH_2)_4CH_3$$

ethyl formate
(rum)

pentyl acetate
(bananas)

$$CH_3CH_2CH_2\overset{\overset{\displaystyle O}{\|}}{C}OCH_2CH_3$$

ethyl butanoate
(pineapple)

methyl salicylate
(oil of wintergreen)

Amides

An **amide** *is also a derivative of a carboxylic acid, where an amine group is substituted for the hydroxyl group of the acid.* Amides are named by dropping the *-oic acid* portion of the carboxylic acid name and substituting *-amide.* Examples of amides are:

$$H-\overset{\overset{\displaystyle O}{\|}}{C}-NH_2 \qquad CH_3-\overset{\overset{\displaystyle O}{\|}}{C}-NH_2 \qquad \langle\!\!\!\bigcirc\!\!\!\rangle-\overset{\overset{\displaystyle O}{\|}}{C}-NH_2$$

methanamide
(formamide)

ethanamide
(acetamide)

benzamide

Amides are made by a reaction similar to that in the preparation of esters. In this case, however, the carboxylic acid reacts with ammonia or an amine instead of alcohol. Again, water is split off from the two reacting molecules, leading to the union of the two remnants.

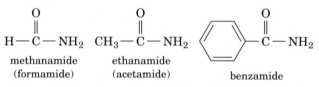

$$H-\overset{\overset{\displaystyle O}{\|}}{C}\boxed{-OH\, +\, H}-NH_2 \xrightarrow{\Delta} H-\overset{\overset{\displaystyle O}{\|}}{C}-NH_2 + H_2O$$

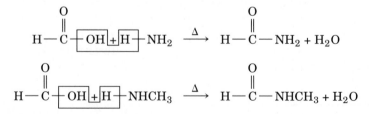

$$H-\overset{\overset{\displaystyle O}{\|}}{C}\boxed{-OH\, +\, H}-NHCH_3 \xrightarrow{\Delta} H-\overset{\overset{\displaystyle O}{\|}}{C}-NHCH_3 + H_2O$$

Uses in Painkillers and Plastics

Carboxylic acids, esters, and amides are frequently present in compounds that have medicinal uses. Salicylic acid has both antipyretic (fever-reducing) and analgesic (pain-relieving) properties. It has the disadvantage, however, of causing severe irritation of the stomach lining. Acetylsalicylic acid (aspirin), which is both an acid and an ester, doesn't irritate the stomach as much. Aspirin is broken up in the small intestine

What do some familiar painkillers have in common?

to form salicylic acid, which is then absorbed. Some people are allergic to aspirin and must take aspirin substitutes. The common aspirin substitutes are acetaminophen and ibuprofen. Acetaminophen is the active ingredient in Tylenol® and Datril®. Ibuprofen is the active ingredient in Advil® and Nuprin®.

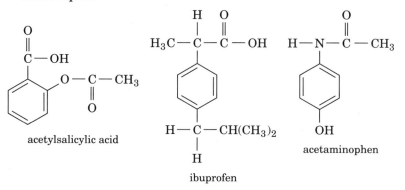

acetylsalicylic acid

ibuprofen

acetaminophen

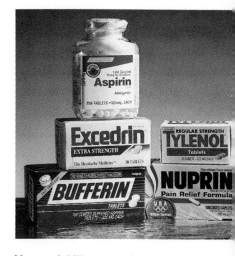

Many painkillers contain carboxylic acid, ester, or amide groups.

One of the biggest uses for carboxylic acids, esters, and amides is in the formation of condensation polymers. *In* **condensation polymers,** *a small molecule (usually H₂O) is given off during formation of the polymers.* These are different from the polymers discussed earlier that resulted from the joining of alkenes (called *addition* polymers). Two of the most widely known condensation polymers are nylon 6,6 (see Figure 16-7) and Dacron®. The first is a polyamide, made from a diacid and a diamine; the second is a polyester, made from a diacid and a dialcohol (diol). Both of these polymers are used to make fibers.

How are many synthetic fibers made?

Figure 16–7
Nylon. Nylon was first synthesized by DuPont chemists. It was one of the first synthetic fibers.

Nylon 6,6

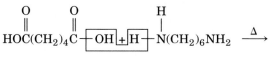

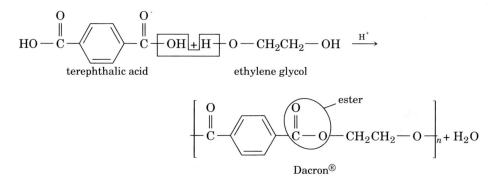

Dacron®

Dacron is used in these sails. It is strong, light, and does not "wet."

Looking Back ▲

The presence of a double or triple bond or a hetero atom has a substantial effect on the nature of a hydrocarbon. These are called functional groups and determine the types of chemical reactions that the compound undergoes. For example, the presence of a carboxyl group makes the compound acidic in water, but the presence of an amine group makes the compound basic in water. Oxygen is the most common hetero atom. Its presence in a compound may produce an alcohol, ether, aldehyde, ketone, carboxylic acid, or ester depending on the environment of the oxygen atom or atoms in the molecule.

Learning Check B

◀**Checking It Out**

B–1. Fill in the blanks.
The functional group in an alcohol is known as a _____ group. In ethers, the oxygen is attached to two _____ groups. The functional group in aldehydes and ketones is a _____ group. In aldehydes, the functional group is attached to at least one _____ atom. The hetero atom in an amine is a _____ atom. The functional group of a carboxylic acid is a _____ group. In esters, a hydrocarbon group replaces a _____ in the acid, whereas in amides an amine group replaces the _____ group of the acid.

B–2. Identify the class of compound from the following functional groups.

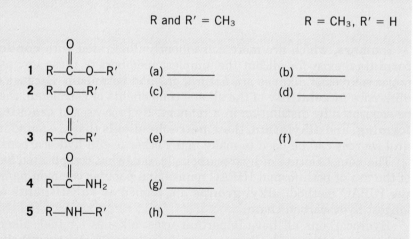

R and R′ = CH₃   R = CH₃, R′ = H

1 R—C(=O)—O—R′	(a) _____	(b) _____
2 R—O—R′	(c) _____	(d) _____
3 R—C(=O)—R′	(e) _____	(f) _____
4 R—C(=O)—NH₂	(g) _____	
5 R—NH—R′	(h) _____	

B–3. Give the common names of each compound in B-2.

Additional Examples: 16-34, 16-36, 16-38, and 16-40.

C H A P T E R   R E V I E W

Chemical compounds have traditionally been divided into two general groups—**organic** and **inorganic.** To introduce the more than ten million registered organic compounds, we need to find common properties or molecular structures so as to classify them into more specific groups. The first classification is the **hydrocarbons,** which can be further subdivided according to the type of bonds in the molecules of the

▲▼**Putting It Together**

various compounds. Each of these groups forms a **homologous series** designated by a general formula. The four hydrocarbon groups are summarized as follows:

Name of Series	General Formula	Example	Name	Comments
Alkane	C_nH_{2n+2}	C_3H_8	propane	Alkanes contain only single bonds and are **saturated.**
Alkene	C_nH_{2n}	C_3H_6	propene propylene	Alkenes contain double bonds and are **unsaturated.**
Alkyne	C_nH_{2n-2}	C_3H_4	propyne	Alkynes contain triple bonds and are unsaturated.
Aromatic	—	C_6H_6	benzene	Aromatic compounds are cyclic, unsaturated hydrocarbons, but are less reactive than alkenes.

Isomers, which are more conveniently illustrated with **condensed formulas,** exist for all but the simplest members of these groups. The major sources of alkanes are natural gas and petroleum. Because of the difference in **volatility** of the alkane components of petroleum, they can be separated by distillation in a refinery. By processes of **cracking, reforming,** and **alkylation,** these **petrochemicals** that are found in natural sources can be used to make many other useful hydrocarbons.

The nomenclature of hydrocarbons is somewhat complicated because of the use of both formal IUPAC names and popular common names. In the IUPAC method, **alkyl groups** are named as **substituents** on the longest hydrocarbon chain.

Hydrocarbons all have important uses: alkanes for fuel, alkenes to make **polymers,** and alkynes to make alkenes for various polymers. Aromatics and **derivatives** of aromatics find application as solvents, drugs, and flavors.

The focal point of chemical reactivity in an organic compound is the **functional group.** This may be a double or triple bond in a hydrocarbon, or a **hetero atom** that is incorporated into the molecule. Functional groups include the **hydroxyl, carbonyl,** and **carboxyl groups.** Derivatives of the carboxyl group include esters and amides. If an ester or amide linkage is made on both sides of a molecule, **condensation polymers,** which are known as polyesters and polyamides, respectively, are formed.

The bonding environment of the hetero atom in a functional group imposes certain common properties on compounds containing that group. Thus the functional group defines a class of compounds. Eight classes of organic compounds and their functional group, formula, example, and name are summarized on the following page:

Name of Class	Functional Group	General Formula*	Example	IUPAC Name (Common Name)
Alcohols	—C—Ö—H	R—OH	CH_3CH_2OH	ethanol (ethyl alcohol)
Ethers	—C—Ö—C—	R—O—R′	$CH_3OCH_2CH_3$	(ethyl methyl ether)
Aldehydes	—C—H (:O: double bond)	R—C—H (O double bond)	CH_3CH_2CHO	propanal (propionaldehyde)
Ketones	—C—C—C— (:O: double bond)	R—C—R′ (O double bond)	$CH_3COCH_2CH_3$	butanone (ethyl methyl ketone)
Amines	—C—N—	R—NH₂	$CH_3CH_2NH_2$	(ethyl amine)
Carboxylic acid	—C—ÖH (:O: double bond)	R—C—OH (O double bond)	$CH_3CH_2CH_2COOH$	butanoic acid (butyric acid)
Esters	—C—Ö—C— (:O: double bond)	R—C—OR′ (O double bond)	CH_3COOCH_3	methyl ethanoate (methyl acetate)
Amides	—C—N— (:O: double bond)	R—C—NR₂ (O double bond)	CH_3CONH_2	ethanamide (acetamide)

*R and R′ stand for hydrocarbon groups (e.g, alkyl). They may be different groups or the same group. R may also represent a hydrogen atom in aldehydes, carboxylic acids, esters, and amides.

E X E R C I S E S

LEWIS STRUCTURES AND ISOMERS

16–1 Draw Lewis structures for each of the following compounds. (More than one structure may be possible.)
  (a) CH_3Br (d) CH_5N
  (b) C_3H_4 (e) C_2H_7N
  (c) C_4H_8 (f) C_2H_8O

16–2 Draw Lewis structures for each of the following compounds. (More than one structure may be possible.)
  (a) C_5H_{10} (c) C_4H_9Cl
  (b) C_2H_4O (d) C_3H_9N

16–3 Write all of the isomers for C_6H_{14}.

16–4 Write all of the isomers for C_6H_{12}.

16–5 Which of the following pairs of compounds are isomers of each other? Why or why not?
  (a) $CH_3CH_2OCH_2CH_3$ and $CH_3CH_2CH_2CH_3$
  (b) $CH_3CH_2CH_2OH$ and $CH_3CH_2OCH_2CH_3$
  (c) $CH_3CH_2NH_2$ and CH_3NHCH_3
  (d) $CH_3CH_2CH_2NHCH_3$ and $CH_3CH_2CH_2NHCH_2CH_3$
  (e) $CH_2CH_2CH_2CH_3$ and $CH_3CH_2CH_2CH_2$
       | |
       OH   OH

16–6 Which of the following pairs of compounds are isomers of each other? Why or why not?
(a) $CH_3(CH_2)_3CH_3$ and $CH_3CH_2CH(CH_3)_2$
(b) CH_3CHCH_3 and $CH_3CH_2CHCH_3$
       | |
      Cl   Cl
(c) $CH_3CH_2CHCH_3$ and $CH_3CHCH_2CH_3$
             | |
          NH_2 NH_2
(d) $(CH_3)_3N$ and $CH_3CH_2CH_2NH_2$

ALKANES

16–7 Which of the following are formulas of open-chain alkanes?
(a) C_4H_{10} (e) C_7H_{15}
(b) C_3H_7 (f) C_7H_{14}
(c) C_8H_{14} (g) $C_{18}H_{38}$
(d) $C_{10}H_{22}$ (h) C_9H_{15}

16–8 The following skeletal structures represent alkanes. Fill in the proper number of hydrogens on each carbon atom.
(a) C—C—C
(b) C—C—C—C—C—C
         | |
         C   C—C—C

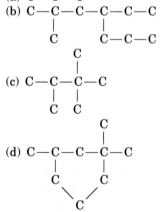
(c) (d)

16–9 The following skeletal structures represent alkanes. Fill in the proper number of hydrogens on each carbon atom.
(a) C—C—C—C
         |
         C
(b) C—C—C
      |
      C
(c)

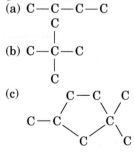

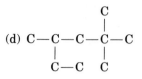
(d) C—C—C—C—C

16–10 Which of the following are formulas of alkyl groups?
(a) C_4H_8 (d) C_6H_6
(b) C_4H_9 (e) C_6H_{13}
(c) C_6H_{10} (f) C_6H_{11}

16–11 Write the formulas of any compounds listed in Problem 16-7 that could be a cyclic alkane.

16–12 Write the formulas of any alkyl groups listed in Problem 16-7.

16–13 Which of the following could be the formula of a cyclic alkane?
(a) C_8H_{18} (c) C_6H_6
(b) C_5H_8 (d) C_5H_{10}

16–14 Write the condensed structure for each of the following compounds.
(a) 3-methylpentane
(b) n-hexane
(c) 2,4,5-trimethyloctane
(d) 4-ethyl-2-methylheptane
(e) 3-isopropylhexane
(f) 2,2-dimethyl-4-t-butylnonane

16–15 Write the condensed structure for each of the following compounds.
(a) cyclopentane
(b) n-pentane
(c) 3,3-dimethylpentane
(d) 3-ethyl-3-methylhexane
(e) 3,5-dimethyl-4-isopropyloctane

16–16 Write the structure and the correct IUPAC name for
(a) 2-ethyl-3-methylpentane
(b) 5-n-propyl-5-isopropylhexane.

16–17 How many carbons are in the longest chain of each of the following compounds? Write the name of the longest alkane chain for each.
                CH_3—CH—CH_3
                      |
(a) $CH_3CHCH_2CHCH_3$
         |
        CH_3
      CH_3 $CH_2CH_2CHCH_3$
       | | |
(b) CH_2CH_2CH CH_3
             |
          CH_2CH_3

(c) 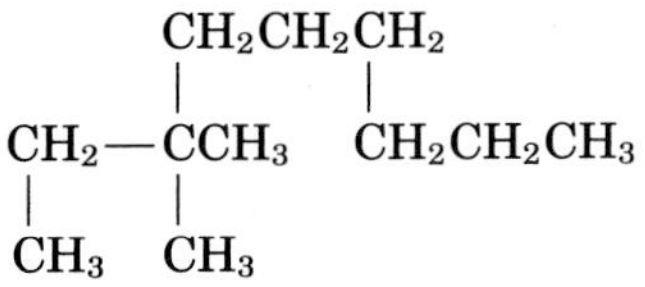

(d) 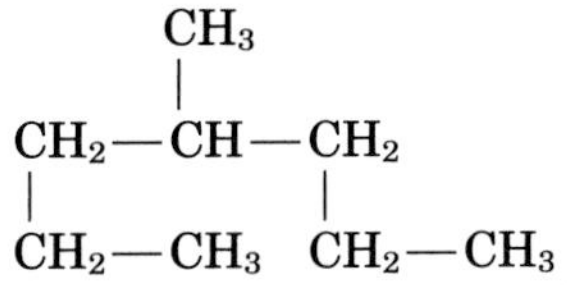

16–18 Name the substituents attached to the longest chains in the previous problem and give the IUPAC names for the compounds.

16–19 Give the longest chain, the substituents, and the proper IUPAC name for the following compounds.

(a) 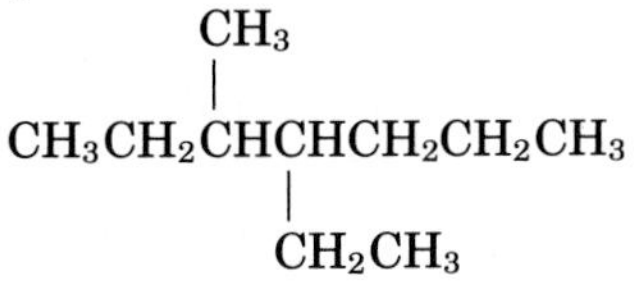

(b) 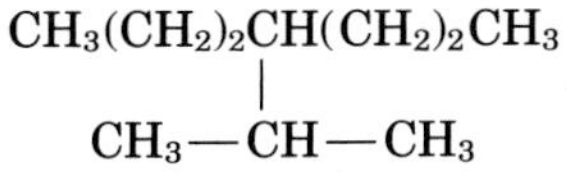

(c) 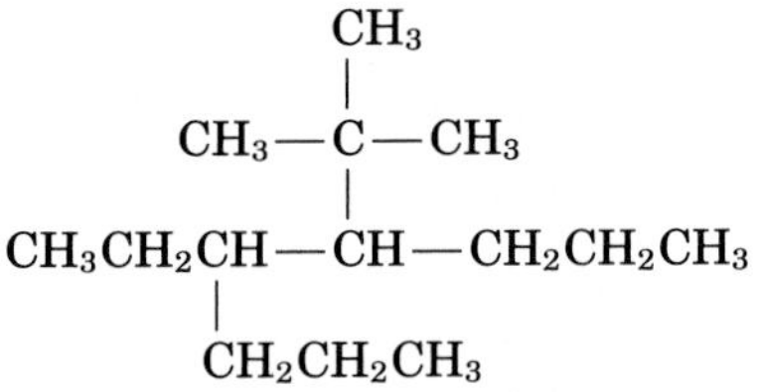

(d) 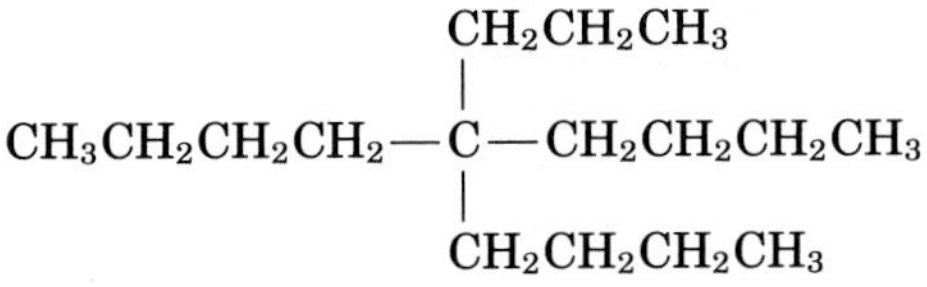

16–20 Give the proper IUPAC names for (a), (b), and (c) in Problem 16-8.

16–21 Give the proper IUPAC names for (a), (b), and (d) in Problem 16-9.

ALKENES AND ALKYNES

16–22 C_nH_{2n} can represent an alkane or an alkene. Explain.

16–23 What is the general formula of a cyclic alkene that contains one double bond?

16–24 The following skeletal structures represent alkenes or alkynes. Fill in the proper number of hydrogens on each carbon.

(a) C—C—C=C

(b) C—C≡C—C

(c) 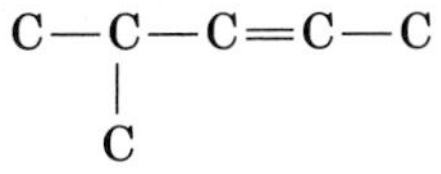

(d) 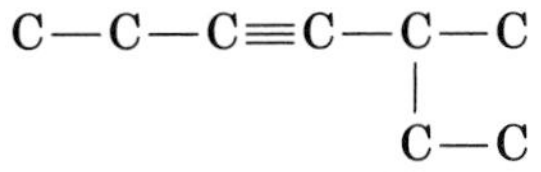

16–25 The following skeletal structures represent alkenes or alkynes. Fill in the proper number of hydrogens on each carbon.

(a) C—C=C—C—C

(b) 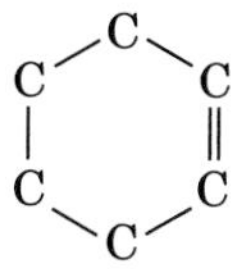

(c) C—C—C—C≡C

(d) 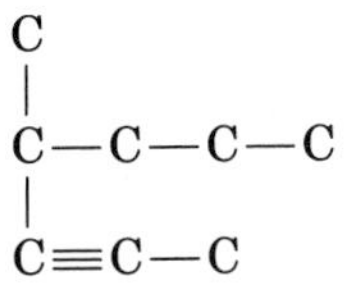

16–26 Write the condensed structure for each of the following compounds.

(a) 2-butyne

(b) 3-methyl-1-butene

(c) 3-propyl-2-hexene

(d) 5-methyl-3-octyne

16–27 Give the proper IUPAC names for the compounds in Problem 16-24.

16–28 Give the proper IUPAC names for the compounds in Problem 16-25.

16–29 Give the abbreviated structure of the polymer that would be formed from each of the following alkenes. Name the polymer.

(a) 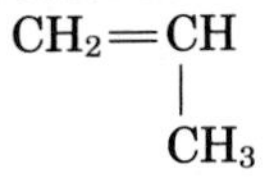 proplyene

(b) 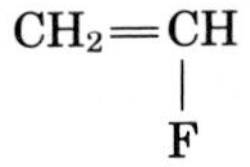 vinyl fluoride

(c) 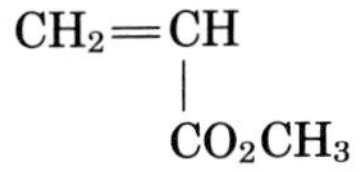 methyl acrylate

16–30 A certain polymer has repeating units represented by the formula

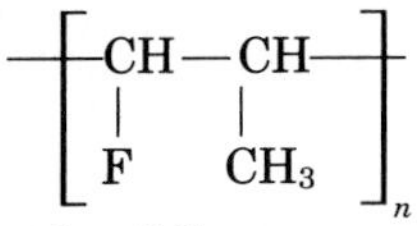

Write the formula of the monomer.

AROMATICS

*16–31   Naphthalene has the formula $C_{10}H_8$ and is composed of two benzene rings that share two carbons. Write equivalent Lewis resonance structures, including hydrogens, for naphthalene.

*16–32   Anthracene and phenanthrene are isomers with the formula $C_{14}H_{10}$. They are both composed of one benzene ring that shares four of its carbons with two other benzene rings. Write Lewis structures for the two isomers including hydrogens.

DERIVATIVES OF HYDROCARBONS

16–33   Which of the eight classes of compounds containing hetero atoms contains a carboxyl group? Which contains a hydroxyl group?

16–34   Tell how the following differ in structure.
(a) alcohols and ethers
(b) aldehydes and ketones
(c) amines and amides
(d) carboxylic acids and esters

16–35   Which two classes of compounds combine to form esters? Which two classes combine to form amides?

16–36   To which class does each of the following compounds belong?
(a) 3-heptanone
(b) 3-nonene
(c) 2-methylpentane
(d) 2-ethylhexanal
(e) ethylbenzene
(f) propanol

16–37   To which class does each of the following compounds belong?
(a) 2-octanol
(b) 2-butyne
(c) ethyl pentanoate
(d) 3-ethylhexene
(e) dimethyl ether
(f) aminobenzene

16–38   Circle and name the functional groups in each of the following compounds.

(a) $CH_3CH{=}CHCH_2CH$
$$\overset{\displaystyle O}{\underset{\displaystyle \|}{}}$$

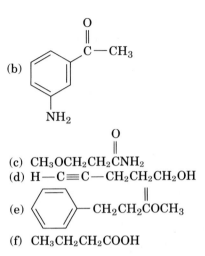

(b)

(c) $CH_3OCH_2CH_2\overset{\displaystyle O}{\overset{\displaystyle \|}{C}}NH_2$
(d) $H{-}C{\equiv}C{-}CH_2CH_2CH_2OH$

(e) 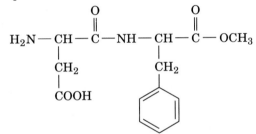 $-CH_2CH_2\overset{\displaystyle O}{\overset{\displaystyle \|}{C}}OCH_3$

(f) $CH_3CH_2CH_2COOH$

16–39   Aspartame is an effective sweetener that is used as a low-calorie substitute for sugar. Identify the functional groups present in an aspartame molecule.

$$H_2N{-}CH{-}\overset{O}{\overset{\|}{C}}{-}NH{-}CH{-}\overset{O}{\overset{\|}{C}}{-}OCH_3$$

16–40   Write condensed structures for the following compounds.
(a) n-butyl alcohol
(b) di-n-propyl ether
(c) trimethylamine
(d) propanal
(e) 3-pentanone
(f) propanoic acid
(g) methyl acetate
(h) propanamide

*16–41   Name the following compounds.
(a) $CH_3\overset{O}{\overset{\|}{C}}CH_2CH_2CH_3$
(b) $CH_3CH_2CH_2{-}\overset{O}{\overset{\|}{C}}{-}OH$
(c) $CH_3{-}\overset{O}{\overset{\|}{C}}{-}NH_2$
(d) $(CH_3)_3C{-}OH$
(e) $CH_3CH_2CH_2CH_2{-}\overset{O}{\overset{\|}{C}}{-}H$

*16-42   Name the following compounds.

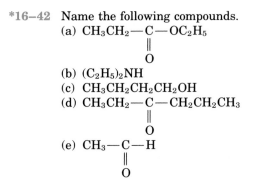

(a) $CH_3CH_2 — \overset{\displaystyle O}{\underset{\displaystyle \|}{C}} — OC_2H_5$

(b) $(C_2H_5)_2NH$

(c) $CH_3CH_2CH_2CH_2OH$

(d) $CH_3CH_2 — \overset{\displaystyle O}{\underset{\displaystyle \|}{C}} — CH_2CH_2CH_3$

(e) $CH_3 — \overset{\displaystyle O}{\underset{\displaystyle \|}{C}} — H$

GENERAL QUESTIONS

16-43   Identify the following as an alkane, alkene, or alkyne. Assume that the compounds are not cyclic.

(a) C_8H_{16} (e) CH_4

(b) C_5H_{12} (f) $C_{10}H_{20}$

(c) C_4H_8 (g) $C_{18}H_{38}$

(d) $C_{20}H_{38}$

16-44   There are three isomers for dichlorobenzene. In these molecules, two chlorines are substituted for two hydrogens on the benzene rings. Write the Lewis structures for the three isomers.

16-45   Write the condensed structures of a compound with seven carbon atoms in each of the four hydrocarbon classes.

16-46   Write the condensed structure of a compound with five carbon atoms containing each of the eight hetero atom functional groups discussed.

16-47   Complete the following equations.

(a) $C_3H_8 + excess\ O_2 \rightarrow$

(b) $CH_2 = CHCH_3 + H_2O \xrightarrow{\ H^+\ }$

(c) $C_2H_5Cl + 2NH_3 \rightarrow$

(d) $NH(CH_3)_2(aq) + HCl(aq) \rightarrow$

(e) $CH_3CH_2COOH + NH(CH_3)_2 \rightarrow$

(f) $HC≡CH + HBr \rightarrow$

(g) $CH_3CH_2COOH + H_2O \rightarrow$

16-48   Complete the following equations.

(a) $C_3H_7OH + excess\ O_2 \rightarrow$

(b) $CH_2 = CHCH_3 + Br_2 \rightarrow$

(c) $NH(CH_3)_2 + H_2O \rightarrow$

(d) $CH_3CH_2CH_2OH + [O] \xrightarrow{catalyst}$

(e) $CH_3CH_2COOH + CH_3OH \rightarrow$

(f) $CH_3CH_2COOH(aq) + NaOH \rightarrow$

16-49   Give a general method for making each of the following classes of compounds.

(a) alcohols   (e) amides

(b) ketones   (f) aldehydes

(c) carboxylic acids (g) amines

(d) esters

16-50   Tell how the following can be obtained from natural sources.

(a) ethanol   (d) methane

(b) acetic acid   (e) gasoline

(c) methanol

16-51   Give one possible use for each of the following compounds or class of compounds.

(a) formaldehyde

(b) acetic acid

(c) ethylene glycol

(d) acetylsalicylic acid

(e) amines

(f) esters

(g) alkenes

(h) alkanes

*16-52   There are three isomers with the formula C_3H_8O, and three isomers with the formula $C_3H_6O_2$. Show the Lewis structures of these six compounds and the common names of each. All contain one of the functional groups discussed.

*16-53   There are two isomers with the formula C_3H_6O, and four isomers with the formula C_3H_9N. Show the Lewis structures of these six compounds and the common names of each. All contain one of the functional groups discussed except hydroxyl.

*16-54   Glycine has the formula $C_2H_5NO_2$. It is a member of a class of compounds called amino acids that combine in long chains to form proteins. Glycine contains both an amine and a carboxylic acid group. Write the structure for glycine. Write an equation illustrating the reaction of glycine as an acid in water, and an equation illustrating its reaction as a base in water. Illustrate how two glycine molecules can combine to form an amide linkage (also called a peptide linkage).

SOLUTIONS TO LEARNING CHECKS

A–1 four, isomers, alkane, natural, petroleum, refinery, volatile, unsaturated, double, polymers, triple, aromatic, derivative

A–2 Pentane = C_5H_{12}
(a) $CH_3—CH_2—CH_2—CH_2—CH_3$
(b) $CH_3—CH_2—CH_2—CH_3$
$\qquad\qquad\qquad |$
$\qquad\qquad\quad CH_3$

$\qquad\quad CH_3$
$\qquad\qquad |$
(c) $CH_3—CH_2—CH_3$
$\qquad\qquad\ |$
$\qquad\quad CH_3$

A–3 (a) *n*-pentane　　　　(b) methylbutane　　　(c) dimethylpropane

A–4 (a) $CH_3—CH{=}CH—CH_2—CH_3$　　　(c)
$\qquad\qquad CH_3$
$\qquad\qquad\ |$
(b) $CH_3—C—CH_2—C{\equiv}CH$
$\qquad\qquad\ |$
$\qquad\quad CH_3$

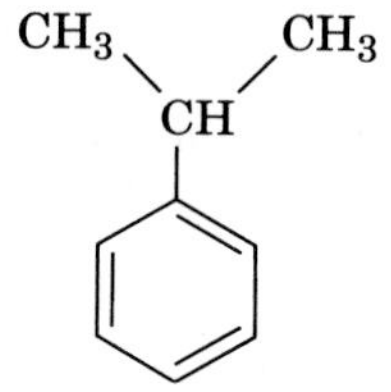

B–1 hydroxyl, hydrocarbon, carbonyl, hydrogen, nitrogen, carboxyl, hydrogen, hydroxyl

B–2 (a) ester　　　　　　　(e) ketone
(b) carboxylic acid　　　(f) aldehyde
(c) ether　　　　　　　(g) amide
(d) alcohol　　　　　　(h) amine

B–3 (a) methyl acetate　　　(e) dimethyl ketone (acetone)
(b) acetic acid　　　　　(f) acetaldehyde
(c) dimethyl ether　　　(g) acetamide
(d) methyl alcohol　　　(h) dimethyl amine

FOREWORD TO THE APPENDIXES

Why do some (one or two, anyway) students seem so self-assured in the study of chemistry and yet others (all the rest) seem so worried? Most likely, it has a lot to do with preparation. Preparation in this case probably does not mean a prior course in chemistry, but it does mean having a solid mathematical background. Most of the students using this text probably are a little rusty on at least some aspects of basic arithmetic, algebra, and scientific notation. There are several reasons for this. Some have not had a good secondary school background in math courses, and others have been away from their high school or college math courses for a number of years. It makes no difference—most students need access to a few reminders, hints, and review exercises to get in shape. The sooner students admit that they have forgotten some math, the faster they do something about it and start to enjoy chemistry. It is really difficult to appreciate the study of this science if math deficiencies get in the way.

Appendix A reviews some of the basic arithmetic concepts such as manipulation of fractions, expressing decimal fractions, and, very importantly, the expression and use of percent. Appendix B reviews the manipulation and solution of simple algebra equations, which are so important in the quantitative aspects of chemistry. Appendix C supplements the discussion of scientific notation in Chapter 1 with more examples and exercises. Appendix D discusses logarithms. Although many students can handle logs on calculators, there is often little understanding of the actual concept of a logarithm and why it is used. Appendix E is a brief discussion of graphing, an important tool of the social sciences as well as the natural sciences. Appendix F contains a discussion of the function of the more common type of calculators used by students in chemistry classes. Appendix G is a glossary of terms, and Appendix H contains answers to about two-thirds of the problems at the ends of the chapters.

BASIC MATHEMATICS

The following is a quick (very quick) refresher of fundamentals of math. This may be sufficient to aid you if you are just a little rusty on some of the basic concepts. For more thorough explanations and practice, however, you are urged to use a more comprehensive math review workbook or consult with your instructor.

1 ADDITION AND SUBTRACTION

Since most calculations in this text use numbers expressed in decimal form, we will emphasize the manipulation of this type of number. In addition and subtraction, it is important to line up the decimal point carefully before doing the math.

Subtraction is simply the addition of a negative number. Remember that subtraction of a negative number changes the sign to a plus (two negatives make a positive). For example, $4 - 7 = -3$, but $4 - (-7) = 4 + 7 = 11$.

EXAMPLE A–1

Addition and Subtraction

Carry out the following calculations:
(a) $16.75 + 13.31 + 175.67$
(b) $11.8 + 13.1 - 6.1$
(c) $47.82 - 111.18 - (-12.17)$

Working It Out ●

570

(a)

$$\downarrow$$

$$16.75$$
$$13.31$$
$$175.67$$
$$\overline{205.73}$$

(b) $11.8 + 13.1 - 6.1$

$$\begin{array}{rr} 11.8 & \to 24.9 \\ +13.1 & -6.1 \\ \overline{24.9} & \overline{18.8} \end{array}$$

(c) $47.82 - 111.18 - (-12.17)$

This is the same as $47.82 - 111.18 + 12.17$.

$$\begin{array}{rr} 47.82 & -111.18 \\ +12.17 & \to +59.99 \\ \overline{59.99} & \overline{-51.19} \end{array}$$

A–1 Carry out the following calculations.

◀ **Exercises**

(a) $47 + 1672$ **(d)** $-97 + 16 - 118$

(b) $11.15 + 190.25$ **(e)** $0.897 + 1.310 - 0.063$

(c) $114 + 26 - 37$ **(f)** $-0.377 - (-0.101) + 0.975$

           (g) $17.489 - 318.112 - (-0.315) + (-3.330)$

Answers: **(a)** 1719 **(b)** 201.40 **(c)** 103 **(d)** -199 **(e)** 2.144
(f) 0.699 **(g)** -303.638

2 MULTIPLICATION

Multiplication is expressed in various ways as follows:

$$13.7 \times 115.35 = 13.7 \cdot 115.35 = (13.7)(115.35) = 13.7(115.35)$$

If it is necessary to carry out the multiplication in longhand, you must be careful to place the decimal point correctly in the answer. Count the *total* number of digits to the right of the decimal point in both multipliers (three in this example). The answer has that number of digits to the right of the decimal point in the answer. Finally, round off to the proper number of significant figures.

$$\begin{array}{r} 13.7 \\ \times\ 2.15 \\ \hline 685 \\ 137 \\ 274 \\ \hline 29455 \end{array} = 29.455 = 29.5^*$$

$13.7 \times \underbrace{2.15}_{} = $ $1 + 2 = \boxed{3}$ (decimal places) $\to \boxed{3}$

When a number (called a *base*) is multiplied by itself one or more times it is said to be raised to a *power*. The power (called the *exponent*) indicates the num-

---

* Rounded off to three significant figures. See Section 1–2.

ber of bases multiplied. For example, the exact values of the following numbers raised to a power are

$$4^2 = 4 \times 4 = 16 \qquad \text{("four squared")}$$
$$2^4 = 2 \times 2 \times 2 \times 2 = 16 \qquad \text{("two to the fourth power")}$$
$$4^3 = 4 \times 4 \times 4 = 64 \qquad \text{("four cubed")}$$

$$(14.1)^2 = 14.1 \times 14.1 = 198.81$$

In the calculations used in this book, most numbers have specific units. In multiplication, the units as well as the numbers are multiplied. For example,

$$3.7 \text{ cm} \times 4.61 \text{ cm} = 17 \text{ (cm} \times \text{cm)} = 17 \text{ cm}^2$$

$$(4.5 \text{ in.})^3 = 91 \text{ in.}^3$$

In the multiplication of a series of numbers, grouping is possible.

$$(a \times b) \times c = a \times (b \times c)$$

$$3.0 \text{ cm} \times 148 \text{ cm} \times 3.0 \text{ cm} = (3.0 \times 3.0) \times 148 \times (\text{cm} \times \text{cm} \times \text{cm})$$

$$= \underline{1300 \text{ cm}^3}$$

When multiplying signs, remember:

$$(+) \times (-) = - \qquad (+) \times (+) = + \qquad (-) \times (-) = +$$

For example, $(-3) \times 2 = -6$; $(-9) \times (-8) = +72$.

Exercises ▶

A–2 Carry out the following calculations. For (a) through (d) carry out the multiplications completely. For (e) through (h) round off the answer to the proper number of significant figures and include units.

(a) $16.2 \times (-118)$ (d) $(-47.8) \times (-9.6)$

(b) $(4 \times 2) \times 879$ (e) $3.0 \text{ ft} \times 18 \text{ lb}$

(c) $(-8) \times (-2) \times (-37)$ (f) $17.7 \text{ in.} \times (13.2 \text{ in.} \times 25.0 \text{ in.})$

(g) What is the area of a circle where the radius is 2.2 cm? (Area $= \pi r^2$, $\pi = 3.14$.)

(h) What is the volume of a cylinder 5.0 in. high with a cross-sectional radius of 0.82 in.? (Volume = area of cross section × height.)

Answers: (a) -1911.6 (b) 7032 (c) -592 (d) 458.88
(e) $54 \text{ ft} \cdot \text{lb}$ (f) 5840 in.^3 (g) 15 cm^2 (h) 11 in.^3

3 ROOTS OF NUMBERS

A root of a number is a fractional exponent. It is expressed as

$$\sqrt[x]{a} = a^{1/x}$$

If x is not shown (on the left), it is assumed to be 2 and is known as the *square root*. The square root is the number that when multiplied by itself gives the base a. For example,

$$\sqrt{4} = 2 \qquad (2 \times 2 = 4)$$

$$\sqrt{9} = 3 \qquad (3 \times 3 = 9)$$

The square root of a number may have either a positive or a negative sign. Generally, however, we are interested only in the positive root in chemistry calculations.

If the square root of a number is not a whole number, it may be computed on a calculator or found in a table. Without these tools available, an educated approximation can come close to the answer. For example, the square root of 54 lies between 7 ($7^2 = 49$) and 8 ($8^2 = 64$) but closer to 7. An educated guess of 7.3 would be excellent.

The cube root of a number is expressed as

$$\sqrt[3]{b} = b^{1/3}$$

It is the number multiplied by itself two times that gives b. For example,

$$\sqrt[3]{27} = 3.0 \qquad (3 \times 3 \times 3 = 27)$$

$$\sqrt[3]{64} = 4.0 \qquad (4 \times 4 \times 4 = 64)$$

A hand calculator (see Appendix F) is the most convenient source of roots of numbers.

A–3 Find the following roots. If necessary, first approximate the answer, then check with a calculator.

◀**Exercises**

    (a) $\sqrt{25}$ **(b)** $\sqrt{36 \text{ cm}^2}$ **(c)** $\sqrt{144 \text{ ft}^4}$ **(d)** $\sqrt{40}$

    (e) $\sqrt{7.0}$ **(f)** $110^{1/2}$ **(g)** $100^{1/3}$ **(h)** $\sqrt[3]{50}$

    (i) What is the radius of a circle that has an area of 150 ft^2? (Area = πr^2)

    (j) What is the radius of the cross section of a cylinder if it has a volume of 320 m^3 and a height of 6.0 m? (Volume = $\pi r^2 \times$ height)

Answers: **(a)** 5.0 **(b)** 6.0 cm **(c)** 12.0 ft^2 **(d)** 6.3 **(e)** 2.6
(f) 10.5 **(g)** 4.64 **(h)** 3.7 **(i)** 6.91 ft **(j)** 4.1 m

4 DIVISION, FRACTIONS, AND DECIMAL NUMBERS

Division problems are expressed in fraction form as follows:

Division form	Common fraction form	Decimal form

$$88.8 \div 2.44 = \frac{88.8}{2.44} = 88.8/2.44 = 36.4$$

Generally, in chemistry, answers are expected in decimal form rather than common fraction form. Therefore, to obtain the answer, the *numerator* (the number on the top) is divided by the *denominator* (the number on the bottom). Before doing the actual calculation, it helps to have a feeling for how the fractional number should look in decimal form. When the numerator is smaller than the denominator, the fraction is known as a *proper fraction*, and the decimal number is less than one. When the numerator is larger than the denominator, the fraction is known as an *improper fraction*, and the decimal number is greater than one.

To carry out the division longhand, it is easier to divide a whole number in the denominator into the numerator. To do this, move the decimal in both numerator and denominator the same number of places. In effect, you are multiplying both numerator and denominator by the same number, which does not change the value of the fraction.

$$\frac{a}{b} = \frac{a \times c}{b \times c}$$

$$\frac{88.8}{2.44} = \frac{88.8 \times 100}{2.44 \times 100} = \frac{8880}{244} = \underline{\underline{36.4}}$$

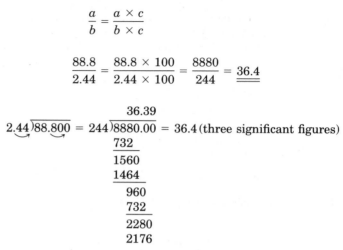

Many divisions can be simplified by cancellation, which is the elimination of common factors in the numerator and denominator. This is possible because a number divided by itself is equal to unity (e.g., $25/25 = 1$). As in multiplication, all units also must be divided. If identical units appear in both numerator and denominator, they also can be canceled.

$$\frac{a \times c}{b \times c} = \frac{a}{b}$$

$$\frac{\overset{1}{\cancel{190}} \times 4 \ \cancel{torr}}{\underset{1}{\cancel{190} \ \cancel{torr}}} = \underline{\underline{4}}$$

$$\frac{2500 \ cm^3}{150 \ cm} = \frac{\overset{1}{\cancel{50}} \times 50 \ \cancel{cm} \times cm \times cm}{\underset{1}{\cancel{50}} \times 3 \ \cancel{cm}} = \frac{50 \ cm^2}{3} = \underline{\underline{17 \ cm^2}}$$

$$\frac{2800 \ mi}{45 \ hr} = \frac{\cancel{5} \times 560 \ mi}{\cancel{5} \times 9 \ hr} = \frac{62 \ mi}{1 \ hr} = \underline{\underline{62 \ mi/hr}}$$

This is read as 62 miles "per" one hour or simply 62 miles per hour. The word "per" implies a fraction or a ratio with the unit after "per" in the denominator. If a number is not written or read in the denominator with a unit, it is assumed that the number is unity and is known to as many significant figures as the number in the numerator (i.e., 62 miles per 1.0 hr).

Exercises ▶

A–4   Express the following in decimal form.

(a) 892 miles ÷ 41 hr

(b) 982.6 ÷ 0.250

(c) 195 ÷ 2650

(d) $\dfrac{67.5 \ g}{15.2 \ mL}$

(e) $\dfrac{1890 \ cm^3}{66 \ cm}$

(f) $\dfrac{146 \ ft \cdot hr}{0.68 \ ft}$

(g) $\dfrac{0.8772 \ ft^3}{0.0023 \ ft^2}$

(h) $\dfrac{37.50 \ ft}{0.455 \ sec}$

Answers: (a) 22 miles/hr   (b) 3930   (c) 0.0736   (d) 4.44 g/mL
(e) 29 cm² (f) 210 hr   (g) 380 ft   (h) 82.4 ft/sec

5 MULTIPLICATION AND DIVISION OF FRACTIONS

When two or more fractions are multiplied, all numbers *and units* in both numerator and denominator can be combined into one fraction.

The division of one fraction by another is the same as the multiplication of the numerator by the *reciprocal* of the denominator. The reciprocal of a fraction is simply the fraction in an inverted form (e.g., $\frac{3}{5}$ is the reciprocal of $\frac{5}{3}$).

$$\frac{a}{b/c} = a \times \frac{c}{b} \qquad \frac{a/b}{c/d} = \frac{a}{b} \times \frac{d}{c} = \frac{a \times d}{b \times c}$$

EXAMPLE A–2

Multiplication and Division

Carry out the following calculations. Round off the answer to two digits.

● **Working It Out**

(a) $\dfrac{3}{5} \times \dfrac{75}{4} \times \dfrac{16}{7} = \dfrac{3 \times 75 \times 16}{5 \times 4 \times 7} = \dfrac{3 \times \overset{15}{\cancel{75}} \times \overset{4}{\cancel{16}}}{\underset{1}{\cancel{5}} \times \underset{1}{\cancel{4}} \times 7} = \dfrac{180}{7} = \underline{\underline{26}}$

(b) $\dfrac{42\ \text{mi}}{\text{hr}} \times \dfrac{3}{7}\text{hr} \times \dfrac{5280\ \text{ft}}{\text{mi}} = \dfrac{\overset{6}{\cancel{42}} \times 3 \times 5290\ \cancel{\text{mi}} \times \cancel{\text{hr}} \times \text{ft}}{\underset{1}{\cancel{7}}\ \cancel{\text{hr}} \times \cancel{\text{mi}}}$

$= \underline{\underline{95{,}000\ \text{ft}}}$

(c) $\dfrac{3}{4}\ \text{mol} \times \dfrac{0.75\ \text{g}}{\text{mol}} \times \dfrac{1\ \text{mL}}{19.3\ \text{g}} = \dfrac{3 \times 0.75 \times 1 \times \cancel{\text{mol}} \times \text{g} \times \text{mL}}{4 \times 1 \times 19.3 \times \cancel{\text{mol}} \times \text{g}}$

$= \underline{\underline{0.029\ \text{mL}}}$

(d) $\dfrac{1650}{3/5} = 1650 \times \dfrac{5}{3} = \underline{\underline{2800}}$

(e) $\dfrac{145\ \text{g}}{7.5\ \text{g/mL}} = 145\ \cancel{\text{g}} \times \dfrac{1\ \text{mL}}{7.5\ \cancel{\text{g}}} = \underline{\underline{19\ \text{mL}}}$

◀ **Exercises**

A–5 Express the following answers in decimal form. If units are not used, round off the answer to three digits. If units are included, round off to the proper number of significant figures and include units in the answer.

(a) $\frac{3}{8} \times \frac{4}{7} \times \frac{21}{20}$

(b) $\frac{250}{273} \times \frac{175}{300} \times (-6)$

(c) $\frac{4}{9} \times \left(-\frac{5}{8}\right) \times \left(-\frac{3}{4}\right)$

(d) $195\ \text{g/mL} \times 47.5\ \text{mL}$

(e) $0.75\ \text{mol} \times 17.3\ \text{g/mol}$

(f) $(3.57\ \text{in.})^2 \times 0.85\ \text{in.} \times \dfrac{16.4\ \text{cm}^3}{\text{in.}^3}$

(g) $\dfrac{\frac{150}{350}}{\frac{25}{42}}$

(h) $\dfrac{\left(-\frac{3}{7}\right)}{\left(-\frac{4}{9}\right)}$

(i) $\dfrac{\left(-\frac{17}{3}\right)}{\frac{8}{9}}$

(j) $\dfrac{\frac{16}{9} \times \frac{10}{14}}{\frac{5}{6}}$

(k) $\dfrac{75.2\ \text{torr}}{760\ \text{torr/atm}}$

(l) $\dfrac{(55.0\ \text{mi/hr}) \times (5280\ \text{ft/mi}) \times (1\ \text{hr/60 min})}{60\ \text{sec/min}}$

(m) $\dfrac{305\ \text{K} \times 62.4\dfrac{\text{L} \cdot \text{torr}}{\text{K} \cdot \text{mol}} \times 0.25\ \text{mol}}{650\ \text{torr}}$

Answers: **(a)** 0.225 **(b)** −3.21 **(c)** 0.208 **(d)** 9260 g **(e)** 13 g
(f) 180 cm³ **(g)** 0.720 **(h)** 0.964 **(i)** −6.38 **(j)** 1.52
(k) 0.0989 atm **(l)** 80.7 ft/sec **(m)** 7.3 L

6 DECIMAL NUMBERS AND PERCENT

In the examples of fractions thus far, we have seen that the units of the numerator can be profoundly different from those of the denominator (e.g., miles/hr, g/mL, etc.). In other problems in chemistry we use fractions to express a component part in the numerator to the total in the denominator. In most cases such fractions are expressed without units and in decimal form.

EXAMPLE A–3

Decimal Numbers

Working It Out ●

(a) A box of nails contains 985 nails; 415 of these are 6-in. nails, 375 are 3-in. nails, and the rest are roofing nails. What is the fraction of roofing nails in decimal form?

Solution

$$\text{Roofing nails} = \text{total} - \text{others} = 985 - (415 + 375) = 195$$

$$\frac{\text{component}}{\text{total}} = \frac{195}{375 + 415 + 195} = \underline{\underline{0.198}}$$

(b) A mixture contains 4.25 mol of N_2, 2.76 mol of O_2, and 1.75 mol of CO_2. What is the fraction of moles of O_2 present in the mixture? (This fraction is known, not surprisingly, as "the mole fraction." The mole is a unit of quantity, like dozen.

Solution

$$\frac{\text{component}}{\text{total}} = \frac{2.76}{4.25 + 2.76 + 1.75} = \underline{\underline{0.315}}$$

Exercises ▶

A–6 A grocer has 195 dozen boxes of fruit; 74 dozen boxes are apples, 62 dozen boxes are peaches, and the rest are oranges. What is the fraction of the boxes that are oranges?

A–7 A mixture contains 9.85 mol of gas. A 3.18-mol quantity of the gas is N_2, 4.69 mol is O_2, and the rest is He. What is the mole fraction of He in the mixture?

A–8 The total pressure of a mixture of two gases, N_2 and O_2, is 0.72 atm. The pressure due to O_2 is 0.41 atm. What is the fraction of the pressure due to O_2?

Answers: **A–6** 0.30   **A–7** 0.201   **A–8** 0.57

The decimal numbers that have just been discussed are frequently expressed as percentages. Percent simply means parts per 100. Percent is obtained by multiplying a fraction in decimal form by 100%.

EXAMPLE A–4

Expressing Percent

Working It Out ●

If 57 out of 180 people at a party are women, what is the percent women?

Solution

The fraction of women in decimal form is

$$\frac{57}{180} = 0.317$$

The percent women is

$$0.317 \times 100\% = \underline{31.7\% \text{ women}}$$

The general method used to obtain percent is

$$\frac{\text{component}}{\text{total}} \times 100\% = \underline{\hspace{2cm}} \% \text{ of component}$$

To change from percent back to a decimal number, divide the percent by 100%, which moves the decimal to the left two places.

$$86.2\% = \frac{86.2\%}{100\%} = 0.862 \quad \text{(fraction in decimal form)}$$

A–9   Express the following fractions or decimal numbers as percents: $\frac{1}{4}, \frac{3}{8}, \frac{9}{8}$, $\frac{55}{25}$, 0.67, 0.13, 1.75, 0.098.

◀Exercises

A–10   A bushel holds 198 apples, 27 of which are green. What is the percent of green apples?

A–11   A basket contains 75 pears, 8 apples, 15 oranges, and 51 grapefruit. What is the percent of each?

Answers: **A–9** $\frac{1}{4} = 25\%$, $\frac{3}{8} = 37.5\%$, $\frac{9}{8} = 112.5\%$, $\frac{55}{25} = 220\%$, $0.67 = 67\%$, $0.13 = 13\%$, $1.75 = 175\%$, $0.098 = 9.8\%$ **A–10** 13.6%,   **A–11** 50.3% pears, 5.4% apples, 10.1% oranges, 34.2% grapefruit

We have seen how the percent is calculated from the total and the component part. We now consider problems where percent is given and we calculate either the component part as in Example A–5(a) or the total as in Example A–5(b). Such problems can be solved in two ways. The method we employ here uses the percent as a conversion factor, and the problems are solved by the factor-label method. (See Section 1–5.) They also can be solved algebraically as is done in Appendix B.

EXAMPLE A–5

Using Percent in Calculations

(a) A crowd at a rock concert was composed of about 87% teenagers. If the crowd totaled 586 people, how many were teenagers?

●Working It Out

Procedure
Remember that percent means "per 100." In this case it means 87 teenagers per 100 people or, in fraction form,

$$\frac{87 \text{ teenagers}}{100 \text{ people}}$$

If this fraction is then multiplied by the number of people, the result is the component part or the number of teenagers.

$$586 \text{ ~~people~~} \times \frac{87 \text{ teenagers}}{100 \text{ ~~people~~}} = (586 \times 0.87) = \underline{510 \text{ teenagers}}$$

(b) A professional baseball player got a hit 28.7% of the times he batted. If he got 246 hits, how many times did he bat?

Procedure

The percent can be written in fraction form and then inverted. It thus relates the total at bats to the number of hits:

$$28.7\% = \frac{28.7 \text{ hits}}{100 \text{ at bats}} \quad \text{or} \quad \frac{100 \text{ at bats}}{28.7 \text{ hits}}$$

If this is now multiplied by the number of hits, the result is the total number of at bats.

$$246 \text{ hits} \times \frac{100 \text{ at bats}}{28.7 \text{ hits}} = \left(\frac{246}{0.287}\right) = \underline{\underline{857 \text{ at bats}}}$$

Exercises ▶

A–12 In a certain audience, 45.9% were men. If there were 196 people in the audience, how many women were present?

A–13 In the alcohol molecule, 34.8% of the mass is due to oxygen. What is the mass of oxygen in 497 g of alcohol?

A–14 The cost of a hamburger in 1994 is 216% of the cost in 1970. If hamburgers cost $0.75 each in 1970, what do they cost in 1994?

A–15 In a certain audience, 46.0% are men. If there are 195 men in the audience, how large is the audience?

A–16 If a solution is 23.3% by mass HCl and it contains 14.8 g of HCl, what is the total mass of the solution?

A–17 An unstable isotope has a mass of 131 amu. This is 104% of the mass of a stable isotope. What is the mass of the stable isotope?

Answers: **A–12** 106 women **A–13** 173 g **A–14** $1.62 **A–15** 424 people
A–16 63.5 g **A–17** 126 amu

BASIC ALGEBRA

1 OPERATIONS ON ALGEBRA EQUATIONS

Many of the quantitative problems of chemistry require the use of basic algebra. As an example of a simple algebra equation we use

$$x = y + 8$$

In any algebraic equation the equality remains valid when identical operations are performed on both sides of the equation. The following operations illustrate this principle.

1 A quantity may be added to or subtracted from both sides of the equation.

(add 8) $\quad x \underline{+ 8} = y + 8 \underline{+ 8} \quad x + 8 = y + 16$

(subtract 8) $\quad x \underline{- 8} = y + 8 \underline{- 8} \quad x - 8 = y$

2 Both sides of the equation may be multiplied or divided by the same quantity.

(multiply by 4) $\quad \underline{4}x = \underline{4}(y + 8) = 4y + 32$

(divide by 2) $\quad \dfrac{x}{2} = \dfrac{(y + 8)}{2} \qquad \dfrac{x}{2} = \dfrac{y}{2} + 4$

3 Both sides of the equation may be raised to a power, or a root of both sides of an equation may be taken.

(equation squared) $\quad x^2 = (y + 8)^2$

(square root taken) $\quad \sqrt{x} = \sqrt{y + 8}$

4 Both sides of an equation may be inverted.

$$\dfrac{1}{x} = \dfrac{1}{y + 8}$$

In addition to operation on both sides of an equation, two other points must be recalled.

1 As in any fraction, identical factors in the numerator and the denominator in an algebraic equation may be canceled.

$$\dfrac{\cancel{4}x}{\cancel{4}} = x = y + 8 \qquad \text{or} \qquad x = \dfrac{\cancel{4}(y + 8)}{\cancel{4}} = y + 8$$

2 Quantities equal to the same quantity are equal to each other. Thus substitutions for equalities may be made in algebraic equations.

$$x = y + 8$$

$$x = 27$$

Therefore, since $x = x$,

$$y + 8 = 27$$

We can use these basic rules to solve algebraic equations. Usually, we need to isolate one variable on the left-hand side of the equation with all other numbers and variables on the right-hand side of the equations. The operations previously listed can be simplified for this purpose in two ways.

In practice, a number or a variable may be moved to the other side of an equation with a change of sign. For example, if

$$x + z = y,$$

then subtracting z from both sides, in effect, gives us

$$x = y - z$$

Also, the numerator of a fraction on the left becomes the denominator on the right. The denominator of a fraction on the left becomes the numerator on the right. For example, consider the following two cases.

If $xz = y$,   If $\dfrac{x}{k + 5} = B$,

then by dividing both   then multiplying both
sides by z gives us, in effect,   sides by $k + 5$ gives us

$xz = y$ $\dfrac{x}{k + 5} = B$

$$x = \frac{y}{z} \qquad\qquad\qquad x = B(k + 5)$$

The following examples illustrate the isolation of one variable (x) on the left-hand side of the equation.

EXAMPLE B–1

Solving Algebra Equations for a Variable

(a) Solve for x in $x + y + 8 = z + 6$.

Solution
Move $+y$ and $+8$ to the right by changing signs.

$$x = z + 6 - y - 8$$

$$= z - y + 6 - 8 = \underline{\underline{z - y - 2}}$$

(b) Solve for x in

$$\frac{x + 8}{y} = z$$

Solution

First, move y to the right by multiplying both sides by y.

$$\cancel{y} \cdot \frac{x+8}{\cancel{y}} = z \cdot y$$

This leaves

$$x + 8 = zy$$

Subtract 8 from both sides to obtain the final answer.

$$\underline{\underline{x = zy - 8}}$$

(c) Solve for x in

$$\frac{4x+2}{3+x} = 7$$

Solution

First, multiply both sides by $(3 + x)$ to clear the fraction.

$$\cancel{(3+x)} \cdot \frac{4x+2}{\cancel{(3+x)}} = 7(3+x)$$

This leaves

$$4x + 2 = 21 + 7x$$

To move integers to the right and the x variable to the left, subtract $7x$ and 2 from both sides of the equation. This leaves

$$-3x = 19$$

Finally, divide both sides by -3 to move the -3 to the right.

$$x = -\frac{19}{3} = \underline{\underline{-6.33}}$$

(d) Solve for T_2 in

$$\frac{P_1 V_1}{T_1} = \frac{P_2 V_2}{T_2}$$

Solution

To move T_2 to the left, multiply both sides by T_2.

$$T_2 \cdot \frac{P_1 V_1}{T_1} = \cancel{T_2} \cdot \frac{P_2 V_2}{\cancel{T_2}} = P_2 V_2$$

Move $P_1 V_1$ to the right by dividing by $P_1 V_1$.

$$\frac{T_2}{\cancel{P_1 V_1}} \cdot \frac{\cancel{P_1 V_1}}{T_1} = \frac{P_2 V_2}{P_1 V_1}$$

Finally, move T_1 to the right by multiplying both sides by T_1.

$$T_2 = \underline{\underline{\frac{T_1 P_2 V_2}{P_1 V_1}}}$$

(e) Solve the following equation for y.

$$\frac{2y}{3} + x = 9z + 4$$

Solution

First, to clear the fraction, multiply both sides by 3. This leaves

$$2y + 3x = 27z + 12$$

Subtract $3x$ from both sides, which leaves

$$2y = 27z + 12 - 3x$$

Finally, divide both sides by 2, which leaves

$$y = \frac{27z + 12 - 3x}{2}$$

Exercises ▶

B–1 Solve for x in $17x = y - 87$.

B–2 Solve for x in

$$\frac{y}{x} + 8 = z + 16$$

B–3 Solve for T in $PV = (\text{mass}/MM)RT$.

B–4 Solve for x in

$$\frac{7x - 3}{6 + 2x} = 3r$$

B–5 Solve for x in $18x - 27 = 2x + 4y - 35$. If $y = 3x$, what is the value of x?

B–6 Solve for x in

$$\frac{x}{4y} + 18 = y + 2$$

B–7 Solve for x in $5x^2 + 12 = x^2 + 37$.

B–8 Solve for r in

$$\frac{80}{2r} + \frac{y}{r} = 11$$

What is the value of r if $y = 14$?

Answers: **B–1** $x = (y - 87)/17$ **B–2** $\dot{x} = y/(8 + z)$ **B–3** $T = PV \cdot MM/\text{mass} \cdot R$ **B–4** $x = 3(6r + 1)/7 - 6r)$ **B–5** $x = (y - 2)/4$. When $y = 3x$, $x = -2$.   **B–6** $x = 4y(y - 16)$ **B–7** $x = \pm 2.5$ **B–8** $r = (40 + y)/11$. When $y = 14$, $r = \frac{54}{11}$

2 WORD PROBLEMS AND ALGEBRA EQUATIONS

Eventually, a necessary skill in chemistry is the ability to translate word problems into algebra equations and then solve. The key is to assign a variable (usually x) to be equal to a certain quantity and then to treat the variable consistently throughout the equation. Again, examples are the best way to illustrate the problems.

EXAMPLE B–2

Solving Abstract Word Equations

Working It Out ●

Translate each of the following to an equation.
(a) A number x is equal to a number that is 4 larger than y.

$$\underline{\underline{x = y + 4}}$$

(b) A number z is equal to three-fourths of u.

$$\underline{\underline{z = \tfrac{3}{4}u}}$$

(c) The square of a number r is equal to 16.9% of the value of w.

$$\underline{r^2 = 0.169w} \quad \text{(change percent to a decimal number)}$$

(d) A number t is equal to 12 plus the square root of q.

$$\underline{\underline{t = 12 + \sqrt{q}}}$$

Write algebraic equations for the following:

B–9   A number n is equal to a number that is 85 smaller than m.

◀ **Exercises**

B–10   A number y is equal to one-fourth of z.

B–11   Fifteen percent of a number k is equal to the square of another number d.

B–12   A number x is equal to 14 more than the square root of v.

B–13   Four times the sum of two numbers, q and w, is equal to 68.

B–14   Five times the product of two variables, s and t, is equal to 16 less than the square of s.

B–15   Five-ninths of a number C is equal to 32 less than a number F.

Answers: **B–9** $n = m - 85$ **B–10** $y = z/4$ **B–11** $0.15k = d^2$
B–12 $x = \sqrt{v} + 14$ **B–13** $4(q + w) = 68$ **B–14** $5st = s^2 - 16$
B–15 $\frac{5}{9}C = F - 32$

We now move from the abstract to the real. In the following examples it is necessary to translate the problem into an algebraic expression, as in the previous examples. There are two types of examples that we will use. The first you will certainly recognize, but the second type may be unfamiliar, especially if you have just begun the study of chemistry. However, it is *not* important that you understand the units of chemistry problems at this time. What *is* important is for you to notice that the problems are worked in the same manner regardless of the units.

EXAMPLE B–3

Solving Concrete Word Equations

(a) John is 2 years more than twice as old as Mary. The sum of their ages is 86. How old is each?

● **Working It Out**

Solution
Let x = age of Mary. Then $2x + 2$ = age of John.

$$x + (2x + 2) = 86$$

$$3x = 84$$

$$x = \underline{\underline{28}} \quad \text{(age of Mary)}$$

$$[2(28) + 2] = \underline{\underline{58}} \quad \text{(age of John)}$$

(b) One mole of SF_6 has a mass 30.0 g less than four times the mass of 1 mol of CO_2. The mass of 1 mol of SF_6 plus the mass of 1 mol of CO_2 is equal to 190 g. What is the mass of 1 mol of each?

Solution

Let x = mass of 1 mol of CO_2. Then $4x - 30$ = mass of 1 mol of SF_6.

$$x + (4x - 30) = 190$$

$$x = \underline{\underline{44 \text{ g}}} \quad \text{(mass of 1 mol of } CO_2\text{)}$$

$$[4(44) - 30] = \underline{\underline{146 \text{ g}}} \quad \text{(mass of 1 mol of } SF_6\text{)}$$

(c) Two students took the same test, and their percent scores differed by 10%. If there were 200 points on the test and the total of their point scores was 260 points, what was each student's percent score?

Procedure

Set up an equation relating each person's percent scores to their total points (260).

Let x = percent score of higher test.

Then $x - 10$ = percent score of lower test.

The points that each person scores is the percent in fraction form multiplied by the points on the test.

$$\frac{\% \text{ grade}}{100 \text{ points}} \times (\text{points on test}) = \text{points scored}$$

Solution

$$\left[\frac{x}{100}(200 \text{ points}) \right] + \left[\frac{x - 10}{100}(200 \text{ points}) \right] = 260 \text{ points}$$

$$200x + 200x - 2000 = 26{,}000$$

$$400x = 28{,}000$$

$$x = 70$$

higher score = $\underline{\underline{70\%}}$ lower score = $70 - 10 = \underline{\underline{60\%}}$

(d) If an 8.75-g quantity of sugar represents 65.7% of a mixture, what is the mass of the mixture?

Solution

Let x = mass of the mixture. Then

$$\frac{65.7}{100}x = 0.657 \, x = 8.75$$

$$x = \frac{8.75}{0.657} = \underline{\underline{13.3 \text{ g}}}$$

(e) A used car dealer has Fords, Chevrolets, and Plymouths. There are 120 Fords, 152 Chevrolets, and the rest are Plymouths. If the fraction of Fords is 0.310, how many Plymouths are on the lot?

Solution

Let x = number of Plymouths.

$$\text{fraction of Fords} = \frac{\text{number of Fords}}{\text{total number of cars}} = 0.310$$

$$\frac{120}{120 + 152 + x} = 0.310$$

$$120 = 0.310(272 + x)$$

$$120 = 84.3 + 0.310x$$

$$x = \underline{\underline{115 \text{ Plymouths}}}$$

(f) There is a 0.605-mol quantity of N_2 present in a mixture of N_2 and O_2. If the mole fraction of N_2 is 0.251, how many moles of O_2 are present?

Solution

Let x = number of moles of O_2. Then

$$\text{mole fraction } N_2 = \frac{\text{mol of } N_2}{\text{total mol present}} = 0.251$$

$$\frac{0.605}{0.605 + x} = 0.251$$

$$0.605 = 0.251(0.605 + x)$$

$$x = \underline{\underline{1.80 \text{ mol}}}$$

In the following exercises, a problem concerning an everyday situation is followed by one or more closely analogous problems concerning a chemistry situation. In both cases the mechanics of the solution are similar. Only the units differ.

◀Exercises

B–16 The total length of two boards is 18.4 ft. If one board is 4.0 ft longer than the other, what is the length of each board?

B–17 An isotope of iodine has a mass 10 amu less than two-thirds the mass of an isotope of thallium. The total mass of the two isotopes is 340 amu. What is the mass of each isotope?

B–18 An isotope of gallium has a mass 22 amu more than one-fourth the mass of an isotope of osmium. The difference in the two masses is 122 amu. What is the mass of each?

B–19 An oil refinery held 175 barrels of oil. When refined, each barrel yields 24 gallons of gasoline. If 3120 gallons of gasoline were produced, what percentage of the original barrels of oil was refined?

B–20 A solution contained 0.856 mol of a substance A_2X. In solution some of the A_2X's break up into A's and X's. (Note that each mole of A_2X yields 2 mol of A.) If 0.224 mol of A is present in the solution, what percentage of the moles of A_2X dissociated (broke apart)?

B–21 In Las Vegas, a dealer starts with 264 decks of cards. If 42.8% of the decks were used in an evening, how many jacks (four per deck) were used?

B–22 A solution originally contains a 1.45-mol quantity of a compound A_3X_2. If 31.5% of the A_3X_2 dissociates (three A's and two X's per A_3X_2), how many moles of A are formed? How many moles of X? How many moles of undissociated A_3X_2 remain? How many moles of particles (A's, X's, and A_3X_2's) are present in the solution?

B–23   The fraction of kerosene that can be recovered from a barrel of crude oil is 0.200. After a certain amount of oil was refined, 8.90 gal of kerosene, some gasoline, and 18.6 gal of other products were produced. How many gallons of gasoline were produced?

B–24   The fraction of moles (mole fraction) of gas A in a mixture is 0.261. If the mixture contains 0.375 mol of gas B and 0.175 mol of gas C as well as gas A, how many moles of gas A are present?

Answers: **B–16** 7.2 ft, 11.2 ft   **B–17** thallium, 210 amu; iodine, 130 amu
B–18 gallium, 70 amu; osmium, 192 amu   **B–19** 74.3%   **B–20** 13.1%
B–21 452 jacks   **B–22** 1.37 mol of A, 0.914 mol of X, 0.99 mol of A_3X_2, 3.27 mol total   **B–23** 17.0 gal   **B–24** 0.195 mol

3 DIRECT AND INVERSE PROPORTIONALITIES

There is one other point that should be included in a review on algebra—direct and inverse proportionalities. We use these often in chemistry.

When a quantity is directly proportional to another, it means that an increase in one variable will cause a corresponding increase of the same percent in the other variable. A direct proportionality is shown as

$$A \propto B \quad (\propto \text{ is the proportionality symbol})$$

which is read "A is directly proportional to B." A proportionality can be easily converted to an algebraic equation by the introduction of a constant (in our examples designated k), called a constant of proportionality. Thus the proportion becomes

$$A = kB$$

or, rearranging,

$$\frac{A}{B} = k$$

Note that k is not a variable but has a certain numerical value that does not change as do A and B under experimental conditions.

A common, direct proportionality that we will study relates Kelvin temperature T and volume V of a gas at constant pressure. This is written as

$$V \propto T \quad V = kT \quad \frac{V}{T} = k$$

(This is known as Charles's law.)

In a hypothetical case, $V = 100$ L and $T = 200$ K. From this information we can calculate the value of the constant k.

$$\frac{V}{T} = \frac{100 \text{ L}}{200 \text{ K}} = 0.50 \text{ L/K}$$

A change in volume or temperature requires a corresponding change in the other *in the same direction*. For example, if the temperature of the gas is changed to 300 K, we can see that a corresponding change in volume is required from the following calculation:

$$V = kT = 0.50 \text{ L/K} \times 300 \text{ K} = \underline{\underline{150 \text{ L}}}$$

When a quantity is inversely proportional to another quantity, an increase in one brings about a corresponding *decrease* in the other. An inverse proportionality between A and B is written as

$$A \propto \frac{1}{B}$$

As before, the proportionality can be written as an equality by the introduction of a constant (which has a value different from the example above).

$$A = \frac{k}{B} \quad \text{or} \quad AB = k$$

A common inverse proportionality that we use relates the volume V of a gas to the pressure P at a constant temperature. This is written as

$$V \propto \frac{1}{P} \quad V = \frac{k}{P} \quad PV = k$$

(This is known as Boyle's law.)

In a hypothetical case, $V = 100$ L and $P = 1.50$ atm. From this information we can calculate the value of the constant k.

$$PV = k = 1.50 \text{ atm} \times 100 \text{ L} = 150 \text{ atm} \cdot \text{L}$$

A change in volume or pressure requires a corresponding change in the other *in the opposite direction*. For example, if the pressure on the gas is changed to 3.00 atm, we can see that a corresponding change in volume is required.

$$V = \frac{k}{P} = \frac{150 \cancel{\text{ atm}} \cdot \text{L}}{3.00 \cancel{\text{ atm}}} = \underline{50.0 \text{ L}}$$

When one variable (e.g., x) is directly proportional to two other variables (e.g., y and z), the proportionality can be written as the product of the two.

$$\text{If} \quad x \propto b \quad \text{and} \quad x \propto z$$

$$\text{then} \quad x \propto yz$$

When one variable, (e.g., a) is directly proportional to one variable (i.e., b) and inversely proportional to another (e.g., c), the proportionality can be written as the ratio of the two.

$$\text{If} \quad a \propto b \quad \text{and} \quad a \propto \frac{1}{c}$$

$$\text{then} \quad a \propto \frac{b}{c}$$

Quantities can be directly or inversely proportional to the square, square root, or any other function of another variable or number, as illustrated by the examples that follow.

EXAMPLE B–4

Solving Equations with Proportionalities

(a) A quantity C is directly proportional to the square of D. Write an equality for this statement and explain how a change in D affects the value of C.

● Working It Out

Solution
The equation is

$$\underline{\underline{C = kD^2}}$$

Note that a change in D will have a significant effect on the value of C. For example,

$$\text{If } D = 1, \text{ then } C = k$$

$$\text{If } D = 2, \text{ then } C = 4k$$

$$\text{If } D = 3, \text{ then } C = 9k$$

Note that when the value of D is doubled, the value of C is increased *fourfold*.

(b) A variable X is directly proportional to the square of the variable Y and inversely proportional to the square of another variable Z. This can be written as two separate equations if it is assumed that Y is constant when Z varies and vice versa.

$$X = k_1 Y^2 \quad (Z \text{ constant})$$

$$X = k_2/Z^2 \quad (Y \text{ constant})$$

(k_1 and k_2 are different constants)

This relationship can be combined into one equation when both Y and Z are variables.

$$X = \frac{k_3 Y^2}{Z^2}$$

k_3 is a third constant that is a combination of k_1 and k_2.

Exercises ▶

Write equalities for the following relations.

B–25 X is inversely proportional to $Y + Z$.

B–26 $[H_3O^+]$ is inversely proportional to $[OH^-]$.

B–27 $[H_2]$ is directly proportional to the square root of r.

B–28 B is directly proportional to the square of y and the cube of z.

B–29 The pressure P of a gas is directly proportional to the number of moles n and the temperature T, and inversely proportional to the volume V.

Answers: **B–25** $X = k/(Y + Z)$ **B–26** $[H_3O^+] = k/[OH^-]$
B–27 $[H_2] = k\sqrt{r}$ **B–28** $B = ky^2z^3$ **B–29** $P = knT/V$

SCIENTIFIC NOTATION

1 Review of Scientific Notation

2 Addition and Subtraction

3 Multiplication and Division

4 Powers and Roots

Although this topic was first introduced in Section 1–3 in this text, we focus on a review of the mathematical manipulation of numbers expressed in scientific notation in this appendix. Specifically, addition, multiplication, division, and taking the roots of numbers expressed in scientific notation are covered.

As mentioned in Chapter 1, scientific notation makes use of powers of 10 to simplify awkward numbers that employ more than two or three zeros that are not significant figures. The exponent of 10 simply indicates how many times we should multiply or divide a number (called the coefficient) by 10 to produce the actual number. For example, $8.9 \times 10^3 = 8.9$ (the coefficient) multiplied by 10 *three* times, or

$$8.9 \times 10 \times 10 \times 10 = 8900$$

Also, $4.7 \times 10^{-3} = 4.7$ (the coefficient) divided by 10 *three* times, or

$$\frac{4.7}{10 \times 10 \times 10} = 0.0047$$

1 EXPRESSING NUMBERS IN STANDARD SCIENTIFIC NOTATION

The method for expressing numbers in scientific notation was explained in Section 1–3. However, to simplify a number or to express it in the standard form with one digit to the left of the decimal point in the coefficient, it is often necessary to change a number already expressed in scientific notation. If this is done in a hurry, errors may result. Thus it is worthwhile to practice moving the decimal point of numbers expressed in scientific notation.

EXAMPLE C–1

Changing Scientific Notation to Normal Numbers

Change the following numbers to the standard form.
(a) 489×10^4 (b) 0.00489×10^8

● **Working It Out**

Procedure

All you need to remember is to raise the power of 10 one unit for each place the decimal point is moved to the left, and lower the power of 10 one unit for each place that the decimal point is moved to the right in the coefficient.

Solution

$$489 \times 10^4 = (4\,8\,9) \times 10^4 = 4.89 \times 10^{4+2} = \underline{\underline{4.89 \times 10^6}}$$

$$0.00489 \times 10^8 = (0.0\,0\,4\,8\,9) \times 10^8 = 4.89 \times 10^{8-3} = 4.89 \times 10^5$$

As an aid to remembering whether you should raise or lower the exponent as you move the decimal point, it is suggested that you write (or at least imagine) the coefficient on a slant. For each place that you move the decimal point *up*, add one to the exponent. For each place that you move the decimal point *down*, subtract one from the exponent. Note that the exponent moves up or down with the decimal point.

EXAMPLE C–2

Changing the Decimal Point in the Coefficient

Working It Out ●

Change the following numbers to the standard form in scientific notation.
(a) 4223×10^{-7} (b) 0.00076×10^{18}

Solution

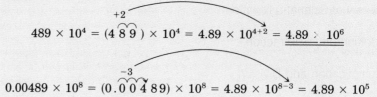

$$4223 \times 10^{-7} = \begin{bmatrix} 4 \\ 2 \\ 2 \\ 3 \end{bmatrix} \begin{matrix} +3 \\ +2 \\ +1 \end{matrix} \times 10^{-7} = 4.223 \times 10^{-7+3} = \underline{\underline{4.223 \times 10^{-4}}}$$

$$0.00076 \times 10^{18} = \begin{bmatrix} 0 \\ 0 \\ 0 \\ 0 \\ 7 \\ 6 \end{bmatrix} \begin{matrix} -1 \\ -2 \\ -3 \\ -4 \end{matrix} \times 10^{18} = 7.6 \times 10^{18-4} = \underline{\underline{7.6 \times 10^{14}}}$$

C–1   Change the following numbers to standard scientific notation with one digit to the left of the decimal point in the coefficient.

Exercises ▶

   (a) 787×10^{-6} (d) 0.0037×10^9

   (b) 43.8×10^{-1} (e) 49.3×10^{15}

   (c) 0.015×10^{-16} (f) 6678×10^{-16}

C–2   Change the following numbers to a number with two digits to the left of the decimal point in the coefficient.

(a) 9554×10^4 (d) 116.5×10^4

(b) 1.6×10^{-5} (e) 0.023×10^{-1}

(c) 1×10^6 (f) 0.005×10^{23}

Answers: **C–1**(a) 7.87×10^{-4} **C–1**(b) 4.38 **C–1**(c) 1.5×10^{-18}
C–1(d) 3.7×10^6 **C–1**(e) 4.93×10^{16} **C–1**(f) 6.678×10^{-13}
C–2(a) 95.54×10^6 **C–2**(b) 16×10^{-6} **C–2**(c) 10×10^5
C–2(d) 11.65×10^5 **C–2**(e) 23×10^{-4} **C–2**(f) 50×10^{19}

2 ADDITION AND SUBTRACTION

Addition or subtraction of numbers in scientific notation can be accomplished only when all coefficients have the same exponent of 10. When all the exponents are the same, the coefficients are added and then multiplied by the power of 10. The correct number of places to the right of the decimal point must be shown as discussed in Section 1–2.

EXAMPLE C–3

Addition of Numbers in Scientific Notation

(a) Add the following numbers:

$$3.67 \times 10^{-4}, 4.879 \times 10^{-4}, \text{ and } 18.2 \times 10^{-4}$$

● **Working It Out**

Solution

$$
\begin{array}{r}
3.67 \times 10^{-4} \\
4.879 \times 10^{-4} \\
\underline{18.2 \times 10^{-4}} \\
26.749 \times 10^{-4} = 26.7 \times 10^{-4} = 2.67 \times 10^{-3}
\end{array}
$$

(b) Add the following numbers:

$$320.4 \times 10^3, 1.2 \times 10^5, \text{ and } 0.0615 \times 10^7$$

Solution
Before adding, change all three numbers to the same exponent of 10.

$$
\begin{array}{r}
320.4 \times 10^3 = 3.204 \times 10^5 \\
1.2 \times 10^5 = 1.2 \times 10^5 \\
\underline{0.0615 \times 10^7 = 6.15 \times 10^5} \\
10.554 \times 10^5 = 10.6 \times 10^5 = 1.06 \times 10^6
\end{array}
$$

C–3 Add the following numbers. Express the answer to the proper decimal place.

◀**Exercises**

(a) $152 + (8.635 \times 10^2) + (0.021 \times 10^3)$

(b) $(10.32 \times 10^5) + (1.1 \times 10^5) + (0.4 \times 10^5)$

(c) $(1.007 \times 10^{-8}) + (118 \times 10^{-11}) + (0.1141 \times 10^{-6})$

(d) $(0.0082) + (2.6 \times 10^{-4}) + (159 \times 10^{-4})$

C–4 Carry out the following calculations. Express your answer to the proper decimal place.

(a) $(18.75 \times 10^{-6}) - (13.8 \times 10^{-8}) + (1.0 \times 10^{-5})$

(b) $(1.52 \times 10^{-11}) + (17.7 \times 10^{-12}) - (7.5 \times 10^{-15})$

(c) $(481 \times 10^6) - (0.113 \times 10^9) + (8.5 \times 10^5)$

(d) $(0.363 \times 10^{-6}) + (71.2 \times 10^{-9}) + (519 \times 10^{-12})$

Answers: **C–3**(a) 1.036×10^3 **C–3**(b) 1.18×10^6 **C–3**(c) 1.254×10^{-7}
C–3(d) 2.44×10^{-2} **C–4**(a) 2.9×10^{-5} **C–4**(b) 3.29×10^{-11}
C–4(c) 3.69×10^8 **C–4**(d) 4.35×10^{-7}

3 MULTIPLICATION AND DIVISION

When numbers expressed in scientific notation are multiplied, the exponents of 10 are *added*. When the numbers are divided, the exponent of 10 in the denominator (the divisor) is subtracted from the exponent of 10 in the numerator (the dividend).

EXAMPLE C–4

Multiplication and Division

Exercises ▶

(a) Carry out the following calculation:
$$(4.75 \times 10^6) \times (3.2 \times 10^5)$$

Solution
In the first step, group the coefficients and the powers of 10. Carry out each step separately.
$$= (4.75 \times 3.2) \times (10^6 \times 10^5)$$

$$= 15.200 \times 10^{6+5} = 15 \times 10^{11} = 1.5 \times 10^{12}$$

(b) Carry out the following calculation:
$$(1.62 \times 10^{-8}) \div (8.55 \times 10^{-3})$$

Solution

$$= \frac{1.62 \times 10^{-8}}{8.55 \times 10^{-3}} = \frac{1.62}{8.55} \times \frac{10^{-8}}{10^{-3}} = 0.189 \times 10^{-8 - (-3)}$$

$$= 0.189 \times 10^{-5} = \underline{\underline{1.89 \times 10^{-6}}}$$

Exercises ▶

C–5 Carry out the following calculations. Express your answer to the proper number of significant figures with one digit to the left of the decimal point.

(a) $(7.8 \times 10^{-6}) \times (1.12 \times 10^{-2})$

(b) $(0.511 \times 10^{-3}) \times (891 \times 10^{-8})$

(c) $(156 \times 10^{-12}) \times (0.010 \times 10^4)$

(d) $(16 \times 10^9) \times (0.112 \times 10^{-3})$

(e) $(2.35 \times 10^3) \times (0.3 \times 10^5) \times (3.75 \times 10^2)$

(f) $(6.02 \times 10^{23}) \times (0.0100)$

C–6 Follow the instructions in Problem C–5.

(a) $(14.6 \times 10^8) \div (2.2 \times 10^8)$

(b) $(6.02 \times 10^{23}) \div (3.01 \times 10^{20})$

(c) $(0.885 \times 10^{-7}) \div (16.5 \times 10^3)$

(d) $(0.0221 \times 10^3) \div (0.57 \times 10^{18})$

(e) $238 \div (6.02 \times 10^{23})$

C–7 Follow the instructions in Problem C–5.

(a) $[(8.70 \times 10^6) \times (3.1 \times 10^8)] \div (5 \times 10^{-3})$

(b) $(47.9 \times 10^{-6}) \div [(0.87 \times 10^6) \times (1.4 \times 10^2)]$

(c) $1 \div [(3 \times 10^6) \times (4 \times 10^{10})]$

(d) $1.00 \times 10^{-14} \div [(6.5 \times 10^5) \times (0.32 \times 10^{-5})]$

(e) $[(147 \times 10^{-6}) \div (154 \times 10^{-6})] \div (3.0 \times 10^{12})$

Answers: **C–5**(a) 8.7×10^{-8} **C–5**(b) 4.55×10^{-9} **C–5**(c) 1.6×10^{-8}
C–5(d) 1.8×10^6 **C–5**(e) 3×10^{10} **C–5**(f) 6.02×10^{21} **C–6**(a) 6.6
C–6(b) 2.00×10^3 **C–6**(c) 5.36×10^{-12} **C–6**(d) 3.9×10^{-17}
C–6(e) 3.95×10^{-22} **C–7**(a) 5×10^{17} **C–7**(b) 3.9×10^{-13}
C–7(c) 8×10^{-18} **C–7**(d) 4.8×10^{-15} **C–7**(e) 3.2×10^{-13}

4 POWERS AND ROOTS

When a number expressed in scientific notation is raised to a power, the coefficient is raised to the power and the exponent of 10 is *multiplied* by the power.

To take the root of a number expressed in scientific notation, take the root of the coefficient and divide the exponent of 10 by the root. Before taking a root, however, adjust the number so that the exponent of 10 divided by the root produces a whole number. Adjust the exponent in the direction that will leave the coefficient as a number greater than 1 (to avoid mistakes).

EXAMPLE C–5

Powers and Roots

● **Working It Out**

(a) Carry out the following calculation:
$$(3.2 \times 10^3)^2 = (3.2)^2 \times 10^{3 \times 2}$$
$$= 10.24 \times 10^6 = \underline{\underline{1.0 \times 10^7}}$$
$[(10^3)^2 = 10^3 \times 10^3 = 10 \times 10 \times 10 \times 10 \times 10 \times 10 = 10^6]$

(b) Carry out the following calculation:
$$(1.5 \times 10^{-3})^3 = (1.5)^3 \times 10^{-3 \times 3}$$
$$= \underline{\underline{3.4 \times 10^{-9}}}$$

(c) Carry out the following calculation:
$$\sqrt{2.9 \times 10^5}$$

Solution
First adjust the number so that the exponent of 10 is divisible by 2.
$$\sqrt{2.9 \times 10^5} = \sqrt{29 \times 10^4} = \sqrt{29} \times \sqrt{10^4} = \sqrt{29} \times 10^{4/2}$$
$$= \underline{\underline{5.4 \times 10^2}}$$

(d) Carry out the following calculation:
$$\sqrt[3]{6.9 \times 10^{-8}}$$

Solution
Adjust the number so that the exponent of 10 is divisible by 3.

$$\sqrt[3]{6.9 \times 10^{-8}} = \sqrt[3]{69 \times 10^{-9}} = \sqrt[3]{69} \times \sqrt[3]{10^{-9}}$$

$$= 4.1 \times 10^{-9/3} = \underline{\underline{4.1 \times 10^{-3}}}$$

Exercises ▶

C–8 Carry out the following operations.

(a) $(6.6 \times 10^4)^2$

(b) $(0.7 \times 10^6)^3$

(c) $(1200 \times 10^{-5})^2$ (It will be easier to square if you change the number to $1.2 \times 10^?$ first.)

(d) $(0.035 \times 10^{-3})^3$

(e) $(0.7 \times 10^7)^4$

C–9 Take the following roots. Approximate the answer if necessary.

(a) $\sqrt{36 \times 10^4}$ (d) $\sqrt[3]{1.6 \times 10^5}$

(b) $\sqrt[3]{27 \times 10^{12}}$ (e) $\sqrt{81 \times 10^{-7}}$

(c) $\sqrt{64 \times 10^9}$ (f) $\sqrt{180 \times 10^{10}}$

Answers: **C–8**(a) 4.4×10^9 **C–8**(b) 3×10^{17} **C–8**(c) 1.4×10^{-4}
C–8(d) 4.3×10^{-14} **C–8**(e) 2×10^{27} **C–9**(a) 6.0×10^2 **C–9**(b) 3.0×10^4
C–9(c) 2.5×10^5 **C–9**(d) 54 **C–9**(e) 2.8×10^{-3} **C–9**(f) 1.3×10^6

LOGARITHMS

In Appendix C we learned that scientific notation is particularly useful in expressing very large or very small numbers. In certain areas of chemistry, however, such as in the expression of H_3O^+ concentration, even the repeated use of scientific notation becomes tedious. In this situation, it is convenient to express the concentration as simply the *exponent of 10. The exponent to which 10 must be raised to give a certain number is called its* **common logarithm.** With common logarithms (or just logs) it is possible to express the coefficient and the exponent of 10 as one number.

Since logarithms are simply exponents of 10, logs of numbers such as 100 can be easily determined. Note that 100 can be expressed as 10^2, so that the log of 100 is simply 2. Other examples of simple logs are

$$1 = 10^0 \qquad \log 1 = 0 \qquad 0.1 = 10^{-1} \qquad \log 0.1 = -1$$

$$10 = 10^1 \qquad \log 10 = 1 \qquad 0.01 = 10^{-2} \qquad \log 0.01 = -2$$

$$100 = 10^2 \qquad \log 100 = 2 \qquad 0.001 = 10^{-3} \qquad \log 0.001 = -3$$

$$1000 = 10^3 \qquad \log 1000 = 3 \qquad 0.0001 = 10^{-4} \qquad \log 0.0001 = -4$$

The same mathematical operations that apply to exponents also apply to logs. For example, consider the multiplication and division of exponents of 10 and logs.

	Exponents	Logarithms
Multiplication	$10^4 \times 10^3 = 10^{4+3}$ $= 10^7$	$\log(10^4 \times 10^3) = \log 10^4 + \log 10^3$ $= 4 + 3 = 7$
Division	$\dfrac{10^{10}}{10^4} = 10^{10-4} = 10^6$	$\log \dfrac{10^{10}}{10^4} = \log 10^{10} - \log 10^4$ $= 10 - 4 = 6$

By definition, the log of a number whose coefficient is exactly one is simply the value of the exponent. When the coefficient is not exactly one (which means the exponent is fractional) then some aid is required to find the value. Most modern hand calculators have a log function, which, of course, is the most convenient method of finding logs. This method is described in Appendix F. If a calculator does not have a log function or a calculator is not available, log tables are available and simple to use if one understands the meaning of a logarithm. The following describes the use of a simple log table to find logs. *It is helpful to work through these sections even if you have an appropriate calculator so as to appreciate the function and meaning of a logarithm.*

1 TAKING THE LOG OF A NUMBER BETWEEN 1 AND 10

Note that log 1 = 0 and log 10 = 1. The log of a number between 1 and 10 is therefore between 0 and 1. The log of two-significant-figure numbers can be found in a two-place log table such as shown in this section. Three-, four-, and even five-place log tables are available for more precise calculations. The two-place table is suitable for our purposes. The log of a number between 1 and 10 can be read directly from the body of the table. The first integer in the number is found in the vertical column, and the second (the tenths) is located in the horizontal column. The log is where the two intersect. Since we are taking logs of two-significant-figure numbers, the log should be expressed to two significant figures.

Exercises ▶

EXAMPLE D–1

Logs of Numbers between 1 and 10

What is the log of (a) 5.8, (b) 3.7, and (c) 1.1?

Solution
The answer is between 0 and 1 and is expressed as

$$0.\text{log}$$

$$\log 5.8 = \underline{0.76} \quad \text{(5 down and 8 across)}$$

$$\log 3.7 = \underline{0.57} \quad \text{(3 down and 7 across)}$$

$$\log 1.1 = \underline{0.04} \quad \text{(1 down and 1 across)}$$

2 TAKING THE LOGS OF A NUMBER LESS THAN 1 AND GREATER THAN 10

To find the log of a number less than 1 or greater than 10, follow these steps:

1 Write the number in standard scientific notation (one digit to the left of the decimal point in the coefficient).

2 Look up the log of the coefficient and write as described in Section 1.

3 The value of the log of the coefficient is added to the value of the exponent of 10. That is a result of the general rule

$$\log(A \times B) = \log A + \log B$$

$$\log(A \times 10^B) = \log A + \log 10^B = (\log A) + B$$

TABLE D–1 Two-Place Logarithms

Ones Column ↓ / Tenths Row →	0.0	0.1	0.2	0.3	0.4	0.5	0.6	0.7	0.8	0.9
1	0.00	0.04	0.08	0.11	0.15	0.18	0.20	0.23	0.26	0.28
2	0.30	0.32	0.34	0.36	0.38	0.40	0.41	0.43	0.45	0.46
3	0.48	0.49	0.51	0.52	0.53	0.54	0.56	0.57	0.58	0.59
4	0.60	0.61	0.62	0.63	0.64	0.65	0.66	0.67	0.68	0.69
5	0.70	0.71	0.72	0.72	0.73	0.74	0.75	0.76	0.76	0.77
6	0.78	0.79	0.79	0.80	0.81	0.81	0.82	0.83	0.83	0.84
7	0.85	0.85	0.86	0.86	0.87	0.88	0.88	0.89	0.89	0.90
8	0.90	0.91	0.91	0.92	0.92	0.93	0.93	0.94	0.94	0.95
9	0.95	0.96	0.96	0.97	0.97	0.98	0.98	0.99	0.99	1.00

The sum of the two numbers (the exponent plus the log of the coefficient) has two parts. *The number to the right of the decimal point is called the* **mantissa.** As mentioned before, the mantissa is expressed to the same number of decimal places as there are significant figures in the coefficient. *The whole number to the left of the decimal point is called the* **characteristic.** For example,

$$\log(5.7 \times 10^6) = \log 5.7 + \log 10^6 = 0.76 + 6$$

$$= \boxed{6}\,.\,\boxed{76}$$

characteristic mantissa

EXAMPLE D–2

Logs of Large and Small Numbers

What are the logs of the following numbers?

(a) 4.7×10^3

● Working It Out

$$\log(4.7 \times 10^3) = \log 4.7 + \log 10^3$$

$$= 0.67 + 3 = \underline{\underline{3.67}}$$

(b) 15

$$\log 15 = \log(1.5 \times 10^1)$$

$$= \log 1.5 + \log 10^1$$

$$= 0.18 + 1 = \underline{\underline{1.18}}$$

(c) 0.066

$$\log 0.066 = \log(6.6 \times 10^{-2})$$

$$= \log 6.6 + \log 10^{-2}$$

$$= 0.82 + (-2) = \underline{-1.18}$$

Note that logs of numbers less than one have negative values.

(d) 8.7×10^{-7}

$$\log(8.7 \times 10^{-7}) = \log 8.7 + \log 10^{-7}$$

$$= 0.94 + (-7)$$

$$= \underline{\underline{-6.06}}$$

D–1 Find the logs of the following numbers:

(a) 7.4	**(f)** 1700	**(k)** 0.057×10^8
(b) 5.2	**(g)** 7.3×10^4	**(l)** 32×10^{-5}
(c) 0.87	**(h)** 7.3×10^{-4}	**(m)** 160×10^6
(d) 0.087	**(i)** 4.1×10^{-12}	**(n)** 0.60×10^{-7}
(e) 85	**(j)** 4.5	

D–2 pH is defined as the negative of the log of a number [e.g., $-\log 10^{-2} = -(-2) = 2$]. Find the negative logs of the following:

(a) 8.5×10^{-6}	**(c)** 71×10^{-10}	**(e)** 1.1
(b) 3.4×10^{-11}	**(d)** 0.217×10^{-3}	

Answers: **D–1**(a) 0.87 **D–1**(b) 0.72 **D–1**(c) −0.06 **D–1**(d) −1.06
D–1(e) 1.93 **D–1**(f) 3.23 **D–1**(g) 4.86 **D–1**(h) −3.14 **D–1**(i) −11.39
D–1(j) 0.65 **D–1**(k) 6.76 **D–1**(l) −3.49 **D–1**(m) 8.20 **D–1**(n) −7.22
D–2(a) 5.07 **D–2**(b) 10.47 **D–2**(c) 8.15 **D–2**(d) 3.66 **D–2**(e) −0.04

3 FINDING THE ANTILOG OF A POSITIVE LOG

Not only should we be able to find the log of a number, but we must be able to find *the number whose log has a certain value.* This is called the **antilog** of a log. The procedure is simply the reverse of finding the log. Remember, we are asking the question, "What is the number whose exponent of 10 is a certain value?" For example, What is the antilog (x) of 2?

$$\log x = 2$$

$$x = \underline{\underline{10^2}} \quad \text{since } \log 10^2 = 2$$

When the log is between 0 and 1, the antilog is between 1 and 10. (This is just the reverse of the statement in Section 1.) To find the antilog, locate the given log in the *body* of the log table. Find the first number on the left margin and the second on the top that correspond to this log. Write the two numbers you found with one digit to the left of the decimal point.

If the log is a number greater than one, divide the number into two parts, the mantissa, which lies to the right of the decimal, and the characteristic, which lies to the left. For example,

$$2.56 = 0.56 + 2$$

$$11.11 = 0.11 + 11$$

Take the antilog of the mantissa and express the number as explained above. This becomes the coefficient of the number. The characteristic then becomes the exponent of 10.

EXAMPLE D–3

Antilogs of Positive Logs

What are the antilogs of the following?

(a) 0.84

$$\text{antilog } 0.84 = \underline{\underline{6.9}}$$

(b) 0.23

$$\text{antilog } 0.23 = \underline{\underline{1.7}}$$

(c) 0.96

$$\text{antilog } 0.96 = \underline{\underline{9.1}}$$

(d) $4.65 = 0.65 + 4$

$$\text{antilog } 0.65 = 4.5$$
$$\text{antilog } 4 = 10^4$$
$$= \underline{\underline{4.5 \times 10^4}}$$

(e) $3.89 = 0.89 + 3$

$$\text{antilog } 0.89 = 7.8$$
$$\text{antilog } 3 = 10^3$$
$$= \underline{\underline{7.8 \times 10^3}}$$

● **Working It Out**

---

4 FINDING THE ANTILOG OF A NEGATIVE LOG

A negative log must be adjusted so that the mantissa is positive without changing the actual value of the log itself. For example, -0.18 is the same number as $(+0.82 - 1)$, and -3.28 is the same number as $(+0.72 - 4)$. To do this, first separate and write the mantissa and the characteristic as described before. *Add* 1 to the mantissa and *subtract* 1 from the characteristic. Note that if you add and subtract the same number from a quantity, you have not actually changed the number (e.g., $15 + 1 - 1 = 15$).

	Separate	Add 1 to mantissa	Subtract 1 from characteristic	
$-3.28 =$	$-0.28 - 3 =$	$(-0.28 + 1) -$	$3 - 1$	$= +0.72 - 4$
$-0.18 =$	$-0.18 - 0 =$	$(-0.18 + 1) -$	$0 - 1$	$= +0.82 - 1$

Locate the mantissa in the body of the log table as described before and locate the corresponding numbers in the left-hand column and the top. The negative characteristic becomes the negative exponent of 10.

Working It Out

EXAMPLE D–4

Antilogs of Negative Logs

Find the following antilogs.

(a) -0.02

$$-0.02 = (1 - 0.02) - 1 = 0.98 - 1$$

$$\text{antilog } 0.98 = 9.5$$

$$\text{antilog } (-1) = 10^{-1}$$

$$= 9.5 \times 10^{-1} = \underline{\underline{0.95}}$$

(b) -4.54

$$-4.54 = (1 - 0.54) - 5 = 0.46 - 5$$

$$\text{antilog } 0.46 = 2.9$$

$$\text{antilog } (-5) = 10^{-5}$$

$$= \underline{\underline{2.9 \times 10^{-5}}}$$

Exercises

D–3 Find the antilogs of the following:

(a) 0.81	**(c)** −0.70	**(e)** 10.94	**(g)** −0.08	**(i)** 0.34
(b) 3.46	**(d)** −5.48	**(f)** −2.60	**(h)** 8.40	**(j)** −5.96

Answers: **D–3**(a) 6.5 **D–3**(b) 2.9×10^3 **D–3**(c) 0.20 **D–3**(d) 3.3×10^{-6} **D–3**(e) 8.7×10^{10} **D–3**(f) 2.5×10^{-3} **D–3**(g) 0.83 **D–3**(h) 2.5×10^8 **D–3**(i) 2.2 **D–3**(j) 1.1×10^{-6}

GRAPHS

1 Direct Linear Relationships

2 Nonlinear Relationships

The graphical representation of data is important in all branches of science, social as well as natural. Trends and cycles in data may be obscure when presented in a table but become immediately apparent in graphical form. For example, consider a graph in which the average monthly temperature (represented on the vertical axis) is plotted as a function of the month (represented on the horizontal axis). The graph is constructed by marking the temperature for each month and drawing a smooth curve through the points. A quick look at the graph tells us what we already knew—the temperature cycles up and down throughout the seasons.

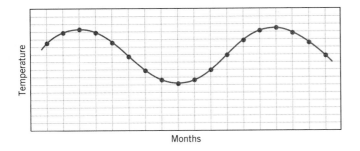

Some useful graphs of this nature are shown in Figures 2-11 and 5-17. In the first, the temperature of a liquid near and at its boiling point is graphed as it varies with time. The graph illustrates the discontinuity in temperature change as the liquid boils. In Figure 5-17, the two graphs illustrate the periodic nature of atomic radius and ionization energy as atomic number increases. These graphs illustrate both short-term and long-term trends.

1 DIRECT LINEAR RELATIONSHIPS

Other types of graphs can be useful in predicting data as well as simply illustrating measurements. For example, in Appendix B we discussed proportionalities. In a proportionality a variable (x) is related to another variable (y) by some mathematical operation (a power, a root, a reciprocal, etc.). A graph translates a proportionality into a line that illustrates the relationship of x to y. There are two types of graphs: linear and nonlinear. A linear graph is represented by a straight line and is a result of a direct, first-power proportionality between the variables x and y. This is expressed as

$$x \propto y$$

Such a graph is illustrated in Figure 9-7, which relates the volume of a gas to the Celsius temperature. Since the data points fall in a straight line, the line can be extended to provide data points beyond the limits of the experiments. (We say the line has been extrapolated.) Using a linear graph, as few as two experiments can be used to predict the results of other measurements.

The relationship between the volume (V) of a gas at constant pressure and the Kelvin temperature (T) can be expressed as

$$V \propto T$$

We will use this direct relationship to construct a linear graph. The following two experiments provide the information necessary to construct such a graph.

Experiment 1 At a temperature of 350°C, the volume was 12.5 L.

Experiment 2 At a temperature of 250°C, the volume was 10.5 L.

Actually, it is always risky to assume that only two experiments provide an accurate graph or give all of the needed information, but for simplicity we will assume that it is valid in this case. We will check the graph with an additional experiment later.

To record this information on a graph, we first need some graph paper with regularly spaced divisions. On the vertical axis (called the ordinate or y axis) we put temperature. Pick divisions on the graph paper that spread out the range of temperatures as much as possible. In our example, the temperatures range between 250 and 350°C, or 100 degrees. Note that the two axes do not need to intersect at $x = y = 0$. We then plot the volume on the horizontal axis (called the abscissa or x axis). The volumes range between 12.5 and 10.5 L, or 2.0 L.

The graph is shown in Figure E-1. An ⊙ is marked at the locations of the two points from the experiments, and a straight line has been drawn between them.

We can now check the accuracy of our graph. A third experiment tells us that at $T = 300$°C, $V = 11.5$ L. Refer back to the graph. Locate 300°C on the ordinate and trace that line to where it intersects the vertical line representing the volume of 11.5 L. Since the point of intersection is right on the line, we have confirmation of the accuracy of the original plot. *Normally, four or five measurements are required to provide enough points on the graph so that a line can be drawn through or as near as many points as possible.*

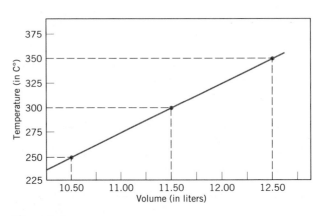

Figure E–1

A straight line graph can be expressed by the linear algebraic equation

$$y = mx + b$$

where y = a value on the ordinate
x = the corresponding value on the abscissa
b = the point on the y axis or ordinate where the straight line intersects the axis
m = the slope of the line

The slope of a line is the ratio of the change on the ordinate to the change on the abscissa. It is determined by choosing two widely spaced points on the line $(x_1, y_1$ and $x_2, y_2)$. The slope is the difference in y divided by the difference in x.

$$m = \frac{y_2 - y_1}{x_2 - x_1}$$

The larger the slope *for a particular graph,* the steeper the straight line. For the graph shown in Figure E–1, if we pick point 1 to be $x_1 = 10.50$, $y_1 = 250$ and point 2 to be $x_2 = 12.50$, $y_2 = 350$, then

$$m = \frac{350 - 250}{12.50 - 10.50} = \frac{100}{2.00} = 50.0$$

The value for b cannot be determined directly from Figure E–1. The value for b is found by determining the point on the y axis where the line intersects when $x = 0$. If we plotted the data on such a graph, we would find that $b = -273°C$. Therefore, the linear equation for this line is

$$y = 50.0x - 273$$

From this equation we can determine the value for y (the temperature) by substitution of any value for x (the volume).

E–1 A sample of water was heated at a constant rate, and its temperature was recorded. The following results were obtained.

◀**Exercises**

Experiment	Temp (°C)	Time (min)
1	10.0	0
2	20.0	4.5
3	30.0	7.5
4	55.0	17.8
5	85.0	30.0

(a) Construct a graph that includes all of the information above with temperature on the ordinate and time on the abscissa.
(b) Calculate the slope of the line and the linear equation $y = mx + b$ for the line.
(c) From the graph find the temperature at 11.0 min and compare it with the value calculated from the equation.
(d) From the graph find the time when the temperature is 65.0°C and compare it with the value calculated from the equation.

Answers: **E–1**(a) Draw the line touching or coming close to as many points as possible. Note that the straight line does not go through all points. **E–1**(b) $m = 2.5$, $y = 2.5x + 10$ **E–1**(c) From the equation, when $x = 11.0$,

$y = 2.5 \cdot 11.0 + 10$, so $y = 37.5°C$. The graph agrees. **E–1**(d) From the equation, when $y = 65$, $65 = 2.5x + 10$, and $x = 22.0$ min. The graph agrees.

2 NONLINEAR RELATIONSHIPS

Nonlinear graphs are represented by a curved line and result from an inverse proportionality, or any type of proportionality other than direct. Examples of such relationships can be expressed as

$$x \propto y^2 \quad \text{and} \quad x \propto \frac{1}{y}$$

An example of an inverse relationship that produces a curve when plotted is the relationship between volume of a gas (V) at constant temperature and pressure (P). The relationship is

$$V \propto \frac{1}{P}$$

In the following experiments the pressure on a volume of gas was varied and the volume was measured at constant temperature.

Experiment	Pressure (torr)	Volume (L)
1	500	15.2
2	600	12.6
3	760	10.0
4	900	8.44
5	1200	6.40
6	1600	4.80

This information has been graphed in Figure E–2. Note that the line has a gradually changing slope, starting with a very steep slope on the left and going to a small slope on the right of the graph. We can interpret this. In the region of the steep slope, the increase in pressure causes little decrease in volume com-

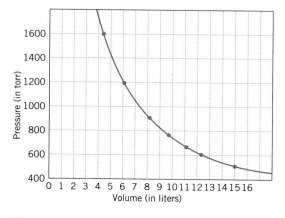

Figure E–2
An Inverse Relationship

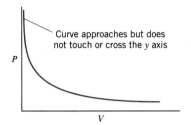

Curve approaches but does not touch or cross the y axis

P

V

Figure E–3

pared with the part of the curve with the small slope to the right. Note that this is a plot of an inverse relationship. That is, the higher the pressure, the lower the volume. If we regraph the information and include much higher pressures and the corresponding volumes, as shown in Figure E–3, we see that *the curve approaches the Y axis but never actually touches it,* no matter how high the pressure. Such a curve is said to approach the axis **asymptotically.** In our example, the curve would eventually appear to change to a vertical line parallel to the y axis.

E–2 Construct a graph from the following experimental information. Plot the time on the abscissa and the concentration on the ordinate. Describe what this graph tells you about the change in concentration of HI as a function of (how it varies with) the time.

◀**Exercises**

Experiment	Time (min)	Concentration of HI (mol/L)
1	0.0	2.50×10^{-2}
2	5.0	1.50×10^{-2}
3	10.0	0.90×10^{-2}
4	15.0	0.65×10^{-2}
5	20.0	0.55×10^{-2}
6	35.0	0.50×10^{-2}

(a) From the graph, find the concentration of HI at 12.0 min and at 25 min.

(b) From the graph, find the times at which the concentration of HI is 2.00×10^{-2} and 0.75×10^{-2}.

Answers: **E–2** The concentration decreases rapidly at the start (the curve has a large slope) but then changes little after about 25 min. At that point the curve appears to change to a straight line parallel to the x axis.   (a) At 12.0 min, the concentration is 0.80×10^{-2} mol/L; at 25 min, the concentration is about 0.52×10^{-2} mol/L.   (b) When the concentration is 2.00×10^{-2} mol/L, the time is 2.4 min; when the concentration is 0.75×10^{-2} mol/L, the time is 13.0 min.

CALCULATORS

1 Addition, Subtraction, Multiplication, and Division

2 Chain Calculations

3 Scientific Notation

4 Powers, Roots, Logs, and Antilogs

An inexpensive hand calculator has become a necessity for almost all chemistry students. The convenience of this handy instrument is undeniable. Calculations are easy, quick, and accurate. In a way, however, use of calculators creates the risk that we may lose the ability to add, multiply, and divide simple numbers on our own. It is sad for an instructor to see students pull out their calculators when the only exercise is to multiply a number by 10. This appendix is included last because it is hoped that students will first review math operations, scientific notation, and logs before using the calculator. Obviously, it is easy to become overly dependent on the calculator, but they are here to stay.

There are two basic types of hand calculators. Almost all inexpensive hand calculators now in use have a logic system known as Algebraic Operating System (AOS). A few students may have calculators that use a system called Reverse Polish Notation (RPN). Calculators using this system have a key labeled "ENTER." In this appendix, we will describe use of the more common AOS system. If your calculator uses the RPN system, consult your instruction book.

Different brands of calculators vary as to the number of digits shown in the answer, the labeling of some functions, and, of course, the number of functions available. You should consult the instruction booklet for more details. In this appendix, we concentrate on the types of calculations that you will encounter in general chemistry courses. This should save time and confusion when reading through an often complicated instruction booklet. Calculations encountered in chemistry include powers, logs, and scientific notation. Recall that the calculator usually gives at least six digits in an answer. This is almost always far too many for the result of a chemical calculation. You must know how to express the answer to the proper number of significant figures (see Section 1-2). A chemistry teacher is not impressed with a student who reports a pH of 8.678346 for a calculation made from a two-significant-figure number. Reporting more numbers than necessary makes an answer look worse, not better.

1 ADDITION, SUBTRACTION, MULTIPLICATION, AND DIVISION

In one-step calculations, you simply enter one number and then tell the calculator what you want it to do with the next number (i.e., $+$, $-$, $\times$, or $\div$). You then enter the next number and ask for the answer (i.e., press the $=$ key). The following examples illustrate simple one-step calculations.

EXAMPLE F–1

One-Step Calculations

(a) 6.368 + 43.852

(b) 0.0835 ÷ 37.2

Enter	Display
6.368	6.368
+	6.368
43.852	43.852
=	50.22

Enter	Display
0.0835	0.0835
÷	0.0835
37.2	37.2
=	2.2446237^{-03}

The answer in (a) should be reported as 50.220 to indicate the proper number of decimal places. A calculator does not include zeros that are to the right of the last digit. Our calculator did not realize that the last zero is significant.

The answer in (b) includes a subscript −03 which is the power of 10. We should report this answer to three significant figures in the form 2.24×10^{-3}.

2 CHAIN CALCULATIONS

One of the most convenient features of the modern hand calculator allows us to do a series of calculations without pausing. For example, many gas law problems require successive calculations as illustrated in the following example.

EXAMPLE F–2

Chain Calculations

(a) Solve the following problem:

$$n = \frac{PV}{RT} = \frac{0.749 \times 7.22}{0.0821 \times 312} = \underline{} \text{ mol}$$

Enter	Display
0.749	0.749
×	0.749
7.22	7.22
÷	5.40778
0.0821	0.0821
÷	65.86821
312	0.211116
Report as 0.211	

You can do this problem in other sequences. For example, you could just as easily carry out the calculation in the following sequence:

$(0.749 \div 0.0821) \times 7.22 \div 312 = 0.211$. One sequence that you *cannot do* is to multiply the numerator (0.749×7.22) and then divide by 0.0821 *times* 312. The final calculation must again be a division by 312. To calculate the formula weight of a compound the atomic masses are multiplied by the number of atoms and the masses totaled. This is illustrated as follows.

(b) Find the formula weight of B_2O_3.

Formula weight $= (2 \times 10.8) + (3 \times 16.0) = $ _____ amu.

Enter	Display
2	2
×	2
10.8	10.8
+	21.6
3	3
×	3
16.0	16.0
=	69.6

One other note on the use of the calculator for successive calculations concerns changing signs. For example, in the calculation (2×-6), after the 6 has been entered, 6 appears on the display. To change it to a -6, type the $+/-$ key. The display now reads -6 and you can proceed with the calculation.

Exercises ▶

F–1 Perform the following calculations.

(a) 0.662×7.99

(b) $73.8 + 0.27 + 45.8$

(c) $43.8 - 178.34$

(d) $7.3 + 8.5 - (-3.4)$

(e) $\dfrac{46.8 \times 23.4}{0.00786}$

(f) $\dfrac{172 + 67 - 89}{55}$

(*Hint:* In problem (f) press the = key after the value of the numerator is calculated. Then carry out the division.)

(g) $(4 \times 14.006) + (3 \times 15.998) + (6 \times 1.008)$

(h) $(0.810 \times 102.99) + (0.190 \times 104.88)$

(i) $\dfrac{0.233 \times 0.0821 \times 323}{0.645}$

Answers: **F–1**(a) 5.29 **F–1**(b) 119.9 **F–1**(c) -134.5 **F–1**(d) 19.2 **F–1**(e) 139000 **F–1**(f) 2.7 **F–1**(g) 110.066 **F–1**(h) 103 **F–1**(i) 9.58

3 SCIENTIFIC NOTATION

Most calculators have the capacity to easily handle scientific notation. This is done with a key marked EE, EXP, or EEX. The coefficient is entered first; then the exponent key is pressed and the exponent entered. If it is a negative exponent, one must press the $+/-$ key to change the sign. The following examples illustrate the use of scientific notation in typical calculations.

EXAMPLE F–3

Scientific Notation

Carry out the following calculations:

(a) $\dfrac{0.250 \times (6.02 \times 10^{23})}{1.55}$

(b) $\dfrac{3.76 \times 10^{-18}}{0.862 \times 10^{6}}$

● **Working It Out**

Enter	Display	
0.250	0.250	
×	0.250	
6.02	6.02	
EXP	6.02	00
23	6.02	23
÷	1.505	23
1.55	1.55	
=	9.71	22
	(9.71×10^{22})	

Enter	Display	
3.76	3.76	
EXP	3.76	00
18	3.76	18
$+/-$	3.76	-18
÷	3.76	-18
0.862	0.862	
EXP	0.862	00
6	0.862	06
=	4.36	-24
	(4.36×10^{-24})	

4 POWERS, ROOTS, LOGS, AND ANTILOGS

These are all one-step calculations that require a special function on the calculator. On some calculators a direct key function is not available. However, notice that there are other functions listed above the key that are accessed by pressing the inverse (INV) or second (2nd) keys. For example, an x^2 key is not available on some calculators, but the function is obtained by pressing the INV key first and then the $\sqrt{x}$ key. The square is the inverse of the square root. On one calculator, the cube root of a number $\sqrt[3]{x}$ is the inverse of the $+/-$ key. On this calculator, the cube of the number must be obtained by first squaring the number, then multiplying the number again. On another calculator, taking the cube root is the same as raising the number to the 1/3 power. There are several variations of how cube functions are presented, and you should experiment with your own calculator in this regard.

Working It Out ●

EXAMPLE F–4

Roots and Powers

Carry out the following calculations:

(a) $(6.2)^2$

Enter	Display
6.2	6.2
INV	6.2
$\sqrt{x}$	38.44

(b) $(6.2)^3$

Enter	Display
6.2	6.2
x^y	6.2
3	3
=	238.328

(c) $\sqrt{6.2}$

Enter	Display
6.2	6.2
$\sqrt{x}$	2.49

(d) $\sqrt[3]{6.2}$

Enter	Display
6.2	6.2
INV*	6.2
x^y	6.2
3	3
=	1.837

*(INV = x^y) is the same as $x^{1/y}$

EXAMPLE F–5

Logs And Antilogs

Find the following logs (common) and antilogs.

(a) 35.6

Enter	Display
35.6	35.6
log	1.55

(b) 4.88×10^{-9}

Enter	Display	
4.88	4.88	
EXP	4.88	00
9	4.88	09
+/−	4.88	−09
log	−8.312	

(c) the antilog of −3.876

Enter	Display
3.876	3.876
+/−	−3.876
INV	−3.876
log	0.000133 or 1.3304 −04

◀**Exercises**

F–2 Perform the following calculations on your calculator.

(a) 0.230^2

(b) $\sqrt{6.80 \times 10^{-3}}$

(c) $\sqrt[3]{0.986}$

(d) $(3.60)^4$

(e) $\sqrt[5]{4.32 \times 10^6}$

(f) the log of 8.33

(g) the log of 0.0257

(h) the log of 567

(i) the antilog of 0.665

(j) the antilog of 2.977

(k) the antilog of −4.005

Answers (given to three significant figures): **F–2**(a) 0.0529 **F–2**(b) 0.0825 **F–2**(c) 0.995 **F–2**(d) 168 **F–2**(e) 21.2 **F–2**(f) 0.921 **F–2**(g) −1.590 **F–2**(h) 2.754 **F–2**(i) 4.62 **F–2**(j) 948 **F–2**(k) 9.89×10^{-5}

GLOSSARY

The numbers in parentheses at the end of each entry refer to the chapter in which the entry is first discussed plus any chapter in which the topic is discussed in detail.

A

absolute zero. The lowest possible temperature. The temperature at which translational motion ceases. Defined as zero on the Kelvin scale. (1, 9)

acid. A compound that increases the H^+ concentration in aqueous solution. In the Brønsted definition, an acid is a proton donor. (4, 8, 12, 14)

acid anhydride. A molecular oxide that dissolves in water to form an oxyacid. (12)

acid ionization constant (K_a). The equilibrium constant specifically for the ionization of a weak acid. The magnitude of the constant relates to the strength of the acid. (14)

acid salt. An ionic compound containing one or more acidic hydrogens on the anion. (12)

acidic solution. A solution with pH < 7 or $[H_3O^+] > 10^{-7} M$. (12)

accuracy. How close a measurement is to the true value. (1)

actinide. One of the fourteen elements between Ac and Unq. An element whose 5f subshell is filling. (4, 5)

activated complex. The transition state formed at the instant of maximum impact between reacting molecules. (14)

activation energy. The minimum energy needed by reactant molecules to form an activated complex so that a reaction may occur. It is the difference in potential energy between the reactants and the activated complex. (14)

actual yield. The measured amount of product in a chemical reaction. (8)

alcohol. An organic compound containing at least one hydroxyl group (i.e., R—-OH). (16)

aldehyde. An organic compound containing a carbonyl group (i.e., C=O) bound to at least one hydrogen (i.e., RCHO). (16)

alkali metal. An element in Group IA in the periodic table (except H). Elements with the electron configuration $[NG]ns^1$. (4, 5)

alkaline earth metal. An element in Group IIA in the periodic table. Elements with the electron configuration $[NG]ns^2$. (4, 5)

alkane. A hydrocarbon containing only single bonds (a saturated hydrocarbon). (16)

alkene. A hydrocarbon containing at least one double bond (an unsaturated hydrocarbon). (16)

alkylation. The formation of large hydrocarbon molecules from small ones in the refining process. (16)

alkyl group. A substituent that is an alkane minus a hydrogen atom [e.g., C_2H_6 (ethane) $- H = C_2H_5$ (ethyl)]. (16)

alkyne. A hydrocarbon containing a triple bond. (16)

allotropes. Different forms of the same element. (6)

alloy. A homogeneous mixture of metallic elements in one solid phase. (2)

alpha (α) particle. A helium nucleus ($^4_2He^{2+}$) emitted from a radioactive nucleus. (3, 15)

amide. A derivative of a carboxylic acid in which an amine group replaces the hydroxyl group of the acid (i.e., $RCONH_2$). (16)

amine. An organic compound containing a nitrogen with a single bond to a hydrocarbon group and a total of two other bonds to hydrocarbon groups or hydrogens (i.e., RNH_2, R_2NH, R_3N). (16)

amorphous solid. A solid without a defined shape. The molecules in such a solid do not occupy regular, symmetrical positions. (10)

amphiprotic. Refers to a molecule or ion that has both a conjugate acid and a conjugate base. (12)

anion. A negatively charged ion. (3)

anode. The electrode at which oxidation takes place. (13)

aromatic compound. A cyclic hydrocarbon (usually six-membered) which can be written with alternating single and double bonds. (16)

atmosphere. The sea of gases above the surface of Earth. Also, the average pressure of the atmosphere at sea level which is defined as the standard pressure and is abbreviated atm. (9)

atom. The smallest fundamental particle of an element that has the properties of that element. (3)

atomic mass. The weighted average of the isotopic masses of all of the naturally occurring isotopes of an element. The mass of an "average" atom of a naturally occurring element compared to ^{12}C. (3)

atomic mass unit (amu). A mass that is exactly one-twelfth of the mass of an atom of ^{12}C, which is defined as exactly 12 amu. (3)

atomic number. The number of protons (positive charge) in the nuclei of the isotopes of a particular element. (3)

atomic radius. The distance from the nucleus of an atom to the outermost electron. (5)

atomic theory. A theory first proposed by John Dalton in 1803 which holds that the basic components of matter are atoms. (3)

aufbau principle. A rule that states that electrons fill the lowest available energy level first. (5)

autoionization. The dissociation of a solvent to produce positive and negative ions. (12)

Avogadro's law. A law which states that the volume of a gas is directly proportional to the number of moles of gas present at constant temperature and pressure (i.e., $V = kn$). (9)

Avogadro's number. The number of objects or particles in one mole, which is 6.02×10^{23}. (7)

B

balanced equation. A chemical equation that has the same number and type of atoms on both sides of the equation. (8)

barometer. A device that measures the atmospheric pressure. (9)

base. A compound that increases the OH^- concentration in aqueous solution. In the Brønsted definition, a base is a proton acceptor. (8, 12)

base anhydride. An ionic oxide that dissolves in water to form an ionic hydroxide. (12)

base ionization constant (K_b). An equilibrium constant specifically for the ionization of a weak base. The magnitude of the constant relates to the strength of the base. (14)

basic solution. A solution with pH > 7 or $[OH^-] > 10^{-7}\ M$. (12)

battery. One or more voltaic cells joined together as a single unit. (13)

beta (β) particle. A high-energy electron emitted from a radioactive nucleus. (15)

binary acid. An acid composed of hydrogen and one other element. (4)

Bohr model. A model of the atom in which electrons orbit around a central nucleus in discrete energy levels. (5)

boiling point. The temperature at which a pure substance changes from the liquid to the gaseous state. (2) It is the temperature at which the vapor pressure equals the restraining pressure. (10)

box diagram. See *orbital diagram.*

Boyle's law. A law which states that the volume of a gas is inversely proportional to the pressure at constant temperature (i.e., $PV = k$). (9)

buffer solution. A solution that resists changes in pH from the addition of limited amounts of strong acid or base. Made from a solution of a weak acid and a salt containing its conjugate base, or a weak base and a salt containing its conjugate acid. (12)

C

calorie. The amount of heat required to raise one gram of water one degree Celsius. Equal to exactly 4.184 joules. (2)

carbonyl group. The functional group of ketones and aldehydes, made up of a carbon with a double bond to an oxygen (i.e., $C=O$). (16)

carboxylic acid. An organic compound containing a carbonyl group attached to a hydroxide group and a hydrocarbon group (i.e., RCOOH). (16)

catalyst. A substance that is not consumed in a reaction but whose presence increases the rate of the reaction. A catalyst lowers the activation energy. (14)

cathode. The electrode at which reduction takes place. (13)

cation. A positively charged ion. (3)

Celsius. A temperature scale with 100 equal divisions between the freezing and boiling points of water at average sea level pressure, with exactly zero assigned to the freezing point of water. (1)

chain reaction. A self-sustaining reaction. In a nuclear chain reaction, one reacting neutron produces between two and three other neutrons that in turn cause reactions. (15)

Charles's law. A law which states that the volume of a gas is directly proportional to the Kelvin temperature at constant pressure (i.e., $P = kT$). (9)

chemical change. A change in a substance to another substance or substances. (2)

chemical equation. The representation of a chemical reaction using the symbols of the elements and the formulas of compounds. (8)

chemical property. The property of a substance relating to its tendency to undergo chemical changes. (2)

chemical thermodynamics. The study of heat and its relationship to chemical changes. (8)

chemistry. The branch of science dealing with the nature, composition, and structure of matter and the changes it undergoes. (2)

coefficient. The number before an element or compound in a balanced chemical equation indicating the number of molecules or moles of that substance. (8)

colligative property. A property that depends only on the relative amounts of solute and solvent present, not on their identity. (11)

collision theory. A theory that states that chemical reactions are brought about by collisions of molecules. (14)

combination reaction. A chemical reaction whereby one compound is formed from two elements and/or compounds. (8)

combined gas law.   A law that relates the pressure, volume, and temperature of a gas (i.e., $PV = kT$). (9)

combustion reaction.   A chemical reaction whereby an element or compound reacts with elemental oxygen. (8)

compound.   A pure substance composed of two or more elements that are chemically combined in fixed proportions. (2)

concentrated solution.   A solution containing a relatively large amount of a specified solute per unit volume. (11)

concentration.   The amount of solute present in a specified amount of solvent or solution. (11)

condensation point.   The temperature at which a pure substance changes from the gas to the liquid or solid state. (2, 10)

conductor.   A substance that allows a flow of electricity. (11)

conjugate acid.   The addition of one H^+ to a base to form the conjugate acid. (12)

conjugate base.   The removal of one H^+ from an acid to form the conjugate base. (12)

conservation of energy.   A law which states that energy is neither created nor destroyed but can be transformed from one form to another. (2)

conservation of mass.   A law which states that matter is neither created nor destroyed in a chemical reaction. (2)

continuous spectrum.   The visible spectrum where one color blends gradually into another. A spectrum containing all wavelengths of visible light. (5)

conversion factor.   A relationship between two units or quantities expressed in fractional form. (1)

covalent bond.   The force that bonds two atoms together by a shared pair or pairs of electrons. (3, 6)

cracking.   The formation of small hydrocarbon molecules from large ones in the refining process. (16)

crystal lattice.   See *lattice*.

crystalline solid.   A solid with a regular, symmetrical shape where the molecules or ions occupy set positions in the crystal lattice. (10)

D

Dalton's law.   A law which states that the total pressure of a gas in a system is the sum of the partial pressures of each component gas. (9)

Daniell cell.   A voltaic cell made up of zinc and copper electrodes each immersed in solutions of their respective ions connected by a salt bridge. (13)

decomposition reaction.   A chemical reaction whereby one compound decomposes to two or more elements and/or compounds. (8)

density.   The ratio of the mass (usually in grams) to the volume (usually in milliliters or liters) of a substance. (2)

derivative.   A compound produced by the substitution of a group or hetero atom on the molecules of the original or parent compound. (16)

diffusion.   The mixing of one substance into others. (9)

dilute solution.   A solution containing a relatively small amount of a specified solute per unit volume. (11)

dilution.   The preparation of a dilute solution from a concentrated solution by the addition of more solvent. (11)

dimensional analysis.   See *factor-label method*.

dipole.   Two poles—one positive and one negative that may exist in a bond or molecule. (6)

dipole-dipole attractions.   The force of attraction between a dipole on one polar molecule and a dipole on another polar molecule. (10)

diprotic acid.   An acid that can produce two H^+ ions per molecule. (12)

discrete spectrum.   A spectrum containing specific wavelengths of light originating from the hot gaseous atoms of an element. (5)

distillation.   A laboratory procedure where a solution is separated into its components by boiling the mixture and condensing the vapor to form a liquid. (2)

double bond.   The sharing of two pairs of electrons between two atoms. (6)

double-replacement reaction.   A chemical reaction whereby the cations and anions in two compounds exchange, leading to either formation of a precipitate or a molecular compound such as water. (8, 11, 12)

dry cell.   A voltaic cell composed of zinc and graphite electrodes immersed in an aqueous paste of NH_4Cl and MnO_2. (13)

dynamic equilibrium.   See *point of equilibrium*.

E

effusion.   The movement of gases through an opening or hole. (9)

electricity.   A flow of electrons through a conductor. (11)

electrode.   A surface in a cell where the oxidation or reduction reaction takes place. (13)

electrolysis.   The process of forcing electrical energy through an electrolytic cell, thereby causing a chemical reaction. (13)

electrolyte.   A solute whose aqueous solution or molten state conducts electricity. (11)

electrolytic cell. A voltaic cell that converts electrical energy into chemical energy. It involves a nonspontaneous reaction. (13)

electron. A negatively charged particle in an atom, with a comparatively very small mass. (3)

electron affinity. The energy released when a gaseous atom adds an electron. (5)

electron capture. The capture of an orbital electron by a radioactive isotope. (15)

electron configuration. The designation of all of the electrons in an atom into a specific shell and subshell. (5)

electronegativity. The ability of an atom of an element to attract electrons to itself in a covalent bond. (6)

electrostatic forces. The forces of attraction between unlike charges, and repulsion between like charges. (3)

element. A pure substance that cannot be broken down into simpler substances. The most basic form of matter existing under ordinary conditions. (2)

empirical formula. The simplest whole-number ratio of atoms in a compound. (7)

endothermic reaction. A chemical reaction that absorbs heat from the surroundings. (2, 8, 14)

energy. The capacity or the ability to do work. It comes in several different forms (e.g., light and heat) and two types (kinetic and potential). (2)

energy level. The discrete orbits available to an electron in a hydrogen atom. (5)

equilibrium. See *point of equilibrium.*

equilibrium constant. A number that defines the position of equilibrium for a particular reaction at a specified temperature. (14)

equilibrium constant expression. The ratio of the concentrations of products to reactants, each raised to the power corresponding to its coefficient in the balanced equation. (14)

equilibrium vapor pressure. The pressure exerted by a vapor above a liquid at a certain temperature. (10)

ester. A derivative of a carboxylic acid in which a hydrocarbon group (R′) replaces the hydrogen on the oxygen in the parent acid (i.e., RCOOR′). (16)

ether. An organic compound containing an oxygen bonded to two hydrocarbon groups (i.e., ROR′). (16)

evaporation. The vaporization of a liquid below its boiling point. (10)

exact number. A number that results from a definition or an actual count. (1)

excited state. An energy level higher than the lowest available energy level. (5)

exothermic reaction. A chemical reaction that occurs with the evolution of heat to the surroundings. (2, 8, 14)

exponent. In scientific notation, it is the power to which ten is raised. (1)

F

factor-label method. A problem-solving technique that converts from one unit to another by use of conversion factors. (1)

Fahrenheit. A temperature scale with 180 divisions between the freezing and boiling points of water, with exactly 32 assigned to the freezing point of water. (1)

family. See *group.*

filtration. A laboratory procedure where solids are removed from liquids by passing the heterogeneous mixture through a filter. (2)

fission. The splitting of a large, unstable nucleus into two smaller nuclei of similar size, resulting in the production of energy. (15)

formula. The symbols of the elements, and the number of atoms of each element that make up a compound. (3)

formula unit. The simplest whole-number ratio of ions in an ionic compound. (3)

formula weight. The mass of a compound (in amu), which is determined from the number of atoms and the atomic mass of each element indicated by the formula. (7)

freezing point. The temperature at which a pure substance changes from the liquid to the solid state. (2)

fuel cell. A voltaic cell that can generate a continuous flow of electricity from the reaction of hydrogen and oxygen to produce water. (13)

functional group. The atom or group of atoms that determines the chemical nature of the molecules of an organic compound. (16)

fusion (nuclear). The combination of two small nuclei to form a larger nucleus, resulting in the production of energy. (16)

G

gamma (γ) ray. A high-energy form of light emitted from a radioactive nucleus. (15)

gas. A physical state that has neither a definite volume nor a definite shape and fills a container uniformly. (2)

gas constant. The constant of proportionality (R) in the ideal gas law. (9)

gas law. A law governing the behavior of gases that is consistent with the kinetic molecular theory as applied to gases. (9)

Gay-Lussac's law. A law which states that the pressure of a gas is directly proportional to the Kelvin temperature at constant volume. (i.e., $P = kT$). (9)

Graham's law. A law that states that the rates of diffusion of gases are inversely proportional to the square root of their molar masses. (9)

ground state. The lowest available energy level in an atom. (5)

group. A vertical column of elements in the periodic table. (4)

H

half-life. The time required for one-half of a given sample of an isotope to undergo radioactive decay. (15)

half-reaction. The oxidation or reduction process in a redox reaction written separately. (13)

halogen. An element in Group VIIA in the periodic table. Elements with the electron configuration $[NG]ns^2np^5$. (4, 5)

heat of fusion. The amount of heat in calories or joules required to melt one gram of a substance. (10)

heat of reaction. The amount of heat energy absorbed or evolved in a specified chemical reaction. (5)

heat of vaporization. The amount of heat in calories or joules required to vaporize one gram of the substance. (10)

heating curve. The graphical representation of the temperature as a solid is heated through two phase changes, plotted as a function of the time of heating. (10)

heterogeneous mixture. A nonuniform mixture containing two or more phases with definite boundaries between phases. (2)

homogeneous mixture. A mixture that is the same throughout and contains only one phase. (2)

homologous series. A series of related compounds in which one member differs from the next by a constant number of atoms. (16)

Hund's rule. A rule that states that electrons occupy separate orbitals with parallel spins if possible. (5)

hydrocarbon. An organic compound containing only carbon and hydrogen. (16)

hydrogen bonding. A force of attraction between a N, O, or F on one molecule and a hydrogen bonded to a N, O, or F on another. (10)

hydrolysis reaction. The reaction of an anion as a base with water or a cation as an acid. (12)

hydrometer. A device that measures the density or specific gravity of a liquid. (2)

hydronium ion. A representation of the hydrogen ion in aqueous solution (H_3O^+). (12)

hydroxyl group. The —OH group as found in an organic compound. It is the functional group of alcohols. (16)

hypothesis. A tentative explanation of related data. It can be used to predict results of more experiments. (Prologue)

I

ideal gas. A hypothetical gas whose molecules are considered to have no volume or interactions with each other. An ideal gas would obey the ideal gas law under all conditions. (9)

ideal gas law. A relationship between the pressure, volume, temperature, and number of moles of gas. (i.e., $PV = nRT$). (9)

immiscible liquids. Two liquids that do not mix and thus form a heterogeneous mixture. (11)

improper fraction. A fraction whose numerator is larger than the denominator and thus has a value greater than one. (Appendix A)

infrared light. Light with wavelengths somewhat longer than those of red light in the visible spectrum. (5)

inner transition element. Either a lanthanide, where the $4f$ subshell is filling, or an actinide, where the $5f$ subshell is filling. (4, 5)

insoluble compound. A compound that does not dissolve to any appreciable amount in a certain solvent. (11)

insulator. See *nonconductor*.

intermolecular forces. The attractive forces between molecules. (10)

ion. An atom or group of covalently bonded atoms that have a net electrical charge. (3)

ion product (K_w). The equilibrium expression of the anion and the cation of water (i.e, $[H_3O^+][OH^-] = K_w$). (12)

ion-dipole force. The force between an ion and the dipole of a polar molecule. (11)

ion-electron method. A method of balancing oxidation-reduction reactions where two half-reactions are balanced separately and then added. (13)

ionic bond. The electrostatic force holding the positive and negative ions together in an ionic compound. (3, 6)

ionic compound. Compounds containing positive and negative ions. (3)

ionic solid. A solid where the crystal lattice positions are occupied by ions. (10)

ionization. The process of forming an ion or ions from a molecule or atom. (5, 12, 15)

ionization energy. The energy required to remove an electron from a gaseous atom or ion. (5)

isotopes. Atoms having the same atomic number but different mass numbers. (3)

isotopic mass. The mass of an isotope compared to ^{12}C, which is defined as having a mass of exactly 12 amu. (3)

J

joule. The SI unit for the measurement of heat energy. (2)

K

Kelvin scale. A temperature scale in which 0 K is the lowest possible temperature. $T(K) = t(°C) + 273$. (1, 9)

ketone. An organic compound containing a carbonyl group bonded to two hydrocarbon groups (i.e., RCOR). (16)

kinetic energy. Energy as a result of motion; equal to $\frac{1}{2}mv^2$ (2, 9)

kinetic molecular theory. A theory advanced in the late 1800s to explain the nature of gases. (9)

L

lanthanide. One of fourteen elements between La and Hf. Elements whose $4f$ subshell is filling. (4, 5)

lattice. A three-dimensional array of ions or molecules in a solid crystal. (6)

law. A concise statement or mathematical relationship that describes some behavior of matter. (Prologue)

law of mass action. The relationship that describes the relative distribution of reactants and products at equilibrium. It is composed of a constant and a ratio called the equilibrium constant expression. (14)

lead-acid battery. A rechargeable voltaic battery composed of lead and lead dioxide electrodes in a sulfuric acid solution. (13)

Le Châtelier's principle. A principle which states that when stress is applied to a system at equilibrium, the system reacts in such a way so as to counteract the stress. (14)

Lewis dot symbols. The representation of an element by its symbol and its valence electrons as dots. (6)

Lewis structure. The representation of a molecule or ion showing the order and arrangement of the atoms as well as the bonded pairs and unshared electrons of all of the atoms. (6)

limiting reactant. The reactant that produces the least amount of product when that reactant is completely consumed. (8)

liquid. A physical state that has a definite volume but not a definite shape. Liquids take the shape of the bottom of the container. (2)

London forces. The instantaneous dipole-induced dipole forces between molecules caused by an instantaneous imbalance of electrical charge in a molecule. The force is dependent on the size of the molecule. (10)

M

mass. The quantity of matter (usually in grams or kilograms) in a sample. (1)

mass number. The number of nucleons (neutrons and protons) in a nucleus. (3)

matter. Anything that has mass and occupies space. (2)

measurement. The quantity, dimensions, or extent of something, usually in comparison to a specific unit. (1)

melting point. The temperature at which a pure substance changes from the solid to the liquid state. (2)

metal. An element with a comparatively low ionization energy that forms positive ions in compounds. Generally, they are hard, lustrous elements that are ductile and malleable. (4, 5)

main group element. See *representative element.*

metallic solid. A solid made of metals where positive metal ions occupy regular positions in the crystal lattice with the valence electrons moving freely among these positive ions. (10)

metalloid. Elements with properties intermediate between metals and nonmetals. Many of the elements on the metal-nonmetal borderline. (4)

metallurgy. The conversion of metal ores into metals. (Prologue)

metric system. A system of measurement based on multiples of 10. (1)

miscible liquids. Two liquids that mix or dissolve in each other to form a solution. (11)

model. A description or analogy used to help visualize a phenomenon. (5)

molality. A unit of concentration that relates the moles of solute to the mass (kg) of solvent. (11)

molar mass. The atomic mass of an element or the formula weight of a compound expressed in grams. (7)

molar volume.   The volume of one mole of a gas at STP, which is 22.4 L. (9)

molarity.   A unit of concentration that relates moles of solute to volume (in liters) of solution. (11)

mole.   A unit of 6.02×10^{23} atoms, molecules, or formula units. It is the same number of particles as there are atoms in exactly 12 grams of ^{12}C. It also represents the atomic mass of an element, or the formula weight of a compound expressed in grams. (7)

mole ratio.   The ratios of moles from a balanced equation that serve as conversion factors in stoichiometry calculations. (8)

molecular compound.   A compound composed of discrete molecules. (3)

molecular dipole.   The combined effect of all of the bond dipoles in a molecule as determined by the molecular geometry. (6)

molecular equation.   A chemical equation showing all reactants and products as neutral compounds. (11)

molecular formula.   See *formula*.

molecular geometry.   The geometry of a molecule or ion described by the bonded atoms. It does not include the unshared pairs of electrons. (6).

molecular solid.   A solid where the individual molecules in the crystal lattice are held together by London forces, dipole-dipole attractions, or hydrogen bonding. (10)

molecular weight.   The formula weight of a molecular compound. (7)

molecule.   The basic unit of a molecular compound, which is two or more atoms held together by covalent bonds. (3)

monoprotic acid.   An acid that can produce only one H^+ ion per molecule. (12)

N

net ionic equation.   A chemical equation shown in ionic form with spectator ions eliminated. (11)

network solid.   A solid where the atoms are covalently bonded to each other throughout the entire crystal. (10)

neutral.   Pure water or a solution with pH = 7. (12)

neutralization reaction.   A reaction whereby an acid reacts with a base to form a salt and water. The reaction of $H^+(aq)$ with $OH^-(aq)$. (8, 12)

neutron.   A particle in the nucleus with a mass of about 1 amu and no charge. (3)

noble gas.   An element with a full outer s and p subshell. Group VIII in the periodic table. (4, 5)

nonconductor.   A substance that does not conduct electricity. (11)

nonelectrolyte.   A solute whose aqueous solution or molten state does not conduct electricity. (11)

nonmetal.   Elements to the right in the periodic table. These elements generally lack metallic properties. They have relatively high ionization energies. (4, 5)

nonpolar bond.   A covalent bond in which electrons are shared equally. (6)

normal boiling point.   The temperature at which the vapor pressure of a liquid is equal to exactly one atmosphere pressure. (10)

nuclear equation.   A symbolic representation of the changes of a nucleus or nuclei into other nuclei and particles. (15)

nuclear reactor.   A device that can maintain a controlled nuclear fission reaction. Used either for research or generation of electrical power. (15)

nucleons.   The protons and neutrons that make up the nucleus of the atom. (3, 15)

nucleus.   The core of the atom containing neutrons, protons, and most of the mass. (3, 15)

O

octet rule.   A rule which states that atoms of representative elements form bonds so as to have access to eight electrons. (6)

orbital.   A region of space where there is the highest probability of finding a particular electron. Each orbital in a subshell has a capacity of two electrons. (5)

orbital diagram.   The representation of specific orbitals of a subshell as boxes and the electrons as arrows in the boxes. (5)

organic chemistry.   The branch of chemistry that deals with most of the compounds of carbon. (16)

osmosis.   The tendency of a solvent to move through a semipermeable membrane from a region of low concentration to a region of high concentration of solutes. (11)

osmotic pressure.   The pressure needed to counteract the movement of solvent through a semipermeable membrane from a region of low concentration of solute to a region of high concentration. (11)

oxidation.   The loss of electrons as indicated by an increase in oxidation state. (13)

oxidation-reduction reaction.   A chemical reaction involving an exchange of electrons. (13)

oxidation state (number).   The charge on an atom in a compound if all atoms were present as monatomic ions. The electrons in bonds are assigned to the more electronegative atom. (13)

oxidation state method.   A method of balancing oxidation-reduction reactions that focuses on the atoms of the elements undergoing a change in oxidation state. (13)

oxidizing agent. The element, compound, or ion that oxidizes another reactant. It is reduced. (12)

oxyacid. An acid composed of hydrogen and an oxyanion. (4)

oxyanions. Anions composed of oxygen and one other element. (4)

P

partial pressure. The pressure of one component in a mixture of gases. (9)

parts per billion (ppb). A unit of concentration obtained by multiplying the ratio of the mass of solute to the mass of solvent by 10^9 ppb. (11)

parts per million (ppm). A unit of concentration obtained by multiplying the ratio of the mass of solute to the mass of solvent by 10^6 ppm. (11)

Pauli exclusion principle. A rule which states that no two electrons can have the same spin in the same orbital. (5)

percent by mass. The mass of solute expressed as a percent of the mass of solution. (11)

percent composition. The mass of each element expressed per 100 mass units of the compound. (7)

percent yield. The actual yield in grams or moles divided by the theoretical yield in grams or moles and multiplied by 100%. (8)

period. A horizontal row of elements between noble gases in the periodic table. (4)

periodic law. A law which states that the properties of elements are periodic functions of their atomic numbers. (4)

periodic table. An arrangement of elements in order of increasing atomic number. Elements with the same number of outer electrons are arranged in vertical columns. (4, 5)

pH. The negative of the common logarithm of the H_3O^+ concentration. (12)

phase. A homogeneous state (solid, liquid, or gas) with distinct boundaries and uniform properties. (2)

physical change. A change in a substance that does not involve a change in composition but a change in physical state or dimensions. (2)

physical properties. Properties that can be observed without changing a substance into another substance.

physical states. The physical condition of matter—solid, liquid, or gas. (2)

pOH. The negative of the common logarithm of the OH^- concentration. (12)

point of equilibrium. The point at which the forward and reverse reactions in a reversible reaction occur at the same rate so that the concentrations of both reactants and products remain constant. (14)

polar covalent bond. A covalent bond that has a partial separation of charge due to the unequal sharing of electrons. (6)

polyatomic ion. A group of atoms covalently bonded to each other that have a net electrical charge. (3)

polymer. A high-molar-mass substance made from repeating units of an alkene (addition) or from combinations of other molecules (condensation). (16)

polyprotic acid. An acid that can produce more than one H^+ ion per molecule. (12)

positron. A positively charged electron emitted from a radioactive isotope. (15)

potential energy. Energy as a result of position or composition. (2, 14)

precipitate. A solid compound formed in a solution. (8, 11)

precipitation reaction. A type of double-replacement reaction in which an insoluble ionic compound is formed by an exchange of ions in the reactants. (8, 11)

precision. The reproducibility of a measurement as indicated by the number of significant figures expressed. (1)

pressure. The force per unit area. (9)

principal quantum number (n). A number that corresponds to a particular shell occupied by the electrons in an atom. (5)

product. An element or compound in an equation that is formed as a result of a chemical reaction. (8)

proper fraction. A fraction whose numerator is smaller than the denominator and thus has a value less than one. (Appendix A)

property. A particular characteristic or trait of a substance. (2)

protons. A particle in the nucleus with a mass of about 1 amu and a charge of +1. (3)

pure substance. A substance that has a definite composition with definite and unchanging properties (i.e., elements and compounds). (2)

Q

quantized energy level. An energy level with a definite and measurable energy. (5)

R

radiation. Particles or high-energy light rays that are emitted by an atom or a nucleus of an atom. (15)

radioactive decay series. A series of elements formed from the successive emission of alpha and beta particles starting from a long-lived isotope and ending with a stable isotope. (15)

radioactivity. The emission of energy or particles from an unstable nucleus. (15)

rate of reaction. A measure of the increase in concentration of a product or the decrease in concentration of a reactant per unit time. (14)

reactant. An element or compound in an equation that undergoes a chemical reaction. (8)

recrystallization. A laboratory procedure whereby a solid compound is purified by saturating a solution at a high temperature and then forming a precipitate at a lower temperature. (11)

redox reaction. See *oxidation-reduction reaction.*

reducing agent. An element, compound, or ion that reduces another reactant. It is oxidized. (13)

reduction. The gain of electrons as indicated by a decrease in the oxidation state. (13)

reforming. The rearrangement of long-chain hydrocarbons to branched hydrocarbons and/or removal of hydrogen from hydrocarbons. (16)

representative element. Elements whose outer *s* and *p* subshells are filling. The A Group elements in the periodic table. (4, 5)

resonance hybrid. The actual structure of the molecule as implied by the resonance structures. (6)

resonance structure. A Lewis structure showing one of two or more possible Lewis structures. (6)

reversible reaction. A reaction where both a forward reaction (forming products) and a reverse reaction (reforming reactants) can occur. (14)

room temperature. The standard reference temperature for physical state, which is exactly 25°C. (4)

S

salt. An ionic compound formed by the combination of most cations and anions. Also, the compound that results from the cation from a base and the anion of an acid. (4, 8, 12)

salt bridge. An aqueous gel that allows anions to migrate between compartments in a voltaic cell. (13)

saturated hydrocarbon. A hydrocarbon containing only single covalent bonds (an alkane). (16)

saturated solution. A solution containing the maximum amount of dissolved solute at a specific temperature. (11)

scientific method. The method whereby modern scientists explain the behavior of nature with hypotheses and theories, or describe the behavior of nature with laws. (Prologue)

scientific notation. A number expressed with one nonzero digit to the left of the decimal point multiplied by 10 raised to a given power. (1, Appendix C)

semimetal. See *metalloid.*

SI unit. An international system of units of measurement. (1)

significant figure. A digit or number in a measurement that is either reliably known or is estimated. (1)

single-replacement reaction. A chemical reaction whereby one free element substitutes for another in a compound. (8)

solid. A physical state with both a definite shape and a definite volume. (2)

solubility. The maximum amount of a solute that dissolves in a specific amount of solvent at a certain temperature. (11)

soluble compound. A compound that dissolves to an appreciable extent in a solvent. (11)

solute. A substance that dissolves in a solvent. (11)

solution. A homogeneous mixture with one phase. It is composed of a solute dissolved in a solvent. (2, 11)

solvent. A medium, usually a liquid, that disperses a solute to form a solution. (11)

specific gravity. The ratio of the mass of a substance to the mass of an equal volume of water under the same conditions. (2)

specific heat. The amount of heat required to raise the temperature of one gram of a substance one degree Celsius. (2)

spectator ion. An ion that is in an identical state on both sides of an equation. (11)

spectrum. The separate color components of a beam of light. (5)

standard temperature and pressure (STP). The defined standard conditions for a gas, which are exactly 0°C and one atmosphere pressure. (9)

Stock method. A method used to name metal-nonmetal or metal-polyatomic ion compounds where the charge on the metal is indicated by Roman numerals enclosed in parentheses. (4)

stoichiometry. The quantitative relationship between reactants and/or products. (8)

strong acid/base. An acid or base that is completely ionized in aqueous solution. (12)

structural formula. Formulas written so that the order and arrangement of specific atoms are shown. (3, 9, 16)

sublimation. The vaporization of a solid. (10)

subshell. One of the subdivisions (*s, p, d,* or *f*) of an energy level in an atom. The orbitals of each subshell have characteristic shapes. (5)

substance. A form of matter. Usually thought of as either an element or a compound. (2)

substituent. An atom or group of atoms that can substitute for a hydrogen atom in an organic compound. (16)

supersaturated solution. A solution containing more than the maximum amount of solute, indicated by the compound's solubility at that temperature. (11)

surface tension. The forces of attraction between molecules that cause a liquid surface to contract. (10)

symbol. The first one or two letters of an element's English or, in some cases, Latin name. (3)

T

temperature. A measure of the intensity of heat of a substance. It relates to the average kinetic energy of the substance. (1, 9)

theoretical yield. The calculated amount of product that would be obtained if all of a reactant were converted to a certain product. (8)

theory. A hypothesis that withstands the test of time and experiments designed to test the hypothesis. (Prologue)

thermochemical equation. A balanced equation that includes the amount of heat energy. (8)

thermometer. A device that measures temperature. (1)

torr. A unit of gas pressure equivalent to the height of one millimeter of mercury. (9)

total ionic equation. A chemical equation showing all soluble compounds that exist as ions in aqueous solution as separate ions. (11)

transition element. Elements whose outer s and d subshells are filling. The B Group elements in the periodic table. (4, 5)

transmutation. The changing of one element into another by a nuclear reaction. (15)

triple bond. The sharing of three pairs of electrons in a bond between two atoms. (6)

triprotic acid. An acid that can produce three H^+ ions per molecule. (12)

U

ultraviolet light. Light with wavelengths somewhat shorter than those of violet light. (7)

unit. A definite quantity adapted as a standard of measurement. (1)

unit factor. A fractional expression that relates a quantity in a certain unit to "one" of another unit. (1)

unsaturated hydrocarbon. A hydrocarbon that contains at least one double or triple bond. (16)

unsaturated solution. A solution that contains less than the maximum amount of solute, indicated by the compound's solubility at that temperature. (11)

V

valence electron. An outer s and p electron in the atom of a representative element. (6)

valence shell electron pair repulsion theory (VSEPR). A theory that predicts that electron pairs either unshared or in a bond repel each other to the maximum extent. (6)

vapor pressure. See *equilibrium vapor pressure.*

viscosity. A measure of the resistance of a liquid to flow. (10)

volatile. Referring to a liquid or solid with a significant vapor pressure. (11, 16)

voltaic cell. A spontaneous oxidation-reduction reaction that can be used to produce electrical energy. (13)

volume. The space that a certain quantity of matter occupies. (1)

W

wavelength (λ). The distance between two adjacent peaks in a wave. (5)

wave mechanics. A complex mathematical approach to the electrons in an atom that considers the electron as having both a particle and a wave nature. (5)

weak acid/base. An acid or base that is only partially ionized in aqueous solution. (12)

weak electrolyte. A solute whose aqueous solution allows only a limited amount of electrical conduction. (11)

weight. A measure of the attraction of gravity for a sample of matter. (1)

ANSWERS TO PROBLEMS

CHAPTER 1

1–1 (b) 74.212 gal (the most significant figures)

1–2 A device used to produce a measurement may provide a reproducible answer to several significant figures, but if the device itself is inaccurate (such as a ruler with the tip broken off) the measurement is inaccurate.

1–4 (a) three (b) two (c) three (d) one (e) four (f) two (g) two (h) three

1–6 (a) 16.0 (b) 1.01 (c) 0.665 (d) 4890 (e) 87,600 (f) 0.0272 (g) 301

1–8 (a) 0.250 (b) 0.800 (c) 1.67 (d) 1.17

1–10 (a) 188 (b) 12.90 (c) 2300 (d) 48 (e) 0.84

1–12 37.9 qt

1–14 (a) 120 cm^2 (b) 394 ft^2 (c) 2 cm (d) 2.3 in.

1–16 (a) 138 (b) 1200 (c) 28

1–18 (a) 1.57×10^2 (b) 1.57×10^{-1} (c) 3.00×10^{-2} (d) 4.0×10^7 (e) 3.49×10^{-2}
(f) 3.2×10^4 (g) 3.2×10^{10} (h) 7.71×10^{-4} (i) 2.34×10^3

1–19 (a) 4.23×10^5 (b) 4.338×10^2 (c) 2.0×10^{-3} (d) 8.8×10^2 (e) 8×10^{-5}
(f) 8.20×10^7 (g) 7.5×10^{13} (h) 1.06×10^{-6}

1–20 (a) 9×10^7 (b) 8.7×10^7 (c) 8.70×10^7

1–22 (a) 0.000476 (b) 6550 (c) 0.00788 (d) 48,900 (e) 4.75 (f) 0.0000034

1–24 (a) 4.89×10^{-4} (b) 4.56×10^{-5} (c) 7.8×10^3 (d) 5.71×10^{-2} (e) 4.975×10^8
(f) 3.0×10^{-4}

1–26 (a) 1.597×10^{-3} (b) 2.30×10^7 (c) 3.5×10^{-5} (d) 2.0×10^{14}

1–28 (a) 3.1×10^{10} (b) 2×10^9 (c) 4×10^{13} (d) 14 (e) 2.56×10^{-14}

1–30 (a) milliliter (mL)   (b) hectogram (hg)   (c) nanojoule (nJ)
(d) centimeter (cm)   (e) microgram (μg)   (f) decipascal (dPa)

1–32 (a) 720 cm, 7.2 m, 7.2×10^{-3} km
(b) 5.64×10^4 mm, 5640 cm, 0.0564 km
(c) 2.50×10^5 mm, 2.50×10^4 cm, 250 m

1–33 (a) 8.9 g, 8.9×10^{-3} kg (b) 2.57×10^4 mg, 0.0257 kg (c) 1.25×10^6 mg, 1250 g

1–35 (a) 12 = 1 doz (c) 3 ft = 1 yd (e) 10^3 m = 1 km

1–36 (a) $\dfrac{1 \text{ g}}{10^3 \text{ mg}}$ (b) $\dfrac{1 \text{ km}}{10^3 \text{ m}}$ (c) $\dfrac{1 \text{ L}}{100 \text{ cL}}$ (d) $\dfrac{1 \text{ m}}{10^3 \text{ mm}}, \dfrac{1 \text{ km}}{10^3 \text{ m}}$

1–38 (a) $\dfrac{1 \text{ ft}}{12 \text{ in.}}$ (b) $\dfrac{2.54 \text{ cm}}{\text{in.}}$ (c) $\dfrac{5280 \text{ ft}}{\text{mi}}$
(d) $\dfrac{1.06 \text{ qt}}{\text{L}}$ (e) $\dfrac{1 \text{ qt}}{2 \text{ pt}}, \dfrac{1 \text{ L}}{1.06 \text{ qt}}$

1–40 (a) 4.8 mi, 2.6×10^4 ft, 7.8 km   (b) 2380 ft, 725 m, 0.725 km
(c) 1.70 mi, 2740 m, 2.74 km   (d) 4.21 mi, 2.22×10^4 ft, 6780 m

1–41 (a) 27.1 qt, 25.6 L   (b) 170 gal, 630 L   (c) 2.03×10^3 gal, 8.12×10^3 qt

1–43 55.4 kg

1–44 $28.0 \ \cancel{m} \ \times \dfrac{10^2 \ \cancel{cm}}{\cancel{m}} \times \dfrac{1 \ \cancel{in.}}{2.54 \ \cancel{cm}} \times \dfrac{1 \ \cancel{ft}}{12 \ \cancel{in.}} \times \dfrac{1 \ yd}{3 \ \cancel{ft}} = \underline{\underline{30.6 \ yd}}$
(New punter is needed.)

1–45 0.354 L

1–46 6 ft 10 1/2 in. = 82.5 in. $82.5 \ \cancel{in.} \times \dfrac{2.54 \ \cancel{cm}}{\cancel{in.}} \times \dfrac{1 \ m}{10^2 \ \cancel{cm}} = \underline{\underline{2.10 \ m}}$

$212 \ \cancel{lb} \times \dfrac{1 \ kg}{2.20 \ \cancel{lb}} = \underline{\underline{96.4 \ kg}}$

1–47 14.6 gal

1–49 $0.200 \ \cancel{gal} \times \dfrac{4 \ qt}{\cancel{gal}} = 0.800 \ qt$ $0.800 \ \cancel{qt} \times \dfrac{1 \ \cancel{L}}{1.06 \ \cancel{qt}} \times \dfrac{1 \ mL}{10^{-3} \ \cancel{L}} = \underline{\underline{755 \ mL}}$
There is slightly more in a "fifth" than in 750 mL.

1–50 105 km/hr

1–53 $\dfrac{\$0.899}{\cancel{gal}} \times \dfrac{1 \ \cancel{gal}}{4 \ \cancel{qt}} \times \dfrac{1.06 \ \cancel{qt}}{L} = \$0.238/L$ $80.0 \ \cancel{L} \times \dfrac{\$0.238}{\cancel{L}} = \underline{\underline{\$19.04}}$

1–54 $23.59 (551 mi) $12.81 (482 km)

1–55 1690 km

1–58 382 nails

1–59 $28.20

1–60 $350 \ \cancel{km} \times \dfrac{1 \ \cancel{mi}}{1.61 \ \cancel{km}} \times \dfrac{1 \ \cancel{gal}}{24.5 \ \cancel{mi}} \times \dfrac{\$1.22}{\cancel{gal}} = \underline{\underline{\$10.83}}$

1–64 2100 s or 0.58 hr

1–65 572°F

1–66 24°C

1–68 −38°F

1–70 95°C

1–71 (a) −98°C   (b) 22°C   (c) 27°C   (d) −48°C   (e) 600°C

1–72 (a) 320 K   (b) 296 K   (c) 200 K   (d) 261 K   (e) 291 K   (f) 244 K

1–74 Since $t(°C) = t(°F)$, substitute $t(°C)$ for $t(°F)$ and set the two equations equal.

$$[t(°C) \times 1.8] + 32 = \dfrac{t(°C) - 32}{1.8} \qquad t(°C) = \underline{\underline{-40°C}}$$

1–75 (a) 3×10^2 (b) 8.26 g · cm   (c) 5.24 g/mL   (d) 19.1

1–76 (a) $\dfrac{1 \ g}{10^3 \ mg}, \dfrac{1 \ lb}{454 \ g}$ (b) $\dfrac{1.06 \ qt}{L}, \dfrac{2 \ pt}{qt}$
(c) $\dfrac{1 \ km}{10 \ hm}, \dfrac{1 \ mi}{1.61 \ km}$ (d) $\dfrac{1 \ in.}{2.54 \ cm}, \dfrac{1 \ ft}{12 \ in.}$

1–77 $5.34 \times 10^{10} \, \cancel{\text{ng}} \times \dfrac{10^{-9} \, \cancel{\text{g}}}{\cancel{\text{ng}}} \times \dfrac{1 \text{ lb}}{454 \, \cancel{\text{g}}} = \underline{\underline{0.118 \text{ lb}}}$

1–79 $11,100

1–80 $\dfrac{247 \text{ lb}}{82.3 \text{ doz}} = \underline{\underline{3.00 \text{ lb/doz}}}$ $\dfrac{82.3 \text{ doz}}{247 \text{ lb}} = \underline{\underline{0.333 \text{ doz/lb}}}$

1–81 $12.0 \, \cancel{\text{fur}} \times \dfrac{1 \, \cancel{\text{mi}}}{8 \, \cancel{\text{fur}}} \times \dfrac{5280 \, \cancel{\text{ft}}}{\cancel{\text{mi}}} \times \dfrac{12 \, \cancel{\text{in.}}}{\cancel{\text{ft.}}} \times \dfrac{1 \text{ hand}}{4 \, \cancel{\text{in.}}} = \underline{\underline{2.38 \times 10^4 \text{ hands}}}$

1–83 1030 packages, 1.41 years

1–84 5.02 L

1–86 $5.4 \times 10^7 \, ^\circ\text{F}$ $3.0 \times 10^7 \, ^\circ\text{C} + 273 = 3.0 \times 10^7 \text{ K}$

CHAPTER 2

2–1 Gases: nitrogen, oxygen, helium, neon, argon, krypton, and xenon. Solids: gold, silver, copper, platinum, palladium, carbon (diamond and graphite), sulfur (in subsurface deposits)

2–3 (a) compound (b) element (c) element (d) element (e) compound (f) compound

2–5 It was found that water could be decomposed into simpler substances (hydrogen and oxygen).

2–6 Not necessarily. For example, the compound calcium carbonate can be decomposed into two other compounds, calcium oxide and carbon dioxide.

2–7 (c) It has a definite volume but not a definite shape.

2–8 The gaseous state is compressible because the basic particles are very far apart and thus the volume of a gas is mostly empty space.

2–10 (a) physical (b) chemical (c) physical (d) chemical (e) chemical
(f) physical (g) physical (h) chemical (i) physical

2–12 (a) chemical (b) physical (c) physical (c) physical (d) chemical
(e) physical

2–14 No. Both compounds and elements have definite and unchanging properties. A compound can be chemically decomposed, however, but an element cannot.

2–16 Physical property: melts at 660°C
Physical change: melting
Chemical property: burns in oxygen
Chemical change: formation of aluminum oxide

2–17 Original substance: green, solid (physical); can be decomposed (chemical). Substance is a compound. Gas: gas, colorless (physical); can be decomposed (chemical). Substance is a compound since it can be decomposed. Solid: shiny, solid (physical); cannot be decomposed (chemical). Substance is an element since it cannot be decomposed.

2–19 2.60 g/mL

2–21 1064 g/657 mL = $\underline{\underline{1.62 \text{ g/mL}}}$ (carbon tetrachloride)

2–23 1450 g

2–24 670 g

2–26 625 mL

2–27 1.74 g/mL (magnesium)

2–28 0.951 g/mL; 4790 mL. Pumice floats in water but sinks in alcohol.

2–30 2080 g

2–31 mass of liquid = 143.5 − 32.5 = 111.0 g $\quad \dfrac{111.0 \text{ g}}{125 \text{ mL}} = \underline{\underline{0.888 \text{ g/mL}}}$

2–33 Water: 1000 g   Gasoline: 670 g

2–36 One needs a conversion factor between mL (cm³) and ft³.

$$\left(\frac{2.54 \text{ cm}}{\text{in.}}\right)^3 = \frac{16.4 \text{ cm}^3}{\text{in.}^3} = \frac{16.4 \text{ mL}}{\text{in.}^3} \qquad \left(\frac{12 \text{ in.}}{\text{ft}}\right)^3 = \frac{1728 \text{ in.}^3}{\text{ft}^3}$$

$$\frac{1.00 \cancel{\text{ g}}}{\cancel{\text{mL}}} \times \frac{1.00 \text{ lb}}{454 \cancel{\text{ g}}} \times \frac{16.4 \cancel{\text{ mL}}}{\cancel{\text{in.}^3}} \times \frac{1728 \cancel{\text{ in.}^3}}{\text{ft}^3} = \underline{\underline{62.4 \text{ lb/ft}^3}}$$

2–37 2.0×10^5 lb (100 tons)

2–38 Carbon dioxide is a compound composed of carbon and oxygen. It can be prepared from a mixture of carbon and oxygen, but the compound is no longer a mixture of the two elements.

2–39 Ocean water is the least pure because it contains a large amount of dissolved compounds. That is why it is not drinkable and cannot be used for crop irrigation. Drinking water also contains chlorine and some dissolved compounds but not nearly as much as ocean water. Rain water is most pure but still contains some dissolved gases from the air.

2–41 (a) liquid only

2–42 (a) gasoline—homogeneous   (b) dirt—heterogeneous   (c) smog—heterogeneous
(d) alcohol—homogeneous   (e) a new nail—homogeneous
(f) vinegar—homogeneous solution   (g) aerosol spray—heterogeneous
(h) air—homogeneous

2–44 (a) liquid   (b) various solid phases   (c) gas and liquid   (d) liquid   (e) solid
(f) liquid   (g) solid and liquid   (h) liquid and gas   (i) gas

2–46 A mixture of all three would have carbon tetrachloride on the bottom, water in the middle, and kerosene on top. Water and kerosene float on carbon tetrachloride; kerosene floats on water.

2–48 Ice is less dense than water. An ice-water mixture is pure but heterogeneous.

2–49 (a) solution (a solid dissolved in a liquid)   (b) heterogeneous mixture (probably a solid suspended in a liquid such as dirty water)   (c) element   (d) compound   (e) solution (two liquids)

2–51 Mass of mixture = 85 + 942 = 1027 g $\quad \dfrac{85 \cancel{\text{ g}}}{1027 \cancel{\text{ g}}} \times 100\% = \underline{\underline{8.3\% \text{ tin}}}$

2–53 64 kg nickel

2–55 $122 \text{ lb } \cancel{\text{iron}} \times \dfrac{100 \text{ lb duriron}}{86 \text{ lb } \cancel{\text{iron}}} = \underline{\underline{140 \text{ lb duriron}}}$

2–57 (a) exothermic   (b) endothermic   (c) endothermic   (d) exothermic   (e) exothermic

2–58 Gasoline is converted into heat energy when it burns. The heat energy moves the pistons, which is mechanical energy. The mechanical energy turns the alternator, which generates electrical energy. The electrical energy is converted into chemical energy in the battery.

2–60 (a) potential   (b) kinetic   (c) potential (It is stored because of its composition.)
(d) kinetic   (e) kinetic

2–63 Kinetic energy is at a maximum nearest the ground when the swing is moving the fastest. Potential energy is at a maximum when the swing has momentarily stopped at the highest point. Assuming no gain or loss of energy, the total of the two energies is constant.

2–64 0.853 J/g · °C

2–65 $\dfrac{56.6 \text{ cal}}{365 \text{ g} \cdot 5.0°C} = \underline{\underline{0.31 \text{ cal/g} \cdot °C}}$ (gold)

2–67 $°C = \dfrac{\text{cal}}{\text{sp. heat} \times \text{g}} = \dfrac{150 \cancel{\text{ cal}}}{0.092 \dfrac{\cancel{\text{cal}}}{\cancel{\text{g}} \cdot °C} \times 50.0 \cancel{\text{ g}}} = 33°C \text{ rise}$ $t(°C) = 25 + 33 = \underline{\underline{58°C}}$

This compares to a 3.0°C rise in temperature for 50.0 g of water.

2–68 506 J

2–69 58 − 25 = 33°C rise in temperature

$g = \dfrac{\text{cal}}{\text{sp. heat} \cdot °C} = \dfrac{16.0 \cancel{\text{ cal}}}{0.106 \dfrac{\cancel{\text{cal}}}{\text{g} \cdot \cancel{°C}} \times 33 \cancel{°C}} = \underline{\underline{4.6 \text{ g}}}$

2–71 Iron: 9.38°C rise   Gold: 32°C rise   Water: 0.997°C rise

2–73 45°C

2–74 2190 J

2–75 140 g

2–77 Heat lost by metal = heat gained by water

100 g × 68.7°C × sp. heat = 100.0 g × 6.3°C × $\dfrac{1.00 \text{ cal}}{\text{g} \cdot °C}$

specific heat = $\underline{\underline{0.092 \text{ cal/g} \cdot °C}}$ The metal is copper.

2–78 Density of A = 0.86 g/mL; density of B = 0.89 g/mL.
Liquid A floats on liquid B.

2–80 15 g of sugar

2–82 50.0 $\cancel{\text{mL gold}}$ × $\dfrac{19.3 \text{ g}}{\cancel{\text{mL gold}}}$ = 965 g gold

50.0 $\cancel{\text{mL alum.}}$ × $\dfrac{2.70 \text{ g}}{\cancel{\text{mL alum.}}}$ = 135 g alum.

$\dfrac{965 \text{ g gold}}{(965 + 135)\text{g alloy}}$ × 100% = $\underline{\underline{87.7\% \text{ gold}}}$

2–83 specific heat = 0.13 $\dfrac{\text{J}}{\text{g} \cdot °C}$ (gold)   25.0 $\cancel{\text{g gold}}$ × $\dfrac{1 \text{ mL}}{19.3 \cancel{\text{ g gold}}}$ = $\underline{\underline{1.30 \text{ mL}}}$

2–85 When a log burns, most of the compounds formed in the combustion are gases that dissipate into the atmosphere. Only some solid residue (ashes) is left. When zinc and sulfur (both solids) combine, the only product is a solid, so there is no weight change. When iron burns, however, its only product is a solid. It weighs more than the original iron because the iron has combined with oxygen gas from the air.

CHAPTER 3

3–1 (a) B—boron   (b) O—oxygen   (c) Pb—lead   (d) Na—sodium   (e) S—sulfur

3–2 cadmium—Cd, calcium—Ca, californium—Cf, carbon—C, cerium—Ce, cesium—Cs, chlorine—Cl, chromium—Cr, cobalt—Co, copper—Cu, curium—Cm

3–5 (a) barium—Ba (b) neon—Ne (c) cesium—Cs
(d) platinum—Pt (e) manganese—Mn (f) tungsten—W

3–7 (a) B—boron (b) Bi—Bismuth (c) Ge—germanium
(d) U—uranium (e) Co—cobalt (f) Hg—mercury
(g) Be—beryllium (h) As—arsenic

3–8 (b) Br_2 (b) S_8 (f) P_4

3–9 P—phosphorus   O—oxygen   Br—bromine
F—fluorine   S—sulfur   Mg—magnesium

3–10 (e) N_2, diatomic element   (b) CO, diatomic compound

3–11 N—nitrogen, O—oxygen, C—carbon, K—potassium
H—hydrogen, S—sulfur

3–12 (a) six carbons, four hydrogens, two chlorines
(b) two carbons, six hydrogens, one oxygen
(c) one copper, one sulfur, 18 hydrogens, 13 oxygens
(d) nine carbons, eight hydrogens, four oxygens
(e) two aluminums, three sulfurs, 12 oxygens
(f) two nitrogens, eight hydrogens, one carbon, three oxygens

3–13 (a) 12 (b) 9 (c) 33 (d) 21 (e) 17 (f) 14

3–14 (a) 8 (b) 7 (c) 4 (d) 3

3–15 (a) SO_2 (b) CO_2 (c) H_2SO_4 (d) C_2H_2

3–17 Neon would appear as individual spheres that are comparatively widely spaced. Chlorine would appear as molecules with two spheres joined together.

3–19 (a) $Ca(ClO_4)_2$ (b) $(NH_4)_3PO_4$ (c) $FeSO_4$

3–21 (c) S^{2-}

3–23 (d) Li^+

3–25 FeS, Li_2SO_3

3–27 (b) and (e)

3–28 (c)

3–31 (a) 21 p, 21 e, 24 n   (b) 90 p, 90 e, 142 n   (c) 87 p, 87 e, 136 n   (d) 38 p, 38 e, 52 n

3–33

Isotope Name	Isotope Symbol	Atomic No.	Mass No.	p	n	e
(a) silver–108	$^{108}_{47}Ag$	47	108	47	61	47
(b) silicon–28	$^{28}_{14}Si$	14	28	14	14	14
(c) potassium–39	$^{39}_{19}K$	19	39	19	20	19
(d) cerium–140	$^{140}_{58}Ce$	58	140	58	82	58
(e) iron–56	$^{56}_{26}Fe$	26	56	26	30	26
(f) tin–110	$^{110}_{50}Sn$	50	110	50	60	50
(g) iodine–118	$^{118}_{53}I$	53	118	53	65	53
(h) mercury–196	$^{196}_{80}Hg$	80	196	80	116	80

3-35 (a) K^+ : 19 p, 18 e (b) Br^- : 35 p, 36 e (c) S^{2-} : 16 p, 18 e
(d) NO_2^- : 7 + 16 = 23 p, 24 e (e) Al^{3+} : 13 p, 10 e
(f) NH_4^+ : 7 + 4 = 11 p, 10 e

3-37 (a) Re: at. no. 75, at. mass 186.2
(b) Co: at. no. 27, at. mass 58.9332
(c) Br: at. no. 35, at. mass 79.904
(d) Si: at. no. 14, at. mass 28.086

3-38 copper (Cu)

3-40 O—at. no. 8, mass no. 16
N—at. no. 7, mass no. 14
Si—at. no. 14, mass no. 28
Ca—at. no. 20, mass no. 40

3-41 $5.81 \times 12.00 = 69.7$ amu. The element is Ga.

3-42 $3.33 \times 12.00 = 40.0$ amu. The element is Ca.

3-43 79.9 amu

3-44 ^{28}Si $0.9221 \times 27.98 = 25.80$
^{29}Si $0.0470 \times 28.98 = 1.362$
^{30}Si $0.0309 \times 29.97 = \underline{0.926}$
$\phantom{^{30}Si   0.0309 \times 29.97 = }\underline{27.088} = \underline{\underline{27.09 \text{ amu}}}$

3-46 Let X = decimal fraction of ^{35}Cl and Y = decimal fraction of ^{37}Cl. Since there are two isotopes present,
X + Y = 1, Y = 1 − X.

$(X \times 35) + (Y \times 37) = 35.5$
$(X \times 35) + [(1 - X) \times 37] = 35.5$
$X = 0.75 \ (\underline{\underline{75\%}} \ ^{35}Cl)$ $Y = 0.25 \ (\underline{\underline{25\%}} \ ^{37}Cl)$

3-49 (a) $^{90}_{38}Sr^{2+}$ (b) $^{52}_{24}Cr^{3+}$ (c) $^{79}_{34}Se^{2-}$ (d) $^{14}_{7}N$ (e) $^{139}_{57}La^{3+}$

3-51 Let x = mass no. of I and y = mass no. of Tl.
Then (1) $x + y = 340$ or $x = 340 - y$.

Also, (2) $x = \frac{2}{3}y - 10$. Substituting for x from (1) and solving for y:

$y = 210$ amu (Tl)
$x = 340 - 210 = \underline{\underline{130 \text{ amu}}}$ (I)

3-53 121.8 (Sb). Sb^{3+} has 51 − 3 = 48 electrons.

^{121}Sb has 121 − 51 = 70 neutrons. ^{123}Sb has 123 − 51 = 72 neutrons.

$^{121}Sb = 57.9\%$ due to neutrons. ^{123}Sb 58.3% due to neutrons.

3-54 118 neutrons and 78 protons [platinum (Pt)]
78 − 2 = 76 electrons for Pt^{2+}

3-56 Mass of other atoms = 16 (oxygen)
$NO^+ = (7 + 8) - 1 = 14$ electrons

3-58 (a) H: $\frac{1}{12} \times 8.00 = 0.667$ (b) N: $\frac{14}{12} \times 8.00 = 9.33$

(c) Na: 15.3 (d) Ca: 26.7

3-59 $43.3/10 = 4.33$ times as heavy as ^{12}C.
4.33×12.0 amu = 52.0 amu. The element is Cr.

REVIEW TEST ON CHAPTERS 1–3

Multiple Choice

1. (a) **2.** (b) **3.** (c) 4. (c) **5.** (a) **6.** (d) **7.** (a) **8.** (a) **9.** (d) **10.** (b) **11.** (d) **12.** (a) **13.** (b)
14. (c) **15.** (c) **16.** (c) **17.** (d) **18.** (c) **19.** (a) **20.** (a) **21.** (c) **22.** (d) **23.** (a) **24.** (a) **25.** (b)

Problems

1. (a) 174 **(b)** 0.00232 **(c)** 18,400 **2.** 17.17 cm, **3.** 69 in.^{2} **4.** 13 g/mL **5.** 2.0×10^7 cm^2
6. 0.014 mL **7.** 1.2×10^{12} **8.** 2×10^{-9} **9.** 1.6×10^{13} **10.** 74.4 in. **11.** 41 L
12. 317 km/hr **13.** 1.00×10^2 g/cm **14.** 4.3 g/cm^3 = 4.3 g/mL **15.** 60.0 mL
16. 102 ml **17.** 44°C **18.** 0.288 J/(g · °C) **19.** 121.8 amu, Sb, antimony **20.** 114.6 amu

CHAPTER 4

4–1 An active metal reacts with water and air. A noble metal is not affected by air, water, or most acids.

4–2 32

4–3 (c) I_2 and (g) Br_2

4–5 (a) Fe and (d) La—transition elements
     (b) Te, (f) H, and (g) In—representative elements
     (e) Xe—noble gas
     (c) Pm—inner transition element

4–7 (b) Ti, (e) Pd, and (g) Ag

4–8 The most common physical state is a solid. Metals are more common than nonmetals.

4–10 (a) Ne (d) Cl (f) N

4–12 (a) Ru (b) Sn (c) Hf (h) W

4–13 (d) Te (f) B

4–14 (a) Ar (b) Hg (c) N_2 (d) Be (e) Po

4–16 Element 118 is in group VIII (noble gas).

4–18 (a) lithium fluoride   (b) barium telluride
      (c) strontium nitride   (d) barium hydride
      (e) aluminum chloride

4–20 (a) Rb_2Se (b) SrH_2 (c) RaO (d) Al_4C_3 (e) BeF_2

4–22 (a) bismuth(V) oxide   (b) tin(II) sulfide
      (c) tin(IV) sulfide   (d) copper(I) telluride
      (e) titanium(IV) carbide

4–24 (a) Cu_2S (b) V_2O_3 (c) AuBr (d) Ni_2C (e) CrO_3

4–26 In 4–22: (a) Bi_2O_5, (c) SnS_2, and (e) TiC
      In 4–24: (e) CrO_3

4–28 (c) ClO_3^-

4–30 ammonium, NH_4^+

4–31 (b) permanganate (MnO_4^-)   (c) perchlorate (ClO_4^-)
      (e) phosphate (PO_4^{3-})   (f) oxalate ($C_2O_4^{2-}$)

4–32 (a) chromium(II) sulfate   (b) aluminum sulfite
     (c) iron(II) cyanide   (d) rubidium hydrogen carbonate
     (e) ammonium carbonate   (f) ammonium nitrate
     (g) bismuth(III) hydroxide

4–34 (a) $Mg(MnO_4)_2$ (b) $Co(CN)_2$ (c) $Sr(OH)_2$
     (d) Tl_2SO_3 (e) $In(HSO_3)_3$ (f) $Fe_2(C_2O_4)_3$
     (g) $(NH_4)_2Cr_2O_7$ (h) $Hg_2(C_2H_3O_2)_2$

4–36

	HSO_3^-	Te^{2-}	PO_4^{3-}
NH_4^+	NH_4HSO_3 ammonium bisulfite	$(NH_4)_2Te$ ammonium telluride	$(NH_4)_3PO_4$ ammonium phosphate
Co^{2+}	$Co(HSO_3)_2$ cobalt(II) bisulfite	*CoTe* cobalt(II) telluride	$Co_3(PO_4)_2$ cobalt(II) phosphate
Al^{3+}	$Al(HSO_3)_3$ aluminum bisulfite	Al_2Te_3 aluminum telluride	$AlPO_4$ *aluminum phosphate*

4–38 (a) sodium chloride   (b) sodium hydrogen carbonate
     (c) calcium carbonate   (d) sodium hydroxide
     (e) sodium nitrate   (f) ammonium chloride
     (g) aluminum oxide   (h) calcium hydroxide
     (i) potassium hydroxide

4–39 (a) Ca_2XeO_6 (b) K_4XeO_6 (c) $Al_4(XeO_6)_3$

4–40 (a) phosphonium fluoride   (b) potassium hypobromite
     (c) cobalt(III) periodate   (d) calcium silicate (actual name is calcium metasilicate)
     (e) aluminum phosphite   (f) chromium(II) molybdate

4–41 (a) Si (b) I (c) H (d) Kr (e) H (f) As

4–43 (a) carbon disulfide   (b) boron trifluoride
     (c) tetraphosphorus decoxide (d) dibromine trioxide
     (e) sulfur trioxide   (f) dichlorine oxide or dichlorine monoxide
     (g) phosphorus pentachloride (h) sulfur hexafluoride

4–45 (a) P_4O_6 (b) CCl_4 (c) IF_3 (d) Cl_2O_7 (e) SF_6 (f) XeO_2

4–47 (a) hydrochloric acid   (b) nitric acid
     (c) hypochlorous acid   (d) permanganic acid
     (e) periodic acid   (f) hydrobromic acid

4–48 (a) HCN (b) H_2S (c) $HClO_2$ (d) H_2CO_3 (e) HI (f) $HC_2H_3O_2$

4–50 (a) hypobromous acid   (b) iodic acid   (c) phosphorous acid
     (d) molybdic acid   (e) perxenic acid

4–51 H_3AsO_3, arsenious acid
     H_3AsO_4, arsenic acid

4–52 A = F, X = Br
     BrF_5 (Br is more metallic.)

4–54 Gas = N_2, and Al forms only +3. Thus AlN, (N^{3-}).
     Ti_3N_2; titanium(II) nitride

4–56 Co^{2+} and Br^-
     $CoBr_2$; cobalt(II) bromide

4–57 Metal = Mg, nonmetal = S
MgH_2; magnesium hydride
H_2S; hydrogen sulfide or hydrosulfuric acid

4–59 NiI_2—nickel(II) iodide H_3PO_4—phosphoric acid
$Sr(ClO_3)_2$—strontium chlorate H_2Te—hydrogen telluride or hydrotelluric acid
As_2O_3—diarsenic trioxide
Sb_2O_3—antimony(III) oxide SnC_2O_4—tin(II) oxalate

4–61 tin(II) hypochlorite—$Sn(ClO)_2$ chromic acid—H_2CrO_4
xenon hexafluoride—XeF_6 barium nitride—Ba_3N_2
hydrofluoric acid—HF   iron(III) carbide—Fe_4C_3
lithium phosphate—Li_3PO_4

CHAPTER 5

5–1 Ultraviolet light has shorter wavelengths but higher energy than visible light. Because of this high energy, ultraviolet light can damage living cells in tissues, thus causing a burn.

5–2 Since these two shells are close in energy, transitions of electrons from these two levels to the $n = 1$ shell have similar energy. Thus the wavelengths of light from the two transitions are very close together.

5–3 Since these two shells are comparatively far apart in energy, transitions from these two levels to the $n = 1$ shell have comparatively different energies. Thus the wavelengths of light from the two transitions are quite different. (The $n = 3$ to $n = 1$ transition has a shorter wavelength than the $n = 2$ to the $n = 1$ transition.)

5–4 $2n^2 = 2(4)^2 = 32$

5–6 $2d$, $1p$, $3f$

5–7 s (2), p (6), d (10), f (14)

5–10 The $n = 5$ shell holds 50 electrons. The s, p, d, and f subshells hold a total of 32 electrons. Therefore, the g subshell must hold $50 - 32 = 18$ electrons.

5–11 (a) three   (b) five   (c) none, because there is no $2d$ subshell
(d) one   (e) seven

5–12 $3s$(one), $3p$(three), $3d$(five), for a total of nine.

5–13 If there are 18 electrons in the $5g$ subshell, there would be 9 orbitals (2 electrons in each).

5–14 A $4p$ orbital is shaped roughly like a two-sided baseball bat with two "lobes" lying along one of the three axes. This shape represents the region of highest probability of finding the $4p$ electrons.

5–16 The $3s$ orbital is spherical in shape. There is an equal probability of finding the electron regardless of the orientation from the nucleus. (In fact, the probability lies in three concentric spheres with the highest probability in the sphere farthest from the nucleus.) The highest probability of finding the electron lies farther from the nucleus in the $3s$ than in the $2s$.

5–17 (a) $6s$ (b) $5p$ (c) $4p$ (d) $4d$

5–19 $4s$, $4p$, $5s$, $4d$, $5p$, $6s$, $4f$

5–20 (a) Mg: $1s^2\ 2s^2\ 2p^6\ 3s^2$
(b) Ge: $1s^2\ 2s^2\ 2p^6\ 3s^2\ 3p^6\ 4s^2\ 3d^{10}\ 4p^2$
(c) Pd: $1s^2\ 2s^2\ 2p^6\ 3s^2\ 3p^6\ 4s^2\ 3d^{10}\ 4p^6\ 5s^2\ 4d^8$
(d) Si: $1s^2\ 2s^2\ 2p^6\ 3s^2\ 3p^2$

5–22 (a) S: [Ne] $3s^2\,3p^4$ (b) Zn: [Ar] $4s^2\,3d^{10}$
(c) Pu: [Rn] $7s^2\,6d^1\,5f^5$ (d) I: [Kr] $5s^2\,4d^{10}\,5p^5$

5–24 (a) This is excluded by Hund's rule since electrons are not shown in separate orbitals of the same subshell with parallel spins.
(b) This is correct.
(c) This is excluded by the Aufbau principle because the $2s$ subshell fills before the $2p$.
(d) This is excluded by the Pauli exclusion principle since the two electrons in the $2s$ orbital cannot have the same spin.

5–25

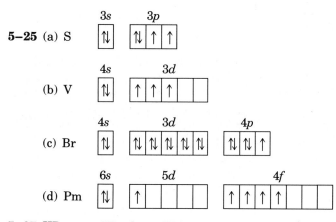

(a) S

(b) V

(c) Br

(d) Pm

5–27 IIB, none; VB, three; VIA, two; VIIA, one; $[ns^1(n-1)d^5]$, six; Pm, five

5–29 IVA and VIA

5–31 (a) In   (b) Y   (c) Ce   (d) Ar

5–33 (a) $3p$ (b) $5p$ (c) $5d$ (d) $5s$

5–35 Both have three valence electrons. The outer electron in IIIA is in a p subshell; in IIIB it is in a d subshell.

5–37 (a) F   (b) Ga   (c) Ba   (d) Gd   (e) Cu

5–39 (a) VIIA   (b) IIIA   (c) IIA   (e) IB

5–40 VA [noble gas] $ns^2\,np^3$ VB [noble gas] $ns^2(n-1)d^3$

5–42 (b) and (e) belong to Group IVA

5–44 (a) IVA   (b) VIII   (c) IB   (d) Pr and Pa

5–45 (a) [NG] ns^2 (b) [NG] $ns^2(n-1)d^{10}$
(c) [NG] $ns^2\,np^4$ or [NG] $ns^2(n-1)d^{10}np^4$ (d) [NG] $ns^2(n-1)d^2$

5–46 Helium does not have a filled p subshell. (There is no $1p$ subshell.)

5–48 [Ne] $3s^2\,3p^6$

5–50 The theoretical order of filling is $6d$, $7p$, $8s$, $5g$. The $6d$ is completed at element #112. The $7p$ and $8s$ fill at element #120. Thus element #121 would theoretically begin the filling of the $5g$ subshell. This assumes filling in the order indicated by Figure 5–14.

5–52 $7s^2\,6d^{10}\,5f^{14}\,7p^6$ Total 32

5–53 $8s^2\,5g^{18}\,7d^{10}\,6f^{14}\,8p^6$ Total 50

5–54 (a) transition   (b) representative
(c) noble gas   (d) inner transition

5–56 Element number 118 (under Rn)

5–58 (a) As (b) Ru (c) Ba (d) I

5–60 Cr, 117 pm; Nb, 134 pm

5–62 (a) V (b) Cl (c) Mg (d) Fe (e) B

5–63 In 558 kJ/mol; Ge, 762 kJ/mol

5–64 Te, 869 kJ/mol; Br, 1140 kJ/mol

5–66 (a) Cs^+ easiest, (b) Rb^{2+} hardest

5–68 (d) I

5–70 (d) Mg^{2+} (e) K^+ (c) S (a) Mg (b) S^{2-} (f) Se^{2-}

5–72 The outer electron in Hf is in a shell higher in energy than Zr. This alone would make Hf a larger atom. However, in between Zr and Hf lie several subshells, including the long $4f$ subshell (Ce through Lu). The filling of these subshells, especially the $4f$, causes a gradual contraction that offsets the higher shell for Hf.

5–73 C^+ (1086 kJ/mol), C^{2+} (3439 kJ/mol), C^{3+} (8059 kJ/mol), C^{4+} (14,282 kJ/mol), C^{5+} (52,112 kJ/mol). Notice that the energy required to form C^+ (1086 kJ) is about twice the energy required to form Ga^+ (579 kJ), which is a metal. Thus it is apparent that metal cations form easier than nonmetal cations.

5–74 (a) B (b) Ar (c) K, Cr, and Cu (d) Ga (e) Hf

5–76 (a) B (b) Br (c) Sb (d) Si

5–77 (a) Nb and Rh (b) Bi (c) Hg (d) Ge

5–78 P

5–80 Z = Sn, X = Zr

5–84 (b) Po. Although its outer subshell can accommodate 2 electrons, Po is a metal.
(c) Cl. Its outer subshell can accommodate only one electron.
(d) H. Its one subshell can accommodate only one additional electron.

5–85 Se and N are nonmetals that can accommodate two and three electrons respectively in their outer subshells. Therefore, Se^{2-} and N^{3-} are likely. F^{2-} and Ar^- are not likely because both would have one electron in the next higher shell; Be^- because metals do not easily form anions.

5–86

s^1	s^2												p^4
1										p^1	p^2	p^3	2
3	4									5	6	7	8
9	10	d^1	d^2	d^3	d^4	d^5	d^6	d^7	d^8	11	12	13	14
15	16	17	18	19	20	21	22	23	24	25	26	27	28
29	30	31	32	33	34	35	36	37	38	39	40	41	42
43	44	45*											

*46–50 would be in a 4f subshell

(a) 6 in second period, 14 in fourth
(b) third period, #14; fourth period, #28
(c) first inner transition element is #46 (assuming an order of filling like that on Earth)
(d) Elements #11 and #17 are most likely to be metals.
(e) Element #12 would have the larger radius in both cases.
(f) Element #7 would have the higher ionization energy in all three cases.
(g) The ions that would be reasonable are 16^{2+} (metal cation), 13^- (nonmetal anion), 15^+ (metal cation), and 1^- (nonmetal anion). The 9^{2+} ion is not likely because the second electron would come from a filled

inner subshell. The 7^+ ion is not likely because it would be a nonmetal cation. The 17^{4+} ion is not likely because #17 has only three electrons in the outer shell.

CHAPTER 6

6–1 (a) Ca· (b) ·S̈b· (c) ·Sn· (d) ·Ï:

(e) :N̈e: (f) ·B̈i· (g) ·V̈IA:

6–2 (a) Group IIIA (b) Group VA (c) Group IIA

6–3 The electrons from filled inner subshells are not involved in bonding.

6–5 (a) Sr and S (c) H and K (d) Al and F

6–7 (b) S^- (c) Cr^{2+} (e) In^+ (f) Pb^{2+} (h) Tl^{3+}

6–8 They have the noble gas configuration of He, which requires only two electrons

6–9 (a) K^+ (b) ·Ö:$^-$ (c) :Ï:$^-$ (d) :P̈:$^{3-}$ (e) Ba^+ (f) ·Ẍe:$^+$ (g) Sc^{3+}

6–11 (b) O^- (e) Ba^+ (f) Xe^+

6–12 (a) Mg^{2+} (b) Ga^{3+} (pseudo-noble gas configuration) (c) Br^-
(d) S^{2-} (e) P^{3-}

6–15 Se^{2-}, Br^-, Rb^+, Sr^{2+}, Y^{3+}

6–16 Tl^{3+} has a full $5d$ subshell. This is a pseudo-noble gas configuration.

6–17 Cs_2S, Cs_3N
$BaBr_2$, BaS, Ba_3N_2
$InBr_3$, In_2S_3, InN

6–18 (a) CaI_2 (b) CaO (c) Ca_3N_2 (d) $CaTe$ (e) CaF_2

6–20 (a) Cr^{3+} (b) Fe^{3+} (c) Mn^{2+} (d) Co^{2+} (e) Ni^{2+} (f) V^{3+}

6–22 (a) H_2S (b) GeH_4 (c) ClF (d) Cl_2O (e) NCl_3 (f) CBr_4

6–24 (b) Cl_3 (d) NBr_4 (e) H_3O

6–25 (a) $SO_4{}^{2-}$ (b) $IO_3{}^-$ (c) $SeO_4{}^{2-}$ (d) $H_2PO_4{}^-$ (e) $C_2{}^{2-}$

6–27 (a)
$$\begin{array}{c} \text{H} \quad \text{H} \\ | \quad\ | \\ \text{H}-\text{C}-\text{C}-\text{H} \\ | \quad\ | \\ \text{H} \quad \text{H} \end{array}$$
(b) $H-\ddot{O}-\ddot{O}-H$
(c)
$$:\ddot{F}-\ddot{N}-\ddot{F}: \\ \quad\quad\ :\ddot{F}:$$

(d) :C̈l—S̈—C̈l:
(e)
$$\begin{array}{c} \text{H} \quad\quad \text{H} \\ | \quad\quad\ | \\ \text{H}-\text{C}-\ddot{\text{O}}-\text{C}-\text{H} \\ | \quad\quad\ | \\ \text{H} \quad\quad \text{H} \end{array}$$
$$\begin{array}{c} \text{H} \quad \text{H} \\ | \quad\ | \\ \text{H}-\text{C}-\text{C}-\ddot{\text{O}}-\text{H} \\ | \quad\ | \\ \text{H} \quad \text{H} \end{array}$$

6–29 (a) :C≡O:
(b)
$$\begin{array}{c} \cdot\dot{O} \qquad \dot{O}\cdot \\ \diagdown\ \diagup \\ \text{S} \\ | \\ :\ddot{O}: \end{array}$$
(c) $K^+[:C{\equiv}N:]^-$
(d)
$$\begin{array}{c} \text{H}-\ddot{\text{O}}: \\ | \\ :\text{S}-\ddot{\text{O}}: \\ | \\ :\ddot{\text{O}}-\text{H} \end{array}$$

6–30 (a) valence electrons $= 4 + 6 = 10$, $N = 2$
$6N + 2 = 14$ $14 - 10 = 4$ (two double bonds or one triple bond)

(b) valence electrons = $(3 \times 6) + 6 = 24$, N = 4
   $6N + 2 = 26$ $26 - 24 = 2$ (one double bond)

(c) valence electrons for CN^-: $4 + 5 + 1 = 10$, N = 2
   $6N + 2 = 14$ $14 - 10 = 4$ (two double bonds or one triple bond)

(d) valence electrons = $(3 \times 6) + 6 + (2 \times 1) = 26$, N = 4
   $6N + 2 = 26$ $26 - 26 = 0$ (no multiple bonds)

6–31 (a) N_2O: valence electrons = $(2 \times 5) + 6 = 16$, N = 3
   $6N + 2 = 20$ $20 - 16 = 4$ (two double bonds or one triple bond)

(b) $Ca(NO_2)_2$: For the NO_2^- ion, valence electrons = $5 + (2 \times 6) + 1 = 18$, N = 3
   $6N + 2 = 20$ $20 - 18 = 2$ (one double bond)

(c) $AsCl_3$: valence electrons $5 + (3 \times 7) = 26$. N = 4
   $6N + 2 = 26$ $26 - 26 = 0$ (no multiple bonds)

(d) H_2S: valence electrons = $(2 \times 1) + 6 = 8$, N = 1
   $6N + 2 = 8$ $8 - 8 = 0$ (no multiple bonds)

(e) CH_2Cl_2: valence electrons $4 + (2 \times 1) + (2 \times 7) = 20$, N = 3
   $6N + 2 = 20$ $20 - 20 = 0$ (no multiple bonds)

(f) NH_4^+: valence electrons = $(4 \times 1) + 5 - 1 = 8$, N = 1
   $6N + 2 = 8$ $8 - 8 = 0$ (no multiple bonds)

6–32 (a) $\ddot{N}=N=\ddot{O}$ (b) Ca^{2+} [structure] (f) [structure]

(c) [structure] (d) [structure] (e) [structure]

6–33 (a) [structure] (b) [structure]

(c) [structure]

(d) [structure] (e) [structure] (f) $:N\equiv O:^+$

6–37 (a) [structures with resonance]

(b) [structures with resonance]

(c) [structure] (only one structure)

6–39

$$\left[\begin{array}{c}H\\|\\H-B-C\overset{\displaystyle \ddot{\cdot}\ddot{O}\cdot}{\underset{\displaystyle \ddot{O}\cdot}{\diagdown\!\!\diagdown}}\\|\\H\end{array}\right]^{2-}\longleftrightarrow\left[\begin{array}{c}H\\|\\H-B-C\overset{\displaystyle \ddot{O}}{\underset{\displaystyle \ddot{O}\cdot}{\diagup\diagup\diagdown}}\\|\\H\end{array}\right]^{2-}$$

6–40 A resonance hybrid is the actual structure of a molecule or ion that is implied by the various resonance structures. Each resonance structure contributes a portion of the actual structure. For example, the two structures shown in Problem 6–39 imply that both C—O bonds have properties that are halfway between those of a single and a double bond.

6–41 :O≡C—Ö: ↔ :Ö—C≡O:

The two resonance structures imply that both C—O bonds are halfway between a single and a triple bond, which is a double bond. This is the same as the one structure shown below.

$$\ddot{O}\!=\!C\!=\!\ddot{O}$$

6–42 Cs, Ba, Be, B, C, Cl, O, F

6–43 (a) $\overset{\delta-}{N}\underset{\leftarrow}{-}\overset{\delta+}{H}$ (b) $\overset{\delta+}{B}\underset{\to}{-}\overset{\delta-}{H}$ (c) $\overset{\delta+}{Li}\underset{\to}{-}\overset{\delta-}{H}$ (d) $\overset{\delta-}{F}\underset{\leftarrow}{-}\overset{\delta+}{O}$ (e) $\overset{\delta-}{O}\underset{\leftarrow}{-}\overset{\delta+}{Cl}$

(f) $\overset{\delta-}{S}\underset{\leftarrow}{-}\overset{\delta+}{Se}$ (g) $\overset{\delta-}{C}\underset{\leftarrow}{-}\overset{\delta+}{B}$ (h) $\overset{\delta+}{Cs}\underset{\to}{-}\overset{\delta-}{N}$ (i) C—S (very low polarity)

6–44 (i) nonpolar, (b) = (f), (d) = (e) = (g), (a), (c), (h)

6–45 Only (d) Al—F is predicted to be ionic on this basis since the electronegativity difference is 2.5.

6–47 (b) I–I

6–48 (a) 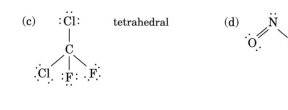 angular (b) :S=C=S: linear

(c) :Cl:
|
C
/ | \
:Cl: :F: :F:
tetrahedral

(d) N double bond O, Cl: angular

(e) Cl
/ \
:O: :O:
angular

6–50 If the molecule has a symmetrical geometry and all bonds are the same, the bond dipoles cancel and the molecule is nonpolar.

6–51 (a) polar (b) nonpolar (c) polar (d) polar (e) polar

6–53 :Ö:
|
S
/ | \
:O: :Cl: Cl: The molecule is polar.

6–54 Since the H—S bond is much less polar than the H—O bond, the resultant molecular dipole is much less. The H_2S molecule is less polar than the H_2O molecule.

6–56 The CHF_3 molecule is more polar than the $CHCl_3$ molecule. The C—F bond is more polar than the C–Cl bond, which means that the resultant molecular dipole is larger for CHF_3.

6–58

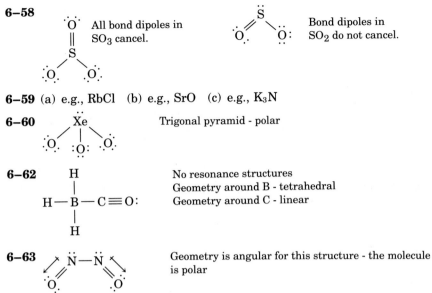

All bond dipoles in SO₃ cancel.

Bond dipoles in SO₂ do not cancel.

6–59 (a) e.g., RbCl (b) e.g., SrO (c) e.g., K₃N

6–60   Xe   Trigonal pyramid - polar

6–62
No resonance structures
Geometry around B - tetrahedral
Geometry around C - linear

H—B—C≡O:

6–63
Geometry is angular for this structure - the molecule is polar

Other resonance structures:

:Ö—N≡N—Ö: ⟷ :O≡N—N̈—Ö: ⟷ :Ö—N̈—N≡O:

6–65 K₃N—potassium nitride
KN₃ = K⁺ N₃⁻
Resonance structures:

N̈=N=N̈ ⟷ :N̈—N≡N: ⟷ :N≡N—N̈:

In all resonance structures, the geometry around N is linear.

6–66 :F̈—N̈=N̈—F̈:

6–68 Oxygen difluoride, OF₂. Lewis structure:

Dioxygen difluoride, O₂F₂. Lewis structure:

Oxygen is less electronegative (more metallic) than fluorine and so is named first. Both molecules are angular; thus they are polar.

6–70 (1) (a) 17 (b) 31 (c) 51₃ (d) 97 (e) 7(13) (f) 10(13)₂ (g)67₂ (h) 3₂6
(2) On Zerk, six electrons fill the outer s and p orbitals to make a noble gas configuration. Therefore, we have a "sextet" rule on Zerk.

(a) 1 – 7̈: (b) 3⁺1⁻(ionic) (c) 1—5—1

(d) 9⁺ :7̈:⁻ (ionic) (e) :7̈—1̈3̈: (f) 10²⁺(:1̈3̈:⁻)₂ (ionic)

(g) :7̈—6̈—7̈: (h) (3⁺)₂:6̈:²⁻ (ionic)

(3) $4N + 2$

REVIEW TEST ON CHAPTERS 4–6

Multiple Choice

1. (e) **2.** (c) **3.** (d) **4.** (b) **5.** (b) **6.** (c) **7.** (d) **8.** (a) **9.** (d) **10.** (c) **11.** (c) **12.** (b) **13.** (c) **14.** (a)
15. (a) **16.** (a) **17.** (b) **18.** (b) **19.** (b) **20.** (a) **21.** (c) **22.** (d)
23. (c) **24.** (c) **25.** (a) **26.** (c) **27.** (c) **28.** (d)

Problems

1. Aluminum (Al): (a) Group IIIA and a representative element   (b) [Ne]$3s^2 3p^1$ (c) solid   (d) metal   (e) +3
(f) neon   (g) Al_2S_3, $AlBr_3$, AlN   (h) aluminum sulfide, aluminum bromide, aluminum nitride

2. Nitrogen (N): (a) Group VA and a representative element   (b) $1s^2 2s^2 2p^3$ (c) three   (d) nonmetal   (e) gas
(f) N_2 (g) :N≡N: (h) −3   (i) neon   (j) oxygen—positive, boron—negative   (k) Mg_3N_2, Li_3N, NF_3
(l) magnesium nitride, lithium nitride, nitrogen trifluoride

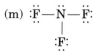

3.

Ionic Form	Name	Lewis Structure
(a) $Mg^{2+}SO_4{}^{2-}$	magnesium sulfate	Mg^{2+} ...
(b) no ions	nitrous acid	$H-\ddot{O}-\ddot{N}=\ddot{O}$
(c) $Li^+NO_3{}^-$	litium nitrate	Li^+ ...
(d) $2(Co^{3+})3(CO_3{}^{2-})$	cobalt(III) carbonate	$2Co^{3+}$ 3 ...
(e) no ions	dichlorine monoxide	$:\ddot{Cl}-\ddot{O}-\ddot{Cl}:$

4.

Formula	Ionic Form	Lewis Structure
(a) N_2O_3	no ions	...

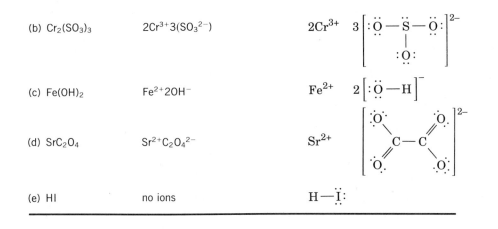

(b) $Cr_2(SO_3)_3$ $2Cr^{3+} 3(SO_3{}^{2-})$

(c) $Fe(OH)_2$ $Fe^{2+} 2OH^-$

(d) SrC_2O_4 $Sr^{2+} C_2O_4{}^{2-}$

(e) HI   no ions

CHAPTER 7

7–1 0.609 lb of pennies

7–3 $145 \text{ g Au} \times \dfrac{108 \text{ g Ag}}{197 \text{ g Au}} = \underline{\underline{79.5 \text{ g Ag}}}$

7–4 94.2 lb C

7–6 71.4 g Cu

7–8 $25.0 \text{ g C} \times \dfrac{x \text{ g}}{12.0 \text{ g C}} = 33.3 \text{ g} \quad x = 16.0 \text{ g (0)}$

The compound is CO.

7–10 40.1 lb S

7–11 $6.02 \times 10^{23} \text{ units} \times \dfrac{1 \text{ sec}}{2 \text{ units}} + \dfrac{1 \text{ min}}{60 \text{ sec}} \times \dfrac{1 \text{ hr}}{60 \text{ min}} \times \dfrac{1 \text{ day}}{24 \text{ hr}} = \dfrac{1 \text{ year}}{365 \text{ day}} =$

$\underline{\underline{9.54 \times 10^{15} \text{ years}}}$ (9.54 quadrillion)

It would take 5.5 billion people $\underline{\underline{1.7 \times 10^6 \text{ years}}}$ (1.7 million).

7–13 6.02×10^{26} (if mass in kg); 6.02×10^{20} (if mass in mg)

7–14 (a) 0.468 mol P, 2.82×10^{23} atoms P
(b) 150 g Rb, 1.05×10^{24} atoms Rb
(c) 27.0 g Al, 1.00 mol Al
(d) Ge, 5.00 mol Ge
(e) 1.66×10^{-24} mol Ti, 7.96×10^{-23} g Ti

7–16 (a) 63.5 g Cu   (b) 16 g S   (c) 40.1 g Ca

7–18 (a) 1.93×10^{25} atoms S   (b) 6.02×10^{23} atoms S
(c) 1.20×10^{24} atoms O

7–20 $50.0 \text{ g Al} \times \dfrac{1 \text{ mol Al}}{27.0 \text{ g Al}} = 1.85 \text{ mol Al}$

$50.0 \text{ g Fe} \times \dfrac{1 \text{ mol Fe}}{55.8 \text{ g Fe}} = 0.896 \text{ mol Fe}$

There are more moles of atoms (more atoms) in 50.0 g of Al.

7-21 $20.0 \text{ g Ni} \times \dfrac{1 \text{ mol Ni}}{58.7 \text{ g Ni}} = 0.341 \text{ mol Ni}$

$2.85 \times 10^{23} \text{ atoms} \times \dfrac{1 \text{ mol Ni}}{6.02 \times 10^{23} \text{ atoms}} = 0.473 \text{ mol Ni}$

The 2.85×10^{23} atoms of Ni contain more atoms than 20.0 g.

7-23 $1.40 \times 10^{21} = 2.33 \times 10^{-3}$ moles
$0.251 \text{ g}/2.33 \times 10^{-3} \text{ mol} = 107.8 \text{ g/mol}$
The element is silver.

7-25 (a) 107 amu   (b) 80.1 amu   (c) 108 amu
(d) 98.1 amu   (e) 106 amu   (f) 60.0 amu
(g) 460 amu

7-27 $Cr_2(SO_4)_3$ $(2 \times 52.0) + (3 \times 32.1) + (12 \times 16.0) = 392$ amu

7-29 (a) 189 g H_2O, 6.32×10^{24} molecules
(b) 0.339 g BF_3, 5.00×10^{-3} mol
(c) 0.218 mol SO_2, 1.31×10^{23} molecules
(d) 0.0209 g K_2SO_4, 7.22×10^{19} formula units
(e) 599 g SO_3, 7.48 mol
(f) 7.61×10^{-3} mol $N(CH_3)_3$, 4.59×10^{21} molecules

7-31 $21.5 \text{ g}/0.0684 \text{ mol} = 314 \text{ g/mol}$

7-33 161 g/mol

7-35 5.10 mol C, 15.3 mol H, 2.55 mol O   Total = 23.0 mol of atoms
61.2 g C, 15.4 g H, 40.8 g O   Total mass = 117.4 g

7-36 0.135 mol Ca $(ClO_3)_2$, 0.135 mol Ca, 0.270 mol Cl, 0.810 mol O
Total = 1.215 mol of atoms

7-38 $1.50 \text{ mol } H_2SO_4 \times \dfrac{2 \text{ mol H}}{\text{mol } H_2SO_4} \times \dfrac{1.01 \text{ g H}}{\text{mol H}} = \underline{\underline{3.03 \text{ g H}}}$
48.2 g S, 72.0 g O

7-40 $1.20 \times 10^{22} \text{ molecules} \times \dfrac{1 \text{ mol } O_2}{6.02 \times 10^{23} \text{ molecules}} = \underline{\underline{0.0199 \text{ mol } O_2}}$

$0.0199 \text{ mol } O_2 \times \dfrac{2 \text{ mol O atoms}}{\text{mol } O_2} = 0.0398 \text{ mol O atoms}$

$0.0199 \text{ mol } O_2 \times \dfrac{32.0 \text{ g } O_2}{\text{mol } O_2} = 0.637 \text{ g } O_2$

The mass is the same.

7-42 Total mass of compound $= 1.375 + 3.935 = 5.310$ g
25.89% N, 74.11% O

7-43 46.7% Si, 53.3% O

7-45 (a) C_2H_6O 52.1% C, 13.1% H, 34.8% O
(b) C_3H_6 85.5% C, 14.5% H
(c) C_9H_{18} 85.7% C, 14.3% H
(b) and (c) are actually the same. The difference comes from rounding off.
(d) Na_2SO_4 32.4% Na, 22.5% S, 45.1% O
(e) $(NH_4)_2CO_3$ 9.1% N, 8.41% H, 12.5% C, 49.9% O

7-47 12.1% Na, 11.3% B, 71.4% O, 5.3% H

7–49 $C_7H_5SNO_3$ Formula weight = $(7 \times 12.0) + (5 \times 1.01) + 32.1 + 14.0 + (3 \times 16.0) = 183$ amu

%C: $\dfrac{84.0 \text{ amu}}{183 \text{ amu}} \times 100\% = \underline{\underline{45.9\%}}$ %H: $\dfrac{5.05 \text{ amu}}{183 \text{ amu}} \times 100\% = \underline{\underline{2.76\%}}$

%S: $\dfrac{32.1 \text{ amu}}{183 \text{ amu}} \times 100\% = \underline{\underline{17.5\%}}$ %N: $\dfrac{14.0 \text{ amu}}{183 \text{ amu}} \times 100\% = \underline{\underline{7.65\%}}$

%O: $100\% - (45.9 + 2.8 + 17.5 + 7.6) = 26.2\%$

7–51 $125 \text{ g Na}_2\text{C}_2\text{O}_4 \times \dfrac{1 \text{ mol Na}_2\text{C}_2\text{O}_4}{134 \text{ g Na}_2\text{C}_2\text{O}_4} \times \dfrac{2 \text{ mol C}}{\text{mol Na}_2\text{C}_2\text{O}_4} \times \dfrac{12.0 \text{ g C}}{\text{mol C}} = \underline{\underline{22.4 \text{ g C}}}$

7–52 4.73 lb P

7–54 1.40×10^3 lb Fe

7–55 (a) N_2O_4 and (d) $H_2C_2O_4$

7–56 (a) FeS   (b) SrI_2 (c) $KClO_3$ (d) I_2O_5
(e) $Fe_2O_{2.66} = Fe_{6/3}O_{8/3} = Fe_6O_8 = Fe_3O_4$ (f) $C_3H_5Cl_3$

7–58 N_2O_3

7–60 KO_2

7–62 MgC_2O_4

7–63 CH_2Cl

7–65 $N_2H_8SO_3$

7–66 $C_{8/3}H_{8/3}O = C_8H_8O_3$

7–68 $C_3H_4Cl_4$ (empirical formula) $C_9H_{12}Cl_{12}$ (molecular formula)

7–70 $B_2C_2H_6O_4$

7–71 Empirical formula = $KC_2NH_3O_2$
Empirical mass = 112 g/emp. unit

$\dfrac{224 \text{ g/mol}}{112 \text{ g/emp. unit}} = 2$ emp. units/mol

$K_2C_4N_2H_6O_4$ (molecular formula)

7–73 I_6C_6

7–75 7.5×10^{-10} mol of pennies

7–76 0.442 g N equals 0.0316 mol N. Therefore, 1.420 g of M also equals 0.0316 mol M.
1.420 g/0.0316 mol = 44.9 g/mol. The metal is scandium (Sc).

7–79 2.77×10^{-3} mol P_4

$2.77 \times 10^{-3} \text{ mol P}_4 \times \dfrac{4 \text{ mol P}}{\text{mol P}_4} = \underline{\underline{0.0111 \text{ mol P atoms}}}$

7–80 100 mol H_2 = 202 g H_2
Therefore, 100 H atoms < 100 H_2 molecules < 100 g H_2 < 100 mol H_2

7–82 Since there is 120 g/mol of compound, there is 64 g/mol of sulfur, which is two moles of atoms. The formula is, therefore, FeS_2.

7–83 (a) $2Na^+$ and $S_4O_6^{2-}$ (b) 27.9 g S   (c) NaS_2O_3 (d) 270 g/mol
(e) 0.0926 mol; 5.57×10^{22} formula units   (f) 35.6% oxygen

7–85 $\dfrac{2N}{2N + x\,O} = 0.368$ $\dfrac{28.0}{28.0 + 16.0\,x} = 0.368$ $x = 3$

N_2O_3—dinitrogen trioxide

7–87 $C_{12}H_4Cl_4O_2$

7–89 Empirical formula: $CrCl_3O_{12}$
Actual formula: $Cr(ClO_4)_3$
Name: chromium(III) perchlorate

7–91 $1.20\,\text{g CO}_2 \times \dfrac{1\,\text{mol CO}_2}{44.0\,\text{g CO}_2} \times \dfrac{1\,\text{mol C}}{\text{mol CO}_2} = 0.0273\,\text{mol C}$

$0.489\,\text{g H}_2\text{O} \times \dfrac{1\,\text{mol H}_2\text{O}}{18.0\,\text{g H}_2\text{O}} \times \dfrac{2\,\text{mol H}}{\text{mol H}_2\text{O}} = 0.0543\,\text{mol H}$

C: $\dfrac{0.0273}{0.0273} = 1.0$ H: $\dfrac{0.0543}{0.0273} = 2.0$

$\underline{\underline{CH_2}}$

CHAPTER 8

8–1 (a) $Cl_2(g)$ (b) $C(s)$ (c) $K_2SO_4(s)$ (d) $H_2O(l)$ (e) $P_4(s)$ (f) $H_2(g)$ (g) $Br_2(l)$ (h) $NaBr(s)$
(i) $S_8(s)$ (j) $Na(s)$ (k) $Hg(l)$ (l) $CO_2(g)$

8–2 (a) $CaCO_3 \longrightarrow CaO + CO_2$
(b) $4Na + O_2 \longrightarrow 2Na_2O$
(c) $H_2SO_4 + 2NaOH \longrightarrow Na_2SO_4 + 2H_2O$
(d) $2H_2O_2 \longrightarrow 2H_2O + O_2$
(e) $2Al + 2H_3PO_4 \longrightarrow 2AlPO_4 + 3H_2$
(f) $Ca(OH)_2 + 2HCl \longrightarrow CaCl_2 + 2H_2O$
(g) $3Mg + N_2 \longrightarrow Mg_3N_2$
(h) $2C_2H_6 + 7O_2 \longrightarrow 4CO_2 + 6H_2O$
(i) $2B_4H_{10} + 11O_2 \longrightarrow 4B_2O_3 + 10H_2O$
(j) $SF_6 + 2SO_3 \longrightarrow 3O_2SF_2$
(k) $CS_2 + 3O_2 \longrightarrow CO_2 + 2SO_2$

8–4 (a) $Mg_3N_2 + 6H_2O \longrightarrow 3Mg(OH)_2 + 2NH_3$
(b) $2H_2S + O_2 \longrightarrow 2S + 2H_2O$
(c) $Si_2H_6 + 8H_2O \longrightarrow 2Si(OH)_4 + 7H_2$
(d) $C_2H_6 + 5Cl_2 \longrightarrow C_2HCl_5 + 5HCl$
(e) $NH_3 + 2Cl_2 \longrightarrow NHCl_2 + 2HCl$
(f) $PBr_3 + 3H_2O \longrightarrow 3HBr + H_3PO_3$

8–6 (a) $2Na(s) + 2H_2O(l) \longrightarrow H_2(g) + 2NaOH(aq)$
(b) $2KClO_3(s) \longrightarrow 2KCl(s) + 3O_2(g)$
(c) $NaCl(aq) + AgNO_3(aq) \longrightarrow AgCl(s) + NaNO_3(aq)$
(d) $2H_3PO_4(aq) + 3Ca(OH)_2(aq) \longrightarrow Ca_3(PO_4)_2(s) + 6H_2O(l)$

8–8 *Problem 8–2*
(a) decomposition   (g) combination
(b) combustion/combination (h) combustion
(c) double–replacement   (i) combustion
(d) decomposition   (j) combination
(e) single–replacement   (k) combustion
(f) double–replacement

Problem 8–6
     (a) single–replacement   (c) double–replacement
     (b) decomposition   (d) double–replacement

8–10 (a) $2K(s) + Cl_2(g) \longrightarrow 2KCl(s)$
      (b) $Ca(s) + 2H_2O(l) \longrightarrow Ca(OH)_2(aq) + H_2(g)$
      (c) $2C_6H_6(l) + 15O_2(g) \longrightarrow 12CO_2(g) + 6H_2O(l)$
      (d) $Na_2S(aq) + Cu(NO_3)_2(aq) \longrightarrow CuS(s) + 2NaNO_3(aq)$
      (e) $2Au_2O_3(s) \longrightarrow 4Au(s) + 3O_2(g)$

8–12 (a) $\dfrac{1 \text{ mol } H_2}{1 \text{ mol Mg}}$ (b) $\dfrac{2 \text{ mol HCl}}{1 \text{ mol Mg}}$ (c) $\dfrac{1 \text{ mol } H_2}{2 \text{ mol HCl}}$ (d) $\dfrac{2 \text{ mol HCl}}{1 \text{ mol } MgCl_2}$

8–13 (a) $\dfrac{2 \text{ mol } C_4H_{10}}{8 \text{ mol } CO_2}$ (b) $\dfrac{2 \text{ mol } C_4H_{10}}{13 \text{ mol } O_2}$ (c) $\dfrac{13 \text{ mol } O_2}{8 \text{ mol } CO_2}$ (d) $\dfrac{10 \text{ mol } H_2O}{13 \text{ mol } O_2}$

8–15 (a) 3.33 mol Al_2O_3, 3.33 mol $AlCl_3$, 10.0 mol NO, 20.0 mol H_2O
      (b) 1.00 mol Al_2O_3, 1.00 mol $AlCl_3$, 3.00 mol NO, 6.00 mol H_2O

8–17 (a) 15.0 mol O_2 and 10.0 mol CH_4 react
      (b) 6.67 mol HCN and 20.0 mol H_2O produced

8–19 (a) 126 g SiF_4 and (b) 43.7 g H_2O produced
      (c) 73.0 g SiO_2 reacts

8–20 (a) mol $H_2O \longrightarrow$ mol H_2

$$0.400 \text{ mol } H_2O \times \frac{2 \text{ mol } H_2}{2 \text{ mol } H_2O} = 0.400 \text{ mol } H_2$$

      (b) g $O_2 \longrightarrow$ mol $O_2 \longrightarrow$ mol H_2O

$$0.640 \text{ g } O_2 \times \frac{1 \text{ mol } O_2}{32.0 \text{ g } O_2} \times \frac{2 \text{ mol } H_2O}{1 \text{ mol } O_2} = 0.0400 \text{ mol } H_2O$$

      (c) g $O_2 \longrightarrow$ mol $O_2 \longrightarrow$ mol H_2

$$0.032 \text{ g } O_2 \times \frac{1 \text{ mol } O_2}{32.0 \text{ g } O_2} \times \frac{2 \text{ mol } H_2}{1 \text{ mol } O_2} = 0.0020 \text{ mol } H_2$$

      (d) g $H_2 \longrightarrow$ mol $H_2 \longrightarrow$ mol $H_2O \longrightarrow$ g H_2O

$$0.400 \text{ g } H_2 \times \frac{1 \text{ mol } H_2}{2.02 \text{ g } H_2} \times \frac{2 \text{ mol } H_2O}{2 \text{ mol } H_2} \times \frac{18.0 \text{ g } H_2O}{\text{mol } H_2O} = \underline{3.56 \text{ g } H_2O}$$

8–21 (a) 1.35 mol CO_2, 1.80 mol H_2O, and 2.25 mol O_2
      (b) 4.80 g H_2O (c) 1.10 g C_3H_8 (d) 44.0 g C_3H_8
      (e) molecules $O_2 \longrightarrow$ mol $O_2 \longrightarrow$ mol $CO_2 \longrightarrow$ g CO_2

$$1.20 \times 10^{23} \text{ molecules} \times \frac{1 \text{ mol } O_2}{6.02 \times 10^{23} \text{ molecules}} \times \frac{3 \text{ mol } CO_2}{5 \text{ mol } O_2} \times \frac{44.0 \text{ g } CO_2}{\text{mol } CO_2} = \underline{5.26 \text{ g } CO_2}$$

      (f) 0.0997 mol H_2O

8–23 47.2 g N_2

8–25 0.730 g HCl

8–27 mol $FeS_2 \longrightarrow$ mol $H_2S \longrightarrow$ molecules H_2S

$$0.520 \text{ mol } FeS_2 \times \frac{1 \text{ mol } H_2S}{1 \text{ mol } FeS_2} \times \frac{6.02 \times 10^{23} \text{ molecules}}{\text{mol } H_2S} = \underline{3.13 \times 10^{23} \text{ molecules}}$$

8–28 16.9 kg HNO_3

8–30 2.30×10^3 g (2.30 kg) C_2H_5OH

8–31 8140 g (8.14 kg) CO

8–32 (a) CuO limiting reactant producing 1.00 mol N_2
   (b) Stoichiometric mixture producing 1.00 mol N_2
   (c) NH_3 limiting reactant producing 0.500 mol N_2
   (d) CuO limiting reactant producing 0.209 mol N_2
   (e) NH_3 limiting reactant producing 1.75 mol N_2

8–33 (a) $3.00 \text{ mol CuO} \times \dfrac{2 \text{ mol } NH_3}{3 \text{ mol CuO}} = 2.00 \text{ mol } NH_3$ used

   $3.00 - 2.00 = 1.00 \text{ mol } NH_3$ in excess
   (c) 1.50 mol CuO in excess

8–36 H_2SO_4 is the limiting reactant, and the yield of H_2 is <u>1.00 mole.</u>

   $1.00 \text{ mol } H_2SO_4 \times \dfrac{2 \text{ mol Al}}{3 \text{ mol } H_2SO_4} = 0.667 \text{ mol of Al used}$

   $0.800 - 0.667 = \underline{\underline{0.133 \text{ mol Al remaining}}}$

8–37 $3.44 \text{ mol } C_5H_6 \times \dfrac{10 \text{ mol } CO_2}{2 \text{ mol } C_5H_6} = 17.7 \text{ mol } CO_2$

   $20.6 \text{ mol } O_2 \times \dfrac{10 \text{ mol } CO_2}{13 \text{ mol } O_2} = 15.8 \text{ mol } CO_2$

   Since O_2 is the limiting reactant:

   $15.8 \text{ mol } CO_2 \times \dfrac{44.0 \text{ g } CO_2}{\text{mol } CO_2} = \underline{\underline{695 \text{ g } CO_2}}$

8–39 The limiting reactant is NH_3 and the yield of N_2 is <u>0.750 mol.</u>

8–40 $20.0 \text{ g } AgNO_3 \times \dfrac{1 \text{ mol } AgNO_3}{170 \text{ g } AgNO_3} \times \dfrac{2 \text{ mol AgCl}}{2 \text{ mol } AgNO_3} = 0.118 \text{ mol AgCl}$

   $10.0 \text{ g } CaCl_2 \times \dfrac{1 \text{ mol } CaCl_2}{111 \text{ g } CaCl_2} \times \dfrac{2 \text{ mol AgCl}}{1 \text{ mol } CaCl_2} = 0.180 \text{ mol AgCl}$

   Since $AgNO_3$ produces the least AgCl, it is the limiting reactant.

   $0.118 \text{ mol AgCl} \times \dfrac{143 \text{ g AgCl}}{\text{mol AgCl}} = \underline{\underline{16.9 \text{ g AgCl}}}$

   Convert moles of AgCl (the limiting reactant) to grams of $CaCl_2$ used.

   $0.118 \text{ mol AgCl} \times \dfrac{1 \text{ mol } CaCl_2}{2 \text{ mol AgCl}} \times \dfrac{111 \text{ g } CaCl_2}{\text{mol } CaCl_2} = 6.55 \text{ g } CaCl_2$ used

   $10.0 \text{ g} - 6.55 \text{ g} = \underline{\underline{3.5 \text{ g } CaCl_2 \text{ remaining}}}$

8–42 Products: 4.72 g S, 2.94 g NO, 4.72 g S

   Reactants remaining: 3.8 g HNO_3

8–43 30.0 g SO_3 theoretical yield, and 70.7% yield

8–46 86.4% yield

8–47 If 86.4% is converted to CO_2, the remainder (13.6%) is converted to CO. Thus $0.136 \times 57.0 \text{ g} = 7.75 \text{ g}$ of C_8H_{18} is converted to CO. Notice that 1 mole of C_8H_{18} forms 8 moles of CO (because of the eight carbons in C_8H_{18}). Thus

   $7.75 \text{ g } C_8H_{18} \times \dfrac{1 \text{ mol } C_8H_{18}}{114 \text{ g } C_8H_{18}} \times \dfrac{8 \text{ mol CO}}{1 \text{ mol } C_8H_{18}} \times \dfrac{28.0 \text{ g CO}}{\text{mol CO}} = \underline{\underline{15.2 \text{ g CO}}}$

8–48 Theoretical yield $\times 0.700 = 250 \text{ g}$ (actual yield)
   Theoretical yield $= 250 \text{ g}/0.700 = 357 \text{ g } N_2$

$g\ N_2 \rightarrow mol\ N_2 \rightarrow mol\ H_2 \rightarrow g\ H_2$

$$357\ g\ N_2 \times \frac{1\ mol\ N_2}{28.0\ g\ N_2} \times \frac{4\ mol\ H_2}{1\ mol\ N_2} \times \frac{2.02\ g\ H_2}{mol\ H_2} = \underline{103\ g\ H_2}$$

8–51 $2Mg(s) + O_2(g) \rightarrow 2MgO(s) + 1204\ kJ$

$2Mg(s) + O_2(g) \rightarrow 2MgO(s)$ $\Delta H = -1204\ kJ$

8–53 $CaCO_3(s) + 176\ kJ \rightarrow CaO(s) + CO_2(g)$

$CaCO_3(s) \rightarrow CaO(s) + CO_2(g)$ $\Delta H = 176\ kJ$

8–54 $1.00\ g\ C_8H_{18} \times \dfrac{1\ mol\ C_8H_{18}}{114\ g\ C_8H_{18}} \times \dfrac{5480\ kJ}{mol\ C_8H_{18}} = \underline{48.1\ kJ}$ (gasoline)

$1.00\ g\ CH_4 \times \dfrac{1\ mol\ CH_4}{16.0\ g\ CH_4} \times \dfrac{890\ kJ}{mol\ CH_4} = \underline{55.6\ kJ}$ (natural gas)

8–56 $kJ \rightarrow mol\ Al \rightarrow g\ Al$

$$35.8\ kJ \times \frac{2\ mol\ Al}{850\ kJ} \times \frac{27.0\ g\ Al}{mol\ Al} = \underline{\underline{2.27\ g\ Al}}$$

8–57 $69.7\ g\ C_6H_{12}O_6$

8–60 $125\ g\ Fe_2O_3 \times \dfrac{1\ mol\ Fe_2O_3}{160\ g\ Fe_2O_3} \times \dfrac{2\ mol\ Fe_3O_4}{3\ mol\ Fe_2O_3} \times \dfrac{3\ mol\ FeO}{1\ mol\ Fe_3O_4} \times \dfrac{1\ mol\ Fe}{1\ mol\ FeO} \times \dfrac{55.8\ g\ Fe}{mol\ Fe} = \underline{\underline{87.2\ g\ Fe}}$

8–61 $2KClO_3 \rightarrow 2KCl + 3O_2$

Find the mass of $KClO_3$ needed to produce $12.0\ g\ O_2$.

$$12.0\ g\ O_2 \times \frac{1\ mol\ O_2}{32.0\ g\ O_2} \times \frac{2\ mol\ KClO_3}{3\ mol\ O_2} \times \frac{123\ g\ KClO_3}{mol\ KClO_3} = 30.8\ g\ KClO_3$$

$$percent\ purity = \frac{30.8\ g}{50.0\ g} \times 100\% = \underline{\underline{61.6\%}}$$

8–62 $4.49\%\ FeS_2$

8–64 NH_3 is the limiting reactant; $141\ g\ NO$ (theoretical yield); $\underline{28.4\%\ yield}$

8–66 H_2O is the limiting reactant producing $11.0\ g$ of hydrate.

8–67 Molecular formula $= CH_2Cl_2$

(a) $CH_4(g) + 2Cl_2(g) \rightarrow CH_2Cl_2(l) + 2HCl(g)$

(b) Cl_2 is the limiting reactant producing $9.01\ g\ CH_2Cl_2$.

REVIEW TEST ON CHAPTERS 4, 7, AND 8

Multiple Choice

1. (c) **2.** (d) **3.** (c) **4.** (d) **5.** (c) **6.** (b) **7.** (d) **8.** (b) **9.** (e) **10.** (d) **11.** (d) **12.** (d) **13.** (c) **14.** (a)
15. (d) **16.** (a) **17.** (b) **18.** (c) **19.** (a) **20.** (c) **21.** (d) **22.** (b) **23.** (e) **24.** (d) **25.** (a) **26.** (d) **27.** (e)
28. (d) **29.** (c) **30.** (c)

Problems

1. $1.50\ mol\ K_2SO_4$ has a mass of $261\ g$ and contains 9.03×10^{23} formula units.
It contains $6.00\ mol\ O$ atoms, $117\ g\ K$, and $1.81 \times 10^{24}\ K$ atoms

2. $C_4H_{10}N_2$ has a molar mass of $86.1\ g/mol$, is $32.5\%\ N$, with an empirical formula of C_2H_5N.

3. Fe_3O_4

4. $C_2H_{14}B_{10}$

5. (a) $6K(s) + Al_2Cl_6 \rightarrow 6KCl(s) + 2Al$
   (b) $2C_2H_6 + 15O_2(g) \rightarrow 12CO_2(g) + 6H_2O$
   (c) $Cl_2O_3 + H_2O \rightarrow 2HClO_2$
   (d) $3K_2S + 2H_3PO_4 \rightarrow 3H_2S(g) + 2K_3PO_4$
   (e) $B_4H_{10} + 12H_2O \rightarrow 4B(OH)_3 + 11H_2$

6. (a) 0.0233 mol SbF_3 (b) 0.826 mol $SbCl_3$
   (c) 1.10×10^3 g CCl_4 (d) 4.75×10^{25} molecules of CCl_4

CHAPTER 9

9–1 The molecules of water are closely packed together and thus offer much more resistance. The molecules in a gas are dispersed into what is mostly empty space.

9–3 Since a gas is mostly empty space, more molecules can be added. In a liquid, the space is mostly occupied by the molecules so no more can be added.

9–5 Gas molecules are in rapid but random motion. When gas molecules collide with a light dust particle suspended in the air, they impart a random motion to the particle.

9–6 1260 mi/hr

9–7 $SF_6(146 \text{ amu}) < SO_2(64.1 \text{ amu}) < N_2O(44.0 \text{ amu}) \cong CO_2(44.0 \text{ amu}) < N_2(28.0 \text{ amu}) < H_2(2.02 \text{ amu})$

9–8 $\text{rate}_{N_2} = 1.19 \text{ rate}_{Ar}$

9–11 127 g/mol

9–12 $\dfrac{r_{(235)}}{r_{(238)}} = \sqrt{\dfrac{352 \text{ g/mol}}{349 \text{ g/mol}}}$ $r_{(235)} = 1.004 r_{(238)}$ (about 0.4% faster)

9–13 When the pressure is high, the gas molecules are forced close together. In a highly compressed gas, the molecules can occupy an appreciable part of the total volume. When the temperature is low, molecules have a lower average velocity. If there is some attraction, they can momentarily stick together when moving slowly.

9–14 (a) 2.17 atm   (b) 0.0266 torr   (c) 9560 torr
   (d) 0.0558 atm   (e) 3.68 lb/in.^{2} (f) 11 kPa

9–15 (a) 768 torr   (b) 2.54×10^4 atm   (c) 8.40 lb/in.^{2} (d) 19 torr

9–17 0.0102 atm

9–19 Assume a column of Hg has a cross–section of 1 cm^2 and is 76.0 cm high.
   weight of Hg = 76.0 cm $\times$ 1 cm^2 $\times$ 13.6 g/cm^3 = 1030 g

If water is substituted, 1030 g of water in the column is required.
   height $\times$ 1 cm^2 $\times$ 1.00 g/cm^3 = 1030 g   height = 1030 cm
   $1030 \text{ cm} \times \dfrac{1 \text{ in.}}{2.54 \text{ cm}} \times \dfrac{1 \text{ ft}}{12 \text{ in.}} = \underline{\underline{33.8 \text{ ft}}}$

If a well is 40 ft deep, the water cannot be raised in one stage by suction since 33.8 ft is the theoretical maximum height that is supported by the atmosphere.

9–20 10.2 L

9–22 978 mL

9–23 67.9 torr

9–26 $V_{final}(V_f) = 15\ V_{initial}(V_i)$

$$\frac{V_f}{V_i} = \frac{P_i}{P_f}; \frac{15\ V_i}{V_i} = \frac{0.950\ atm}{P_f} \qquad P_f = 0.950\ atm \times 15 = \underline{\underline{14.2\ atm}}$$

9–27

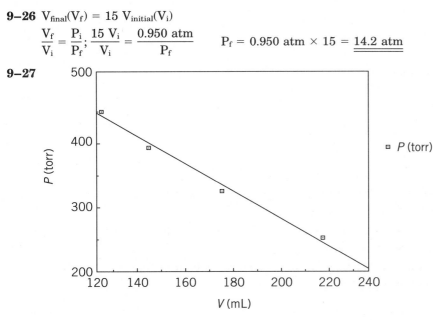

$$k_{av} = 5.61 \times 10^4\ mL \cdot torr$$

9–28 1.94 L

9–30 77°C

9–32 2.60×10^4 L

9–33 341 K (68°C)

9–35 2.94 atm

9–36 191 K − 273 = −82°C

9–38 596 K (323°C)

9–39 30.2 lb/in.^{2}

9–42 1.70 L

9–43 Let X = total moles needed in the expanded balloon

$$188\ L \times \frac{X}{8.40\ mol} = 275\ L \qquad X = 12.3\ mol$$

$12.3 - 8.4 = \underline{3.9\ moles\ must\ be\ added.}$

9–45 $n_2 = 2.50 \times 10^{-3}\ mol \times \dfrac{164\ \cancel{mL}}{75.0\ \cancel{mL}} = 5.47 \times 10^{-3}\ mol$

$(5.47 \times 10^{-3}) - (2.50 \times 10^{-3}) = 2.97 \times 10^{-3}\ mol\ added$

$2.97 \times 10^{-3}\ \cancel{mol\ N_2} \times \dfrac{28.0\ g\ N_2}{\cancel{mol\ N_2}} = \underline{\underline{0.0832\ g\ N_2}}$

9–47 (a) and (d)

9–49 Exp. 1, T increases   Exp. 2, V decreases
Exp. 3, P increases   Exp. 4, T increases

9–51 1.24 atm

9–52 76.8 L

9–54 88.5 K = −185°C

9–57 258 K (−15°C)

9–58 409 mL

9–59 98°C

9–61 13.0 g

9–62 589 torr

9–64 3.35 g

9–65 0.0148 g

9–66 $n = 1.04 \times 10^6$ mol of gas in the balloon.
9200 lb He or 66,000 lb air
Lifting power with He = 66,000 − 9200 = <u>57,000 lb</u>
Lifting power with H_2 = 66,000 − 5000 = <u>61,000 lb</u>
Helium is a noncombustible gas whereas hydrogen forms an explosive mixture with O_2.

9–68 762 torr

9–70 6.8 torr

9–73 $P(N_2)$ = 756 torr; $P(O_2)$ = 84.0 torr; $P(SO_2)$ = 210 torr

9–74 730 torr

9–75 fraction N_2 = mol N_2/total mol = 0.250/(0.250 + 0.427) = 0.369
$P(N_2)$ = 1.02 atm; $P(CO_2)$ = 1.73 atm

9–76 $P(N_2)$ = 300 torr $P(O_2)$ = 85 torr $\times \dfrac{4.00 \; L}{2.00 \; L}$ = 170 torr

$P(CO_2)$ = 225 torr P_T = 695 torr

9–77 P_A = 0.550 atm; P_B = 0.300 atm

9–78 $n_{total} = \dfrac{6.25 \; g \; N_2}{28.0 \; g/mol \; N_2} + \dfrac{12.6 \; g \; CO_2}{44.0 \; g/mol \; CO_2}$ = 0.510 mol

$P = \dfrac{nRT}{V} = \dfrac{0.510 \; mol \times 0.0821 \; \dfrac{L \cdot atm}{K \cdot mol} \times 306 \; K}{0.825 \; L}$ = <u>15.5 atm</u>

7–79 7.64 L

9–81 112 L

9–83 1.39×10^{-3} g

9–84 1.24 g/L

9–86 34.0 g/mol

9–88 1.00 L $\times \dfrac{273 \; K}{298 \; K} \times \dfrac{1.20 \; atm}{1.00 \; atm}$ = 1.10 L (STP)

$\dfrac{3.60 \; g}{1.10 \; L}$ = <u>3.27 g/L (STP)</u>

9–89 Find moles of N_2 in 1 L at 500 torr and 22°C using the ideal gas law.
n = 0.272 mol N_2; mass = 0.762 g N_2
density = 0.762 g/L (500 torr and 22°C)

9–91 25.8 L CO_2 (STP)

9–92 vol. O_2 → mol O_2 → mol Mg → g Mg

$$5.80 \text{ L O}_2 \times \frac{1 \text{ mol O}_2}{22.4 \text{ L O}_2} \times \frac{2 \text{ mol Mg}}{1 \text{ mol O}_2} \times \frac{24.3 \text{ g Mg}}{\text{mol Mg}} = \underline{\underline{12.6 \text{ g Mg}}}$$

9–94 3.47 L

9–96 (a) g C_4H_{10} → mol C_4H_{10} → mol CO_2 → vol CO_2

$$85.0 \text{ g C}_4\text{H}_{10} \times \frac{1 \text{ mol C}_4\text{H}_{10}}{58.0 \text{ g C}_4\text{H}_{10}} \times \frac{8 \text{ mol CO}_2}{2 \text{ mol C}_4\text{H}_{10}} \times \frac{22.4 \text{ L}}{\text{mol CO}_2} = \underline{\underline{131 \text{ L}}}$$

(b) 96.3 L O_2 (c) 124 L CO_2

9–97 2080 kg Zr (2.29 tons)

9–99 $n(O_2)$ = 1.45 mol O_2

$$1.45 \text{ mol O}_2 \times \frac{1 \text{ mol CO}_2}{2 \text{ mol O}_2} = 0.725 \text{ mol CO}_2$$

$$V = \underline{\underline{11.9 \text{ L CO}_2}}$$

9–100 force = $12.0 \text{ cm}^2 \times 15.0 \text{ cm} \times \dfrac{13.6 \text{ g}}{\text{cm}^3} = 2450 \text{ g}$

$$P = \frac{2450 \text{ g}}{12.0 \text{ cm}^2} = 204 \text{ g/cm}^2 \quad 1 \text{ atm} = 76.0 \text{ cm} \times \frac{13.6 \text{ g}}{\text{cm}^3} = 1030 \text{ g/cm}^2$$

$$204 \text{ g/cm}^2 \times \frac{1 \text{ atm}}{1030 \text{ g/cm}^2} = \underline{\underline{0.198 \text{ atm}}}$$

9–102 n = 0.0573 mol $\dfrac{8.37 \text{ g}}{0.0573 \text{ mol}} = \underline{\underline{146 \text{ g/mol}}}$

9–103 empirical formula = CH_2; molar mass = 41.9 g/mol; molecular formula = C_3H_6

9–105 n_T = 0.265 mol O_2 + 0.353 mol N_2 + 0.160 mol CO_2 = 0.778 mol of gas
V = 6.92 L

$$P(O_2) = \frac{0.265}{0.778} \times 2.86 \text{ atm} = 0.974 \text{ atm}; \ P(N_2) = 1.30 \text{ atm}; \ P(CO_2) = 0.59 \text{ atm}$$

9–106 $n = 5 \times 10^{-20}$ mol; $P = 4 \times 10^{-17}$ atm

9–108 19.6 L/mol; density of CO_2 = 2.24 g/L

9–109 density = $\dfrac{\text{mass}}{V} = \dfrac{P \times MM}{RT} = 0.525 \text{ g/L}$ (hot air)

Density at STP = 1.29 g/L 0.525/1.29 = 0.41
(Hot air is less than half as dense as air at STP.)

9–111 0.0103 mol of H_3BCO produces 0.0412 mol of gaseous products; V = 1.12 L (products at 25°C and 0.900 atm)

9–113 $2Al(s) + 3F_2(g) \rightarrow 2AlF_3(s)$
(original F_2) = 0.310 mol F_2 (leftover F_2) = 0.0921 mol
0.310 − 0.092 = 0.218 mol F_2 reacts forming 12.2 g AlF_3

CHAPTER 10

10–1 Since gas molecules are far apart, they move a comparatively large distance between collisions. Liquid molecules, on the other hand, are close together and so do not move far between collisions. The farther molecules move, the faster they mix.

10–3 Both the liquid and food coloring molecules are in motion. Through constant motion and collisions, the food coloring molecules eventually become dispersed.

10–5 Generally, solids have greater densities than liquids. (Ice and water are notable exceptions.) Since the molecules of a solid are held in fixed positions, more of them usually fit into the same volume compared to the liquid state. This is similar to being able to get more people into a room if they are standing still than if they are moving around.

10–6 $CH_4 < CCl_4 < GeCl_4$

10–8 All are nonpolar molecules with only London forces between molecules. The higher the molar mass, the greater the London forces and the more likely that the compound is a solid. I_2 is the heaviest and is a solid; Cl_2 is the lightest and is a gas.

10–10 If H_2O were linear, the two equal bond dipoles would be exactly opposite and would therefore cancel. Hydrogen bonding can occur only when the molecule is polar.

10–12

NH$_3$ molecules interact with hydrogen bonding.

10–14 (a) HBr   (b) SO_2 (f) CO

10–17 (a) HF   (c) H_2NCl (d) H_2O (f) HCOOH

10–19 (a) ion–ion   (b) hydrogen bonding plus London
(c) dipole–dipole plus London   (d) London only

10–20 $F_2 < HCl < HF < KF$

10–23 CO_2 is a nonpolar molecular compound. SiO_2 is a network solid.

10–25 $PbCl_2$ is most likely ionic (Pb^{2+}, $2Cl^-$), while the melting point of $PbCl_4$ indicates that it is a molecular compound.

10–26 Motor oil is composed of large molecules that increase viscosity. Motor oil also has a higher surface tension.

10–28 Water evaporates quickly on a hot day, which lowers the air temperature.

10–29 Alcohol is more volatile (i.e., has a higher vapor pressure) than water. It thus evaporates more rapidly and produces a pronounced cooling effect.

10–32 The liquid other than water. The higher the vapor pressure, the faster the liquid evaporates and the liquid cools.

10–33 Equilibrium refers to a state where opposing forces are balanced. In the case of a liquid in equilibrium with its vapor, it means that a molecule escaping to the vapor is replaced by one condensing to the liquid.

10–35 The substance is a gas at one atm and at 75°C. It would boil at a temperature below 75°C.

10–38 Ethyl alcohol boils at about 52°C at that altitude. At 10°C ethyl ether is a gas at that altitude.

10–39 Figure 10–15 is not very precise but at 10°C the actual vapor pressure of water is 9.2 torr. Liquid water could theoretically exist, but it would rapidly evaporate and change to ice due to the cooling effect.

10–40 The liquid with the higher vapor pressure (420 torr) has the lower boiling point. It also probably has the lower heat of vaporization.

10–42 Molecular O_2 is heavier than Ne atoms and has higher intermolecular forces (London). CH_3OH has hydrogen bonding, which would indicate a much higher heat of vaporization than the other two.

10–43 This is a comparatively high heat of fusion, so the melting point is probably also comparatively high.

10–45 This is a comparatively high boiling point, so the compound probably has a high melting point.

10–46

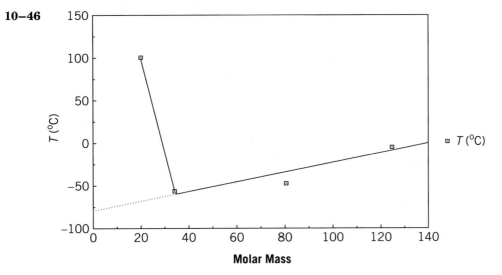

The boiling point of H_2O would be about $-75°C$ without hydrogen bonding.

10–49 7.07×10^3 J

10–51 2.54 g H_2O, 1.64 g NaCl, 6.69 g benzene

10–53 555 cal (ether)
2000 cal (H_2O)
Water would be more effective.
1.30×10^4 J (13.0 kJ)

10–54 NH_3: 6.12×10^5 J
Freon: 7.25×10^4 J
On the basis of mass, ammonia is the more effective refrigerant.

10–55 Condensation: 62.2×10^4 J
Cooling: 8.6×10^4 J
Total = 7.08×10^5 J (708 kJ)

10–57 2.83×10^4 cal

10–58 heat ice: $132 \ \cancel{g} \times 20°C \times \dfrac{0.492 \ \text{cal}}{\cancel{g} \cdot \cancel{°C}} = 1300$ cal

melt ice: $132 \ \cancel{g} \times \dfrac{79.8 \ \text{cal}}{\cancel{g}} = 10,500$ cal

heat H_2O: $132 \ \cancel{g} \times 100°\cancel{C} \times \dfrac{1.00 \ \text{cal}}{\cancel{g} \cdot \cancel{°C}} = 13,200$ cal

vap. H_2O: $132 \ \cancel{g} \times \dfrac{540 \ \text{cal}}{\cancel{g}} = 71,300$ cal

Total = $\underline{\underline{96,300 \ \text{cal} \ (96.3 \ \text{kcal})}}$

10–60 Let Y = the mass of the sample in grams. Then

$$\left(\frac{2260 \text{ J}}{\text{g}} \times Y\right) + \left(25°\text{C} \times Y \times \frac{4.18 \text{ J}}{\text{g} \cdot °\text{C}}\right) = 28{,}400 \text{ J} \quad Y = \underline{\underline{12.0 \text{ g}}}$$

10–62 18°C

10–63 (b) melting   (c) boiling

10–64 The average kinetic energy of all molecules of water at the same temperature is the same regardless of the physical state.

10–66 Because H_2O molecules have an attraction for each other, moving them apart increases the potential energy. Since the molecules in a gas at 100°C are farther apart than in a liquid at the same temperature, the potential energy of the gas molecules is greater.

10–67 The water does not become hotter, but it will boil faster.

10–68 $t(°\text{C})$

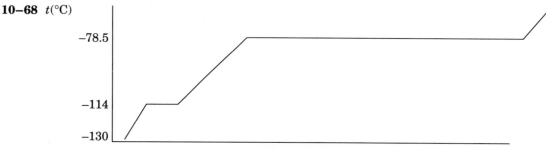

time coordinate

10–70 $C_2H_5NH_2$: 17°C, hydrogen bonding
CH_3OCH_3: −25°C, dipole–dipole (polar)
CO_2: −78°C, London forces only (nonpolar)

10–71 SF_6 is a molecular compound and SnO is an ionic compound (i.e., Sn^{2+}, O^{2-}). All ionic compounds are found as solids at room temperature because of the strong ion–ion forces.

10–73 There is hydrogen bonding in CH_3OH. Hydrogen bonding is a considerably stronger interaction than ordinary dipole–dipole forces. The stronger the interactions, the higher the boiling point.

10–74 SiH_4 is nonpolar. PH_3 and H_2S are both polar, but H_2S is more polar with more and stronger dipole–dipole interactions.

10–76 Carbon monoxide is polar whereas nitrogen is not. The added dipole–dipole interaction in CO may account for the slightly higher boiling point.

10–77 $2000 \text{ lb} \times \dfrac{454 \text{ g}}{\text{lb}} \times \dfrac{266 \text{ J}}{\text{g}} = \underline{\underline{2.42 \times 10^8 \text{ J}}}$

Heating 1 g of H_2O from 25.0°C to 100.0°C and then vaporizing the water requires
$(75.0°\text{C} \times 4.184 \text{ J/°C}) + 2{,}260 \text{ J} = 2.57 \times 10^3 \text{ J/g } H_2O$

$\dfrac{2.42 \times 10^8 \text{ J}}{2.57 \times 10^3 \text{ J/g } H_2O} = 9.42 \times 10^4 \text{ g } H_2O = \underline{\underline{94.2 \text{ kg } H_2O}}$

10–78 $V = 100 \text{ L}$, $T = 34 + 273 = 307 \text{ K}$
$P(H_2O) = 0.70 \times 39.0 \text{ torr} = 27.3 \text{ torr}$
Use the ideal gas law to find moles of water.

$$n = \frac{PV}{RT} = \frac{27.3 \text{ torr} \times 100 \text{ L}}{62.4 \dfrac{\text{L} \cdot \text{torr}}{\text{K} \cdot \text{mol}} \times 307 \text{ K}} = 0.143 \text{ mol}$$

$0.143 \text{ mol } H_2O \times \dfrac{18.0 \text{ g } H_2O}{\text{mol } H_2O} = \underline{\underline{2.57 \text{ g } H_2O}}$

CHAPTER 11

11–1 (a) $Na_2S \rightarrow 2Na^+(aq) + S^{2-}(aq)$
   (b) $Li_2SO_4 \rightarrow 2Li^+(aq) + SO_4{}^{2-}(aq)$
   (c) $K_2Cr_2O_7 \rightarrow 2K^+(aq) + Cr_2O_7{}^{2-}(aq)$
   (d) $CaS \rightarrow Ca^{2+}(aq) + S^{2-}(aq)$
   (e) $2(NH_4)_2S \rightarrow 4NH_4{}^+(aq) + 2S^{2-}(aq)$
   (f) $4Ba(OH)_2 \rightarrow 4Ba^{2+}(aq) + 8OH^-(aq)$

11–3 (a) 2 mol of Al^{3+} and 3 mol of $SO_4{}^{2-}$
   (b) 6 mol of Mg^{2+} and 4 mol of $PO_4{}^{3-}$

11–4

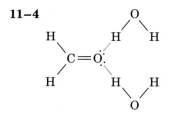

There is hydrogen bonding between solute and solvent.

11–6 (b) $PbSO_4$,   (c) $MgSO_3$,   (d) Ag_2O

11–8 At 10°C, Li_2SO_4 is the most soluble; at 70°C, KCl is the most soluble.

11–10 (a) unsaturated   (b) supersaturated
   (c) saturated   (d) unsaturated

11–12 A specific amount of Li_2SO_4 will precipitate unless the solution becomes supersaturated.

$$\left(500 \text{ g } H_2O \times \frac{35 \text{ g}}{100 \text{ g } H_2O}\right) - \left(500 \text{ g } H_2O \times \frac{28 \text{ g}}{100 \text{ g } H_2O}\right) = \underline{\underline{35 \text{ g } Li_2SO_4 \text{ (precipitate)}}}$$

11–13 (a) $2KI(aq) + Pb(C_2H_3O_2)_2(aq) \rightarrow PbI_2(s) + 2KC_2H_3O_2(aq)$
   (b) and (c) no reaction occurs
   (d) $BaS(aq) + Hg_2(NO_3)_2(aq) \rightarrow Hg_2S(s) + Ba(NO_3)_2(aq)$
   (e) $FeCl_3(aq) + 3KOH(aq) \rightarrow Fe(OH)_3(s) + 3KCl(aq)$

11–15 (a) $2K^+(aq) + S^{2-}(aq) + Pb^{2+}(aq) + 2NO_3{}^-(aq) \rightarrow PbS(s) + 2K^+(aq) + 2NO_3{}^-(aq)$
   $S^{2-}(aq) + Pb^{2+} \rightarrow PbS(s)$
   (b) $2NH_4{}^+(aq) + CO_3{}^{2-}(aq) + Ca^{2+}(aq) + 2Cl^-(aq) \rightarrow CaCO_3(s) + 2NH_4{}^+(aq) + 2Cl^-(aq)$
   $CO_3{}^{2-}(aq) + Ca^{2+}(aq) \rightarrow CaCO_3(s)$
   (c) $2Ag^+(aq) + 2ClO_4{}^-(aq) + 2Na^+(aq) + CrO_4{}^{2-}(aq) \rightarrow Ag_2CrO_4(s) + 2Na^+(aq) + 2ClO_4{}^-(aq)$
   $2Ag^+(aq) + CrO_4{}^{2-}(aq) \rightarrow Ag_2CrO_4(s)$

11–16 (a) $2K^+(aq) + 2I^-(aq) + Pb^{2+}(aq) + 2C_2H_3O_2{}^-(aq) \rightarrow PbI_2(s) + 2K^+(aq) + 2C_2H_3O_2{}^-(aq)$
   (d) $Ba^{2+}(aq) + S^{2-}(aq) + Hg_2{}^{2+}(aq) + 2NO_3{}^-(aq) \rightarrow Hg_2S(s) + Ba^{2+}(aq) + 2NO_3{}^-(aq)$
   (e) $Fe^{3+}(aq) + 3Cl^-(aq) + 3K^+(aq) + 3OH^-(aq) \rightarrow Fe(OH)_3(s) + 3K^+(aq) + 3Cl^-(aq)$

11–18 (a) $Pb^{2+}(aq) + 2I^-(aq) \rightarrow PbI_2(s)$
   (d) $Hg_2{}^{2+}(aq) + S^{2-}(aq) \rightarrow Hg_2S(s)$
   (e) $Fe^{3+}(aq) + 3OH^-(aq) \rightarrow Fe(OH)_3(s)$

11–20 (a) $CuCl_2(aq) + Na_2CO_3(aq) \rightarrow CuCO_3(s) + 2NaCl(aq)$
   Filter the solid $CuCO_3$.
   (b) $(NH_4)_2SO_3(aq) + Pb(NO_3)_2(aq) \rightarrow PbSO_3(s) + 2NH_4NO_3(aq)$
   Filter the solid $PbSO_3$.
   (c) $2KI(aq) + Hg_2(NO_3)_2(aq) \rightarrow Hg_2I_2(s) + 2KNO_3(aq)$
   Filter the solid Hg_2I_2.
   (d) $NH_4Cl(aq) + AgNO_3(aq) \rightarrow AgCl(s) + NH_4NO_3(aq)$
   Filter the solid AgCl; the desired product remains after water is removed by boiling.

(e) $Ca(C_2H_3O_2)_2(aq) + K_2CO_3(aq) \rightarrow CaCO_3(s) + KC_2H_3O_2(aq)$
Filter the solid $CaCO_3$ the desired product remains after the water is removed by boiling.

11–21 1.49%

11–22 15.77%

11–23 0.375 mol NaOH

11–26 8.81% NaOH

11–27 100 ppm

11–29 0.542 M

11–30 (a) 0.873 M (b) 1.40 M (c) 12.4 L   (d) 41.3 g
(e) 0.294 M (f) 0.024 mol   (g) 310 mL   (h) 49.0 g
(i) 2.00 mL

11–34 $1.17 \times 10^{-3}\ M$

11–35 0.166 M (in Ba^{2+}), 0.332 M (in OH^-)

11–36 1.84 M

11–38 1.32 g/mL

11–39 0.833 L

11–41 Slowly add <u>313 mL</u> of the 0.800 M NaOH to about 500 mL of water in a 1–L volumetric flask. Dilute to the 1–L mark with water.

11–43 0.140 M

11–44 720 mL

11–46 2.86 mL

11–47 Find the total moles and the total volume.
$n = V \times M = 0.150\ \cancel{L} \times 0.250\ mol/\cancel{L} = 0.0375$ mol (solution 1)
$n = V \times M = 0.450\ \cancel{L} \times 0.375\ mol/\cancel{L} = 0.169$ mol (solution 2)
$V_T = 0.600$ L; $n_T = 0.206$ mol $\dfrac{n}{V} = \dfrac{0.206\ mol}{0.600\ L} = \underline{\underline{0.343\ M}}$

11–48 $0.500\ \cancel{L} \times \dfrac{0.200\ \cancel{mol\ KOH}}{\cancel{L}} \times \dfrac{1\ \cancel{mol\ Cr(OH)_3}}{3\ \cancel{mol\ KOH}} \times \dfrac{103\ g\ Cr(OH)_3}{\cancel{mol\ Cr(OH)_3}} = \underline{\underline{4.29\ g\ Cr(OH)_3}}$

11–50 145 g $BaSO_4$

11–51 4.08 L

11–53 0.653 L

11–55 NaOH is the limiting reactant producing 1.75 g $Mg(OH)_2$.

11–57 An aqueous solution of AB is a good conductor of electricity. AB is dissociated into ions such as A^+ and B^-. A solution of AC is a weak conductor of electricity, which means that AC is only partially dissociated into ions: i.e.,

$$AC \rightleftharpoons A^+ + C^-$$

A solution of AD is a nonconductor of electricity because it is present as undissociated molecules in solution.

11–59 2.50 m (The molality is the same in both solvents since the mass of solvent is the same.)

11–61 15.8 g NaOH

11–63 −0.37°C

11–65 The salty water removes water from the cells of the skin by osmosis. After a prolonged period, a person would dehydrate and become thirsty.

11–68 We can concentrate a dilute solution by boiling away some solvent if the solute is not volatile. Reverse osmosis can also be used to concentrate a solution if pressure greater than the osmotic pressure is applied on a concentrated solution separated from the solvent by a semipermeable membrane. As the solution becomes more concentrated, the osmotic pressure becomes greater, and the corresponding pressure that is applied must be increased.

11–69 1.00 m

11–70 894 g H_2O

11–71 834 g glycol

11–72 101.4°C

11–74 2.93 m

11–75 80.5°C

11–77 −14.5°C

11–79 (a) −5.8°C (b) −6.4°C (c) −5.0°C

11–80 Dissolve the mixture in 100 g of H_2O and heat to over 45°C. Cool to 0°C where the solution is saturated with about 10 g of KNO_3 and about 25 g of KCl. 50 − 25 = 25 g of KCl precipitates.

11–82 $Fe^{2+}(aq) + 2NO_3^-(aq) + 2K^+(aq) + S^{2-}(aq) \rightarrow FeS(s) + 2K^+(aq) + 2NO_3^-(aq)$

$$Fe(NO_3)_2(aq) + K_2S(aq) \rightarrow FeS(s) + 2KNO_3(aq)$$

11–83 90.9% H_2O; 3.03% salt; 6.06% sugar

11–85 0.10 mol of Ag^+ reacts with 0.10 mol of Cl^- leaving 0.05 mol of Cl^- in 1.0 L of solution. $\underline{[Cl-] = 0.05\ M}$.

11–87 $0.225\ \cancel{L} \times \dfrac{0.196\ \cancel{\text{mol HCl}}}{\cancel{L}} \times \dfrac{1\ \text{mol M}}{2\ \cancel{\text{mol HCl}}} = 0.0221\ \text{mol}\ M$

$\dfrac{1.44\ \text{g}}{0.0221\ \text{mol}} = \underline{\underline{65.2\ \text{g/mol (Zn)}}}$

11–89 molality = 0.426 m
molar mass = 60.1 g/mol
empirical formula = CH_4N
molecular formula = $C_2H_8N_2$

REVIEW TEST ON CHAPTERS 9–11

Multiple Choice

1. (c) **2.** (b) **3.** (a) **4.** (c) **5.** (e) **6.** (e) **7.** (c) **8.** (c) **9.** (c) **10.** (d) **11.** (b) **12.** (b) **13.** (b) **14.** (a) **15.** (c) **16.** (d) **17.** (d) **18.** (b) **19.** (a) **20.** (a) **21.** (c) **22.** (e) **23.** (d) **24.** (b) **25.** (c) **26.** (d) **27.** (e) **28.** (a) **29.** (c) **30.** (c)

Problems

1. 0.446 mol of O_2, 2.69×10^{23} molecules of O_2, 8.06 L at −53°C, 13.1 L at 580 torr, 1.10 atm at 27°C, 21.2 L if 0.500 mol of a gas is added

2. Molar mass = 147 g/mol, empirical formula = C_3H_2Cl, molecular formula = $C_6H_4Cl_2$

3. BeF$_2$ (800°C) Ion–ion forces are strongest
C$_2$H$_5$OH (−117°C) A molecular compound with hydrogen bonding
(CH$_3$)$_2$O (−138°C) A polar covalent compound with ordinary dipole–dipole forces
C$_3$H$_8$ (−189°C) A nonpolar covalent compound with only London forces between molecules

4. 1.82×10^4 cal

5. (a) 0.480 M (b) 0.0873 M (c) 625 mL

6. (a) $Pb(NO_3)_2(aq) + 2NaI(aq) \rightarrow PbI_2(s) + 2NaNO_3(aq)$ (b) 17.1 L

CHAPTER 12

12–1 (a) HNO$_3$, nitrous acid (b) HNO$_2$ nitrous acid
(c) HClO$_3$, chloric acid (d) H$_2$SO$_3$, sulfurous acid

12–2 (a) CsOH, cesium hydroxide
(b) Sr(OH)$_2$, strontium hydroxide
(c) Al(OH)$_3$, aluminum hydroxide
(d) Mn(OH)$_3$, manganese(III) hydroxide

12–4 (a) $HNO_3 + H_2O \rightarrow H_3O^+ + NO_3^-$
(b) $HNO_2 + H_2O \rightarrow H_3O^+ + NO_2^-$
(c) $HClO_3 + H_2O \rightarrow H_3O^+ + ClO_3^-$
(d) $H_2SO_3 + H_2O \rightarrow H_3O^+ + HSO_3^-$

12–6 (a) acid (b) normal salt (c) acid (d) acid salt
(e) normal salt (f) base

12–7 (a) $HC_2H_3O_2 + KOH \rightarrow KC_2H_3O_2 + H_2O$
(b) $2HI + Ca(OH)_2 \rightarrow CaI_2 + 2H_2O$
(c) $H_2SO_4 + Ca(OH)_2 \rightarrow CaSO_4 + 2H_2O$

12–9 (b) $2H^+(aq) + 2I^-(aq) + Ca^{2+}(aq) + 2OH^-(aq) \rightarrow Ca^{2+}(aq) + 2I^-(aq) + 2H_2O$
$H^+(aq) + OH^-(aq) \rightarrow H_2O$ (net ionic)
(c) $2H^+(aq) + SO_4^{2-}(aq) + Ca^{2+}(aq) + 2OH^-(aq) \rightarrow CaSO_4(s) + 2H_2O$
Net ionic equation is the same as the total ionic equation since CaSO$_4$ precipitates.

12–10 (a) $2HBr + Ca(OH)_2 \rightarrow CaBr_2 + 2H_2O$
(b) $2HClO_2 + Sr(OH)_2 \rightarrow Sr(ClO_2)_2 + 2H_2O$
(c) $2H_2S + Ba(OH)_2 \rightarrow Ba(HS)_2 + 2H_2O$
(d) $H_2S + 2LiOH \rightarrow Li_2S + 2H_2O$

12–12 $LiOH + H_2S \rightarrow LiHS + H_2O$
$LiOH + LiHS \rightarrow Li_2S + H_2O$

12–14 $H_2S + NaOH \rightarrow NaHS + H_2O$

12–17 $HN(CH_3)_2 + H_2O \leftrightarrows H_2N(CH_3)_2^+ + OH^-$

12–18 HX is a weak acid. The concentration of H$_3$O$^+$ must equal the concentration of the HX that ionized.

$$\frac{0.010}{0.100} \times 100\% = \underline{\underline{10\% \text{ ionized}}}$$

12–20 [H$_3$O$^+$] = 0.55 M

12–21 [H$_3$O$^+$] = $0.030 \times 0.55 = 0.016$ M

12–23 From the first ionization: [H$_3$O$^+$] = 0.354 M.
Of that, 25% undergoes further ionization.

$0.25 \times 0.354 = 0.088 \ M \ [H_3O^+]$ from the second ionization.
The total $[H_3O^+] = 0.354 + 0.088 = \underline{\underline{0.442 \ M}}$

12-25 The system would not be at equilibrium if $[H_3O^+] = [OH^-] = 10^{-2} \ M$. Therefore, H_3O^+ reacts with OH^- until the concentration of each is reduced to $10^{-7} \ M$. This is a neutralization reaction—i.e.,

$$H_3O^+ + OH^- \rightarrow H_2O$$

12-26 (a) $[H_3O^+] = \dfrac{K_w}{[OH^-]} = \dfrac{10^{-14}}{10^{-12}} = 10^{-2} \ M$

(b) $[H_3O^+] = 10^{-15} \ M$

(c) $[OH^-] = 5.0 \times 10^{-10} \ M$

12-28 $[H_3O^+] = 0.0250 \ M$, $[OH^-] = 4.00 \times 10^{-13} \ M$

12-29 In lye: $[H_3O^+] = 3.92 \times 10^{-15} \ M$
In ammonia: $[H_3O^+] = 2.5 \times 10^{-12} \ M$

12-30 (a) acidic (b) basic (c) acidic

12-32 (a) acidic (b) basic (c) acidic (d) basic

12-34 (a) 6.00 (b) 9.00 (c) 12.00 (d) 9.40 (e) 10.19

12-36 (a) $1.0 \times 10^{-3} \ M$ (b) $2.9 \times 10^{-4} \ M$ (c) $1.0 \times 10^{-6} \ M$
(d) $2.4 \times 10^{-8} \ M$ (e) $2.0 \times 10^{-13} \ M$

12-38 For Problem 12–34: (a) acidic (b) basic (c) basic (d) basic (e) basic
For Problem 12–36: (a) acidic (b) acidic (c) acidic (d) basic (e) basic

12-40 pH = 1.12

12-42 pH = 12.56

12-43 $[H_3O^+] = 0.100 \times 0.10 = 0.010 = 1.0 \times 10^{-2} \ M$ pH = $\underline{\underline{2.00}}$

12-45 (a) strongly acidic   (b) strongly acidic   (c) weakly acidic
(d) strongly basic   (e) neutral   (f) weakly basic
(g) weakly acidic   (h) strongly acidic

12-46 ammonia (pH = 11.4), eggs (pH = 7.8), rain water ($[H_3O^+] = 2.0 \times 10^{-6} \ M$, pH = 5.70), grape juice ($[OH^-] = 1.0 \times 10^{-10} \ M$, pH = 4.0), vinegar ($[H_3O^+] = 2.5 \times 10^{-3} \ M$, pH = 2.60), sulfuric acid (pOH = 13.6, pH = 0.4)

12-49 (a) NO_3^- (b) HSO_4^- (c) PO_4^{3-}
(d) CH_3^- (e) OH^- (f) NH_2^-

12-51 (a) $HClO_4$, ClO_4^- and H_2O, OH^- (b) HSO_4^-, SO_4^{2-} and $HClO$, ClO^-
(c) H_2O, OH^- and NH_3, NH_2^- (d) NH_4^+, NH_3 and H_3O^+, H_2O

12-54 (a) $NH_3 + H_2O \longrightarrow NH_4^+ + OH^-$
B_1 ⎣___ A_1 ___⎦
A_2 B_2

(b) $N_2H_4 + H_2O \longrightarrow N_2H_5^+ + OH^-$
B_1 ⎣___ A_1 ___⎦
A_2 B_2

(c) $HS^- + H_2O \longrightarrow H_2S + OH^-$
B_1 ⎣___ A_1 ___⎦
A_2 B_2

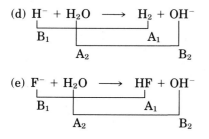

(e) F⁻ + H₂O ⟶ HF + OH⁻

12–55 Base: $HS^- + H_3O^+ \rightleftharpoons H_2S + H_2O$
Acid: $HS^- + OH^- \rightleftharpoons S^{2-} + H_2O$

12–57 (a) $H_2S + H_2O \rightleftharpoons HS^- + H_3O^+$
(b) no reaction
(c) $HBr + H_2O \rightarrow H_3O^+ + Br^-$
(d) $NH_4^+ + H_2O \rightleftharpoons NH_3 + H_3O^+$

12–59 (c) H^- cannot exist in water.

12–61 (a) HS^- (b) H_3O^+ (c) OH^- (d) $HN(CH_3)_2$

12–63 (a) $F^- + H_2O \rightleftharpoons HF + OH^-$
(b) $SO_3^{2-} + H_2O \rightleftharpoons HSO_3^- + OH^-$
(c) $H_2N(CH_3)_2^+ + H_2O \rightleftharpoons HN(CH_3)_2 + H_3O^+$
(d) $HPO_4^{2-} + H_2O \rightleftharpoons H_2PO_4^- + OH^-$
(e) $CN^- + H_2O \rightleftharpoons HCN + OH^-$
(f) no hydrolysis (cation of a strong base)

12–67 (a) neutral (neither ion hydrolizes)   (b) acidic (cation hydrolysis)
(c) basic (anion hydrolysis)   (d) neutral (neither ion hydrolyzes)
(e) acidic (cation hydrolysis)   (f) basic (anion hydrolysis)

12–69 (b) and (d) are favorable.

12–71 (a) $H_3O^+ + HS^- \rightarrow H_2S + H_2O$
(b) $HClO + CN^- \rightarrow HCN + ClO^-$
(c) $H_2SO_3 + OH^- \rightarrow HSO_3^- + H_2O$
(d) $HBr + CH_3^- \rightarrow CH_4 + Br^-$

12–73 $CaC_2(s) + 2H_2O(l) \rightarrow C_2H_2(g) + Ca^{2+}(aq) + 2OH^-(aq)$

12–74 Cation: $NH_4^+ + H_2O \rightleftharpoons NH_3 + H_3O^+$
Anion: $CN^- + H_2O \rightleftharpoons HCN + OH^-$
Since the solution is basic, the anion hydrolysis reaction must take place to a greater extent than the cation hydrolysis.

12–76 (a), (b), (e), (h), and (i) are buffer solutions.

12–77 There is no equilibrium when HCl dissolves in water. A reservoir of un–ionized acid must be present to react with any strong base that is added. Likewise, the Cl^- ion does not exhibit base behavior in water, so it cannot react with any H_3O^+ added to the solution.

12–78 $N_2H_4(aq) + H_2O \rightleftharpoons N_2H_5^+(aq) + OH^-(aq)$
Added H_3O^+: $H_3O^+ + N_2H_4 \rightarrow N_2H_5^+ + H_2O$
Added OH^-: $OH^- + N_2H_5^+ \rightarrow N_2H_4 + H_2O$

12–80 $HPO_4^{2-}(aq) + H_2O \rightleftharpoons PO_4^{3-}(aq) + H_3O^+(aq)$
Added H_3O^+: $H_3O^+ + PO_4^{3-} \rightarrow HPO_4^{2-} + H_2O$
Added OH^-: $OH^- + HPO_4^{2-} \rightarrow PO_4^{3-} + H_2O$

12–83 (a) $Sr(OH)_2$ (b) H_2SeO_4 (c) H_3PO_4
(d) $CsOH$ (e) HNO_2 (f) $HClO_3$

12–85 $CO_2(g) + LiOH(aq) \rightarrow LiHCO_3(aq)$

12–86 $2LiNO_3(s)$

12–87 (a) $HCN + NH_3 \rightarrow NH_4^+ + CN^-$
(b) $NH_3 + H^- \rightarrow H_2 + NH_2^-$
(c) $HCO_3^- + NH_3 \rightarrow NH_4^+ + CO_3^{2-}$
(d) $NH_4^+Cl^- + Na^+NH_2^- \rightarrow Na^+Cl^- + 2NH_3$

12–88 (a) $HCl + CH_3OH \rightarrow CH_3OH_2^+ + Cl^-$
(b) $CH_3OH + NH_2^- \rightarrow NH_3 + CH_3O^-$

12–90 (a) Acidic: $H_2S + H_2O \rightleftharpoons H_3O^+ + HS^-$
(b) Basic: $ClO^- + H_2O \rightleftharpoons HClO + OH^-$
(c) Neutral
(d) Basic: $NH_3 + H_2O \rightleftharpoons NH_4^+ + OH^-$
(e) Acidic: $N_2H_5^+ + H_2O \rightleftharpoons N_2H_4 + H_3O^+$
(f) Basic: $Ba(OH)_2 \rightarrow Ba^{2+} + 2OH^-$
(g) Neutral
(h) Basic: $NO_2^- + H_2O \rightleftharpoons HNO_2 + OH^-$
(i) Acidic: $H_2SO_3 + H_2O \rightleftharpoons H_3O^+ + HSO_3^-$
(j) Acidic: $Cl_2O_3 + H_2O \rightarrow HClO_2$; $HClO_2 + H_2O \rightleftharpoons H_3O^+ + ClO_2^-$

12–92 $LiOH$, strongly basic, pH = 13.0; $SrBr_2$, neutral, pH = 7.0; $KClO$, weakly basic, pH = 10.2; NH_4Cl, weakly acidic, pH = 5.2; HI, strongly acidic, pH = 1.0. When pH = 1.0, $[H_3O^+] = 0.10\ M$. If HI is completely ionized, its initial concentration must be $0.10\ M$.

12–94 8.2 kg of H_2SO_4

12–95 0.024 mol of HCl remains after neutralization. pH = 1.62

12–97 $2HNO_3 + Ca(OH)_2 \rightarrow Ca(NO_3)_2 + 2H_2O$
0.050 mol of HNO_3 reacts with 0.025 mol of $Ca(OH)_2$.
0.025 mol of $Ca(OH)_2$ remains. $0.025 \times 2 = 0.05$ mol of OH^- remains.
$[OH^-] = 0.050$ pOH = 1.30   pH = 12.70

CHAPTER 13

13–1 (a) Pb +4, O −2   (b) P +5, O −2   (c) C −1, H +1
(d) N −2, H +1   (e) Li +1, H −1   (f) B +3, Cl −1
(g) Rb +1, Se −2   (h) Bi +3, S −2

13–3 (a) Li, (e) K, and (g) Rb

13–5 +3

13–6 (a) P = +5   (b) C = +3   (c) Cl = +7   (d) Cr = +6
(e) S = +6   (f) N = +5   (g) Mn = +7

13–7 (a) S = +6   (b) Co = +3   (c) U = +6
(d) N = +5   (e) Cr = +6   (f) Mn = +6

13–9 (a), (c), and (f)

13–10 (a) oxidation   (b) reduction   (c) oxidation
(d) reduction   (e) reduction

13–12

Reactant Oxidized	Product of Oxidation	Reactant Reduced	Product of Reduction	Oxidizing Agent	Reducing Agent
Br^-	Br_2	MnO_2	Mn^{2+}	MnO_2	Br^-
CH_4	CO_2	O_2	CO_2, H_2O	O_2	CH_4
Fe^{2+}	Fe^{3+}	MnO_4^-	Mn^{2+}	MnO_4^-	Fe^{2+}

13–13

Reactant Oxidized	Product of Oxidation	Reactant Reduced	Product of Reduction	Oxidizing Agent	Reducing Agent
Al	AlO_2^-	H_2O	H_2	H_2O	Al
Mn^{2+}	MnO_4^-	$Cr_2O_7^{2-}$	Cr^{3+}	$Cr_2O_7^{2-}$	Mn^{2+}

$$-5e^- \times \mathbf{4} = -20e^-$$

13–16 (a)
$$\overset{-3}{4NH_3} + 5O_2 \longrightarrow \overset{+2}{4NO} + 6H_2O$$
$$\underset{0 \qquad\qquad -2}{}$$
$$+4e^- \times \mathbf{5} = +20e^-$$

$$-4e^- \times \mathbf{1} = -4e^-$$

(b)
$$\overset{0}{Sn} + 4H\overset{+5}{N}O_3 \longrightarrow \overset{+4}{Sn}O_2 + 4\overset{+4}{N}O_2 + 2H_2O$$
$$+1e^- \times \mathbf{4} = +4e^-$$

(c) Before the number of electrons lost is calculated, notice that a temporary coefficient of "2" is needed for the Na_2CrO_4 in the products since there are 2 Cr's in Cr_2O_3 in the reactants.

$$+2e^- \times \mathbf{3} = +6e^-$$

$$\overset{+5}{Cr_2O_3} + 2Na_2CO_3 + 3K\overset{+5}{N}O_3 \longrightarrow 2CO_2 + 2Na_2\overset{+3}{Cr}O_4 + 3K\overset{+3}{N}O_2$$
$$\overset{+6}{}\qquad\qquad\qquad\qquad\qquad\qquad\qquad\qquad\overset{+12}{}$$
$$-6e^- \times \mathbf{1} = -6e^-$$
$$-4e^- \times \mathbf{3} = -12e^-$$

(d)
$$\overset{0}{3Se} + 2Br\overset{+5}{O_3^-} + 3H_2O \longrightarrow 3H_2\overset{+4}{Se}O_3 + 2\overset{-1}{Br^-}$$
$$+6e^- \times \mathbf{2} = +12e^-$$

13–18 (a) $2H_2O + Sn^{2+} \rightarrow SnO_2 + 4H^+ + 2e^-$
(b) $2H_2O + CH_4 \rightarrow CO_2 + 8H^+ + 8e^-$
(c) $e^- + Fe^{3+} \rightarrow Fe^{2+}$
(d) $6H_2O + I_2 \rightarrow 2IO_3^- + 12H^+ + 10e^-$
(e) $e^- + 2H^+ + NO_3^- \rightarrow NO_2 + H_2O$

13-20 (a)

$$S^{2-} \longrightarrow S + 2e^- \quad\bigg|\, \times 3$$
$$3e^- + 4H^+ + NO_3^- \longrightarrow NO + 2H_2O \quad\bigg|\, \times 2$$
$$\overline{3S^{2-} + 8H^+ + 2NO_3^- \longrightarrow 3S + 2NO + 4H_2O}$$

(b) $2S_2O_3^{2-} + I_2 \longrightarrow S_4O_6^{2-} + 2I^-$

(c)

$$H_2O + SO_3^{2-} \longrightarrow SO_4^{2-} + 2H^+ + 2e^- \quad\bigg|\, \times 3$$
$$6e^- + 6H^+ + ClO_3^- \longrightarrow Cl^- + 3H_2O \quad\bigg|\, \times 1$$
$$\overline{3SO_3^{2-} + ClO_3^- \longrightarrow Cl^- + 3SO_4^{2-}}$$

(d) $2H^+ + 2Fe^{2+} + H_2O_2 \longrightarrow 2Fe^{3+} + 2H_2O$

(e) $AsO_4^{3-} + 2I^- + 2H^+ \longrightarrow I_2 + AsO_3^{3-} + H_2O$

(f) $4Zn + NO_3^- + 10H^+ \longrightarrow 4Zn^{2+} + NH_4^+ + 3H_2O$

13-22 (a) $2OH^- + SnO_2^{2-} \longrightarrow SnO_3^{2-} + H_2O + 2e^-$

(b) $6e^- + 4H_2O + 2ClO_2^- \longrightarrow Cl_2 + 8\,OH^-$

(c) $6\,OH^- + Si \longrightarrow SiO_3^{2-} + 3H_2O + 4e^-$

(d) $8e^- + 6H_2O + NO_3^- \longrightarrow NH_3 + 9\,OH^-$

13-24 (a)

$$8OH^- + S^{2-} \longrightarrow SO_4^{2-} + 4H_2O + 8e^- \quad\bigg|\, \times 1$$
$$I_2 + 2e^- \longrightarrow 2I^- \quad\bigg|\, \times 4$$
$$\overline{S^{2-} + 8OH^- + 4I_2 \longrightarrow SO_4^{2-} + 8I^- + 4H_2O}$$

(b) $8OH^- + I^- + 8MnO_4^- \longrightarrow 8MnO_4^{2-} + IO_4^- + 4H_2O$

(c) $2H_2O + SnO_2^{2-} + BiO_3^- \longrightarrow SnO_3^{2-} + Bi(OH)_3 + OH^-$

(d)

$$32OH^- + CrI_3 \longrightarrow CrO_4^{2-} + 3IO_4^- + 16H_2O + 27e^- \quad\bigg|\, \times 2$$
$$2e^- + Cl_2 \longrightarrow 2Cl^- \quad\bigg|\, \times 27$$
$$\overline{2CrI_3 + 64OH^- + 27Cl_2 \longrightarrow 2CrO_4^{2-} + 6IO_4^- + 32H_2O + 54Cl^-}$$

13-26 (a) $2H_2 + O_2 \longrightarrow 2H_2O$

(b)

$$H_2O_2 \longrightarrow O_2 + 2H^+ + 2e^- \quad\bigg|\, \times 1$$
$$2e^- + 2H^+ + H_2O_2 \longrightarrow 2H_2O \quad\bigg|\, \times 1$$
$$\overline{2H_2O_2 \longrightarrow O_2 + 2H_2O}$$

$$2OH^- + H_2O_2 \longrightarrow O_2 + 2H_2O + 2e^- \quad\bigg|\, \times 1$$
$$2e^- + H_2O_2 \longrightarrow 2\,OH^- \quad\bigg|\, \times 1$$
$$\overline{2H_2O_2 \longrightarrow O_2 + 2H_2O}$$

13-28 $Zn(s)$ reacts at the anode and $MnO_2(s)$ reacts at the cathode. The total reaction is

$$Zn(s) + 2MnO_2(s) + 2H_2O(l) \longrightarrow Zn(OH)_2(s) + 2MnO(OH)(s)$$

13-30 anode: $Cd + 2OH^- \rightarrow Cd(OH)_2 + 2e^-$

cathode: $NiO_2 + 2H_2O + 2e^- \rightarrow Ni(OH)_2 + 2OH^-$

13-32 Reactions (a), (c), (e), and (f) are predicted to be favorable.

13-34 (a) $CuCl_2(aq) + Fe(s) \rightarrow FeCl_2(aq) + Cu(s)$

(A coating of Cu forms on the nails.)

(d) $Ni(s) + Pb(NO_3)_2(aq) \rightarrow Ni(NO_3)_2(aq) + Pb(s)$

(e) $2Al(s) + 6H_2O(l) \rightarrow 2Al(OH)_3(s) + 3H_2(g)$

(h) $2Cr^{3+} + 3Zn \rightarrow 3Zn^{2+} + 2Cr$

No reactions occur in (b), (c), (f), and (g).

13-36 (c) $2F_2 + 2H_2O \rightarrow 4F^- + 4H^+ + O_2$

(e) $Mg + 2H_2O \rightarrow Mg(OH)_2(s) + H_2$

13-38 From Table 13-1:

$Fe + 2H_2O \rightarrow$ no reaction

$Fe + 2H^+ \rightarrow Fe^{2+} + H_2$

Acid rain has a higher $H^+(aq)$ concentration, thus making the second reaction more likely.

13–40 The spontaneous reaction is

$$Pb(NO_3)_2(aq) + Fe(s) \rightarrow Fe(NO_3)_2(aq) + Pb(s)$$

anode: $Fe \rightarrow Fe^{2+} + 2e^-$
cathode: $Pb^{2+} + 2e^- \rightarrow Pb$

13–41 The Daniell cell is more powerful. The greater the separation between oxidizing and reducing agent as shown in Table 13–1, the more powerful is the cell.

13–43 $3Fe(s) + 2Cr^{3+}(aq) \rightarrow 3Fe^{2+}(aq) + 2Cr(s)$
Zinc would spontaneously form a chromium coating as illustrated by the following equation:
$$3Zn(s) + 2Cr^{3+}(aq) \rightarrow 3Zn^{2+}(aq) + 2Cr(s)$$

13–44 Sodium reacts spontaneously with water since it is an active metal. The actual cathode reaction is:
$$2H_2O + 2e^- \rightarrow H_2 + 2OH^-$$
Elemental sodium is produced by electrolysis of the molten salt NaCl.

13–47 The strongest oxidizing agent is reduced the easiest. Thus the reduction of Ag^+ to Ag occurs first. This procedure can be used to purify silver.

13–48 K_3N (-3), N_2H_4 (-2), NH_2OH, (-1), N_2 (0), N_2O $(+1)$, NO $(+2)$, N_2O_3 $(+3)$, N_2O_4 $(+4)$, $Ca(NO_3)_2$ $(+5)$

13–49 $Cd + NiCl_2(aq) \rightarrow Ni + CdCl_2(aq)$
$Zn + CdCl_2(aq) \rightarrow Cd + ZnCl_2(aq)$
These reactions indicate that Cd^{2+} is a stronger oxidizing agent than Zn^{2+}, but is weaker than Ni^{2+}. It appears about the same as Fe^{2+}.

13–51 The reactions that occur are

$$4Au^{3+} + 6H_2O \rightarrow 4Au + 12H^+ + 3O_2$$
$$2Au + 3F_2 \rightarrow 2AuF_3$$

These reactions and the fact that Au does not react with Cl_2 ranks Au^{3+} above H^+ and Cl_2 but below F_2.

13–52 $12H^+(aq) + 5Zn(s) + 2NO_3^-(aq) \rightarrow 5Zn^{2+}(aq) + N_2(g) + 6H_2O(l)$
0.658 g N_2 requires <u>7.68 g Zn.</u>

13–54 $14H^+(aq) + 2NO_3^-(aq) + 3Cu_2O(s) \rightarrow 6Cu^{2+}(aq) + 2NO(g) + 7H_2O(l)$
10.0 g of Cu_2O produces <u>1.04 L of NO</u> measured at STP.

13–56 $7OH^-(aq) + 4Zn(s) + 6H_2O(l) + NO_3^-(aq) \rightarrow NH_3(g) + 4Zn(OH)_4^{2-}(aq)$
6.54 g of Zn produces <u>0.493</u> L of NH_3 at 27.0°C and 1.25 atm pressure.

CHAPTER 14

14-1 Colliding molecules must have the proper orientation relative to each other at the time of the collision, and the colliding molecules must have the minimum kinetic energy for the particular reaction.

14–2 Most likely, the collision would involve the carbon end of the CO molecule colliding with one oxygen on the NO_2 as follows:

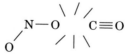

14–3 (a) As the temperature increases, the frequency of collisions between molecules increases, as does the average energy of the collisions. Both contribute to the increased rate of reaction.
(b) The cooking of eggs initiates a chemical reaction that occurs more slowly at lower temperatures.
(c) The average energy of colliding molecules at room temperature is not sufficient to initiate a reaction between H_2 and O_2.
(d) A higher concentration of oxygen increases the rate of combustion.

(e) When a solid is finely divided, a greater surface area is available for collisions with oxygen molecules. Thus it burns faster.

(f) The souring of milk is a chemical reaction that slows as the temperature drops. It takes several days in a refrigerator.

(g) The platinum is a catalyst. Since the activation energy in the presence of a catalyst is lower, the reaction can occur at a lower temperature.

14-6 The rate of the forward reaction was at a maximum at the beginning of the reaction; the rate of the reverse reaction was at a maximum at the point of equilibrium.

14-7 In many cases, reactions do not proceed directly to the right because other products are formed between the same reactants. For example, combustion may produce carbon monoxide as well as carbon dioxide.

14-8 Products are easier to form because the activation energy for the forward reaction is less than for the reverse reaction. This is true of all exothermic reactions. The system should come to equilibrium faster starting with pure reactants.

14-10 (a) right (b) left (c) left (d) right (e) right (f) has no effect (g) yield decreases
(h) yield increases but rate of formation decreases

14-12 (a) increase (b) increase (c) decrease (d) decrease (e) decrease (f) no effect

14-14 (a) Since there are the same number of moles of gas on both sides of the equation, pressure (or volume) has no effect on the point of equilibrium.
(b) decrease the amount of NO
(c) decrease the amount of NO

14-15 (a) $K_{eq} = \dfrac{[COCl_2]}{[CO][Cl_2]}$ (b) $K_{eq} = \dfrac{[CO_2][H_2]^4}{[CH_4][H_2O]^2}$

(c) $K_{eq} = \dfrac{[Cl_2]^2[H_2O]^2}{[HCl]^4[O_2]}$ (d) $K_{eq} = \dfrac{[CH_3Cl][HCl]}{[CH_4][Cl_2]}$

14-17 products

14-19 There will be an appreciable concentration of both reactants and products at equilibrium.

14-20 $K_{eq} = 0.34$

14-21 $K_{eq} = 49.2$

14-23 $K_{eq} = \dfrac{[CO_2][H_2]^4}{[CH_4][H_2O]^2} = \dfrac{(2.20/30.0)(4.00/30.0)^4}{(6.20/30.0)(3.00/30.0)^2} = \underline{\underline{0.0112}}$

14-25 (a) $[H_2] = [I_2] = 0.30$ mol/L   (b) $[H_2] = 0.30$ mol/L; $[I_2] = 0.50$ mol/L
(c) $[HI] = 0.40$ mol/L; $[I_2] = [H_2] = 0.10$ mol/L   (d) $K_{eq} = 0.063$
(e) $K_{eq} = 16$. This is a smaller value than that used in Table 14-1. This indicates that the equilibrium in this problem was established at a different temperature than that of Table 14-1.

14-27 (a) $[N_2]_{reacts} = 0.50 - 0.40 = 0.10$ mol/L

$0.10 \text{ mol } N_2 \times \dfrac{3 \text{ mol } H_2}{1 \text{ mol } N_2} = 0.30$ mol/L H_2 (reacts)

$[H_2]_{eq} = 0.50 - 0.30 = \underline{0.20 \text{ mol/L}}$

$0.10 \text{ mol } N_2 \times \dfrac{2 \text{ mol } NH_3}{1 \text{ mol } N_2} = 0.20$ mol/L NH_3 formed

$[NH_3]_{eq} = 0.50 + 0.20 = \underline{0.70 \text{ mol/L}}$

(b) $K_{eq} = \dfrac{(0.70)^2}{(0.20)^3(0.40)} = \underline{\underline{150}}$

14-28 (a) The concentration of O_2 that reacts is

$0.25 \text{ mol } NH_3 \times \dfrac{5 \text{ mol } O_2}{4 \text{ mol } NH_3} = 0.31$ mol/L O_2

(b) $[NH_3] = 0.75$ mol/L; $[O_2] = 0.69$ mol/L;
$[NO] = 0.25$ mol/L NO (formed)
$[H_2O] = 0.38$ mol/L H_2O

(c) $K_{eq} = \dfrac{[NO]^4[H_2O]^6}{[NH_3]^4[O_2]^5} = \dfrac{(0.25)^4(0.38)^6}{(0.75)^4(0.69)^5}$

14–30 $[PCl_5] = 0.28$ mol/L

14–32 $[H_2O] = 0.26$ mol/L

14–33 Let $X = [HCl] = [CH_3Cl]$ $\quad K_{eq} = \dfrac{[HCl][CH_3Cl]}{[CH_4][Cl_2]} = \dfrac{X^2}{(0.20)(0.40)} = 56$

$X^2 = 4.5 \quad X = \underline{\underline{2.1 \text{ mol/L}}}$

14–35 (a) $K_a = \dfrac{[H_3O^+][BrO^-]}{[HBrO]}$ (d) $K_a = \dfrac{[\mathbf{H_3O^+}][SO_3{}^{2-}]}{[HSO_3{}^-]}$

(b) $K_b = \dfrac{[NH_4{}^+][OH^-]}{[NH_3]}$ (e) $K_a = \dfrac{[H_3O^+][\mathbf{H_2PO_4{}^-}]}{[H_3PO_4]}$

(c) $K_a = \dfrac{[H_3O^+][HSO_3{}^-]}{[H_2SO_3]}$ (f) $K_b = \dfrac{[H_2N(CH_3)_2{}^+][\mathbf{OH^-}]}{[HN(CH_3)_2]}$

14–37 The acid HB is weaker because it produces a smaller concentration (higher pH) at the same initial concentration. The stronger acid HX has the large value of K_a.

14–38 The strongest acid (with the largest K_a) produces the lowest pH. HZ, $HC_2H_3O_2$, HClO.

14–39 (a) $[HOCN]_{eq} = 0.20 - 0.0062 = 0.19$

(b) $K_a = \dfrac{[H_3O^+][OCN^-]}{[HOCN]} = \dfrac{(6.2 \times 10^{-3})(6.2 \times 10^{-3})}{(0.19)}$

(c) pH = 2.21

14–40 (a) $HX + H_2O \rightleftharpoons H_3O^+ + X^- \quad K_a = \dfrac{[H_3O^+][X^-]}{[HX]}$

(b) From the equation $[H_3O^+] = [X^-] = 0.100 \times 0.58$, $[HX] = 0.58 - 0.058 = 0.52$.

(c) $K_a = 6.5 \times 10^{-3}$ (d) pH = 1.24

14–43 $Nv + H_2O \rightleftharpoons NvH^+ + OH^- \quad K_b = 6.6 \times 10^{-6}$

14–44 $[H_3O^+] = 0.300 - 0.277 = 0.023$; pH = 1.64; $K_a = 1.9 \times 10^{-3}$

14–45 pH = 4.43

14–47 $[OH^-] = 3.2 \times 10^{-3}$

14–49 pH = 12.43

14–51 pH = 9.40

14–53 pH = 8.20

14–54 In a buffer solution, when the acid and its conjugate base are present in equimolar amounts,

$K_a = \dfrac{[H_3O^+][\text{anion}]}{[\text{acid}]}$ so $[H_3O^+] = K_a$

Likewise, $[OH^-] = K_b$ for a weak base.
When pH = 7.50, $[H_3O^+] = 3.2 \times 10^{-8}$ and $[OH^-] = 3.1 \times 10^{-7}$.
Since $K_a = 3.2 \times 10^{-8}$ for HClO, an equimolar mixture of HClO and KClO produces the required buffer.

14–56 (a) $3Cl_2(g) + NH_3(g) \rightarrow NCl_3(g) + 3HCl(g)$

(b) $K_{eq} = \dfrac{[NCl_3][HCl]^3}{[NH_3][Cl_2]^3}$ (c) no effect   (d) decreases $[NH_3]$

(e) reactants   (f) $[NH_3] = 0.083$ mol/L

14–58 (a) $2N_2O(g) \to 2N_2(g) + O_2(g)$

(b) $K_{eq} = \dfrac{[N_2]^2[O_2]}{[N_2O]^2}$ (c) decreases $[N_2]$ (d) decreases $[N_2O]$

(e) $[N_2O]_{eq} = 0.10 - (0.015 \times 0.10) = 0.10$ mol/L; $[N_2] = 0.015 \times 0.10 = 1.5 \times 10^{-3}$ mol/L;

$[O_2] = \dfrac{1.5 \times 10^{-3}}{2} = 7.5 \times 10^{-4}$ mol/L. $K_{eq} = 1.7 \times 10^{-7}$

14–60 (a) A solution of $NaHCO_3$ is slightly basic because the hydrolysis reaction occurs to a greater extent than the acid ionization reaction. That is, K_b is larger than K_a.

(b) It is used as an antacid to counteract excess stomach acidity.

(c) $HCl + H_2O \to H_3O^+ + Cl^-$ (HCl is a strong acid.)

$H_3O^+(aq) + HCO_3^-(aq) \to H_2CO_3(aq) + H_2O(l)$

$H_2CO_3(aq) \to H_2O(l) + CO_2(g)$

(b) No. The HSO_4^- does not react with H_3O^+ because it would form H_2SO_4, which is a strong acid. The molecular form of a strong acid does not exist in water.

14–63 The addition of the NaOH neutralizes part of the acetic acid to produce sodium acetate.

$$HC_2H_3O_2 + NaOH \to NaC_2H_3O_2 + H_2O$$

$1.00 - 0.20 = 0.80$ mol of $HC_2H_3O_2$ remains and 0.20 mol of $C_2H_3O_2^-$ is formed after 0.20 mol of OH^- is added. pH $= 4.14$

REVIEW TEST ON CHAPTERS 12-14

Multiple Choice

1. (e) **2.** (d) **3.** (b) **4.** (b) **5.** (b) **6.** (e) **7.** (c) **8.** (c) **9.** (b) **10.** (c) **11.** (b) **12.** (c) **13.** (a) **14.** (d)
15. (e) **16.** (d) **17.** (b) **18.** (c) **19.** (b) **20.** (c)

Problems

1. (a) $H_2SO_4(aq) + 2LiOH(aq) \to Li_2SO_4(aq) + 2H_2O(l)$ (b) 150 mL

2. (a) $HClO + H_2O \leftrightharpoons H_3O^+ + ClO^-$
(b) $NH_3 + H_2O \leftrightharpoons NH_4^+ + OH^-$
(c) $KOH + HNO_2 \to KNO_2 + H_2O$
(d) $H_3AsO_4 + 2NaOH \to Na_2AsO_4 + 2H_2O$

3. (a) N.R.   (b) $CaO + H_2O \to Ca(OH)_2)$
(c) $NH_4^+ + H_2O \leftrightharpoons NH_3 + H_3O^+$ (d) $CO_2 + H_2O \to H_2CO_3$
(e) $ClO^- + H_2O \leftrightharpoons HClO + OH^-$ (f) N.R.

4. (a) $[H_3O^+] = 1.5 \times 10^{-11}$ (b) pH $= 10.82$ (c) pOH $= 3.18$

5. (a) $Sn(s) + 4HNO_3(aq) \to SnO_2(s) + 4NO_2(g) + 2H_2O(l)$
(b) 3.2 g NO_2

6. (a) Sn   (b) HNO_3 (c) SnO_2 (d) NO_2 (e) HNO_3 (f) Sn

7. (a) nonspontaneous
(b) $2Cr^{3+} + 3Mn \to 2Cr + 3Mn^{2+}$
(c) yes: $3I_2 + 2Cr \to 2Cr^{3+} + 6I^-$

8. (a) $K_{eq} = \dfrac{[NOBr]^2}{[NO]^2[Br_2]}$ (b) $K_{eq} = 20$ (c) $[NOBr] = 0.94$ mol/L

9. $B + H_2O \leftrightharpoons BH^+ + OH^-$ $K_b = 1.7 \times 10^{-6}$

10. $H_3BO_3 + H_2O \leftrightharpoons H_3O^+ + H_2BO_3^-$ pH $= 4.72$

CHAPTER 15

15–1 (a) $^{210}_{84}$Po (b) $^{84}_{38}$Sr (c) $^{257}_{100}$Fm (d) $^{206}_{82}$Pb (e) $^{233}_{92}$U

15–3 (a) $^{4}_{2}$He (b) $^{0}_{-1}e$ (c) $^{1}_{0}n$

15–4 $^{2}_{1}$H

15–6 (a) $^{206}_{82}$Pb (b) $^{148}_{62}$Sm (c) $^{248}_{98}$Cf (d) $^{262}_{107}$Uns

15–8 (a) $^{3}_{2}$He (b) $^{153}_{65}$Tb (c) $^{59}_{27}$Co (d) $^{24}_{12}$Mg

15–11 $^{51}_{25}$Mn $\rightarrow$ $^{51}_{24}$Cr + $^{0}_{+1}e$

15–14 $^{68}_{32}$Ge + $^{0}_{-1}e$ $\rightarrow$ $^{68}_{31}$Ga

15–16 (a) $^{0}_{-1}e$ (b) $^{90}_{38}$Sr (c) $^{26}_{13}$Al (d) $^{231}_{90}$Th (e) $^{179}_{72}$Hf
(f) $^{41}_{20}$Ca (g) $^{210}_{81}$Tl

15–18 (a) $^{230}_{90}$Th $\rightarrow$ $^{226}_{88}$Ra + $^{4}_{2}$He
(b) $^{214}_{84}$Po $\rightarrow$ $^{210}_{82}$Pb + $^{4}_{2}$He
(c) $^{210}_{84}$Po $\rightarrow$ $^{210}_{85}$At + $^{0}_{-1}e$
(d) $^{218}_{84}$Rn $\rightarrow$ $^{214}_{82}$Pb + $^{4}_{2}$He
(e) $^{14}_{6}$C $\rightarrow$ $^{14}_{7}$N + $^{0}_{-1}e$
(f) $^{50}_{25}$Mn $\rightarrow$ $^{50}_{24}$Cr + $^{0}_{+1}e$

15–19 $^{234}_{90}$Th, $^{234}_{91}$Pa, $^{234}_{92}$U, $^{230}_{90}$Th, $^{226}_{88}$Ra, $^{222}_{86}$Rn, $^{218}_{84}$Po, $^{214}_{82}$Pb, $^{214}_{83}$Bi, $^{210}_{81}$Tl, $^{210}_{82}$Pb, $^{210}_{83}$Bi, $^{210}_{84}$Po, $^{206}_{82}$Pb

15–20 $\dfrac{1}{16}$

15–22 (a) 5 mg (b) 1.25 mg (c) 1.25 mg

15–23 2.50 g

15–25 about 11,500 years old

15–27 The energy from the radiation causes an electron in an atom or a molecule to be expelled, leaving behind a positive ion. When a molecule in a cell is ionized, it is damaged and may die or mutate.

15–29 Radiation enters a chamber causing ionization. The electrons formed migrate to the central electrode (positive) and cause a burst of current that can be detected and amplified. In a scintillation counter, the radiation is detected by phosphors that glow when radiation is absorbed.

15–30 (a) $^{35}_{16}$S (b) $^{2}_{1}$H (c) $^{30}_{15}$P (d) 8 $^{0}_{-1}e$ (e) $^{254}_{102}$No (f) $^{237}_{92}$U (g) 4 $^{1}_{0}n$

15–32 one neutron

15–34 $^{206}_{82}$Pb

15–35 $^{262}_{107}$Uns

15–36 (a) $^{87}_{34}$Se (b) $^{97}_{38}$Sr

15–38 $^{2}_{1}$H + $^{2}_{1}$H $\longrightarrow$ $^{4}_{2}$He $\nearrow$ $^{3}_{2}$He + $^{1}_{0}n$
$\searrow$ $^{3}_{1}$H + $^{1}_{1}$H

15–39 (a) $^{106}_{46}$Pd + $^{4}_{2}$He $\longrightarrow$ $^{109}_{47}$Ag + $^{1}_{1}$H
(b) $^{266}_{109}$Une $\longrightarrow$ $^{262}_{107}$Uns + $^{4}_{2}$He
$^{262}_{107}$Uns $\longrightarrow$ $^{258}_{105}$Unp + $^{4}_{2}$He
(c) $^{212}_{83}$Bi $\longrightarrow$ $^{212}_{84}$Po + $^{0}_{-1}e$

(d) $^{60}_{30}Zn \longrightarrow \, ^{60}_{29}Cu + \, ^{0}_{+1}e$

(e) $^{239}_{94}Pu + \, ^{1}_{0}n \longrightarrow \, ^{140}_{55}Cs + \, ^{97}_{39}Y + 3\, ^{1}_{0}n$

(f) $^{206}_{82}Pb + \, ^{54}_{24}Cr \longrightarrow \, ^{257}_{106}Unh + 3\, ^{1}_{0}n$

(g) $^{93}_{42}Mo + \, ^{0}_{-1}e \longrightarrow \, ^{93}_{41}Nb$

15–41 (a) $^{90}_{39}Y$ (b) 50 years

Chapter 16

16–1 (a)   (b)

(c)

(d)

(e)

(f)

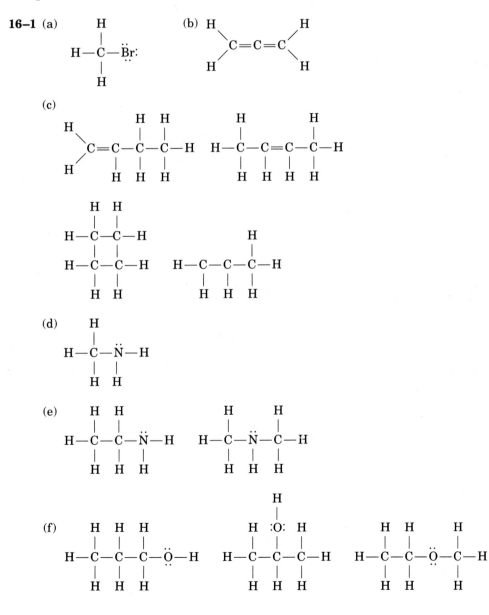

16-3

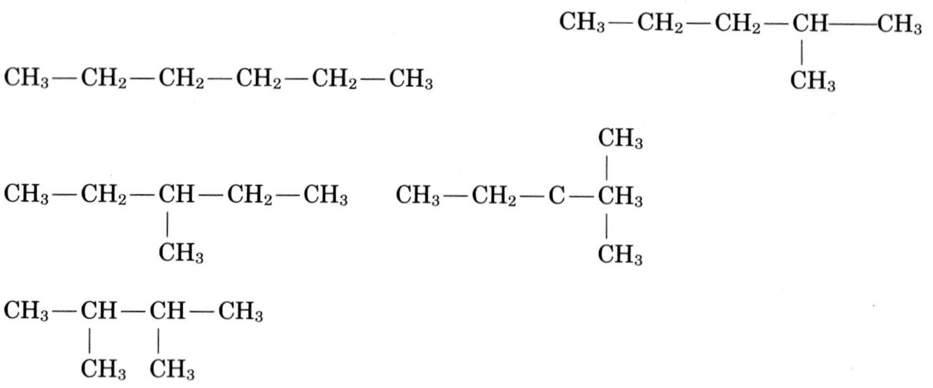

$$CH_3-CH_2-CH_2-CH_2-CH_2-CH_3$$

$$CH_3-CH_2-CH_2-\underset{\underset{CH_3}{|}}{CH}-CH_3$$

$$CH_3-CH_2-\underset{\underset{CH_3}{|}}{CH}-CH_2-CH_3 \qquad CH_3-CH_2-\underset{\underset{CH_3}{|}}{\overset{\overset{CH_3}{|}}{C}}-CH_3$$

$$CH_3-\underset{\underset{CH_3}{|}}{CH}-\underset{\underset{CH_3}{|}}{CH}-CH_3$$

16-5 (a) No. The second compound does not have oxygen.
(b) No. The compounds have unequal numbers of carbon atoms.
(c) Yes. Both compounds have the formula C_2H_7N.
(d) No. The compounds have unequal numbers of carbon atoms.
(e) Yes. The compounds are identical.

16-7 (a) C_4H_{10}, (d) $C_{10}H_{22}$, and (g) $C_{18}H_{38}$

16-8 (a)

$$CH_3-CH_2-CH_3$$

(b) 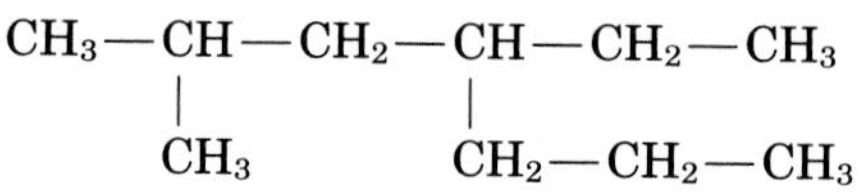 $CH_3-\underset{\underset{CH_3}{|}}{CH}-CH_2-\underset{\underset{CH_2-CH_2-CH_3}{|}}{CH}-CH_2-CH_3$

(c)

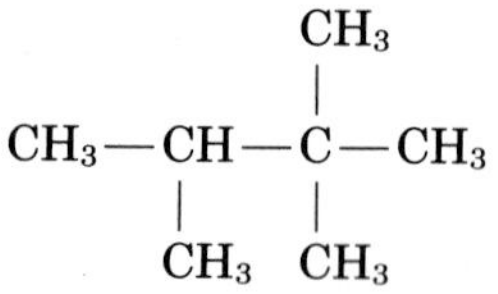

$$CH_3-\underset{\underset{CH_3}{|}}{CH}-\underset{\underset{CH_3}{|}}{\overset{\overset{CH_3}{|}}{C}}-CH_3$$

(d)

$$CH_3-\underset{\underset{CH_2}{|}}{CH}-\underset{\underset{CH_2}{||}}{CH_2}-\underset{\overset{\overset{CH_3}{|}}{}}{C}-CH_3$$
$$\underset{CH_2}{\diagdown \diagup}$$

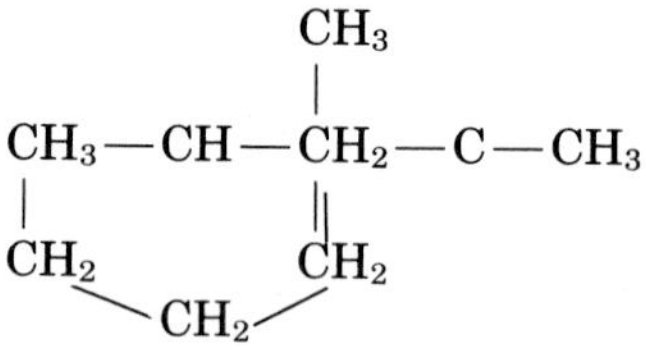

16-11 (f) C_7H_{14}

16-12 (b) C_3H_7, and (e) C_7H_{15}

16-14 (a) 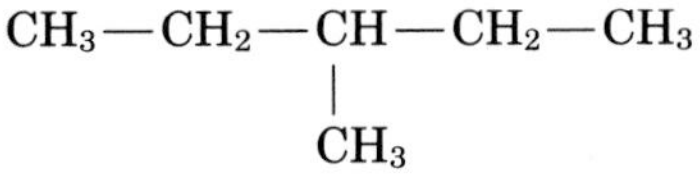 $CH_3-CH_2-\underset{\underset{CH_3}{|}}{CH}-CH_2-CH_3$

(b) $CH_3CH_2CH_2CH_2CH_2CH_3$

(c) 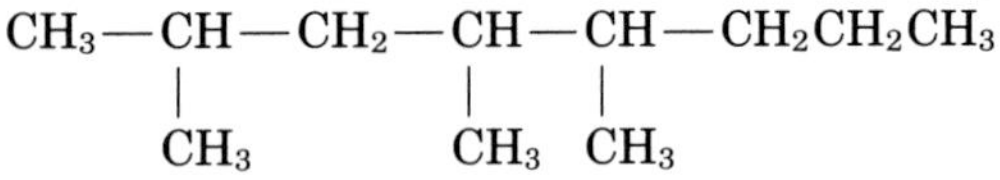 $CH_3-\underset{\underset{CH_3}{|}}{CH}-CH_2-\underset{\underset{CH_3}{|}}{CH}-\underset{\underset{CH_3}{|}}{CH}-CH_2CH_2CH_3$

(d) 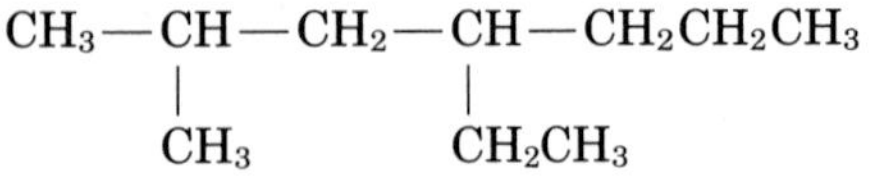 $CH_3-\underset{\underset{CH_3}{|}}{CH}-CH_2-\underset{\underset{CH_2CH_3}{|}}{CH}-CH_2CH_2CH_3$

(e) 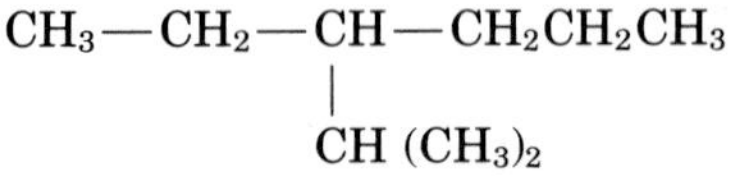 $CH_3-CH_2-\underset{\underset{CH(CH_3)_2}{|}}{CH}-CH_2CH_2CH_3$

(f) 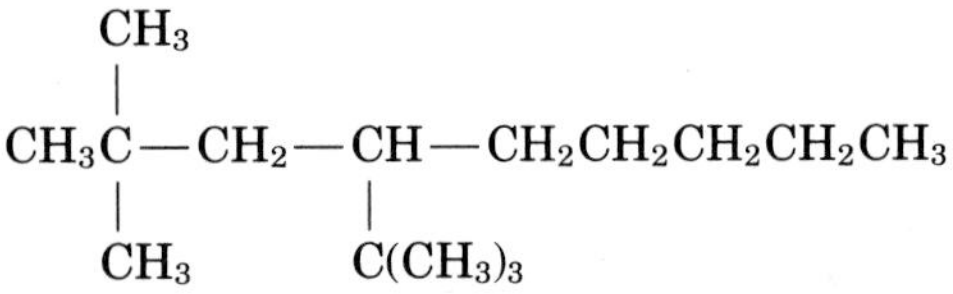 $CH_3\underset{\underset{CH_3}{|}}{\overset{\overset{CH_3}{|}}{C}}-CH_2-\underset{\underset{C(CH_3)_3}{|}}{CH}-CH_2CH_2CH_2CH_2CH_3$

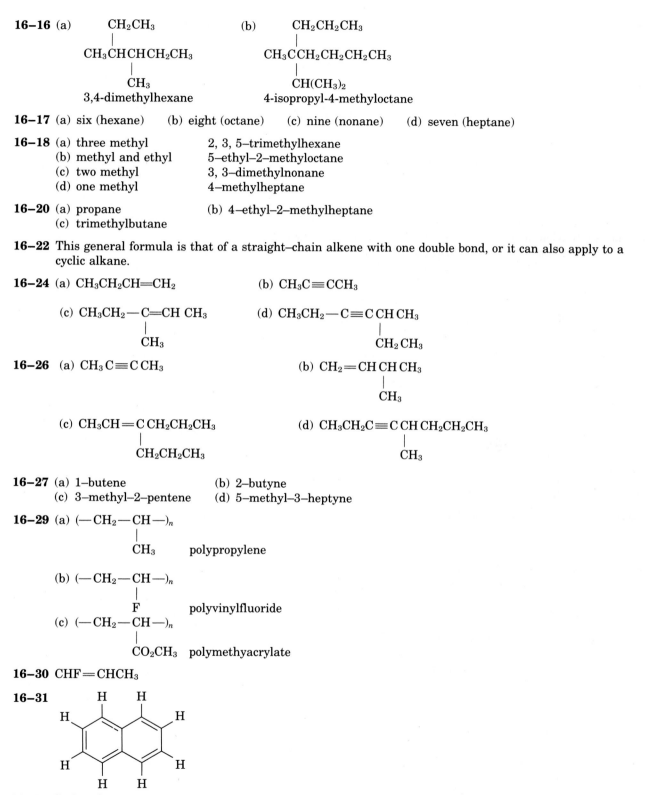

16–16 (a)

$$CH_2CH_3$$
$$|$$
$$CH_3CHCHCH_2CH_3$$
$$|$$
$$CH_3$$
3,4-dimethylhexane

(b)

$$CH_2CH_2CH_3$$
$$|$$
$$CH_3CCH_2CH_2CH_2CH_3$$
$$|$$
$$CH(CH_3)_2$$
4-isopropyl-4-methyloctane

16–17 (a) six (hexane)   (b) eight (octane)   (c) nine (nonane)   (d) seven (heptane)

16–18 (a) three methyl   2, 3, 5–trimethylhexane
(b) methyl and ethyl   5–ethyl–2–methyloctane
(c) two methyl   3, 3–dimethylnonane
(d) one methyl   4–methylheptane

16–20 (a) propane   (b) 4–ethyl–2–methylheptane
(c) trimethylbutane

16–22 This general formula is that of a straight–chain alkene with one double bond, or it can also apply to a cyclic alkane.

16–24 (a) $CH_3CH_2CH{=}CH_2$ (b) $CH_3C{\equiv}CCH_3$

(c)

$$CH_3CH_2{-}C{=}CH\ CH_3$$
$$|$$
$$CH_3$$

(d)

$$CH_3CH_2{-}C{\equiv}C\ CH\ CH_3$$
$$|$$
$$CH_2\ CH_3$$

16–26 (a) $CH_3\ C{\equiv}C\ CH_3$

(b)

$$CH_2{=}CH\ CH\ CH_3$$
$$|$$
$$CH_3$$

(c)

$$CH_3CH{=}C\ CH_2CH_2CH_3$$
$$|$$
$$CH_2CH_2CH_3$$

(d)

$$CH_3CH_2C{\equiv}C\ CH\ CH_2CH_2CH_3$$
$$|$$
$$CH_3$$

16–27 (a) 1–butene   (b) 2–butyne
(c) 3–methyl–2–pentene   (d) 5–methyl–3–heptyne

16–29 (a)

$$({-}CH_2{-}CH{-})_n$$
$$|$$
$$CH_3$$ polypropylene

(b)

$$({-}CH_2{-}CH{-})_n$$
$$|$$
$$F$$ polyvinylfluoride

(c)

$$({-}CH_2{-}CH{-})_n$$
$$|$$
$$CO_2CH_3$$ polymethyacrylate

16–30 $CHF{=}CHCH_3$

16–31

16–33 Carboxylic acids contain a carboxyl group, and alcohols contain only a hydroxyl group.

16–34 (a) In an ether, the oxygen is between two carbons; in an alcohol, the oxygen is between a carbon and a hydrogen.

(b) In an aldehyde, the carbonyl (i.e., C=O) is between a carbon and a hydrogen; in a ketone, the carbonyl is between the two carbons.

(c) In an amine, the NH_2 group is attached to a hydrocarbon group; in an amide, the NH_2 group is attached to a carbonyl group.

(d) Carboxylic acids have a hydrogen attached to an oxygen; in esters the hydrogen is replaced by a hydrocarbon group.

16–36 (a) ketone   (b) alkene   (c) alkane
      (d) aldehyde   (e) aromatic   (f) alcohol

16–38 (a) alkene and aldehyde   (b) ketone and amine
      (c) ether and amide   (d) alkyne and alcohol
      (e) ester   (f) carboxylic acid

16–40 (a) $CH_3CH_2CH_2CH_2OH$ (b) $CH_3CH_2CH_2OCH_2CH_2CH_3$
      (c) $(CH_3)_3N$ (d) CH_3CH_2CHO
      (e) $CH_3CH_2CCH_2CH_3$ (f) CH_2CH_2COOH
             ‖   (g) CH_3COCH_3
             O   ‖
      (h) $CH_3CH_2CNH_2$ O
               ‖
               O

16–41 (a) 2–pentanone   (b) butanoic acid   (c) acetamide   (d) *t*–butyl alcohol   (e) pentanal

16–43 (a) alkene   (b) alkane   (c) alkene   (d) alkyne   (e) alkane   (f) alkene   (g) alkane

16–44

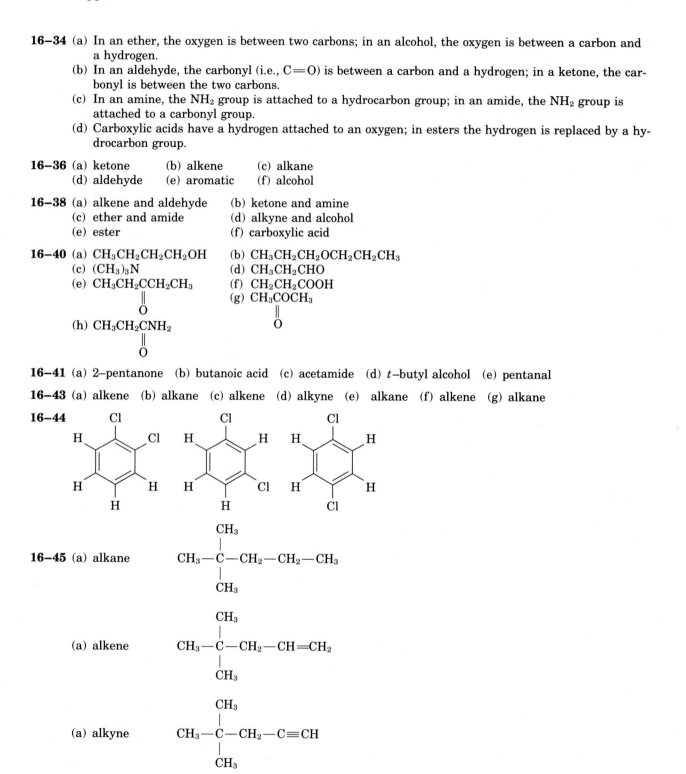

16–45 (a) alkane

(a) alkene

(a) alkyne

(d) aromatic

16–47 (a) $C_3H_8 + 5O_2 \longrightarrow 3CO_2 + 4H_2O$

(b) $CH_2{=}CHCH_3 + H_2O \longrightarrow CH_3CH(OH)CH_3$

(c) $C_2H_5Cl + 2NH_3 \longrightarrow C_2H_5NH_2 + NH_4^+ Cl^-$

(d) $(CH_3)_2NH(aq) + HCl(aq) \longrightarrow (CH_3)_2NH_2^+(aq) + Cl^-(aq)$

(e) $CH_3CH_2COOH + (CH_3)_2NH \longrightarrow CH_3CH_2CON(CH_3)_2 + H_2O$

(f) $HC{\equiv}CH + HBr \longrightarrow CH_2{=}CHBr$

(g) $CH_3CH_2COOH + H_2O \leftrightharpoons CH_3CH_2COO^- + H_3O^+$

16–52 (a) C_3H_8O: $CH_3{-}CH_2{-}CH_2{-}OH$ n–propyl alcohol

$$\begin{array}{c} CH_3 \\ {\diagdown} \\ \quad CH{-}OH \\ {\diagup} \\ CH_3 \end{array}$$ isopropyl alcohol

$CH_3{-}CH_2{-}O{-}CH_3$ ethyl methyl ether

(b) $C_3H_6O_2$:

$$CH_3{-}\overset{\displaystyle O}{\overset{\|}{C}}{-}O{-}CH_3$$ methyl acetate (ester)

$$H{-}\overset{\displaystyle O}{\overset{\|}{C}}{-}O{-}CH_2{-}CH_3$$ ethyl formate (ester)

$CH_3{-}CH_2{-}COOH$ propionic acid

PHOTO CREDITS

Chapter 9   Opener: Clyde H. Smith/Peter Arnold. **Page 277:** Andy Washnik. **Page 278:** OPC, Inc. **Page 280:** Walter Geirsperger/The Stock Market. **Page 285:** R. Rowan/Photo Researchers. **Page 287:** Michael Dalton/Fundamental Photographs. **Page 295:** Patrick Morrow. **Page 299:** Tom Young/The Stock Market.

Chapter 10   Opener: Comstock, Inc. **Figure 10–7a):** Fred Ward/Black Star. **Figure 10–7b):** Herve Berthoule/Photo Researchers. **Figure 10–7c):** M. Claye/Photo Researchers. **Page 325:** Kristian Hilsen/Tony Stone Images. **Page 326 (top):** Paul Silverman/Fundamental Photographs. **Page 326 (bottom):** Richard Megna/Fundamental Photographs. **Page 327:** Yoav Levy/Phototake. **Figure 10–11b):** Courtesy of S. Wilcox, SUNY Binghampton. **Figure 10–11c):** Ken Karp/Fundamental Photographs. **Figure 10–12:** Pat LaCroix/The Image Bank. **Page 332:** Tony Stone Images. **Page 333:** Satellite image data processing by the Environmental Research Institute of Michigan (ERIM), Ann Arbor, Michigan.

Chapter 11   Opener: Luis Castaneda/The Image Bank. **Figure 11–1:** Andy Washnik. **Page 356 (left and right):** Andy Washnik. **Page 356 (margin):** Jacana-American Collection/The Image Bank. **Page 364:** Andy Washnik. **Figure 11–9 and Page 366:** Michael Watson. **Page 372:** Dan McCoy/Rainbow. **Page 374 (top):** Aneal Vohra/Unicorn Stock Photo. **Page 374 (bottom):** Courtesy of Recovery Engineering.

Chapter 12   Opener: Marti Pie/The Image Bank. **Figure 12–1:** Andy Washnik. **Page 391:** Robert Capece. **Page 392:** Ken Karp. **Page 396:** Andy Washnik. **Page 398:** Robert Capece. **Page 402:** Andy Washnik. **Figure 12–8:** Courtesy of Fisher Scientific. **Page 416:** Andy Washnik. **Figure 12–11:** Spencer Grant/Photo Researchers.

Chapter 13   Opener: Don Landerwehrle/The Image Bank. **Figure 13–1a):** OPC, Inc. **Figure 13–1b):** Courtesy of USDA. **Page 438:** Andy Washnik. **Page 439 and Figure 13–3:** Michael Watson. **Figure 13–11:** Bill Gallery/Viesti Associates, Inc. **Figure 13–5:** Courtesy of Delco Remy. **Figure 13–6:** Courtesy of Eveready. **Page 452:** Michael Watson.

Chapter 14   Opener: Gary Ladd Photography. **Page 468:** Craig Hammell/The Stock Market. **Page 469:** Duomo Photography, Inc. **Page 470:** Andy Washnik. **Page 473 (top):** Courtesy of USDA. **Page 473 (bottom):** David Sailors/The Stock Market. **Figure 14–9:** Courtesy of AC Spark Plug. **Figure 14–13:** Andy Washnik. **Figure 14–14:** Courtesy of Monsanto Enviro-Chem.

Chapter 15   Opener: Courtesy of NASA. **Figure 15–4a) and b):** Courtesy of Siemens Medical Systems. **Figure 15–4c):** Courtesy of Alpha Spectra, Inc. **Page 509:** Courtesy of Lead Industries Associates, Inc. **Figure 15–5:** Hank Morgan/Science Photo Library/Photo Researchers. **Figure 15–7:** Courtesy of Superconducting Super Collider Laboratory. **Figure 15–8:** Courtesy of Council for Energy Awareness, Washington, D.C. **Figure 15–9:** Kelly Culpepper/Transparencies, Inc. **Figure 15–10:** Hank Morgan/Rainbow. **Page 522 (top):** Sipa Press. **Figure 15–14:** IFA/Bruce Coleman, Inc. **Page 524:** Phototake.

Chapter 16   Opener: Gregory Scott/Photo Researchers. **Figure 16–1:** Andy Washnik. **Figure 16–2:** Brett Froomer/The Image Bank. **Page 541 (top):** Gary Braasch/AllStock, Inc. **Page 543:** Peter Lerman. **Figure 16–4 (left):** Trevor Wood Picture Library/The Image Bank. **Figure 16–4 (center):** Robert Tringali/Sports Chrome Inc. **Figure 16–4 (right):** Michael Ventura/Bruce Coleman, Inc. **Figures 16–5 and 16–6:** Andy Washnik. **Page 555:** Courtesy Heinz. **Figure 16–7:** Stewart Halperin Photo. **Page 557 (top right):** Andy Washnik. **Page 558:** Presse-Sports/Picture Group.

Appendix   Opener: Chris Harvey/Tony Stone Images.

INDEX